Springer Series in Statistics

Advisors
J. Berger, S. Fienberg, J. Gani,
K. Krickeberg, B. Singer

Springer Series in Statistics

Mike West   Jeff Harrison

Bayesian Forecasting and Dynamic Models

With 108 illustrations

Springer-Verlag
New York Berlin Heidelberg
London Paris Tokyo Hong Kong

Mike West
Institute of Statistics and
    Decision Sciences
Duke University
Durham, NC 27706, USA

Jeff Harrison
Department of Statistics
University of Warwick
Coventry CV4 7AL
United Kingdom

Mathematical Subject Classification Codes: 62A15, 62F15, 62M10, 62M20.

Library of Congress Cataloging-in-Publication Data
West, Mike, 1959–
    Bayesian forecasting and dynamic models / Mike West and Jeff
Harrison.
        p. cm.—(Springer series in statistics)
    Bibliography: p.
    ISBN 0-387-97025-8 (U.S. : alk. paper)
    1. Bayesian statistical decision theory. 2. Linear models
(Statistics) I. Harrison, Jeff. II. Title. III. Series.
QA279.5.W47   1989
519.5′42—dc20   89-11497

Printed on acid-free paper.

Camera-ready copy prepared by the authors using T_EX.
Printed and bound by Edwards Brothers, Incorporated, Ann Arbor, Michigan.
Printed in the United States of America.

9 8 7 6 5 4 3 2 1

ISBN 0-387-97025-8 Springer-Verlag New York Berlin Heidelberg
ISBN 3-540-97025-8 Springer-Verlag Berlin Heidelberg New York

PREFACE

In this book we are concerned with Bayesian learning and forecasting in dynamic environments. We describe the structure and theory of classes of dynamic models, and their uses in Bayesian forecasting.

The principles, models and methods of Bayesian forecasting have been developed extensively during the last twenty years. This development has involved thorough investigation of mathematical and statistical aspects of forecasting models and related techniques. With this has come experience with application in a variety of areas in commercial and industrial, scientific and socio-economic fields. Indeed much of the technical development has been driven by the needs of forecasting practitioners. As a result, there now exists a relatively complete statistical and mathematical framework, although much of this is either not properly documented or not easily accessible. Our primary goals in writing this book have been to present our view of this approach to modelling and forecasting, and to provide a reasonably complete text for advanced university students and research workers.

The text is primarily intended for advanced undergraduate and postgraduate students in statistics and mathematics. In line with this objective we present thorough discussion of mathematical and statistical features of Bayesian analyses of dynamic models, with illustrations, examples and exercises in each Chapter. On the less mathematical side, we have attempted to include sufficient in the way of practical problems, motivation, modelling and data analysis in order that the ideas and techniques of Bayesian forecasting be accessible to students, research workers and practitioners in business, economic and scientific disciplines.

Prerequisites for the technical material in the book include a knowledge of undergraduate calculus and linear algebra, and a working knowledge of probability and statistics such as provided in first and second year undergraduate statistics programs. The exercises are a mixture of drill, mathematical and statistical calculations, generalisations of text material and more practically orientated problems that will involve the use of computers and access to software. It is fair to say that much insight into the practical issues of model construction and usage can be gained by students involved in writing their own software, at least for the simpler models. Computer demonstrations, particularly using graphical displays, and use of suitable software by students, should be an integral part of any university course on advanced statistical modelling and forecasting. Two microcomputer software packages, interactive, menu-driven and highly graphically based, have been developed and written over the last two or three years by the authors. The first, **FAB**,[†] is specifically designed for teaching and training in Bayesian forecasting, concentrating on models and methods contained in the first half of the book. The second, **BATS**,[‡] is more orientated towards application and may be usfully used in gaining experience in the use of an important class of models, and associated intervention and monitoring techniques. Both have been used in teaching and training of students from a variety of backgrounds, and also of practitioners, and may be usefully studied in connection with the book.

The material in the book can be very loosely grouped into Chapters taken three at a time. There are sixteen Chapters, so consider five groups of three with the final Chapter 16 comprising a summary Appendix of mathematical and statistical theory relevant to the first fifteen.

A.   Introduction

The first three Chapters provide a broad introduction to the basic principles, modelling ideas and practice of Bayesian forecasting and dynamic models. In Chapter 1 we discuss general principles of modelling, learning and forecasting, aspects of the role of forecasters within decision systems, and introduce basic elements of dynamic modelling and Bayesian forecasting. Chapter 2 is devoted to the simplest, and most widely used, dynamic model — the first-order polynomial model, or steady model. In this setting, the simplest mathematical framework, we introduce the approach to sequential learning and forecasting, describe important theoretical model features, consider practical issues of model choice and intervention, and relate the approach to well-known alternatives. Chapter 3 continues the introduction to dynamic modelling through simple dynamic

---

[†]Harrison, West and Pole (1987). **FAB**, a training package for Bayesian forecasting, *Warwick Research Report* **122**, Department of Statistics, University of Warwick

[‡]West, Harrison and Pole, (1987). **BATS**: Bayesian Analysis of Time Series, *The Professional Statistician* **6**, (43—46)

regression models. Readers will be familiar with standard regression concepts, so that the rather simple extension of straight line regression models to dynamic regression will be easily appreciated.

B.   Dynamic linear model theory and structure

Chapters 4,5 and 6 provide a comprehensive coverage of the theoretical structure of the class of Dynamic Linear Models (DLMs) and Bayesian analyses within the class. Chapter 4 is key. Here we introduce the general framework and notation, and derive the major theoretical results for learning and forecasting. Chapter 5 is concerned with a special subclass, referred to as Time Series Models, that relate naturally to most existing methods for time series forecasting. Chapter 6 focusses on aspects of model design and specification, developing, in particular, the concepts of models built up from basic components, and discounting.

C.   Classes of dynamic models

Chapters 7, 8 and 9 describe in greater detail the structure of important special classes of dynamic models, and their analyses. Chapter 7 is devoted to Time Series Models for polynomial trends, particularly important cases being first-order polynomials of Chapter 2, and second-order polynomials, or linear trend models. Chapter 8 concerns dynamic linear models for seasonal time series, describing approaches through seasonal factor representations and harmonic models based on Fourier representations. Chapter 9 concerns relationships between time series modelled through dynamic regressions, extending Chapter 3, models for transfer effects of independent variables, and dynamic linear model representations of classical ARIMA type noise models.

D.   DLMs in practice, intervention and monitoring

Chapter 10 illustrates the application of standard classes of dynamic models for analysis and forecasting of time series with polynomial trends, seasonal and regression components. Also discussed are various practical model modifications and data analytic considerations. Chapter 11 focusses on intervention as a key feature of complete forecasting systems. We describe modes of subjective intervention in dynamic models, concepts and techniques of forecast model monitoring and assessment, and methods of feed-forward and feed-back control. Chapter 12 is concerned with multi-process models by which a forecaster may combine several basic DLMs together for a variety of purposes. These include model identification, approximation of more complex models, and modelling of highly irregular behaviour in time series, such as outlying observations and abrupt changes in pattern.

E. Advanced topics

Chapters 13, 14 and 15 are concerned with more advanced and recently developed models. In Chapters 13 and 14 we consider approaches to learning and forecasting in dynamic, non-linear models, where the neat theory of linear models does not directly apply. Chapter 13 describes some standard methods of analytic and numerical approximations, and also some more advanced approaches based on numerical integration. Chapter 14 demonstrates analyses in the class of dynamic generalised linear models. In Chapter 15, we return to linear models but consider aspects of modelling and forecasting in multivariate settings.

Acknowledgements

Section 1.4 of Chapter 1 briefly reviews historical developments and influences on our own work. Over the years many people and organisations have contributed, directly and indirectly, to material presented in this book and to our own efforts in developing the material. In particular, we would mention Colin Stevens, Roy Johnson, Alan Scott, Mike Quinn, Jim Smith and our past research students Mike Green, Rei Souza, Jamal Ameen, Helio Migon, Muhammed Akram, Tomek Brus, Dani Gamerman and Jose Quintana. Special thanks go to Andy Pole, our SERC Research Fellow and colleague over the last three years. In addition to joint work on non-linear models, Andy's computer experience has contributed a great deal to the development of computer software which has been used for many of the examples and case studies presented in the book.

Among the companies who have supported the development of Bayesian forecasting we must single out Imperial Chemical Industries plc. The initial work on forecasting began there in 1957 and took its Bayesian flavour in the late 1960's. We would particularly like to acknowledge the help and encouragement of Richard Munton, Steve Smith and Mike Taylor of ICI. Other companies have supported our work and aided software development. Among these are British Gas Corporation, through the guiding efforts of Chris Burston and Paul Smith, Information Services International (formerly Mars Group Services), and IBM.

We would also like to thank all our colleagues, past and present, for their interest and inputs, in particular Ewart Shaw, Jim Smith, Tony O'Hagan, Tom Leonard, Dennis Lindley and Adrian Smith. In addition, we acknowledge the support and facilities of the Department of Statistics at Warwick University, and, more recently, the Institute of Statistics and Decision Sciences at Duke University.

Mike West & Jeff Harrison
January 1989

CONTENTS

CHAPTER 1

INTRODUCTION

1.1 MODELLING, LEARNING AND FORECASTING

1.1.1 Perspective

This book is concerned with modelling, learning and forecasting. A basic view of scientific modelling is that a model is any *"simplified description of a system (etc.) that assists calculations and predictions"* (Oxford English Dictionary). More broadly, a model is any scheme of description and explanation that organises information and experiences providing a means of learning and forecasting. The prime reason for modelling is to provide efficient learning processes which will enhance understanding and enable wise decisions.

In one way the whole operation of any organisation can be viewed as comprising a sequence of decisions based upon a continual stream of information over time. Consequently there is an accumulation of knowledge which, in principle, should lead to improvements in understanding and better decision making. Suitably formulated and utilised, descriptive models provide vehicles for such learning. The foundation for learning is the Scientific Method. It is often assumed that the scientific investigation and learning is concerned with the pursuit and identification of a single "true" model. This is certainly not our position. Models do not represent truth. Rather they are ways of viewing a system, its problems and their contexts. In modelling we are concerned with ensuring that at all times we have a way of viewing which enables good decisions and enhances performance both in the long and short term. Within a model framework, the scientific learning process enables routine and coherent processing of received information that is used to revise views about the future and guide decision making.

A forecast is an hypothesis, conjecture, or speculative view about something future. Unlike the learning process, forecasting is not itself a science but may be guided by model descriptions. Models do not represent truth, but are ways of seeing that are highly conditional since they are based upon past experience. Consequently they are largely relevant to historic conditions and any extension or extrapolation of these conditions is highly speculative and is quite likely to be far from the mark. Too often forecasting systems ignore the wider aspects of learning and settle for myopic models capable of very restrictive prediction and incapable of much development. One desirable property of a forecasting and learning system is that the way of viewing should not change radically too frequently. Otherwise confidence is impaired, communication breaks down, and performance deteriorates. Hence the fundamentals of a operational model should remain constant for considerable periods of time and regular change should only affect small detail. That is, there should be a routine way of learning during phases when predictions and decisions appear adequate, and an "exceptional" way when they seem unsatisfactory. The routine adjustment will generally be concerned with small improvements in the estimates of model quantities whereas the exceptions may involve basic model revision. Such systems operate according to the important principle of *Management by Exception*, an integral part of the Scientific Method. Essentially, information is routinely processed, usually within an accepted conceptual and qualitative framework. Generally the major features of an operational model will remain unchanged, but minor modifications will be accommodated and the relative merits of any rival models noted. Exceptions can arise in two main ways. The first occurs when "non-routine" information or events happen which lead to anticipation of a future major change which will not be properly reflected by routine learning. The second occurs when performance monitoring identifies deficiencies in the view described by the existing model, questioning model adequacy.

1.1.2 Model structure

Model structure is critical to performance. A good structure will provide model properties which include

- Description
- Control
- Robustness

Consequently we view the structuring of a model through a triple

$$\mathcal{M}: \quad \{C, F, Q\}.$$

The component C describes the **conceptual** basis, F the model **form**, and Q the **quantified** form.

C : The concepts C provide an abstract view of a model. They may express scientific or socio-economic laws; objectives of decision centres; behavioural characteristics etc. As such they are expected to be very durable and rarely changed. Further, at any particular moment in time, rival models representing alternative views may be founded upon the same conceptual base C (although this is not always the case).

F : The qualitative form F represents the conceptual in descriptive terms, selecting appropriate variables and defining relationships. For example a government may be seen as a decision centre wishing to retain power. This may be a part of a general conceptual view of an economic sector, which helps to express the type of decisions to be taken as circumstances evolve. At any time there may be choices about the way in which general objectives are accomplished. Under one policy the relevant control variables will be from, say, set A, and at a time of policy revision, they may suddenly change to a set B. Form defines the relevant sets of variables and their relationships, and may involve algebraic, geometric, and flow sheet representations.

Q : Often many, if not all, rival models will have a common qualitative parametric form and differ only at the quantitative level Q in the values given to the parameters. Then a single form is often durable for reasonable periods of time. It is at the quantitative level that frequent change occurs. Here the merits of rival parametric values are continually changing as new information is received. Generally these changes are small and in accord with the uncertainty conditional upon the adequacy of the concept and form.

Description aims at providing meaning and explanation in an acceptable and communicative way. This is necessary for unifying all concerned with a decision process and its effects. It brings confidence from the fact that all are working and learning together with a well defined view. In particular, it encourages growth in understanding, participation, and progressive change. In most decision situations anticipation of major change is the critical factor upon which the life of an organisation may depend. Hence it is vital to promote creative thinking about the basic concepts and form, and to improve intervention at all levels. Two important aspects of an effective description are parsimony and perspective. Parsimony means simplicity. It excludes the irrelevant and uses familiar canonical concepts, forms, and learning procedures, bringing all the power of past experience to bear. Perspective is concerned with the relative importance of the various model characteristics and what they do and do not affect.

Control is usually taken to mean actually influencing the behaviour of the system being modelled. Given such control there is great opportunity for effective learning by experimenting. In process control this may mean carrying out dynamic experimentation, as embodied in the principles and practice of Evolutionary Operation (Box and Draper, 1969). All the principles and power of the statistical design of experiments can then be utilised. Another aspect of control occurs when the system being modelled cannot be controlled or can only be partially controlled. Then the decision makers may still have control in the sense of having freedom to respond wisely to predictions about systems which they cannot influence. For example a farmer who assesses the weather in order to make farming decisions cannot directly influence the weather but can utilise his forecast to control his actions, perhaps waiting until "the time is ripe". Currently there is much power in the hands of remote centralised organisations who can bring disaster to those they control simply because of their lack of understanding the systems they control and their selfish desires. It is then vital that the true nature of the "controlled" system is understood along with the motivations and reactive responses of the power centre. One look at the past control of European agriculture and industry can illustrate this point only too well.

Robustness is a key property for a learning system. Essentially the aim is to structure so that, at "exceptional" times, intervention is efficiently and economically accomplished. This means only changing that which needs changing. Thus the objective is to extract the maximum from history so that all relevant information is retained whilst accommodating the new. The structure $\{C, F, Q\}$ provides a major source of robustness. It offers the opportunity of carrying out major changes at the quantitative level Q whilst retaining the model form F. It provides a way of retaining the conceptual C whilst drastically altering aspects of form F. However it is also crucial to structure within each of these levels so that major changes affect only the relevant aspects and do not damage others. At the quantitative level, when operating within a parametrised form, modelling component features of a system through distinct though related *model components* is recommended. Each component describes a particular aspect such as a relationship, a seasonal effect, or a trend. Then if intervention necessitates a major change concerning any particular component this can be accomplished without affecting other components.

1.1.3 The role of mathematics

The role of mathematics and statistics is as a language. It is a very powerful language since far reaching implications can often be deduced from really quite simple statements. Nevertheless, the function of mathematics must be seen in perspective. It expresses a view in a way analogous to that of paint on canvas. In this sense it is only as good as the artist who uses the materials and the audience who see the result. Like all sources of power, mathematics can be well used or abused. Selective interpretation is the key weapon of the deceiver. Any modeller, just like an artist, must of necessity select a view of the context under study and is thus, either innocently or deliberately, likely to mislead. Choosing an appropriate view is often very hard work. It may have nothing to do with mathematics, although of course it can involve some data analysis. Many modellers pay scant regard to this vital preliminary effort in their eagerness to play with computers and equations. Consequently so many mathematical models are inappropriate and misleading. With today's computing power it is not far from the truth to say that if a system can be coherently described then it can be expressed mathematically and modelled. To summarise, our position is that modelling is an art; that the first task is to define objectives; the second to select a consistent view of the system; and only later, and if appropriate, to use a mathematical description.

1.1.4 Dynamic models

Learning is dynamic. At any particular time a model describes a routine way of viewing a context, with possible competing views described through alternative models. However, because of uncertainty, the routine view itself is likely to comprise a set of models. For example, consider the view that an output variable Y is related to an input variable X according to a parametrised form

$$Y = X\theta + \epsilon.$$

Here θ is an uncertain parameter and ϵ a (similarly uncertain) random error term. Further suppose that the beliefs of a forecaster about the parameter θ are expressed through a probability distribution $P(\theta)$. Then the view may be described as comprising a set of models each with measurable support $P(.)$. This embodies one form of model uncertainty. The dynamic nature of processes and systems demands also that we recognise uncertainty due to the passage of time. Thus we take the stance that, because the form is only locally appropriate in time, it is necessary to think of θ as slowly changing with time. Further it might be recognised that, at some future time, the whole model form may change and even involve quite different input variables. Hence modelling may be phrased in terms of **Dynamic Models** defined generally as *"sequences of sets of models"*.

At any given time, such a dynamic model $\mathcal{M}$ will comprise member models M, with the forecaster's uncertainty described through a *prior* distribution $P(M), (M \in \mathcal{M})$. In producing a forecast from the dynamic model for an output Y, each member model M will provide a conditional forecast in terms of a probability distribution $P(Y|M)$. In the above example M relates directly to the parametrisation θ and thus to the component Q of the model. This is typical although, more widely, M may involve uncertain aspects of form F and even conceptual descriptions C. The forecast from the dynamic model $\mathcal{M}$ is then simply defined by the marginal probability distribution, namely

$$P(Y) = \int_{M \in \mathcal{M}} P(Y|M)dP(M).$$

Alternative dynamic models are rivals in the sense that they compete with the routine model for its prime position. They provide a means of performing model monitoring and assessment. A dynamic model is not restricted to members which have the same form or conceptual base (although this will often be the case), and so the composition is entirely general.

1.1.5 Routine Learning

Bayesian methodology offers a comprehensive way of routine learning. It is not dependent upon any particular assumptions. For simplicity consider a dynamic model $\mathcal{M}$ with member models M. Before receiving an observation of uncertain quantities Y, the forecaster has a *prior* probability distribution $P(M)$ describing uncertainty about $\mathcal{M}$. Each member model M provides a means of forecasting Y through a probability distribution $P(Y|M)$. This determines conditional views about the future possible values of Y, conditional upon particular possible descriptions M.

By the laws of probability these two sets of probabilities provides a joint probability distribution. In terms of densities rather than distributions,

$$p(Y, M) = p(Y|M)p(M).$$

When Y is observed to take a value Y^*, say, the updated probability distribution for M given $Y = Y^*$ is defined through the density $p(M|Y^*)$, expressed via

$$p(M|Y^*) \propto p(Y^*, M),$$

or equivalently via

$$p(M|Y^*) \propto p(Y^*|M)p(M).$$

This is often expressed as

Posterior $\propto$ Observed likelihood $\times$ Prior.

The proportionality constant is simply the normalising quantity $p(Y^*)$, the prior density for Y at the observed value Y^*, which ensures that the "posterior" density over $M \in \mathcal{M}$ is normalised to unit probability. Hence the usual representation of Bayes' Theorem:

$$p(M|Y) = p(Y|M)p(M)/p(Y), \qquad (M \in \mathcal{M}),$$

for any observed value of Y. Given the dynamic model $\mathcal{M}$, all the routine information contained in the observation Y is expressed through the likelihoods $p(Y|M)$.

1.1.6 Model construction

Some forecasters approach model building in an unsatisfactory way, often resorting to processing historic output data by computer and accepting whatever mathematical expression emerges. Such people are always asking us where we get our models from? The answer is very simple. We apply the natural scientific approach to model building. The first step is, as we have stated previously, to clarify objectives. If both macro and micro decisions are to be made using the model, think about the value of structuring the model hierarchically and initially modelling within hierarchical levels whilst not losing sight of the relationships between levels. The next step is to decide upon the conceptual basis, and then to consider significant factors and relationships. In particular, the aim is to explain as much of the significant variation as possible. This may involve identifying "sure thing" relations. These may arise because of accepted physical laws, relationships and constraints. Often "lead" variables are involved. For example, if you are predicting beef supply at the butchers, it is clear that there is a lag of three years from impregnation of a cow and the resulting meat reaching the shop. Within a "closed" community it is obvious that the decision point for expansion at retail level is three years previously. It is also very clear, from biological considerations, that "in order to expand you must first contract". That is, more heiffers will have to be taken back into the breeding herd at the expense of next year's meat supply. Germinal decision points and these delayed response dynamics appear in many systems only to be completely ignored by many modellers and "fire brigade" decision makers. Hence the reason for so many problems which arise, not only in agriculture (Harrison and Quinn, 1978), but in many other economic sectors. Another key step is to assess the nature of the process being studied. Is it purposeful? It is probably no use studying the detailed flow of a river in forecasting its path; the main thing is to recognise that all rivers follow the principle of steepest descent. Similarly, if you are studying a process dependent upon other decision makers, it may be critical to assess their purposes, what effect the environment will have on them, how their perceptions match reality, and how their statements match their actions. In selecting model factors, control factors are to be prized. When, and only when, any significant structure has been modelled should the modeller resort to time series. Essentially a time series is a confession of ignorance, generally describing situations statistically without relating them to explanatory variables. That is not to say that pure time series models are useless. For example, time series models involving polynomial trends and seasonal components may be very useful in, say short-term sales forecasting, where experience, perhaps across many similar products has led to empirical growth laws and defensible arguments for seasonality. The danger arises when mathematical expressions, such as general stationary noise models models, are adopted without any substantial foundation. Then historic peculiarities are likely to suggest totally inappropriate models. The key message for the modeller is *"THINK, and do not sacrifice yourself to mathematical magic."* This is not to rule out exploratory data analysis. History and data may be used both to assess contending models and to stimulate creative thought, but data analysis should not replace preliminary contextual thinking, nor should it promote mathematical formulae which have no defensible explanations. All analytic methods have some contribution to offer but they must be seen as servants of explanatory thought and not its usurper.

1.2 FORECAST AND DECISION SYSTEMS

1.2.1 Integration

As Dr Johnson said *"People need to be reminded more often than they need to be instructed"*. This is true in modelling especially when dealing with open decision systems. There is much that we know or are able to recognise but that we are not anxious to see or practise. Good modelling demands hard thinking, and good forecasting requires an integrated view of the role of forecasting within decision systems.

The consequences of failing to harmonize forecasts and decisions is sharply (and rather humorously) exemplified in the following incident that occurred over twenty years ago in the brewery trade. At the end of the summer of 1965, one of us received a call from a person who we will call Jack. Jack had just moved companies and, amongst other things, had been given the resposibility for short-term forecasting and was seeking advice. Upon visiting us, we learned that he was to produce beer forecasts for a couple of weeks ahead. He did not appear to be too enthralled about this, and when we enquired what decisions rested on the forecasts his reply was "I am not interested in the decisions. My job is simply to forecast." Pressed further he said that the forecasts would be used in production planning. However he would not see the importance of integrating forecasting and control.We discussed some of his data, which we had examined with the aid of the ICI MULDO package (Harrison and Scott, 1965; Harrison, 1965). As is evident from the typical history displayed in Figure 1.1, there was a marked seasonal effect with a summer peak, a

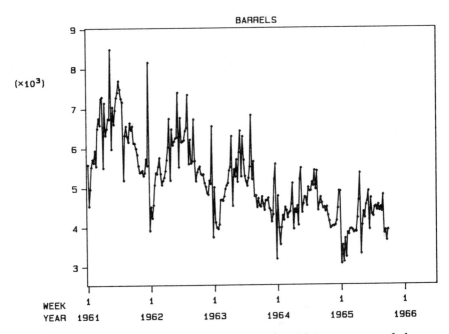

Figure 1.1.   Barrels of beer (in thousands) sold in a sector of the
               UK market.

very marked festive component, particularly at Christmas, and also
a general decline in trend.

A few months later Jack telephoned to announce that he was in
trouble. He did not volunteer the extent of his misfortunes. "No, I
have not been following the methods you advised. I simply estimated
based upon the last few weeks. This seemed to give as good forecasts
as yours and to begin with I did fine." Only six weeks later did it
become clear just how his misfortunes had grown. On the front
page of a number of national daily newspapers was the story. The
colourful Daily Mirror version is reproduced below in Figure 1.2.
We leave the reader to speculate how a computer came to be blamed
when apparently Jack had never used one.

1.2.2   Choice of information

Often there is a choice concerning the observed information on which
forecasts are based. Modellers are generally well aware that they may
need to carry out such operations as allowing for differences in numbers of working days, seasonal holidays and feasts, deflating prices,

Computer sends the beer for a Burton

It happened, of all places, in Burton-upon-Trent, the town made famous
by beer.

First, they found they had **TOO MUCH** beer.

Thousands of gallons too much — all because a computer went wrong.

The computer over-estimated how much thirsty revellers could swallow
over the Christmas and New Year holidays — and now the beer is too
old for use.

Brewery officials in the Staffordshire beer "capital" ordered: "Down the
drain with it ..."

Secret

The throwaway beer — more than 9,000 casks of best bitter and pale ale
produced by the Bass-Worthington group — is worth about £100,000.

Tankers are pouring it down the drain at a secret spot.

But now the brewery officials are feeling sour once again ...

Some pubs in the town yesterday reported a beer **SHORTAGE**.

Production of fresh supplies has been disrupted — again because of the
Christmas and New Year holidays.

A brewery spokesman said: "We were caught on the hop"

Daily Mirror, January 12th 1966
(Journalist: William Daniels)

Figure 1.2.   Caught on the hop.

and so on. However there are other considerations. It is clear that
in controlling a system, information which is relatively independent
of the performance of that system is preferable to that which is dependent upon it. In many cases little thought is given to the matter
or its consequences. As an example in short-term sales forecasting
for stock control and production planning, modellers may, without
question accept sales statistics as their routine observations. The real
objective is to forecast customer requirements. Clearly sales statistics represent what is sold. As such they reflect the ability of the

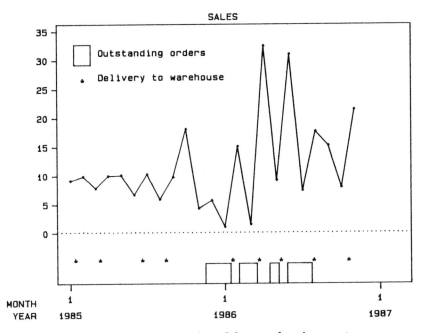

Figure 1.3. Sales variation induced by production system.

system to meet requirements and not necessarily the actual requirements themselves. The use of sales statistics may result in excessive
variation, particularly when products compete for manufacturing capacity, as illustrated by the graph in Figure 1.3. There can also be
an insensitivity to increases in demand, particularly when manufac-
ture is continuous. Other problems include recording delays, due to
waiting for suitable transport, and the possibility of negative figures
when returns occur. In general it is preferable to collect order or demand statistics although even these can mislead on requirements in
periods of shortage when customers deliberately request more than
they need, knowing that they will only receive a partial allocation.
However the short-term problem is then one of allocation.

Another example of care in treating observations comes from forecasting the price of a vegetable oil commodity. Some time ago, a
bumper harvest had been predicted and without thinking had been
converted into its oil equivalent. Unfortunately this proved to be a
significant over estimate of the available oil since there was a limit
on crushing capacity. Thus, although the abundant harvest materi-
alised, the price did not fall nearly as much as anticipated. These

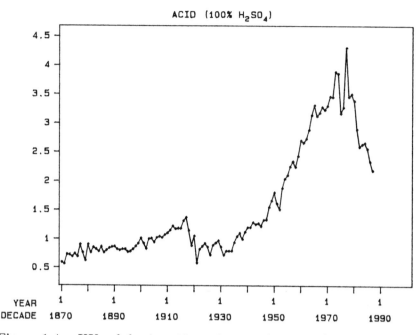

Figure 1.4.   UK sulphuric acid production (tons $\times 10^6$).

examples should suffice to remind modellers to consider both the
direct relevance and the quality of their observations.

1.2.3 Prospective Intervention

A major problem for many mathematically based learning systems
has been that of accommodating subjective information. This is particularly important at times of major change. Consider the example
of the history of UK Sulphuric Acid production. Production statistics from 1870 to 1987 are shown in Figure 1.4 (Source: *Monthly
Digest of Statistics*). A glance shows that there have been major
changes, which, if not anticipated, would have been extremely costly
to producers. The feature of these changes is the sharp decline and
demise of previously significant end uses due to technological and
social change. The first end use to suffer was the Leblanc process
for the production of soda. The ammonia-soda process uses no sulphuric acid and finally superceded Leblanc around the end of 1920.
More recently, hydrochloric acid has challenged in steel pickling since
it offers a faster process and is recoverable. The importance of an
early awareness and assessment of the effect of these developments

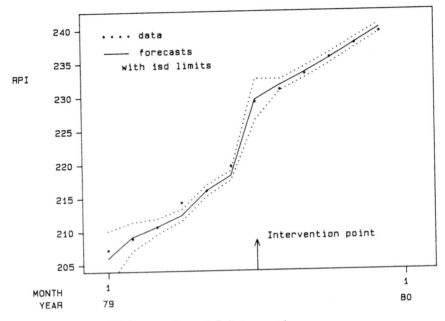

Figure 1.5.   RPI forecasting with intervention.

and consequent prospective intervention is self evident (Harrison and
Pearce, 1972).

We have already discussed the vital role of model structure. Usually, at times of intervention, there is additional uncertainty with an
accompanying sharp change in beliefs about the future. Since the
Bayesian approach to handling uncertainty and synthesising information is phrased in terms of probabilities, it offers natural ways of
intervening. This is one of the prime reasons for its adoption as the
routine learning method within Management by Exception systems.
Often the probabilities are neatly described in meaningful terms thus
facilitating intervention. As an example, consider short-term forecasting of the retail price index (RPI, Figure 1.5) with a routine
view based essentially only on local linear extrapolation. The announcement in the UK in June 1979 that VAT (value added tax)
was to be raised from 10% to 15% makes intervention essential to
sustained short-term forecasting accuracy. Prior to the announcement the July price level might, for example, have been assessed
by a forecaster as 221.5 with an associated standard deviation of
about 1.5 representing uncertainty about the point forecast value.
Upon announcement of the change, one possible response by the

forecaster might have been to subjectively change to a revised estimate of 229.5 with an increase in uncertainty to standard deviation of about 3. The actual mechanics of the change can, and should, be designed with the intervener in mind. Direct communication in terms of statistical terminology is not at all necessary provided the chosen intervention method can be interpreted in such terms. Figure 1.5 further illustrates the incident just described, representing simple forecasts based on such a view. The forecasts are one-step ahead, that for each monthly value of the RPI being made the previous month, and are plotted with associated uncertainties represented simply by intervals of one standard deviation (sd) either side of the point forecasts. Some people object to intervention on the grounds of subjectivity. Let us be quite clear that as soon as anyone starts to model they are being subjective! Of course, any facility which enables subjective communication is open to abuse, but the answer is definitely not to prohibit it. Rather it needs monitoring and control. This may be accomplished within the routine monitoring system or otherwise. It is important that such monitoring is performed as part of the learning process for upon first encountering complete learning and forecasting systems decision makers are tempted to intervene on ill-founded hunches or on situations which are already accommodated in the routine model.

1.2.4 Monitoring

Within the framework of a forecasting system incorporating routine learning and prospective intervention, continual assessment of model adequacy is vital to control and sustain forecasting performance. Model monitoring is concerned with detecting inadequacies in any currently adopted routine model M, and, in particular, in signalling major unanticipated changes in the processes of interest. Practitioners are familiar with standard monitoring schemes, such as used in quality control and inventory management, that operate with a standard model and associated "routine" forecasts unless exceptional circumstances arise. Feed-forward interventions to adapt in the light of information about forthcoming events are always the preferred management action. However, in the absence of anticipatory information, monitoring systems will continually assess the routine model and signal doubtful performance based on the occurrence of

unexpectedly large forecast errors. At such times, relevant explanatory external information may then be sought, or automatic procedures designed to correct for model deficiencies applied. Within the
Bayesian approach, the extent to which observations accord with pre-
dictions is measured through predictive probabilities. With $P(Y|M)$
the predictive probability distribution, based on model M, for a future quantity Y, model adequacy may be questioned if the eventually
observed value of Y does not accord with $P(.|M)$. The initial task in
designing any monitoring system is to decide what constitutes bad
forecasting ie. just how to measure accord with forecast distributions in the context of decisions that may be dependent on future
forecasts.

1.2.5   Retrospective Assessment

Apart from routine learning and prediction, it is often useful to
make retrospective assessments. Typically this is called a *"What
happened?"* analysis. The objective is, in the light of all current
information, to estimate what happened in the past in order to improve understanding. A typical example in marketing is when a novel
market campaign is carried out. Initially, prospective intervention is
undertaken to communicate the predicted effects which are likely to
involve increased sales estimates and associated uncertainties over
the campaign period. This is necessary in order to ensure that appropriate stocks will be available. The progress of the campaign's
effect will be suitably monitored. Then, some time after completion,
the campaign will be retrospectively assessed in order to increase
the store of market knowledge. This involves properly attributing
effects to the many contributing sources of variation such as price,
seasonality, competitive action etc. Retrospective analysis is particularly useful for examining what happened at times of major change
especially where there is debate or ignorance about possible explanations for the change. As such it is an integral part of a good learning
system.

1.2.6   Utilities and Decisions

A statistician, economist or management scientist usually looks at a
decision as comprising a forecast, or belief, and a utility, or reward,
function. In a simple horse racing situation, you may have beliefs
about the relative merits of the runners which lead you to suppose

that horse A is the most likely to win and horse B the least likely to win. However, the bookmaker's odds may be such that you decide to bet on horse B. More formally, in a simple situation you may have a view about the outcome of a future random quantity Y conditional on your decision a expressed through a forecast or probability distribution function $P(Y|a)$, possibly depending on a. A reward function $U(Y, a)$ expresses your gain or loss if outcome Y happens when you take decision a. Thus for each decision a it is possible to calculate the merit as reflected in the expected reward $r(a) = \int U(Y, a) dP(Y|a)$. The optimal decision is that a which maximises $r(a)$. This is the formalised Bayesian view of optimal decision making (DeGroot, 1971; Berger, 1985).

Given that the spirit of this formulation is accepted, it is clear that, in general, for a given forecast distribution, provided the utility favours a particular decision strongly enough, that decision will be chosen. Also, for a given utility function, if the probability distribution or forecast relating to decisions sufficiently favours a particular decision, that decision will be chosen.

A decision maker occupies a powerful situation and often has strong self interests which lead to utilities quite different from those of others. On the otherhand, the decision maker often cannot openly express his selfish utilities and is therefore forced to articulate in terms of a utility function which is acceptable to others who may later authorise the decision or participate in implementation. In such a situation forecasting offers a major source of power, which can be used to manipulate decisions. Furthermore, the extent of forecast manipulation in business, government etc. should not be underestimated. As Bertrand Russell remarks *"It becomes necessary to entertain whole systems of false beliefs in order to hide the nature of what is desired"* (Russell, 1921). It is common for practitioners to be asked to undertake forecasting exercises where the direction is that a given decision should to result. For example, "you must not show that this plant should not be built," and "we must show that the total market is *not* affected by television advertising, but only the shares between competing brands." The former case involving extended capacity is quite common. On one occasion one of us was very strongly reprimanded when a check was carried out which revealed that the current market of an organisation was over 40% less than the figures being used to justify extended capacity. The reasons for such exhortations are clear. Unless the plant is built "I will be out of a job," "the banks cannot lend money," "unemployment

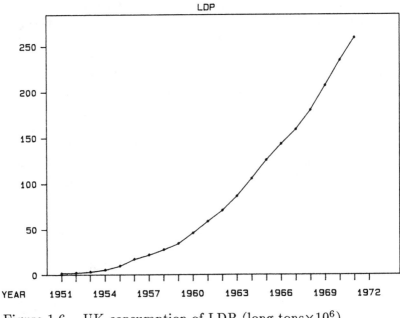

Figure 1.6.   UK consumption of LDP (long tons$\times 10^6$).

will be higher," etc., so that many powerful groups have very strong
and similar utility functions. The key weapon of the manipulator is
selection. This may be data selection, the selection of a particular
set of circumstances to illustrate an argument, the reference to forecasts from other, similarly interested organisations, the selection of
consultants and forecasters who need business, the selection of an
inexperienced project team, etc.

There is, as readers will be well aware, a recent history in gross
over-forecasting across a host of sectors by business and government,
the following examples being typical.

EXAMPLE 1.1. Consider a forecast of UK consumption of Low Density Polyethylene, denoted LDP, made in 1971 (Harrison and Pearce,
1972). The history to that date is shown in Figure 1.6. Historically,
trend curves backed by end use analysis had performed exceedingly
well in estimating UK consumption and the predicted fast growth
had supported past decisions to extend production capacity. In 1971,
the same approach estimated a 1980 market of about 470 thousand
long tons, which did not suit the utility of managers. The only new
significant use was in plastic refuse sacks and no end use argument

could be proposed for increasing this forecast other than surprise or the imposition of unfounded growth constraints. The basic use of trend curves is as a control mechanism. Historically, various empirical rules have been derived such as the percentage growth in demand decreases with time, real production costs decrease with cumulative output etc. These form a standard by which proposals can be judged. Basically, if a case conforms to these rules, it may be safe to sanction extension proposals. If however, the case rests on a vastly different forecast then there must be a very sound and strong reason why the future will evolve in such an unexpected way. In the LDP situation the growth rate had already fallen to 10% and could be expected to fall further to about 6% by 1980. Thus all empirical evidence pointed to a struggle for the market to exceed 500 thousand long tons by 1980. However, the suppliers of the raw material ethylene had produced an incredible 1980 forecast of 906 thousand long tons. Furthermore the board of the plastics company for whom the forecasts were produced decided to plump for an unjustified figure of 860 thousand long tons. To come to fruition both these forecasts needed an average growth rate of about 13% over the next ten years. To say the least, such a growth rate recovery would have been most remarkable and any proposal based upon this would certainly have had to have been very well-founded. The market researcher who did the end use analysis was retired. Extensions were sanctioned, and throughout the next decade, regretted. From the mid 1970's onwards capacity in Western Europe was generally between 30 and 40% over capacity and in 1979 estimated forecasts continually lowered, finally reaching the region of the 470 mark.

EXAMPLE 1.2. The above example is not mentioned as an isolated case but as typical of decision makers. For example, Figure 1.7 shows how official forecasts for ethylene production were continually reduced over the years throughout the 1970's (Source: *Financial Times*, August 9th 1978).

EXAMPLE 1.3. The profits figures and official forecasts shown in Figure 1.8 relate to a period ending around 1970 and should cause no embarrassment now. At that time never, in this particular company's history, had profits been under-forecast. Usually at a boom time, which then regularly occurred with the business cycle at about five yearly intervals, the forecast profits over the next three "recession" years would be about fifty percent too high. It was said that, even if improved forecasts could be obtained, they could not be issued and used because of the consequences in the "City"! Hence totally

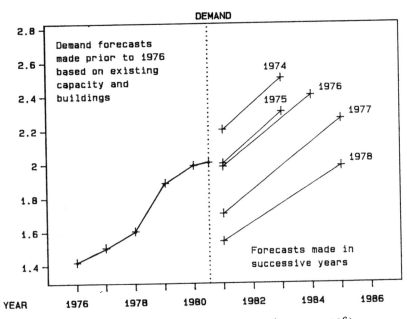

Figure 1.7. Official UK ethylene forecasts (tonnes×10^6).

inappropriate forecasts were adopted with great loss in efficiency as projects were cancelled, workers laid off, etc. It is evident that the profits figures show a cyclic behaviour typical of the second order dynamics generated by reactive control trying to force a system to violate its natural characteristics. Essentially, over this period, the UK government was operating on a deflation/reflation policy governed by the state of UK unemployment and trade balance, unions would take advantage of boom periods to negotiate inappropriate wage rates and "golden" shifts, and companies would take actions to expand activities and profits.

Figure 1.9 (from Harrison and Smith, 1980) gives an insight into what was happening. In 1971 this view led to the forecast that the mid 1970's would see the deepest depression since the 1930's. At the time a bet was made that UK unemployment would exceed 2 million before 1980. Though this figure was not officially achieved until later, it was conceded that we had won in 1980 since, in order to depress the unemployment satistics, the government had introduced new methods of compilation which omitted previously included groups. However, we would not accept that we had won since we argued that a good forecaster would have anticipated manipulation of the figures!

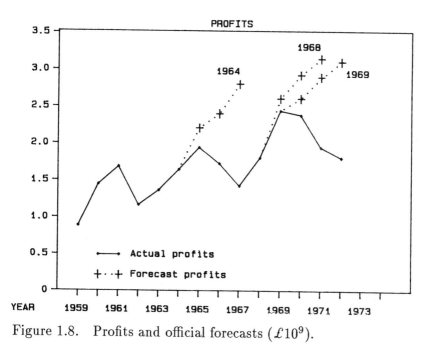

Figure 1.8. Profits and official forecasts ($£10^9$).

1.2.7 Macro and micro forecasts

The production of the above profit figures is interesting and is illustrative of a major point. The process was conducted from the bottom up over hundreds if not thousands of products. Individual point forecasts would be produced, generally judged on their merits as sales targets and summed through a hierarchy into a single figure, which could be amended by the Treasurer but was rarely altered in any significant way. The merits of this system are that it will almost guarantee that top management get a forecast which satisfies their desires and for which they can shelve a good measure of responsibility. The properties of the forecast system are that it focusses on point forecasts which are sales targets, and as such are nearly always greater than the immediate previous achievement, and that it completely ignores the major factors causing profit variation.

This last point is critical for a forecaster. For example, in 1970 within this company there was much argument about the effect of the business cycle. One camp would argue that because "it didn't affect product A, or product B, or any other individual products, then it did not exist." Others would say that it quite clearly existed if one

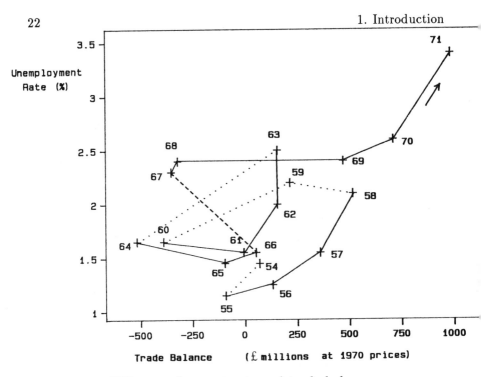

Figure 1.9. UK unemployment rate and trade balance.

looked at total business. The difference of opinion arises because of
a lack of appreciation of the law of large numbers and was simply
demonstrated as follows. Suppose at time t that there are 1000
products with sales Y_{it}, $(i = 1 \ldots, 1000)$, modelled simply as

$$Y_{it} = \mu_{it} + B_t + \epsilon_{it}$$

where the μ's are known means, the ϵ_{it} are independent error terms
with zero means and variances 99, denoted by $\epsilon_{it} \sim [0, 99]$, and
$B_t \sim [0, 1]$ is a common and independent source of variation. Clearly
each Y_{it} has variance 100, $V[Y_{it}] = 100$, and the common factor B_t is
negligible accounting only for 1% of the variance of the individuals
or 0.5% of the standard deviation. Now consider the total sales S_t,
given by

$$S_t = \mu + 1000 B_t + \sum_{i=1}^{1000} \epsilon_{it}.$$

Then the variance of S_t is

$$V[S_t] = 10^6 V[B_t] + \sum_{i=1}^{1000} V[\epsilon_{it}] = 1099 \times 10^3.$$

Now it is seen that the common factor B_t accounts for 91% of the total variance and over 95% of the total standard deviation. The example can be generalised but the key point is quite clear and vital for a modeller, namely that factors which dominate a system at one level of aggregation may be insignificant at other levels and *vice-versa* (Green and Harrison, 1972; Harrison, 1985).

One important message to be drawn from this is that a model should focus on a limited range of questions and not try to answer both macro and micro questions unless it is particularly well structured hierarchically with models within models. On the otherhand forecasts from a range of models for parts, subtotals and totals etc. often need to be consistent. Appreciating how to combine and constrain forecasts is a necessary skill for a forecaster. In the above example, stock control forecasts for the thousand individual products may be produced individually using simple time series model. However, for macro stock decisions a model of the total business needs to be made which can then anticipate major swings in product demand which in turn can effect the total stock policy by their translation into action at the individual product level.

1.3   BAYESIAN MODELLING AND FORECASTING

1.3.1   Preliminaries

We are primarily concerned with modelling and forecasting single time series of observations deriving from a system or process of interest. Much of the book is devoted to the study of the mathematical and statistical structure of classes of dynamic models and their analyses. Whilst the focus is of necessity on detailed features of particular models and their analyses, readers should not lose sight of the wider considerations relevant to real life modelling and forecasting as discussed above. This noted, a full understanding of detailed structure is necessary in order to appreciate the implications for application.

Bayesian statistics is founded on the fundamental premise that all uncertainties be represented and measured by probabilities. The primary justification for this lies in the formal, axiomatic development of the normative framework for rational (or coherent) behaviour of individuals in the face of uncertainty. Any individual desiring to behave (at least in principle) in this way is led to act as if his/her uncertain knowledge is represented using subjective probability. The Bayesian paradigm, in addition to ensuring coherence, then provides

simple rules for the management of uncertainties, based on the laws
of probability. Analyses of complex problems, with possibly many
different but interacting sources of uncertainty, become problems of
mathematical manipulation, and so are, in principle, well-defined.
The laws of probability apply to produce probabilistic inferences
about any quantity, or collection of quantities, of interest. In forecasting over time, such quantities may be future values of time series
and quantities involved in models of the processes generating the
time series. Related forecasts are the predictive probability distributions derived from such models. Thus Bayesian forecasting involves
the provision of forecast information in terms of such distributions
that represent and summarise current uncertain knowledge (or belief). *All* probabilities are subjective. The beliefs represented are
those of an individual, the forecaster or modeller responsible for the
provision of forecast information.

Many underlying principles and features of the Bayesian approach
are identified and described throughout the book in the contexts of
various forecasting models and problems. Routine manipulation of
collections of probability distributions to identify those relevant to
forecasting and related inferences are performed repeatedly. This
provides a useful introduction to Bayesian ideas more generally for
the inexperienced reader, though is no substitute for more substantial introductions such as can be found in the works of Lindley (1965),
De Groot (1971) and Berger (1985).

1.3.2   Basic notation

Notation is introduced in the context of modelling a series of real-
valued quantities observed over time, a generic value being denoted
by Y. The time index t is used as a suffix for the time series, so
that Y_t denotes the tth value of the series. We adopt the convention
that observations begin at $t = 1$, the series developing as $Y_1, Y_2, \ldots$,
or $Y_t, (t = 1, 2, \ldots,)$. There is no need for the series to be equally
spaced in time although for convenience Y_t will often be referred to
as the observation at time t. We do not distinguish between random quantities and their outcomes or realised values. Thus, prior
to observing the value of the series at time t, Y_t denotes the unknown, or, rather more appropriately, *uncertain* random quantity,
which becomes known, or certain, when observed. The context and
notations used in the specification of models and probability distributions make the distinctions as necessary.

Throughout the book we assume that all probability distributions, discrete or continuous, have densities defined with respect to Lebesgue or counting measures as appropriate, preferring to work in terms of density functions. In the main we use continuous distributions with continuous densities, models being primarily based on standard and familiar parametric forms such as normal, Student T, and so forth. Fuller notational considerations are given in the first Section of Chapter 16, a general mathematical and statistical Appendix to the book. Some basic notation is as follows. Density functions are denoted by $p(.)$, and labelled by their arguments. Conditioning events, information sets and quantities are identified as necessary in the argument, following a vertical line. For example, the distribution of the random quantity Y conditional on an information set D has density $p(Y|D)$, and that given D plus additional information denoted H is simply $p(Y|D,H)$ or $p(Y|H,D)$. The joint density of two random quantities Y and X given D is $p(Y,X|D)$ or $p(X,Y|D)$. Conditional upon X taking any specified value (either hypothetical or actually observed), the conditional density of Y given both D and X is simply $p(Y|X,D)$ or $p(Y|D,X)$. Note here that the distinction between X as an uncertain random quantity and as a realised value is clear from the notation. In referring to the distributions of random quantities, we use this notation without the p. Thus we talk of the distributions of Y, $(Y|D)$, $(Y,X|D)$ and $(Y|X,D)$, for example. Expectations of functions of random quantities are denoted by $E[.]$; thus $E[Y|D]$ is the conditional expectation, or mean, of $(Y|D)$. Variances and covariances are represented using $V[.]$ and $C[.,.]$ respectively; thus $V[Y|D]$ is the variance of $(Y|D)$ and $C[Y,X|D]$ the covariance of $(Y,X|D)$.

We generally follow the convention of using Roman characters, both upper and lower case, for quantities that are either known or uncertain but observable. Thus, at any time point t, the observed values of the time series $Y_1, Y_2, \ldots, Y_t$ will usually be known, those of $Y_{t+1}, Y_{t+2}, \ldots$, being uncertain but (at least potentially) observable at some future time. Random quantities that are assumed uncertain and unobservable are denoted by Greek characters, and referred to as unknown model parameters. For example, the mean and variance of a random quantity will be denoted using Roman characters if assumed known, Greek if assumed to be unknown parameters of the distribution. In the latter case, we will usually make the uncertain nature of the parameters explicit by including them in the conditioning of distributions. Thus $p(Y|\mu, D)$ is the density of Y given the information set D and, in addition, the value of an uncertain random quantity μ. On the other hand, we do not usually explictly recognise all assumed quantities in the conditioning.

Notation for specific distributions is introduced and referenced as necessary. Distributions related to the multivariate normal play a central role throughout the book, so the notation is mentioned here (refer to Sections 16.2 and 16.3 for further details). In particular, $(Y|D) \sim N[m, V]$ when $(Y|D)$ is normally distributed with (known) mean m and (known) variance V. Then $X = (Y - m)/\sqrt{V}$ has a standard normal distribution, $(X|D) \sim N[0, 1]$. Similarly, $(Y|D) \sim T_k[m, V]$ when $(Y|D)$ has a Student T distribution on k degrees of freedom, with mode m and scale V. If $k > 1$ then $E[Y|D] = m$; if $k > 2$ then $V[Y|D] = Vk/(k-2)$. $X = (Y - m)/\sqrt{V}$ has a standard T distribution, $(X|D) \sim T_k[0, 1]$. Similar notation applies to other distributions specified by a small number of parameters. In addition, we use $(Y|D) \sim [m, V]$ when $E[Y|D] = m$ and $V[Y|D] = V$, whatever the distributional form may be.

Vector and matrix quantities are denoted by bold typeface, whether Roman or Greek. Matrices are always in upper case, vectors may or may not be depending on context. The above notation for distributions has the following multivariate counterparts (see Section 16.2 for further details). Let $\mathbf{Y}$ be a vector of n random quantities having some joint distribution, $\mathbf{m}$ a known n−dimensional vector and $\mathbf{V}$ a known $n \times n$ non-negative definite matrix. Then $(\mathbf{Y}|D) \sim N[\mathbf{m}, \mathbf{V}]$ when $(\mathbf{Y}|D)$ has a multivariate normal distribution with (known) mean $\mathbf{m}$ and (known) variance matrix $\mathbf{V}$. Similarly $(\mathbf{Y}|D) \sim T_k[\mathbf{m}, \mathbf{V}]$ when $(\mathbf{Y}|D)$ has a multivariate Student T distribution on k degrees of freedom, with mode $\mathbf{m}$ and scale matrix $\mathbf{V}$. If $k > 1$ then $E[\mathbf{Y}|D] = \mathbf{m}$; if $k > 2$ then $V[\mathbf{Y}|D] = \mathbf{V}k/(k-2)$. Finally, $(\mathbf{Y}|D) \sim [\mathbf{m}, \mathbf{V}]$ when $E[\mathbf{Y}|D] = \mathbf{m}$ and $V[\mathbf{Y}|D] = \mathbf{V}$, with the distributional form otherwise unspecified.

1.3.3 Dynamic models

Mathematical and statistical modelling of time series processes is based on classes of *Dynamic Models*, the term *dynamic* relating to changes in such processes due to the passage of time as a fundamental motive force. The most widely known and used sublass is that of *Normal Dynamic Linear Models*, referred to more simply as *Dynamic Linear Models*, or *DLMs*, when the normality is understood. This class of models forms the basis for much of the development in the book.

The fundamental principles used by a Bayesian forecaster in structuring forecasting problems through dynamic modelling comprise

- sequential model definition;
- structuring using parametric models;
- probabilistic representation of information about parameters;
- forecasts derived as probability distributions.

A sequential model definition and structuring is natural in the time series context. As time evolves, information relevant to forecasting the future is received and may be used by a forecaster in revising his/her views, whether this revision be at the quantitative, form or conceptual levels in the general model structure $\mathcal{M}$ of Section 1.1. The sequential approach focuses attention on statements about the future development of a time series conditional on existing information. Suppose, with no loss of generality, that the time origin $t = 0$ represents the current time, and that existing information available to, and recognised by, a forecaster is denoted by the

$$\textit{Initial information set:} \qquad D_0.$$

This represents all the available relevant starting information which is used to form initial views about the future, including history and all defining model quantities. In forecasting ahead to any time $t > 0$, the primary objective is calculation of the forecast distribution for $(Y_t|D_0)$. Similarly, at any time t, statements made concerning the future are conditional on existing information at that time. Generally, we have the

$$\textit{Information set at time } t: \qquad D_t.$$

Thus statements made at time t about any random quantities of interest are based on D_t; in particular, forecasting ahead to time $s > t$ involves consideration of the forecast distribution for $(Y_s|D_t)$. As time evolves, so does the forecaster's information. Observing the value of Y_t at time t implies that D_t includes both the previous information set D_{t-1} and the observation Y_t. If this is all the relevant information, then $D_t = \{Y_t, D_{t-1}\}$, although rather often, as discussed earlier, further relevant information will be incorporated by the forecaster in revising or updating his/her view of the future. Denoting all additional relevant information by I_t at time t, we have the

$$\textit{Information updating:} \qquad D_t = \{I_t, D_{t-1}\}, \qquad (t = 1, 2, \dots)$$

The sequential focus is emphasised through the use of statistical models for the development of the series into the future described via distributions for $Y_t, Y_{t+1}, \ldots$, conditional on past information D_{t-1}. Usually such distributions depend on defining parameters determining moments of distributions, functional relationships, forms of distributions, and so forth. Focusing on one-step ahead, the beliefs of the forecaster are then structured in terms of a parametric model,

$$p(Y_t \mid \boldsymbol{\theta}_t, D_{t-1}),$$

where $\boldsymbol{\theta}_t$ is a defining parameter vector at time t. This mathematical and statistical representation is the language that provides communication between the forecaster, model and decision makers. As such, the parameters must represent constructs meaningful in the context of the forecasting problem. Indexing $\boldsymbol{\theta}_t$ by t indicates that the parametrisation may be dynamic. In addition, although often the number and meaning of the elements of $\boldsymbol{\theta}_t$ will be stable, there are occasions on which $\boldsymbol{\theta}_t$ will be expanded, contracted or changed in meaning according to the forecaster's existing view of the time series. This is true in particular of *open systems*, such as typically arise in social, economic and biological environments, where influential factors affecting the time series process are themselves subject to variation based on the state of the system generating the process. In such cases changes in $\boldsymbol{\theta}_t$ may be required to reflect system learning and the exercise of purposeful control. Such events, although recognisable when they happen, may be difficult to anticipate and so will not typically be included in the model until occurrence.

The model parameters $\boldsymbol{\theta}_t$ provide the means by which information relevant to forecasting the future is summarised and used in forming forecast distributions, and the learning process involves sequentially revising the state of knowledge about such parameters. Probabilistic representation of all uncertain knowledge is the keystone of the Bayesian approach, whether such knowledge relates to future, potentially observable quantities or unobservable model parameters. At time t, historical information D_{t-1} is summarised through a *prior* distribution for future model parameters (prior, that is, to observing Y_t, but *posterior* of course, to learning D_{t-1}). The prior density $p(\boldsymbol{\theta}_t \mid D_{t-1})$ (and the posterior $p(\boldsymbol{\theta}_t \mid D_t)$), provides a concise, coherent and effective transfer of information on the time series process through time. In addition to simply processing the information deriving from observations and feeding it forward to forecast future development, this probabilistic encoding allows new information from external sources to be formally incorporated in the system. It also naturally extends to allow for expansion or contraction of the parameter vector in open systems, with varying degress of uncertainty associated with the effects of such external interventions and changes. Further, inferences about system development and change are directly drawn from components of these distributions in a standard statistical manner.

Two very simple models, the subjects of Chapters 2 and 3, exemplify the class of dynamic models generally, and give concrete settings for the above general ideas.

EXAMPLE 1.4. An archetype statistical model assumes that observations are independent and identically normally distributed, denoted by $(Y_t|\mu) \sim N[\mu, V]$, $(t = 1, 2, \dots,)$. Changes over time in the mean (and sometimes the variance) of this model are natural considerations and typically unavoidable features when the observations are made on a process or system that is itself continually evolving. Time variation may be slow and gradual, reflecting continuous changes in environmental conditions, and more abrupt due to marked changes in major influential factors. For example, a normal model may be assumed as a basically suitable representation of the random variation in "steady" consumer demand for a product, but the *level* of demand will rarely remain absolutely constant over time. A simple extension of the archetype to have a time varying mean provides a considerable degree of flexibility. Subscripting μ by t we could then write $(Y_t|\mu_t) \sim N[\mu_t, V]$, or

$$Y_t = \mu_t + \nu_t,$$

where μ_t represents the level of the series at time t, and $\nu_t \sim N[0, V]$ random, *observational* error or noise about the underlying level. One of the simplest, non-degerate ways in which the level can be modelled as changing through time is according to a *random walk*. In such a case, time evolution of the level, defining the process model, is represented as

$$\mu_t = \mu_{t-1} + \omega_t,$$

where ω_t represents purely random, unpredictable changes in level between time $t - 1$ and t. A standard DLM assumes, for example, a normal random walk model with $\omega_t \sim N[0, W]$ for some variance W. Chapter 2 is devoted to the study of such models, which find wide application in short term forecasting. In product demand forecasting

and inventory management, for instance, underlying levels may be
assumed to be *locally* constant, but expected to change rather more
significantly over longer periods of time. The zero-mean and independence assumptions for the w_t series are consistent with a view
that this longer term variation cannot be predicted, being described
as purely stochastic.

With reference to the general discussion above, if the variances V
and W are assumed known then the model parameter θ_t at time t
represents the uncertain level μ_t alone. Otherwise θ_t may include
uncertain variances for either or both of v_t and w_t. With known
variances, their values are included in the initial information set D_0
and then, with $\theta_t = \mu_t$ for all t, $p(Y_t|\theta_t, D_{t-1}) = p(Y_t|\mu_t, D_0)$ is the
normal density of $(Y_t|\mu_t, D_0) \sim N[\mu_t, V]$. Historical information at
time t leads to the prior distribution for $\theta_t = \mu_t$ given D_{t-1}, the
information set including all past values of the Y series. In demand
forecasting, for example, historical values of the Y series may lead
the forecaster to believe that next month's demand level, μ_t, is most
likely to be in the region of about 250, but is unlikely to be below
230 or to exceed 270. One possible representation of this prior view
is that

$$(\mu_t|D_{t-1}) \sim N[250, 100],$$

having an expected and most likely value of 250, and about 95%
probability of lying between 230 and 270. This sort of model structuring is developed extensively in Chapter 2.

EXAMPLE 1.5. Regression modelling is central to much of statistical practice. Regression is concerned with the construction of
quantitative description of relationships between observables, such
as between two time series. Consider a second time series represented by observations X_t, $(t = 1, 2, \ldots,)$, observed contempora-
neously with Y_t. Regression modelling often focuses on the extent
to which changes in the mean μ_t of Y_t can be explained through
X_t, and possibly past values, X_s for $s < t$. Common terminology
refers to Y_t as the response or dependent variable series, and X_t as
the regressor or independent variable series. Then μ_t is the mean
response, related to the regressor variable through a mean response
function $\mu_t = \mu_t(X_t, X_{t-1}, \ldots)$ defining the regression. For example, a simple linear model for the effect of the current X_t on the
current mean is $\mu_t = \alpha + \beta X_t$, where the defining parameters α and

β take suitable values. Models of this sort find wide use in prediction, interpolation, estimation and control contexts. Construction of models in practice is guided by certain objectives specific to the problem area, such as short-term forecasting of the response series. For some such purposes, simple linear models as above may well be satisfactory *locally* but, as in the previous example, are typically unlikely to adequately describe the relationship globally ie. as time evolves *and* as X_t varies. Flexibility in modelling such changes can be provided by simply allowing for the possibility of time variation in the coefficients, viz

$$\mu_t = \alpha_t + \beta_t X_t.$$

Thus although the *form* of regression model is linear in X_t for all t, the quantified model may have different defining parameters at different times. The distinction between an appropriate *local model form* and an appropriate *quantified local model* is critical and is a fundamental concept in dynamic modelling in general. Often the values of the independent variable X_t change only slowly in time and an appropriate local model description is that above, $\mu_t = \alpha_t + \beta_t X_t$, where the parameters vary only slightly from one time point to the next. This may be modelled as in the previous example using random walk type evolutions for the defining parameters, as in

$$\alpha_t = \alpha_{t-1} + \delta\alpha_t,$$
$$\beta_t = \beta_{t-1} + \delta\beta_t,$$

where $\delta\alpha_t$ and $\delta\beta_t$ are zero-mean error terms. These evolution equations express the concept of *local constancy* of the parameters, subject to variation controlled and modelled through the distribution of the evolution error terms $\delta\alpha_t$ and $\delta\beta_t$. Small degrees of variation here imply a stable linear regression function over time, larger values modelling greater volatility and suggesting caution in extrapolating or forecasting too far ahead in time based on the quantified linear model at the current time. The usual static regression model, so commonly used, is obviously obtained as a special case of this dynamic regression when both evolution errors are identically zero for all time.

Finally, the primary goals of the forecaster are attained by directly applying probability laws. The above components provide one representation of a joint distribution for the observations and parameters, namely that defined via

$$p(Y_t, \boldsymbol{\theta}_t \mid D_{t-1}) = p(Y_t \mid \boldsymbol{\theta}_t, D_{t-1}) p(\boldsymbol{\theta}_t \mid D_{t-1}),$$

from which the relevant one-step forecast may be deduced as the marginal

$$p(Y_t \mid D_{t-1}) = \int p(Y_t, \boldsymbol{\theta}_t \mid D_{t-1}) d\boldsymbol{\theta}_t.$$

Inference for the future Y_t is now a standard statistical problem of summarising the information encoded in the forecast distribution. In connection with end use, loss or utility structures may be imposed to derive coherent optimal decision policies in activities dependent on the unknown future values of the series.

1.4 HISTORICAL PERSPECTIVE AND BIBLIOGRAPHIC COMMENTS

Our approach to modelling and forecasting synthesises concepts, models and methods whose development has been influenced by work in many fields. It would be impossible to fully document historical influences. As we have already discussed, much is involved in modelling that cannot be routinely described using formal, mathematical structures, particularly in the stages of model formulation, choice and criticism. However, in line with the fundamental concepts of scientific method, we identify the Bayesian approach as the framework for routine learning and organisation of uncertain knowledge within complete forecasting systems. Over the last fifty years there has been rapidly increasing support for the Bayesian approach to scientific learning and decision making, with notable recent acceptance by practitioners driven by increased perception outside academia of the commonsense principles on which it is founded. Axiomatic foundations notwithstanding, decision makers find it natural to phrase beliefs as normed or probability measures (as do academics, though some prefer not to recognise the fact). This seems to have been done ever since gambling started — and what else is decision making except compulsory gambling? We all face one-off decisions that have to be made with no chance of repeating the experience, so the value of classical statistics is very limited for decision makers.

Important influences on current Bayesian thinking and practice may be found, in particular, in the books of Box and Tiao (1973), De Finetti (1974/75), De Groot (1971), Jeffreys (1961), Lindley

(1965), and Savage (1954). A notable recent contribution is Berger (1985), which includes a comprehensive bibliography.

Concerning approaches to practical forecasting, little had been done in industry and government in the way of mathematical or socio-economic modelling before the arrival of computers in the 1950's. Exponentially weighted moving averages (EWMAs) and Holt's linear growth and seasonal model (1957) began to find use in the mid to late 1950's in forecasting for stock control and production planning with one of us (PJH) involved in early use at Imperial Chemical Industries Ltd. (ICI). In this company, Holt's method became established, and apart from taking over the routine of forecasting, led to great improvements in the accuracy of forecasts. In fact the methods were used in conjunction with interventions so that a complete forecasting system was in operation at that time. This involved the Product Sales Control Department in adjusting the computer forecasts whenever they felt it necessary, though these adjustments did not always improve the forecasts, probably because hunches were included as well as definite information. Methods developed then are still in use now in sales forecasting and stock control.

The work of Brown in the late 1950's (Brown, 1959, 1962), promoting the use of discount techniques in forecasting, was and still is a major influence for practitioners. A parsimonious concept anchored to a fundamental familiar principle of discounting is very attractive. Work at ICI at the start of the 1960's used the principle of discounting in models for series with trend and seasonality, but applied two discount factors, one for trend and one for seasonality. As far as we know this was the first development of multiple discounting, incorporated in two forecasting programs, Seatrend and Doubts, in 1963, presented at the Royal Statistical Society's Annual conference held at Cardiff in 1964 and later published in Harrison (1965). However, what will not be known by a literature search is what really happened in development. The Society had asked for a paper to be circulated in advance and this presented the multiple discount methods but concluded that, because of simplicity (parsimony was known about then and before, as Occam knew), the single discount factor employed in Brown's Exponentially Weighted Regression (EWR) approach was preferable. However, because of the difficulties in programming, numerical comparisons were unavailable at the time of writing. Almost as soon as the paper was sent to the RSS the results were ready, and surprisingly indicated the enormous improvements to be obtained using multiple discounting. This lead to a reversal of the issued conclusions and a quick seven page addendum was written and circulated at the conference, though the published paper shows no trace of this amusing history. Basically, the conclusion was that different model components need discounting at different rates — a view which dominates our current approach to structure and discounting in Bayesian Forecasting.

Further developments at ICI were described in a paper presented to the Operational Research Society Conference in 1965, the manuscript never being published but available in its original form as a Warwick Research Report (Harrison and Scott, 1965). The contents will surprise many people; they include discussion of Complete Forecasting Systems emphasising intervention and monitoring and present several specific dynamic models. Some of the modelling work was based on developments by Muth (1960), Nerlove and Wage (1964), Thiel and Wage (1964), and Whittle (1965), though a major concern was with more general models with an emphasis on sensitivity to departures from optimal adaptive parameters, partially motivated by unfounded inexperienced opposition to discounting from certain academics. The monitoring work in that paper was largely based on the Backward Cusum controller developed initially in 1961 from Cusums in quality control and published in Harrison and Davies (1964). Much of this followed developments by Page (1954) and Barnard (1959), and the subsequent work of the ICI through Woodward and Goldsmith (1964) and Ewan and Kemp (1960).

So by 1964, simple parametrised, structural models were in use, employing double discounting, and emphasising the approach through complete forecasting systems operating according to the principle of Management by Exception, enabling prospective and retrospective interventions. By the end of the 1960's, the basic models had been extended and generalised to a wide class of dynamic linear models (though not then referred to as such), and to multi-process models which were to model outliers and sudden changes in structure. Some indication of the extent of the developments, both theoretical and practical, was then reported in Harrison and Stevens (1971). Implemented forecasting systems based on this took full advantage of the intervention facilities of the Bayesian approach and also performed automatic model discrimination using multi-process models. Important also for the use of subjective and quantitative information is the application in the mail order business reported rather later in Green and Harrison (1973).

At about the same time in 1969, it becamce clear that some of the mathematical models were similar to those used in engineering control. It is now well-known that, in normal DLMs with known variances, the recurrence relationships for sequential updating of posterior distributions are essentially equivalent to the Kalman Filter equations, based on the early work of Kalman (1960, 1963) in engineering control, using a minimum variance approach. It was clearly not, as many people appear to believe, that Bayesian Forecasting is founded upon Kalman Filtering (see Harrison and Stevens, 1976a, and discussion; and reply to the discussion by Davis of West, Harrison and Migon, 1985). To say that "Bayesian forecasting is Kalman Filtering" is akin to saying that statistical inference is regression!

The late 1970's and 1980's have seen much further development and application of Bayesian forecasting models and methods. Notable amongst these are procedures for variance learning (West, 1982; Smith and West, 1983), discounting (Ameen and Harrison, 1984, 1985a and b), monitoring and intervention (West, 1986; West and Harrison, 1986; West, Harrison and Pole, 1987), non-normal and non-linear model structures (Souza, 1981; Smith, 1979; Migon, 1984; Migon and Harrison, 1985; West, Harrison and Migon, 1985), reference analyses (Pole and West, 1987), and many others, with applications in a variety of areas. There are various continuing developments. Computing developments have led to wider usage, easing communication with less technically orientated practitioners. We do hold the view that modelling is an art and particularly so is Bayesian Forecasting. Any software package is just that; a package of specific, selected facilities, and the limitations of such software are too easily seen by some as the limitations of the whole approach. Indeed, the early Bayesian Forecasting package **SHAFT**, produced in the 1970's by Colin Stevens, led to a widespread view that Bayesian Forecasting was the single model discussed in Harrison and Stevens (1971)! This is like defining an animal as a fox! However, it is now abundantly clear that that to treat the armoury of the Bayesian approach as a paintbox without even a numbered canvas is not going to get many pictures painted. Some results of our approaches to the problem using the computer language APL are two packages **FAB** and **BATS**, with some further developments underway on others. **FAB**, (Harrison, West and Pole, 1987), is a training package specifically designed for teaching and training in Bayesian forecasting. **BATS**, (West, Harrison and Pole, 1987a, b) is more orientated towards application and experimentation with a subset of DLMs — although already people are misinterpreting this class of models as Bayesian Forecasting!

The above discussion concentrates rather closely on what we identify as direct influences on Bayesian forecasting and dynamic modelling as we present the subject in this book. There has, of course, been considerable development of dynamic modelling and related forecasting techniques outside the Bayesian framework, particularly in control engineering. Good references include Anderson and Moore (1979), Jazwinski (1970), Sage and Melsa (1971), for example. Details of work by statisticians, econometricians and others may be found, with useful references, in Duncan and Horne (1972), Harvey (1981), Spall (1988), Thiel (1971), and Young (1984), for example.

CHAPTER 2

INTRODUCTION TO THE DLM:
THE FIRST-ORDER POLYNOMIAL MODEL

2.1 INTRODUCTION

Many important underlying concepts and analytic features of dynamic linear models are apparent in the simplest and most widely used case of the **first-order polynomial model**. By way of introduction to DLMs, this case is described and examined in detail in this Chapter. The first-order polynomial model is the simple, yet non-trivial, time series model in which the observation series Y_t is represented as $Y_t = \mu_t + \nu_t$, μ_t being the current **level** of the series at time t, and $\nu_t \sim N[0, V_t]$ the **observational error** or noise term. The **time evolution** of the level of the series is a simple random walk $\mu_t = \mu_{t-1} + \omega_t$, with **evolution error** $\omega_t \sim N[0, W_t]$. This latter equation describes what is often referred to as a *locally constant mean model*. Note the assumption that the two error terms, observational and evolution errors, are normally distributed for each t. In addition we adopt the assumptions that the error sequences are independent over time and mutually independent. Thus, for all t and all s with $t \neq s$, ν_t and ν_s are independent, ω_t and ω_s are independent, and ν_t and ω_s are independent. Further assumptions at this stage are that the variances V_t and W_t are known for each time t. Figure 2.1 shows two examples of such Y_t series together with their underlying μ_t processes. In each the starting value is $\mu_0 = 25$, and the variances defining the model are constant in time, $V_t = V$ and $W_t = W$, having values $V = 1$ in both cases and evolution variances (a) $W = 0.05$, (b) $W = 0.5$. Thus in (a) the movement in the level over time is small compared to the observational variance, $W = V/20$, leading to a typical locally constant realisation, whereas in (b) the larger value of W leads to greater variation over time in the level of the series.

The notation for this model in terms of observational and evolution equations is as follows. For each $t = 1, 2, \ldots,$

$$(Y_t \mid \mu_t) \sim N[\mu_t, V_t]$$
$$(\mu_t \mid \mu_{t-1}) \sim N[\mu_{t-1}, W_t].$$

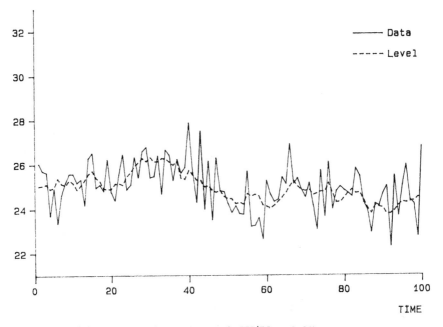

Figure 2.1(a).   Example series with $W/V = 0.05$.

This model is used effectively in numerous applications, particularly in short term forecasting for production planning and stock control. In modelling market demand for a product, for example, μ_t represents *true* underlying market demand at time t. It is supposed that *locally* in time, that is a few periods forward or backwards, the demand level is roughly constant. Significant changes over longer periods of time are expected but the zero-mean and independent nature of the ω_t series imply that the modeller does not wish to anticipate the form of this longer term variation, merely describing it as a purely stochastic process. It is often useful to think of the underlying demand process as a smooth function of time $\mu(t)$ with an associated Taylor series representation $\mu(t + \delta t) = \mu(t) +$ higher order terms. The model then simply describes the higher order terms as zero-mean noise. This is the genesis of the **first-order polynomial** terminology; the model is a local linear (= first-order polynomial) proxy for an underlying smooth evolution. The model is also sometimes, although perhaps a little misleadingly, referred to as a *steady* model. A basic feature of the model, that provides a guide as to the suitability of the model for particular applications, is that, projecting k steps into the future from time t, the expected value of the

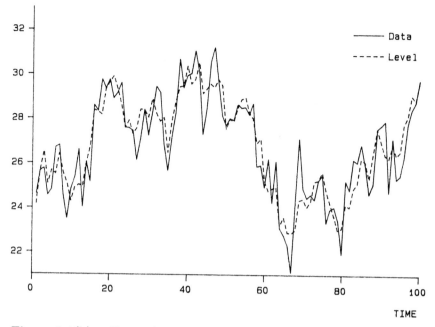

Figure 2.1(b).   Example series with $W/V = 0.5$.

series at that time conditional on the current level is simply

$$E[Y_{t+k} \mid \mu_t] = E[\mu_{t+k} \mid \mu_t] = \mu_t.$$

In practice the forecaster's posterior distribution for μ_t given D_t, the existing information at time t including the history of the series up to and including time t, will have a mean m_t depending on past data so that the **forecast function** $f_t(k)$ is given by

$$f_t(k) = E[Y_{t+k} \mid D_t] = E[\mu_t \mid D_t] = m_t$$

for all $k > 0$ and the expected future level of the series is constant. Consequently such a model is only useful for short-term application, and particularly in cases in which the sampling variation, as measured by V_t, is considerably greater than the systematic variation in the level, measured by W_t. We return to such design considerations later in Section 2.3.

2.2   THE DLM AND RECURRENCE RELATIONSHIPS

2.2.1   Definition

The formal definition of the model for each time t is as follows. Recall that, for all t, including $t = 0$, D_t represents all the available information, including past values of the series.

Definition 2.1. *For each t, the model is defined by:*

Observation equation: $Y_t = \mu_t + \nu_t,$ $\nu_t \sim N[0, V_t],$

System equation: $\mu_t = \mu_{t-1} + \omega_t,$ $\omega_t \sim N[0, W_t],$

Initial information: $(\mu_0 \,|\, D_0) \sim N[m_0, C_0],$

where the error sequences ν_t and ω_t are independent, and mutually independent. In addition, they are independent of $(\mu_0 \mid D_0)$.

The third component here is the probabilistic representation of the forecaster's beliefs and information about the level at time $t = 0$ given information at the time, D_0. The mean m_0 is an estimate of the level and the variance C_0 a measure of uncertainty about the mean. Concerning the information sets, we assume here that D_{t-1} comprises D_0, the values of the variances V_t, and W_t for all t, and the values of the observations $Y_{t-1}, Y_{t-2}, \ldots, Y_1$. Thus the only new information becoming available at any time is the value of the time series observation so that $D_t = \{Y_t, D_{t-1}\}$.

2.2.2   Updating equations

For forecasting, control and learning, the key model components are the distributions summarised in the following Theorem. These are based on the sequential updating equations that define the evolution of information about the model and the series over time.

Theorem 2.1. *For each t we have the following one-step forecast and posterior distributions.*

(a) *Posterior for μ_{t-1} :*
$$(\mu_{t-1} \mid D_{t-1}) \sim N[m_{t-1}, C_{t-1}]$$
for some mean m_{t-1} and variance C_{t-1};

(b) *Prior for μ_t :*
$$(\mu_t \mid D_{t-1}) \sim N[m_{t-1}, R_t]$$
where $R_t = C_{t-1} + W_t$;

(c) *1-step forecast:*
$$(Y_t \mid D_{t-1}) \sim N[f_t, Q_t],$$
where $f_t = m_{t-1}$ and $Q_t = R_t + V_t$;

(d) *Posterior for μ_t :*
$$(\mu_t \mid D_t) \sim N[m_t, C_t],$$
with $m_t = m_{t-1} + A_t e_t$ and $C_t = A_t V_t$,
where $A_t = R_t / Q_t$, and $e_t = Y_t - f_t$.

Proof: The results may be derived in different ways. We consider two methods of deriving (d), each of which is instructive and illuminating in its own right. The first, and most important, uses the generally applicable Bayes' Theorem and standard Bayesian calculations. The second method, appropriate for all DLMs, provides an elegant and instructive derivation using the additivity, linearity and distributional closure properties of the normal linear structure. This latter method extends to all normal DLMs in later chapters although the former is required as a general approach for non-normal models. These results are standard in Bayesian normal theory as detailed in some generality in Section 16.2 of Chapter 16. In the proofs here we appeal to standard properties of the normal distribution also detailed there.

Proof is by induction. Let us suppose the distribution in (a). Then conditional on D_{t-1}, μ_t is the sum of two independent normal random quantities μ_{t-1} and ω_t, and so is itself normal. The mean and variance are obtained by adding means and variances of the summands, leading to

$$(\mu_t \mid D_{t-1}) \sim N[m_{t-1}, R_t]$$

where $R_t = C_{t-1} + W_t$. This is (b) as required. Similarly, and again conditional upon D_{t-1}, Y_t is the sum of the independent normal quantities μ_t and ν_t and so is normal,

$$(Y_t \mid D_{t-1}) \sim N[m_{t-1}, Q_t]$$

where $Q_t = R_t + V_t$, as stated in (c).

As mentioned above we derive (d) twice, using two different techniques.

(1) **Updating via Bayes' Theorem**

The Bayesian method is, of course, general, applying to all models no matter what distributional assumptions are involved. In this case, the sampling model provides the observational distribution with density

$$p(Y_t \mid \mu_t, D_{t-1}) = \exp[-(Y_t - \mu_t)^2/(2V_t)]/(2\pi V_t)^{1/2}.$$

The prior for μ_t given D_{t-1} has, from (b), the density

$$p(\mu_t \mid D_{t-1}) = \exp[-(\mu_t - m_{t-1})^2/(2R_t)]/(2\pi R_t)^{1/2}.$$

On observing Y_t, the *likelihood* for μ_t is provided as proportional to the observational density viewed as a function of μ_t, and, directly from Bayes' Theorem, the posterior for μ_t is simply given by

$$\begin{aligned}
p(\mu_t \mid D_t) &= p(\mu_t \mid D_{t-1}, Y_t) \\
&= p(\mu_t \mid D_{t-1})p(Y_t \mid \mu_t, D_{t-1})/p(Y_t \mid D_{t-1}),
\end{aligned}$$

where the denominator is the one-step forecast density evaluated at the actual observation. This posterior is of a standard Bayesian form being derived from a normal observational model with a normal prior, and the result is well-known. The calculation is simplified by concentrating on the form as a function of μ_t alone, ignoring multiplicative factors depending on the now fixed value Y_t and other constants. Doing this leads to the proportional form of Bayes' Theorem

$$\begin{aligned}
p(\mu_t \mid D_t) &\propto p(\mu_t \mid D_{t-1})p(Y_t \mid \mu_t, D_{t-1}) \\
&\propto \exp\big[-(Y_t - \mu_t)^2/(2V_t) - (\mu_t - m_{t-1})^2/(2R_t)\big].
\end{aligned}$$

As in most applications of Bayes' Theorem the natural logarithmic scale provides a simpler additive expression

$$\begin{aligned}
\ln[p(\mu_t \mid D_t)] = \ \text{constant} &- 0.5[(Y_t - \mu_t)^2/V_t \\
&+ (\mu_t - m_{t-1})^2/R_t],
\end{aligned}$$

which on rearranging gives

$$
\begin{aligned}
&= \text{constant} - 0.5[\mu_t^2/C_t - 2(m_{t-1}/R_t + Y_t/V_t)\mu_t] \\
&= \text{constant} - 0.5C_t^{-1}(\mu_t - m_t)^2,
\end{aligned}
$$

where:

(i) $C_t = 1/[1/R_t + 1/V_t] = R_t V_t/Q_t = A_t V_t$, with $A_t = R_t/Q_t = R_t/(R_t + V_t)$; and

(ii) $m_t = C_t(m_{t-1}/R_t + Y_t/V_t) = m_{t-1} + A_t e_t$, with $e_t = Y_t - m_{t-1}$.

Hence, finally,

$$
p(\mu_t \mid D_t) \propto \exp[-(\mu_t - m_t)^2/(2C_t)],
$$

and renormalisation of this density leads to a normalisation constant $1/(2\pi C_t)^{1/2}$.

(2) **Proof based on standard normal theory.**

Within this special normal, linear framework, a more specific proof based on the bivariate normal distribution (and extending to more general normal DLMs) is derived by: (i) calculating the joint distribution of Y_t and μ_t given D_{t-1}; and (ii) directly deducing the conditional distribution for μ_t given Y_t.

Any linear function of Y_t and μ_t will be a linear combination of the independent normal quantities ν_t, ω_t, and μ_{t-1}, and so, conditional on D_{t-1}, will be normally distributed. By definition then, $(Y_t, \mu_t \mid D_{t-1})$ is bivariate normal. To identify the mean vector and variance matrix note that we already have the appropriate marginal distributions in (b) and (c), $(Y_t \mid D_{t-1}) \sim N[f_t, Q_t]$ and $(\mu_t \mid D_{t-1}) \sim N[m_{t-1}, R_t]$, and the covariance is simply

$$
\begin{aligned}
C[Y_t, \mu_t \mid D_{t-1}] &= C[\mu_t + \nu_t, \mu_t \mid D_{t-1}] \\
&= V[\mu_t \mid D_{t-1}] = R_t,
\end{aligned}
$$

by additivity and using the independence of μ_t and ν_t. So the joint distribution is given by

$$
\begin{pmatrix} Y_t \\ \mu_t \end{pmatrix} D_{t-1} \sim N\left[\begin{pmatrix} m_{t-1} \\ m_{t-1} \end{pmatrix}, \begin{pmatrix} Q_t & R_t \\ R_t & R_t \end{pmatrix} \right].
$$

General multivariate normal theory (Section 16.2, Chapter 16) may be applied to obtain the required distribution conditional on Y_t. Alternatively the particular case of the bivariate normal is directly applicable. The correlation $\rho_t = R_t/(R_tQ_t)^{1/2}$ is clearly positive with $\rho_t^2 = R_t/Q_t = A_t$. Using the referenced material, the conditional distribution of $(\mu_t \mid Y_t, D_{t-1})$ is normal with mean

$$m_{t-1} + \rho_t^2(Y_t - m_{t-1}) = m_t,$$

and variance

$$(1 - \rho_t^2)R_t = R_tV_t/Q_t = A_tV_t = C_t,$$

as required. Note the interpretation of A_t, in this case, as both the regression coefficient and squared correlation.

The result (d) is now established, completing the proof when (a) is assumed to hold. The complete proof now follows by induction since (a) is true for $t = 0$ from the model definition in Definition 2.1.

$$\diamond$$

Some discussion of the various elements of the distributions is in order. e_t is the one step ahead forecast error, the difference between the observed value Y_t and the expectation f_t. A_t is the prior regression coefficient of μ_t upon Y_t and, in this particular case, is the square of their correlation coefficient; clearly $0 \leq A_t \leq 1$. The results are computationally simple and elegant due to the use of normal distributions for each model component. Using these results, sequential updating and revision of forecasts is direct. It is worth noting that an alternative representation for m_t is

$$m_t = A_tY_t + (1 - A_t)m_{t-1},$$

showing that m_t is a weighted average of the prior estimate of level m_{t-1} and the observation on the level Y_t. The weight, or **adaptive coefficient**, A_t, defining this combination lies between 0 and 1, being closer to 0 when the prior distribution is more concentrated than the likelihood with $R_t < V_t$, and closer to 1 when the prior is more diffuse, or less informative, than the likelihood. In addition, the posterior is less diffuse than the prior since $C_t < R_t$, representing an increase in information about μ_t due to the additional observation Y_t.

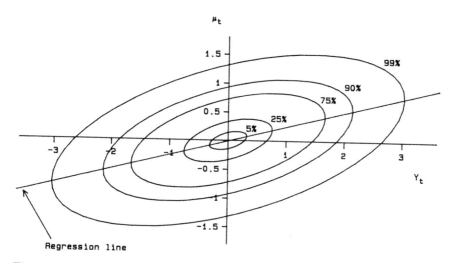

Figure 2.2. Probability contours of bivariate normal: $R_t = 0.25$ and $V_t = 0.75$.

A simple case with $m_{t-1} = 0$, $R_t = 0.25$ and $V_t = 0.75$ is illustrated in Figure 2.2 which provides a contour plot of the joint density. Here $A_t = 0.25$ and the regression line, simply m_t as a function of Y_t, is given by $m_t = Y_t/4$, plotted in the Figure.

2.2.3 Forecast distributions

At time t, the two main distributions required by the forecaster are the marginals of $(Y_{t+k} \mid D_t)$ and $(X_t(k) \mid D_t)$ where $X_t(k) = Y_{t+1} + Y_{t+2} + \cdots + Y_{t+k}$, for $k > 0$. The former is the **k-step ahead forecast**, the latter the **k-step lead time forecast**. These distributions are as follows.

Theorem 2.2. For $k > 0$, the following distributions exist:

(a) k-step ahead: $(Y_{t+k} \mid D_t) \sim \mathrm{N}[m_t, Q_t(k)]$

(b) k-step lead-time: $(X_{t+k} \mid D_t) \sim \mathrm{N}[km_t, L_t(k)]$

where

$$Q_t(k) = C_t + \sum_{j=1}^{k} W_{t+j} + V_{t+k}$$

and

$$L_t(k) = k^2 C_t + \sum_{j=1}^{k} V_{t+j} + \sum_{j=1}^{k} j^2 W_{t+k+1-j}.$$

Proof: From the evolution equation for μ_t we have, for $k > 1$,

$$\mu_{t+k} = \mu_t + \sum_{j=1}^{k} \omega_{t+j};$$

from the observational equation we therefore have

$$Y_{t+k} = \mu_t + \sum_{j=1}^{k} \omega_{t+j} + \nu_{t+k}.$$

Since all terms are normal and mutually independent, $(Y_{t+k} \mid D_t)$ is normal and the mean and variance follow directly. For the lead time, note that

$$X_t(k) = k\mu_t + \sum_{j=1}^{k} j\omega_{t+k+1-j} + \sum_{j=1}^{k} \nu_{t+j},$$

and so is obviously normal with mean $k\mu_t$. Using the independence structure of the error terms,

$$V[X_t(k) \mid D_t] = k^2 C_t + \sum_{j=1}^{k} j^2 W_{t+k+1-j} + \sum_{j=1}^{k} V_{t+j},$$

and the stated form of $L_t(k)$ follows.

◇

2.3   THE CONSTANT MODEL

2.3.1   Introduction

The special case of a model in which the observational and evolution variances are constant in time is referred to as a **constant** model. In addition, since the information set $D_t = \{Y_t, D_{t-1}\}$ for each $t \geq 1$ contains no information external to the time series, the model is called **closed** (to external information). The closed, constant, first-order polynomial model is a practically useful, although somewhat restricted, case that is of interest since it allows the derivation of important limiting results that give insight into the structure of more general DLMs, some of which are related to classical time series models and point forecasting methods in common use.

The closed, constant DLM has V and W both positive, finite and constant, a prior at the time origin $t = 0$ given by $(\mu_0 \mid D_0) \sim N[m_0, C_0]$, and, for $t \geq 1$, $D_t = \{Y_t, D_{t-1}\}$. An important constant is the ratio $r = W/V$, which plays a role related to the concept of *signal to noise* ratios in engineering terminology, measuring relative variation of the state or system evolution equation to the observation equation. Given V and W, the model can always be reparametrised in terms of V and r by setting $W = rV$, and this is adopted for the constant model.

EXAMPLE 2.1. A pharmaceutical company markets an ethical drug KURIT which currently sells an average of 100 units per month. Medical advice leads to a change in drug formulation that is expected to lead to a wider market demand for the product. It is agreed that from January, $t = 1$, the new formulation with new packaging will replace the current product, but the price and brandname KURIT remains unchanged. In order to plan production, stocks and raw material supplies, short-term forecasts of future demand are required. The drug is used regularly by individual patients so that demand tends to be locally constant in time, thus a first-order polynomial model is assumed for the total sales of KURIT per month. Sales fluctuations and observational variation about demand level are expected to considerably exceed month to month variation in the demand level so that ω_t is small compared to ν_t. In accord with this, the DLM which operated successfully on the old formulation was constant, with $V = 100$ and $W = 5$, and these values are retained for the new formulation.

In December, $t = 0$, the view of the market for the new product is that demand is most likely to have expanded by about 30% to 130 units per month. It is believed that demand is unlikely to have fallen by more than 10 units or to have increased by more than 70. This range of 80 units is taken as representing 4 standard deviations for μ_0, thus the initial view of the company prior to launch is described by $m_0 = 130$, $C_0 = 400$, so that

$$(\mu_0 \mid D_0) \sim N[130, 400].$$

Consequently, the operational model for sales Y_t in month t is:

$$Y_t = \mu_t + \nu_t, \qquad \nu_t \sim N[0, 100],$$

$$\mu_t = \mu_{t-1} + \omega_t, \qquad \omega_t \sim N[0, 5],$$

$$(\mu_0 \mid D_0) \sim N[130, 400].$$

Here $r = 0.05$, a low signal to noise ratio typical in this sort of application.

Observations over the next few months and the various components of the one-step forecasting and updating recurrence relationships are given in Table 2.1. Figure 2.3a provides a plot of the observations and one-step forecasts over time and Figure 2.4a is a plot of the adaptive coefficient A_t against time, both up to September, $t = 9$.

Initially, at $t = 0$, the company's prior view of market demand is vitally important in making decisions about production and stocks. Subsequently, however, the value of this particular subjective prior diminishes rapidly as data is received. For example, the adaptive coefficient A_1 is 0.8 so that $m_1 = m_0 + 0.8e_1 = (4Y_1 + m_0)/5$. Thus the January observation is given 4 times the weight of the prior mean m_0 in calculating the posterior mean m_1. At $t = 2$, $A_2 = 0.46$ and $m_2 = m_1 + 0.46e_2 = 0.46Y_2 + 0.43Y_1 + 0.11m_0$, so that Y_2 is also relatively highly weighted and m_0 contributes only

Table 2.1. KURIT example

Month	Forecast Distribution		Adaptive Coefficient	Observation	Error	Posterior Information	
t	Q_t	f_t	A_t	Y_t	e_t	m_t	C_t
0						130.0	400
1	505	130.0	0.80	150	20.0	146.0	80
2	185	146.0	0.46	136	−10.0	141.4	46
3	151	141.4	0.34	143	1.6	141.9	34
4	139	141.9	0.28	154	12.1	145.3	28
5	133	145.3	0.25	135	−10.3	142.6	25
6	130	142.6	0.23	148	5.3	143.9	23
7	128	143.9	0.22	128	−15.9	140.4	22
8	127	140.4	0.21	149	8.6	142.2	21
9	126	142.2	0.21	146	3.8	143.0	20
10	125	143.0	0.20				

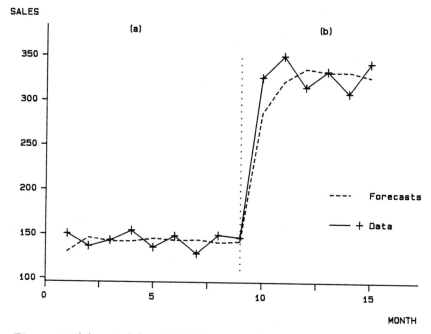

Figure 2.3(a) and (b).   KURIT examples.

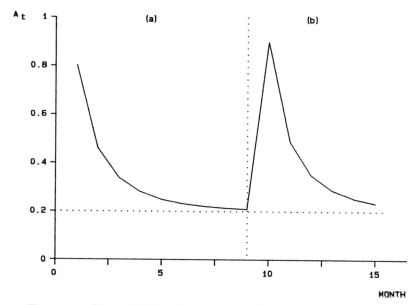

Figure 2.4(a) and (b).   Adaptive coefficient.

11% of the weight of information incorporated in m_2. As t increases, A_t appears to decay rapidly to a limiting value near 0.2. In fact the next section shows that, in this example, A_t does converge to exactly 0.2. Finally note that the coefficient of m_0 in m_t is given by the product $(1 - A_t)(1 - A_{t-1}) \ldots (1 - A_1)$ so that, at $t = 10$, m_0 contributes only 1% of the information used to calculate m_{10}, and as t increases the relevance of this subjective prior decays to zero.

2.3.2 Intervention to incorporate external information

The assumption of model closure precludes the use of extra external information relevant to the forecasting problem and so is generally not acceptable in practice. One of the major advantages of the Bayesian approach lies in the ease with which subjective information and model/data information combine. In the example, at $t = 9$, $(\mu_9 \mid D_9) \sim N[143, 20]$ with one-step forecast distribution $(Y_{10} \mid D_9) \sim N[143, 125]$. Suppose that, at $t = 9$, information denoted by S_9 is received concerning the pending withdrawal from the market of a major competitive drug BURNIT due to suspected side effects. This will occur at $t = 10$ when patients who were prescribed BURNIT will switch to a competitor. It is known that BURNIT currently accounts for roughly 50% of the market which leads the company to estimate a 100% increase in KURIT demand, $E[\mu_{10} \mid D_9, S_9] = 286$. However uncertainty about this figure is high, with estimated increased demand ranging within the company's marketing department from a pessimistic value of 80 units to an optimistic 200. After discussion it is agreed to model the change in demand at time 10 as $(\omega_{10} \mid D_9, S_9) \sim N[143, 900]$ leading to the prior $(\mu_{10} \mid D_9, S_9) \sim N[286, 920]$. Thus the revised forecast distribution is $(Y_{10} \mid D_9, S_9) \sim N[286, 1020]$. This raises A_{10} from 0.2 to 0.9 and leads to a far greater degree of adaptation to further data than would be the case without intervention. Observing $Y_{10} = 326$ now gives $e_{10} = 40$ and, with a revised information set $D_{10} = \{Y_{10}, D_9, S_9\}$, a posterior $(\mu_{10} \mid D_{10}) \sim N[322, 90]$. Figures 2.3b and 2.4b give the continued development over the next few months.

When unexpected and relevant external information of this sort becomes available, intervention is obviously of paramount importance for good decision making. It is clearly unsatisfactory to simply alter the mean estimate of demand in this example, although this is unfortunately an approach adopted by some forecasters. The associated increase in uncertainty leads to greater adaptation to further forecast errors and highly reliable forecasts after the change point in the series.

2.3.3 Limiting behaviour and convergence

In the closed model the rate of adaptation to new data, as measured by the adaptive coefficient A_t, rapidly converges to a constant value. This is proven as follows.

Theorem 2.3. As $t \to \infty$, $A_t \to A$ and $C_t \to C = AV$ where

$$A = r(\sqrt{1 + 4/r} - 1)/2.$$

Proof: Since $C_t = A_t V$ and $0 < A_t < 1$, then $0 < C_t \le V$ for all t. Then, using the recursions $C_t^{-1} = R_t^{-1} + V^{-1}$ and $R_t = C_{t-1} + W$, we have

$$C_t^{-1} - C_{t-1}^{-1} = R_t^{-1} - R_{t-1}^{-1} = K_t(C_{t-1}^{-1} - C_{t-2}^{-1})$$

where $K_t = C_{t-1}C_{t-2}/(R_t R_{t-1}) > 0$. So C_t is a monotonic and bounded sequence with a limit C, say, and $R_t = C_{t-1} + W$ implies that R_t converges to $R = C + W$. Also, using $C_t = R_t V/(R_t + V)$, it follows that $C = RV/(R+V)$ which implies $C^2 + CW - VW = 0$. This quadratic has one positive root given by $C = rV(\sqrt{1 + 4/r} - 1)/2$. It follows that, since $A_t = C_t/V$, then A_t converges to $A = C/V$ which is the required function of r. A useful inversion of this relationship leads to $r = A^2/(1 - A)$.

$\diamond$

A is a function of $r = W/V$ alone, the relationship being illustrated in the following table of values of $1/r$ and A:

$1/r$	9,900	380	90	20	6	0.75	0.01
A	0.01	0.05	0.10	0.20	0.33	0.67	0.99

Summarising the above results, and some easily deduced consequences:

(i) $A_t \to A = r(\sqrt{1 + 4/r} - 1)/2$ with $r = A^2/(1 - A)$

(ii) $C_t \to C = AV$

(iii) $R_t \to R = C/(1 - A) = AV/(1 - A)$

(iv) $Q_t \to Q = V/(1 - A)$ and $V = (1 - A)Q$

(v) $W = A^2 Q$.

Concerning the form of convergence, C_t is monotonically increas-
ing/decreasing according to whether the initial variance C_0 is less/
greater than the limiting value C. Similar comments apply to the
sequence A_t.

To obtain further insight into the behaviour of this latter sequence,
an exact expression in terms of the limiting value A and the starting
value A_1 can be obtained. The derivation of this result appears in the
Appendix, Section 2.7, at the end of this chapter. Define $\delta = 1 - A$.
Clearly $0 < \delta < 1$, and, for large t, $m_t \approx \delta Y_t + (1 - \delta)m_{t-1}$. Given
C_0, the initial adaptive coefficient is $A_1 = (C_0 + W)/(C_0 + W + V)$.
It is shown in Harrison (1985a) that A_t has the form

$$A_t = A \left[\frac{(1 - \delta^{2t-2})A + (\delta + \delta^{2t-2})A_1}{(1 + \delta^{2t-1})A + (\delta - \delta^{2t-1})A_1} \right]$$

from which we can deduce the following:

(a) In the case of a very vague initial prior in which C_0^{-1} is close
to zero, $A_1 \approx 1$, and A_t is monotonically decreasing with

$$A_t \approx A(1 + \delta^{2t-1})/(1 - \delta^{2t}),$$

or, writing $\delta_t = 1 - A_t$,

$$\delta_t \approx \delta(1 - \delta^{2t-2})/(1 - \delta^{2t}).$$

(b) At the other extreme when the initial prior is very precise
with C_0 close to zero, then $A_1 \approx 0$ and A_t is monotonically
increasing with

$$A_t \approx A(1 - \delta^{2t-2})/(1 + \delta^{2t-1}),$$

or

$$\delta_t \approx \delta(1 + \delta^{2t-3})/(1 + \delta^{2t-1}).$$

(c) Generally convergence is extremely fast, being monotonically increasing for $A_1 < A$, and monotonically decreasing for $A_1 > A$.

2.3.4   General comments

The case $A = 1$ corresponds to $V = 0$ hence $A_t = 1$ for $t \geq 2$. This is implied in practical cases by $r \to \infty$ and reduces the model to a pure random walk for the observation series, $Y_t = Y_{t-1} + \omega_t$. In this case $m_t = Y_t$ and the model is of little use for prediction. Often, when applied to series such as daily share, stock and commodity prices, the appropriate values of V in a first-order model may appear to be close to zero or very small compared to the systematic variance W. In other words the DLM appears to infer that all movement between daily prices may be attributed to movement in the underlying price level μ_t. This feature has led many economic forecasters to conclude that such series are purely random and that no better forecast of future prices than $(Y_{t+k} \mid D_t) \sim N[Y_t, kW]$, $(k \geq 0)$, is available. Such conclusions are quite erroneous and reflect a very narrow view of modelling; clearly the model is restricted and inappropriate as far as forecasting is concerned. Accepting the above conclusion is rather like restricting a description of lamp-posts to that relating to men. Within this way of looking at lamp-posts the best description is that of a tall, thin, one-legged man with a large head and one gleaming eye, but this does not mean that this is the best or only description of a lamp-post. Rather it infers that other models, descriptions and ways of seeing are required.

The major applications of the constant model are in short term forecasting and control with, typically, $5 \leq 1/r \leq 1000$, where the main benefit is derived from data smoothing. Denoting the forecasting horizon by L sampling periods, a rough guide to what is meant by short term is that $2rL < 1$ and that generally $1 \leq L \leq 6$. In the limit as $t \to \infty$, the point predictor of future values is $m_t = AY_t + (1 - A)m_{t-1}$. Clearly if $A = 1$ then $m_t = Y_t$ and any changes in Y_t are fully reflected in the predictor. On the other hand if A is close to zero then $m_t \approx m_{t-1}$ and none of the changes in the series will be captured by the predictor. The larger the value of A the more sensitive is the predictor to the latest values of the observation series. However there is a dilemma arising from the conflict between the requirements of sensitivity and robustness here. Note that $Y_t = \mu_{t-1} + \omega_t + \nu_t$. On the one hand, it is desirable to have a large value of A so that any sustained changes which occur through the signal ω_t are fully incorporated in the estimate m_t; on the other hand a small value is desirable in order that the random noise ν_t corrupting the signal is excluded from the estimate. The value of A employed in the estimate reflects the relative expected variation in ω_t and ν_t through the ratio r, a routine value of which is typically used in estimation. However, practical applications should embed the routine model within a complete forecasting system which operates according to the principle of management by exception. That is, relevant subjective information is combined with data to anticipate major movements in the series ν_t and ω_t, and control systems monitor model predictive performance to highlight suspected major changes in the data series. Further details of such important topics are left until later chapters.

2.3.5   Limiting predictors and alternative methods

The limiting form of the one- (and $k > 1$) step ahead point forecast given by the mean $f_{t+1} = m_t = \mathrm{E}[Y_{t+1} \mid D_t]$ as $t \to \infty$ has been shown to satisfy

$$m_t = m_{t-1} + Ae_t = m_{t-1} + A(Y_t - m_{t-1}),$$

in the formal sense that $\lim_{t\to\infty}(m_t - m_{t-1} - Ae_t) = 0$. Consideration of this form provides interpretations for several commonly used methods of calculating point forecasts, described below. In addition, a *limiting* representation of the first-order model as an ARIMA process is described.

(a) **Holt's point predictor.**

Holt (1957) introduced what is now a widely used point forecasting method based on one-step estimates $M_t = \alpha Y_t + (1-\alpha)M_{t-1}$, $(t > 1)$. Clearly this is equivalent to the asymptotic form of m_t with $\alpha = A$. For finite t, however, there are important differences due to the time dependence of the actual adaptive coefficient A_t in the DLM. In particular, with $\alpha = A$ and $\delta = 1 - A$, Holt's predictor has the form

$$M_t = A \sum_{j=0}^{t-1} \delta^j Y_{t-j} + \delta^t M_0,$$

whereas this form is approached only asymptotically by m_t.

(b) **Exponentially weighted moving averages (EWMA).**

Given the sequence $Y_t, \ldots, Y_1$, the EWMA with parameter $\delta, (0 < \delta < 1)$, is defined as

$$M_t = \frac{(1 - \delta)}{(1 - \delta^t)} \sum_{j=0}^{t-1} \delta^j Y_{t-j}.$$

As $t \to \infty$, the definition takes the limiting form

$$M_t = (1 - \delta)Y_t + \delta M_{t-1},$$

clearly equivalent to Holt's predictor. Thus the limiting form of m_t is an EWMA of the observations with $\delta = 1 - A$. More formally, given a finite m_0 in the closed, constant model, and finite M_0, the difference $m_t - M_t$ converges in probability to zero, with $\lim_{t \to \infty} P(|m_t - M_t| > \epsilon) = 0$ for any $\epsilon > 0$.

(c) **Brown's exponentially weighted regression (EWR).**

The EWR estimator (Brown, 1962) for a locally constant mean of $Y_t, \ldots, Y_1$, is defined as the value $M_t = \mu$ that minimises the discounted sum of squares

$$V_t(\mu) = \sum_{j=0}^{t-1} \delta^j (Y_{t-j} - \mu)^2$$

for some given discount factor $\delta, (0 < \delta < 1)$. It is easily shown that $V_t(\mu)$ has a unique minimum as a function of μ at the EWMA for M_t above. Thus EWR is also obtained as a limiting case of the predictor in the DLM.

In each of these cases then, the classical point predictors are derived as limiting forms of m_t from the standard closed, constant model. Unlike the DLM however, these forecasting methods are essentially *ad hoc* devices which, although simple to apply, have no coherent basis. They do not derive for finite t from a time series model and so are difficult to interpret; they have none of the flexibility, such as intervention, associated with the DLM; and, with the exception of EWR, they do not easily generalise.

(d) **An alternative ARIMA(0,1,1) limiting representation of the model.**

The class of ARIMA models of Box and Jenkins (1976) provide probably the most well-known and widely used time series models. An important case, the ARIMA(0,1,1) process, is derived here as a special limiting case of the first-order polynomial DLM.

Given the two relationships $m_{t-1} = Y_t - e_t$ and $m_t = m_{t-1} + A_t e_t$, it follows that

$$Y_t - Y_{t-1} = e_t - (1 - A_t)e_{t-1}.$$

This implies that

$$\lim_{t \to \infty} [Y_t - Y_{t-1} - e_t + (1 - A)e_{t-1}] = 0$$

or, loosely, $Y_t \approx Y_{t-1} + e_t - \delta e_{t-1}$ where $\delta = 1 - A$, as above defined. Now on the basis of the model, the sequence of random quantities $(e_t \mid D_{t-1})$ are zero-mean, independent, and, as $t \to \infty$, the variance Q_t converges to $Q = V/(1 - A)$. Thus, in terms of A and Q rather than V and W, an alternative limiting model for the series is

$$Y_t = Y_{t-1} + a_t - \delta a_{t-1}$$

where the a_t are independent with $a_t \sim N[0, Q]$. This form is that of an ARIMA(0,1,1) process. Note, however, that it is again only the limiting form of the updating recurrence equations in the DLM that has this representation, and also that the moving average error terms a_t are actually the limiting forecast errors.

2.3.6   Forecast distributions

At time t, the step ahead and lead time forecast distributions are special cases of those of the general model in Theorem 2.2. Using that result we obtain:

(a)   k-step ahead: $(Y_{t+k} \mid D_t) \sim N[m_t, Q_t(k)]$

(b)   k-step lead-time: $(X_t(k) \mid D_t) \sim N[k m_t, L_t(k)]$

where

$$Q_t(k) = C_t + kW + V$$

and

$$L_t(k) = k^2 C_t + kV + k(k+1)(2k+1)W/6.$$

It is of interest to note that the lead-time forecast *coefficient of variation*, defined as $L_t(k)^{1/2}/(km_t)$, can be shown to be monotonically decreasing for k less than or equal to the integer part of $k_0 = \sqrt{(0.5 + 3/r)}$, and monotonically increasing for $k > k_0$. Note that k_0 is independent of C_t, being a function only of r. For $r = 0.05$ corresponding to $A = 0.2$, for example, $k_0 = 8$. In this case, with the limiting value of variance $C = 0.2V$ substituted for C_t, it follows that $L_t(k)/(64L_t(1)) = (0.62)^2$ so that the limiting coefficient of variation for the one-step ahead forecast is 61% greater than that for the 8-step lead time forecast.

2.4 SPECIFICATION OF EVOLUTION VARIANCE W_t

The forecasting performance of a first-order polynomial model, constant or otherwise, depends heavily on choosing appropriate values for the variances V_t and W_t. The problem of the former variance is considered in Section 2.5 below; here we examine the choice of the latter.

2.4.1 Robustness to values of W in the constant model

Suppose that a forecaster applies the constant model with observational variance V and signal to noise ratio r when the data are truly generated by a model with values V_0 and r_0. Of course this example is purely hypothetical since no mathematical model can exactly represent a 'true' process, but it provides important insight into the questions of robustness and choice of variances. Convergence is fast so consider the limiting form of the model

$$Y_t - Y_{t-1} = e_t - \delta e_{t-1},$$

where $\delta = 1 - A$ and the forecaster assumes $(e_t \mid D_{t-1}) \sim N[0, Q]$ with $Q = V/\delta$. The true limiting model is

$$Y_t - Y_{t-1} = a_t - \delta_0 a_{t-1},$$

where $\delta_0 = 1 - A_0$, and, in fact, $(a_t \mid D_{t-1}) \sim N[0, Q_0]$ with $Q_0 = V_0/\delta_0$. Equating these two expressions for the series gives $e_t - \delta e_{t-1} = a_t - \delta_0 a_{t-1}$, so that

$$e_t = a_t + (\delta - \delta_0) \sum_{j=0}^{t} \delta^j a_{t-j-1}.$$

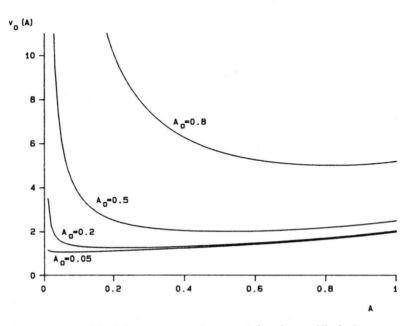

Figure 2.5.   Limiting error variance with misspecified A.

Given the independence of the true error sequence a_t, it follows that the true initial prior mean of the model errors is $E[e_t \mid D_0] = 0$ and, as $t \to \infty$, the corresponding variances are

$$V[e_t \mid D_0] = Q_0[1 + (\delta - \delta_0)^2/(1 - \delta^2)].$$

It is immediately apparent that a gross penalty is incurred if the value of r, and hence W, is too low. As $r \to 0$ then $\delta \to 1$ and the true variance of the error sequence above becomes infinite whereas, of course, the forecaster uses the value V from the model! At the other extreme, as $r \to \infty$ then $\delta \to 0$ and the true variance tends to $Q_0(1 + \delta_0^2)$. Hence choosing r too large is generally much to be preferred, allowing for over adaptation to new data, since this at least limits the margin for error. The case $r = 0$ implies a constant level $\mu_t = \mu$, a model that has been much used in practice but the pitfalls are now clear. Figure 2.5 illustrates this feature by plotting the function $v_0(A) = [1 + (\delta - \delta_0)^2/(1 - \delta^2)]/\delta_0$ as a function of the assumed value of $A = 1 - \delta$ for various true values $A_0 = 0.05, 0.2, 0.5, 0.8$.

Additionally, misspecification leads to correlated errors. Clearly the true covariance between the model forecast errors e_t and e_{t-k} is

$$C[e_t, e_{t-k} \mid D_0] = Q_0(\delta - \delta_0)\delta^{k-1}(1 - \delta\delta_0)/(1 - \delta^2),$$

giving correlations

$$C_0(k) = (\delta - \delta_0)(1 - \delta\delta_0)\delta^{k-1}/(1 - 2\delta\delta_0 + \delta_0^2),$$

instead of zero as assumed in the model. If $\delta > \delta_0$ so that $A < A_0$, the e_t are positively correlated, otherwise they are negatively correlated. In the limit, as $\delta_0 \to 1$, so that μ_t is actually constant, then

$$0 > C_0(k) = -\delta^{k-1}(1 - \delta)/2;$$

and as $\delta_0 \to 0$, so that Y_t is a pure random walk, then

$$0 < C_0(k) = \delta^k.$$

Thus an under-adaptive model ($\delta > \delta_0$) leads to positively correlated forecast errors, and an over-adaptive model ($\delta < \delta_0$) leads to negatively correlated errors. Further details and discussion of these ideas appears in Harrison (1967), and Roberts and Harrison (1984).

2.4.2 Discount factors as an aid to choosing W_t

By definition for the constant model, $R_t = C_{t-1} + W$, and, in the limit, $R = C + W = C/(1 - A)$. Thus $W = AC/(1 - A)$ so that W is a fixed proportion of C. This is a natural way of thinking about the system variance; between observations the addition of the error ω_t leads to an additive increase of $W = 100A/(1 - A)\%$ of the initial uncertainty C. Since $\delta = 1 - A$, then, simply, $R = C/\delta$. Choice of δ by reference to the limiting rate of adaptation to data then guides the choice of W. For $A = 0.1$ we have $\delta = 0.9$ and W is roughly 11% of C. This increases to 25% for $\delta = 0.8$. Bearing in mind that the limiting behaviour is very rapidly achieved in this model it is convenient and natural to adopt a constant *rate* of increase of uncertainty, or decay of information, for all t rather than just in the limiting case. Thus, for a given **discount factor** δ, say 0.8 or 0.9 as typical values, choosing $\omega_t \sim N[0, W_t]$, where $W_t = C_{t-1}(1 - \delta)/\delta$ leads to $R_t = C_{t-1}/\delta$. Clearly this is not now a constant model but the limiting behaviour is that of the constant model with $A = 1 - \delta$. This is most easily seen by noting that in this model, $\{1, 1, V, W_t\}$,

$$C_t^{-1} = V^{-1} + R_t^{-1} = V^{-1} + \delta C_{t-1}^{-1} = V^{-1}[1 + \delta + \delta^2 + \cdots + \delta^t C_0^{-1}],$$

which, as $t \to \infty$, converges to $C^{-1} = V^{-1}/(1-\delta)$. Thus the limiting value of C_t is simply $C = (1 - \delta)V = AV$, as required. Further discussion of this and more general **discount** models appears in later chapters. Here the concept is introduced as a simple and natural way of structuring and assigning values to the W_t sequence in the first-order polynomial model. Note that it applies directly to the general model with variances V_t and W_t since W_t as defined above depends only on the values of C_{t-1} and δ which are known at time at $t - 1$. Furthermore, with the assumption of the known values of the V_t sequence into the initial information set D_0 then clearly W_t is also known at $t = 0$, calculated from the above recursion as a function of δ, C_0 and $V_{t-1}, \dots, V_1$.

2.5   UNKNOWN OBSERVATIONAL VARIANCES

2.5.1   Introduction

In many practical applications the forecaster will not wish to state precise values for the observational variance sequence V_t although he may have beliefs about some particular features. Examples include $V_t = V$ as in the constant model but with V unknown; weighted variances $V_t = Vk_t$ where the weight sequence k_t is known but V is unknown; $V_t = Vk_t(\mu_t)$ where $k_t(.)$ is a possibly time dependent variance function of the mean, such as the common case of a power function $k_t(\mu_t) = k(\mu_t) = \mu_t^p$ for some power p, and V is unknown; and so on. In each case the forecaster may also suspect that the nominally constant scale factor V does, in fact, vary slowly in time, reflecting natural, second-order dynamics of the time series observational process as well as allowing further for errors of approximation in the model. All of these cases are important in practice and are developed further in later chapters. Here we concentrate only on the simplest case of the first-order polynomial model in which $V_t = V$ is constant but unknown. Denote by ϕ the unknown reciprocal variance, or **precision**, $\phi = 1/V$. We will often work in terms of ϕ in place of V^{-1} since the discussion can sometimes appear clearer in terms of the precision. Refer to Section 16.3 of Chapter 16 for wider theoretical discussion of normal linear models with unknown variance, whose analysis provides the basis of that here.

A simple, closed form Bayesian analysis of the DLM with unknown, constant variance V is available if a particular structure is imposed on the W_t sequence and the initial prior for μ_0. With such structure, described below, a *conjugate* sequential updating procedure for V, equivalently for ϕ, is derived, in addition to that for μ_t. To motivate the required structuring, recall that in the constant model it was natural to specify the evolution variance as a multiple of the observational variance for all time. Applying this to the general model suggests that, for known V,

$$(\omega_t \mid V, D_{t-1}) \sim N[0, V W_t^*]$$

for each t. Thus W_t^* is now the variance of ω_t if $V = 1$, equivalently $\phi = 1$. With this structure, suppose that C_0 is very large compared with V. In particular, as C_0 increases, we obtain $C_1 \approx V$, and, for $t > 1$, C_t is given by V multiplied by that value obtained from the standard model in which $V = 1$. Indeed this always happens eventually, whatever the value of C_0, since the effect of the initial prior rapidly decays. Thus based on the form of the C_t sequence as deduced from the data, we are led to scaling by V just as with W_t^* above. In particular, if C_0 is scaled by V directly, then so is C_t for all t. This feature is the final component of the model structuring required for the closed, conjugate analysis that is now described.

2.5.2   The case of a constant unknown variance

The analysis for V follows standard Bayesian theory as detailed in Section 16.3 of Chapter 16 to which the reader can refer. The conjugate analysis is based on gamma distributions for ϕ, and thus, equivalently, inverse gamma distributions for V, for all time, and derives from the following model definition.

Definition 2.2. *For each t, the model is defined by:*

Observation equation:	$Y_t = \mu_t + \nu_t,$	$\nu_t \sim N[0, V],$
System equation:	$\mu_t = \mu_{t-1} + \omega_t,$	$\omega_t \sim N[0, V W_t^*],$
Initial information:	$(\mu_0 \mid D_0, V) \sim N[m_0, V C_0^*],$	
	$(\phi \mid D_0) \sim G[n_0/2, d_0/2],$	

for some m_0, C_0^*, W_t^*, n_0 and d_0. In addition, the usual independence assumptions hold, now conditional on V.

Here the final component is a gamma prior with density, for $\phi > 0$, given by

$$p(\phi \mid D_0) = \left[\frac{(d_0/2)^{n_0/2}}{\Gamma(n_0/2)} \right] \phi^{n_0/2-1} e^{-\phi d_0/2},$$

The defining parameters n_0 and d_0 are both positive and known. Ignoring the constant of normalisation, we have

$$p(\phi \mid D_0) \propto \phi^{n_0/2-1} e^{-\phi d_0/2}.$$

The mean of this prior is $E[\phi \mid D_{t-1}] = n_0/d_0 = 1/S_0$ where S_0 is a prior point estimate of the observational variance V. Alternative expressions of the prior are that, when n_0 is integral, the multiple d_0 of ϕ has a chi-squared distribution with n_0 degrees of freedom,

$$(d_0\phi \mid D_{t-1}) \sim \chi^2_{n_0},$$

or that $(V/d_0 \mid D_0)$ has an inverse chi-squared distribution with the same degrees of freedom. Notice that, in specifying the prior, the forecaster must choose the prior estimate S_0 and the associated degrees of freedom parameter n_0, in addition to m_0 and C_0^*. The starred variances C_0^* and the sequence W_t^*, assumed known, are multiplied by V to provide the actual variances when V is known. Thus they are referred to as *observation-scale free variances*, explicitly recognising that they are independent of the scale of the observations determined by V.

The results obtained below are based on Student T distributions for the level parameters and forecasts. Again details are given in Section 16.3 of Chapter 16 but, for convenience, a short summary is in order here. A real random quantity μ has a Student T distribution with n degrees of freedom, mode m and scale C, denoted

$$\mu \sim T_n[m, C],$$

if and only if the density of μ is given by

$$p(\mu) \propto \left[n + (\mu - m)^2/C \right]^{-(n+1)/2}.$$

The quantity $(\mu - m)/C^{1/2}$ has a standard Student T distribution on n degrees of freedom, $T_n[0, 1]$. As n increases, this form approaches a normal distribution and the notation is chosen by analogy with that for the limiting normal case. The analogy is important; the standard, known variance model is based on normal distributions which are replaced by Student T forms when V is unknown. Note that, in the limit, C is the variance of the distribution. Also, m is the mean whenever $n > 1$, and the variance is $Cn/(n-2)$ when $n > 2$.

It follows from the model definition that, unconditionally (with respect to V),

$$(\mu_0 \mid D_o) \sim T_{n_0}[m_0, C_0]$$

where $C_0 = S_0 C_0^*$ is the scale of the marginal prior T distribution. In specifying the initial prior, the forecaster provides n_0 and d_0 for V, and m_0 and C_0 for μ_0, and these may be considered separately. Note that, from the T prior, $V[\mu_0 \mid D_0] = C_0 n_0/(n_0 - 2)$ if $n_0 > 2$. Also, for large n_0, V is essentially equal to S_0 and the prior for μ_0 is approximately normal with mean m_0 and variance C_0.

Theorem 2.4. *With the model as specified above, the following distributional results obtain at each time $t \geq 1$.*

(a) *Conditional on V:*

$$(\mu_{t-1} \mid D_{t-1}, V) \sim N[m_{t-1}, VC_{t-1}^*],$$
$$(\mu_t \mid D_{t-1}, V) \sim N[m_{t-1}, VR_t^*],$$
$$(Y_t \mid D_{t-1}, V) \sim N[f_t, VQ_t^*],$$
$$(\mu_t \mid D_t, V) \sim N[m_t, VC_t^*],$$

with $R_t^* = C_{t-1}^* + W_t^*$, $f_t = m_{t-1}$, and $Q_t^* = 1 + R_t^*$, the defining components being updated via

$$m_t = m_{t-1} + A_t e_t,$$
$$C_t^* = R_t^* - A_t^2 Q_t^* = A_t,$$

where $e_t = Y_t - f_t$ and $A_t = R_t^*/Q_t^*$.

(b) *For precision $\phi = V^{-1}$:*

$$(\phi \mid D_{t-1}) \sim G[n_{t-1}/2, d_{t-1}/2],$$
$$(\phi \mid D_t) \sim G[n_t/2, d_t/2],$$

where $n_t = n_{t-1} + 1$ and $d_{t-1} = d_{t-1} + e_t^2/Q_t^*$.

(c) *Unconditional on V :*

$$(\mu_{t-1} \mid D_{t-1}) \sim T_{n_{t-1}}[m_{t-1}, C_{t-1}],$$
$$(\mu_t \mid D_{t-1}) \sim T_{n_{t-1}}[m_{t-1}, R_t],$$
$$(Y_t \mid D_{t-1}) \sim T_{n_{t-1}}[f_t, Q_t],$$
$$(\mu_t \mid D_t) \sim T_{n_t}[m_t, C_t],$$

where $C_{t-1} = S_{t-1}C_{t-1}^*$, $R_t = S_{t-1}R_t^*$, $Q_t = S_{t-1}Q_t^*$ and $C_t = S_t C_t^*$, with $S_{t-1} = d_{t-1}/n_{t-1}$ and $S_t = d_t/n_t$.

(d) *Operational definition of updating equations:*

$$m_t = a_t + A_t e_t,$$
$$C_t = (S_t/S_{t-1})\,[R_t - A_t^2 Q_t] = A_t S_t,$$
$$S_t = d_t/n_t,$$

and

$$n_t = n_{t-1} + 1 \quad \text{and} \quad d_t = d_{t-1} + S_{t-1}e_t^2/Q_t,$$

where $Q_t = S_{t-1} + R_t$ and $A_t = R_t/Q_t$.

Proof: Given the model definition, the results in (a) follow directly from Theorem 2.1. They are the standard, known variance results.

The rest of the proof is by induction. Assume that the prior in (b) holds so that, prior to observing Y_t, $(\phi \mid D_{t-1}) \sim G[n_{t-1}/2, d_{t-1}/2]$. From (a) we have $(Y_t \mid D_{t-1}, \phi) \sim N[f_t, Q_t^*/\phi]$ so that

$$p(Y_t \mid D_{t-1}, \phi) \propto \phi^{1/2} \exp(-0.5\phi e_t^2/Q_t^*).$$

Now, by Bayes' Theorem, we have posterior for ϕ given by

$$p(\phi \mid D_t) \propto p(\phi \mid D_{t-1})\,p(Y_t \mid D_{t-1}, \phi).$$

Using the prior from (b) and the likelihood above, we have

$$p(\phi \mid D_t) \propto \phi^{(n_{t-1}+1)/2 - 1} \exp[-(d_{t-1} + e_t^2/Q_t^*)\phi/2]$$

and deduce that $(\phi \mid D_t) \sim G[n_t/2, d_t/2]$ where the updated parameters are $n_t = n_{t-1} + 1$ and $d_t = d_{t-1} + e_t^2/Q_t^*$. This establishes the posterior for ϕ in (b).

Results (c) follow directly from the normal/gamma – Student T theory mentioned earlier in the Section, and reviewed in detail in Sections 16.3.1 and 16.3.2 of Chapter 16. They follow on simply integrating the conditionally normal distributions in (a) with respect to the appropriate prior/posterior gamma distribution for ϕ from (b).

The inductive proof is now completed by noting that the results hold at $t = 0$ by using the same arguments and the initial distributions in Definition 2.2.

$\diamond$

This Theorem provides the key results. At time t, the prior mean of ϕ is $E[\phi \mid D_{t-1}] = n_{t-1}/d_{t-1} = 1/S_{t-1}$ where $S_{t-1} = d_{t-1}/n_{t-1}$ is a prior point estimate of $V = 1/\phi$. Similarly, the posterior estimate is $S_t = d_t/n_t$. Note that the updating equations for the parameters defining the T prior/posterior and forecast distributions are essentially the same as the standard, known variance equations with the estimate S_{t-1} appearing as the variance. The only differences lie in the scaling by S_t/S_{t-1} in the update for C_t and the scaling of e_t^2/Q_t by S_{t-1} in the update for d_t, both to correct for the updated estimate of V. The equations in (d) are those used in practice, the starred, scale free versions appearing only to communicate the theoretical structure.

2.5.3 Summary

These results are summarised in tabular form below for easy reference.

2.5.4 General comments

The estimate S_t of $1/\phi$ may be written in the recursive form $S_t = S_{t-1} + [S_{t-1}/(n_t Q_t)](e_t^2 - Q_t)$. Some early EWMA forecasting systems used a similar point estimate based on the recursion $S_t = S_{t-1} + h(e_t^2 - S_{t-1})$, for some h, for the *one-step forecast* variance rather than the *observational* variance $1/\phi$, an important distinction. Such a method is obviously suspect, particularly when t and/or n_t are small, and the fully coherent learning procedure should be adopted. This prediction variance is actually given, from the Student T forecast distribution, as $V[Y_{t+1} \mid D_t] = Q_{t+1} n_t/(n_t - 2)$, when $n_t > 2$, being infinite otherwise. In the case of large n_t, this variance is approximately equal to Q_{t+1}. If the model is constant, then $Q_{t+1} \approx S_t/\delta$ as t, and hence n_t, increases in which case the one-step variance is approximately given by S_t/δ. Hence, in this special case, the *limiting* prediction variance may be written in the above mentioned *ad hoc* form, as

$$V[Y_{t+1} \mid D_t] \approx V[Y_t \mid D_{t-1}] + (\delta/n_t)(e_t^2 - V[Y_t \mid D_{t-1}]),$$

as $t \to \infty$.

First-Order Polynomial DLM, with constant variance V		
Observation:	$Y_t = \mu_t + \nu_t$	$\nu_t \sim N[0, V]$
System:	$\mu_t = \mu_{t-1} + \omega_t$	$\omega_t \sim T_{n_{t-1}}[0, W_t]$
Information:	$(\mu_{t-1} \mid D_{t-1}) \sim T_{n_{t-1}}[m_{t-1}, C_{t-1}]$	
	$(\mu_t \mid D_{t-1}) \sim T_{n_{t-1}}[m_{t-1}, R_t]$	$R_t = C_{t-1} + W_t$
	$(\phi \mid D_{t-1}) \sim G[n_{t-1}/2, d_{t-1}/2]$	$S_{t-1} = d_{t-1}/n_{t-1}$
Forecast:	$(Y_t \mid D_{t-1}) \sim T_{n_{t-1}}[f_t, Q_t]$	$f_t = m_{t-1}$
		$Q_t = R_t + S_{t-1}$

Updating Recurrence Relationships
$(\mu_t \mid D_t) \sim T_{n_t}[m_t, C_t]$
$(\phi \mid D_t) \sim G[n_t/2, d_t/2]$
with
$m_t = m_{t-1} + A_t e_t,$
$C_t = (S_t/S_{t-1})[R_t - A_t^2 Q_t] = A_t S_t,$
$n_t = n_{t-1} + 1,\ d_t = d_{t-1} + S_{t-1} e_t^2/Q_t,\ \text{and } S_t = d_t/n_t,$
where $e_t = Y_t - f_t,$ and $A_t = R_t/Q_t.$

Forecast Distributions
For $k \geq 1,$
$(Y_{t+k} \mid D_t) \sim T_{n_t}[m_t, Q_t(k)]$
$(X_t(k) \mid D_t) \sim T_{n_t}[km_t, L_t(k)]$
with
$Q_t(k) = C_t + \sum_{j=1}^{k} W_{t+j} + S_t,$
$L_t(k) = k^2 C_t + \sum_{j=1}^{k} j^2 W_{t+k+1-j} + k S_t,$
where, for $j > 0,$
$W_{t+j} = S_t W_{t+j}^*$
for scale free variances $W_{t+j}^*.$

2.6   ILLUSTRATION

The series titled Index plotted in Figure 2.6, and given in Table 2.2, represents the first differences of the logged monthly USA/UK exchange rate \$/£ over the period from January 1975 to July 1984.

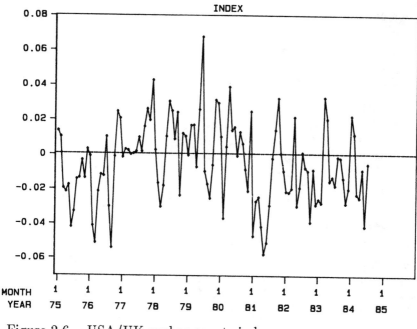

Figure 2.6.   USA/UK exchange rate index.

Table 2.2. USA/UK exchange rate index (×100)

Year	Month											
	1	2	3	4	5	6	7	8	9	10	11	12
75	1.35	1.00	−1.96	−2.17	−1.78	−4.21	−3.30	−1.43	−1.35	−0.34	−1.38	0.30
76	−0.10	−4.13	−5.12	−2.13	−1.17	−1.24	1.01	−3.02	−5.40	−0.12	2.47	2.06
77	−0.18	0.29	0.23	0.00	0.06	0.17	0.98	0.17	1.59	2.62	1.96	4.28
78	0.26	−1.66	−3.03	−1.80	1.04	3.06	2.50	0.87	2.42	−2.37	1.22	1.05
79	−0.05	1.68	1.70	−0.73	2.59	6.77	−0.98	−1.71	−2.53	−0.61	3.14	2.96
80	1.01	−3.69	0.45	3.89	1.38	1.57	−0.08	1.30	0.62	−0.87	−2.11	2.48
81	−4.73	−2.70	−2.45	−4.17	−5.76	−5.09	−2.92	−0.22	1.42	3.26	0.05	−0.95
82	−2.14	−2.19	−1.96	2.18	−2.97	−1.89	0.12	−0.76	−0.94	−3.90	−0.86	−2.88
83	−2.58	−2.78	3.30	2.06	−1.54	−1.30	−1.78	−0.13	−0.20	−1.35	−2.82	−1.97
84	2.25	1.17	−2.29	−2.49	−0.87	−4.15	−0.53					

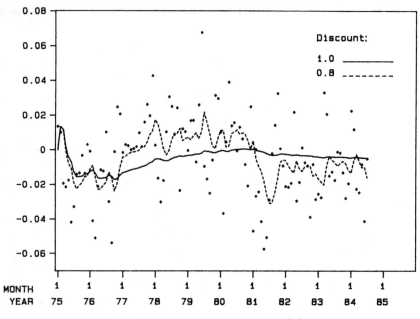

Figure 2.7.   USA/UK exchange rate index and forecasts.

There is clearly a relatively high degree of random variation about a changing level over the period so an initial examination of the data using a first-order polynomial model is performed. The purpose of this here is to simply demonstrate the automatic analysis using several models that differ only in the degree of movement allowed in the level via different W_t sequences. The evolution variances are defined, as we have recommended in Section 2.4, by the discount factor δ, the four models examined being defined by discount values of 0.7, 0.8, 0.9 and 1.0, the latter corresponding to the degenerate static model, $W_t = 0$ with the observations a simple normal random sample. In each case the initial prior distributions are defined by $m_0 = 0, C_0 = 1, n_0 = 1$ and $d_0 = 0.01$. This is a rather vague, or uninformative, joint prior specification that implies, for example, that given D_0, μ_0 lies between -0.1 and 0.1 with probability 0.5, and between -0.63 and 0.63 with probability 0.9. The data and sequences of one-step point forecasts from the two models with $\delta = 0.8$ and $\delta = 1.0$ appear in Figure 2.7. Evidently, as expected, the degree of adaptation to new data increases as δ decreases leading to more erratic forecast sequences.

To compare the models, and hence the suitability of the different discount factors, Table 2.3 displays various summary quantities. The first two are commonly used rules-of-thumb for assessing forecast accuracy, namely the mean absolute deviation, MAD $= \sum_{t=1}^{115} |e_t|/115$, and mean square error, MSE $= \sum_{t=1}^{115} e_t^2/115$, for the entire series. The third summary is based on the observed predictive density

$$p(Y_{115}, Y_{114}, \ldots, Y_1 \mid D_0) = \prod_{t=1}^{t=115} p(Y_t \mid D_{t-1}),$$

the product of the sequence of one-step forecast densities evaluated at the actual observation. By our convention, D_0 contains the value of the discount factor δ used in the model so that, for each such value, the above quantity provides a measure of predictive performance of the model that is actually a likelihood for δ, since it is only through δ that the models differ. Viewing δ as an uncertain quantity we could now introduce a prior distribution and hence calculate a posterior for δ. However, for our purposes here we simply examine the likelihoods above to obtain a rough data-based guide. For convenience we use a log scale and define the measures relative to the model with $\delta = 1$. Thus LLR is the log-likelihood ratio for the values 0.9, 0.8, 0.7 relative to 1.0, larger values indicating a higher degree of support from the data.

Table 2.3. Exchange rate Index example

Discount	115 MAD	115 MSE	LLR	90% Interval		SD
1.0	2.232	0.064	0.00	−0.009	−0.001	0.025
0.9	2.092	0.058	3.62	−0.024	0.000	0.023
0.8	2.061	0.058	2.89	−0.030	0.002	0.022
0.7	2.073	0.059	0.96	−0.035	0.003	0.021

From the table, the MSE and MAD measures indicate better predictive performance for δ between 0.8 and 0.9 than otherwise, the latter more in favour of the smaller value 0.8. The LLR measure, however, favours 0.9 due to the fact that it takes into account the variances of forecast distributions that are ignored by the MAD and MSE. In this measure, there is a balance between forecast accuracy as measured by the e_t sequence, and forecast precision as measured by the spread of the predictive distributions. The value of 0.8 for δ leads to more diffuse distributions than that of 0.9 and this counts against the model in the LLR measure of predictive performance. Note also, in particular, that the static model is clearly unsatisfactory by comparison with the others on all three measures of performance.

Further information provided in the table indicates final values of some of the quantities of interest. In particular we quote 90% posterior probability intervals for the final level μ_{115} based on the posterior Student T distributions. Also quoted are the values of the estimated standard deviation SD $= \sqrt{S_{115}}$. Note that, as expected, smaller values of δ lead to: (i) faster decay of information about the level between observations and so wider posterior intervals; and (ii) smaller estimates of observational variance as indicated by the standard deviations. In this example the differences in the latter are not really large. In some applications, as is illustrated in an example in the next Chapter, the observational variance can be markedly over-estimated by models with discount factors too close to 1, leading to forecast distributions that are much more diffuse than those from models with more suitable, lower discount values.

2.7 APPENDIX

The exact expression for A_t in the constant model is derived here.

In the closed, constant model we have $C_t = A_t V$, $R_t = A_{t-1}V + W$, and $Q_t = R_t + V$ which imply that

$$A_{t+1} = (C_t + W)/(C_t + W + V) = (A_t + r)/(A_t + r + 1)$$

where, as usual, $r = W/V$. Proceeding to the limit we have $A = (A + r)/(A + r + 1)$ and, defining

$$u_t = 1/(A_t - A)$$

for each t, we have by subtraction

$$u_{t+1} = u_t(A_t + r + 1)(A + r + 1).$$

Now, since $\delta = 1 - A$ and $r = A^2/(1 - A)$, we have $A + r + 1 = 1/\delta$ and $A_t + r + 1 = 1/u_t + 1/\delta$, so that $\delta^2 u_{t+1} = u_t + \delta$ and

$$u_t = \frac{[\delta(1 - \delta^{2(t-1)}) + u_1(1 - \delta^2)]}{[(1 - \delta^2)\delta^{2(t-1)}]}.$$

Noting that $1 - \delta^2 = A(1 + \delta)$ and substituting for $u_1 = 1/(A_1 - A)$, it is found that

$$A_t - A = \frac{A(A_1 - A)(1 + \delta)\delta^{2(t-1)}}{[(1 + \delta^{2t-1})A + (\delta - \delta^{2t-1})A_1]}.$$

This in turn implies, after rearrangement, the general solution

$$A_t = A \left[\frac{(1 - \delta^{2t-2})A + (\delta + \delta^{2t-2})A_1}{(1 + \delta^{2t-1})A + (\delta - \delta^{2t-1})A_1} \right].$$

2.8 EXERCISES

(1) Write a computer program to simulate and graph 100 observations from the model $\{1, 1, 1, W\}$ starting with $\mu_0 = 25$. Simulate several series for each value of $W = 0.05$ and 0.5. From these simulations, become familiar with the forms of behaviour such series can display.

(2) Become familiar with the two versions of the proof of Theorem 2.1.

(3) Generalise Theorem 2.1 to models in which v_t and ω_t may have non-zero means.

(4) Derive the alternative forms for the updating equations given by

$$m_t = C_t(R_t^{-1}m_{t-1} + V_t^{-1}Y_t) \quad \text{and} \quad C_t^{-1} = R_t^{-1} + V_t^{-1}.$$

(5) In the model $\{1, 1, V, 0\}$, suppose that C_0 is very large relative to V. Show that

(a) $\qquad m_1 \approx Y_1 \quad$ and $\quad C_1 \approx V;$

(b) $\qquad m_t \approx t^{-1} \sum_{j=1}^{t} Y_j \quad$ and $\quad C_t \approx V/t.$

Comment on these results.

(6) Suppose that Y_t is a missing observation so that $D_t = D_{t-1}$. Given m_{t-1} and C_{t-1} as usual, calculate $p(\mu_t|D_t)$ and $p(Y_{t+1}|D_t)$.

(7) In the known variance model $\{1,1,V_t,W_t\}$, identify the joint distribution of Y_t and ν_t given D_{t-1}. Use this to calculate the posterior for the observational error, $p(\nu_t|D_t)$; verify that this is given by $(\nu_t|D_t) \sim N[(1 - A_t)e_t, A_tV_t]$.

(8) Sometimes it is of interest to look back in time to make inferences about historical levels of a time series based on more recent data (referred to as *retrospective* analysis). As a simple case of this, suppose a forecaster is interested in revising inferences about μ_{t-1} based on Y_t in addition to D_{t-1} in the known variance model. Identify the joint distribution of Y_t and μ_{t-1} given D_{t-1} and deduce the posterior $p(\mu_{t-1}|D_t)$. Verify that this is normal, with moments given by

$$E[\mu_{t-1}|D_t] = m_{t-1} + B_{t-1}(m_t - m_{t-1})$$

and

$$V[\mu_{t-1}|D_t] = C_{t-1} - B_{t-1}^2(R_t - C_t),$$

where $B_{t-1} = C_{t-1}/R_t$. Show how these equations simplify in the case of a discount model in which $R_t = C_{t-1}/\delta$ for some discount factor δ lying between 0 and 1.

(9) In the constant model with m_{t-1} and C_{t-1} as usual, suppose that the data recording procedure at times t and $t + 1$ is such that Y_t and Y_{t+1} cannot be separately observed, but $X = Y_t + Y_{t+1}$ is observed at $t + 1$. Hence $D_t = D_{t-1}$ and $D_{t+1} = \{D_{t-1}, X\}$. Calculate $p(X|D_{t-1})$ and $p(\mu_{t+1}|D_{t+1})$. Generalise this result to the case when only the sum of the next $k > 1$ (rather than just $k = 2$) observations is available. These results are useful in applications where the lengths of sampling interval are not constant. This arises, for example, in companies who divide the year into 12 cost periods, 8 of length 4 weeks and 4 of length 5 weeks, for accounting purposes. A model based on weekly data can then be applied using $k = 4$ or $k = 5$ as appropriate.

(10) Re-analyse the KURIT series of Example 2.1. In place of the assumptions there, suppose that $V_t = V = 64$, $m_0 = 140$ and $C_0 = 16$. Redo the analysis illustrated in Table 2.1, producing the corresponding table from your analysis. Continue the analysis to incorporate the intervention at $t = 9$ describe in Section 2.3.2, and plot the adaptive coefficient A_t as a function of t as in Figure 2.3. Comment on the differences between this and the original analysis, and explain why they arise.

(11) In the framework of Section 2.2.3, for integers $k \geq j \geq 1$, calculate the forecast distribution $p(Z_t(j,k)|D_t)$ where $Z_t(j,k) = \sum_{r=j}^{k} Y_{t+r}$.

(12) In the closed, constant model $\{1,1,V,W\}$, verify the limiting identities $R = AV/(1-A)$, $Q = V/(1-A)$ and $W = A^2 Q$.

(13) In the closed, constant model, with limiting values A, C, R, etc., prove that the sequence C_t decreases/increases as t increases according to whether C_0 is greater/less than the limiting value C. Show also that the sequence A_t behaves similarly.

(14) Verify that $M_t = (1-\delta)Y_t + \delta M_{t-1}$ provides the unique solution to the problem of minimising the discounted sum of squares of Brown's method of EWR in Section 2.3.5(c).

(15) Consider the lead-time forecast variance $L_t(k)$ in Section 2.3.6.
   (a) Show that the value of k minimising the lead-time coefficient of variation is independent of k. What is this value when $V = 97$ and $W = 6$?
   (b) Supposing that $C_t = C$, the limiting value, show that the corresponding value of $L_t(k)/V$ depends only on k and $r = W/V$. For each value of $r = 0.05$ and $r = 0.2$, plot the ratio $L_t(k)/V$ as a function of k over $k = 1,\ldots,20$. Comment on the form of the plots and the differences between the two cases.

(16) In the discount model of Section 2.4.2 with $V_t = V$ constant and known, suppose that C_0 is very large relative to V. Show that

(a) $C_t \approx V(1-\delta)/(1-\delta^t),$ for all $t \geq 1$;

(b) $m_t \approx \dfrac{(1-\delta)}{(1-\delta^t)} \sum_{j=0}^{t-1} \delta^j Y_{t-j}.$

(17) In the constant model $\{1, 1, V, W\}$, show that $R_t = C_{t-1}/\delta_t$ where δ_t lies between 0 and 1. Thus the constant model updating equations are equivalent to those in a discount model with discount factors δ_t changing over time. Find the limiting value of δ_t as t increases, and verify that δ_t increases/decreases with t according to whether the initial variance C_0 lies below/above the limiting value C.

(18) Become familiar with just how heavy-tailed Student T distributions with small and moderate degrees of freedom are relative to normal distributions. To do this use statistical tables (eg. Lindley and Scott, 1984, page 45) to find the upper 90%, 95%, 97.5% and 99% points of the $T_n[0, 1]$ distribution for $n = 2, 5, 10$ and 20 degrees of freedom, and compare these with those of the $N[0, 1]$ distribution.

(19) Perform analyses of the USA/UK exchange rate index series along the lines of those in Section 2.6, one for each value of the discount factor $\delta = 0.6, 0.65, \ldots, 0.95, 1$. Relative to the model with $\delta = 1$, plot the MSE, MAD and LLR measures as functions of δ, and comment on these plots. Sensitivity analyses involve exploring how inferences change with respect to model assumptions. Explore just how sensitive this model is to values of δ in terms, at $t = 115$, of (a) inferences about the final level μ_{115}; (b) inferences about the observational variance V; (c) inferences about the next observation Y_{116}.

CHAPTER 3

INTRODUCTION TO THE DLM:
THE DYNAMIC REGRESSION MODEL

3.1 INTRODUCTION

In this chapter some basic concepts that underlie the general DLM theory are introduced and developed in the context of dynamic linear regression. Although we introduce the general multiple regression model, details of analysis and examples are considered only for the very special case of straight line regression through the origin. Although seemingly trivial, this particular case effectively illustrates the important messages at this introductory stage without the technical complications of larger and more practically important models.

Regression modelling concerns the construction of a mathematical and statistical description of the effect of *independent* or *regressor* variables on the *response* time series Y_t. Considering a single such variable, represented by a time series of observations X_t, regression modelling often begins by relating the mean response function μ_t of the original series to X_t, and possibly X_s, for $s < t$, via a particular *regression function*. For example, a simple linear model for the effect of the current X_t on the current mean is

$$\mu_t = \alpha + \beta X_t,$$

where the defining parameters α and β take suitable values. Models of this sort may be used in a variety of prediction, interpolation, estimation and control contexts, such as:

(i) Prediction using a *lead* variable, or indicator. For example, X_t is the number of new housing starts nine months ago, at time $t - 9$, when μ_t is the current monthly level of demand for roofing tiles and Y_t the current monthly sales;

(ii) Prediction using a *proxy* variable. For example, $X_t = t$, time itself in predicting growth in time of a population;

(iii) Control using a *control variable*. For example, the temperature level μ_t of water from a shower can be related to the setting X_t of the tap mixing incoming hot and cold water flows;

(iv) Interpolation. Suppose that the response Y_t represents a measurement of the latitude of a satellite at time t, when interest lies in describing, and estimating, the trajectory up to that time from $t = 0$, described by $\mu_t = \alpha + \beta t$.

In each of the above cases both Y_t and X_t are continuous measurements. However, practically important and interesting models often include *categorical* regressor variables that classify the response into groups according to type, presence or absence of a control, and so on. Classificatory regressor variables such as these are treated just like measured values of continuous variables. Concerning the Y_t series, however, in some applications this will be a series of discrete, categorical variables too so that the normal models here are clearly inappropriate. Non-normal models for this sort of problem are developed in a later chapter.

Construction of models in practice is guided by certain objectives specific to the problem area. Rarely is a modeller seeking to establish an all embracing model purporting to represent a "true" relationship between the response and regressor. Rather, the model is a way of looking at the problem that is required to capture those features of importance in answering specific questions about the relationship. A frequent objective, for example, is short-term forecasting of the response series. Suppose that there really is an underlying, unknown and complex relationship $f(\mu_t, X_t, t) = c$ between the level of the series μ_t, the regressor variable X_t, and time itself. If the modeller believes that this relationship is sufficiently *smooth* and well-behaved locally as a function of both X_t and t, then for short-term prediction, that is *local* inference, it may well be satisfactory to have a local model, as an approximation that satisfies the specific need, of the form $\mu_t = \alpha_t + \beta_t X_t$. Note the word *form* used here and the indexing of the coefficients α and β by t. The form of a linear model may adequately describe the *qualitative* local characteristics of the relationship for all X_t and t, but the *quantification* of this form may well have to change according to the locality. For illustration suppose that the relationship between the mean response and the regressor looks like that in Figure 3.1 for all t.

It can be seen that the general form $\mu = \alpha + \beta X$ is always locally appropriate as represented by the tangent at X. However, in region 1, β must be negative whereas in region 2 it is clearly positive. Similarly, the intercept coefficient α obviously differs between the two regions.

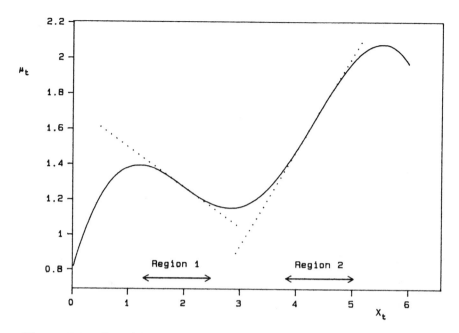

Figure 3.1. Local linearity of mean response function.

The distinction between an appropriate local model form and an appropriate quantified local model is critical and is a fundamental concept in dynamic modelling in general. Often in practice the values of the independent variable X_t change only slowly in time and an appropriate local model description is that above,

$$\mu_t = \alpha_t + \beta_t X_t,$$

where the parameters vary only slightly from one time point to the next. In modelling this time variation in the quantification of the local linear regression, the modeller develops a dynamic model. For example, a simple but often appropriate representation of a slowly evolving process has as a key component a random walk evolution of the parameters, with

$$E\left[\alpha_t \mid \alpha_{t-1}, \beta_{t-1}\right] = \alpha_{t-1},$$

and

$$E\left[\beta_t \mid \alpha_{t-1}, \beta_{t-1}\right] = \beta_{t-1},$$

Temperature X

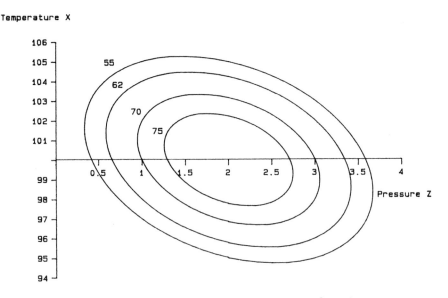

Figure 3.2.   Contours of quadratic mean response function.

together with a statement about potential variation of the expected
changes. This forms the basis of the regression DLM.

Modellers must continually be aware that model building is always
a selective process. Often there are several, if not many, independent
variables that may be considered as useful and important predictors
of Y_t. The modeller typically identifies just a few of these variables which are judged to be of greatest importance. Those which
are judged unimportant, or, indeed, of which the modeller is not
conscious, are omitted from the model with the result that their
effects are either carried via those regressors in the model or, commonly, lumped together into error terms with some broad statistical
description. It is important that this selectivity be identified as potentially contributing to inadequacies in the chosen model. Suppose
as a simple example that Y_t is the percentage yield of a chemical
process that operates under different temperature and pressure controls. Over the operating conditions, the mean yield is related to
temperature X_t and pressure Z_t according to the approximate description

$$\mu_t = 80 - (X_t - 100)^2 - 2(X_t - 100)(Z_t - 2) - 10(Z_t - 2)^2. \quad (3.1)$$

Such a relationship, as in Figure 3.2, is typical of the elliptical nature
of a yield response function of temperature and pressure in the region

of the optimum operating conditions maximising mean yield at, in this case, $X_t = 100$ and $Z_t = 2$.

Consider now two chemical plants running the same process. The first operates with constant pressure $Z_t = 1$ for which, from above,

$$\mu_t = 70 - (X_t - 100)^2 + 2(X_t - 100), \qquad (3.2)$$

so that the effect of raising temperature from 98° to 102° is to *increase* the mean yield by about 13% from 62 to 70. The second plant operates with constant pressure $Z_t = 3$ so that

$$\mu_t = 70 - (X_t - 100)^2 - 2(X_t - 100) \qquad (3.3)$$

and here the effect of the above temperature increase is to *decrease* the mean yield by about 11% from 70 to 62! Suppose that the plants are currently both operating with $X_t = 100$ so that they both have mean yield of 70%. Ignoring the difference in pressure settings will lead to a sharp difference of opinion between plant managers on the effect of raising temperature; one will claim it to be beneficial, the other detrimental. Considering pressure as a contributory factor will clarify the problem, clearly identifying the source of confusion and conflict. However, the general point of note is that all models are conditional although the conditions under which they are constructed are often *not* explicit. Thus, quite frequently, various conflicting models may be proposed for equivalent situations, each supported by empirical evidence and associated statistical tests. The truth may be that, although each model is *conditionally* correct, the conditioning may be so restrictive as to render the models practically useless.

Further study of the example provides valuable insight and a pointer to possible useful modification of simple conditional models. Consider the operating model for the first plant (3.2). If, due to uncontrollable circumstances, pressure begins to increase moving from $Z_t = 1$ to $Z_t = 3$, then the model becomes invalid and practically misleading. In this and more complex situations, and particularly in modelling open systems, the lack of awareness of the variation in, and the interaction with, excluded or unidentified variables causes confusion and misleading inferences. Simple modifications of conditional models to provide a small degree of flexibility of response to possibly changing external conditions and variables are possible. In the example, the operating model may be rewritten as

$$\mu_t = \alpha(Z_t) + \beta(Z_t)X_t + \gamma X_t^2$$

where the coefficients $\alpha(.)$ and $\beta(.)$, (and in more complex situations γ too) are functions of pressure Z_t. If it is assumed that pressure changes only slowly in time then the concept of local modelling described earlier suggests that to account for some of the variability due to the unidentified pressure variable (and possibly others too), a simple local model would be that above with coefficients $\alpha(.)$ and $\beta(.)$ replaced by time-varying, or dynamic, quantities to give

$$\mu_t = \alpha_t + \beta_t X_t + \gamma X_t^2.$$

A simple dynamic model for the coefficients, such as the earlier random walk type of evolution, will now provide a degree of flexibility in the response of the model to changes in underlying conditions and related variables.

It is worth exploring the distinction between the qualitative, that is the local model form, and the quantitative by reference to Taylor series expansions. Suppose that, temporarily suppressing the dependence on time, there exists some unknown, complex but smooth underlying relationship between the mean response function μ and an independent variable X, of the form $\mu = f(X, Z)$ where Z represents a set of possibly many related but omitted variables. For any given Z, the functional dependence on X can be represented locally in a neighbourhood of any point X_0 by the form

$$\mu = f(X_0, Z) + \frac{\delta f(X_0, Z)}{\delta X_0}(X - X_0),$$

or

$$\mu = \alpha_0(Z) + \beta_0(Z)X,$$

say. In considering the use of a linear regression on X for the response function over time, it should now be clear that there are two components important in assessing the adequacy of such a model. Firstly, the adequacy of the linear approximation as a function of X for any given Z, as above; and, secondly, the adequacy of the implicit assumption of constancy as a function of Z. Situations in which X and Z vary only slowly over time, or in which Z varies only slowly and the response is close to linear in X for given Z, provide the most satisfactory cases. Here the response function can be adequately represented as

$$\mu_t = \alpha_t + \beta_t X_t$$

with the coefficients changing slowly over time according to a simple random walk evolution. The form of such an evolution,

$$\begin{pmatrix} \alpha_t \\ \beta_t \end{pmatrix} = \begin{pmatrix} \alpha_{t-1} \\ \beta_{t-1} \end{pmatrix} + \omega_t$$

where ω_t is a zero-mean vector error term, expresses the concept of *local constancy* of the parameters, subject to variation controlled by the variance matrix of ω_t, $V[\omega_t] = \mathbf{W}_t$, say. Clearly small values of $\mathbf{W}_t$ imply a stable linear function over time, larger values leading to greater volatility and suggesting caution in extrapolating or forecasting too far ahead in time based on the quantified linear model at the current time. The usual static regression model, so commonly used, is obviously obtained as a special case of this dynamic regression in which $\omega_t = \mathbf{0}$ for all t. The foregoing discussion has identified some of the potential dangers in employing simple static models and also reveals reasons why they often prove inadequate in practice.

A final point concerns the suggestion that the above type of dynamic model is likely to break down if there happens to be a large, abrupt change in either X or Z, or both. In the chemical plant example, a sudden change in pressure from 1 to 3 leads to an enormous shift in the locally appropriate pair of coefficients α and β with the latter even reversing its sign. In spite of this marked quantitative change, however, it is still the case that the qualitative form is durable with respect to large changes in either X or Z. Furthermore, the changes in parameter values may be estimated and adapted to by suitably designing the variances $\mathbf{W}_t$ to allow for larger values of ω_t at abrupt change points. Models incorporating this important feature of coping with what are essentially discontinuities in an otherwise smooth process are an important feature of our practical approaches in later chapters.

3.2 The Multiple Regression DLM

For reference we specify the structure of the general dynamic regression model before considering the case of a single regressor variable in detail. Suppose that n regressor variables are identified and labelled $X_1, \ldots, X_n$. The value of the ith variable X_i at time t is denoted by X_{ti} and we adopt the convention that if a constant term is included then it is represented by $X_{t1} = 1$, for all t. The regression DLM is now defined.

Definition 3.1. *For each t, the model is defined by:*

Observation equation: $Y_t = \mathbf{F}'_t\,\boldsymbol{\theta}_t + \nu_t\,,$ $\nu_t \sim \mathrm{N}[0, V_t],$

System equation: $\boldsymbol{\theta}_t = \boldsymbol{\theta}_{t-1} + \boldsymbol{\omega}_t\,,$ $\boldsymbol{\omega}_t \sim \mathrm{N}[\mathbf{0}, \mathbf{W}_t],$

where the **regression vector** *at time t is $\mathbf{F}'_t = (X_{t1}, \ldots, X_{tn})$, $\boldsymbol{\theta}_t$ is the $n \times 1$* **regression parameter vector** *at time t, and $\mathbf{W}_t$ the* **evolution variance matrix** *for $\boldsymbol{\theta}_t$.*

The standard, static regression model has the above form with $\mathbf{W}_t = \mathbf{0}$ for all t, so that $\boldsymbol{\theta}_t = \boldsymbol{\theta}$, constant in time. Following the discussion in the introduction, the regression DLM assumes that the linear regression form is only locally appropriate in time with the regression parameter vector varying according to a random walk. The evolution error term $\boldsymbol{\omega}_t$ describes the changes in the elements of the parameter vector between times $t - 1$ and t. The zero mean vector here reflects the belief that $\boldsymbol{\theta}_t$ is *expected* to be constant over the interval whilst the variance matrix $\mathbf{W}_t$ governs the extent of the movements in $\boldsymbol{\theta}_t$ and hence the extent of the time period over which the assumption of local constancy is reasonable. Finally, the error sequences ν_t and $\boldsymbol{\omega}_t$ are assumed to be independent sequences with, in addition, v_t independent of $\boldsymbol{\omega}_s$ for all t and s.

An important special case obtains when $n = 1$ and a constant term is included in the model. The result is a straight line regression on $X = X_2$ specified by $\mathbf{F}_t = (1, X_t)'$, and $\boldsymbol{\theta}_t = (\alpha_t, \beta_t)'$. We then have

$$Y_t = \alpha_t + \beta_t X_t + \nu_t, \qquad \nu_t \sim \mathrm{N}[0, V_t],$$
$$\alpha_t = \alpha_{t-1} + \omega_{t1},$$
$$\beta_t = \beta_{t-1} + \omega_{t2},$$

where $\boldsymbol{\omega}_t = (\omega_{t1}, \omega_{t2})' \sim \mathrm{N}[0, \mathbf{W}_t]$.

3.3   DYNAMIC STRAIGHT LINE THROUGH THE ORIGIN

3.3.1   Introduction and definition

For illustrative purposes the simple dynamic straight line through the origin is considered. Formally this is a special case of the straight line model for which $\alpha_t = 0$ and $\boldsymbol{\theta}_t = \theta_t = \beta_t$ for each t. Thus

it is assumed that a straight line passing through the origin models
the local relationship but that in different localities the appropriate
slope values differ.

For illustration, three datasets appear in Tables 3.1, 3.2 and 3.3.
In the first, the response series Y_t represents the total annual milk

Table 3.1. Annual milk production and milk cows

Year t	1970	1971	1972	1973	1974	1975	1976	1977	1978	1979	1980	1981	1982
Milk: Y_t (lbs $\times 10^9$)	117.0	118.6	120.0	115.5	115.6	115.4	120.2	122.7	121.5	123.4	128.5	130.0	135.8
Cows: F_t ($\times 10^6$)	12.0	11.8	11.7	11.4	11.2	11.1	11.0	11.0	10.8	10.7	10.8	10.9	11.0

Table 3.2. Change in annual sales and industrial production

Year t	1	2	3	4	5	6	7	8	9	10	11	12	13	14
Y_t	12	11	9	5	3	0	−5	−7	−6	−3	7	10	13	12
F_t	4	4	3	2	1	−1	−3	−4	−3	−1	2	3	4	4

Table 3.3. Company Sales/ Total market data

Year	Company Sales Y				Total Market F			
	Quarter				Quarter			
	1	2	3	4	1	2	3	4
1975	71.2	52.7	44.0	64.5	161.7	126.4	105.5	150.7
1976	70.2	52.3	45.2	66.8	162.1	124.2	107.2	156.0
1977	72.4	55.1	48.9	64.8	165.8	130.8	114.3	152.4
1978	73.3	56.5	50.0	66.8	166.7	132.8	115.8	155.6
1979	80.2	58.8	51.1	67.9	183.0	138.3	119.1	157.3
1980	73.8	55.9	49.8	66.6	169.1	128.6	112.2	149.5
1981	70.0	54.8	48.7	67.7	156.9	123.4	108.8	153.3
1982	70.4	52.7	49.1	64.8	158.3	119.5	107.7	145.0
1983	70.0	55.3	50.1	65.6	155.3	123.1	109.2	144.8
1984	72.7	55.2	51.5	66.2	160.6	119.1	109.5	144.8
1985	75.5	58.5			165.8	127.4		

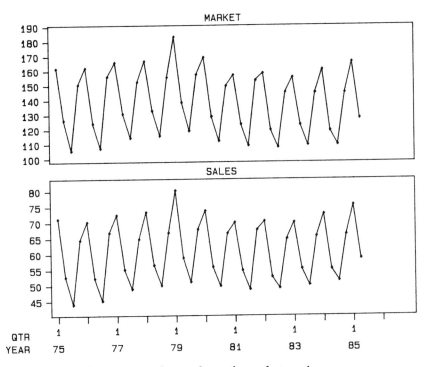

Figure 3.3.   Company sales and total market series.

production in the U.S.A. over the years $t = 1970$ to $t = 1982$, whilst F_t is the total number of milk cows. In this case, θ_t represents the average annual milk output per cow in year t. The model is not of primary interest for forecasting the Y_t series but rather for assessing the changing productivity over time. In the second dataset, a leading indicator series is to be used in forecasting the annual sales of a given product. Here Y_t is the change in annual sales between years $t - 1$ and t, and F_t is the change in an index of industrial production, the leading indicator, between the years $t - 2$ and $t - 1$. The final dataset concerns quarterly sales, Y_t in standardised units, of a company in a market whose total sales comprise the series F_t, over the years 1975 to mid-1985. Primary interest lies in assessing the changing relationship between the two series over time, and in short-term (say, up to one year ahead) forecasting. A major feature of this dataset is the marked annual seasonal pattern exhibited by each of the series over the year. The two series are plotted over time in Figure 3.3. A simple scatter plot of Y_t versus F_t, given in Figure 3.4, removes this

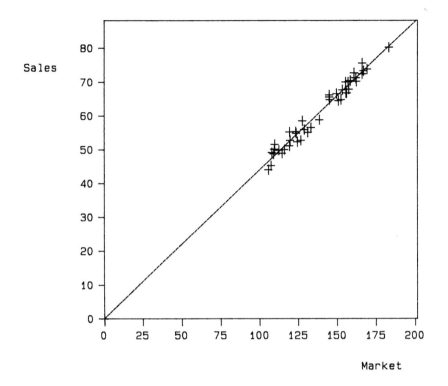

Figure 3.4.   Company sales versus total market sales.

seasonality and seems to support a simple, essentially static straight line regression with, from the nature of the data, zero origin.

Definition 3.2. *For each t, the form of model is given as the special case of Definition 3.1. The full model specification requires the addition of the initial prior, as follows:*

Observation equation:	$Y_t = F_t\theta_t + \nu_t,$	$\nu_t \sim N[0, V_t],$
System equation:	$\theta_t = \theta_{t-1} + \omega_t,$	$\omega_t \sim N[0, W_t],$
Initial prior:	$(\theta_0 \mid D_0) \sim N[m_0, C_0]$	

for some mean m_0 and variances C_0, V_t and W_t.

The sequential model description for the series requires that the defining quantities at time t be known at that time. Similarly, when forecasting more than one step ahead to time $t + k$ at time t, the corresponding quantities F_{t+k}, V_{t+k}, and W_{t+k} must belong to the current information set D_t. In general for this chapter, and unless otherwise specified, it will be assumed that the information set D_0 contains all the future values of F_t, V_t, and W_t. Thus we again have have $D_t = \{Y_t, D_{t-1}\}$ for each t. Finally we have the usual assumptions of mutual independence of the error sequences and θ_0.

3.3.2 Updating and forecasting equations

Theorem 3.1. *One-step forecast and posterior distributions are given, for each t, as follows:*

(a) *Posterior for θ_{t-1}* : $(\theta_{t-1} \mid D_{t-1}) \sim N[m_{t-1}, C_{t-1}]$

for some mean m_{t-1} and variance C_{t-1};

(b) *Prior for θ_t* : $(\theta_t \mid D_{t-1}) \sim N[m_{t-1}, R_t]$

where $R_t = C_{t-1} + W_t$;

(c) *1-step forecast* : $(Y_t \mid D_{t-1}) \sim N[f_t, Q_t]$,

where $f_t = F_t m_{t-1}$ and $Q_t = F_t^2 R_t + V_t$;

(d) *Posterior for θ_t* : $(\theta_t \mid D_t) \sim N[m_t, C_t]$,

with $m_t = m_{t-1} + A_t e_t$ and $C_t = R_t V_t / Q_t$,

where $A_t = F_t R_t / Q_t$, and $e_t = Y_t - f_t$.

Proof: The technique of proof is essentially the same as used in Theorem 2.1. Assume that (a) holds at time t. Then writing Y_t and θ_t in terms of linear functions of θ_{t-1}, ν_t and ω_t it follows immediately that Y_t and θ_t have a bivariate normal distribution conditional on D_{t-1}, with means and variances as stated above, establishing (b) and (c). For the covariance we have

$$\begin{aligned} C[Y_t, \theta_t \mid D_{t-1}] &= C[F_t \theta_t + \nu_t, \theta_t \mid D_{t-1}] \\ &= C[F_t \theta_t, \theta_t \mid D_{t-1}] + C[\theta_t, \nu_t \mid D_{t-1}] \\ &= F_t V[\theta_t \mid D_{t-1}] + 0 = F_t R_t, \end{aligned}$$

so that

$$\left(\begin{matrix} Y_t \\ \theta_t \end{matrix} \middle| D_{t-1} \right) \sim N\left[\left(\begin{matrix} m_{t-1} \\ f_t \end{matrix} \right), \left(\begin{matrix} Q_t & F_t R_t \\ F_t R_t & R_t \end{matrix} \right) \right].$$

From this, the regression coefficient of θ_t on Y_t is just $A_t = F_t R_t / Q_t$ and, using normal theory from Section 16.2 of Chapter 16, we have

$$(\theta_t \mid Y_t, D_{t-1}) \sim N[m_t, C_t],$$

where

$$m_t = m_{t-1} + A_t(Y_t - f_t),$$

and

$$C_t = R_t - (R_t F_t)^2 / Q_t.$$

This latter equation reduces to $C_t = R_t V_t / Q_t$ and (d) follows.

$\diamond$

For forecasting at time t, the main distributions required by the forecaster are the k-step ahead marginals, $p(Y_{t+k} \mid D_t)$ for $k > 0$. These distributions are as follows.

Theorem 3.2. For $k > 0$, the k-step ahead forecast distribution is:

$$(Y_{t+k} \mid D_t) \sim N[f_t(k), Q_t(k)]$$

where $f_t(k) = F_{t+k} m_t$, and $Q_t(k) = F_{t+k}^2 R_t(k) + V_{t+k}$, with $R_t(k) = C_t + \sum_{r=1}^{k} W_{t+r}$.

Proof: From the evolution equation for θ_t we have, for $k \geq 1$,

$$\theta_{t+k} = \theta_t + \sum_{r=1}^{k} \omega_{t+r};$$

from the observational equation we therefore have

$$Y_{t+k} = F_{t+k}\{\theta_t + \sum_{r=1}^{k} \omega_{t+r}\} + \nu_{t+k}.$$

Since all terms are normal and mutually independent, $(Y_{t+k} \mid D_t)$ is normal and the mean and variance follow directly. Note in passing that we also have $(\theta_{t+k} \mid D_t) \sim N[m_t, R_t(k)]$.

$\diamond$

3.3.3.   General comments

The following points are noteworthy.

(i) The posterior mean m_t is obtained by correcting the prior mean m_{t-1} with a term proportional to the forecast error e_t. The coefficient $A_t = R_t F_t / Q_t$ scales the correction term according to the relative precisions of the prior and likelihood, as measured by R_t/Q_t, and by the regressor value F_t. So the correction always takes the sign of F_t and is, in general, unbounded.

(ii) The posterior precision C_t^{-1} is

$$C_t^{-1} = Q_t/(R_t V_t) = R_t^{-1} + F_t^2 V_t^{-1},$$

so that, for $F_t \neq 0$, it always exceeds the prior precision R_t^{-1}. Thus the posterior for θ_t is never more diffuse than the prior. Further, the precision increases with $|F_t|$. If, however, $F_t = 0$, then Y_t provides no information and $C_t = R_t$. If this is true for a sequence of observations then the sequence C_t continues to increase by the addition of further W_t terms reflecting an increasingly diffuse posterior. Thus, although there may exist an appropriate regression relationship changing in time, information relevant to this relationship is not being received. Although seemingly trivial, this point is critical when considering the case of multiple regression. The chemical plant operation of Section 3.1 is a case in point. In view of concern about falling short of set production targets, management of individual plants may be wary of varying operating conditions away from standard, well-used controlling conditions. In such cases, the absence of planned variation in operating conditions means that no new information is obtained about the effect on, say, chemical yield, of variable factors such as temperature. As described in Section 3.1, the yield relationships with controlling factors may change in time and thus actual optimum operating conditions can move away from those initially identified. Hence the lack of information about the effects of changing conditions means that there is little prospect of estimating the new ideal conditions. The need for a continual flow of information on the effects of changes in operating conditions was recognised by G.E.P. Box when at Imperial Chemical Industries. This led to the development of Evolutionary Operation, which advocates continued small variation in conditions near the currently identified optimum so that movements away from this can be identified (Box and Draper, 1969).

(iii) Consider the special case of constant variances, $V_t = V$ and $W_t = W$ for each t. In general the sequence C_t is neither monotonic nor convergent. However, in the degenerate constant model given by $F_t = F \neq 0$, the corresponding model for the scaled series Y_t/F is a constant, first-order polynomial model with observational variance V/F. It follows from Theorem 2.3 that

$$\lim_{t \to \infty} A_t = A(F),$$

and

$$\lim_{t \to \infty} C_t = C(F),$$

where, if $r(F) = WF^2/V$,

$$A(F) = Fr(F)[\sqrt{(1 + 4r(F))} - 1]/2,$$

and

$$C(F) = FA(F)V.$$

Consequently, in the more general case of bounded regressor values, $a < F_t < b$ for all t, then, as t increases, A_t will lie in the interval $[A(a), A(b)]$ and C_t in the interval $[C(c), C(d)]$, where $c = \max(|a|, |b|)$ and $d = \min\{|u| : u \in [a, b]\}$.

(iv) For the static model in which $W_t = 0$ for all t, θ is assumed constant and it follows that

$$C_t^{-1} = C_0^{-1} + \sum_{r=1}^{t} F_r^2 V_r^{-1},$$

and

$$m_t = C_t[C_0^{-1}m_0 + \sum_{r=1}^{t} F_r^2 V_r^{-1} Y_r].$$

This is the standard posterior distribution derived from a non-sequential analysis of the constant regression model. In the particular cases of either large t, or relatively little prior knowledge as represented by small C_0^{-1}, m_t is approximately the usual maximum likelihood point estimate of θ, and C_t the associated variance.

3.3.4 Illustrations

The model is now illustrated by application to the first two datasets introduced above. The objective is simply to demonstrate the analysis and highlight practically interesting features. Reasonably uninformative priors are chosen at $t = 0$ for the purpose of illustration. EXAMPLE 3.1: U.S. Milk production. The years of this dataset, in Table 3.4, are renumbered 1 to 13 for convenience. The constant variance model with $V = 1$ and $W = 0.05$ is applied. θ_t may be interpreted as the level, in thousands of pounds, of milk per cow for year t. Initially we assume $m_0 = 10$ and $C_0 = 100$, representing a high degree of uncertainty about θ_0. Table 3.4a gives the values of m_t, C_t and A_t calculated sequentially as in (d) of Theorem 3.1.

The m_t series is seen to increase except at $t = 4$. This reflects the increasing efficiency in general dairy production over time through better management and improving breeds of cow. In practice a modeller would wish to incorporate this feature by modelling "growth" in θ_t, the current, constant model being obviously deficient. However, this particular simple model illustrates how, even though it is unsatisfactory for long term prediction, the dynamic nature of θ_t leads to reasonable short term forecasts. For example at $t = 11$ and $t = 12$ the one-step ahead forecasts are $(Y_{12} \mid D_{11}) \sim N[129.2, 7.8]$ and $(Y_{13} \mid D_{12}) \sim N[131.1, 7.9]$. The actual observations $Y_{12} = 130.0$ and

Table 3.4. Analyses for Milk data: (a) Dynamic model, (b) Static model.

t	1	2	3	4	5	6	7	8	9	10	11	12	13
F_t	12.0	11.8	11.7	11.4	11.2	11.1	11.0	11.0	10.8	10.7	10.8	10.9	11.0
Y_t	117.0	118.6	120.0	115.5	115.6	115.4	120.2	122.7	121.5	123.4	128.5	130	135.8
(a)													
m_t	9.75	10.01	10.23	10.14	10.30	10.38	10.86	11.12	11.23	11.49	11.85	11.92	12.29
$1000C_t$	6.94	6.38	6.47	6.77	6.99	7.10	7.22	7.22	7.46	7.58	7.46	7.34	7.22
$100A_t$	8.30	7.50	7.60	7.70	7.80	7.90	7.90	7.90	8.10	8.10	8.10	8.00	7.90
(b)													
m_t	9.75	9.90	10.01	10.04	10.09	10.14	10.24	10.35	10.44	10.54	10.65	10.75	10.87
$1000C_t$	6.94	3.53	2.38	1.82	1.48	1.25	1.09	0.96	0.86	0.79	0.72	0.66	0.61
$100A_t$	8.30	4.20	2.80	2.10	1.70	1.40	1.20	1.10	0.90	0.80	0.80	0.70	0.70

$Y_{13} = 135.8$ are well within acceptable forecast limits. By comparison, Table 3.4b gives results from the standard static model with $W_t = 0$. It is clear that forecasts from this model are totally unsatisfactory with, for example, $(Y_{12} \mid D_{11}) \sim N[116.1, 1.09]$ and $(Y_{13} \mid D_{12}) \sim N[118.2, 1.08]$. The beneficial effects of assuming a model form holding only locally rather than globally is clearly highlighted. The dynamic assumption leads to greater robustness and helps to compensate for model inadequacies which, at first, may not be anticipated or noticed. Since the independent variable F_t varies between 10.7 and 12.0 it is not surprising that in the dynamic model the values of C_t and A_t settle down to vary in narrow regions. By contrast in the static model they both decay to zero so that the model responds less and less to the most recent data points and any movements in milk productivity.

EXAMPLE 3.2: Company Sales. The sales data in Table 3.5 is analysed with the constant model in which $V = 1$ and $W = 0.01$, with initial, relatively diffuse prior having $m_0 = 2$ and $C_0 = 0.81$. Table 3.5 gives the values of m_t, C_t and A_t. Neither the variance nor the adaptive coefficient are monotonic, and there is no convergence. Here F_t takes negative and positive values and A_t takes the sign of the current regressor value. When the latter is near zero, C_t tends to increase since the information provided by Y_t is not compensating for increased uncertainty about the regression in the movement from θ_{t-1} to θ_t. On the other hand, when F_t increases in absolute value, the observations are very informative leading to decreases in C_t and increases in the absolute value of A_t.

Looking at m_t, it begins in the region of 3 then drops to near 2 before rising again to 3, apparently completing a cycle. Since F_t varies in a similar manner, there is a suggestion that a large part of this variation in m_t could be accounted for by relating it to F_t, and

Table 3.5. Analysis of company sales data

t	1	2	3	4	5	6	7	8	9	10	11	12	13	14
F_t	4	4	3	2	1	−1	−3	−4	−3	−1	2	3	4	4
Y_t	12	11	9	5	3	0	−5	−7	−6	−3	7	10	13	12
m_t	2.93	2.84	2.88	2.83	2.84	2.69	2.33	2.08	2.06	2.09	2.31	2.63	2.88	2.93
$100C_t$	5.80	3.30	3.10	3.50	4.30	5.00	3.90	2.80	2.80	3.70	3.90	3.40	2.60	2.30
$10A_t$	2.32	1.30	0.92	0.70	0.43	−0.50	−1.17	−0.84	−0.84	−0.37	0.79	1.02	1.03	0.91

a more elaborate model is worth considering. A simple, tentative example would be to extend the regression to include a quadratic term in F_t thus implying a multiple regression rather than the simple straight line here.

One further point that may be queried is that this simple model deals with *changes* in both response and regressor variables, rather than their original values. Typically this will lead to time series that are not conditionally independent as the model implies. If, for example, $Y_t = U_t - U_{t-1}$, then Y_t and Y_{t-1} have U_{t-1} in common and so will be dependent. The reason why many difference models are used in classical forecasting approaches is that there is a desire to use static, stationary models that would be more reasonable for the differenced series Y_t than for U_t, on the basis that dynamic changes in the latter which cannot be handled by static models can be partially eliminated by differencing. Our view, however, is that, although this reasoning may be validated in some cases, it is typically more natural and sounder to develop dynamic, stochastic models directly for the original, undifferenced data U_t.

3.4   MODEL VARIANCES AND SUMMARY

3.4.1   Summary of updating and forecasting equations

Estimation of the observational variance and assignment of values to the evolution variance series are problems discussed in detail for the general DLM in later chapters. With a constant observational variance, the analysis is a minor generalisation of that for the first-order polynomial model as developed in Section 2.5 and the corresponding results are simply summarised here. The observational variance, if assumed unknown but constant, is estimated using the fully conjugate Bayesian analysis based on gamma prior/posterior distributions for the precision parameter. The evolution variance sequence may be assigned values based on the use of a single discount factor. In the regression setting this is particularly important. From the defining equations in Definition 3.2 it is immediately clear that W_t is not invariant to the scale of measurement of the independent variable F_t. Because of this, and the fact that the amount of information about θ_t conveyed by an observation varies with $|F_t|$, it is difficult in practice to assign suitable values to the evolution variance sequence without reference to the prior variances. If a discount approach is used as in the first-order polynomial model in Section 2.4.2, a discount factor

$\delta, (0 < \delta < 1)$, leads to W_t being naturally defined as a multiple of the variance C_{t-1} at time $t - 1$, namely $W_t = C_{t-1}(\delta^{-1} - 1)$ with the result that $R_t = W_t/\delta$.

The table below summarises the updating components of the analysis. Again, reference can be made to Section 4.5 of Chapter 4 for details in the general DLM of which this is a particular case, the full summary, including forecast distributions, being easily deduced.

Again this is completely analogous to the results for the first order polynomial model; n_t is the degree of freedom parameter for the posterior distribution of the scale ϕ, increasing by one for each observation, and d_t is the "sum of squares" parameter which is updated by the addition of a scale multiple of the square of the standardised one-step ahead forecast error $e_t/\sqrt{Q_t}$.

3.4.2 Example

The company sales/total market dataset of Table 3.3 and Figure 3.3 is used for illustration of the sequential analysis summarised above and also to illuminate some further practical points. Sales are plotted against total market data in Figure 3.4, demonstrating an apparently stable linear relationship. Here θ_t is the expected market share of the company in month t. Initially, the prior estimate of θ_0 at $t = 0$ is taken as $m_0 = 0.45$, in line with pre-1975 information, and uncertainty about θ_0 is represented by $C_0 = 0.0025$. A relatively

Regression DLM with zero origin and constant variance		
Observation:	$Y_t = F_t\theta_t + \nu_t$	$\nu_t \sim N[0, V]$
System:	$\theta_t = \theta_{t-1} + \omega_t$	$\omega_t \sim T_{n_{t-1}}[0, W_t]$
Information:	$(\theta_{t-1} \mid D_{t-1}) \sim T_{n_{t-1}}[m_{t-1}, C_{t-1}]$ $(\theta_t \mid D_{t-1}) \sim T_{n_{t-1}}[m_{t-1}, R_t]$ $(\phi \mid D_{t-1}) \sim G[n_{t-1}/2, d_{t-1}/2]$	$R_t = C_{t-1} + W_t$ $S_{t-1} = d_{t-1}/n_{t-1}$
Forecast:	$(Y_t \mid D_{t-1}) \sim T_{n_{t-1}}[f_t, Q_t]$	$f_t = F_t m_{t-1}$ $Q_t = F_t^2 R_t + S_{t-1}$

Updating Recurrence Relationships
$(\theta_t \mid D_t) \sim T_{n_t}[m_t, C_t]$ $(\phi \mid D_t) \sim G[n_t/2, d_t/2]$ with $m_t = m_{t-1} + A_t e_t,$ $C_t = (S_t/S_{t-1})[R_t - A_t^2 Q_t] = R_t S_t/Q_t,$ $n_t = n_{t-1} + 1, \; d_t = d_{t-1} + S_{t-1}e_t^2/Q_t, \text{ and } S_t = d_t/n_t,$ where $e_t = Y_t - f_t$, and $A_t = F_t R_t/Q_t.$

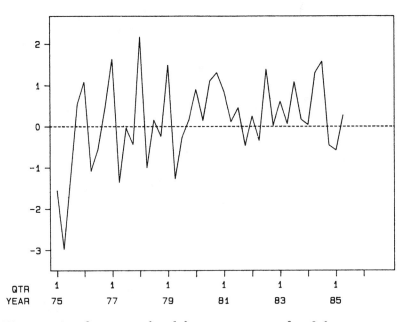

Figure 3.5.   One-step ahead forecast errors : $\delta = 0.6$.

vague prior for the observational precision is assigned using $n_0 = d_0 = 1$ so that the point estimate is $S_0 = 1$. This implies the prior $(\theta_0 \mid D_0) \sim T_1[0.45, 0.0025]$ with, in particular, 90% prior probability that θ_0 lies between 0.13 and 0.77, symmetrically about the mode 0.45.

It is of interest to consider several analyses of the data using this model and initial prior, differing only through the value of the discount factor δ. Consider first an analysis with $\delta = 0.6$ corresponding to a view that θ_t may vary rather markedly over time. Given the initial prior, sequential updating and one-step ahead forecasting proceeds directly. Figure 3.5 displays a plot over time of the raw one-step ahead forecast errors e_t. Note that there is a preponderance of positive errors during the latter half of the time period, years 1980 onwards, indicating that the model is generally under-forecasting the data series Y_t during that time. Figure 3.6 provides a plot over time of the on-line estimated value m_t of the regression parameter θ_t, referred to as the estimated *trajectory* of the coefficient. This is the full line in the figure. An indication of uncertainty about the value at time t is indicated by the dashed lines symmetrically located either side of the estimate. These limits provide 90% posterior probability intervals (in fact, HPD intervals) for the corresponding values of

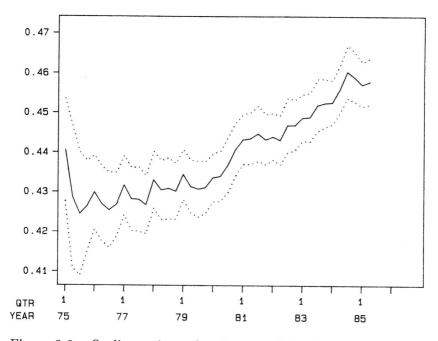

Figure 3.6.   On-line estimated trajectory of θ_t: $\delta = 0.6$.

θ_t, calculated from the Student T posterior distributions $p(\theta_t|D_t)$. From this figure it is clear that θ_t drifts upwards over time, particularly over the last six or seven years of data. The model, though not predicting such positive drift, adapts as data is processed, sequentially adjusting the posterior to favour higher values consistent with the data. The fact that the model is very adaptive (with δ rather low at 0.6), leads to the resulting marked degree of inferred change in θ_t. However, since the model implies that θ_t undergoes a random walk, the positive drift cannot be anticipated and so, generally, under-forecasting results as evidenced in the forecast errors in Figure 3.5.

Consider now an analysis with $\delta = 1$. Here the model is a standard, static regression with $\theta_t = \theta_0$, constant over time. The sequentially calculated one-step ahead errors e_t from this analysis appear in Figure 3.7. The preponderance of positive errors is evident here as with the previous analysis, but the effect is profound. From 1980 onwards, all the errors are positive, and tending to increase, indicating continual deterioration in model adequacy. Figure 3.8 displays the on-line estimates m_t with 90% HPD intervals. Again the model is adapting to higher values of θ_t as time progresses, but the rate

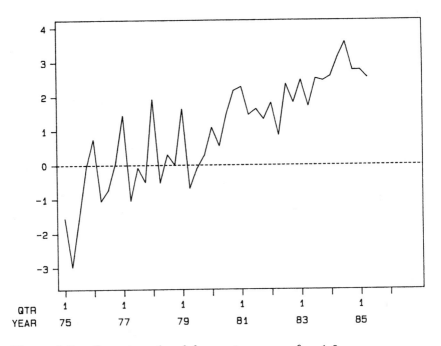

Figure 3.7.   One-step ahead forecast errors : $\delta = 1.0$.

of adaptation is much lower than the previous, very adaptive model with $\delta = 0.6$. This under-adaptation leads to increasingly poor forecasts as time progresses.

Obviously the regression model form with no expected drift in θ_t is deficient for this dataset. Despite this, the model with $\delta = 0.6$ adapts sufficiently to new data that the forecast errors, though tending to be positive, are relatively small. By comparison, the static regression model with $\delta = 1$ performs extremely poorly in one-step ahead forecasting due to its under-adaptivity. Thus, interpreting Figure 3.4 as suggesting a simple static regression is certain to mislead. This plot masks the effects of time on the relationship between the two series. Many readers will immediately note that a simple plot of the ratio Y_t/F_t (empirical estimates of the θ_t values) over time would indicate the increasing nature of θ_t and suggest a more appropriate model form; Figure 3.9 provides such a plot, the ratio values appearing as crosses. Also plotted are the sequences of on-line estimates m_t from each of the two analyses described above; the full line is that from the adaptive analysis with $\delta = 0.6$, the dashed line from the static model. That the adaptive model tracks changes in θ_t rather well is further evidenced here.

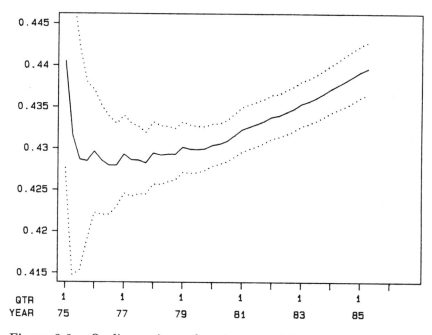

Figure 3.8.   On-line estimated trajectory of θ_t: $\delta = 1.0$.

To further compare the two analyses, some details are given below. Tables 3.6 and 3.7 display the data and some components of the prior/posterior distributions from the two analyses for the final 12 quarters, $t = 31, 32, \ldots, 42$.

The major features of these tables are as follows.

(a) As already noted, the increasing market share is identified most clearly by the model with the lower discount of 0.6 where m_t approaches 0.46 at the end of the data, though it is not continually increasing. The one-step forecasts are good, although the increasing θ_t leads to a dominance of positive forecast errors e_t, albeit very small compared to their distributions. Note that $n_t = t + 1$ so that the Student T forecast distributions are close to normality and Q_t is then roughly the variance associated with e_t.

In contrast, the static regression model with discount 1.0 is far less adaptive, the m_t sequence slowly increasing to 0.44 at the end of the data. This is well below the more reasonable values near 0.46 as is evident from the sustained, and significant, under-forecasting of the Y_t series. The key point here is that, although the model form is really inappropriate with,

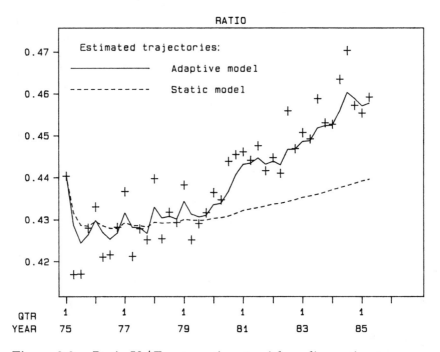

Figure 3.9. Ratio Y_t/F_t versus time t, with on-line estimates.

Table 3.6. Sales data example: discount factor = 0.6.

t	F_t	Y_t	f_t	$Q_t^{1/2}$	$S_t^{1/2}$	m_t	$C_t^{1/2}$
31	107.7	49.1	47.71	0.94	0.80	0.447	0.0040
32	145.0	64.8	64.79	1.10	0.79	0.447	0.0037
33	155.3	70.0	69.39	1.09	0.78	0.449	0.0035
34	123.1	55.3	55.23	0.96	0.77	0.449	0.0036
35	109.2	50.1	49.02	0.92	0.77	0.452	0.0039
36	144.8	65.6	65.43	1.06	0.76	0.452	0.0036
37	160.6	72.7	72.66	1.07	0.75	0.453	0.0033
38	119.1	55.2	53.90	0.91	0.76	0.456	0.0036
39	109.5	51.5	49.93	0.92	0.78	0.460	0.0039
40	144.8	66.2	66.65	1.07	0.77	0.459	0.0037
41	165.8	75.5	76.08	1.10	0.77	0.457	0.0033
42	127.4	58.5	58.23	0.94	0.76	0.458	0.0034

for example, a growth term needed in θ_t, the lower discount model adapts over time to the changes and the resulting short-term forecast are acceptable. The standard, static model is clearly extremely poor by comparison, and the message extends to more general models, and regressions in particular.

(b) The more adaptive model allows for a much greater decay of information about θ_t over time and this results in a larger posterior variance. At the end of the data, for example, the posterior standard deviation of θ_{42} in the adaptive model is almost 80% greater than that in the static model. This is a large difference due to a very small discount at 0.6. Typically, if the form of the model is generally appropriate, as is clearly not the case here, then discount factors for regression will exceed 0.9. Concerning the static model, note that, in addition to very poor forecasts, the posteriors for θ_t are overly precise, being highly concentrated about a mode that is far from suitable as an estimate of θ_t !

(c) The adaptive model correctly attributes a high degree of variation in the Y_t series to movement in θ_t and so much less than the static model to observational variation about level. The final estimate of observational standard deviation in the adaptive model is 0.76 compared with 1.69 in the static case.

Table 3.7. Sales data example: discount factor = 1.0.

t	F_t	Y_t	f_t	$Q_t^{1/2}$	$S_t^{1/2}$	m_t	$C_t^{1/2}$
31	107.7	49.1	46.74	1.19	1.23	0.434	0.0016
32	145.0	64.8	62.98	1.26	1.25	0.435	0.0016
33	155.3	70.0	67.52	1.28	1.30	0.435	0.0016
34	123.1	55.3	53.59	1.32	1.32	0.436	0.0016
35	109.2	50.1	47.58	1.33	1.36	0.436	0.0016
36	144.8	65.6	63.14	1.38	1.40	0.437	0.0017
37	160.6	72.7	70.11	1.43	1.44	0.437	0.0017
38	119.1	55.2	52.06	1.46	1.51	0.438	0.0017
39	109.5	51.5	47.92	1.52	1.59	0.438	0.0018
40	144.8	66.2	63.44	1.62	1.63	0.439	0.0018
41	165.8	75.5	72.73	1.66	1.66	0.439	0.0019
42	127.4	58.5	55.96	1.68	1.69	0.440	0.0019

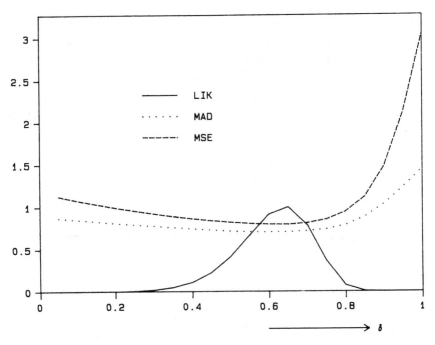

Figure 3.10.   MSE, MAD and LIK measures as functions of δ.

(d) The one-step forecast variances in the adaptive model are much smaller than those in the static model. The final one-step forecast standard deviation in the former is 0.94 compared to 1.68 in the latter. This is due to (c); the observational variance is heavily over-estimated in the static model. Thus, in addition to providing much more accurate point forecasts than the static model, the adaptive model produces much more concentrated forecast distributions. By any measures of comparative performance, the adaptive model is way ahead of the standard form.

Further, similar analyses with different values of δ can be assessed and compared using the MSE, MAD and LLR (log likelihood ratio) criteria as demonstrated in the example of Section 2.7. Each of these measures is calculated from analyses with $\delta = 0.05, 0.1, \ldots, 0.95, 1.0$. MSE and MAD measures are graphed as functions of δ over this range in Figure 3.10. Also plotted is the actual model likelihood LIK=exp(LLR). This figure indicates very clearly that a static model ($\delta = 1.0$) is highly inappropriate. Both MSE and MAD measures, as functions of δ, decay rapidly from their maxima at $\delta = 1.0$ to minima near $\delta = 0.6$, rising only slightly thereafter as δ approaches 0. Also, the curves are very flat between $\delta = 0.4$ and $\delta = 0.8$ indicating the usual marked robustness to particular values within a suitable range. The likelihood function LIK for δ peaks between 0.6 and 0.7, being neglible above $\delta = 0.85$. Also, very adaptive models with δ less than 0.5 have low likelihood, being penalised since the corresponding one-step forecast distributions are very diffuse.

3.5 EXERCISES

(1) One measure of predictability of Y_{t+1} at time t is the recipro-cal of the coefficient of variation, given by $f_t/Q_t^{1/2}$ (sometimes referred to as a signal to noise ratio). Explore the dependency of this measure as a function of F_t when $Q_t = 1 + R_t F_t^2$.

(2) In the model $\{F_t, 1, V, W\}$ suppose that the regressor variable F_t is actually a sequence of independent normal random quan-tities, $F_t \sim N[X, U]$ independently, with known mean X and variance U. Suppose also that F_t is independent of ω_s for all t and s. Show that, given θ_t, Y_t and F_t have a bivariate nor-mal distribution and identify the mean vector and covariance matrix. What is the correlation between Y_t and F_t?

(3) In the model $\{F_t, 1, V_t, W_t\}$, suppose that $F_t \neq 0$ for all t.
  (a) Show that the series $X_t = Y_t/F_t$ follows a first-order poly-nomial DLM and identify the defining observational and evolution variance sequences.
  (b) Verify that the updating equations for the regression DLM can then be deduced from those of the simpler, first-order polynomial model.
  (c) What happens if $F_t = 0$?

(4) Suppose that V_t is a known function of F_t and that F_t is con-trollable. In particular, let $V_t = V(a + |F_t|^p)$ for some known quantities V, a and p.
  (a) How should F_t be chosen in order to maximise the posterior precision C_t subject to $|F_t| < k$ for some $k > 0$? What is the optimal design value of F_t in the case $a = 8$, $p = 3$ and $k = 10$?
  (b) How does this problem change when V is unknown and estimated as usual?

(5) In the regression DLM with constant V and W, suppose that F_t is controlled to be $F_t = (-1)^t k$ for some constant $k > 0$. Prove that the updating equations have a stable limiting form as t increases, in the sense that each of the series C_t, R_t, Q_t and A_t

converge to limiting values. Find the limiting value of C_t as a function of k, and verify that this limit is W if $k = \sqrt{0.5V/W}$.

(6) Suppose that $V_t = Vk_t$ where $V = 1/\phi$ is unknown and k_t is a known variance multiplier. Show how the analysis summarised in the table in Section 3.4.1 is modified.

(7) Consider the discount regression DLM in which V_t is known for all t and $R_t = C_{t-1}/\delta$. Show that the updating equations can be written as

$$m_t = C_t(\delta^t C_0^{-1} m_0 + \sum_{j=0}^{t-1} \delta^j F_{t-j} V_{t-j}^{-1} Y_{t-j})$$

and

$$C_t^{-1} = \delta^t C_0^{-1} + \sum_{j=0}^{t-1} \delta^j F_{t-j}^2 V_{t-j}^{-1}.$$

Deduce that, when C_0 is very large, m_t is approximately given by the exponentially weighted regression (EWR) form

$$m_t \approx (\sum_{j=0}^{t-1} \delta^j F_{t-j} V_{t-j}^{-1} Y_{t-j}) / \sum_{j=0}^{t-1} \delta^j F_{t-j}^2 V_{t-j}^{-1}.$$

(8) Consider the company sales/total market series in the example of Section 3.4.2. Perform similar analyses of this data using the same DLM but varying the discount factor over the range $0.6, 0.65, \ldots, 1$. Explore the sensitivity to inferences about the time trajectory of θ_t as the discount factor varies, in the following ways.

(a) Plot m_t versus t, with intervals based on $C_t^{1/2}$ to represent uncertainty, for each value of δ and comment on differences with respect to δ.

(b) Compare the final estimates of observational variance S_{42} as δ varies. Do the same for prediction variances Q_{42}. Discuss the patterns of behaviour.

(c) Use MSE, MAD and LLR measures to assess the predictive performance of the models relative to the static model with $\delta = 1$.

(9) Consider the problem of looking back over time to make inferences about historical values of θ_t based on more recent data (*retrospective* analysis). For any t, prove the following, *one-step back* updating results for the model $\{F_t, 1, V_t, W_t\}$ with V_t known.

(a) Use the system equation directly to show that

$$C[\theta_t, \theta_{t-1} | D_{t-1}] = B_{t-1} V[\theta_t | D_{t-1}]$$

for some B_{t-1} lying between 0 and 1, and identify B_{t-1}.

(b) Deduce that

$$C[\theta_{t-1}, Y_t | D_{t-1}] = B_{t-1} C[\theta_t, Y_t | D_{t-1}].$$

(c) Hence identify the moments of the joint normal distribution of $(\theta_{t-1}, \theta_t, Y_t | D_{t-1})$, and from this those of the conditional distribution of $(\theta_{t-1} | D_t)$ (by conditioning on Y_t in addition to D_{t-1}). Verify that the regression coefficient of θ_{t-1} on Y_t is $B_{t-1} A_t$ where A_t is the usual regression coefficient (adaptive coefficient) of θ_t on Y_t.

(d) Deduce that $(\theta_{t-1} | D_t)$ is normal with moments that can be written as

$$E[\theta_{t-1} | D_t] = m_{t-1} + B_{t-1}(E[\theta_t | D_t] - E[\theta_t | D_{t-1}])$$

and

$$V[\theta_{t-1} | D_t] = C_{t-1} - B_{t-1}^2 (V[\theta_t | D_{t-1}] - V[\theta_t | D_t]).$$

(10) Generalise the results of the previous exercise to allow retrospection back over time more than one step, calculating the distribution of $(\theta_{t-k} | D_t)$ for any k, $(0 \le k \le t)$. Do this as follows.

(a) Using the observation and evolution equations directly, show that, for any $r \ge 1$,

$$C[\theta_{t-k}, Y_{t-k+r} | D_{t-k}] = B_{t-k} C[\theta_{t-k+1}, Y_{t-k+r} | D_{t-k}]$$

where, for any s, $B_s = C_s / R_{s+1}$ lies between 0 and 1.

(b) Writing $\mathbf{X}_t(k) = (Y_{t-k+1}, \ldots, Y_t)'$, deduce from (a) that

$$C[\theta_{t-k}, \mathbf{X}_t(k) | D_{t-k}] = B_{t-k} C[\theta_{t-k+1}, \mathbf{X}_t(k) | D_{t-k}].$$

(c) Hence identify the moments of the joint normal distribution of

$$(\theta_{t-k}, \theta_{t-k+1}, \mathbf{X}_t(k) | D_{t-k}),$$

and from this those of the conditional distributions of
$(\theta_{t-k}|D_t)$ and $(\theta_{t-k+1}|D_t)$ (by conditioning on $\mathbf{X}_t(k)$ in
addition to D_{t-k}, and noting that $D_t = \{\mathbf{X}_t(k), D_{t-k}\}$).
Verify, using (b), that the regression coefficient vector of
θ_{t-k} on $\mathbf{X}_t(k)$ is B_{t-k} times that of θ_{t-k+1} on $\mathbf{X}_t(k)$.

(d) Deduce that $(\theta_{t-k}|D_t)$ is normal with moments that can
be written as

$$E[\theta_{t-k}|D_t] = m_{t-k} + B_{t-k}(E[\theta_{t-k+1}|D_t] - E[\theta_{t-k+1}|D_{t-k}])$$

and

$$V[\theta_{t-k}|D_t] = C_{t-k} - B_{t-k}^2(V[\theta_{t-k+1}|D_{t-k}] - V[\theta_{t-k+1}|D_t]).$$

(e) By way of notation, let the above moments be denoted by
$a_t(-k)$ and $R_t(-k)$ so that the result here is $(\theta_{t-k}|D_t) \sim$
$N[a_t(-k), R_t(-k)]$. Verify that the above, retrospective updating equations provide these moments backwards over
time for $k = t - 1, t - 2, \ldots, 0$ via

$$a_t(-k) = m_{t-k} + B_{t-k}[a_t(-k+1) - a_{t-k+1}]$$

and

$$R_t(-k) = C_{t-k} - B_{t-k}^2[R_t(-k+1) - R_{t-k+1}],$$

with $a_s = m_{s-1}$ and $R_s = C_{s-1} + W_s$ for all s as usual,
$a_t(0) = m_t$ and $R_t(0) = C_t$.

(f) In the discount model with $W_t = C_{t-1}(\delta^{-1} - 1)$ for all t,
verify that $B_{t-1} = \delta$ for all t.

CHAPTER 4

THE DYNAMIC LINEAR MODEL

4.1 OVERVIEW

The first-order polynomial and simple regression models of the preceding two chapters illustrate many basic concepts and important features of the general class of Normal Dynamic Linear Models, referred to as Dynamic Linear Models (DLMs) when the normality is understood. This class of models is described and analysed here, providing a basis for the special cases in later chapters and for further generalisations to follow. The principles used by a Bayesian forecaster in structuring forecasting problems, as introduced in Section 1.3 of Chapter 1, are reaffirmed here. The approach of Bayesian forecasting and dynamic modelling comprises, fundamentally,

(i) a sequential model definition;

(ii) structuring using parametric models with meaningful parametrisation;

(iii) probabilistic representation of information about parameters;

(iv) forecasts derived as probability distributions.

The sequential approach focusses attention on statements about the future development of a time series conditional on existing information. Thus, if interest lies in the scalar series Y_t, statements made at time $t-1$ are based on the existing information set D_{t-1}. These statements are derived from a representation of the relevant information obtained by structuring the beliefs of the forecaster in terms of a parametric model defining $p(Y_t \mid \boldsymbol{\theta}_t, D_{t-1})$, $\boldsymbol{\theta}_t$ being the defining parameter vector at time t. This mathematical and statistical representation is the language that provides communication between the forecaster, model and decision makers. Clearly the parameters must represent constructs meaningful in the context of the forecasting problem. Indexing $\boldsymbol{\theta}_t$ by t indicates that the parametrisation may be dynamic. In addition, although often the number and meaning of the elements of $\boldsymbol{\theta}_t$ will be stable, there are occasions on which $\boldsymbol{\theta}_t$ will be expanded, contracted or changed in meaning according to the forecaster's existing view of the time series. This is true in particular of *open systems*, such as typically arise in social, economic and biological environments, where influential factors affecting the time series process are themselves subject to variation based on the state of the system generating the process. In such cases changes in $\boldsymbol{\theta}_t$ may be required to reflect system learning and the exercise of purposeful control. Such events, although recognisable when they occur, may be difficult to identify initially and so will not typically be included in the model until occurrence.

The second component of the structuring of forecast information, the prior distribution with density $p(\boldsymbol{\theta}_t \mid D_{t-1})$ (and the posterior $p(\boldsymbol{\theta}_t \mid D_t)$), provides a concise, coherent and effective transfer of information on the time series process through time. In addition to simply processing the information deriving from observations and feeding it forward to forecast future development, this probabilistic encoding allows new information from external sources to be formally incorporated in the system. It also naturally extends to allow for expansion or contraction of the parameter vector in open systems, with varying degrees of uncertainty associated with the effects of such external interventions and changes. Further, inferences about system development and change are directly drawn from components of these distributions in a standard statistical manner. Finally, an ultimate goal of the forecaster is attained by directly applying probability laws. The above components provide one representation of a joint distribution for the observations and parameters, namely that defined via

$$p(Y_t, \boldsymbol{\theta}_t \mid D_{t-1}) = p(Y_t \mid \boldsymbol{\theta}_t, D_{t-1}) p(\boldsymbol{\theta}_t \mid D_{t-1}),$$

from which the relevant one-step forecast may be deduced as the marginal $p(Y_t \mid D_{t-1})$. Inference for the future Y_t is again a standard statistical problem of summarising the information encoded in the forecast distribution. In connection with end use, loss or utility structures may be imposed to derive coherent optimal decision policies in activities dependent on the unknown future values of the series.

It should be clear that there are important conceptual and practical differences between our view of time series modelling and that of the classical, "stationary" schools. The major difference is that in the latter a fixed process model is adopted for all time, the forecaster being committed to that model be it for 55BC or 2001AD. In addition, the models are typically closed to other information sources, and so are unable to adapt to unforeseen external events impinging on the system. By contrast, the operational Bayesian models specifically aim to incorporate information from any relevant source as it arises, including, most importantly, subjective views, leading to amended and updated model structures. In practice Bayesian models generally operate on the principle of *Management by Exception*. This involves the use of a proposed model and associated "routine" forecasts unless exceptional circumstances arise. These occur in two main, and distinct, ways. The first is when relevant information, from a source external to the system, is received. Examples in consumer demand forecasting include information such as the notification of changes in regulatory legislation, for instance licensing laws, or advance warning that a major consuming industry is about to become strike-bound, and so on. Such information, which is often feed-forward and anticipatory, is naturally included in the existing system by formally combining it with existing probabilistic information. By contrast, the second type of exception is feedback and occurs when a monitoring system assessing the routine model signals breakdown based on the occurrence of unexpectedly large forecast errors. At such a time, relevant explanatory external information may be sought, or, perhaps, automatic default procedures designed to correct for model deficiencies are initiated. Such considerations, discussed in later chapters, are important practical components of Bayesian forecasting with the models described below.

4.2 DEFINITIONS AND NOTATION

For reference we define the general normal DLM for a vector observation $\mathbf{Y}_t$ rather than the more usual scalar special case. Much of the discussion, in this chapter and elsewhere, is, however, restricted to the scalar case. Let $\mathbf{Y}_t$ be an $(r \times 1)$ vector observation on the time series $\mathbf{Y}_1, \mathbf{Y}_2, \dots$.

Definition 4.1. *The general Normal Dynamic Linear Model is characterised by a quadruple*

$$\{\mathbf{F}, \mathbf{G}, \mathbf{V}, \mathbf{W}\}_t = \{\mathbf{F}_t, \mathbf{G}_t, \mathbf{V}_t, \mathbf{W}_t\}$$

for each time t, where:

(a) $\mathbf{F}_t$ *is a known* $(n \times r)$ *matrix;*
(b) $\mathbf{G}_t$ *is a known* $(n \times n)$ *matrix;*
(c) $\mathbf{V}_t$ *is a known* $(r \times r)$ *variance matrix;*
(d) $\mathbf{W}_t$ *is a known* $(n \times n)$ *variance matrix.*

This quadruple defines the model relating $\mathbf{Y}_t$ *to the* $(n \times 1)$ *parameter vector* $\boldsymbol{\theta}_t$ *at time t, and the* $\boldsymbol{\theta}_t$ *sequence through time, via the sequential specification of distributions:*

$$(\mathbf{Y}_t \mid \boldsymbol{\theta}_t) \sim N[\mathbf{F}_t'\boldsymbol{\theta}_t, \mathbf{V}_t], \tag{4.1a}$$

$$(\boldsymbol{\theta}_t \mid \boldsymbol{\theta}_{t-1}) \sim N[\mathbf{G}_t\boldsymbol{\theta}_{t-1}, \mathbf{W}_t]. \tag{4.1b}$$

Equations (4.1) are implicitly also conditional on D_{t-1}, the information set available prior to time t. This includes, in particular, the values of the defining variances $\mathbf{V}_t$ and $\mathbf{W}_t$ and the past observations $\mathbf{Y}_{t-1}, \mathbf{Y}_{t-2}, \dots$, as well as the initial information set D_0 from $t = 0$. D_{t-1} is not explicitly recognised in the conditioning in equations (4.1) for notational simplicity, but it should be remembered that it is always conditioned upon. An alternative representation of these defining equations is:

$$\mathbf{Y}_t = \mathbf{F}_t'\boldsymbol{\theta}_t + \boldsymbol{\nu}_t, \qquad \boldsymbol{\nu}_t \sim N[\mathbf{0}, \mathbf{V}_t], \tag{4.2a}$$

$$\boldsymbol{\theta}_t = \mathbf{G}_t\boldsymbol{\theta}_{t-1} + \boldsymbol{\omega}_t, \qquad \boldsymbol{\omega}_t \sim N[\mathbf{0}, \mathbf{W}_t], \tag{4.2b}$$

where the errors sequences $\boldsymbol{\nu}_t$ and $\boldsymbol{\omega}_t$ are independent and mutually independent.

Equation (4.2a) is the **observation equation** for the model, defining the sampling distribution for $\mathbf{Y}_t$ conditional on the quantity $\boldsymbol{\theta}_t$. It is assumed that, given this vector, $\mathbf{Y}_t$ is conditionally independent of the past values of the series. This equation relates the observation to $\boldsymbol{\theta}_t$ via a dynamic linear regression with a multivariate normal error structure having known, though possibly time varying, observational variance matrix $\mathbf{V}_t$. The matrix $\mathbf{F}_t$ plays the role of the **regression matrix** of known values of independent variables, and $\boldsymbol{\theta}_t$ is the dynamic vector of regression parameters, referred to as the **state vector**, or sometimes **system vector**, of the model. The **mean response** at t is $\mu_t = \mathbf{F}_t'\boldsymbol{\theta}_t$, simply the expected value of $\mathbf{Y}_t$ in (4.2a), which defines the **level** of the series at time t. Finally in the observation equation, $\boldsymbol{\nu}_t$ is the **observational error** at time t. Equation (4.2b) is the state or system **evolution equation**, defining the time evolution of the state vector. A one-step Markov evolution is evident; given $\boldsymbol{\theta}_{t-1}$, and the known values of $\mathbf{G}_t$ and $\mathbf{W}_t$, the distribution of $\boldsymbol{\theta}_t$ is fully determined independently of values of the state vector and data prior to time $t - 1$. The deterministic, or systematic, component of the evolution is the transition from state $\boldsymbol{\theta}_{t-1}$ to $\mathbf{G}_t\boldsymbol{\theta}_{t-1}$, a simple linear transformation via the system or state **evolution transfer matrix** $\mathbf{G}_t$. The evolution is completed with the addition of the **evolution error** term $\boldsymbol{\omega}_t$, with known evolution variance $\mathbf{W}_t$. Finally note that the defining quadruple, assumed known throughout, does not appear in the conditioning of the distributions. This is a convention that is followed throughout the book; known quantities will not be made explicit in conditioning distributions for notational clarity unless explicity required. Some further discussion of this appears below in Section 4.3.

Definition 4.2. *Of special interest are the following two subsets of the general class of DLMs.*

   (i) *If the pair* $\{\mathbf{F}, \mathbf{G}\}_t$ *is constant for all* t *then the model is referred to as a* **Time Series DLM***, or* **TSDLM***.*

   (ii) *A TSDLM whose observational and evolution variances are constant for all* t *is referred to as a* **Constant DLM***.*

Thus a constant model is characterised by a time independent, or static, quadruple

$$\{\mathbf{F}, \mathbf{G}, \mathbf{V}, \mathbf{W}\}.$$

It will be seen that this important subset of DLMs includes essentially all classical linear time series models.

The general Univariate DLM is defined by Definition 4.1 with $r = 1$ and is therefore characterised by a quadruple

$$\{\mathbf{F}_t, \mathbf{G}_t, V_t, \mathbf{W}_t\},$$

leading to:

$$(Y_t \mid \boldsymbol{\theta}_t) \sim N[\mathbf{F}'_t\boldsymbol{\theta}_t, V_t],$$
$$(\boldsymbol{\theta}_t \mid \boldsymbol{\theta}_{t-1}) \sim N[\mathbf{G}_t\boldsymbol{\theta}_{t-1}, \mathbf{W}_t].$$

These equations, together with the initial prior at time 0, provide the full definition as follows.

Definition 4.3. *For each* t*, the general Univariate DLM is defined by:*

Observation equation:	$Y_t = \mathbf{F}'_t\boldsymbol{\theta}_t + \nu_t,$	$\nu_t \sim N[0, V_t],$
System equation:	$\boldsymbol{\theta}_t = \mathbf{G}_t\boldsymbol{\theta}_{t-1} + \boldsymbol{\omega}_t,$	$\boldsymbol{\omega}_t \sim N[\mathbf{0}, \mathbf{W}_t],$
Initial prior:	$(\boldsymbol{\theta}_0 \mid D_0) \sim N[\mathbf{m}_0, \mathbf{C}_0],$	

for some prior moments $\mathbf{m}_0$ *and* $\mathbf{C}_0$*. The observational and evolution error sequences are assumed to be independent and mutually independent, and are independent of* $(\boldsymbol{\theta}_0 \mid D_0)$*.*

Some comments about slightly different model definitions are in order. Firstly, note that this definition, with initial prior specified for time 0, applies in particular when the data $Y_1, Y_2, \ldots$, represent the continuation of a series observed previously, the time origin $t = 0$ being, essentially, an arbitrary label. In such cases, the initial prior is viewed as sufficiently summarising the information from the past, θ_0 having the concrete interpretation of the final state vector for the historical data. Otherwise, θ_0 has no such interpretation and the model may be initialised, essentially equivalently, by specifying the initial prior for the first state vector, providing a normal prior for $(\theta_1|D_0)$. The results derived below follow directly with no changes. Secondly, apparently more general models could be obtained by allowing the error sequences $\{\nu_t\}$ and $\{\omega_t\}$ to be both autocorrelated and cross-correlated, and some definitions of dynamic linear models would allow for such structure. It is always possible, however, to rephrase such a correlated model in terms of one that satisfies the independence assumptions. Thus we lose nothing by imposing such restrictions which lead to the simplest and most easily analysed mathematical form. Further, the independence model is typically more natural and meaningful. The ν_t error is simply a random perturbation in the measurement process that affects the observation Y_t but has no further influence on the series. ω_t by contrast influences the development of the system into the future. The independence assumption clearly separates these two sources of stochastic input to the system and clarifies their roles in the model. Finally, the model can be simply generalised to allow for known, non-zero means for either of the noise terms ν_t or ω_t. In addition, some or all of the noise components can be assumed to be known by taking the appropriate variances (and covariances) to be zero. These features are not central to the model theory here but do appear in particular models in later chapters where they are discussed as required.

4.3   UPDATING EQUATIONS: THE UNIVARIATE DLM

We consider univariate DLMs that are closed to inputs of external information at times $t \geq 1$. Thus, given initial prior information D_0 at $t = 0$, the information set available at any time t is simply

$$D_t = \{Y_t, D_{t-1}\}$$

where Y_t is the observed value of the series at t. To formally incorporate the known values of the defining quadruples $\{\mathbf{F}, \mathbf{G}, V, \mathbf{W}\}_t$

for each t, it is assumed that D_0 includes these values. This convention is purely for notational convenience and economy in explanation since, in fact, only those values which are to be used in calculating forecast distributions need to be known at any particular time.

The central characteristic of the normal model is that, at any time, existing information about the system is represented and sufficiently summarised by the posterior distribution for the current state vector. The key results are as follows.

Theorem 4.1. *In the univariate DLM of Definition 4.3, one-step forecast and posterior distributions are given, for each t, as follows.*

(a) *Posterior at $t-1$:*

For some mean $\mathbf{m}_{t-1}$ and variance matrix $\mathbf{C}_{t-1}$,

$$(\boldsymbol{\theta}_{t-1} \mid D_{t-1}) \sim \mathrm{N}[\mathbf{m}_{t-1}, \mathbf{C}_{t-1}].$$

(b) *Prior at t:*

$$(\boldsymbol{\theta}_t \mid D_{t-1}) \sim \mathrm{N}[\mathbf{a}_t, \mathbf{R}_t],$$

where

$$\mathbf{a}_t = \mathbf{G}_t \mathbf{m}_{t-1} \qquad and \qquad \mathbf{R}_t = \mathbf{G}_t \mathbf{C}_{t-1} \mathbf{G}_t' + \mathbf{W}_t.$$

(c) *One-step forecast:*

$$(Y_t \mid D_{t-1}) \sim \mathrm{N}[f_t, Q_t],$$

where

$$f_t = \mathbf{F}_t' \mathbf{a}_t \qquad and \qquad Q_t = \mathbf{F}_t' \mathbf{R}_t \mathbf{F}_t + V_t.$$

(d) *Posterior at t:*

$$(\boldsymbol{\theta}_t \mid D_t) \sim \mathrm{N}[\mathbf{m}_t, \mathbf{C}_t],$$

with

$$\mathbf{m}_t = \mathbf{a}_t + \mathbf{A}_t e_t \qquad and \qquad \mathbf{C}_t = \mathbf{R}_t - \mathbf{A}_t \mathbf{A}_t' Q_t,$$

where

$$\mathbf{A}_t = \mathbf{R}_t \mathbf{F}_t Q_t^{-1} \qquad and \qquad e_t = Y_t - f_t.$$

Proof: Proof is by induction and relies on the theory of the multi-variate normal distribution of Section 16.2 in Chapter 16. Suppose (a) to hold. Two proofs of (d) are given. The first is simple and informative but based on the normality of all distributions involved. The second is a particular application of Bayes' Theorem that is, of course, quite general.

The linear, normal evolution equation coupled with (a) lead directly to the prior in (b). Clearly $(\boldsymbol{\theta}_t \mid D_{t-1})$ is normally distributed since it is a linear function of $\boldsymbol{\theta}_{t-1}$ and $\boldsymbol{\omega}_t$ which are independently normally distributed. The mean $\mathbf{a}_t$ and variance $\mathbf{R}_t$ are easily deduced. Now the observation equation provides the conditional normal density

$$p(Y_t|\boldsymbol{\theta}_t) = p(Y_t|\boldsymbol{\theta}_t, D_{t-1}).$$

Hence, using (b) together with the evolution equation, we have Y_t and $\boldsymbol{\theta}_t$ jointly normally distributed conditional on D_{t-1}, with covariance

$$\begin{aligned}
C[Y_t, \boldsymbol{\theta}_t \mid D_{t-1}] &= C[\mathbf{F}'_t\boldsymbol{\theta}_t + \nu_t, \boldsymbol{\theta}_t \mid D_{t-1}] \\
&= \mathbf{F}'_t V[\boldsymbol{\theta}_t \mid D_{t-1}] + \mathbf{0}' = \mathbf{F}'_t\mathbf{R}_t = \mathbf{A}'_t Q_t,
\end{aligned}$$

where $\mathbf{A}_t = \mathbf{R}_t\mathbf{F}_t/Q_t$. In addition, the mean f_t and variance Q_t of Y_t follow easily from the observation equation and (b), establishing (c), and therefore

$$\begin{pmatrix} Y_t \\ \boldsymbol{\theta}_t \end{pmatrix} D_{t-1} \sim N\left[\begin{pmatrix} f_t \\ \mathbf{a}_t \end{pmatrix}, \begin{pmatrix} Q_t & Q_t\mathbf{A}'_t \\ \mathbf{A}_t Q_t & \mathbf{R}_t \end{pmatrix} \right].$$

(1) **Proof of (d) using normal theory.**

Then, using standard normal theory as in Section 16.2 of Chapter 16, the conditional distribution of $\boldsymbol{\theta}_t$ given $D_t = \{Y_t, D_{t-1}\}$ is directly obtained. The regression vector for the regression of $\boldsymbol{\theta}_t$ on Y_t is just $\mathbf{A}_t$, and (d) follows.

(2) **Proof using Bayes' Theorem.**

Bayes' Theorem implies directly that, as a function of $\boldsymbol{\theta}_t$,

$$p(\boldsymbol{\theta}_t \mid D_t) \propto p(\boldsymbol{\theta}_t \mid D_{t-1}) p(Y_t \mid \boldsymbol{\theta}_t),$$

the second term derived directly from the observation equation as

$$p(Y_t \mid \boldsymbol{\theta}_t) \propto \exp\{-0.5(Y_t - \mathbf{F}'_t\boldsymbol{\theta}_t)^2/V_t\},$$

and the first from (b) as

$$p(\boldsymbol{\theta}_t \mid D_{t-1}) \propto \exp\{-0.5(\boldsymbol{\theta}_t - \mathbf{a}_t)'\mathbf{R}_t^{-1}(\boldsymbol{\theta}_t - \mathbf{a}_t)\}.$$

Thus, taking natural logarithms and multiplying by -2 we have

$$-2\ln[p(\boldsymbol{\theta}_t \mid D_t)] = (\boldsymbol{\theta}_t - \mathbf{a}_t)'\mathbf{R}_t^{-1}(\boldsymbol{\theta}_t - \mathbf{a}_t)$$
$$+ (Y_t - \mathbf{F}_t'\boldsymbol{\theta}_t)^2 V_t^{-1} + \text{constant},$$

where the constant term does not involve $\boldsymbol{\theta}_t$. This quadratic function of $\boldsymbol{\theta}_t$ can be expanded and rearranged with a new constant as

$$\boldsymbol{\theta}_t'(\mathbf{R}_t^{-1} + \mathbf{F}_t\mathbf{F}_t'V_t^{-1})\boldsymbol{\theta}_t - 2\boldsymbol{\theta}_t'(\mathbf{R}_t^{-1}\mathbf{a}_t + \mathbf{F}_tY_tV_t^{-1}) + \text{constant}.$$

Now, using the definition of $\mathbf{C}_t$ in the statement of the Theorem, it is easily verified that

$$(\mathbf{R}_t^{-1} + \mathbf{F}_t\mathbf{F}_t'V_t^{-1})\mathbf{C}_t = \mathbf{I},$$

the $n \times n$ identity matrix, so that

$$(\mathbf{R}_t^{-1} + \mathbf{F}_t\mathbf{F}_t'V_t^{-1}) = \mathbf{C}_t^{-1}.$$

Using this form for $\mathbf{C}_t^{-1}$ and the expression for $\mathbf{m}_t$ in the Theorem statement, it follows that

$$\mathbf{C}_t^{-1}\mathbf{m}_t = (\mathbf{R}_t^{-1}\mathbf{a}_t + \mathbf{F}_tY_tV_t^{-1}).$$

As a consequence we then have

$$-2\ln[p(\boldsymbol{\theta}_t \mid D_t)] = \boldsymbol{\theta}_t'\mathbf{C}_t^{-1}\boldsymbol{\theta}_t - 2\boldsymbol{\theta}_t'\mathbf{C}_t^{-1}\mathbf{m}_t + \text{constant},$$
$$= (\boldsymbol{\theta}_t - \mathbf{m}_t)'\mathbf{C}_t^{-1}(\boldsymbol{\theta}_t - \mathbf{m}_t) + \text{constant},$$

or

$$p(\boldsymbol{\theta}_t \mid D_t) \propto \exp\{-0.5(\boldsymbol{\theta}_t - \mathbf{m}_t)'\mathbf{C}_t^{-1}(\boldsymbol{\theta}_t - \mathbf{m}_t)\}.$$

The result (d) follows on normalising this density.

In deriving the proofs the following identities, that may be used elsewhere, have been produced:

(a) $\mathbf{A}_t \quad = \mathbf{R}_t \mathbf{F}_t Q_t^{-1} = \mathbf{C}_t \mathbf{F}_t V_t^{-1};$

(b) $\mathbf{C}_t \quad = \mathbf{R}_t - \mathbf{A}_t \mathbf{A}_t' Q_t = \mathbf{R}_t (\mathbf{I} - \mathbf{F}_t \mathbf{A}_t');$

(c) $\mathbf{C}_t^{-1} = \mathbf{R}_t^{-1} + \mathbf{F}_t \mathbf{F}_t' V_t^{-1};$

(d) $Q_t \quad = (1 - \mathbf{F}_t' \mathbf{A}_t)^{-1} V_t;$

(e) $\mathbf{F}_t' \mathbf{A}_t = 1 - V_t Q_t^{-1}.$

By way of terminology, e_t is the one-step forecast error and $\mathbf{A}_t$ the adaptive coefficient vector at time t.

As noted earlier, the model definition may be marginally extended to incorporate known, non-zero means for the observational or evolution noise terms. It is trivial to derive the appropriate corrections to the above results in such cases. Generally, if $E[\nu_t]$ and/or $E[\omega_t]$ are known means for these terms, the above results apply with the modifications that $\mathbf{a}_t = \mathbf{G}_t \mathbf{m}_{t-1} + E[\omega_t]$ and $f_t = \mathbf{F}_t' \mathbf{a}_t + E[\nu_t]$. The verification of these modifications is left to the reader.

4.4 FORECAST DISTRIBUTIONS

Definition 4.4. *For any current time t, the **forecast function** $f_t(k)$ is defined for all integers $k \geq 0$ as*

$$f_t(k) = E[\mu_{t+k} | D_t] = E[\mathbf{F}_{t+k}' \boldsymbol{\theta}_{t+k} \mid D_t],$$

where

$$\mu_{t+k} = \mathbf{F}_{t+k}' \boldsymbol{\theta}_{t+k}$$

*is the **mean response function**.*

For k strictly greater than 0, the forecast function provides the expected values of future observations given current information,

$$f_t(k) = E[Y_{t+k} \mid D_t], \qquad (k \geq 1).$$

For completeness, however, the definition is given in terms of the expected values of the mean response μ_{t+k} rather than $Y_{t+k} = \mu_{t+k}+$

ν_{t+k}. In doing so, the definition includes the case $k = 0$, providing $f_t(0) = E[\mu_t \mid D_t]$, a posterior point estimate of the current level of the series.

In designing DLMs we focus on the form of the forecast function in k as a major guide to the suitability of particular models in any application. The following results provide the full forecast distributions of which the forecast functions are central components.

Theorem 4.2. *For each time t and $k \geq 1$, the $k-$step ahead distributions for $\boldsymbol{\theta}_{t+k}$ and Y_{t+k} given D_t are given by:*

    (a) State distribution: $(\boldsymbol{\theta}_{t+k} \mid D_t) \sim N[\mathbf{a}_t(k), \mathbf{R}_t(k)],$

    (b) Forecast distribution : $(Y_{t+k} \mid D_t) \sim N[f_t(k), Q_t(k)],$

with moments defined recursively by:

$$f_t(k) = \mathbf{F}_t' \mathbf{a}_t(k)$$

and

$$Q_t(k) = \mathbf{F}_t' \mathbf{R}_t(k) \mathbf{F}_t + V_{t+k},$$

where

$$\mathbf{a}_t(k) = \mathbf{G}_{t+k} \mathbf{a}_t(k-1)$$

and

$$\mathbf{R}_t(k) = \mathbf{G}_{t+k} \mathbf{R}_t(k-1) \mathbf{G}_{t+k}' + \mathbf{W}_{t+k},$$

with starting values $\mathbf{a}_t(0) = \mathbf{m}_t$, and $\mathbf{R}_t(0) = \mathbf{C}_t$.

Proof: For all t and integer $r < t$, define the $n \times n$ matrices $\mathbf{H}_t(r) = \mathbf{G}_t \mathbf{G}_{t-1} \ldots \mathbf{G}_{t-r+1}$, with $\mathbf{H}_t(0) = \mathbf{I}$. From repeat application of the state evolution equation we then have

$$\boldsymbol{\theta}_{t+k} = \mathbf{H}_{t+k}(k) \boldsymbol{\theta}_t + \sum_{r=1}^{k} \mathbf{H}_{t+k}(k-r) \boldsymbol{\omega}_{t+r}.$$

Thus, by linearity and independence of the normal summands,

$$(\boldsymbol{\theta}_{t+k} \mid D_t) \sim N[\mathbf{a}_t(k), \mathbf{R}_t(k)],$$

where

$$\mathbf{a}_t(k) = \mathbf{H}_{t+k}(k) \mathbf{m}_t = \mathbf{G}_{t+k} \mathbf{a}_t(k-1)$$

and

$$\mathbf{R}_t(k) = \mathbf{H}_{t+k}(k)\mathbf{C}_t\mathbf{H}_{t+k}(k)' + \sum_{r=1}^{k}\mathbf{H}_{t+k}(k-r)\mathbf{W}_{t+r}\mathbf{H}_{t+k}(k-r)'$$
$$= \mathbf{G}_{t+k}\mathbf{R}_t(k-1)\mathbf{G}'_{t+k} + \mathbf{W}_{t+k},$$

with starting values $\mathbf{a}_t(0) = \mathbf{m}_t$, and $\mathbf{R}_t(0) = \mathbf{C}_t$.

This establishes (a). The forecast distribution is deduced using the observation equation at time $t + k$, as

$$(Y_{t+k} \mid D_t) \sim N[f_t(k), Q_t(k)],$$

where

$$f_t(k) = \mathbf{F}'_{t+k}\mathbf{a}_t(k),$$

and

$$Q_t(k) = \mathbf{F}'_{t+k}\mathbf{R}_t(k)\mathbf{F}_{t+k} + V_{t+k},$$

as stated.

$$\diamond$$

These marginal distributions form the basis for forecasting individual values of the series. For other purposes, such as forecasting cumulative totals or averages of future observations, joint distributions are necessary. Since all distributions involved are multivariate normal, joint forecasts are completely determined by the above marginals and the pairwise covariances. The relevant covariances are given in the following corollary, the proofs of these, and related, results being left to the exercises.

Corollary 4.1. *For any integers j and k with $1 \leq j < k$,*

$$C[\boldsymbol{\theta}_{t+k}, \boldsymbol{\theta}_{t+j} \mid D_t] = \mathbf{C}_t(k, j)$$

and

$$C[Y_{t+k}, Y_{t+j} \mid D_t] = \mathbf{F}'_{t+k}\mathbf{C}_t(k, j)\mathbf{F}_{t+j},$$

where the $\mathbf{C}_t(k, j)$ are recursively defined via

$$\mathbf{C}_t(r, j) = \mathbf{G}_{t+r}\mathbf{C}_t(r - 1, j), \qquad (r = j + 1, \ldots, k),$$

with starting value $\mathbf{C}_t(j, j) = \mathbf{R}_t(j)$, for all t.

In the special but highly important cases when $\mathbf{G}_t$ and/or $\mathbf{F}_t$ are constant, the above results simplify considerably. In particular, the form of the forecast function is of such importance that it is detailed here for such cases.

Corollary 4.2. *In the special case that the evolution matrix* $\mathbf{G}_t$ *is constant,* $\mathbf{G}_t = \mathbf{G}$ *for all* t, *then, for* $k \geq 0$,

$$\mathbf{a}_t(k) = \mathbf{G}^k \mathbf{m}_t$$

so that

$$f_t(k) = \mathbf{F}'_{t+k} \mathbf{G}^k \mathbf{m}_t.$$

If, in addition, $\mathbf{F}_t = \mathbf{F}$ *for all* t *so that the model is a TSDLM, then the forecast function has the form*

$$f_t(k) = \mathbf{F}' \mathbf{G}^k \mathbf{m}_t.$$

In the latter case, the form of the forecast function, as a function of the step ahead k, is entirely determined by the powers of the system matrix $\mathbf{G}$. This is a fundamental guiding feature in time series model design investigated extensively in Chapters 5 and 6.

4.5   OBSERVATIONAL VARIANCES

So far this Chapter has assumed the defining quadruple of the univariate DLM to be completely known for all time. The regression and evolution matrices are defined by the modeller in accordance with model design principles related to desired forecast function form explored in following chapters. The evolution variance matrix is also chosen by the modeller, generally in accordance with the principles of discounting explored in later chapters. The final element of the quadruple, the observational variance V_t, is, however, quite typically subject to uncertainty. Accordingly, we present analyses that provide Bayesian learning procedures for unknown observational variances. Attention is restricted in this chapter to the special cases in which the variance is constant but unknown, $V_t = V$, given by the reciprocal of the precision ϕ as in earlier chapters. Practically important generalisations, that include both stochastically changing variances and mean response/variance laws, are introduced later where the basic analysis of this section is modified appropriately.

The analysis here is a fully conjugate Bayesian analysis that has already been covered in essence in Sections 2.5 and 3.4. We will often work in terms of the unknown **precision** parameter $\phi = V^{-1}$. The key feature of the analysis is that all variances in the model definition and analysis at each time are scaled by V.

Definition 4.5. *For each* t, *the model is defined by:*

Observation equation: $\qquad\qquad Y_t = \mathbf{F}_t'\boldsymbol{\theta}_t + \nu_t, \qquad \nu_t \sim \mathrm{N}[0, V],$

System equation: $\qquad\qquad\quad \boldsymbol{\theta}_t = \mathbf{G}_t\boldsymbol{\theta}_{t-1} + \boldsymbol{\omega}_t, \quad \boldsymbol{\omega}_t \sim \mathrm{N}[0, V\, W_t^*],$

Initial information: $\quad (\boldsymbol{\theta}_0 \mid D_0, \phi) \sim \mathrm{N}[\mathbf{m}_0, V\, \mathbf{C}_0^*],$

$$(\phi \mid D_0) \sim \mathrm{G}[n_0/2, d_0/2],$$

where $\phi = V^{-1}$.

Here the initial quantities $\mathbf{m}_0$, $\mathbf{C}_0^*$, n_0 and d_0 are to be specified, as are the matrices $\mathbf{W}_t^*$. Notice that all variances and covariances in the model have V as a multiplier, or scale factor, providing a *scale-free* model in terms of the *scale-free* variances that are starred, $\mathbf{C}_0^*$ and the $\mathbf{W}_t^*$. We lose no generality here; for V fixed, the model coincides with the original Definition 4.3 with the scale factor V simply being absorbed into these matrices.

The usual independence assumptions hold, now conditional on V, equivalently ϕ. As in Chapters 2 and 3, the prior for ϕ has mean $\mathrm{E}[\phi \mid D_{t-1}] = n_0/d_0 = 1/S_0$ where S_0 is a prior point estimate of the observational variance V.

The results obtained below are based on multivariate T distributions for the state vector at all times. Details are given in Section 16.3.3 of Chapter 16 but, for convenience, a short summary is in order here. The notation for Student T distributions is extended, by analogy with that for normal distributions, to multivariate cases. In particular, if the $n \times 1$ random vector $\boldsymbol{\theta}$ has a multivariate T distribution with h degrees of freedom, mode $\mathbf{m}$ and positive definite scale matrix $\mathbf{C}$, then the density is

$$p(\boldsymbol{\theta}) \propto [h + (\boldsymbol{\theta} - \mathbf{m})'\mathbf{C}^{-1}(\boldsymbol{\theta} - \mathbf{m})]^{-(n+h)/2}.$$

The notation is simply $\boldsymbol{\theta} \sim \mathrm{T}_h[\mathbf{m}, \mathbf{C}]$ as in the univariate case. Note that, as h tends to infinity, the distribution approaches normality, with $\boldsymbol{\theta} \sim \mathrm{N}[\mathbf{m}, \mathbf{C}]$.

Theorem 4.3. *With the model as specified above, the following distributional results obtain at each time* $t \geq 1$.

(a) *Conditional on* V:

$$(\boldsymbol{\theta}_{t-1} \mid D_{t-1}, V) \sim \mathrm{N}[\mathbf{m}_{t-1}, V\mathbf{C}_{t-1}^*],$$
$$(\boldsymbol{\theta}_t \mid D_{t-1}, V) \sim \mathrm{N}[\mathbf{a}_t, V\mathbf{R}_t^*],$$
$$(Y_t \mid D_{t-1}, V) \sim \mathrm{N}[f_t, VQ_t^*],$$
$$(\boldsymbol{\theta}_t \mid D_t, V) \sim \mathrm{N}[\mathbf{m}_t, V\mathbf{C}_t^*],$$

with $a_t = G_t m_{t-1}$, and $R_t^* = G_t C_{t-1}^* G_t' + W_t^*$, $f_t = F_t' a_t$, and $Q_t^* = 1 + F_t' R_t^* F_t$, the defining components being updated via

$$m_t = a_t + A_t e_t,$$
$$C_t^* = R_t^* - A_t A_t' Q_t^*,$$

where $e_t = Y_t - f_t$ and $A_t = R_t^* F_t / Q_t^*$.

(b) For precision $\phi = V^{-1}$:

$$(\phi \mid D_{t-1}) \sim G[n_{t-1}/2, d_{t-1}/2],$$
$$(\phi \mid D_t) \sim G[n_t/2, d_t/2],$$

where $n_t = n_{t-1} + 1$ and $d_{t-1} = d_{t-1} + e_t^2/Q_t^*$.

(c) Unconditional on V :

$$(\theta_{t-1} \mid D_{t-1}) \sim T_{n_{t-1}}[m_{t-1}, C_{t-1}],$$
$$(\theta_t \mid D_{t-1}) \sim T_{n_{t-1}}[a_t, R_t],$$
$$(Y_t \mid D_{t-1}) \sim T_{n_{t-1}}[f_t, Q_t],$$
$$(\theta_t \mid D_t) \sim T_{n_t}[m_t, C_t],$$

where $C_{t-1} = S_{t-1} C_{t-1}^*$, $R_t = S_{t-1} R_t^*$, $Q_t = S_{t-1} Q_t^*$ and $C_t = S_t C_t^*$, with $S_{t-1} = d_{t-1}/n_{t-1}$ and $S_t = d_t/n_t$.

(d) Operational definition of updating equations:

$$m_t = a_t + A_t e_t,$$
$$C_t = (S_t/S_{t-1}) [R_t - A_t A_t' Q_t],$$
$$S_t = d_t/n_t,$$

and

$$n_t = n_{t-1} + 1 \quad \text{and} \quad d_t = d_{t-1} + S_{t-1} e_t^2/Q_t,$$

where $Q_t = S_{t-1} + F_t' R_t F_t$ and $A_t = R_t F_t / Q_t$.

Proof: Given the model definition, the results in (a) follow directly from Theorem 4.1. They are simply the standard, known variance results.

The rest of the proof is by induction, and uses standard results from normal/gamma theory as detailed in Section 16.3 of Chapter

16. Assume that the prior for precision ϕ in (b) holds so that $(\phi \mid D_{t-1}) \sim \mathrm{G}[n_{t-1}/2, d_{t-1}/2]$. From (a) we have $(Y_t \mid D_{t-1}, \phi) \sim \mathrm{N}[f_t, Q_t^*/\phi]$ so that

$$p(Y_t \mid D_{t-1}, \phi) \propto \phi^{1/2} \exp(-0.5\phi e_t^2/Q_t^*).$$

Now, by Bayes' Theorem, we have the posterior for ϕ given by

$$p(\phi \mid D_t) \propto p(\phi \mid D_{t-1}) \, p(Y_t \mid D_{t-1}, \phi).$$

Using the prior from (b) and the likelihood above, we have

$$p(\phi \mid D_t) \propto \phi^{(n_{t-1}+1)/2-1} \exp\left[-(d_{t-1} + e_t^2/Q_t^*)\phi/2\right]$$

and deduce that $(\phi \mid D_t) \sim \mathrm{G}[n_t/2, d_t/2]$ where the updated parameters are $n_t = n_{t-1} + 1$ and $d_t = d_{t-1} + e_t^2/Q_t^*$. This establishes the posterior for ϕ in (b).

To establish (c), recall from Section 16.3 of Chapter 16 that, if the $n \times 1$ random vector $\boldsymbol{\theta}$ has distribution $(\boldsymbol{\theta} \mid \phi) \sim \mathrm{N}[\mathbf{m}, \mathbf{C}^*/\phi]$ with $\phi \sim \mathrm{G}[n/2, d/2]$, then, unconditionally on ϕ, $\boldsymbol{\theta} \sim \mathrm{T}_n[\mathbf{m}, \mathbf{C}^*S]$ where $S = d/n$. The results in (c) now follow by marginalisation of the distributions in (a) with respect to the appropriate prior/posterior gamma distribution for ϕ. The summary updating equations in (d) simply follow from those in (c), the difference being that the variances and covariances in the T distributions (all now unstarred) include the relevant estimate of V.

The inductive proof is now completed by noting that the results hold at $t = 0$ by using the same arguments and the initial distributions in Definition 4.2.

$\diamond$

This Theorem provides the key results. At time t, the prior mean of ϕ is $\mathrm{E}[\phi \mid D_{t-1}] = n_{t-1}/d_{t-1} = 1/S_{t-1}$ where $S_{t-1} = d_{t-1}/n_{t-1}$ is a prior point estimate of $V = 1/\phi$. Similarly, the posterior estimate is $S_t = d_t/n_t$. Note that the updating equations for the parameters defining the T prior/posterior and forecast distributions are essentially the same as the standard, known variance equations with the estimate S_{t-1} appearing as the variance. The only differences lie in the scaling by S_t/S_{t-1} in the update for $\mathbf{C}_t$ and the scaling of e_t^2/Q_t by S_{t-1} in the update for d_t, both to correct for the updated estimate of V. The equations in (d) are those used in practice, the starred, scale free versions appearing only to communicate the theoretical structure.

4.6 SUMMARY

These results are summarised in tabular form for easy reference. In the univariate DLM, the results above are summarised below.

Univariate DLM: unknown, constant variance $V = \phi^{-1}$		
Observation:	$Y_t = \mathbf{F}'_t\boldsymbol{\theta}_t + \nu_t$	$\nu_t \sim N[0, V]$
System:	$\boldsymbol{\theta}_t = \mathbf{G}_t\boldsymbol{\theta}_{t-1} + \boldsymbol{\omega}_t$	$\boldsymbol{\omega}_t \sim T_{n_{t-1}}[\mathbf{0}, \mathbf{W}_t]$
Information:	$(\boldsymbol{\theta}_{t-1} \mid D_{t-1}) \sim T_{n_{t-1}}[\mathbf{m}_{t-1}, \mathbf{C}_{t-1}]$	
	$(\boldsymbol{\theta}_t \mid D_{t-1}) \sim T_{n_{t-1}}[\mathbf{a}_t, \mathbf{R}_t]$	$\mathbf{a}_t = \mathbf{G}_t\mathbf{m}_{t-1}$
		$\mathbf{R}_t = \mathbf{G}_t\mathbf{C}_{t-1}\mathbf{G}'_t + \mathbf{W}_t$
	$(\phi \mid D_{t-1}) \sim G[n_{t-1}/2, d_{t-1}/2]$	$S_{t-1} = d_{t-1}/n_{t-1}$
Forecast:	$(Y_t \mid D_{t-1}) \sim T_{n_{t-1}}[f_t, Q_t]$	$f_t = \mathbf{F}'_t\mathbf{a}_t$
		$Q_t = \mathbf{F}'_t\mathbf{R}_t\mathbf{F}_t + S_{t-1}$

Updating Recurrence Relationships
$(\boldsymbol{\theta}_t \mid D_t) \sim T_{n_t}[\mathbf{m}_t, \mathbf{C}_t]$
$(\phi \mid D_t) \sim G[n_t/2, d_t/2]$
with
$\mathbf{m}_t = \mathbf{a}_t + \mathbf{A}_t e_t,$
$\mathbf{C}_t = (S_t/S_{t-1})[\mathbf{R}_t - \mathbf{A}_t\mathbf{A}'_t Q_t],$
$n_t = n_{t-1} + 1, \; d_t = d_{t-1} + S_{t-1}e_t^2/Q_t, \; \text{and} \; S_t = d_t/n_t,$
where $e_t = Y_t - f_t$, and $\mathbf{A}_t = \mathbf{R}_t\mathbf{F}_t/Q_t$.

Forecast Distributions
For $k \geq 1$,
$(\boldsymbol{\theta}_{t+k} \mid D_t) \sim T_{n_t}[\mathbf{a}_t(k), \mathbf{R}_t(k)]$
$(Y_{t+k} \mid D_t) \sim T_{n_t}[f_t(k), Q_t(k)]$
with moments as defined in Theorem 4.2, V_t replaced by the estimate S_{t-1}.

4.7 FILTERING RECURRENCES

The sequential updating and forecasting components of the DLM are geared to inferences about the state of the time series process and about observations at time $t, t+1, \ldots, t+k$, $(k \geq 0)$ based on information available at the current time t. Often there are complimentary requirements for inferences about the state of the process in the *past*, at times $t - k$ for $k > 0$. In process monitoring and control, for example, it is typically of interest to compare the current level of the series μ_t with that of previous observation stages, μ_{t-k}, in order to assess changes in level over time. In monthly series with a seasonal pattern over the year, the seasonal components of a DLM over the past twelve months provide information typically of interest in reporting seasonal form, whilst historical values of the remaining model components provide a "deseasonalised" series of interest in assessing underlying trends. In this case, inferences are required about θ_{t-k} for $k = 0, 1, \ldots, 11$, conditional on current information D_t.

By way of terminology, the distribution of $(\theta_{t-k} \mid D_t)$, for $k \geq 1$ and any fixed t, is called the **k-step filtered** distribution for the state vector at that time, analogous to the k-step ahead forecast distribution. The act of using recent data to revise inferences about previous values of the state vector is called **filtering**; the information recently obtained is filtered back to previous time points. A related concept is that of **smoothing** a time series. The retrospective estimation of the historical development of a time series mean response function μ_t using the filtered distributions $(\mu_{t-k} \mid D_t)$ for $k \geq 1$ is called **smoothing** the series.

At any given time t, the filtered distributions are derived recursively backwards in time using relationships, proven below, that are similar in structure to the standard sequential updating equations. For $k \geq 1$ we extend the definition of the k-step ahead state forecast distributions with moments $\mathbf{a}_t(k)$ and $\mathbf{R}_t(k)$ to negative arguments $\mathbf{a}_t(-k)$ and $\mathbf{R}_t(-k)$ and prove the following results.

Theorem 4.4. *In the univariate DLM* $\{\mathbf{F}_t, \mathbf{G}_t, V_t, \mathbf{W}_t\}$, *define*

$$\mathbf{B}_t = \mathbf{C}_t \mathbf{G}'_{t+1} \mathbf{R}_{t+1}^{-1}$$

for all t. Then, for all k, $(1 \leq k < t)$, the filtered distributions are defined by

$$(\theta_{t-k} \mid D_t) \sim \mathrm{N}[\mathbf{a}_t(-k), \mathbf{R}_t(-k)],$$

where:

$$\mathbf{a}_t(-k) = \mathbf{m}_{t-k} + \mathbf{B}_{t-k}[\mathbf{a}_t(-k+1) - \mathbf{a}_{t-k+1}]$$

and

$$\mathbf{R}_t(-k) = \mathbf{C}_{t-k} - \mathbf{B}_{t-k}[\mathbf{R}_{t-k+1} - \mathbf{R}_t(-k+1)]\mathbf{B}'_{t-k},$$

with starting values given by $\mathbf{a}_t(0) = \mathbf{m}_t$, and $\mathbf{R}_t(0) = \mathbf{C}_t$. Also, as already defined, $\mathbf{a}_{t-k+1} = \mathbf{a}_{t-k}(1)$ and $\mathbf{R}_{t-k+1} = \mathbf{R}_{t-k}(1)$.

Proof: The filtered densities are defined recursively via

$$p(\theta_{t-k} \mid D_t) = \int p(\theta_{t-k} \mid \theta_{t-k+1}, D_t)\, p(\theta_{t-k+1} \mid D_t)\, d\theta_{t-k+1}$$

$$(4.3)$$

which suggests proof by induction. Assume therefore that the stated result is true with k replaced by $k-1$ so that it applies to the second term in the integrand of (4.3),

$$(\boldsymbol{\theta}_{t-k+1} \mid D_t) \sim N[\mathbf{a}_t(-k+1), \mathbf{R}_t(-k+1)].$$

The first term in the integrand is, by Bayes' Theorem,

$$p(\boldsymbol{\theta}_{t-k} \mid \boldsymbol{\theta}_{t-k+1}, D_t) =$$
$$\frac{p(\boldsymbol{\theta}_{t-k} \mid \boldsymbol{\theta}_{t-k+1}, D_{t-k}) \, p(Y_{t-k+1}, \ldots, Y_t \mid \boldsymbol{\theta}_{t-k}, \boldsymbol{\theta}_{t-k+1}, D_{t-k})}{p(Y_{t-k+1}, \ldots, Y_t \mid \boldsymbol{\theta}_{t-k+1}, D_{t-k})}.$$

Now, given $\boldsymbol{\theta}_{t-k+1}$, the data $\{Y_{t-k+1}, \ldots, Y_t\}$ is independent of the previous value $\boldsymbol{\theta}_{t-k}$ so that the final two terms here cancel. The first is, again by Bayes' Theorem, given by

$$p(\boldsymbol{\theta}_{t-k} \mid \boldsymbol{\theta}_{t-k+1}, D_{t-k}) \propto p(\boldsymbol{\theta}_{t-k} \mid D_{t-k}) \, p(\boldsymbol{\theta}_{t-k+1} \mid \boldsymbol{\theta}_{t-k}, D_{t-k}). \tag{4.4}$$

But

$$(\boldsymbol{\theta}_{t-k+1} \mid \boldsymbol{\theta}_{t-k}, D_{t-k}) \sim N[\mathbf{G}_{t-k+1}\boldsymbol{\theta}_{t-k}, \mathbf{W}_{t-k+1}]$$

and

$$(\boldsymbol{\theta}_{t-k} \mid D_{t-k}) \sim N[\mathbf{m}_{t-k}, \mathbf{C}_{t-k}]$$

which define a joint distribution for the two state vectors having conditional (4.4) given directly by

$$(\boldsymbol{\theta}_{t-k} \mid \boldsymbol{\theta}_{t-k+1}, D_{t-k}) \sim N[\mathbf{h}_t(k), \mathbf{H}_t(k)], \tag{4.5}$$

where

$$\mathbf{h}_t(k) = \mathbf{m}_{t-k} + \mathbf{C}_{t-k}\mathbf{G}'_{t-k+1}\mathbf{R}^{-1}_{t-k+1}[\boldsymbol{\theta}_{t-k+1} - \mathbf{a}_{t-k+1}]$$

and

$$\mathbf{H}_t(k) = \mathbf{C}_{t-k} - \mathbf{C}_{t-k}\mathbf{G}'_{t-k+1}\mathbf{R}^{-1}_{t-k+1}\mathbf{G}_{t-k+1}\mathbf{C}_{t-k}.$$

With $\mathbf{B}_t$ as defined in the theorem statement, it follows that

$$\mathbf{h}_t(k) = \mathbf{m}_{t-k} + \mathbf{B}_{t-k}[\boldsymbol{\theta}_{t-k+1} - \mathbf{a}_{t-k+1}]$$

and

$$\mathbf{H}_t(k) = \mathbf{C}_{t-k} - \mathbf{B}_{t-k}\mathbf{R}_{t-k+1}\mathbf{B}'_{t-k}.$$

Returning then to (4.3), the required density $p(\boldsymbol{\theta}_{t-k} \mid D_t)$ is the expectation of (4.5) with respect to $(\boldsymbol{\theta}_{t-k+1} \mid D_t)$ which, by assumption, has the form stated in the Theorem with k replaced by $k-1$. Hence, directly,

$$(\boldsymbol{\theta}_{t-k} \mid D_t) \sim N[\mathbf{a}_t(-k), \mathbf{R}_t(-k)],$$

where

$$
\begin{aligned}
\mathbf{a}_t(-k) &= E[\mathbf{h}_t(k) \mid D_t] \\
&= \mathbf{m}_{t-k} + \mathbf{B}_{t-k}[\mathbf{a}_t(-k+1) - \mathbf{a}_{t-k+1}]
\end{aligned}
$$

and

$$
\begin{aligned}
\mathbf{R}_t(-k) &= E[\mathbf{H}_t(k) \mid D_t] + V[\mathbf{h}_t(k) \mid D_t] \\
&= \mathbf{C}_{t-k} - \mathbf{B}_{t-k}[\mathbf{R}_{t-k+1} - \mathbf{R}_t(-k+1)]\mathbf{B}'_{t-k},
\end{aligned}
$$

which are the values in the Theorem statement.

To complete the inductive proof consider the case $k = 1$ so that $t-k+1 = t$. The above derivation applies with $\mathbf{a}_t(-k+1) = \mathbf{a}_t(0) = \mathbf{m}_t$ and $\mathbf{R}_t(-k+1) = \mathbf{R}_t(0) = \mathbf{C}_t$, and the Theorem follows.

$\diamond$

Corollary 4.3. *If $V_t = V = \phi^{-1}$ is unknown and the conjugate analysis of Section 4.5 applied, then*

$$(\boldsymbol{\theta}_{t-k} \mid D_t) \sim T_{n_t}[\mathbf{a}_t(-k), (S_t/S_{t-k})\mathbf{R}_t(-k)].$$

Note that, as with the sequential updating equations, a change of scale is implied when the observational variance is unknown, estimated by the conjugate normal/gamma analysis.

Corollary 4.4. *The corresponding smoothed distributions for the mean response of the series are given by*

$$(\mu_{t-k} \mid D_t) \sim T_{n_t}[f_t(-k), (S_t/S_{t-k})\mathbf{F}'_{t-k}\mathbf{R}_t(-k)\mathbf{F}_{t-k}],$$

where, in an extension of the notation for the forecast function to negative arguments,

$$f_t(-k) = \mathbf{F}'_{t-k}\mathbf{a}_t(-k).$$

4.8   REFERENCE ANALYSIS OF THE DLM[†]

4.8.1   Introductory comments

The specification of proper, possibly highly informative, prior distributions at $t = 0$ to initialise models can be particularly beneficial if relevent prior information exists and can be satisfactorily articulated. Whilst this emphasis on informative priors and interventionist ideas is primary to Bayesian forecasting, we believe that there is also an important role to be played by reference analyses based on standard vague or uninformative priors (Bernardo, 1979) used to initialise models without further inputs from the user.

Such analyses are developed in this section. The reference analysis of a DLM (including learning about unknown observational variances) has a structure differing from standard analysis and is of interest from a purely theoretical standpoint. More importantly, and as the name suggests, a reference prior based analysis provides a reference level, or benchmark, against which alternative analyses may be compared. In particular, the implications for inference of various informative priors may be gauged and highlighted by comparison with the reference analysis. Also, it is our experience that users unfamiliar with the full complexities of a proper Bayesian analysis can learn rapidly through practice, and a default, reference initialisation of a model has been found to aid this learning.

In our opinion, in spite of the extensive research into the development of reference priors, there is no way to represent a state of complete "ignorance" within the Bayesian framework. In multiparameter problems in particular, the various approaches suggested typically lead to different reference priors, no single approach being obviously preferable (see Bernardo, 1979, and the discussion). These comments notwithstanding, there has emerged a consensus on the problem of normal, linear regression type models, and in this area there is what we may term a *standard* analysis, the implications of the standard reference prior being well investigated and understood. See Box and Tiao (1973), and DeGroot (1970) for further details of linear models. The DLM has the same basic, linear regression structure and thus the standard reference prior is adopted.

The relevant theoretical results for the DLM are stated here and some are proven. All results in this section are given, with full proofs

---

[†]This section may be omitted on a first reading.

and further discussion and related theory, in Pole and West (1989a) to which reference may be made.

4.8.2 Updating equations in reference analysis

Sequential updating equations defining the distributions in the DLM subject to a reference initial prior specification are given here. The results in this Section, summarised in Theorem 4.5 below, are general, applying to all univariate DLMs in which the system evolution variance matrices $\mathbf{W}_t$ are non-singular for all t. The cases of observational variances either known for all time or unknown but constant are considered together. Following this discussion, Section 4.8.3 develops similar ideas for the very important special case of $\mathbf{W}_t = \mathbf{0}$.

The reference analysis of the DLM is based on one of the following reference prior specifications for time $t = 1$ (see Box and Tiao, 1973, for example):

(1) In the model of Definition 4.3, with observational variances known, the reference initial prior specification is defined via

$$p(\boldsymbol{\theta}_1|D_0) \propto constant.$$

(2) In the model of Definition 4.5, with $V_t = V$ unknown, the initial prior information is represented by the reference form

$$p(\boldsymbol{\theta}_1, V|D_0) \propto V^{-1}.$$

Definitions 4.6. *In the models of Definitions 4.3 and 4.5 (observational variance known and unknown respectively), sequentially define the following quantities:*

$$\mathbf{H}_t = \mathbf{W}_t^{-1} - \mathbf{W}_t^{-1}\mathbf{G}_t\mathbf{P}_t^{-1}\mathbf{G}_t'\mathbf{W}_t^{-1},$$
$$\mathbf{P}_t = \mathbf{G}_t'\mathbf{W}_t^{-1}\mathbf{G}_t + \mathbf{K}_{t-1},$$
$$\mathbf{h}_t = \mathbf{W}_t^{-1}\mathbf{G}_t\mathbf{P}_t^{-1}\mathbf{k}_{t-1},$$

and

$$\mathbf{K}_t = \begin{cases} \mathbf{H}_t + \mathbf{F}_t\mathbf{F}_t' & \text{if } V_t = V \text{ is unknown,} \\ \mathbf{H}_t + \mathbf{F}_t\mathbf{F}_t'/V_t & \text{if } V_t \text{ is known,} \end{cases}$$

with

$$\mathbf{k}_t = \begin{cases} \mathbf{h}_t + \mathbf{F}_t Y_t & \text{if } V_t = V \text{ is unknown,} \\ \mathbf{h}_t + \mathbf{F}_t Y_t / V_t & \text{if } V_t \text{ is known,} \end{cases}$$

having initial values $\mathbf{H}_1 = \mathbf{0}$ *and* $\mathbf{h}_1 = \mathbf{0}$.

In the case of $V_t = V$, *unknown,* $\mathbf{W}_t$ *in the above definitions is replaced by the scale-free matrix* $\mathbf{W}_t^*$. *In addition in this case, define*

$$\gamma_t = \gamma_{t-1} + 1,$$
$$\lambda_t = \delta_{t-1} - \mathbf{k}_{t-1}' \mathbf{P}_t^{-1} \mathbf{k}_{t-1},$$
$$\delta_t = \lambda_t + Y_t^2,$$

with initial values $\lambda_1 = 0$ *and* $\gamma_0 = 0$.

Theorem 4.5.

    (1) **Case of variance known**

        In the model of Definition 4.3, with the reference prior

$$p(\boldsymbol{\theta}_1 | D_0) \propto constant,$$

        the prior and posterior distributions of the state vector at time t are given by (the possibly improper forms)

$$p(\boldsymbol{\theta}_t | D_{t-1}) \propto \exp\{-\tfrac{1}{2}(\boldsymbol{\theta}_t' \mathbf{H}_t \boldsymbol{\theta}_t - 2\boldsymbol{\theta}_t' \mathbf{h}_t)\},$$
$$p(\boldsymbol{\theta}_t | D_t) \propto \exp\{-\tfrac{1}{2}(\boldsymbol{\theta}_t' \mathbf{K}_t \boldsymbol{\theta}_t - 2\boldsymbol{\theta}_t' \mathbf{k}_t)\},$$

        with defining parameters given in Definition 4.5.

    (2) **Case of variance unknown**

        In the model of Definition 4.5, with

$$p(\boldsymbol{\theta}_1, V | D_t) \propto V^{-1},$$

        the joint prior and posterior distributions of the state vector and the observation variance at time $t = 1, 2, \ldots$ are given by

$$p(\boldsymbol{\theta}_t, V | D_{t-1}) \propto V^{-(1 + \frac{\gamma_{t-1}}{2})} \exp\{-\tfrac{1}{2}V^{-1}(\boldsymbol{\theta}_t' \mathbf{H}_t \boldsymbol{\theta}_t - 2\boldsymbol{\theta}_t' \mathbf{h}_t + \lambda_t)\},$$
$$p(\boldsymbol{\theta}_t, V | D_t) \propto V^{-(1 + \frac{\gamma_t}{2})} \exp\{-\tfrac{1}{2}V^{-1}(\boldsymbol{\theta}_t' \mathbf{K}_t \boldsymbol{\theta}_t - 2\boldsymbol{\theta}_t' \mathbf{k}_t + \delta_t)\}.$$

Proof: The proof of the results in the variance unknown case are given, the variance known case having essentially the same structure and being left as an exercise for the reader. In this case, then, $(\omega_t|V) \sim N[\mathbf{0}, V\mathbf{W}_t^*]$ and $\mathbf{W}_t^*$ replaces $\mathbf{W}_t$ in Definition 4.6.

The proof is by induction. When $V_t = V$ is unknown then, assume that the prior for $(\boldsymbol{\theta}_t, V|D_{t-1})$ is as stated,

$$p(\boldsymbol{\theta}_t, V|D_{t-1}) \propto V^{-(1+\frac{\gamma_{t-1}}{2})} \exp\{-\tfrac{1}{2}V^{-1}(\boldsymbol{\theta}_t'\mathbf{H}_t\boldsymbol{\theta}_t - 2\boldsymbol{\theta}_t'\mathbf{h}_t + \lambda_t)\}.$$

The likelihood from the observation Y_t is

$$p(Y_t|\boldsymbol{\theta}_t, V, D_{t-1}) \propto V^{-\frac{1}{2}} \exp\{-\tfrac{1}{2}V^{-1}(Y_t - \mathbf{F}_t'\boldsymbol{\theta}_t)^2\}$$

and it follows that the posterior for time t is of the form stated in the theorem. Consider now the implied prior for $t+1$, specifically

$$p(\boldsymbol{\theta}_{t+1}, V|D_t) = \int p(\boldsymbol{\theta}_{t+1}, V|\boldsymbol{\theta}_t, D_t)p(\boldsymbol{\theta}_t|D_t)d\boldsymbol{\theta}_t$$

$$= \int p(\boldsymbol{\theta}_{t+1}|\boldsymbol{\theta}_t, V, D_t)p(\boldsymbol{\theta}_t, V|D_t)d\boldsymbol{\theta}_t.$$

The first term in the integral is the normal distribution of $\boldsymbol{\theta}_{t+1} \sim N[\mathbf{G}_{t+1}\boldsymbol{\theta}_t, V\mathbf{W}_{t+1}^*]$ so that

$$p(\boldsymbol{\theta}_{t+1}, V|D_t) \propto \int V^{-\frac{n}{2}} \exp\{-\tfrac{1}{2}V^{-1}(\boldsymbol{\theta}_{t+1} - \mathbf{G}_{t+1}\boldsymbol{\theta}_t)'$$

$$\times \mathbf{W}_{t+1}^{*}{}^{-1}(\boldsymbol{\theta}_{t+1} - \mathbf{G}_{t+1}\boldsymbol{\theta}_t)\}$$

$$\times V^{-(1+\frac{\gamma_t}{2})} \exp\{-\tfrac{1}{2}V^{-1}(\boldsymbol{\theta}_t'\mathbf{K}_t\boldsymbol{\theta}_t - 2\boldsymbol{\theta}_t'\mathbf{k}_t + \delta_t)\}d\boldsymbol{\theta}_t$$

$$\propto V^{-(1+\frac{\gamma_t+n}{2})} \int \exp\{-\tfrac{1}{2}V^{-1}[(\boldsymbol{\theta}_t - \boldsymbol{\alpha}_{t+1})'\mathbf{P}_{t+1}(\boldsymbol{\theta}_t - \boldsymbol{\alpha}_{t+1})$$

$$+ R_{t+1}]\}d\boldsymbol{\theta}_t,$$

where

$$\mathbf{P}_{t+1} = \mathbf{K}_t + \mathbf{G}_{t+1}'\mathbf{W}_{t+1}^{*}{}^{-1}\mathbf{G}_{t+1},$$

$$\boldsymbol{\alpha}_{t+1} = \mathbf{P}_{t+1}^{-1}(\mathbf{k}_t + \mathbf{G}_{t+1}'\mathbf{W}_{t+1}^{*}{}^{-1}\boldsymbol{\theta}_{t+1}),$$

$$R_{t+1} = \boldsymbol{\theta}_{t+1}'\mathbf{W}_{t+1}^{*}{}^{-1}\boldsymbol{\theta}_{t+1} + \delta_t - \boldsymbol{\alpha}_{t+1}'\mathbf{P}_{t+1}\boldsymbol{\alpha}_{t+1}.$$

Using the well known normal integral result it is straightforward to obtain

$$p(\boldsymbol{\theta}_{t+1}, V|D_t) \propto V^{-(1+\frac{\gamma_t}{2})} \exp\{-\tfrac{1}{2}V^{-1}R_{t+1}\}.$$

Expanding $\mathbf{R}_{t+1}$ gives

$$\mathbf{R}_{t+1} = \boldsymbol{\theta}'_{t+1}\mathbf{H}_{t+1}\boldsymbol{\theta}_{t+1} - 2\boldsymbol{\theta}'_{t+1}\mathbf{h}_{t+1} + \lambda_{t+1}$$

where $\mathbf{H}_{t+1}$, $\mathbf{h}_{t+1}$, and λ_{t+1} are defined as stated in the theorem. It remains to validate the theorem for $t = 1$. Setting $\mathbf{H}_1 = 0$, $\mathbf{h}_1 = 0$, $\lambda_1 = 0$ and $\gamma_0 = 0$ provides this validation directly.

$\diamond$

Recall that the model has n parameters in the state vector at each time t and so, starting from the reference prior, the posteriors will be improper until at least time n in the observation known case, time $n + 1$ in the observation unknown case. After sufficient observations have been processed the improper posterior distributions characterising the model become proper. Although the above recursions remain valid it is more usual to revert to the standard forms involving directly updating the state vector posterior mean and covariances since these do not require any matrix inversions. The number of observations required to achieve proper distributions depends on the form of the model and the data. As mentioned above, the least number which will suffice is the number of unknown parameters in the model, including 1 for the observational variance if unknown. This requires that there are no missing data in these first observations and also that there are no problems of collinearity if the model includes regressors. In practice, more than the minimum will be rarely required. For generality, however, define

$$[n] = \min\{t : \text{posterior distributions are proper}\}.$$

The necessary relationships between the quantities defining the posterior distributions as represented in Theorem 4.5 and those in the original representation are easily obtained as follows, the proof being left as an exercise for the reader.

Corollary 4.5.

(1) **Case of variance known**
For $t \geq [n]$, the posterior distribution of $(\boldsymbol{\theta}_t | D_t)$ is as in Section 4.3, with

$$\mathbf{C}_t = \mathbf{K}_t^{-1} \qquad \text{and} \qquad \mathbf{m}_t = \mathbf{K}_t^{-1}\mathbf{k}_t.$$

(2) **Case of variance unknown**

For $t \geq [n]$, the posterior distributions of $(\boldsymbol{\theta}_t, V | D_t)$ is as in Section 4.6, with

$$\mathbf{C}_t = S_t \mathbf{K}_t^{-1} \qquad \text{and} \qquad \mathbf{m}_t = \mathbf{K}_t^{-1} \mathbf{k}_t,$$

where $S_t = d_t / n_t$ as usual, with $n_t = \gamma_t - n$ and $d_t = \delta_t - \mathbf{k}_t' \mathbf{m}_t$. In the usual case that $[n] = n + 1$, then $n_{n+1} = 1$ and it is easily shown that $d_{n+1} = S_{n+1} = e_{n+1}^2 / Q_{n+1}^*$.

4.8.3 Important special case of $\mathbf{W}_t = 0$

Consider now the case of models with deterministic evolution equations, that is, those in which $\mathbf{W}_t = \mathbf{W}_t^* = \mathbf{0}$. Whilst of interest from the point of view of theoretical completion and generality, this special case is discussed primarily for practical reasons. The basic motivation derives from the need to specify $\mathbf{W}_t$ in the recursions detailed above. As has been mentioned elsewhere this problem has led to much resistance to the use of these models by practitioners and the introduction of discount techniques in the usual conjugate prior analysis. Unfortunately these methods do not apply in the reference analysis for $t < [n]$ because the posterior covariances do not then exist. Hence, an alternative approach is required, and the practical use of zero covariances initially, $\mathbf{W}_t = \mathbf{0}$ for $t = 1, 2, \ldots, [n]$ is recommended. The rational behind this is as follows.

In the reference analysis with $n + 1$ model parameters (including V), we need $[n]$ (at least $n + 1$) observations to obtain a fully specified joint posterior distribution: one observation for each parameter. More generally, at time $t = [n]$ we have essentially only one observation's worth of information for each parameter. Thus it is *impossible* to detect or estimate any changes in parameters during the first $(n + 1)$ observation time points over which the reference analysis is performed. Consequently, use of non-zero $\mathbf{W}_t$ matrices is irrelevant since they basically allow for changes that cannot be estimated, and so we lose nothing by setting them to zero for $t = 1, 2, \ldots, [n]$. At time $t = [n]$, the posteriors are fully specified and future parametric changes can be identified. Thus, at this time, we revert to a full dynamic model with suitable, non-zero evolution covariance matrices.

Theorem 4.6. *In the framework of Theorem 4.5, suppose that $\mathbf{G}_t$ is non-singular and $\mathbf{W}_t = \mathbf{W}_t^* = \mathbf{0}$. Then the prior and posterior distributions of $\boldsymbol{\theta}_t$ and V have the forms of Theorem 4.5 with recursions defined as follows.*

(1) **Case of variance known**

$$\mathbf{H}_t = \mathbf{G}_t^{-1'}\mathbf{K}_{t-1}\mathbf{G}_t^{-1}$$
$$\mathbf{h}_t = \mathbf{G}_t^{-1'}\mathbf{k}_{t-1}.$$

(2) **Case of variance unknown**

$$\mathbf{H}_t = \mathbf{G}_t^{-1'}\mathbf{K}_{t-1}\mathbf{G}_t^{-1}$$
$$\mathbf{h}_t = \mathbf{G}_t^{-1'}\mathbf{k}_{t-1}$$
$$\lambda_t = \delta_{t-1}.$$

Proof:

(1) **Case of variance known**

Again the proof is inductive. Suppose first that $p(\boldsymbol{\theta}_{t-1}|D_{t-1})$ has the stated form. Now the system equation in this case is $\boldsymbol{\theta}_t = \mathbf{G}_t\boldsymbol{\theta}_{t-1}$ which may be inverted when $\mathbf{G}_t$ is non-singular so that $\boldsymbol{\theta}_{t-1} = \mathbf{G}_t^{-1}\boldsymbol{\theta}_t$. Applying this linear transformation, which has a constant Jacobian, to $p(\boldsymbol{\theta}_{t-1}|D_{t-1})$ immediately results in the prior

$$p(\boldsymbol{\theta}_t|D_{t-1}) \propto \exp\{-\tfrac{1}{2}(\boldsymbol{\theta}_t'\mathbf{G}_t^{-1'}\mathbf{K}_{t-1}\mathbf{G}_t^{-1}\boldsymbol{\theta}_t - 2\boldsymbol{\theta}_t'\mathbf{G}_t^{-1'}\mathbf{k}_{t-1})\}.$$

Hence, multiplying by the likelihood, the posterior is seen to be

$$p(\boldsymbol{\theta}_t|D_t) \propto \exp\{-\tfrac{1}{2}[\boldsymbol{\theta}_t'\mathbf{H}_t\boldsymbol{\theta}_t - 2\boldsymbol{\theta}_t'\mathbf{h}_t + V_t^{-1}(Y_t - \mathbf{F}_t'\boldsymbol{\theta}_t)'(Y_t - \mathbf{F}_t'\boldsymbol{\theta}_t)]\}$$
$$\propto \exp\{-\tfrac{1}{2}(\boldsymbol{\theta}_t'\mathbf{K}_t\boldsymbol{\theta}_t - 2\boldsymbol{\theta}_t'\mathbf{k}_t)\}$$

where $\mathbf{K}_t$, $\mathbf{k}_t$ are as defined in Theorem 4.5, $\mathbf{H}_t$, and $\mathbf{h}_t$ as above. Initially, for $t = 1$,

$$p(\boldsymbol{\theta}_1|D_1) \propto \exp\{-\tfrac{1}{2}V_1^{-1}(\boldsymbol{\theta}_1'\mathbf{F}_1\mathbf{F}_1'\boldsymbol{\theta}_1 - \boldsymbol{\theta}_1'\mathbf{F}_1 Y_1)\}$$

and the result follows by induction.

(2) **Case of variance unknown**

The proof follows as a simple extension of that for part (1) and is left as an exercise.

◇

The updating equations derived in Theorem 4.6 are of key practical importance. The assumption that the system matrices G_t are nonsingular is obviously crucial to the results. In practical models this is typically satisfied. This is true, in particular, of the important class of Time Series models.

4.8.4   Filtering

Filtering in the case of reference initial priors uses exactly the same results as in the conjugate (proper) priors case for times $t \geq [n]$ since all the distributions in this range are proper. In particular, the usual recursions of Section 4.7 are valid. However, for $t < [n]$ these recursions do not apply since the on-line posterior means and covariances required do not exist. The following Theorem provides the solutions for this case.

Theorem 4.7. *In the framework of Theorem 4.5, the filtered distributions in the DLM for times* $t - r$, $r = 0, 1, 2, ..., [n] - 1$ *are defined as follows.*

   (1) **Case of variance known**

$$p(\boldsymbol{\theta}_{t-r}|D_t) \propto \exp\{-\tfrac{1}{2}[\boldsymbol{\theta}'_{t-r}\mathbf{K}_t(-r)\boldsymbol{\theta}_{t-r} - 2\boldsymbol{\theta}'_{t-r}\mathbf{k}_t(-r)]\}$$

   (2) **Case of variance unknown**

$$p(\boldsymbol{\theta}_{t-r}, V|D_t) \propto V^{-(1+\frac{t}{2})} \exp\{-\tfrac{1}{2}V^{-1}[\boldsymbol{\theta}'_{t-r}\mathbf{K}_t(-r)\boldsymbol{\theta}_{t-r}$$
$$- 2\boldsymbol{\theta}'_{t-r}\mathbf{k}_t(-r) + \delta_t(-r)]\}$$

where the defining quantities are calculated recursively backwards in time according to:

$$\mathbf{K}_t(-r) = \mathbf{G}'_{t-r+1}\mathbf{W}^{-1}_{t-r+1}\mathbf{G}_{t-r+1}$$
$$- \mathbf{G}'_{t-r+1}\mathbf{W}^{-1}_{t-r+1}\mathbf{P}^{-1}_t(-r+1)\mathbf{W}^{-1}_{t-r+1}\mathbf{G}_{t-r+1} + \mathbf{K}_{t-r}$$
$$\mathbf{P}_t(-r+1) = \mathbf{W}^{-1}_{t-r+1} + \mathbf{K}_t(-r+1) - \mathbf{H}_{t-r+1}$$
$$\mathbf{k}_t(-r) = \mathbf{k}_{t-r} + \mathbf{G}'_{t-r+1}\mathbf{W}^{-1}_{t-r+1}\mathbf{P}^{-1}_t(-r+1)$$
$$\times [\mathbf{k}_t(-r+1) - \mathbf{h}_{t-r+1}]$$
$$\delta_t(-r) = \delta_t(-r+1) - \lambda_{t-r+1}$$
$$- [\mathbf{k}_t(-r+1) - \mathbf{h}_{t-r+1}]'\mathbf{P}^{-1}_t(-r+1)$$
$$\times [\mathbf{k}_t(-r+1) - \mathbf{h}_{t-r+1}]$$

and $\mathbf{H}_t$, $\mathbf{h}_t$, $\mathbf{K}_t$, $\mathbf{k}_t$, λ_t, δ_t are as in Definition 4.6 (Again note that, in the case of V unknown, $\mathbf{W}_t^*$ replaces $\mathbf{W}_t$ throughout, for all t). Starting values for these recursions are $\mathbf{K}_t(0) = \mathbf{K}_t$, $\mathbf{k}_t(0) = \mathbf{k}_t$, $\delta_t(0) = \delta_t$.

Proof: An exercise for the reader — complete proofs of these result, as the other in this Section, appearing in Pole and West (1989a).

$\diamond$

For theoretical completion, note that the recursions in Theorem 4.7 also apply for $t > [n]$. Although the standard filtering equations should be used, those of the Theorem provide the relevant distributions, as follows.

Corollary 4.6. For $t > [n]$, the distributions defined in Theorem 4.7 are proper, as given in Section 4.7, with

(1) **Case of variance known**

$$(\boldsymbol{\theta}_{t-r}|D_t) \sim \mathrm{N}[\mathbf{a}_t(-r), \mathbf{R}_t(-r)]$$

where $\mathbf{a}_t(-r) = \mathbf{K}_t(-r)^{-1}\mathbf{k}_t(-r)$ and $\mathbf{R}_t(-r) = \mathbf{K}_t(-r)^{-1}$.

(2) **Case of variance unknown**

$$(\boldsymbol{\theta}_{t-r}|D_t) \sim \mathrm{T}_{\gamma_t - [n]}[\mathbf{a}_t(-r), \mathbf{R}_t(-r)]$$

where $\mathbf{a}_t(-r) = \mathbf{K}_t(-r)^{-1}\mathbf{k}_t(-r)$ and $\mathbf{R}_t(-r) = S_t\mathbf{K}_t(-r)^{-1}$.

Theorem 4.8. In the case of zero evolution disturbance variances as in Theorem 4.6, the results of Theorem 4.7 are still valid but with the following changes to the recursions.

$$\mathbf{K}_t(-r) = \mathbf{G}'_{t-r+1}\mathbf{K}_t(-r + 1)\mathbf{G}_{t-r+1}$$
$$\mathbf{k}_t(-r) = \mathbf{G}'_{t-r+1}\mathbf{k}_t(-r + 1)$$
$$\delta_t(-r) = \delta_t(-r + 1)$$

Proof: Again left as an exercise.

$\diamond$

4.9 LINEAR BAYES' OPTIMALITY

4.9.1 Introduction

The results of Theorem 4.1 provide the central updating equations for dynamic models based on the assumed normality of the observational and evolution error sequences and the prior at time 0. The recurrences for $\mathbf{m}_t$ and $\mathbf{C}_t$ do, however, possess certain strong optimality properties that do not in any way depend on normality assumptions. These properties, essentially decision theoretic in nature, are derived from a general approach to Bayesian estimation when distributions are only partially specified in terms of their means and variances. The general theory is that of **Linear Bayes' Estimation** described below. In part this parallels the non-Bayesian techniques of minimum variance/least squares estimation that some readers may be familar with in connection with Kalman filtering as used by many authors; see, for example, Anderson and Moore (1979), Harvey (1981), Jazwinski (1970), Kalman (1960, 1963), and Sage and Melsa (1971).

The linear Bayes' approach to the DLM assumes the standard model but drops the normality assumptions. Thus in the univariate case the model equations become

$$
\begin{aligned}
Y_t &= \mathbf{F}'_t\boldsymbol{\theta}_t + \nu_t, & \nu_t &\sim [0, V_t], \\
\boldsymbol{\theta}_t &= \mathbf{G}_t\boldsymbol{\theta}_{t-1} + \boldsymbol{\omega}_t, & \boldsymbol{\omega}_t &\sim [\mathbf{0}, \mathbf{W}_t],
\end{aligned}
\tag{4.6}
$$

with, initially, $(\boldsymbol{\theta}_0 \mid D_0) \sim [\mathbf{m}_0, \mathbf{C}_0]$. All distributions here are unspecified apart from their first and second order moments which are just as in the normal model.

4.9.2 Linear Bayes' Estimation

The theory underlying linear Bayes' estimation is detailed, for completeness, in a general setting with a vector observation. Further discussion of the underlying principles and theoretical details, as well as related applications, can be found in Hartigan (1969), and Goldstein (1976).

Suppose that $\mathbf{Y}$, a $p \times 1$ vector, and $\boldsymbol{\theta}$, an unobservable $n \times 1$ vector, have some joint distribution, and that $\mathbf{Y}$ is to be observed and used to make inferences about $\boldsymbol{\theta}$. Estimation of $\boldsymbol{\theta}$ is derived within the following decision theoretic framework.

Let $\mathbf{d}$ be any estimate of $\boldsymbol{\theta}$ and suppose that accuracy in estimation is measured by a loss function $L(\boldsymbol{\theta}, \mathbf{d})$. Then the estimate $\mathbf{d} = \mathbf{m} = \mathbf{m}(\mathbf{Y})$ is optimal with respect to the loss function if the posterior risk, or expected loss, function $r(\mathbf{d}) = E[L(\boldsymbol{\theta}, \mathbf{d}) \mid \mathbf{Y}]$ is minimised as a function of $\mathbf{d}$ when $\mathbf{d} = \mathbf{m}$. In particular, if the loss function is quadratic,

$$L(\boldsymbol{\theta}, \mathbf{d}) = (\boldsymbol{\theta} - \mathbf{d})'(\boldsymbol{\theta} - \mathbf{d}) = \text{trace } (\boldsymbol{\theta} - \mathbf{d})(\boldsymbol{\theta} - \mathbf{d})',$$

then the posterior risk is minimised at the posterior mean, namely, $\mathbf{m} = E[\boldsymbol{\theta} \mid \mathbf{Y}]$, and the minimum risk is the trace of the posterior variance matrix, equivalently the sum of posterior variances of the elements of $\boldsymbol{\theta}$, namely $r(\mathbf{m}) = \text{trace} V[\boldsymbol{\theta} \mid \mathbf{Y}]$. This is a standard result in Bayesian decision theory; see, for example, DeGroot (1971), or Berger (1985).

Suppose now, however, that the decision maker has specified the joint distribution of $\mathbf{Y}$ and $\boldsymbol{\theta}$ only partially by providing joint first and second order moments, the mean and variance matrix

$$\begin{pmatrix} \mathbf{Y} \\ \boldsymbol{\theta} \end{pmatrix} \sim \left[\begin{pmatrix} \mathbf{f} \\ \mathbf{a} \end{pmatrix}, \begin{pmatrix} \mathbf{Q} & \mathbf{S}' \\ \mathbf{S} & \mathbf{R} \end{pmatrix} \right]. \tag{4.7}$$

With a quadratic loss function, or indeed any other, this specification does not provide enough information to identify the optimal estimate, the posterior mean, and the associated posterior variance. They are, in fact, undefined. The linear Bayes' method is designed effectively in order to side step this problem by providing an alternative estimate that may be viewed as an approximation to the optimal procedure. Since the posterior risk function cannot be calculated, the overall risk

$$r(\mathbf{d}) = \text{trace } E[(\boldsymbol{\theta} - \mathbf{d})(\boldsymbol{\theta} - \mathbf{d})'], \tag{4.8}$$

unconditional on $\mathbf{Y}$, is used instead. Furthermore, estimates $\mathbf{d} = \mathbf{d}(\mathbf{Y})$ of $\boldsymbol{\theta}$ are restricted to linear functions of $\mathbf{Y}$, of the form

$$\mathbf{d}(\mathbf{Y}) = \mathbf{h} + \mathbf{HY}, \tag{4.9}$$

for some $n \times 1$ vector $\mathbf{h}$ and $n \times p$ matrix $\mathbf{H}$. Clearly this may be viewed as a standard application of linear regression; the unknown regression function $E[\boldsymbol{\theta} \mid \mathbf{Y}]$ is approximated by the linear model above.

Definition 4.7. *A Linear Bayes' Estimate (LBE) of θ is a linear form (4.9) that is optimal in the sense of minimising the overall risk (4.8).*

The basic result is as follows.

Theorem 4.9.
$$m = a + SQ^{-1}(Y - f)$$

is the unique LBE of θ in the above framework. The associated Risk Matrix (RM), given by

$$C = R - SQ^{-1}S',$$

is the value of $E[(\theta - m)(\theta - m)']$ so that the minumum risk is simply $r(m) = \text{trace}(C)$.

Proof: For $d = h + HY$, define $R(d) = E[(\theta - d)(\theta - d)']$. Then

$$R(d) = R + HQH' - SH' - HS' + (a - h - Hf)(a - h - Hf)'.$$

Using the identity

$$(H - SQ^{-1})Q(H - SQ^{-1})' = SQ^{-1}S' + HQH' - SH' - HS',$$

we then have

$$R(d) = [R - SQ^{-1}S'] + (H - SQ^{-1})Q(H - SQ^{-1})'$$
$$+ (a - h - Hf)(a - h - Hf)'.$$

Hence the risk $r(d) = \text{trace}R(d)$ is the sum of three terms:
  (1) $\text{trace}(R - SQ^{-1}S')$ which is independent of d;
  (2) $\text{trace}(H - SQ^{-1})Q(H - SQ^{-1})'$ which has a minimum value of 0 at $H = SQ^{-1}$; and
  (3) $(a - h - Hf)'(a - h - Hf)$ which has a minimum value of 0 at $h + Hf = a$.

Thus $r(d)$ is minimised at $H = SQ^{-1}$ and $h = a - SQ^{-1}$ so that

$$d(Y) = a + SQ^{-1}(Y - f) = m,$$

as required. Also, at $d = m$, the risk matrix is

$$E[(\theta - m)(\theta - m)'] = R - SQ^{-1}S' = C,$$

as required.

$\diamond$

Two important, and obvious, properties are as follows.

Corollary 4.7. *If θ_t is any subvector of θ, then the LBE of θ and the associated RM from the marginal distribution of $\mathbf{Y}$ and θ coincide with the relevant marginal terms of $\mathbf{m}$ and $\mathbf{C}$.*

Corollary 4.8. $\mathbf{m}$ *and* $\mathbf{C}$ *are equivalent to the posterior mean and variance matrix of* $(\theta \mid \mathbf{Y})$ *in the case of joint normality of $\mathbf{Y}$ and θ.*

The use and interpretation of $\mathbf{m}$ and $\mathbf{C}$ is as approximations to posterior moments. Given the restriction to linearity of the estimates $\mathbf{d}$ in $\mathbf{Y}$, $\mathbf{m}$ is sometimes called the **Linear Posterior Mean** and $\mathbf{C}$ the **Linear Posterior Variance** of $(\theta \mid \mathbf{Y})$. This terminology is due to Hartigan (1969).

4.9.3 Linear Bayes' estimation in the DLM

Consider the DLM as specified, without further distributional assumptions, by equations (4.6). The observational and evolution error sequences, assumed independent in the normal framework, may, in fact, be dependent so long as they remain uncorrelated. Under normality assumptions this implies independence although this is not generally true. Suppose, in addition, the initial prior is partially specified in terms of moments as

$$(\theta_0 \mid D_0) \sim [\mathbf{m}_0, \mathbf{C}_0],$$

being uncorrelated with the error sequences. Finally, as usual, $D_t = \{D_{t-1}, Y_t\}$ with D_0 containing the known values of the error variance sequences.

Theorem 4.10. *The moments $\mathbf{m}_t$ and $\mathbf{C}_t$ as defined in the normal DLM of Theorem 4.1 are the linear posterior mean and variance matrix of* $(\theta_t \mid D_t)$.

Proof: For any t, let $\mathbf{Y}$ and θ be defined respectively as the $t \times 1$ and $n(t+1) \times 1$ vectors

$$\mathbf{Y} = (Y_t, Y_{t-1}, \ldots, Y_1)'$$

and

$$\theta = (\theta'_t, \theta'_{t-1}, \ldots, \theta'_0)'.$$

The linear structure of the DLM implies that the first and second order moments of the initial forecast distribution for $\mathbf{Y}$ and $\boldsymbol{\theta}$ are then

$$\begin{pmatrix} \mathbf{Y} \\ \boldsymbol{\theta} \end{pmatrix} \Big| D_0 \sim \left[\begin{pmatrix} \mathbf{f} \\ \mathbf{a} \end{pmatrix}, \begin{pmatrix} \mathbf{Q} & \mathbf{S}' \\ \mathbf{S} & \mathbf{R} \end{pmatrix} \right]$$

where the component means, variances and covariances are precisely those in the special case of normality. The actual values of these moments are not important here; the feature of interest is that they are the same whatever the full joint distribution may be.

Now, from Corollary 4.7, the first $n \times 1$ subvector $\boldsymbol{\theta}_t$ of $\boldsymbol{\theta}$ has LBE given by the corresponding subvector of the LBE $\mathbf{m}$ of $\boldsymbol{\theta}$. But, from Corollary 4.8, $\mathbf{m}$ is just the posterior mean for $\boldsymbol{\theta}$ in the normal case so that the required subvector is $\mathbf{m}_t$. Similarly, the corresponding RM is the relevant submatrix $\mathbf{C}_t$.

$\diamond$

Corollary 4.9. *For integers $k \geq 1$, the normal DLM moments*

$$[f_t(k), Q_t(k)] \qquad \text{and} \qquad [\mathbf{a}_t(k), \mathbf{R}_t(k)]$$

of k-step ahead forecast distributions, and, for $k > 0$, the corresponding moments

$$[f_t(-k), \mathbf{F}'_{t-k}\mathbf{R}_t(-k)\mathbf{F}_{t-k}] \qquad \text{and} \qquad [\mathbf{a}_t(-k), \mathbf{R}_t(-k)]$$

of the k-step smoothed and filtered distributions, are also the linear posterior means and variances of the corresponding random quantities.

These results justify the use of the sequential updating, forecasting and filtering recurrences outside the strict assumptions of normality. They have been presented in the case of a known variance DLM but it should be noted that, as in the normal model, the recurrences apply when the observational variances are known up to a constant scale factor, when all variances are simply scaled by this factor as in Section 4.5. Estimation of the unknown scale then proceeds using the updated estimates S_t of Section 4.5. This procedure may also be justified within a modified Linear Bayes' framework although this is not detailed here.

4.10 EXERCISES

(1) Become familiar with the methods of proof of the updating and forecasting results in Theorems 4.1, 4.2 and 4.3.

(2) Verify the identities listed in points (a) to (e) following the proof of Theorem 4.1.

(3) Generalise the updating and forecasting equations of Theorem 4.1 to models in which $E[\nu_t]$ and $E[\omega_t]$ are both non-zero (cf. discussion following proof of Theorem 4.1).

(4) The defining moments of the posterior for the mean response at time t, $(\mu_t|D_t) \sim N[f_t(0), Q_t(0)]$, are easily seen to be given by

$$E[\mu_t|D_t] = f_t(0) = \mathbf{F}'_t\mathbf{m}_t$$

and

$$V[\mu_t|D_t] = Q_t(0) = \mathbf{F}'_t\mathbf{C}_t\mathbf{F}_t.$$

Use the updating equations for $\mathbf{m}_t$ and $\mathbf{C}_t$ to derive the following results.

(a) Show that the mean can be simply updated via

$$E[\mu_t|D_t] = E[\mu_t|D_{t-1}] + (1 - b_t)e_t,$$

or via the similar expression

$$f_t(0) = (1 - b_t)Y_t + b_tf_t.$$

where $b_t = V_t/Q_t$. Interpret the second expression here as a weighted average of two estimates of μ_t.

(b) Show that the variance can be similarly updated using

$$V[\mu_t|D_t] = b_tV[\mu_t|D_{t-1}],$$

or, equivalently, using

$$Q_t(0) = (1 - b_t)V_t.$$

(5) Identify the posterior distribution of the observational and evolution errors at time t in the standard DLM of Definition 4.3. With $b_t = V_t/Q_t$, prove the following results.

(a) $(\nu_t|D_t) \sim N[b_te_t, (1 - b_t)V_t]$.

(b) $(\omega_t|D_t) \sim N[\mathbf{W}_t\mathbf{F}_te_t/Q_t, \mathbf{W}_t - \mathbf{W}_t\mathbf{F}_t\mathbf{F}'_t\mathbf{W}_t/Q_t]$.

(c) ν_t and ω_t are jointly normally distributed given D_t, having covariance vector given by $C[\omega_t, \nu_t | D_t] = -b_t \mathbf{W}_t \mathbf{F}_t$.

(6) Show how the results in Theorem 4.2 simplify in the case of a TSDLM with constant defining quadruple $\{\mathbf{F}, \mathbf{G}, V, \mathbf{W}\}$.

(7) Prove Corollary 4.1, and show how the results simplify in the case of a TSDLM.

(8) At time t, let $X_t(k) = \sum_{r=1}^{k} Y_{t+r}$, the cumulative total over the next $k > 0$ sampling intervals. Calculate the lead-time forecast distribution $p(X_t(k)|D_t)$. Show how this simplifies in a TSDLM.

(9) Generalise the updating and forecasting eqautions summarised in Section 4.6 to the DLM in which the observational errors are heteroscedastic, ie. $\nu_t \sim N[0, V k_t]$, where V is, as usual, an unknown variance but now k_t is a known, positive variance multiplier.

(10) Consider the closed, constant model $\{1, \lambda, V, W\}$ with λ, V and W known, and $-1 < \lambda < 1$. For all t, identify the mean and variance of the $k-$step ahead forecast distribution for $(Y_{t+k}|D_t)$ as functions of m_t, C_t and g. Show that $\lim_{k \to \infty} f_t(k) = 0$ and $\lim_{k \to \infty} Q_t(k) = V + W/(1 - \lambda^2)$.

(11) Generalise Theorem 4.1 to a DLM whose observation and evolution noise are instantaneously correlated. Specifically, suppose that $C[\omega_t, \nu_t] = \mathbf{c}_t$, a known $n-$vector of covariance terms, but that all other assumptions remain valid. Show now that Theorem 4.1 applies with the modifications given by

$$Q_t = (\mathbf{F}_t' \mathbf{R}_t \mathbf{F}_t + V_t) + \mathbf{F}_t' \mathbf{c}_t$$

and

$$\mathbf{A}_t = (\mathbf{R}_t \mathbf{F}_t + \mathbf{c}_t)/Q_t.$$

(12) Prove the filtering results of Theorem 4.4 an alternative way using mulivariate normal theory directly (rather than indirectly via Bayes' Theorem as used in the proof given). To do this, let $\mathbf{X}_t(k) = (Y_{t-k+1}, \dots, Y_t)'$, the vector of observations obtained between times $t-k+1$ and t, inclusive, and prove the following.

(a) Using the observation and evolution equations directly, show that, for any $r \geq 1$,

$$C[\theta_{t-k}, Y_{t-k+r} | D_{t-k}] = \mathbf{B}_{t-k} C[\theta_{t-k+1}, Y_{t-k+r} | D_{t-k}]$$

where, for any s,

$$\mathbf{B}_s = \mathbf{C}_s \mathbf{G}_{s+1}' \mathbf{R}_{s+1}^{-1}.$$

(b) Deduce that

$$C[\boldsymbol{\theta}_{t-k}, \mathbf{X}_t(k)|D_{t-k}] = \mathbf{B}_{t-k}C[\boldsymbol{\theta}_{t-k+1}, \mathbf{X}_t(k)|D_{t-k}].$$

(c) Hence identify the moments of the joint normal distribution of

$$(\boldsymbol{\theta}_{t-k}, \boldsymbol{\theta}_{t-k+1}, \mathbf{X}_t(k)|D_{t-k}),$$

and from this those of the conditional distributions of $(\boldsymbol{\theta}_{t-k}|D_t)$ and $(\boldsymbol{\theta}_{t-k+1}|D_t)$. (Note that $D_t = \{\mathbf{X}_t(k), D_{t-k}\}$). Verify, using (b), that the regression coefficient matrix of $\boldsymbol{\theta}_{t-k}$ on $\mathbf{X}_t(k)$ is $\mathbf{B}_{t-k}$ times that of $\boldsymbol{\theta}_{t-k+1}$ on $\mathbf{X}_t(k)$.

(d) Deduce that $(\boldsymbol{\theta}_{t-k}|D_t)$ is normal with moments that can be written as

$$E[\boldsymbol{\theta}_{t-k}|D_t] = \mathbf{m}_{t-k} + \mathbf{B}_{t-k}(E[\boldsymbol{\theta}_{t-k+1}|D_t] - E[\boldsymbol{\theta}_{t-k+1}|D_{t-k}])$$

and

$$V[\boldsymbol{\theta}_{t-k}|D_t] = \mathbf{C}_{t-k} - \mathbf{B}_{t-k}(V[\boldsymbol{\theta}_{t-k+1}|D_{t-k}] - V[\boldsymbol{\theta}_{t-k+1}|D_t])\mathbf{B}'_{t-k}.$$

Verify that these are as originally stated in Theorem 4.4.

(13) Verify the filtering and smoothing results in Corollaries 4.3 and 4.4.

(14) Prove the results in part (1) of Theorem 4.5 and part (2) of Theorem 4.6.

(15) Prove the results stated in Corollary 4.5, providing the conversion from reference analysis updating to the usual recurrence equations when posteriors become proper.

(16) Consider the first-order polynomial model $\{1, 1, V, W\}$, with $n = 1$ and $\boldsymbol{\theta}_t = \mu_t$, in the reference analysis updating of Section 4.8.

(a) Assume that V is known. Using the known variance results of Theorem 4.5 and Corollary 4.5, show that $(\mu_1|D_1) \sim N[Y_1, V]$.

(b) Assume that $V = 1/\phi$ is unknown so that the results of Theorem 4.5 and Corollary 4.5 apply. Show that the posterior for μ_t and V becomes proper and of standard form at $t = 2$, and identify the defining quantities m_2, C_2, n_2 and S_2.

CHAPTER 5

UNIVARIATE TIME SERIES DLM THEORY

5.1 UNIVARIATE TIME SERIES DLMS

The class of univariate Time Series DLMs, denoted TSDLMs, was introduced in Section 4.2 as defined by quadruples of the form

$$\{\mathbf{F}, \mathbf{G}, V_t, \mathbf{W}_t\},$$

for any V_t and $\mathbf{W}_t$. When this is understood, we refer to the quadruple simply as

$$\{\mathbf{F}, \mathbf{G}, ., .\}.$$

The theoretical structure of this highly important class of models is explored in this Chapter. Much of classical time series analysis concerns itself with models of *stationary* processes (Box and Jenkins, 1976), otherwise referred to as processes exhibiting *stationarity*. It will be shown in this Chapter that such models can essentially be expressed as Constant TSDLMs, those with $V_t = V$ and $\mathbf{W}_t = \mathbf{W}$ for all t. This is, in practice, obviously a very restrictive assumption, particularly since the variances V_t and $\mathbf{W}_t$ can often be expected to vary both randomly and as a function of the level of the series, and also should be open to intervention by the forecaster. Consequently, we do not restrict attention to constant models, but consider processes that cannot typically be reduced to stationarity.

Of central interest are the mean response function, denoted by $\mu_{t+k} = \mathrm{E}[Y_{t+k} \mid \boldsymbol{\theta}_{t+k}] = \mathbf{F}'\boldsymbol{\theta}_{t+k}$, and the forecast function $f_t(k) = \mathrm{E}[\mu_{t+k} \mid D_t] = \mathbf{F}'\mathbf{G}^k\mathbf{m}_t$. The structure of the forecast function in k is central in describing the implications of a given DLM and in designing DLMs consistent with a forecaster's view of the future development of a series. A further ingredient in this design activity is the concept of *observability* of DLMs, which is where we start.

5.2 OBSERVABILITY

5.2.1 Introduction and definition

Observability is a fundamental concept in linear systems theory that has important ramifications for TSDLMs. It relates essentially to the parametrisation θ_t of DLMs. To introduce the concept, consider the evolution error free case with $\mathbf{W}_t = \mathbf{0}$ for all t so that $\theta_t = \mathbf{G}\theta_{t-1}$. Thus $\mu_{t+k} = \mathbf{F}'\mathbf{G}^k\theta_t$. The state vector, comprising n elements, is chosen to reflect identifiable features of the time series and, accordingly, knowledge of the mean response over time provides knowledge of the state vector. Clearly, at least n distinct values of the mean response are required for such identification, with parametric parsimony suggesting that no more than n be necessary. The n distinct values starting at t, denoted by $\boldsymbol{\mu}_t = (\mu_t, \mu_{t+1}, \ldots, \mu_{t+n-1})'$, are related to the state vector via

$$\boldsymbol{\mu}_t = \mathbf{T}\theta_t,$$

where $\mathbf{T}$ is the $n \times n$ matrix

$$\mathbf{T} = \begin{pmatrix} \mathbf{F}' \\ \mathbf{F}'\mathbf{G} \\ \vdots \\ \mathbf{F}'\mathbf{G}^{n-1} \end{pmatrix}. \tag{5.1}$$

Thus, to precisely determine the state vector from the necessary minimum of n consecutive values of the mean response function, we require that $\mathbf{T}$ be non-singular and then

$$\theta_t = \mathbf{T}^{-1}\boldsymbol{\mu}_t.$$

These ideas of economy and identifiability of parameters in the case of purely deterministic evolution motivates the formal definition of observability in the general case.

Definition 5.1. *Any TSDLM* $\{\mathbf{F}, \mathbf{G}, ., .\}$ *is observable if and only if the $n \times n$ observability matrix $\mathbf{T}$ in (5.1) has full rank n.*

5.2.2 Examples

EXAMPLE 5.1. The model

$$\left\{ \begin{pmatrix} 1 \\ 0 \end{pmatrix}, \begin{pmatrix} 1 & 1 \\ 0 & 1 \end{pmatrix}, \ldots \right\}$$

is observable since

$$\mathbf{T} = \begin{pmatrix} 1 & 0 \\ 1 & 1 \end{pmatrix}$$

has rank 2. If $\mathbf{W}_t = \mathbf{0}$, then $\theta_t = \mathbf{T}^{-1}\boldsymbol{\mu}_t$ where $\boldsymbol{\mu}'_t = (\mu_t, \mu_{t+1})$. Thus $\theta'_t = (\mu_t, \mu_{t+1} - \mu_t)$ comprises $\theta_{t1} = \mu_t$, the current level of the series, and $\theta_{t2} = \mu_{t+1} - \mu_t$, the growth in level between times t and $t+1$.

EXAMPLE 5.2. The model

$$\left\{ \begin{pmatrix} 1 \\ 1 \end{pmatrix}, \begin{pmatrix} 1 & 0 \\ 0 & 1 \end{pmatrix}, \dots \right\}$$

is unobservable since

$$\mathbf{T} = \begin{pmatrix} 1 & 1 \\ 1 & 1 \end{pmatrix}$$

has rank 1. Here we have $\theta_t' = (\theta_{t1}, \theta_{t2})$ and $\omega_t' = (\omega_{t1}, \omega_{t2})$, with observation and evolution equations

$$Y_t = \theta_{t1} + \theta_{t2} + \nu_t,$$
$$\theta_{t1} = \theta_{t-1,1} + \omega_{t1},$$
$$\theta_{t2} = \theta_{t-1,2} + \omega_{t2}.$$

Noting that $\mu_t = \theta_{t1} + \theta_{t2}$ and defining $\psi_t = \theta_{t1} - \theta_{t2}$, the model becomes

$$Y_t = \mu_t + \nu_t,$$
$$\mu_t = \mu_{t-1} + \delta_{t1},$$
$$\psi_t = \psi_{t-1} + \delta_{t2},$$

where $\delta_{t1} = \omega_{t1} + \omega_{t2}$ and $\delta_{t2} = \omega_{t1} - \omega_{t2}$. The first two equations here completely define the DLM as, essentially, a first-order polynomial model. The random quantity ψ_t has no influence on the mean response and is in this sense redundant. The model is over-parametrised.

The conclusion in Example 5.2 is a special case of the general result that if $\mathbf{T}$ has rank $n - r$ for some r, $(1 \le r < n)$, then there exists a reparametrisation, based on linearly transforming the state vector, to an observable model of dimension $n - r$ that has the same mean response function.

EXAMPLE 5.3. Consider the reparametrised model in Example 5.2 and suppose that

$$\begin{pmatrix} \mu_0 \\ \psi_0 \end{pmatrix} \Big| D_0 \sim N \left[\begin{pmatrix} m_{1,0} \\ m_{2,0} \end{pmatrix}, \begin{pmatrix} C_{1,0} & 0 \\ 0 & C_{2,0} \end{pmatrix} \right],$$

with, for all t,

$$\mathbf{W}_t = \begin{pmatrix} W_1 & 0 \\ 0 & W_2 \end{pmatrix}.$$

In this case there is no correlation between μ_t, or Y_t, and ψ_t so the observations provide no information about the latter. Applying the updating equations leads to $\mathbf{C}_t$ diagonal for all t and the marginal posterior

$$(\psi_t \mid D_t) \sim \mathrm{N}[m_{2,0}, C_{2,0} + t W_2],$$

with increasingly large variance.

EXAMPLE 5.4. Example 5.3 suggests that, in the case of unobservability, the observations provide no information about some of the state parameters, or some linear functions of them. In the specific case there, this will always be true unless $\mathbf{W}_t$ has non-zero off diagonal elements introducing correlations between the state vector elements. Consider, for example, the unobservable model

$$\left\{ \begin{pmatrix} 1 \\ 0 \end{pmatrix}, \begin{pmatrix} 1 & 0 \\ 0 & 0 \end{pmatrix}, 2, \begin{pmatrix} 1 & W \\ W & W_2 \end{pmatrix} \right\}.$$

Clearly θ_{t2} is redundant, an equivalent mean response function being provided by the reduced model $\{1, 1, 2, 1\}$. This is a first-order polynomial model with, from Section 2.3 of Chapter 2, limiting distribution $(\theta_{t1} \mid D_t) \sim \mathrm{N}[m_{t1}, 1]$ as $t \to \infty$. In the 2-dimensional model, however, it is easily verified that the updating equations lead to the limiting form

$$(\theta_{t2} \mid D_{t-1}) \sim \mathrm{N}[W e_t / 4, 3W^2 / 2],$$

as $t \to \infty$. Thus, unless $W = 0$, Y_t is informative about the parameter that, for forecasting purposes, is redundant.

EXAMPLE 5.5. In the purely deterministic model $\{\mathbf{F}, \mathbf{G}, 0, \mathbf{0}\}$, $Y_{t+k} = \mu_{t+k}$ for all k so that, defining

$$\mathbf{Y}_t = (Y_t, Y_{t+1}, \dots, Y_{t+n-1})',$$

then $\mathbf{Y}_t = \boldsymbol{\mu}_t = \mathbf{T}\boldsymbol{\theta}_t$ and, in the case of observability,

$$\boldsymbol{\theta}_t = \mathbf{T}^{-1} \mathbf{Y}_t.$$

Hence the values of the state vector elements are precisely determined, or observed, as linear functions of any n values of the series.

If, however, $\mathbf{T}$ has rank $n - r$, then every Y_t is a linear combination of any $n - r$ observations, say $Y_1, Y_2, \ldots, Y_{n-r}$, and $\boldsymbol{\theta}_t$ is *not* fully determinable from the observations. This corresponds closely to the concept of observability as applied in linear systems theory and related fields.

5.2.3 Observability and the forecast function

For a TSDLM with $f_t(k) = \mathbf{F}'\mathbf{G}^k\mathbf{m}_t$ for $k \geq 0$, we have

$$\mathbf{T}\mathbf{m}_t = \begin{pmatrix} f_t(0) \\ f_t(1) \\ \vdots \\ f_t(n-1) \end{pmatrix}$$

and

$$\mathbf{T}\mathbf{G}^k\mathbf{m}_t = \begin{pmatrix} f_t(k) \\ f_t(k+1) \\ \vdots \\ f_t(n+k-1) \end{pmatrix}.$$

So observability implies that the first n consecutive terms of the forecast function are linearly independent, and vice versa. Any further values of the forecast function are then linear combinations of the first n values. If, however, $\mathbf{T}$ has rank $n - r$, then all values of the forecast function are linear functions of the first $n - r$.

EXAMPLE 5.6. The model

$$\left\{ \begin{pmatrix} 1 \\ 0 \end{pmatrix}, \begin{pmatrix} 1 & 1 \\ 0 & 1 \end{pmatrix}, \ldots \right\}$$

has forecast function

$$f_t(k) = m_{t1} + k m_{t2} = f_t(0) + k[f_t(1) - f_t(0)],$$

a polynomial of order 2, a straight line.

EXAMPLE 5.7. The model

$$\left\{ \begin{pmatrix} 0 \\ 1 \end{pmatrix}, \begin{pmatrix} 1 & 1 \\ 0 & 1 \end{pmatrix}, \ldots \right\},$$

is unobservable with $\mathbf{T}$ of rank 1 and

$$f_t(k) = m_{t2} = f_t(0),$$

simply that of a reduced, first-order polynomial model.

5.2.4 Limiting behaviour of constant models

One of the distinguishing features of constant models is that the components of the updating equations defining posterior and forecast distributions eventually, and typically rather rapidly, converge to stable values. This convergence allows an investigation of the links between constant DLMs and classical point forecasting methods based on the limiting form of the forecast distributions in the former. Such links are explored in various special models in later chapters. Here, based on Harrison (1985b), we state and prove general convergence results that underlie the later discussions.

Related (though more restricted) results can be found in Anderson and Moore (1979, Chapter 4), and some matrix analysis results used in this Theorem are found there.

Theorem 5.1. *Every observable constant DLM* $\{\mathbf{F}, \mathbf{G}, V, \mathbf{W}\}$, *for which* V *and* $\mathbf{W}$ *are finite, is such that, given fixed* D_0 *and* $D_t = \{Y_t, D_{t-1}\}$,

$$\lim_{t \to \infty} \{\mathbf{A}_t, \mathbf{C}_t, \mathbf{R}_t, Q_t\} = \{\mathbf{A}, \mathbf{C}, \mathbf{R}, Q\}$$

exists with finite elements.

Proof: It is convenient to present the proof in several parts.

(a) Consider initially the case of a reference, uninformative initial distribution for $\boldsymbol{\theta}_0$, so that formally, $\mathbf{C}_0^{-1} = \mathbf{0}$. In the notation, represent this initial condition by a superscript $*$, so that posterior distributions are denoted $(\boldsymbol{\theta}_t | D_t^*)$. Here $D_t^* = \{Y_t, \dots, Y_1, D_0^*\}$ where D_0^* simply indicates the vague initial distribution. From Section 4.8.2 of Chapter 4, it follows that, due to the assumed observability of the model, the posterior for $(\boldsymbol{\theta}_t | D_t^*)$ is proper, with a positive definite variance matrix $\mathbf{C}_t^*$, for $t \geq n$.

Now, from the evolution equation we have, for $t > n$,

$$\boldsymbol{\theta}_t = \mathbf{G}^{n-1} \boldsymbol{\theta}_{t-n+1} + \boldsymbol{\eta}_t$$

where $\boldsymbol{\eta}_t$ is a linear combination of $n-1$ terms of the evolution error sequence. In addition, $\mathbf{Y}_{t-n+1} = (Y_{t-n+1}, \dots, Y_t)'$ is such that

$$\mathbf{Y}_{t-n+1} = \mathbf{T} \boldsymbol{\theta}_{t-n+1} + \boldsymbol{\xi}_t$$

where $\boldsymbol{\xi}_t$ is a linear combination of the most recent n terms in the observation and evolution error sequences up to time t. Therefore, again using observability so that $\mathbf{T}^{-1}$ exists,

$$\boldsymbol{\theta}_{t-n+1} = \mathbf{T}^{-1}(\mathbf{Y}_{t-n+1} - \boldsymbol{\xi}_t)$$

whence

$$\boldsymbol{\theta}_t = \mathbf{G}^{n-1}\boldsymbol{\theta}_{t-n+1} + \boldsymbol{\eta}_t = \mathbf{G}^{n-1}\mathbf{T}^{-1}\mathbf{Y}_{t-n+1} + \boldsymbol{\epsilon}_t,$$

where $\boldsymbol{\epsilon}_t$ is a linear function of ν_{t-k} and ω_{t-k} for $0 \le k \le n-1$. Hence $\boldsymbol{\epsilon}_t$ has zero mean and a constant, finite variance matrix and, therefore, so has $\boldsymbol{\theta}_t$ *conditional* on $\mathbf{Y}_{t-n+1}$. Denote the latter by $\mathbf{U}$, so that, for all $t > n$,

$$(\boldsymbol{\theta}_t \mid \mathbf{Y}_{t-n+1}, D_0^*) \sim \mathrm{N}[\mathbf{G}^{n-1}\mathbf{T}^{-1}\mathbf{Y}_{t-n+1}, \mathbf{U}].$$

Now, since $\boldsymbol{\theta}_t, Y_t, Y_{t-1}, \ldots, Y_1$ are jointly normally distributed given D_0^* and $\mathbf{Y}_{t-n+1}$ is a subset of D_t^*, then for any, non-zero vector $\mathbf{a}$,

$$0 \le \mathrm{V}[\mathbf{a}'\boldsymbol{\theta}_t \mid D_t^*] \le \mathrm{V}[\mathbf{a}'\boldsymbol{\theta}_t \mid \mathbf{Y}_{t-n+1}, D_0^*] = \mathbf{a}'\mathbf{U}\mathbf{a}.$$

So the sequence $\mathrm{V}[\mathbf{a}'\boldsymbol{\theta}_t \mid D_t^*]$ is bounded above and below. Note that the lower bound is automatic. The upper bound is of little interest in its own right, what is important is that $\mathrm{V}[\mathbf{a}'\boldsymbol{\theta}_t \mid D_t^*]$ eventually (ie. for $t \ge n$) is finite for any $\mathbf{a}$. The upper bound is a sufficient condition for this that derives from the assumption of observability.

(b) The symmetry of the model with respect to the time subscript combines with the completely uninformative assumption regarding D_0^* to imply that, for $t > n$,

$$\mathrm{V}[\boldsymbol{\theta}_t | Y_t, Y_{t-1}, \ldots, Y_2, D_0^*] = \mathrm{V}[\boldsymbol{\theta}_{t-1} | Y_{t-1}, \ldots, Y_1, D_0^*] = \mathbf{C}_{t-1}^*.$$

But, $\mathbf{C}_t^* \le \mathrm{V}[\boldsymbol{\theta}_t | Y_t, \ldots, Y_2, D_0^*]$ in the sense that, for all $\mathbf{a}$,

$$\mathbf{a}'\mathbf{C}_t^*\mathbf{a} \le \mathbf{a}'\mathrm{V}[\boldsymbol{\theta}_t | Y_t, Y_{t-1}, \ldots, Y_2, D_0^*]\mathbf{a} = \mathbf{a}'\mathbf{C}_{t-1}^*\mathbf{a}.$$

Thus the sequence $\mathbf{a}'\mathbf{C}_t^*\mathbf{a}$ is non-increasing (and decreasing in non-degenerate models).

The boundedness and monotonicity of the sequence $V[\mathbf{a}'\boldsymbol{\theta}_t|D_t^*]$ implies convergence to a limit. Since this is true for any $\mathbf{a}$, it implies that $\mathbf{C}_t^*$ converges to a finite limit $\mathbf{C}^*$, say.

(c) The result in part (b) shows that the posterior variance matrix for $\boldsymbol{\theta}_t$ at time t converges to a finite, limiting value as t increases in the case of reference initial conditions, $\mathbf{C}_0^{-1} = \mathbf{0}$. Now $\mathbf{C}_t^*$ is the posterior variance of $\boldsymbol{\theta}_t$ based on the data alone in this reference case. Since the model is normal, this is the same as the variance matrix of the maximum likelihood estimator of $\boldsymbol{\theta}_t$ based on the data alone. In moving to a general, non-reference initial variance matrix $\mathbf{C}_0$, the effect is to increases the precision of $\boldsymbol{\theta}_t$ to

$$V[\boldsymbol{\theta}_t|D_t]^{-1} = \mathbf{C}_t^{*-1} + V[\boldsymbol{\theta}_t|D_0]^{-1},$$

or

$$\mathbf{C}_t^{-1} = \mathbf{C}_t^{*-1} + \mathbf{K}_t^{-1},$$

say, where $\mathbf{K}_t = V[\boldsymbol{\theta}_t|D_0]$, the step-ahead variance of $\boldsymbol{\theta}_t$ from $t = 0$ in the non-reference case.

For any non-zero vector $\mathbf{a}$, the above equation for precision matrices has the analogue

$$V[\mathbf{a}'\boldsymbol{\theta}_t|D_t]^{-1} = (\mathbf{a}'\mathbf{C}_t^*\mathbf{a})^{-1} + a_t^{-1},$$

where

$$a_t = V[\mathbf{a}'\boldsymbol{\theta}_t|D_0] = \mathbf{a}'\mathbf{K}_t\mathbf{a}.$$

Since $\mathbf{C}_t^*$ converges to a finite limit, it can be deduced that $V[\mathbf{a}'\boldsymbol{\theta}_t|D_t]$ does too if the sequence a_t^{-1} converges to a finite limit. Thus attention switches to this latter sequence.

(d) From the evolution equation we have

$$\mathbf{K}_t = \mathbf{G}\mathbf{K}_{t-1}\mathbf{G}' + \mathbf{W}$$

for all $t > 0$, having starting value $\mathbf{K}_0 = \mathbf{C}_0$, so that a_t is defined via

$$a_t = \mathbf{a}'\mathbf{G}\mathbf{K}_{t-1}\mathbf{G}'\mathbf{a} + \mathbf{a}'\mathbf{W}\mathbf{a}$$

This recurrence equation for $\mathbf{K}_t$ is such that, as t increases, the limiting behaviour of a_t takes one of three forms:

Case (i): a_t can tend to infinity, the addition of further evolution errors as t increases leading to greater and greater uncertainty about $\mathbf{a}'\boldsymbol{\theta}_t$ given only the initial information D_0.

Case (ii): a_t can converge (not necessarily monotonically) to a finite, positive limit a, say. This will occur for *all* vectors $\mathbf{a}$ when the eigenvalues of $\mathbf{G}$ are all less than unity, in which case the model is referred to as *stationary* and *stable* (Anderson and Moore, 1979, Chapter 4). If this condition is not met, then convergence may be obtained for only some $\mathbf{a}$.

Case (iii): If $\mathbf{W}_t = \mathbf{0}$ and $\mathbf{G}$ has complex eigenvalues of modulus unity, a_t may enter a limit cycle, asymptotically oscillating through a fixed and finite set of limit points. In such cases however, it can be shown that the corresponding quantity $\mathbf{a}'\mathbf{C}_t^*\mathbf{a}$ will converge to zero, so that $(\mathbf{a}'\mathbf{C}_t^*\mathbf{a})^{-1}$ diverges to infinity.

In each case it follows that, for any non-zero $\mathbf{a}$, $\mathbf{C}_t = V[\mathbf{a}'\boldsymbol{\theta}_t|D_t]$ converges to a finite limit (that is non-zero except in the rare Case (iii).) Since this holds for any non-zero $\mathbf{a}$ then $\mathbf{C}_t$ converges to a finite, stable value $\mathbf{C}$. The limits for $\mathbf{R}$, $\mathbf{A}$ and Q are then a simple consequence, and the proof is complete.

$\diamond$

In the limit, we have defining relationships

$$\mathbf{R} = \mathbf{G}\mathbf{C}\mathbf{G}' + \mathbf{W},$$
$$Q = \mathbf{F}'\mathbf{R}\mathbf{F} + V,$$
$$\mathbf{A} = \mathbf{R}\mathbf{F}/Q.$$

The limiting form of the updating equation for $\mathbf{m}_t$ is then given by

$$\mathbf{m}_t \approx \mathbf{G}\mathbf{m}_{t-1} + \mathbf{A}e_t.$$

Based on this limiting result, a limiting representation of the observation series in terms of the one-step forecast errors may be obtained. Let B be the *backshift* operator such that, for any time series X_t, we have $BX_t = X_{t-1}$ and $B^p X_t = X_{t-p}$ for any integer p. Let $\mathbf{H}$ be the $n \times n$ matrix defined by $\mathbf{H} = (\mathbf{I} - \mathbf{A}\mathbf{F}')\mathbf{G} = \mathbf{C}\mathbf{R}^{-1}\mathbf{G}$. We have the following result.

Theorem 5.2. *In the DLM of Theorem 5.1, denote the eigenvalues of $\mathbf{G}$ by λ_i, and those of $\mathbf{H}$ by ρ_i, $(i = 1, \ldots, n)$. Then*

$$\lim_{t \to \infty} \left\{ \prod_{i=1}^{n}(1 - \lambda_i B)Y_t - \prod_{i=1}^{n}(1 - \rho_i B)e_t \right\} = 0.$$

Proof: The proof, based on application of the Cayley-Hamilton Theorem (Section 16.4.2 of Chapter 16), is given in Ameen and Harrison (1985a), and omitted here.

◇

It should be noted that Theorems 5.1 and 5.2 are based solely on the forms of the updating equations in TSDLMs, with absolutely no assumptions about a "true" data generating process. They obviously apply outside the strict confines of normality since the updating equations are also obtained via the linear Bayesian (or Bayesian least squares) approach in Section 4.9 of Chapter 4. In addition, they apply even if the observational variance V is unknown and subject to the usual variance learning. In such cases, the posterior distribution for V will concentrate about its mode as t increases, asymptotically degenerating, and the model therefore converges to a known variance model.

From Theorem 5.2, the limiting representation of the observation series in terms of forecast errors is given by

$$Y_t = \sum_{j=1}^{n} \alpha_j Y_{t-j} + e_t + \sum_{j=1}^{n} \beta_j e_{t-j}$$

with coefficients given by

$$\alpha_1 = \sum_{i=1}^{n} \lambda_i,$$

$$\alpha_2 = -\sum_{i=1}^{n} \sum_{k=i+1}^{n} \lambda_i \lambda_k,$$

$$\vdots$$

$$\alpha_n = (-1)^n \lambda_1 \lambda_2 \ldots \lambda_n,$$

and

$$\beta_1 = -\sum_{i=1}^{n} \rho_i,$$

$$\beta_2 = \sum_{i=1}^{n} \sum_{k=i+1}^{n} \rho_i \rho_k,$$

$$\vdots$$

$$\beta_n = (-1)^{n+1} \rho_1 \rho_2 \ldots \rho_n.$$

This representation provides a link with familiar techniques of forecasting based on ARIMA models (Box and Jenkins, 1976), and with alternative methods including exponentially weighted regression, or exponential smoothing, (McKenzie, 1976). The following comments on the relationship with ARIMA modelling are pertinent.

(1) The ARIMA form is a limiting result in the DLM and therefore primarily of theoretical interest. In practice the use of non-constant variances and interventions will mean that the limiting forms are rarely utilised.

(2) No reference is made to an assumed "true" data generating process. The limiting representions of Y_t above *always* holds, whether or not the DLM is an adequate model for the series. If the data actually follow the DLM (impossible in practice, of course), then the e_t are zero-mean and uncorrelated, asympotically distributed as $N[0, Q]$.

(3) Using a generalisation of the notation of Box and Jenkins (1976) in which both stationary and corresponding non-stationary, or "explosive" processes are encompassed, the limiting representation is that of a generalised $ARMA(n, n)$ process. By suitably choosing $\mathbf{F}$, $\mathbf{G}$ and $\mathbf{W}$ it can be arranged that, for any integer q, $(0 \leq q \leq n)$, $\beta_{q+1} = \ldots = \beta_n = 0$ so that the limiting form is that of an $ARMA(n, q)$ process with $q \leq n$. Similarly, $ARMA(p, n)$ processes, for $p \leq n$, can be obtained as special cases. Thus all ARIMA processes can be modelled as limiting versions of the Time Series subset of DLMs.

5.2.5 Constrained observability

It will be assumed from now on that, because of indeterminism, parametric redundancy and the reducibility of unobservable to observable models, working TSDLMs will be observable unless otherwise stated. In some cases we do use models that, although having a non-singular observability matrix, are subject to additional structure that leads to observability in a wider sense.

Consider, for example, the model

$$\left\{ \begin{pmatrix} 1 \\ 1 \\ 0 \end{pmatrix}, \begin{pmatrix} 1 & 0 & 0 \\ 0 & 0 & 1 \\ 0 & 1 & 0 \end{pmatrix}, .,., \right\}$$

having

$$\mathbf{T} = \begin{pmatrix} 1 & 1 & 0 \\ 1 & 0 & 1 \\ 1 & 1 & 0 \end{pmatrix}$$

which has rank 2 so the model is unobservable. If, however, θ_{t2} and θ_{t3} represent the effects of a factor variable, say a seasonal cycle of periodicity 2, then the modeller will typically apply a *constraint* of the form, for all t,

$$\theta_{t1} + \theta_{t2} = 0.$$

This additional structure ensures that, for all $t \geq 2$, $\boldsymbol{\theta}_t$ is determinable from the mean response function via

$$\mu_t = \theta_{t1} + \theta_{t2},$$
$$\mu_{t+1} = \theta_{t1} + \theta_{t3},$$

and the constraint

$$0 = \theta_{t1} + \theta_{t2},$$

since then $\theta_{t1} = (\mu_t + \mu_{t+1})/2$ and $\theta_{t2} = -\theta_{t3} = (\mu_t - \mu_{t+1})/2$.

Such models for *effects* are common in statistics, relating to classifying factors such as seasonal period, blocking variables, treatment regimes and so on. Rather different constraints arise in other ways such as, for example, in studies of compositional data. Here it may be that some of the elements of the state vector represent proportions that must sum to 1, implying a constraint on, again, a linear function of $\boldsymbol{\theta}_t$.

Clearly the unobservable model above can be reduced to the observable model

$$\left\{ \begin{pmatrix} 1 \\ 1 \end{pmatrix}, \begin{pmatrix} 1 & 0 \\ 0 & -1 \end{pmatrix}, \ldots \right\}$$

producing the same forecast function. For interpretation and communication in practice, however, it is often desirable to retain the full unobservable model. Hence a wider definition of observability is required to cover models subject to linear constraints.

Definition 5.2. *Suppose the unobservable model* $\{\mathbf{F}, \mathbf{G}, V_t, \mathbf{W}_t\}$ *of dimension* n *is subject to constraints on the state vector of the form*

$$\mathbf{C}\boldsymbol{\theta}_t = \mathbf{c}$$

for some known, constant matrix $\mathbf{C}$ *and vector* $\mathbf{c}$. *Then the DLM is said to be* **Constrained Observable** *if and only if the* **extended observability matrix**

$$\begin{pmatrix} \mathbf{T} \\ \mathbf{C} \end{pmatrix}$$

has full rank n.

5.3 SIMILAR AND EQUIVALENT MODEL

5.3.1 Introduction

The concept of observability allows a modeller to restrict attention to a subclass of DLMs that are parsimoniously parametrised but provide the full range of forecast functions. This subclass is still large, however, and any given form of forecast function may typically be derived from many observable models. For designing models, therefore, we require further guidelines in order to identify small, practically meaningful collections of suitable models and, generally, a preferred single model for any given forecast function. The two key concepts that aid us here are those of **Similarity** and **Equivalence** of TS-DLMs. Similarity identifies and groups together all observable models consistent with any given forecast function. Any two such models are called **Similar models**. Equivalence strengthens this relationship by requiring that, in addition to having the same *qualitative form* of forecast function, the precise **quantitative** specification of the full forecast distributions for future observations be the same. Any two models producing the same forecasts are called **Equivalent models**. Similarity and equivalence are essentially related to the **reparametrisation** of a model via a linear map of the state vector. This is particularly useful in model building when a simpler identified, **canonical** model may be reparametrised to provide an equivalent model that is operationally more meaningful, efficient and easily understood.

5.3.2 Similar models

Consider any two observable TSDLMs, denoted by M and M_1, characterised by quadruples

$$M : \qquad \{\mathbf{F}, \mathbf{G}, V_t, \mathbf{W}_t\},$$
$$M_1 : \qquad \{\mathbf{F}_1, \mathbf{G}_1, V_{1t}, \mathbf{W}_{1t}\},$$

having forecast functions $f_t(k)$ and $f_{1t}(k)$, and observability matrices $\mathbf{T}$ and $\mathbf{T}_1$ respectively. The formal definition of similarity is as follows.

Definition 5.3. M and M_1 are **Similar** models, denoted by $M \sim M_1$, if and only if the system matrices $\mathbf{G}$ and $\mathbf{G}_1$ have identical eigenvalues.

The implications of similarity are best appreciated in the special case when the system matrices both have n *distinct* eigenvalues $\lambda_1, \ldots, \lambda_n$. Here $\mathbf{G}$ is diagonalisable. If $\boldsymbol{\Lambda} = \operatorname{diag}(\lambda_1, \ldots, \lambda_n)$, then there exists a non-singular $n \times n$ matrix $\mathbf{E}$ such that

$$\mathbf{G} = \mathbf{E}\boldsymbol{\Lambda}\mathbf{E}^{-1} \qquad \text{and} \qquad \mathbf{G}^k = \mathbf{E}\boldsymbol{\Lambda}^k\mathbf{E}^{-1}$$

for all $k \geq 0$. By definition, the forecast function of M is

$$f_t(k) = \mathbf{F}'\mathbf{G}^k\mathbf{m}_t = \mathbf{F}'\mathbf{E}\boldsymbol{\Lambda}^k\mathbf{E}^{-1}\mathbf{m}_t = \sum_{r=1}^{n} a_{tr}\lambda_r^k$$

for some coefficients $a_{t1}, \ldots, a_{tn}$ which do not depend on $\boldsymbol{\Lambda}$. Since $M \sim M_1$, we have, similarly,

$$f_{1t}(k) = \sum_{r=1}^{n} b_{tr}\lambda_r^k$$

for some coefficients $b_{t1}, \ldots, b_{tn}$ not involving $\boldsymbol{\Lambda}$. Thus M and M_1 have forecast functions of precisely the same *algebraic form* as functions of the step ahead integer k. This is the key to understanding similarity. If a modeller has a specific forecast function form in mind, then any two similar models are, for forecasting purposes, essentially the same. Although the above example concerns the special case of distinct eigenvalues, it is *always* the case that two observable models have the same form of forecast function if and only if they are similar as defined. An equivalent definition is that, again noting the necessity of observability, M and M_1 are similar models if and only if the system matrices $\mathbf{G}$ and $\mathbf{G}_1$ are *similar matrices;* that is, if and only if there exists a non-singular $n \times n$ **similarity matrix H** such that

$$\mathbf{G} = \mathbf{H}\mathbf{G}_1\mathbf{H}^{-1}.$$

Hence similarity of observable models is defined via similarity of system matrices. Further discussion, and justification of these points, is given in Section 5.4 below where the eigenstructure of system matrices is thoroughly explored.

5.3.3   Equivalent models and reparametrisation

The observable model M_1 has the usual form

$$Y_t = \mathbf{F}_1' \boldsymbol{\theta}_{1t} + \nu_{1t}, \qquad \nu_{1t} \sim N[0, V_{1t}];$$
$$\boldsymbol{\theta}_{1t} = \mathbf{G}_1 \boldsymbol{\theta}_{1t-1} + \boldsymbol{\omega}_{1t}, \qquad \boldsymbol{\omega}_{1t} \sim N[\mathbf{0}, \mathbf{W}_{1t}].$$

Given any $n \times n$ nonsingular matrix $\mathbf{H}$, M_1 may be **reparametrised** by linearly transforming the state vector $\boldsymbol{\theta}_{1t}$ to obtain

$$\boldsymbol{\theta}_t = \mathbf{H} \boldsymbol{\theta}_{1t} \qquad (5.2)$$

or

$$\boldsymbol{\theta}_{1t} = \mathbf{H}^{-1} \boldsymbol{\theta}_t$$

for all t. On doing this we have

$$Y_t = \mathbf{F}_1' \mathbf{H}^{-1} \boldsymbol{\theta}_t + \nu_{1t},$$
$$\boldsymbol{\theta}_t = \mathbf{H} \mathbf{G}_1 \mathbf{H}^{-1} \boldsymbol{\theta}_{t-1} + \mathbf{H} \boldsymbol{\omega}_{1t}.$$

Defining $\mathbf{F}$ and $\mathbf{G}$ via the equations

$$\begin{aligned} \mathbf{F}' &= \mathbf{F}_1' \mathbf{H}^{-1} \\ \mathbf{G} &= \mathbf{H} \mathbf{G}_1 \mathbf{H}^{-1}, \end{aligned} \qquad (5.3)$$

we can write

$$f_{1t}(k) = \mathbf{F}_1' \mathbf{G}_1^k \mathbf{m}_{1t} = \mathbf{F}_1' \mathbf{H}^{-1} \mathbf{H} \mathbf{G}_1^k \mathbf{H}^{-1} \mathbf{H} \mathbf{m}_{1t} = \mathbf{F}' \mathbf{G}^k \mathbf{m}_t,$$

where $\mathbf{m}_t = \mathbf{H} \mathbf{m}_{1t}$. It follows that *any* model M with $\mathbf{F}$ and $\mathbf{G}$ given by (5.3) for some $\mathbf{H}$ is similar to M_1. In addition, the matrix $\mathbf{H}$ is defined as follows.

Theorem 5.3. *If M and M_1 are such that $\mathbf{F}' = \mathbf{F}_1' \mathbf{H}^{-1}$ and $\mathbf{G} = \mathbf{H} \mathbf{G}_1 \mathbf{H}^{-1}$ for some nonsingular matrix $\mathbf{H}$, then:*

(i) *$M \sim M_1$; and*
(ii) *$\mathbf{H} = \mathbf{T}^{-1} \mathbf{T}_1$ where $\mathbf{T}$ and $\mathbf{T}_1$ are the observability matrices of M and M_1 respectively.*

Proof: (i) follows from the definition of similarity since the system matrices are similar, having similarity matrix $\mathbf{H}$. For (ii), note that, by definition of $\mathbf{T}$ and $\mathbf{T}_1$,

$$\mathbf{T}_1 = \mathbf{T}_1\mathbf{H}^{-1}\mathbf{H} = \mathbf{T}\mathbf{H}$$

so that $\mathbf{H} = \mathbf{T}^{-1}\mathbf{T}_1$.

$\diamond$

$\mathbf{H}$ is called the **Similarity** matrix of the transformation from M_1 to M, since it is that of the similar transformation from $\mathbf{G}_1$ to $\mathbf{G}$.

EXAMPLE 5.8. Let

$$\mathbf{F} = \begin{pmatrix} 1 \\ 0 \end{pmatrix}, \qquad \mathbf{G} = \begin{pmatrix} 1 & 1 \\ 0 & 1 \end{pmatrix};$$
$$\mathbf{F}_1 = \begin{pmatrix} -6 \\ 5 \end{pmatrix}, \qquad \mathbf{G}_1 = \begin{pmatrix} 9 & 16 \\ -4 & -7 \end{pmatrix}$$

M and M_1 are observable with

$$\mathbf{T} = \begin{pmatrix} 1 & 0 \\ 1 & 1 \end{pmatrix}, \qquad \mathbf{T}_1 = \begin{pmatrix} -6 & 5 \\ -74 & -131 \end{pmatrix}.$$

$\mathbf{G}$ and $\mathbf{G}_1$ each have a single eigenvalue 1 of multiplicity 2 so that $M \sim M_1$. Also

$$\mathbf{H} = \mathbf{T}^{-1}\mathbf{T}_1 = \begin{pmatrix} -6 & 5 \\ -68 & -136 \end{pmatrix},$$

and it is easily verified that $\mathbf{F}_1 = \mathbf{H}'\mathbf{F}$ and $\mathbf{G}_1 = \mathbf{H}^{-1}\mathbf{G}\mathbf{H}$. The forecast function has the form

$$f_t(k) = f_t(0) + k[f_t(1) - f_t(0)],$$

although, of course, the precise numerical values may differ between models.

Some further general features of the reparametrisation defined by (5.2) to (5.3) are as follows.

(a) In addition to equivalent forecast function forms, the defining state vector $\boldsymbol{\theta}_t$ of M may be obtained as a linear transformation of that $\boldsymbol{\theta}_{1t}$ in M_1 via

$$\boldsymbol{\theta}_t = \mathbf{H}\boldsymbol{\theta}_{1t},$$

and vice versa. One model may thus be reparametrised to obtain the other as far as the structural components $\mathbf{F}$, $\mathbf{G}$ and $\boldsymbol{\theta}_t$ are concerned.

(b) The full defining quadruple of M, $\{\mathbf{F}, \mathbf{G}, V_t, \mathbf{W}_t\}$ is obtained via this reparametrisation from that of M_1, $\{\mathbf{F}_1, \mathbf{G}_1, V_{1t}, \mathbf{W}_{1t}\}$ if, in addition to (a), the variances are related via

$$V_t = V_{1t}, \qquad \mathbf{W}_t = \mathbf{H}\mathbf{W}_{1t}\mathbf{H}'.$$

(c) If, in addition to (a) and (b), the information models $(\boldsymbol{\theta}_{t-1} \mid D_{t-1}) \sim \mathrm{N}[\mathbf{m}_{t-1}, \mathbf{C}_{t-1}]$ and $(\boldsymbol{\theta}_{1t-1} \mid D_{t-1}) \sim \mathrm{N}[\mathbf{m}_{1t-1}, \mathbf{C}_{1t-1}]$ are related via

$$\mathbf{m}_{t-1} = \mathbf{H}\mathbf{m}_{1t-1}, \qquad \mathbf{C}_{t-1} = \mathbf{H}\mathbf{C}_{1t-1}\mathbf{H}',$$

then the entire quantitative specification of M is obtained from that of M_1 by the reparametrisation. This obtains, in particular, if these relationships hold between the initial priors at $t = 0$.

These final comments motivate the concept of model equivalence.

Definition 5.4. *The observable models M and M_1 are **equivalent** models, denoted by $M \equiv M_1$, if and only if the similarity matrix $\mathbf{H}$ used to reparametrise M_1 according to $\boldsymbol{\theta}_t = \mathbf{H}\boldsymbol{\theta}_{1t}$ also leads to the relationships in (b) and (c) above, thus resulting in the model M with the same forecast distributions for the series.*

EXAMPLE 5.9. Consider the models

$$M: \quad \left\{ \begin{pmatrix} 1 \\ 0 \end{pmatrix}, \begin{pmatrix} \lambda & 1 \\ 0 & \rho \end{pmatrix}, V, \begin{pmatrix} 2 & \rho - \lambda \\ \rho - \lambda & (\rho - \lambda)^2 \end{pmatrix} \right\};$$

$$M_1: \quad \left\{ \begin{pmatrix} 1 \\ 1 \end{pmatrix}, \begin{pmatrix} \lambda & 0 \\ 0 & \rho \end{pmatrix}, V, \begin{pmatrix} 1 & 0 \\ 0 & 1 \end{pmatrix} \right\},$$

where λ and ρ are real and distinct. Then

$$\mathbf{T} = \begin{pmatrix} 1 & 0 \\ \lambda & 1 \end{pmatrix}, \qquad \mathbf{T}_1 = \begin{pmatrix} 1 & 1 \\ \lambda & \rho \end{pmatrix},$$

so that both models are observable. Thus, since both system matrices have eigenvalues λ and ρ, $M \sim M_1$. Further

$$\mathbf{H} = \begin{pmatrix} 1 & 1 \\ 0 & \rho - \lambda \end{pmatrix},$$

and it can be verified that $\mathbf{F}' = \mathbf{F}_1' \mathbf{H}^{-1}$ and $\mathbf{W} = \mathbf{H}\mathbf{W}_1\mathbf{H}'$. So, if the initial priors conform in the sense described in (c) above, $M \equiv M_1$. Finally notice that, if $\rho = \lambda$, the models cannot even be similar, let alone equivalent, since M_1 is not then observable.

5.3.4 General equivalence[†]

The definition of equivalence is a key concept in DLM design. It fails to apply, however, in an essentially degenerate case when two, or more, similar models that produce precisely the same forecast distributions cannot be linked by linear transformations. It is of interest to note such cases although, from a practical viewpoint, they are of little consequence. The anomaly arises for $n \geq 2$ when, given any model M as above, there is an uncountable set of models which differ from M only through the system evolution variance matrix $\mathbf{W}_t$ that produce exactly the same forecast distributions.

For the general case with M defined by the standard quadruple $\{\mathbf{F}, \mathbf{G}, V_t, \mathbf{W}_t\}$, let ϵ_t be an independent sequence of random $n-$vectors with

$$(\epsilon_t \mid D_{t-1}) \sim \mathrm{N}[\mathbf{0}, \mathbf{U}_t].$$

If, for each t, $\mathbf{U}_t$ is a non-negative definite matrix such that $\mathbf{F}'\mathbf{U}_t\mathbf{F} = 0$, then $\mathbf{F}'\epsilon_t = 0$ with probability one for all t. Consider now the model defined by adding the term $\mathbf{G}\epsilon_{t-1}$ to the system evolution equation M to get

$$Y_t = \mathbf{F}'\boldsymbol{\theta}_t + \nu_t$$

and

$$\boldsymbol{\theta}_t = \mathbf{G}\boldsymbol{\theta}_{t-1} + (\boldsymbol{\omega}_t + \mathbf{G}\epsilon_{t-1}).$$

[†]This Section is of rather theoretical interest and may be omitted without loss on a first reading.

Defining the new state vector $\boldsymbol{\psi}_t$ via

$$\boldsymbol{\psi}_t = \boldsymbol{\theta}_t + \boldsymbol{\epsilon}_t,$$

we then have

$$Y_t = \mathbf{F}'(\boldsymbol{\psi}_t - \boldsymbol{\epsilon}_t) + \nu_t = \mathbf{F}'\boldsymbol{\psi}_t + \nu_t;$$

and

$$\boldsymbol{\psi}_t = \mathbf{G}\boldsymbol{\psi}_{t-1} + \boldsymbol{\omega}_{1t}$$

where $\boldsymbol{\omega}_{1t} = \boldsymbol{\omega}_t + \boldsymbol{\epsilon}_t$ is an independent evolution error sequence with variance

$$\mathbf{W}_{1t} = \mathbf{W}_t + \mathbf{U}_t.$$

The *stochastic* shift from $\boldsymbol{\theta}_t$ to $\boldsymbol{\psi}_t$ has transformed M to the model M_1 given by $\{\mathbf{F}, \mathbf{G}, V_t, \mathbf{W}_{1t}\}$. Thus, for *any* specified $\mathbf{W}_{1t}$, the model $\{\mathbf{F}, \mathbf{G}, V_t, \mathbf{W}_{1t}\}$ can be written in an uncountable number of ways by choosing $\mathbf{W}_t$ and $\mathbf{U}_t$ to give $\mathbf{W}_{1t}$ as above. Clearly, although reparametrisations are involved, they are stochastic and cannot be linked to a simple, deterministic linear transformation of the state vector.

EXAMPLE 5.10. To show that this can in fact be done, consider the model with $\mathbf{F}' = (1,1)$ and $\mathbf{G} = \mathbf{I}$, the 2×2 identity matrix. In order to satisfy $\mathbf{F}'\boldsymbol{\epsilon}_t = 0$ for all t, we require $\boldsymbol{\epsilon}_t$ be of the form $\boldsymbol{\epsilon}_t = (\epsilon_{t1}, -\epsilon_{t1})'$. Then, with $\epsilon_{t1} \sim N[0, U_t]$ for some $U_t > 0$,

$$\mathbf{U}_t = U_t \begin{pmatrix} 1 & -1 \\ -1 & 1 \end{pmatrix}.$$

Thus, given any variance matrix

$$\mathbf{W}_{1t} = \begin{pmatrix} W_{t1} & W_{t3} \\ W_{t3} & W_{t2} \end{pmatrix},$$

to obtain $\mathbf{W}_{1t} = \mathbf{W}_t + \mathbf{U}_t$ we simply choose $\mathbf{W}_t = \mathbf{W}_{1t} - \mathbf{U}_t$ and this is a valid variance matrix whenever

$$0 < U_t < (W_{t1}W_{t2} - W_{t3}^2)/(W_{t1} + W_{t2} + 2W_{t3}).$$

This can always be satisfied since, as $\mathbf{W}_{1t}$ is positive definite, the numerator of this upper bound is the determinant $|\mathbf{W}_{1t}| > 0$, and the denominator is the positive term $\mathbf{F}'\mathbf{W}_{1t}\mathbf{F}$.

These stochastic shift models seem to suggest that any specified model can be transformed to one with a particularly simple form, based on *diagonal* evolution variance matrix. This is certainly an appealing simplification for practical purposes and, in a large number of practical cases, can be done. There are cases, however, when this is not possible. Consider Example 5.10. Here, $\mathbf{W}_t$ will be diagonal only when $W_{t3} = -U_t$ which will never be possible if $W_{t3} > 0$. More commonly, as will become clear later, we use evolution matrices that are *block-diagonal*.

This discussion motivates the following definition.

Definition 5.5. *The models M and M_1, which are either observable or constrained observable, are said to be* **generally equivalent** *if and only if they produce the same forecast distributions for all time.*

5.4   CANONICAL MODELS

5.4.1   Introduction

Similarity groups together observable models having similar forecast functions. Within each such group, we identify particular models with specific, simple structure that provide *canonical* DLMs consistent with the required forecast function form. Similarity is, as described, essentially related to the eigenstructure of system evolution matrices and therefore, in this section, we explore the various possible eigenvalue configurations that may, and do, arise. In relation to simple, canonical matrices that are similar to any given $\mathbf{G}$, the focus is naturally on diagonal, block diagonal and **Jordan** forms. Supporting material on the linear algebra associated with these matrices appears in Section 16.4 of Chapter 16. We begin naturally with the simplest case when the $n \times n$ system matrix has a single real eigenvalue of multiplicity n, in which system matrices in the forms of simple Jordan blocks are fundamental.

Definition 5.6. *The $n \times n$* **Jordan block** *is defined, for real or complex λ, as the $n \times n$ upper diagonal matrix*

$$
\mathbf{J}_n(\lambda) = \begin{pmatrix}
\lambda & 1 & 0 & 0 & \cdots & 0 \\
0 & \lambda & 1 & 0 & \cdots & 0 \\
0 & 0 & \lambda & 1 & \cdots & 0 \\
\vdots & \vdots & \vdots & \vdots & \ddots & \vdots \\
0 & 0 & 0 & 0 & \cdots & 1 \\
0 & 0 & 0 & 0 & \cdots & \lambda
\end{pmatrix}.
$$

Thus the diagonal elements are all equal to λ, those on the super-diagonal are 1, and the remaining elements are 0.

5.4.2 System matrix with one real eigenvalue

Suppose that $\mathbf{G}$ has a single real eigenvalue λ of multiplicity n. There are, clearly, an infinite number of such matrices, the simplest being $\mathbf{G} = \lambda \mathbf{I}$, a multiple of the $n \times n$ identity matrix. The first result, of great importance, shows that the class of observable models is restricted to those whose system matrix is similar to the Jordan block $\mathbf{J}_n(\lambda)$.

Theorem 5.4. *If $\mathbf{G}$ has one eigenvalue λ of multiplicity n but is not similar to $\mathbf{J}_n(\lambda)$, then any TSDLM $\{\mathbf{F}, \mathbf{G}, ., .\}$ is unobservable.*

Proof: The proof uses properties of similar matrices and Jordan forms for which reference may be made to Section 16.4.3 of Chapter 16. Since $\mathbf{G}$ is not similar to the Jordan block, then it must be similar to a Jordan form

$$\mathbf{J}_s = \text{block diag}\,[\mathbf{J}_{n_1}(\lambda), \mathbf{J}_{n_2}(\lambda), \ldots, \mathbf{J}_{n_s}(\lambda)],$$

for some $s \geq 2$ and $n_r \geq 1$, $(r = 1, \ldots, s)$, with $n_1 + n_2 + \cdots + n_s = n$. For each r, let $\mathbf{f}_r$ be any n_r dimensional vector, and define $\mathbf{F}_s$ via $\mathbf{F}'_s = (\mathbf{f}'_1, \mathbf{f}'_2, \ldots, \mathbf{f}'_s)$. Then the observability matrix $\mathbf{T}$ of any model with regression vector $\mathbf{F}_s$ and system matrix $\mathbf{J}_s$ has rows

$$\mathbf{t}'_{k+1} = \mathbf{F}'_s \mathbf{J}^k_s(\lambda) = [\mathbf{f}'_1 \mathbf{J}^k_{n_1}(\lambda), \ldots, \mathbf{f}'_s \mathbf{J}^k_{n_s}(\lambda)], \quad (k = 0, \ldots, n-1).$$

Define $m = \max_{r=1,\ldots,s}\{n_r\}$, so that $m \leq n - s + 1$. Then using the above referenced appendix, it follows that for $k > m$,

$$\sum_{r=1}^{k} \binom{k}{r} (-\lambda)^r \mathbf{t}_{k-r+1} = \mathbf{0}.$$

This implies that $\mathbf{T}$ has at most $m < n$ linearly independent rows and so is of less than full rank. The result follows.

$\diamond$

Corollary 5.1. *If* **G** *and* **G**$_1$ *each have a single eigenvalue* λ *of multiplicity* n *and, for some* **F** *and* **F**$_1$*, the models* $\{\mathbf{F}, \mathbf{G}, ., .\}$ *and* $\{\mathbf{F}_1, \mathbf{G}_1, ., .\}$ *are observable, then* **G** *is similar to* **G**$_1$ *and the models are similar.*

These results establish the similarity of models with a common, single eigenvalue to observable models whose system matrix has the simple, canonical form of a Jordan block. There are still many such models. The following result identifies some further structure that then leads to the identification of a unique, simple canonical model within the class.

Theorem 5.5. *Any TSDLM* $\{\mathbf{F}, \mathbf{J}_n(\lambda), ., .\}$ *is observable if and only if the first element of* **F** *is non-zero.*

Proof: Let $\mathbf{F} = (f_1, \ldots, f_n)'$ and denote the rows of the observability matrix **T** by

$$\mathbf{t}'_{r+1} = \mathbf{F}' \mathbf{J}_n^r(\lambda), \qquad (r = 0, \ldots, n-1).$$

Let **A** be the $n \times n$ matrix whose rows are

$$\mathbf{a}'_{k+1} = \sum_{r=0}^{k} \binom{k}{r} (-\lambda)^{k-r} \mathbf{t}'_{r+1}, \qquad (k = 0, \ldots, n-1).$$

Then

$$\mathbf{a}'_{k+1} = \mathbf{F}' [\mathbf{J}_n(\lambda) - \lambda \mathbf{I}_n]^k = (0, \ldots, 0, f_1, \ldots, f_{n-k}),$$

having k leading zeros. Thus **A** is an upper triangular matrix with leading diagonal $(f_1, \ldots, f_1)'$ and so has determinant f_1^n. Therefore **A** is non-singular if and only if f_1 is non-zero. The rows of **A** have been constructed, however, as linearly independent linear combinations of those of **T** which implies that **A** and **T** have the same rank so the result follows.

◇

Notice that even in the degenerate case of $\lambda = 0$, when $\mathbf{J}_n(0)$ has rank $n - 1$, the model is still observable. Thus, for observability, it is not necessary in general that the system matrix has full rank. It is necessary, however, that the system matrix has rank at least $n - 1$.

Within the class of models identified by the above results, the model having the simplest mathematical form is that with $\mathbf{F}' = \mathbf{E}'_n = (1, 0, \ldots, 0)$, the only non-zero element of which is the leading 1. This specific form is adopted as the basic model with a single real eigenvalue. The $\mathbf{E}_n$ notation is used throughout.

Definition 5.7. *Let* $\{\mathbf{F}, \mathbf{G}, V_t, \mathbf{W}_t\}$ *be any observable TSDLM in which the system matrix* $\mathbf{G}$ *has a single real eigenvalue* λ *of multiplicity* n. *Let* $\mathbf{T}$ *be the observability matrix of this model and define*

$$\mathbf{E}_n = (1, 0, \dots, 0)'.$$

Then:

(i) *any model* $\{\mathbf{E}_n, \mathbf{J}_n(\lambda), ., .\}$ *with observability matrix* $\mathbf{T}_0$ *is defined as a* **canonical similar model**; *and*

(ii) *the model* $\{\mathbf{E}_n, \mathbf{J}_n(\lambda), V_t, \mathbf{H}\mathbf{W}_t\mathbf{H}'\}$, *where* $\mathbf{H} = \mathbf{T}_0^{-1}\mathbf{T}$, *is defined as the* **canonical equivalent model**.

5.4.3 Multiple real eigenvalues

Suppose $\mathbf{G}$ has s distinct eigenvalues $\lambda_1, \dots, \lambda_s$ with λ_i having multiplicity $r_i \geq 1$, so that $n = r_1 + \cdots + r_s$. Again using Section 16.4.3 of Chapter 16, it follows that $\mathbf{G}$ is similar to the block diagonal **Jordan form** matrix

$$\mathbf{J} = \text{block diag}[\mathbf{J}_{r_1}(\lambda_1), \dots, \mathbf{J}_{r_s}(\lambda_s)],$$

defined by the superposition of s Jordan blocks, one for each of the distinct eigenvalues and having dimension given by the multiplicity of that eigenvalue. In such cases it can be shown, in a generalisation of Theorem 5.5, that the DLM is observable if and only if it is similar to models of the form

$$\{\mathbf{E}, \mathbf{J}, ., .\},$$

where

$$\mathbf{E}' = (\mathbf{E}'_{r_1}, \dots, \mathbf{E}'_{r_s})$$

is constructed as the corresponding catenation of s vectors of the form

$$\mathbf{E}'_r = (1, 0, \dots, 0)$$

of dimension r, for $r = r_1, \dots, r_s$. The form of $\mathbf{E}$ and $\mathbf{J}$ provide the algebraically simplest similar models for this case and so are adopted.

Definition 5.8. *Let* $\{\mathbf{F}, \mathbf{G}, V_t, \mathbf{W}_t\}$ *be any observable TSDLM in which the system matrix* $\mathbf{G}$ *has a s distinct real eigenvalues* $\lambda_1, \dots, \lambda_s$ *with multiplicities* $r_1, \dots, r_s$ *respectively. Let* $\mathbf{T}$ *be the observability matrix of this model and define*

$$\mathbf{E} = (\mathbf{E}_{r_1}, \dots, \mathbf{E}_{r_s})',$$

and

$$\mathbf{J} = block\ diag[\mathbf{J}_{r_1}(\lambda_1), \ldots, \mathbf{J}_{r_s}(\lambda_s)].$$

Then:

(i) *any model* $\{\mathbf{E}, \mathbf{J}, ., .\}$ *with observability matrix* $\mathbf{T}_0$ *is defined as a* **canonical similar model**; *and*

(ii) *the model* $\{\mathbf{E}, \mathbf{J}, V_t, \mathbf{H}\mathbf{W}_t\mathbf{H}'\}$, *where* $\mathbf{H} = \mathbf{T}_0^{-1}\mathbf{T}$, *is defined as the* **canonical equivalent model**.

5.4.4   Complex eigenvalues when $n = 2$

Suppose that $n = 2$ and the 2×2 system matrix $\mathbf{G}$ has complex eigenvalues. Since $\mathbf{G}$ is real valued, it must have a pair of complex conjugate eigenvalues of the form

$$\lambda_1 = \lambda e^{i\omega} \quad \text{and} \quad \lambda_2 = \lambda e^{-i\omega},$$

for some real λ and ω, i being the imaginary square root of -1. Thus $\mathbf{G}$ is similar to $\text{diag}(\lambda_1, \lambda_2)$ and, as in the case of distinct real eigenvalues of Section 5.4.3, the model is similar to any model of the form $\{(1,1)', \text{diag}(\lambda_1, \lambda_2), ., .\}$. In this case we do not, however, use this canonical similar model. Since the eigenvalues are complex, this would imply a complex parametrisation which would horrify practitioners and is clearly to be avoided. Instead we identify a *real canonical form* of $\mathbf{G}$ and the associated similar, **real** models. To proceed, note that, again following Section 16.4.3 of Chapter 16, if

$$\mathbf{H} = \begin{pmatrix} 1 & 1 \\ i & -i \end{pmatrix},$$

then

$$\mathbf{H} \begin{pmatrix} \lambda_1 & 0 \\ 0 & \lambda_2 \end{pmatrix} \mathbf{H}^{-1} = \lambda \begin{pmatrix} \cos(\omega) & \sin(\omega) \\ -\sin(\omega) & \cos(\omega) \end{pmatrix}$$

and

$$\mathbf{H}' \begin{pmatrix} 1 \\ 0 \end{pmatrix} = \begin{pmatrix} 1 \\ 1 \end{pmatrix},$$

from which it follows that the model is similar to any model with regression vector $(1,0)'$ and system matrix with the above cos/sin form.

Definition 5.9. *Let* $\{\mathbf{F}, \mathbf{G}, V_t, \mathbf{W}_t\}$ *be any 2-dimensional observable TSDLM in which the system matrix* $\mathbf{G}$ *has a pair of distinct, complex conjugate eigenvalues*

$$\lambda_1 = \lambda e^{i\omega} \qquad and \qquad \lambda_2 = \lambda e^{-i\omega}$$

for real, non-zero λ *and* ω. *Let* $\mathbf{T}$ *be the* 2×2 *observability matrix of this model and define*

$$\mathbf{J}_2(\lambda, \omega) = \lambda \begin{pmatrix} \cos(\omega) & \sin(\omega) \\ -\sin(\omega) & \cos(\omega) \end{pmatrix}$$

Then:

(i) *any model* $\{\mathbf{E}_2, \mathbf{J}_2(\lambda, \omega), ., .\}$ *with observability matrix* $\mathbf{T}_0$ *is defined as a* **real canonical similar model;** *and*

(ii) *the model* $\{\mathbf{E}_2, \mathbf{J}_2(\lambda, \omega)V_t, \mathbf{H}\mathbf{W}_t\mathbf{H}'\}$, *where* $\mathbf{H} = \mathbf{T}_0^{-1}\mathbf{T}$, *is defined as the* **real canonical equivalent model.**

It is easily checked that the observability matrix of the real canonical model is simply

$$\mathbf{T}_0 = \begin{pmatrix} 1 & 0 \\ \lambda\cos(\omega) & \lambda\sin(\omega) \end{pmatrix}.$$

5.4.5 Multiple complex eigenvalues[†]

Following Section 16.4 of Chapter 16 once again, we directly define the real canonical models for cases in which the system matrix has multiple complex eigenvalues. Although rare in practice, there are instances on which such models may be used and the canonical forms defined here provide the simplest possible construction. Note that, since complex eigenvalues must occur in conjugate pairs as $\mathbf{G}$ is real valued, then for multiple complex eigenvalues is is necessary that the dimension n of the model is even.

Definition 5.10. *Let* $\{\mathbf{F}, \mathbf{G}, V_t, \mathbf{W}_t\}$ *be any n-dimensional observable TSDLM in which* $n = 2v$ *for some positive integer* v. *Suppose the system matrix* $\mathbf{G}$ *has* v *multiples of a pair of distinct, complex conjugate eigenvalues*

$$\lambda_1 = \lambda e^{i\omega} \qquad and \qquad \lambda_2 = \lambda e^{-i\omega},$$

[†]The remaining sections of this Chapter are of rather theoretical interest and may be omitted without loss on a first reading.

for real, non-zero λ and ω. Let $\mathbf{T}$ be the $n \times n$ observability matrix of this model and define

$$\mathbf{J}_{2,v}(\lambda,\omega) = \lambda \begin{pmatrix} \mathbf{J}_2(\lambda,\omega) & \mathbf{I} & \mathbf{0} & \cdots & \mathbf{0} \\ \mathbf{0} & \mathbf{J}_2(\lambda,\omega) & \mathbf{I} & \cdots & \mathbf{0} \\ \vdots & \vdots & \vdots & \ddots & \vdots \\ \mathbf{0} & \mathbf{0} & \mathbf{0} & \cdots & \mathbf{I} \\ \mathbf{0} & \mathbf{0} & \mathbf{0} & \cdots & \mathbf{J}_2(\lambda,\omega) \end{pmatrix}.$$

Thus $\mathbf{J}_{2,v}(\lambda,\omega)$ is a $2v \times 2v$ block matrix entirely composed of 2×2 submatrices. The v diagonal blocks are the basic cos/sin 2×2 blocks $\mathbf{J}_2(\lambda,\omega)$, the super-diagonal blocks are the 2×2 identity $\mathbf{I}$, and the remaining elements are zero. Finally, define the $2v \times 1$ vector $\mathbf{E}_{2,v}$ via

$$\mathbf{E}_{2,v} = (\mathbf{E}_2', \ldots, \mathbf{E}_2')'.$$

Then:

(i) any model $\{\mathbf{E}_{2,v}, \mathbf{J}_{2,v}(\lambda,\omega), ., .\}$ with observability matrix $\mathbf{T}_0$ is defined as a **real canonical similar model**; and

(ii) the model $\{\mathbf{E}_{2,v}, \mathbf{J}_{2,v}(\lambda,\omega), V_t, \mathbf{H}\mathbf{W}_t\mathbf{H}'\}$, where $\mathbf{H} = \mathbf{T}_0^{-1}\mathbf{T}$, is defined as the **real canonical equivalent model**.

Note in particular the special case of Section 5.4.4 when $v = 1$, $\mathbf{E}_{2,1} = \mathbf{E}_2$, and $\mathbf{J}_{2,1}(\lambda,\omega) = \mathbf{J}_2(\lambda,\omega)$.

5.4.6   General case

In the most general case, $\mathbf{G}$ has s real and distinct eigenvalues $\lambda_1, \ldots, \lambda_s$ of multiplicities $r_1, \ldots, r_s$ respectively, and v pairs of complex conjugate eigenvalues

$$\lambda_{s+k}e^{i\omega_k} \qquad \text{and} \qquad \lambda_{s+k}e^{-i\omega_k}, \qquad\qquad (k = 1, \ldots, v),$$

for some real and distinct $\lambda_{s+1}, \ldots, \lambda_{s+v}$, some real and distinct $\omega_1, \ldots, \omega_v$, with the kth pair having multiplicity r_{s+k}, $(k = 1, \ldots, v)$, respectively. Note that the dimension of the model is now

$$n = \sum_{k=1}^{s} r_k + 2 \sum_{k=1}^{v} r_{s+k}.$$

The simplest real similar model is based on a similar system matrix formed of the superposition of diagonal blocks corresponding to the individual real and complex conjugate eigenvalues defined in previous sections. Again the reader is referred to Section 16.4 of Chapter 16.

Definition 5.11. *Let* $\{\mathbf{F}, \mathbf{G}, V_t, \mathbf{W}_t\}$ *be any n-dimensional observable TSDLM with the eigenvalue structure as detailed above. Let* $\mathbf{T}$ *be the $n \times n$ observability matrix of this model and define the $n \times n$ block diagonal matrix* $\mathbf{J}$ *as*

$$\mathbf{J} = block\ diag[\mathbf{J}_{r_1}(\lambda_1), \mathbf{J}_{r_2}(\lambda_2), \ldots, \mathbf{J}_{r_s}(\lambda_s);$$
$$\mathbf{J}_{2,r_s+1}(\lambda_{s+1}, \omega_1), \mathbf{J}_{2,r_s+2}(\lambda_{s+2}, \omega_2),$$
$$\ldots, \mathbf{J}_{2,r_s+v}(\lambda_{s+v}, \omega_v)].$$

Finally, define the $n \times 1$ vector $\mathbf{E}$ *via*

$$\mathbf{E} = (\mathbf{E}'_{r_1}, \mathbf{E}'_{r_2}, \ldots, \mathbf{E}_{r_s}; \mathbf{E}_{2,r_s+1}, \mathbf{E}'_{2,r_s+2}, \ldots, \mathbf{E}'_{2,r_s+v})'.$$

Then:

(i) *any model* $\{\mathbf{E}, \mathbf{J}, \cdot, \cdot\}$ *with observability matrix* $\mathbf{T}_0$ *is defined as a* **real canonical similar model**; *and*

(ii) *the model* $\{\mathbf{E}, \mathbf{J}, V_t, \mathbf{HW}_t\mathbf{H}'\}$, *where* $\mathbf{H} = \mathbf{T}_0^{-1}\mathbf{T}$, *is defined as the* **real canonical equivalent model.**

This general canonical form is simply constructed from the simpler canonical models for each of the real and complex pairs of eigenvalues. The individual sytem matrices are simply superposed to form an overall block diagonal $\mathbf{J}$, and the corresponding individual regression vectors are catenated in the same order to provide the general vector $\mathbf{E}$. This construction of a very general model from the component building blocks provided by simpler models is a key concept in model design, the subject of the next chapter.

5.5 EXERCISES

(1) Consider the model $\{1, \lambda, V, W\}$ where λ is real. Apply Theorems 5.1 and 5.2 to deduce
  (a) the limiting values of C, R, A and Q; and
  (b) the limiting representation of Y_t as a linear function of Y_{t-1}, e_t and e_{t-1} (with coefficients depending on the limiting value C). Distinguish between, and comment upon, the cases $|\lambda| < 1$, $\lambda = 1$, and $|\lambda| > 1$.

(2) The eigenvalues of the matrix $\mathbf{H} = (\mathbf{I} - \mathbf{AF}')\mathbf{G}$ determine the limiting representation of Y_t in Theorem 5.2. Verify the matrix identity $\mathbf{H} = \mathbf{CR}^{-1}\mathbf{G}$.

(3) (A harder question). Apply Theorem 5.2 to obtain limiting representations of the observation series in terms of past observations and one-step forecast errors in the model

$$\left\{ \begin{pmatrix} 1 \\ 0 \end{pmatrix}, \begin{pmatrix} 1 & \lambda \\ 0 & \lambda \end{pmatrix}, V, V \begin{pmatrix} U + \lambda^2 W & \lambda^2 W \\ \lambda^2 W & \lambda^2 W \end{pmatrix} \right\}.$$

Here U and W are any two variances and $|\lambda| \leq 1$.

(4) Verify the stated posterior distributions for the parameters ψ_t in Examples 5.3, and θ_{t2} in Example 5.4.

(5) Determine which of the following models is observable. The defining components $\mathbf{F}$ and $\mathbf{G}$ are given.
  (a) $\mathbf{F} = 1$ and $\mathbf{G} = \lambda$ for some real λ.
  (b) $\mathbf{F}' = (1,1)$ and $\mathbf{G} = \mathrm{diag}(\lambda_1, \lambda_2)$ for some real λ_1 and λ_2.
  (c) $\mathbf{F}' = (1,0)$ and

$$\mathbf{G} = \begin{pmatrix} 0 & 1 \\ 0 & 0 \end{pmatrix}.$$

  (d) $\mathbf{F}' = (1,1,1)$ and

$$\mathbf{G} = \begin{pmatrix} 4 & -1 & 2 \\ 3 & 9 & 3 \\ 1 & 5 & 5 \end{pmatrix}.$$

  (e) $\mathbf{F}' = (0,1)$ and $\mathbf{G} = \mathbf{J}_2(1,\omega)$ for some real ω.
  (f) $\mathbf{F}' = \mathbf{E}'_n$ and $\mathbf{G} = \mathbf{J}_n(\lambda)$ for $n \geq 2$.

(6) Give an example of an observable, n−dimensional model whose $\mathbf{G}$ matrix has rank $n - 1$. Show that it is necessary for observability that the rank of $\mathbf{G}$ in such models is either $n - 1$ or n.

(7) Transform the following observable models to canonical forms, identifying the corresponding similarity matrices $\mathbf{H}$ (Theorem 5.2).
  (a) $\mathbf{F}' = (1,0)$ and $\mathbf{G} = \mathrm{diag}(\lambda_1, \lambda_2)$ for some real and distinct λ_1 and λ_2.
  (b) $\mathbf{F}' = (1,1,1)$ and

$$\mathbf{G} = \begin{pmatrix} 1 & 1 & 1 \\ 0 & 1 & 1 \\ 0 & 0 & 0.5 \end{pmatrix}.$$

(c) $\mathbf{F}' = (1,1)$ and

$$\mathbf{G} = \begin{pmatrix} 0.5e^{i\omega} & 0 \\ 0 & 0.5e^{-i\omega} \end{pmatrix},$$

where ω is not an integer multiple of π.

(8) Obtain the forecast functions of models with the following $\mathbf{F}$ and $\mathbf{G}$, noting whether or not the models are observable (nb. all models have real parameters).

(a) $\mathbf{F}' = 1$ and $\mathbf{G} = \lambda$. Investigate all possible cases $\lambda < -1$, $\lambda = -1$, $-1 < \lambda < 0$, $\lambda = 0$, $0 < \lambda < 1$, $\lambda = 1$ and $\lambda > 1$.

(b) $\mathbf{F}' = (1,0,0)$ and $\mathbf{G} = \mathbf{J}_3(\lambda)$ for $0 < \lambda < 1$. Examine the form of $f_t(k)$ as a function of k, determining, in particular, the turning points of the forecast function.

(c) $\mathbf{F}' = (1,0,1,0,1)$ and

$$\mathbf{G} = \text{block diag}\left\{ \begin{pmatrix} 1 & 1 \\ 0 & 1 \end{pmatrix}, \lambda \begin{pmatrix} \cos(\omega) & \sin(\omega) \\ -\sin(\omega) & \cos(\omega) \end{pmatrix}, \phi \right\},$$

with $\lambda > 0$, $1 > \phi > 0$ and ω is not an integer multiple of π.

(9) Consider a model with

$$\mathbf{F} = \begin{pmatrix} 1 \\ 0 \\ 0 \end{pmatrix} \quad \text{and} \quad \mathbf{G} = \begin{pmatrix} 0 & 1 & 0 \\ 0 & 0 & 1 \\ 1 & 0 & 0 \end{pmatrix}.$$

Show that $f_t(k) = f_t(k + 3n)$ for all positive integers k and n, so that the forecast function is *cyclical* of *period* 3. Show that the model is observable and transform it to canonical form.

(10) Generalise the previous example to models that are cyclical of period $n > 1$, having $\mathbf{F} = \mathbf{E}_n = (1,0,\ldots,0)'$ and

$$\mathbf{G} = \begin{pmatrix} \mathbf{0} & \mathbf{I} \\ 1 & \mathbf{0}' \end{pmatrix},$$

where $\mathbf{I}$ is the $n \times n$ identity matrix. Distinguish between the cases of even and odd values of n.

(11) For some integer $n > 1$, suppose that $\mathbf{F}' = (1, \mathbf{E}'_n)$ and

$$\mathbf{G} = \begin{pmatrix} 1 & 0 & \mathbf{0}' \\ \mathbf{0} & 0 & \mathbf{I} \\ \mathbf{0} & 1 & \mathbf{0}' \end{pmatrix},$$

where $\mathbf{I}$ is the $n \times n$ identity matrix. Show that the model is unobservable. Let ϕ_t denote the final $n-1$ elements of the state vector at time t, so that $\boldsymbol{\theta}'_t = (\theta_t, \phi'_t)$, say. If $\mathbf{1}'\phi_t = 1$ show that the model is constrained observable.

(12) Consider any two DLMs M and M_1 characterised by quadruples

$$M : \qquad \{\mathbf{E}_3, \mathbf{J}_3(1), V_t, \mathbf{W}_t\},$$
$$M_1 : \qquad \{\mathbf{E}_3, \mathbf{L}_3, V_t, \mathbf{W}_{1t}\},$$

where $\mathbf{L}_3$ is the 3×3 upper triangular matrix of unities,

$$\mathbf{L}_3 = \begin{pmatrix} 1 & 1 & 1 \\ 0 & 1 & 1 \\ 0 & 0 & 1 \end{pmatrix}.$$

(a) Calculate the observability matrices $\mathbf{T}$ and $\mathbf{T}_1$ for M and M_1 respectively, and deduce that both models are observable TSDLMs.

(b) Show that the models are similar, and that the similarity matrix $\mathbf{H} = \mathbf{T}^{-1}\mathbf{T}_1$ is given by

$$\mathbf{H} = \begin{pmatrix} 1 & 0 & 0 \\ 0 & 1 & 1 \\ 0 & 0 & 1 \end{pmatrix}.$$

(c) Identify the common form of the forecast function, and interpret the meaning of the state parameters in each model.

(c) Under what conditions are M and M_1 equivalent?

CHAPTER 6

MODEL SPECIFICATION AND DESIGN

6.1 BASIC FORECAST FUNCTIONS

The mean response function μ_{t+k} of any given model defines, as a function of the step ahead index k, the implied form of the time series and is thus of fundamental interest in model design. The expectation of this, the forecast function $f_t(k)$, provides the forecaster's view of the expected development of the series and we focus on this, rather than the mean response itself, as the central guide to constructing appropriate models. This is purely convention on our part; equivalent discussion could be based on the mean response function instead. We begin with discussion of forecast functions derived from the various TSDLMs of the previous chapter. Together with complementary forms from models for the effects of independent variables, these provide essentially all practically important dynamic linear models.

6.1.1 Real Jordan block system matrices

The first, and essentially the simplest, observable class of models is that in which the system matrices have a single real eigenvalue.

Theorem 6.1. For real λ, any canonical model $\{\mathbf{E}_n, \mathbf{J}_n(\lambda), . , \mathbf{W}_t\}$, and hence any similar model, has a forecast function $f_t(k)$, $(k \geq 0)$, of the following form:

(1) If λ is non-zero, as with most practical models, then

$$f_t(k) = \lambda^k \sum_{r=0}^{n-1} a_{tr} k^r,$$

for some coefficients $a_{t0}, \ldots, a_{tn-1}$ that do not depend on k;

(2) In the irregular case $\lambda = 0$ then

$$f_t(k) = m_{tk+1}, \qquad\qquad (0 \leq k < n),$$

with $f_t(k) = 0$ for $k \geq n$.

Proof: Given $E[\boldsymbol{\theta}_t \mid D_t] = \mathbf{m}_t$, as usual, we have, by definition,

$$f_t(k) = \mathbf{E}'_n \mathbf{J}_n(\lambda)^k \mathbf{m}_t.$$

Thus when λ is non-zero we have (Section 16.4.3 of Chapter 16)

$$f_t(k) = \lambda^k \sum_{r=0}^{n-1} a_{tr} k^r,$$

where the coefficients $a_{t0}, \ldots, a_{tn-1}$ do not depend on k although they do depend on λ. For any similar model it follows from Chapter 5 that the associated forecast function has the same form in k, though generally with different coefficients.

If $\lambda = 0$ then $\mathbf{J}_n(\lambda)^k$ has zero elements everywhere but for the kth super-diagonal which is comprised of elements all equal to 1. Thus it follows that $f_t(k) = m_{tk+1}$, the $(k+1)^{st}$ element of $\mathbf{m}_t$, when $0 \leq k < n$. Also $\mathbf{J}_n(\lambda)^k = \mathbf{0}$ for $k \geq n$ and the result follows.

$\diamond$

The practically important cases have λ non-zero, and thus $f_t(k)$ has the form of the kth power of the eigenvalue λ multiplying a polynomial of order n, degree $n - 1$, in the step-ahead index k.

EXAMPLE 6.1. Consider the case $n = 1$ so that the canonical model has $\mathbf{F} = 1$ and $\mathbf{G} = \lambda$, with $\boldsymbol{\theta}_t = \mu_t$ and $\mathbf{m}_t = m_t$, all scalars. Here then $f_t(k) = a_{t0} \lambda^k$ with $a_{t0} = m_t$. This special case is important since it illustrates the nature of the contribution of a single eigenvalue of multiplicity one to any observable DLM. The value of λ clearly determines the behaviour of the forecast function. The various possible cases, illustrated in Figure 6.1 with $m_t = 1$, are described.

    (a) $\lambda = 0$.

        The model is simply $Y_t = \omega_t + \nu_t$ and m_t is a point estimate of $\mu_t = \omega_t$. Here $f_t(0) = m_t$, and $f_t(k) = 0$ for $k > 0$.

    (b) $\lambda = 1$.

        Here $f_t(k) = m_t$ for all $k \geq 1$. This is the case of the first-order polynomial model discussed in Chapter 2.

    (c) $0 < \lambda < 1$.

        Here we have $f_t(k) = \lambda^k m_t$ which decays to zero exponentially in k.

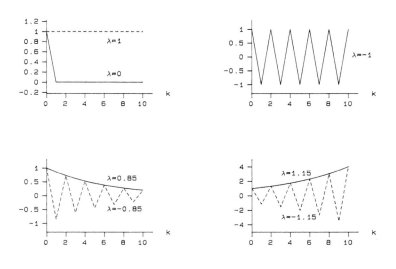

Figure 6.1.   Forecast functions $f_t(k) = \lambda^k$.

(d) $-1 < \lambda < 0$.

Here we have $f_t(k) = (-\lambda)^k m_t$, which oscillates in an exponentially decaying fashion to zero as k increases.

(e) $\lambda = -1$.

Here we have $f_t(k) = (-1)^k m_t$ which describes a periodic function with period 2, oscillating perfectly between m_t and $-m_t$. This period 2 DLM is called a **Nyquist harmonic,** and appears in models for cyclical or seasonal series in Chapter 8.

(f) $\lambda > 1$.

Here $f_t(k) = \lambda^k m_t$, which, for non-zero m_t, explodes exponentially, and monotonically, to ∞ if $m_t > 0$, and to $-\infty$ if $m_t < 0$.

(g) $\lambda < -1$.

Here $f_t(k) = (-\lambda)^k m_t$ oscillates explosively when m_t is non-zero.

EXAMPLE 6.2. Consider the case $n = 2$ so that the canonical model has the form

$$\left\{ \begin{pmatrix} 1 \\ 0 \end{pmatrix}, \begin{pmatrix} \lambda & 1 \\ 0 & \lambda \end{pmatrix}, \ldots, \right\}.$$

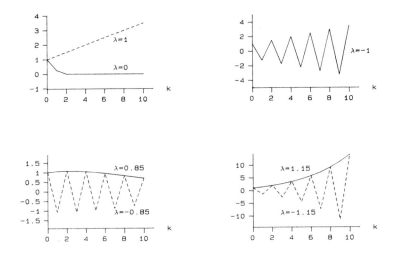

Figure 6.2.   Forecast functions $f_t(k) = (1 + 0.25k/\lambda)\lambda^k$.

With state vector and evolution error $\boldsymbol{\theta}_t = (\theta_{t1}, \theta_{t2})'$ and $\boldsymbol{\omega}_t = (\omega_{t1}, \omega_{t2})'$ respectively, we have

$$Y_t = \theta_{t1} + \nu_t,$$
$$\theta_{t1} = \lambda\theta_{t-1,1} + \theta_{t-1,2} + \omega_{t1},$$
$$\theta_{t2} = \lambda\theta_{t-1,2} + \omega_{t2}.$$

With $\mathbf{m}_t = (m_{t1}, m_{t2})'$ we have, for $k \geq 0$, and λ non-zero,

$$f_t(k) = (m_{t1} + km_{t2}/\lambda)\,\lambda^k.$$

The various possible cases, determined by the value of λ, are described below and illustrated in Figure 6.2 with $m_{t1} = 1$ and $m_{t2} = 0.25$.

(a) $\lambda = 0$.
The model is simply $Y_t = \omega_{t1} + \omega_{t-1,2} + \nu_t$ and $f_t(0) = m_{t1}$, $f_t(1) = m_{t2}$ and $f_t(k) = 0$ for $k > 1$. If the DLM is a constant model, the Y_t series can be expressed as a moving average process of order 2, MA(2).

(b) $\lambda = 1$.
Here $f_t(k) = m_{t1} + km_{t2}$ for all $k \geq 0$. This is an extremely

important model for short term forecasting. The forecast function is a straight line, or polynomial of order 2. The model may be viewed as representing a "locally linear" development of the mean response function over time.

(c) $0 < \lambda < 1$.

In this case the forecast function eventually decays exponentially to zero as k increases. The initial behaviour depends on the actual values of m_{t1} and m_{t2}; the forecast function is either monotonically decreasing or, alternatively, it may rise to a maximum for some $k > 0$ before decaying.

(d) $-1 < \lambda < 0$.

Here the forecast function oscillates between positive and negative values whilst $|f_t(k)|$ behaves essentially as described in (c).

(e) $\lambda = -1$.

Here we have $f_t(k) = (-1)^k(m_{t1} - km_{t2})$. This oscillates between points on two straight lines, tending to infinity as k increases when m_{t2} is non-zero. This case is not of real practical interest.

(f) $\lambda > 1$.

$f_t(k)$ explodes exponentially to $\pm\infty$ if $\mathbf{m}_t$ is non-zero.

(g) $\lambda < -1$.

Here $f_t(k)$ oscillates between positive and negative values whilst $|f_t(k)|$ behaves as in (f).

EXAMPLE 6.3. For any n, if $\lambda = 1$ we have

$$f_t(k) = a_{t0} + a_{t1}k + a_{t2}k^2 + \cdots + a_{t,n-1}k^{n-1}$$

which is a polynomial model of order $n-1$, degree n. For all n, these models provide the important class of **Polynomial DLMs** in which, essentially, the expected behaviour of the series is viewed as being *locally* described by a polynomial. Typically this can be seen as an approximation, usually using polynomial forms of low order 1,2 or 3, say, to an unknown but essentially smooth mean response function. Chapter 2 was devoted to the case $n = 1$. Chapter 7 describes the general case but is largely concerned with the case $n = 2$, including the important class of linear growth models.

EXAMPLE 6.4. In the special case that $\lambda = 0$, the model $\{\mathbf{E}_n, \mathbf{J}_n(0),$ $V_t, \mathbf{W}_t\}$ has the form

$$Y_t = \theta_{t1} + \nu_t,$$
$$\theta_{tr} = \theta_{t-1,r+1} + \omega_{tr}, \qquad (r = 1, \ldots, n-1),$$
$$\theta_{tn} = \omega_{tn},$$

so that

$$Y_t = \nu_t + \sum_{r=1}^{n} \omega_{t+1-r,i}.$$

From Theorem 6.1 with $\mathbf{m}_t = (m_{t1}, \ldots, m_{tn})'$, we have

$$f_t(k) = \begin{cases} \mathrm{E}[\theta_{t,k+1} \mid D_t] = m_{t,k+1}, & \text{for } 0 \le k < n; \\ 0, & \text{for } k \ge n. \end{cases}$$

The forecast function takes irregular values in the first n steps, $k = 0, 1, \ldots, n-1$, and then takes the value zero. If the model is constant with $V_t = V$ and $\mathbf{W}_t = \mathbf{W}$ for all t, then Y_t has a moving average $MA(n-1)$ representation

$$Y_t = \sum_{r=0}^{n-1} \psi_r \epsilon_{t-r}$$

where $\epsilon_t \sim N[0,1]$, $(t = 1, 2, \ldots)$ is a sequence of independent random quantities. This representation may be useful to those readers familiar with standard linear, stationary time series modelling (eg. Box and Jenkins, 1976). Note that it is derived as the very special case of zero eigenvalues. Also note that we can make this link with classical models even more apparent by transforming to the equivalent model $\{\mathbf{E}_n, \mathbf{J}_n(0), 0, \mathbf{W}_{1t}\}$ where

$$\mathbf{W}_{1t} = \mathbf{W}_t + \begin{bmatrix} V_t & \mathbf{0}' \\ \mathbf{0} & \mathbf{0}_{n-1} \end{bmatrix},$$

with $\mathbf{0}_r$ being the $(r \times r)$ zero matrix. Thus, whenever zero eigenvalues occur, V_t can be set as zero by suitably amending $\mathbf{W}_t$.

6.1.2 Single complex block system matrices

Complex eigenvalues in the system matrix lead to sinusoidal components in the forecast function. The simplest case is that of a single sin/cosine wave of that which corresponds to a pair of complex conjugate eigenvalues in the case $n = 2$.

Theorem 6.2. *In the 2-dimensional real canonical model*

$$\{\mathbf{E}_2, \mathbf{J}_2(\lambda,\omega), V_t, \mathbf{W}_t\}$$

with real, non-zero λ, *and real* ω, $(0 < \omega < 2\pi)$, *the forecast function has the form*

$$f_t(k) = [m_{t1}\cos(k\omega) + m_{t2}\sin(k\omega)]\lambda^k,$$

for real m_{t1} *and* m_{t2}.

Proof: By induction, and using standard trigonometric identities, it is easily shown that

$$\mathbf{J}_2(\lambda,\omega)^k = \lambda^k \begin{pmatrix} \cos(k\omega) & \sin(k\omega) \\ -\sin(k\omega) & \cos(k\omega) \end{pmatrix} = \mathbf{J}_2(\lambda^k, k\omega),$$

for all integers k. Thus, with $\mathbf{m}_t = (m_{t1}, m_{t2})'$,

$$f_t(k) = \mathbf{E}_2' \mathbf{J}_2(\lambda,\omega)^k \mathbf{m}_t = [m_{t1}\cos(k\omega) + m_{t2}\sin(k\omega)]\lambda^k.$$

$$\diamond$$

An alternative expression for the forecast function is

$$f_t(k) = \lambda^k r_t \cos(k\omega + \phi_t),$$

where:

(a) $r_t^2 = m_{t1}^2 + m_{t2}^2$. r_t is the **amplitude** of the **periodic**, or **harmonic**, component $m_{t1}\cos(k\omega) + m_{t2}\sin(k\omega)$; and

(b) $\phi_t = \arctan(-m_{t2}/m_{t1})$ is the **phase-angle**, or just the **phase**, of the periodic component. The quantity ω is the **frequency** of the periodic component that defines the number of time intervals over which the harmonic completes a full cycle, this number being simply $2\pi/\omega$. This follows since, for all integer $h \geq 0$,

$$f_t(k) = f_t(k + 2\pi h/\omega).$$

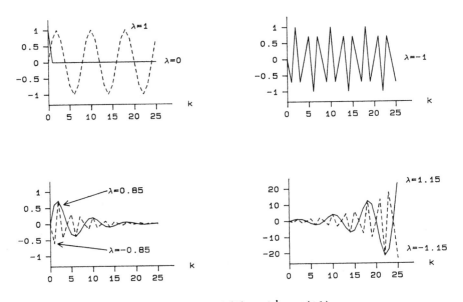

Figure 6.3.   Forecast functions $f_t(k) = \lambda^k \cos(8k)$.

The forecast function thus has the form of a sin/cosine wave, whose form is determined essentially by ω, that is modified by the multiplicative term λ^k determined by λ. The effect of this latter term is to either *dampen* or *explode* the periodic component. If $|\lambda| < 1$, the sinusoidal form is damped, decaying asymptotically to zero; if $|\lambda| > 1$ it is exploded, diverging as k increases. Negative values of λ lead to the forecast function oscillating between positive and negative values for consecutive values of k whilst either decaying to zero or diverging. Figure 6.3 illustrates the various possibilities in the particular case with coefficients $m_{t1} = 1$, $m_{t2} = 0$, and frequency $\omega = \pi/4$ so that a full cycle of the periodic component takes exactly 8 time intervals. Of greatest practical importance are the cases with $0 < \lambda \leq 1$. In particular, $\lambda = 1$ leads to a pure cosine wave of *period* $2\pi/\omega$, or *frequency* ω, that is often used to model simple, constant wave forms in seasonal time series.

6.1.3   Models with multiple complex eigenvalues[†]

Models with forecast functions of basic periodic forms discussed in the previous section provide for essentially all the cyclical/seasonal behaviour associated with linear models in practice. Only very rarely will a model with multiple complex eigenvalues be required. The forecast functions of such models are discussed here for completeness, although practitioners may ignore this section without loss on a first reading.

Refer to Definition 5.10 for the specification of the real canonical model in the case when a system matrix has a pair of complex conjugate eigenvalues $\lambda e^{i\omega}$ and $\lambda e^{-i\omega}$ with multiplicity v.

Theorem 6.3. *In the real canonical Jordan form model*

$$\{\mathbf{E}_{2,v}, \mathbf{J}_{2,v}(\lambda,\omega), V_t, \mathbf{W}_t\}$$

with λ and ω real, and $\lambda\omega$ non-zero, the forecast function is given by

$$f_t(k) = \left[\sum_{j=0}^{v-1} a_{tj}k^j\right] \lambda^k \cos(k\omega) + \left[\sum_{j=0}^{v-1} b_{tj}k^j\right] \lambda^k \sin(k\omega)$$

for some coefficients a_{tj} and b_{tj}, $(j = 0, \ldots, v-1)$. A neater expression is

$$f_t(k) = \lambda^k \sum_{j=0}^{v-1} r_{tj}k^j \cos(k\omega + \phi_{tj})$$

for some amplitudes r_{tj} and phase angles ϕ_{tj}, $(j = 0, \ldots, v-1)$.

Proof: The proof, an exercise in linear algebra, is left to the reader.

$\diamond$

As mentioned above, this case is rarely used in practice in any generality. Of some interest are the particular models in which $\lambda = 1$ and $\phi_{tj} = 0$, $(j = 0, \ldots, v-1)$, so that the forecast function

---

[†]This Section is of rather theoretical interest and may be omitted without loss on a first reading.

represents a cosine wave whose amplitude is a polynomial in k. For example, with $\lambda = 1$, $v = 2$, and $\phi_{tj} = 0$ for each j, we have

$$f_t(k) = (r_{t0} + r_{t1}k)\cos(k\omega).$$

Such forms can be useful in representing increasing seasonal patterns in some applications.

6.2 SPECIFICATION OF F_t AND G_t

Models for real series illustrating a variety of features are constructed by collecting together several simple component DLMs that model these individual features. We refer to this construction of complex models from components as **Superposition**, and the reverse process, that of identifying components of a given model, as **Decomposition**. We discuss these two important modelling concepts in detail here.

6.2.1 Superposition

Some examples introduce the basic ideas.

EXAMPLE 6.5. Consider the two special models M_1, of dimension n, and M_2, of dimension 1, specified by quadruples

$$M_1 : \quad \{\mathbf{F}, \mathbf{G}, 0, \mathbf{W}\};$$
$$M_2 : \quad \{0, 0, V, 0\}.$$

M_1 is a standard DLM whose observation series is purely determined by $\boldsymbol{\theta}_t$ through the mean response μ_t. If M_1 generates a series Y_{1t}, then $Y_{1t} = \mu_t = \mathbf{F}'\boldsymbol{\theta}_t$, with probability 1, where $\boldsymbol{\theta}_t$ is the state vector in this model. M_2 has no state vector, generating a simple and independent noise series, $Y_{2t} = \nu_t \sim N[0, V]$ for all t. Now, defining the series $Y_t = Y_{1t} + Y_{2t}$ it follows that Y_t is modelled by $\{\mathbf{F}, \mathbf{G}, V, \mathbf{W}\}$. Thus in adding the series defined by two DLMs we create a more complex DLM defined by combining the quadruples in a certain way. This is a simple example of superposition of DLMs.

EXAMPLE 6.6. Consider two purely deterministic models M_1 and M_2, with state vectors $\boldsymbol{\theta}_{1t}$ and $\boldsymbol{\theta}_{2t}$ respectively, defined via

$$M_1 : \quad \{\mathbf{F}_1, \mathbf{G}_1, 0, \mathbf{0}\};$$
$$M_2 : \quad \{\mathbf{F}_2, \mathbf{G}_2, 0, \mathbf{0}\}.$$

Adding the observations generated by these two models gives a series that obeys the model $\{\mathbf{F}, \mathbf{G}, 0, \mathbf{0}\}$ with state vector given by

$$\boldsymbol{\theta}'_t = (\boldsymbol{\theta}'_{1t}, \boldsymbol{\theta}'_{2t}),$$

where the regression vector and system matrix are given by

$$\mathbf{F}' = (\mathbf{F}'_1, \mathbf{F}'_2)$$

and

$$\mathbf{G} = \text{block diag}[\mathbf{G}_1, \mathbf{G}_2].$$

More generally, the linear combination of observations from any h deterministic models $M_1, \ldots, M_h$, in an obvious extension of the notation, follows a DLM with

$$\mathbf{F}' = (\mathbf{F}'_1, \ldots, \mathbf{F}'_h)$$

and

$$\mathbf{G} = \text{block diag}[\mathbf{G}_1, \ldots, \mathbf{G}_h].$$

These examples illustrate, in particular cases, the construction of DLMs from a collection of component DLMs by *superposition* of the corresponding state vectors, regression vectors and system matrices. (Of course if the two component state vectors have common elements, the model formed by this superposition may be reparametrised to one of lower dimension.) In general, we have the following result.

Theorem 6.4. *For integer $h \geq 2$, consider the h time series Y_{it} generated by DLMs*

$$M_i : \quad \{\mathbf{F}_i, \mathbf{G}_i, V_i, \mathbf{W}_i\}_t,$$

with state vectors $\boldsymbol{\theta}_{it}$ of dimensions n_i, for $i = 1, \ldots, h$. Denote the observation and evolution error series in M_i by ν_{it} and $\boldsymbol{\omega}_{it}$ respectively. Assume that, for all distinct i and j, $(1 < i, j < h)$, the series ν_{it} and $\boldsymbol{\omega}_{it}$ are mutually independent of the series ν_{jt} and $\boldsymbol{\omega}_{jt}$.
    Then the series defined by

$$Y_t = \sum_{i=1}^{h} Y_{it}$$

follows the DLM $\{\mathbf{F}, \mathbf{G}, V, \mathbf{W}\}_t$ *with state vector given by*

$$\boldsymbol{\theta}_t' = (\boldsymbol{\theta}_{1t}', \dots, \boldsymbol{\theta}_{ht}'),$$

of dimension $n = n_1 + \cdots + n_h$, *and quadruple defined via:*

$$\mathbf{F}_t' = (\mathbf{F}_{1t}', \dots, \mathbf{F}_{ht}');$$

$$\mathbf{G}_t = \text{block diag}[\mathbf{G}_{1t}, \dots, \mathbf{G}_{ht}];$$

$$V_t = \sum_{i=1}^{h} V_{it};$$

and

$$\mathbf{W}_t = \text{block diag}[\mathbf{W}_{1t}, \dots, \mathbf{W}_{ht}].$$

Proof: Directly, we have $Y_t = \mathbf{F}_t' \boldsymbol{\theta}_t + \nu_t$ where $\nu_t = \sum_{i=1}^{h} \nu_{it}$. Clearly ν_t is normally distributed with mean zero and the independence assumptions imply a variance V_t as required. Next, $\boldsymbol{\theta}_t = \mathbf{G}_t \boldsymbol{\theta}_{t-1} + \boldsymbol{\omega}_t$ where $\boldsymbol{\omega}_t' = (\boldsymbol{\omega}_{1t}', \dots, \boldsymbol{\omega}_{ht}')$. Again it is clear that $\boldsymbol{\omega}_t \sim N[\mathbf{0}, \mathbf{W}_t]$ independently of ν_t and the result follows.

$\diamond$

This obvious, but highly important, result is termed the **Principle of Superposition**. It simply states that the linear combination of series generated by independent DLMs follows a DLM that is defined via the superposition of the corresponding model components. The principle is heavily dependent on the additivity properties associated with linear normal models familiar in other areas of statistical modelling. The independence of models assumed here is not crucial to the structure, however, and can be relaxed if required. A most general statement of superposition would simply have to assume that $\nu_{1t}, \dots, \nu_{ht}$ are jointly normally distributed and, similarly, $\boldsymbol{\omega}_{1t}, \dots, \boldsymbol{\omega}_{ht}$ are jointly normally distributed, for each t. (Marginal normality of the terms within each model does not necessarily imply joint normality across models although the practical circumstances in which joint normality is violated are rare and of little importance). In any case, as is explored in depth below, practical utilisation of the superposition principle naturally and appropriately adopts the independence assumptions of the Theorem.

Design implications of superposition do not depend on the independence structure. As far as the forecast function form is concerned, the additivity property is itself sufficient to determine the following result. The trivial proof is left to the reader.

Theorem 6.5. *Consider the models in Theorem 6.4 and assume the error sequences are jointly normal as described above. Denote the forecast function of M_i by $f_{it}(k)$, $(k \geq 0)$, for $i = 1, \ldots, h$. Then the forecast function for the Y_t series generated by the superposition of the k component models is given by*

$$f_t(k) = \sum_{i=1}^{h} f_{it}(k)$$

for $k \geq 0$.

6.2.2 Decomposition and model design

The practical value of the superposition principle lies in the construction of models for complex problems by combining simpler components for easily identified features of the process. The technique used in doing this involves the reversal of superposition to **decompose** models with complex forecast functions into simple, canonical components. These components are familiar and easily understood and allow a modeller to structure complex problems component by component. Superposition then provides the overall model for the series by simply aggregating the individual building blocks.

The starting point for model design is the form of the forecast function. Given this, we construct observable models by identifying the component canonical forms. This effectively solves the design problem so far as choice of regression vector $\mathbf{F}_t$ and system matrix $\mathbf{G}_t$ is concerned. The practically important canonical components are related to forecast functions as follows.

(1) Suppose that, for all $t \geq 0$ and $k \geq 0$ the forecast function has the form

$$f_t(k) = \lambda^k \sum_{r=0}^{n-1} a_{tr} k^r,$$

for some real, non-zero λ, integer $n \geq 1$, and real coefficients $a_{t0}, \ldots, a_{tn-1}$ not depending on k. From Theorem 6.1 the canonical model $\{\mathbf{E}_n, \mathbf{J}_n(\lambda), ., .\}$ can be identified immediately. Then an observable TSDLM has the required forecast function form if and only if it is similar to the canonical model.

(2) The generalisation to several real eigenvalues is as follows. Suppose that it is desired to have a forecast function of the form

$$f_t(k) = \sum_{i=1}^{s} \left[\lambda_i^k \sum_{r=0}^{n_i-1} a_{tr}(i)k^r \right],$$

where $s > 1$ is integral, $\lambda_1, \ldots, \lambda_s$ are real, non-zero and distinct, and the real coefficients $a_{tr}(i)$ do not depend on k. Immediately we can write

$$f_t(k) = \sum_{i=1}^{s} f_{it}(k)$$

where, as in (1) above, $f_{it}(k)$ is the forecast function of any observable model similar to the canonical form $\{\mathbf{E}_{n_i}, \mathbf{J}_{n_i}(\lambda_i), ., .\}$, for $i = 1, \ldots, s$. Now, using Theorem 6.5, this form is that of a model obtained from the superpostion of these s canonical models. Thus the required forcast function is provided by any observable model that is similar to the **block form** $\{\mathbf{E}, \mathbf{J}, ., .\}$ where

$$\mathbf{E}' = (\mathbf{E}'_{n_1}, \ldots, \mathbf{E}'_{n_s}),$$

and

$$\mathbf{J} = \text{block diag}[\mathbf{J}_{n_1}(\lambda_1), \ldots, \mathbf{J}_{n_s}(\lambda_s)].$$

(3) Suppose

$$f_t(k) = \lambda^k \sum_{r=0}^{v-1} a_{tr} k^r \cos(k\omega + \phi_{tr})$$

for some real, non-zero λ and ω, integer $v > 0$, and real coefficients a_{tr} and ϕ_{tr} not depending on k. Immediately, from Theorem 6.3, it follows that this form of forecast function is provided by any model similar to the real canonical form $\{\mathbf{E}_{2,v}, \mathbf{J}_{2,v}(\lambda, \omega), ., .\}$.

(4) **(General TSDLM)**
A general forecast function of an observable Time Series Model has the following form. For some non-negative integers s and v such that $s + v > 0$, define $f_t(k)$ as

$$f_t(k) = \sum_{i=1}^{s+v} f_{it}(k)$$

where:

$$f_{it}(k) = \lambda_i^k \sum_{r=0}^{n_i-1} a_{tr}(i)k^r, \qquad\qquad (i = 1, \dots, s)$$

$$f_{it}(k) = \lambda_i^k \sum_{r=0}^{n_i-1} a_{tr}(i)k^r \cos(k\omega_r + \phi_{tr}(i)),$$
$$(i = s+1, \dots, s+v).$$

Here n_i, $(i = 1, \dots, s+v)$, are non-negative integers and, for each i and r, the real, non-zero quantities λ_i, ω_i, $a_{tr}(i)$ and $\phi_{tr}(i)$ do not depend on k.

Following Theorem 6.5 and using the results of Section 5.4.6, this forecast function form is provided by any TSDLM similar to the real canonical model of Definition 5.11. This caters for all real, non-zero eigenvalues of the system matrix for $i = 1, \dots, s$, and complex pairs for $i = s+1, \dots, s+v$. The most general model would also allow for zero eigenvalues as in Example 6.4. This would then add a further component to the forecast function, namely

$$f_{s+v+1,t}(k) = \begin{cases} b_{tk}, & \text{for } 0 \le k < n_{s+v}; \\ 0, & \text{for } l \ge n, \end{cases}$$

where n_{s+v+1} is the multiplicity of the zero eigenvalue, and the b_{tk} are known contants.

(5) The above cases cover the forms of forecast function encountered in TSDLMs. The complementary cases of models incorporating regression effects of independent variables, considered in simple cases in Chapter 3, are rather simple in form. Suppose a related regressor variable gives rise to a time series X_t with X_{t+k} known at t for $k \ge 0$. The X_t may be raw or transformed values of a related series, or values filtered through a known, possibly non-linear, transfer function to provide a constructed *effect* variable that depends on past, or lagged, values of the related series. From Chapter 3, a simple regression DLM whose forecast function has the form $f_t(k) = m_{t0} + m_{t1}X_{t+k}$, is the straight line model with regression vector $\mathbf{F}_t = (1, X_t)'$ and system matrix $\mathbf{G}_t = \mathbf{I}$ respectively. This is obtained as the superposition of the first-order polynomial model $\{1, 1, .,.\}$, and the simple dynamic straight line through the origin $\{X_t, 1, .,.\}$.

Generalising to multiple linear regression DLMs, consider a collection of h possible regressor variables $X_{1t}, \ldots, X_{ht}$. By superposition of the corresponding h simple models and a first-order polynomial we obtain a multiple regression model with $\mathbf{F}'_t = (1, X_{1t}, \ldots, X_{ht})$ and $\mathbf{G}_t = \mathbf{I}$.

EXAMPLE 6.7. For the pure polynomial forecast function $f_t(k) = \sum_{r=0}^{n-1} a_{tr} k^r$ a unit eigenvalue of multiplicity n is required and the canonical model is $\{\mathbf{E}_n, \mathbf{J}_n(1), ., .\}$. Any similar model is called a **n^{th}-order polynomial DLM**.

EXAMPLE 6.8. Suppose a modeller requires a forecast function that represents a single harmonic oscillation of period p about a linear trend. Such forms are fundamental in short-term forecasting of seasonal series. The linear trend is provided by Example 6.7 with $n = 1$. The cyclical term with frequency $\omega = 2\pi/p$ is provided by the 2×2 real canonical model for a single pair of complex conjugate eigenvalues with modulus $\lambda = 1$ and argument ω. Thus superposition provides the required form in the model with regression vector $(\mathbf{E}'_2, \mathbf{E}'_2)'$, and system matrix block $\mathrm{diag}[\mathbf{J}_2(1), \mathbf{J}_2(1, \omega)]$. This is referred to as a **second-order polynomial/seasonal** model in which the seasonal pattern has the form of a simple cosine wave.

EXAMPLE 6.9. Suppose the forecast function required is as in Example 6.8 with the additional demand that: (i) the seasonal pattern is more complex, being modelled by adding another harmonic of frequency $\omega^* = 2\pi/p^*$; and (ii) there is an exponential progression from $t = 0$ to the specified forecast function by a term of the form $a_t \lambda^k$, where $0 < \lambda < 1$. Then the canonical model is obtained by superposition as having regression vector $(\mathbf{E}'_2, 1, \mathbf{E}'_2, \mathbf{E}'_2)'$ and system matrix block $\mathrm{diag}[\mathbf{J}_2(1), \lambda, \mathbf{J}_2(1, \omega), \mathbf{J}_2(1, \omega^*)]$.

In each of these examples we have identified the canonical models for a given forecast function. They may be transformed, if required, to alternative similar models via reparametrisation or time shifts affecting the interpretation of the state vector. The final example illustrates the the reverse problem of finding the forecast function of a given DLM. In general observability is checked directly by examining the observability matrix. Then the eigenvalues of the system matrix are all that is required to assess the component models and hence deduce the form of the forecast function.

EXAMPLE 6.10. Consider the non-trivial TSDLM with $\mathbf{F} = \mathbf{E}_n$ and $\mathbf{G} = \mathbf{P}_n$ where, for some integer $q > 0$, $n = 2q + 1$ and

$$\mathbf{P}_n = \begin{pmatrix} \mathbf{0} & \mathbf{I} \\ 1 & \mathbf{0}' \end{pmatrix},$$

with $\mathbf{I}$ the $(2q) \times (2q)$ identity matrix. It can be easily verified that the observability matrix is simply the identity, and thus the model is observable. It can also be verified that $\mathbf{G}$ has roots given by the nth roots of 1, namely $e^{iv\omega}$ where $\omega = 2\pi/n$, for $v = 0, \dots, n - 1$. When $v = 0$, the root is just 1 so that $\mathbf{G}$ has an eigenvalue 1 and q distinct pairs of complex conjugates $(e^{iv\omega}, e^{-iv\omega})$, $(v = 1, \dots, q)$. Thus the model decomposes into one whose forecast function comprises the sum of a first-order polynomial term a_{t0}, and q harmonics, or cosine waves, of the form $r_t(v) \cos(kv\omega + \phi_t(v))$ for $v = 1, \dots, q$. The model is known as a **seasonal effects** model with period n, where n is, of course, an odd, positive integer. The canonical similar block model has regression vector $(1, \mathbf{E}_2', \dots, \mathbf{E}_2')'$, and system matrix

$$\text{block diag}\,[1, \mathbf{J}_2(1, \omega), \mathbf{J}_2(1, 2\omega), \dots, \mathbf{J}_2(1, q\omega)],$$

with dimension $n = 2q + 1$.

6.3 DISCOUNT FACTORS AND COMPONENT MODEL SPECIFICATION

6.3.1 Component models

The above design principles lead naturally to DLMs structures in **block** or **component** form. The system matrix is block diagonal with individual sub-matrices providing contributions from simple component models. The regression vector is partitioned into the catenation of corresponding sub-vectors. To complete the model specification we need three further components. These are the sequence of state evolution variance matrices $\mathbf{W}_t$, $(t = 1, \dots)$; the observational variance sequence V_t, $(t = 1, \dots)$; and the parameters determining the initial priors for state vector, and/or observational error variance, at $t = 0$ given D_0. Estimation of constant observational variance has already been considered in Chapter 4 and will be extended to more general, time dependent variances in a later Chapter. The initial prior settings, and related questions concerning representation of subjective information of the forecaster in terms of probability distributions, are also covered extensively in later chapters and application specific contexts. A general point here is that, typically, the component structure of DLMs naturally leads to these initial priors being specified in terms of a collection of priors, one for each of the sub-vectors of $\boldsymbol{\theta}_0$ corresponding to the individual component models, with independence between components. In this Section we consider the third remaining requirement, the specification of the sequence of evolution variance matrices $\mathbf{W}_t$. We refer throughout to the general model of Definition 4.3, with minor, purely technical changes to the case of unknown observational variance as in Definition 4.5.

The specification of suitable structure and magnitude of $\mathbf{W}_t$ is crucially important for successful modelling and forecasting. The values control the extent of the stochastic variation in the evolution of the model and hence determine the stability over time. In the system equation $\mathbf{W}_t$ leads to an increase in uncertainty or, equivalently, a loss of information, about the state vector between times $t-1$ and t. More precisely, consider the sequential information updating equations summarised in Section 4.5. At $t-1$ the posterior for the current state vector has variance $V[\boldsymbol{\theta}_{t-1} \mid D_{t-1}] = \mathbf{C}_{t-1}$ which leads, via the evolution equation, to prior variance for $\boldsymbol{\theta}_t$ given by $V[\boldsymbol{\theta}_t \mid D_{t-1}] = \mathbf{G}_t \mathbf{C}_{t-1} \mathbf{G}_t' + \mathbf{W}_t$. Let $\mathbf{P}_t$ denote the first term here, that is

$$\mathbf{P}_t = \mathbf{G}_t \mathbf{C}_{t-1} \mathbf{G}_t' = V[\mathbf{G}_t \boldsymbol{\theta}_{t-1} \mid D_{t-1}].$$

$\mathbf{P}_t$ may be viewed as the appropriate prior variance in the model $\{\mathbf{F}_t, \mathbf{G}_t, V_t, \mathbf{0}\}$, that is the standard model with no evolution error at time t. This would therefore be the required prior variance of an ideal, stable state vector with no stochastic changes. Adding the evolution error $\boldsymbol{\omega}_t$ to $\mathbf{G}_t \boldsymbol{\theta}_{t-1}$ to give the true state vector $\boldsymbol{\theta}_t$ simply has the effect of increasing the uncertainty from the ideal $\mathbf{P}_t$ to the actual $\mathbf{R}_t = \mathbf{P}_t + \mathbf{W}_t$.

The evolution error term thus models an *additive* increase in uncertainty, or *loss of information*, about the state vector between observations. To achieve a suitable degree of information decay over time, however, it is clear that the relative magnitudes of $\mathbf{W}_t$ and $\mathbf{P}_t$ are important. This leads to thinking in terms of a natural *rate* of decay of information that suggests a *multiplicative*, rather than additive, increase in uncertainty. We have already met and explored this idea in Chapters 2 and 3 where all the quantities, in particular the variances, are scalars, with $\mathbf{P}_t = C_{t-1}$, $\mathbf{W}_t = W_t$, $\mathbf{R}_t = C_{t-1} + W_t$ and $\mathbf{C}_t = C_t$. There the use of a discount factor $\delta, (0 < \delta \leq 1)$, linked

W_t to $P_t = C_{t-1}$ via $W_t = P_t(1-\delta)/\delta$ so that $R_t = P_t/\delta$. This implies an increase in variance, or loss of information, of $100(1-\delta)/\delta\%$. There are various ways in which this idea can be generalised to the multiparameter case to aid the choice of $\mathbf{W}_t$. Here we discuss the most important practical approach based on *component discounting*.

The basic idea is a simple extension of that for the scalar models of Chapters 2 and 3. Suppose initially that the DLM is essentially a single, canonical component consistent with a specified forecast function. This may be based, for example, on a system matrix with one real eigenvalue of multiplicity n such as in Example 6.2, or a periodic component such as in Theorem 6.2. A parsimonious form for the evolution variance sequence is obtained by directly extending the scalar discounting to the matrix case, defining

$$\mathbf{R}_t = \mathbf{P}_t/\delta$$

for some scalar discount factor δ, as usual. This implies that

$$\mathbf{R}_t = \mathbf{P}_t + \mathbf{W}_t,$$

where

$$\mathbf{W}_t = \mathbf{P}_t(1-\delta)/\delta.$$

Thus $\mathbf{W}_t$ has precisely the same internal structure, in terms of correlations, as $\mathbf{P}_t$. The magnitude of variances and covariances is controlled by the discount factor in just the same way as described in Chapter 2 in the scalar case. The implication is that information decays at the same rate for each of the elements of the state vector. This is appropriate, in particular, when the quantification of the *entire* forecast function of the model is viewed as subject to change at a constant rate without reference to components of that function. Very often this is a suitable assumption in practice. Note, however, that any such discount approach, including those to be described in following sections, is not appropriate in cases when some or all of the state parameters are known at time t. In such circumstances, the corresponding variances and covariances of $\mathbf{P}_t$ will be zero so that the discount construction cannot apply.

6.3.2 Component discounting

Turning to a general DLM comprising the superposition of several components, the idea extends naturally suggesting one discount factor for each component. As in Theorem 6.4, consider a DLM comprising the superposition of $h \geq 1$ sub-models

$$M_i : \quad \{\mathbf{F}_i, \mathbf{G}_i, V_i, \mathbf{W}_i\}_t,$$

with state vectors $\boldsymbol{\theta}_{it}$, and evolution errors $\boldsymbol{\omega}_{it}$, of dimensions n_i, for $i = 1, \ldots, h$. The DLM is thus specified by the quadruple

$$\{\mathbf{F}, \mathbf{G}, V, \mathbf{W}\}_t$$

with state vector, of dimension $n = n_1 + \cdots + n_h$, given by

$$\boldsymbol{\theta}_t' = (\boldsymbol{\theta}_{1t}', \ldots, \boldsymbol{\theta}_{ht}'),$$

where:

$$\mathbf{F}_t' = (\mathbf{F}_{1t}', \ldots, \mathbf{F}_{ht}'),$$

$$\mathbf{G}_t = \text{block diag}[\mathbf{G}_{1t}, \ldots, \mathbf{G}_{ht}]$$

$$= \begin{pmatrix} \mathbf{G}_{1t} & \mathbf{0} & \mathbf{0} & \cdots & \mathbf{0} \\ \mathbf{0} & \mathbf{G}_{2t} & \mathbf{0} & \cdots & \mathbf{0} \\ \mathbf{0} & \mathbf{0} & \mathbf{G}_{3t} & \cdots & \mathbf{0} \\ \vdots & \vdots & \vdots & \ddots & \vdots \\ \mathbf{0} & \mathbf{0} & \mathbf{0} & \cdots & \mathbf{G}_{ht} \end{pmatrix},$$

and

$$\mathbf{W}_t = \text{block diag}[\mathbf{W}_{1t}, \ldots, \mathbf{W}_{ht}]$$

$$= \begin{pmatrix} \mathbf{W}_{1t} & \mathbf{0} & \mathbf{0} & \cdots & \mathbf{0} \\ \mathbf{0} & \mathbf{W}_{2t} & \mathbf{0} & \cdots & \mathbf{0} \\ \mathbf{0} & \mathbf{0} & \mathbf{W}_{3t} & \cdots & \mathbf{0} \\ \vdots & \vdots & \vdots & \ddots & \vdots \\ \mathbf{0} & \mathbf{0} & \mathbf{0} & \cdots & \mathbf{W}_{ht} \end{pmatrix}.$$

At time t, the variance matrix

$$\mathbf{P}_t = V[\boldsymbol{\theta}_t \mid D_{t-1}] = \mathbf{G}_t \mathbf{C}_{t-1} \mathbf{G}_t'$$

represents uncertainty about $\mathbf{G}_t \boldsymbol{\theta}_t$ *before* the addition of the evolution noise term. Denote the corresponding block diagonal components by $\mathbf{P}_{it}$, namely

$$\mathbf{P}_{it} = V[\mathbf{G}_{it} \boldsymbol{\theta}_{it} \mid D_{t-1}] \qquad (i = 1, \ldots, h).$$

Note that $\mathbf{P}_t$ will not be a block diagonal (in other than degenerate models), but these components provide measures of information about the corresponding components of $\mathbf{G}_t\boldsymbol{\theta}_t$. Now, adding the evolution noise $\boldsymbol{\omega}_t$ with block diagonal variance matrix $\mathbf{W}_t$ above, the prior variance matrix $\mathbf{R}_t$ for $\boldsymbol{\theta}_t$ has off-diagonal components the same as those of $\mathbf{P}_t$, but diagonal blocks

$$\mathbf{R}_{it} = \mathbf{P}_{it} + \mathbf{W}_{it}$$

for each i. The discount concept applies naturally to the model structured through components in this manner, as follows.

Definition 6.1. *In the above framework, let $\delta_1,\ldots,\delta_h$ be any h discount factors, $(0 < \delta_i \leq 1;\ i = 1,\ldots,h)$, with δ_i being the discount factor associated with the component model M_i. Suppose that the component evolution variance matrices $\mathbf{W}_{it}$ are defined as in Section 6.3.1 above, via*

$$\mathbf{W}_{it} = \mathbf{P}_{it}(1 - \delta_i)/\delta_i, \qquad (i = 1,\ldots,h).$$

Then the model is referred to as a **component discount DLM**.

The effect of component discounting is to model decay of information over time at a possibly different rate for each component model. The modeller chooses the discount factors, some of which may, of course, be equal, to reflect his beliefs about the stability over time of the individual components. Note that, from an operational point of view in updating, the evolution from $\mathbf{P}_t$ to $\mathbf{R}_t$ need not make reference to the constructed $\mathbf{W}_t$ sequence. It is simply achieved by taking the component covariances as unchanged and dividing the block diagonal elements by the appropriate discount factors,

$$\mathbf{R}_{it} = \mathbf{P}_{it}/\delta_i,$$

for each i. Block discounting is the recommended approach to structuring the evolution variance sequence in almost all applications. The approach is parsimonious, applying a single discount factor to each component of a larger model, naturally interpretable and robust. Sometimes a single discount factor applied to an entire model viewed as a single component will be adequate, but the flexibility remains to model up to n separate components, each with individual (though not necessarily distinct) discounts. Also, and importantly, the derived $\mathbf{W}_t$ matrix is naturally scaled, the discount factors being dimensionless quantities on a standardised scale. With or without variance learning, the discount construction applies directly. The following Section describes some practical features of the use of component discount models. The basic ideas underlying discounting have a long history, the specific approach here being introduced in Ameen and Harrison (1985), described and developed in practical detail in Harrison and West (1986, 1987), Harrison (1988), and implemented in the **BATS** package of West, Harrison and Pole (1987). Some theoretical variations are considered in Section 6.4 below although they are of restricted practical interest.

6.3.3 Practical discount strategy

Discounting should be viewed as a technique to be used to structure the evolution variance just *one-step ahead*. Given C_{t-1}, hence P_t, and a notion of the relative durability of components quantified in terms of the discount factors, the derived W_t is identified as an appropriate variance matrix for the evolution noise at time t. Looking further ahead than this one time point, it is *not* the case that repeat application of discounting will produce the relevant sequence of variance matrices. For example, with a single component model having discount factor δ, repeat application would lead to the use of δ^k as a discount factor k–steps ahead. The implied *exponential* decay of information into the future is obviously not consistent with the DLM in which the information decays *arithmetically* through the addition of future evolution error variance matrices. Hence, though perfectly coherent one-step ahead, the discount approach must be applied with thought in extrapolating ahead (and also, therefore, when encountering missing values in the time series). Ameen and Harrison (1985) discuss this point, and use one-step discounting from $t = 0$ to determine the implied sequence W_t for all future times t. This is possible since, given the other model components, these matrices are simply functions of quantities assumed known initially. It can be seen that this is also possible in models where the observational variance is being estimated.

From a practical viewpoint, an alternative, more flexible and less computationally demanding approach is suggested in Harrison and West (1986). This simply assumes that, at time t, the implied one-step evolution variance matrix is appropriate for extrapolation into the future, determining a *constant* step-ahead variance matrix. The resulting practical discount strategy is as follows.

(1) Consider the position posterior to D_t, with the usual posterior variance $\mathbf{C}_t$, and the one-step prior variance $\mathbf{R}_{t+1} = \mathbf{P}_{t+1} + \mathbf{W}_{t+1}$. Here $\mathbf{W}_{t+1}$ is constructed through discounting as in the previous Section (but with t replaced by $t+1$, of course).

(2) Forecasting k–steps ahead, it will usually be adequate to adopt a conditionally constant variance matrix, taking

$$V[\omega_{t+k}|D_t] = \mathbf{W}_t(k) = \mathbf{W}_{t+1}, \qquad (k = 1, \dots).$$

Thus step-ahead forecast distributions will be based on the addition of evolution errors with the same variance matrix $\mathbf{W}_{t+1}$ for all k.

(3) Once Y_{t+1} is observed, the posterior for $\boldsymbol{\theta}_{t+1}$ is calculated, hence so is $\mathbf{P}_{t+2}$ and $\mathbf{W}_{t+2}$ deduced using discounting. Thus forecasting ahead from time $t+1$, we have

$$V[\omega_{t+k+1}|D_{t+1}] = \mathbf{W}_{t+1}(k) = \mathbf{W}_{t+2}, \qquad (k = 1, \dots).$$

(4) Proceed in the manner at time $t+2$, and into the future.

The computational simplicity of this strategy is evident; at any time, a single evolution variance matrix is calculated and used k–step-ahead for any desired k. Note an important modification of the standard DLM analysis implied here. Hithertofore, the evolution errors were assumed to have variance matrices known for all time, and also independent of the history of the series. With the discount strategy this assumption has been weakened and modified to allow the variance matrices in the future to depend on the current state of information. Mathematically, the assumption that, for any $k = 1, \dots$, $V[\omega_{t+k}|D_t] = V[\omega_{t+k}|D_0]$ has been revised; it is now the case that $V[\omega_{t+k}|D_t] = \mathbf{W}_t(k)$ depends on t in addition to $t+k$. For example, at time t, the 2–step ahead variance matrix is

$$V[\omega_{t+2}|D_t] = \mathbf{W}_t(2) = \mathbf{W}_{t+1}.$$

Obtaining a further observation, this is revised to

$$V[\omega_{t+2}|D_{t+1}] = \mathbf{W}_{t+1}(1) = \mathbf{W}_{t+2}.$$

This modification is straightforward, and has no complicating consequences in practice. In updating (and also in filtering backwards over time), the actual one-step value $\mathbf{W}_t$ used at time t is used in the relevant updating (and filtering) equations. As time progesses, the current assumptions about the evolution variances into the future are revised. This may be intepreted as successive intervention to change the future values of the variance sequence.

6.4   Further Comments on Discount Models[†]

From an applied viewpoint, the above framework provides a complete operational approach to structuring the evolution variance matrices of all DLMs. The use of single discount ideas to structure forecasting models based on TSDLMs is discussed in Brown (1962), Harrison (1965), Godolphin and Harrison (1975), and Harrison and Akram (1983). The first extension to multiple discount factors is to be found in Harrison (1965) and is discussed in Whittle (1965). The general extension to multiple discount factors for components described above is generally appropriate outside the restricted class of TSDLMs. Applications can be found in Ameen and Harrison (1985), West and Harrison (1986), Harrison and West (1986, 1987), and implementation in West, Harrison and Pole (1987). It can be seen that these discount factors play a role analogous to those used in non-Bayesian point forecasting methods, in particular to exponential smoothing techniques (Ledolter and Abraham, 1983, Chapters 3 and 4, for example), providing interpretation and meaning within the DLM framework. Some further theoretical discussion of discount models in general is now given.

When concerned with TSDLMs having constant $\mathbf{F}$ and $\mathbf{G}$, questions then arise about the limiting behaviour in discount models. Some features of this are discussed in particular models in later Chapters, in particular Chapter 7, and have already been noted in the first-order polynomial model in Chapter 2. In more complex models, using multiple discount factors applied to components as in the preceding sections, the updating equations may observed in practice to to converge to stable, limiting forms, although theoretical results for such models are (at time of writing) unavailable. It is conjectured that, in any closed model $\{\mathbf{F}, \mathbf{G}, V, \mathbf{W}_t\}$ with $\mathbf{W}_t$ structured in block diagonal form as in Definition 6.1, the updating equations have stable limiting forms, with $\mathbf{C}_t$, $\mathbf{R}_t$, $\mathbf{A}_t$ and Q_t converging (rapidly) to finite limits $\mathbf{C}$, $\mathbf{R}$, $\mathbf{A}$ and Q. If this is so, then $\mathbf{W}_t$ also has a stable limiting form being based on $\mathbf{C}$ and the fixed

---

[†]This Section is of theoretical interest only, and may be omitted without loss on a first reading.

discount factors. Thus, in the limiting forms, component discount TSDLMs are essentially standard DLMs with constant, block diagonal evolution variance matrices. The limiting representations of the observation series in generalised ARIMA form are then deducable from Theorem 5.2 of Chapter 5.

Some support for this conjectured limiting behaviour is provided in the work of Ameen and Harrison (1985) who consider alternative approaches to multiple discounting. The various alternatives are basically very similar to component discounting, defining matrices $\mathbf{W}_t$ based on possible several discount factors. Consider the n-dimensional DLM $\{\mathbf{F}_t, \mathbf{G}_t, V_t, \mathbf{W}_t\}$. The approaches of Ameen and Harrison (1985) involves, generally, n discount factors $\delta_1, \ldots, \delta_n$, one for each element of the state vector. Note that some or all of the discount factors may coincide. Introduce the $n \times n$ diagonal matrix $\boldsymbol{\Delta}$, defined by

$$\boldsymbol{\Delta} = \operatorname{diag}(\delta_1^{-1/2}, \ldots, \delta_n^{-1/2}).$$

Then, given the posterior variance matrix $\mathbf{C}_{t-1}$ at time $t-1$, two possible approaches (alternatives to component discounting) to multiple discounting are to define the prior variance matrix $\mathbf{R}_t$ at time t by either of the forms

(a) $\mathbf{R}_t = \boldsymbol{\Delta}\mathbf{G}_t\mathbf{C}_{t-1}\mathbf{G}_t'\boldsymbol{\Delta}$,
(b) $\mathbf{R}_t = \mathbf{G}\boldsymbol{\Delta}\mathbf{C}_{t-1}\boldsymbol{\Delta}\mathbf{G}_t$.

Ameen and Harrison (1985) discuss, in particular, scheme (a), and derive theoretical results about the stable limiting forms of updating equations when $\mathbf{F}$, $\mathbf{G}$ and V are constant over time. In particular, it is shown that $\mathbf{C}_t$, $\mathbf{R}_t$, $\mathbf{A}_t$, Q_t and $\mathbf{W}_t$ typically converge rapidly to finite limits $\mathbf{C}$, $\mathbf{R}$, $\mathbf{A}$, Q and $\mathbf{W}$, so that the limiting form of the model is just that of a standard, constant TSDLM $\{\mathbf{F}, \mathbf{G}, V, \mathbf{W}\}$. Theorem 5.2 of Chapter 5 then applies directly to deduce limiting representations of the observation series in generalised ARIMA form. These models thus provide Bayesian analogues of standard point forecasting techniques based on the use of multiple discount factors, such as multiple exponential smoothing (Abraham and Ledholter, 1983, Chapter 7; McKenzie, 1974, 1976). Using either (a) or (b), note the following.

(1) Each method implies $\mathbf{W}_t = \mathbf{R}_t - \mathbf{G}_t\mathbf{C}_{t-1}\mathbf{G}_t'$. Whilst obviously symmetric, this matrix is not guaranteed to be non-negative definite, and so may not define a valid evolution variance matrix. To be used, $\mathbf{W}_t$ must be a valid variance matrix. Hence these approaches are not generally applicable.

In certain models, however, the $\mathbf{W}_t$ thus defined will always be valid. Note that component discounting has no such problem.

(2) Unlike component discounting, $\mathbf{W}_t$ will not typically be block diagonal in a model with several components.

(3) If the discount factors coincide, $\delta_i = \delta$ for $i = 1, \dots, n$, then $\mathbf{R}_t = \mathbf{G}_t \mathbf{C}_{t-1} \mathbf{G}_t'/\delta$ in either case. Here, then, the model is viewed as a single component model with just one discount factor.

6.5 EXERCISES

(1) Identify $\mathbf{F}$ and $\mathbf{G}$ of observable models with forecast functions of the following forms.

(a) $f_t(k) = a_{t1}\lambda_1^k + a_{t2}\lambda_2^k$ for some, non-zero and distinct values of λ_1 and λ_2.

(b) $f_t(k) = a_{t1} + a_{t2}k + a_{t3}k^2 + a_{t4}k^3$.

(c) $f_t(k) = a_{tj}$ where $j = k|4$, $(j = 1, \dots, 4)$.

(d) $f_t(k) = a_{tk}$ for $k = 0, 1, 2$ and 3, but $f_t(k) = 0$ for $k > 3$.

(e) $f_t(k) = a_{t1} + a_{t2}k + a_{t3}\lambda^k + a_{t4}\lambda\cos(k\omega) + a_{t5}\lambda\sin(k\omega)$ for some λ and ω, where ω is not an integer multiple of π.

(f) $f_t(k) = a_{t1} + a_{t2}k + a_{t3}\lambda_1^k + a_{t4}\lambda_2^k\cos(k\omega) + a_{t5}\lambda_2^k\sin(k\omega)$ for some λ_1, λ_2 and ω, where ω is not an integer multiple of π.

(2) In a general DLM, suppose that $\mathbf{G}_t$ is non-singular for all t. Suppose further that $\mathbf{W}_t$ is defined using a single discount factor δ so that $\mathbf{R}_t = \mathbf{G}_t \mathbf{C}_{t-1} \mathbf{G}_t'/\delta$ for all t, $(0 < \delta < 1)$.

(a) Show that the filtering recurrence equations in Theorem 4.4 of Chapter 4 simplify to

$$\mathbf{a}_t(-k) = (1 - \delta)\mathbf{m}_{t-k} + \delta\mathbf{G}_{t-k+1}^{-1}\mathbf{a}_t(-k+1),$$

and

$$\mathbf{R}_t(-k) = (1 - \delta)\mathbf{C}_{t-k} + \delta^2\mathbf{G}_{t-k+1}^{-1}\mathbf{R}_t(-k+1)(\mathbf{G}_{t-k+1}')^{-1}.$$

(b) Comment on the forms of these equations, with particular reference to the implied computational demands relative to those of the original, general recurrences.

(3) In the framework of the previous example, suppose that $\mathbf{G}_t = \mathbf{I}$, the $n \times n$ identity matrix, for all t, so that the model is a multiple regression DLM as in Definition 3.1 of Chapter 3.

(a) Show how the filtering equations simplify in this special case.

(b) Deduce that

$$\mathbf{a}_{t+1}(-k-1) - \mathbf{a}_t(-k) = \delta[\mathbf{a}_{t+1}(-k) - \mathbf{a}_t(-k+1)]$$

and

$$\mathbf{R}_{t+1}(-k-1) - \mathbf{R}_t(-k) = \delta^2[\mathbf{R}_{t+1}(-k) - \mathbf{R}_t(-k+1)].$$

Viewing $s = t - k$ as fixed, intepret these equations in terms of the extent of change in filtered distributions for time s given D_t as t increases. Show further that, for $k \geq -1$, the equations may be written in the computationally convenient forms

$$\mathbf{a}_{t+1}(-k-1) = \mathbf{a}_t(-k) + \delta^{k+1}\mathbf{A}_{t+1}e_{t+1}$$

and

$$\mathbf{R}_{t+1}(-k-1) = \mathbf{R}_t(-k) + \delta^{2(k+1)}\mathbf{A}_{t+1}\mathbf{A}'_{t+1}Q_{t+1}.$$

(4) Consider the discount model $\{1, \lambda, 1, W_t\}$ with $W_t = C_{t-1}(\delta^{-1} - 1)$.

(a) Prove that $\lim_{t \to \infty} C_t$ exists if, and only if, $\delta < \lambda^2$. Under this condition, show that the limiting value is

$$C = 1 - \delta/\lambda^2.$$

Identify the corresponding limiting values of R_t, Q_t and A_t.

(b) Using the limiting form of the updating equation $m_t = \lambda m_{t-1} + Ae_t$, show that

$$Y_t - \lambda Y_{t-1} = e_t - \lambda(1 - \delta/\lambda^2)e_{t-1}.$$

(c) Compare the result in (b) with the corresponding limiting result in the model $\{1, \lambda, 1, W\}$ obtained by applying Theorem 5.2 of Chapter 5. In particular, show how the same limiting form can be obtained by appropriately choosing the discount factor δ as a function of λ, W and the limiting adaptive coefficient A.

(d) Consider the implications of this result for setting component discount factors in relationship to the eigenvalues of the components.

(5) Consider the model

$$\left\{ \begin{pmatrix} 1 \\ 1 \end{pmatrix}, \begin{pmatrix} \lambda_1 & 0 \\ 0 & \lambda_2 \end{pmatrix}, 1, \mathbf{W}_t \right\}$$

for any distinct, non-zero values λ_1 and λ_2. Suppose a single discount model so that $\mathbf{W}_t = \mathbf{G}\mathbf{C}_{t-1}\mathbf{G}'(\delta^{-1}-1)$. Let $\mathbf{K}_t = \mathbf{C}_t^{-1}$ for all t.

(a) Show that the updating equation for $\mathbf{C}_t$ can be written in terms of precision matrices as

$$\mathbf{K}_t = \delta \mathbf{G}^{-1}\mathbf{K}_{t-1}\mathbf{G}^{-1} + \mathbf{F}\mathbf{F}'.$$

(b) Writing

$$\mathbf{K}_t = \begin{pmatrix} K_{t1} & K_{t3} \\ K_{t3} & K_{t2} \end{pmatrix},$$

deduce the recurrence equations

$$K_{t1} = 1 + \delta K_{t-1,1}/\lambda_1^2,$$
$$K_{t2} = 1 + \delta K_{t-1,2}/\lambda_2^2,$$
$$K_{t3} = 1 + \delta K_{t-1,3}/\lambda_1\lambda_2.$$

(c) Deduce that, as t increases, $\mathbf{K}_t$ converges to a limit if $\delta < \min\{\lambda_1^2, \lambda_2^2\}$.

(d) Assuming this to hold, deduce expressions for the elements of the limiting matrix, $\mathbf{K}$, say, as functions of δ, λ_1 and λ_2. Deduce the limiting variance matrix $\mathbf{C} = \mathbf{K}^{-1}$.

(e) Suppose that $\delta = 0.7$, $\lambda_1 = 0.8$ and $\lambda_2 = 0.9$. Calculate $\mathbf{K}$ and deduce the limiting values of $\mathbf{C}_t$, $\mathbf{R}_t$, Q_t and $\mathbf{A}_t$.

(6) Show that the results of the previous example apply in the case of complex eigenvalues

$$\lambda_1 = e^{i\omega} \qquad \text{and} \qquad \lambda_2 = e^{-i\omega}$$

for some, real ω, (not an integer multiple of π).

CHAPTER 7

POLYNOMIAL TREND MODELS

7.1 INTRODUCTION

Polynomial models find wide use in time series and forecasting as they do in other branches of applied statistics, such as static regression and experimental design. Their use in time series modelling is to provide simple yet flexible forms to describe the local trend components, where by trend we mean smooth variation in time. Relative to the sampling interval of the series and the required forecast horizons, such trends can frequently be well approximated by low order polynomial functions of time. Indeed a first or second-order polynomial model alone is often quite adequate for short-term forecasting, and very widely applicable when combined via superposition with other components such as seasonality and regression. In Chapter 2 we introduced polynomial DLMs with the simplest, yet most widely used, case of the first-order models. The next in terms of both complexity and applicability are the second-order models, also sometimes referred to as linear growth models, that much of this chapter is concerned with. Higher order polynomial models are also discussed for completeness, although it is rare that polynomials of order greater than three are required for practice. The structure of polynomial models was discussed in Harrison (1965, 1967), and theoretical aspects explored in Godolphin and Harrison (1975). See also Abraham and Ledolter (1983, Chapter 3).

Polynomial DLMs are a subset of the class of Time Series DLMs, or TSDLMs, defined in Chapter 4 as those models whose regression vector $\mathbf{F}$ and system matrix $\mathbf{G}$ are constant for all time. Let n be a positive integer.

Definition 7.1. *Any observable TSDLM with a forecast function of the form*

$$f_t(k) = a_{t0} + a_{t1}k + \cdots + a_{t,n-1}k^{n-1},$$

for all $t \geq 0$ and $k \geq 0$, is defined as an nth**-order polynomial DLM.**

From the model theory concerning forecast function form in Section 5.3, it follows immediately that the system matrix $\mathbf{G}$ of any such model has a single eigenvalue of unity with multiplicity n. We stress that, with this definition, $\mathbf{G}$ has no other eigenvalues, in particular, no zero eigenvalues. As a consequence of the definition of canonical models in Section 5.4 we have the following comments.

(1) A DLM is an nth-order polynomial model if and only if it is similar to the canonical model

$$\{\mathbf{E}_n, \mathbf{J}_n(1), \cdot, \cdot\}.$$

(2) Any DLM equivalent to the constant model $\{\mathbf{E}_n, \mathbf{J}_n(1), V, \mathbf{W}\}$ is a constant n^{th}-order polynomial DLM.

In the main, the development of the rest of this Chapter concentrates on the highly important, second-order model, elaborating on the structure and features of analysis in isolation from other components. The primary objectives are to provide familiarity with the nature of the sequential updating equations, as was done in Chapter 2 for the first-order model, and to link with other models and concepts of forecasting, such as discounting, used by practitioners. Constant models, and in particular constant linear growth models, are central to this discussion and are considered in depth. Throughout this chapter we ignore the possibility that the observational variance sequence is unknown since this would detract from the primary goals. The general variance learning procedure of Section 4.5 applies to any model without affecting the primary features of the updating equations for model state parameters and forecasts. Thus, from here on, we suppose that V_t is known for all t.

Consider the canonical model $\{\mathbf{E}_n, \mathbf{J}_n(1), \cdot, \cdot\}$. For reasons which will be made clear below, denote the state vector of this model by $\boldsymbol{\lambda}_t$ rather than the usual $\boldsymbol{\theta}_t$, and let the evolution error be denoted by $\boldsymbol{\delta\lambda}_t$. Then, with

$$\boldsymbol{\lambda}'_t = (\lambda_{t1}, \dots, \lambda_{tn})$$

and

$$\boldsymbol{\delta\lambda}'_t = (\delta\lambda_{t1}, \dots, \delta\lambda_{tn}),$$

the model equations can be written as

Observation equation: $Y_t = \lambda_{t1} + \nu_t,$

System equation: $\lambda_{tj} = \lambda_{t-1,j} + \lambda_{t-1,j-1} + \delta\lambda_{tj},$

$$(j = 1, \dots, n-1),$$

$$\lambda_{tn} = \lambda_{t-1,n} + \delta\lambda_{tn}.$$

Here $\mu_t = \lambda_{t1}$ is the level at time t, which between times $t - 1$ and t changes by the addition of $\lambda_{t-1,2}$ plus the noise $\delta\lambda_{t1}$. $\lambda_{t-1,2}$ represents a systematic change in level, which itself changes by the addition of $\lambda_{t-1,3}$ plus noise. Proceeding through the state parameters, for $j = 1,\dots,n-1$ each λ_{tj} changes systematically via the increment $\lambda_{t-1,j-1}$, and also by the addition of the noise term $\delta\lambda_{tj}$. The nth component λ_{tn} changes only stochastically. Although this is the canonical model form, from the point of view of intepretation of the model parameters, it is convenient to work in terms of an alternative, similar model, defined as follows.

Let $\mathbf{L}_n$ denote the $n \times n$, upper triangular matrix of unities,

$$\mathbf{L}_n = \begin{pmatrix} 1 & 1 & 1 & \cdots & 1 \\ 0 & 1 & 1 & \cdots & 1 \\ 0 & 0 & 1 & \cdots & 1 \\ \vdots & \vdots & \vdots & \ddots & \vdots \\ 0 & 0 & 0 & \cdots & 1 \end{pmatrix}.$$

Consider the model

$$\{\mathbf{E}_n, \mathbf{L}_n, . , .\}.$$

It can be easily verified (the verification being left to the exercises for the reader) that this model is an nth-order polynomial model similar to the canonical model above. Let the state vector in this model be denoted, as usual, by

$$\boldsymbol{\theta}_t' = (\theta_{t1},\dots,\theta_{tn})$$

and the evolution error by

$$\boldsymbol{\omega}_t = (\omega_{t1},\dots,\omega_{tn})'.$$

Then the model equations are

Observation equation: $\quad Y_t = \theta_{t1} + \nu_t,$

System equation: $\quad \theta_{tj} = \theta_{t-1,j} + \sum_{r=j-1}^{n} \theta_{t-1,r} + \omega_{tj},$

$$(j = 1,\dots,n-1),$$

$$\theta_{tn} = \theta_{t-1,n} + \omega_{tn}.$$

Ignoring the complications due to the evolution noise terms, the state parameters in this non-canonical representation can be thought of as "derivatives" of the mean response function $\mu_t = \theta_{t1}$. For each $j = 1,\dots,n$, θ_{tj} represents the jth derivative of the mean response. At time t, the expected future trajectory of θ_{tj} is a polynomial of degree $n - j$.

The difference between this and the canonical model lies simply in a time shift in the definition of the elements of the state vector. In this representation, the state parameters have the natural interpretation as derivatives at time t. In the canonical model, the higher order terms in the state vector play the same role but are shifted back to time $t - 1$. This is most easily seen by ignoring the complications due to evolution noise, setting the evolution errors to zero in each model. If this is done, then the parameters in the canonical model relate to those in the alternative representation via

$$\lambda_{t1} = \theta_{t1},$$
$$\lambda_{tj} = \theta_{tj} + \theta_{t,j-1}, \qquad (j = 2,\dots,n-1),$$
$$\lambda_{tn} = \theta_{tn}.$$

The evolution error vector ω_t confuses the interpretation since, in general, the elements may be arbitarily correlated through the variance matrix $\mathbf{W}_t$, although the basic notion of the elements of the state vector as derivatives of the mean response is sound. A class of models with naturally interpretable and appropriate variance structure are defined as follows.

Definition 7.2. *Let* $\delta\theta_t' = (\delta\theta_{t1},\dots,\delta\theta_{tn})$ *be a zero-mean evolution noise term with a diagonal variance matrix,*

$$\delta\theta_t \sim \text{N}[\mathbf{0}, \text{diag}(W_{t1},\dots,W_{tn})].$$

Suppose that

$$\omega_t = \mathbf{L}_n \delta\theta_t$$

so that $\omega_t \sim \text{N}[\mathbf{0}, \mathbf{W}_t]$ *where*

$$\mathbf{W}_t = \mathbf{L}_n \text{diag}(W_{t1},\dots,W_{tn})\mathbf{L}_n'.$$

Then the model $\{\mathbf{E}_n, \mathbf{L}_n, . , \mathbf{W}_t\}$ *is an* $\mathbf{n^{th}}$**-order polynomial growth DLM.**

The distinguishing features of *polynomial growth*, rather than simply *polynomial*, models are that

    (a) the system matrix is not in canonical form, the alternative $\mathbf{L}_n$ matrix being used; and

(b) the evolution variance matrix has the special form consistent with the definition of $\boldsymbol{\omega}_t = \mathbf{L}_n \boldsymbol{\delta\theta}_t$ where the elements of $\boldsymbol{\delta\theta}_t$ are uncorrelated.

In such a model, the defining equations simplify as follows. We can write

Observation equation: $Y_t = \theta_{t1} + \nu_t,$

System equation: $\quad \theta_{tj} = \theta_{t-1,j} + \theta_{t,j-1} + \delta\theta_{tj},$

$$(j = 1, \ldots, n-1),$$

$$\theta_{tn} = \theta_{t-1,n} + \delta\theta_{tn}.$$

In such a model, therefore, the "derivative" θ_{tj}, for any $(j > 1)$, changes stochastically between times $t - 1$ and t via the increment θ_{tj} and also by those affecting the higher order derivatives $\delta\theta_{tk}$, $(j < k \leq n)$. We proceed to explore the special cases $n = 2$ and $n = 3$ in detail.

7.2 SECOND-ORDER POLYNOMIAL MODELS

7.2.1 The general model form

By definition, the models of interest have forecast functions of the linear form

$$f_t(k) = a_{t0} + a_{t1}k, \tag{7.1}$$

for all $t \geq 0$ and $k \geq 0$. Alternatively,

$$f_t(k) = f_t(0) + [f_t(1) - f_t(0)]k,$$

identifying the coefficients of the linear forecast function in terms of initial values at times $k = 0$ and $k = 1$. The canonical model has the form

$$\left\{ \begin{pmatrix} 1 \\ 0 \end{pmatrix}, \begin{pmatrix} 1 & 1 \\ 0 & 1 \end{pmatrix}, \ldots \right\}. \tag{7.2}$$

Notice that, in this special case of $n = 2$, $\mathbf{J}_2(1) = \mathbf{L}_2$ so that the similar model with system matrix $\mathbf{L}_2$ coincides with the canonical model. With parameterisation

$$\boldsymbol{\theta}_t = \begin{pmatrix} \theta_{t1} \\ \theta_{t2} \end{pmatrix} = \begin{pmatrix} \mu_t \\ \beta_t \end{pmatrix},$$

we can write the model in terms of the usual equations as

$$\text{Observation equation:} \quad Y_t = \mu_t + \nu_t, \quad\quad\quad (7.3a)$$

$$\text{System equation:} \quad \mu_t = \mu_{t-1} + \beta_{t-1} + \omega_{t1}, \quad (7.3b)$$

$$\beta_t = \beta_{t-1} + \omega_{t2}, \quad\quad\quad (7.3c)$$

where, with $\omega_t = (\omega_{t1}, \omega_{t2})'$, we have $\nu_t \sim N[0, V_t]$ and $\omega_t \sim N[0, \mathbf{W}_t]$, for some V_t and $\mathbf{W}_t$, subject to the usual independence assumptions. As usual, μ_t is the level of the series, and now β_{t-1} represents change in level. Standard DLM theory supplies the third component at time t, namely the information model specifying the prior at $t-1$ as

$$(\boldsymbol{\theta}_{t-1} \mid D_{t-1}) \sim N[\mathbf{m}_{t-1}, \mathbf{C}_{t-1}],$$

where

$$\mathbf{m}_{t-1} = \begin{pmatrix} m_{t-1} \\ b_{t-1} \end{pmatrix}$$

and

$$\mathbf{C}_{t-1} = \begin{pmatrix} C_{t-1,1} & C_{t-1,3} \\ C_{t-1,3} & C_{t-1,2} \end{pmatrix}.$$

It easily follows that, in (7.1), $a_{t0} = m_{t-1}$ and $a_{t1} = b_{t-1}$. As mentioned in the introduction, it is assumed here that the observational variance in the model is known. Otherwise the normal posterior distributions here would, of course, be replaced by Student T forms.

7.2.2 Updating equations

We use the general theory of Section 4.4 to provide the updating equations for this model in explicit terms rather than vector and matrix form. Write the evolution variance matrix as

$$\mathbf{W}_t = \begin{pmatrix} W_{t1} & W_{t3} \\ W_{t3} & W_{t2} \end{pmatrix}.$$

The sequential analysis has the following components.

(1) $(\boldsymbol{\theta}_t \mid D_{t-1}) \sim N[\mathbf{a}_t, \mathbf{R}_t]$ where

$$\mathbf{a}_t = \begin{pmatrix} m_{t-1} + b_{t-1} \\ b_{t-1} \end{pmatrix}, \quad \mathbf{R}_t = \begin{pmatrix} R_{t1} & R_{t3} \\ R_{t3} & R_{t2} \end{pmatrix},$$

with $R_{t1} = (C_{t-1,1} + 2C_{t-1,3} + C_{t-1,2}) + W_{t1}$, $R_{t2} = C_{t-1,2} + W_{t2}$ and $R_{t3} = (C_{t-1,2} + C_{t-1,3}) + W_{t3}$.

(2) The one-step forecast distribution is $(Y_t \mid D_{t-1}) \sim N[f_t, Q_t]$ where $f_t = f_{t-1}(1) = m_{t-1} + b_{t-1}$ and $Q_t = R_{t1} + V_t$.

(3) The adaptive vector is given by

$$\mathbf{A}_t = \begin{pmatrix} A_{t1} \\ A_{t2} \end{pmatrix} = \begin{pmatrix} R_{t1}/Q_t \\ R_{t3}/Q_t \end{pmatrix}.$$

(4) The updating to posterior at time t leads to posterior moments

$$\mathbf{m}_t = \begin{pmatrix} m_t \\ b_t \end{pmatrix} \quad \text{and} \quad \mathbf{C}_t = \begin{pmatrix} C_{t1} & C_{t3} \\ C_{t3} & C_{t2} \end{pmatrix},$$

with components as follows:

$$
\begin{aligned}
m_t &= m_{t-1} + b_{t-1} + A_{t1}e_t, \\
b_t &= b_{t-1} + A_{t2}e_t, \\
C_{t1} &= A_{t1}V_t, \\
C_{t2} &= R_{t2} - A_{t2}R_{t3}, \\
C_{t3} &= A_{t2}V_t,
\end{aligned}
$$

with $e_t = Y_t - f_t$ as usual.

(5) It is of some interest to note the relationships $Q_t = V_t/(1 - A_{t1})$ and

$$\mathbf{R}_t = \begin{pmatrix} A_{t1} & A_{t2} \\ A_{t2} & c_t \end{pmatrix} V_t,$$

where, if $r_t = (C_{t-1,2} + W_{t2})/Q_t$, we have $c_t = (r_t - A_{t2}^2)/(1 - A_{t1})$.

(6) The updating equations lead to an alternative representation of the observation series in terms of past observations and forecast errors. Use of this representation provides easy comparison of the Bayesian model with alternative forecasting techniques. To derive the representation, note that the three identities

$$
\begin{aligned}
Y_t &= m_{t-1} + b_{t-1} + e_t, \\
m_t &= m_{t-1} + b_{t-1} + A_{t1}e_t, \\
b_t &= b_{t-1} + A_{t2}e_t
\end{aligned}
$$

lead directly to the second difference equation

$$Y_t - 2Y_{t-1} + Y_{t-2} = e_t + \beta_{t1}e_{t-1} + \beta_{t2}e_{t-2},$$

where $\beta_{t1} = -(2 - A_{t1} - A_{t2})$ and $\beta_{t2} = 1 - A_{t1}$. Note that this representation is a consequence of the model however inappropriate it may be for any particular series.

7.2.3   Constant models and limiting behaviour

Denote by M the constant model

$$\left\{ \begin{pmatrix} 1 \\ 0 \end{pmatrix}, \begin{pmatrix} 1 & 1 \\ 0 & 1 \end{pmatrix}, 1, \mathbf{W} = \begin{pmatrix} W_1 & W_3 \\ W_3 & W_2 \end{pmatrix} \right\}.$$

Note that we assume the observational variance is unity for all time in M, losing no generality since otherwise $\mathbf{W}$, and all other variances in the model, may be scaled by a constant factor V and, if this factor is unknown, the standard variance learning procedure may be applied. Now, from Section 5.2, it follows that the updating equations have a stable limiting form. It is also the case that this limiting form is typically rapidly approached. For M, the limiting behaviour is as follows.

Theorem 7.1. *In the model M,*

$$\lim_{t \to \infty} \{ \mathbf{A}_t, \mathbf{C}_t, \mathbf{R}_t, Q_t \} = \{ \mathbf{A}, \mathbf{C}, \mathbf{R}, Q \}$$

exist. Writing $\mathbf{A} = (A_1, A_2)'$, the positive scalars A_1, A_2 and Q satisfy the equations

(i) $$(1 - A_1)Q = 1,$$
(ii) $$A_2^2 Q = W_2,$$
(iii) $$(A_1^2 + A_1 A_2 - 2A_2)Q = W_1 - W_3.$$

In addition,

(iv) $$\mathbf{R} = \begin{pmatrix} A_1 Q & A_2 Q \\ A_2 Q & A_1 A_2 Q - W_3 + W_2 \end{pmatrix},$$

and the feasible region for the limiting adaptive coefficient vector is given by

(v) $$0 < A_1 < 1, \qquad 0 < A_2 < 4 - 2A_1 - 4(1 - A_1)^{1/2} < 2.$$

Proof: Theorem 5.1 shows that the limit exists. Write

$$\mathbf{F} = \mathbf{E}_2, \qquad \mathbf{G} = \begin{pmatrix} 1 & 1 \\ 0 & 1 \end{pmatrix}, \qquad \mathbf{R} = \begin{pmatrix} R_1 & R_3 \\ R_3 & R_1 \end{pmatrix}.$$

Then, by definition in the limiting form of the updating equations, $\mathbf{A} = \mathbf{R}\mathbf{F}Q^{-1}$ so that $R_1 = A_1 Q$ and $R_3 = A_2 Q$ establishing the values of R_1 and R_3 in part (iv) of the Theorem. Next, $Q = \mathbf{F}'\mathbf{R}\mathbf{F} + 1 = A_1 Q + 1$ and (i) follows. Noting that $\mathbf{R}$ is involved in the two matrix equations $\mathbf{R} = \mathbf{G}\mathbf{C}\mathbf{G}' + \mathbf{W}$ and $\mathbf{C} = \mathbf{R} - \mathbf{A}\mathbf{A}'Q$, it follows by eliminating $\mathbf{C}$ that

$$\mathbf{R} - \mathbf{G}^{-1}\mathbf{R}(\mathbf{G}')^{-1} = \mathbf{A}\mathbf{A}'Q - \mathbf{G}^{-1}\mathbf{W}(\mathbf{G}')^{-1},$$

leading to

$$\begin{pmatrix} 2R_3 - R_2 & R_2 \\ R_2 & 0 \end{pmatrix} = \begin{pmatrix} A_1^2 & A_1 A_2 \\ A_1 A_2 & A_2^2 \end{pmatrix} Q - \begin{pmatrix} W_1 - 2W_3 + W_2 & W_3 - W_2 \\ W_3 - W_2 & W_2 \end{pmatrix}.$$

Equating components and rearranging leads to $A_2^2 Q = W_2$, $R_2 = A_1 A_2 Q - W_3 + W_2$ and $(A_1^2 + A_1 A_2 - 2A_2)Q = W_1 - W_3$.

This completes the proofs of statements (i) to (iv).

For the feasible region, note firstly that $A_1 = R_1/Q > 0$ and $1 - A_1 = Q^{-1} > 0$ so that $0 < A_1 < 1$. For A_2 proceed as follows. Define $a = Q(A_1^2 + A_1 A_2 - 2A_2)$. Then $|\mathbf{W}| \geq 0$ implies $W_1 W_2 - W_3^2 \geq 0$. Substituting $W_2 = A_2^2 Q$ from (ii) and $W_3 = W_1 - a$ from (iii) leads to $W_1^2 - W_1(A_2^2 Q + 2a) + a^2 \leq 0$. At the boundary, the roots of this quadratic in W_1 must be real valued, and thus $(A_2^2 Q + 2a)^2 - 4a^2 \geq 0$. This reduces to $A_2^2 Q + 4a \geq 0$ and, on substituting for a, we obtain the quadratic $A_2^2 - 4A_2(2 - A_1) + 4A_1^2 \geq 0$. For this to be true, A_2 must lie below the lower root of this quadratic and it follows that $A_2 \leq 4 - 2A_1 - 4(1 - A_1)^{1/2}$ as stated.

$\diamond$

Note that, following note (6) of the previous Section, we have a limiting representation of the observations in terms of forecast errors given by

$$Y_t - 2Y_{t-1} + Y_{t-2} = e_t + \beta_1 e_{t-1} + \beta_2 e_{t-2},$$

where $\beta_1 = \lim_{t \to \infty} \beta_{t1} = -(2 - A_1 - A_2)$ and $\beta_2 = \lim_{t \to \infty} \beta_{t2} = 1 - A_1$. We comment further on this limiting result below.

7.2.4 Single discount models

Often in practice the trend model will be a component, perhaps the only component in some cases, of a larger model. Generally, then, the discount concept will be used with such component models to structure the evolution variance matrices, as described for general block models in Section 6.3. This, of course, leads to a non-constant model although the practical differences are small. The following definition is simply a special case of the class of single discount model of Section 6.3 in Chapter 6.

Definition 7.3. *For any discount factor δ, $0 < \delta \leq 1$, a single discount, second-order polynomial growth DLM is any second-order model in which, for all t,*

$$\mathbf{W}_t = \mathbf{L}_2 \mathbf{C}_{t-1} \mathbf{L}_2'(1 - \delta)/\delta. \tag{7.4}$$

From a practical viewpoint, this is the most important class of second-order models. Note that a single discount factor is applied to the trend model as a whole, consistent with the ideas underlying component models. This does not, however, mean that different discount factors cannot be applied to the two elements μ_t and β_t of the model separately if desired, or indeed that some other form of evolution variance sequence be used. There are cases when such alternative approaches may be suitable, but, for the vast majority of applications, a single discount factor for the entire trend model will be satisfactory with alternatives leading to little, practically unimportant, difference. Under this single discount model, it can be shown that the updating equations have stable, limiting forms just as in the constant DLM in Theorem 7.1. The proof is left as an exercise for the reader at the end of the Chapter. Eventually and typically rapidly, the discount model behaves just like a constant model in which, from the definition of the former, $\mathbf{W} = \mathbf{L}_2 \mathbf{C} \mathbf{L}_2'(1 - \delta)/\delta$. The particular limiting values of the elements in the updating equations are given by

$$\mathbf{C} = \begin{pmatrix} 1 - \delta^2 & (1 - \delta)^2 \\ (1 - \delta)^2 & (1 - \delta)^3/\delta \end{pmatrix},$$

$$\mathbf{R} = \mathbf{L}_2 \mathbf{C} \mathbf{L}_2'/\delta,$$

$$\mathbf{A} = \begin{pmatrix} A_1 \\ A_2 \end{pmatrix},$$

$$Q = 1 + R_1 = \delta^{-2},$$

where

$$A_1 = 1 - \delta^2 \text{ and } A_2 = (1 - \delta)^2,$$

and R_1 is the upper-left-hand element of $\mathbf{R}$.

Note that the limiting adaptation is much greater for the first component μ_t than for the second β_t. Typical values, for example, are given in the case $\delta = 0.9$ when $A_1 = 0.19$ and $A_2 = 0.01$. Note also that any chosen value of either A_1 or A_2 suffices to determine both δ and the remaining limiting coefficient. In particular, choosing A_1 implies that $\delta = (1 - A_1)^{1/2}$ and $A_2 = 2 - A_1 - 2(1 - A_1)^{1/2}$.

7.2.5 Double discount models[†]

From a practical point of view, it is usually the case that the entire polynomial trend is best viewed as a component to be discounted as a block using a single discount factor. Approaches using two discount factors, one for the trend and one for the growth, are, of course, possible, and, though of restricted practical interest, provide Bayesian analogues of standard double exponential smoothing techniques (Abraham and Ledolter, 1983, Chapters 3 and 7; McKenzie, 1976). We restrict comment here to one of the approach described in Section 6.4 of Chapter 6. Approach (b) of that Section concerns the use of two discount factors δ_1 and δ_2, the former for the level and the latter for the growth. Let $\mathbf{\Delta} = \text{diag}(\delta_1^{-1/2}, \delta_2^{-1/2})$.

In this approach, $\mathbf{R}_t$ is given as a function of $\mathbf{C}_{t-1}$ by $\mathbf{R}_t = \mathbf{G}\mathbf{\Delta}\mathbf{C}_{t-1}\mathbf{\Delta}\mathbf{G}'$ so that, inverting, we have

$$\mathbf{C}_t^{-1} = \mathbf{R}_t^{-1} + \mathbf{F}\mathbf{F}'$$
$$= \mathbf{G}'^{-1}\mathbf{\Delta}^{-1}\mathbf{C}_{t-1}^{-1}\mathbf{\Delta}^{-1}\mathbf{G}^{-1} + \mathbf{F}\mathbf{F}'.$$

[†]This Section can be omitted without loss on a first reading.

Now, it may be shown that $\mathbf{C}_t$ has a limiting value $\mathbf{C}$. Proceeding to the limit in the above equations, we thus deduce that

$$\mathbf{C}^{-1} = \mathbf{G}'^{-1}\mathbf{\Delta}^{-1}\mathbf{C}^{-1}\mathbf{\Delta}^{-1}\mathbf{G}^{-1} + \mathbf{F}\mathbf{F}'.$$

Write

$$\mathbf{C} = \begin{pmatrix} C_1 & C_3 \\ C_3 & C_2 \end{pmatrix}$$

and $d = |\mathbf{C}|$ so that

$$\mathbf{C}^{-1} = d^{-1}\begin{pmatrix} C_2 & -C_3 \\ -C_3 & C_1 \end{pmatrix}.$$

Noting also that

$$\mathbf{G}^{-1} = \begin{pmatrix} 1 & -1 \\ 0 & 1 \end{pmatrix},$$

it follows that

$$\begin{pmatrix} C_2 & -C_3 \\ -C_3 & C_1 \end{pmatrix} = \begin{pmatrix} 1 & 0 \\ -1 & 1 \end{pmatrix}\begin{pmatrix} \delta_1 C_2 & -(\delta_1\delta_2)^{1/2}C_3 \\ -(\delta_1\delta_2)^{1/2}C_3 & \delta_2 C_1 \end{pmatrix}\begin{pmatrix} 1 & -1 \\ 0 & 1 \end{pmatrix} + \begin{pmatrix} d & 0 \\ 0 & 0 \end{pmatrix}.$$

Multiplying out and matching elements, it can be deduced that

$$C_1 = 1 - \delta_1\delta_2,$$
$$C_2 = (1 - \delta_2)[1 - (\delta_1\delta_2)^{1/2}]^2/\delta_1,$$
$$C_3 = (1 - \delta_2)[1 - (\delta_1\delta_2)^{1/2}].$$

The proof is left an an exercise for the reader. The corresponding limiting value for $\mathbf{R}_t$ can be calculated from $\mathbf{R} = \mathbf{G}\mathbf{\Delta}\mathbf{C}\mathbf{\Delta}\mathbf{G}'$, and, with $Q = 1 + \mathbf{F}'\mathbf{R}\mathbf{F}$, the corresponding limiting value for the adaptive vector $\mathbf{A} = \mathbf{R}\mathbf{F}/Q$ deduced. Again this is an exercise. The reader can verify that $\mathbf{A} = (A_1, A_2)'$ where

$$A_1 = 1 - \delta_1\delta_2 \qquad \text{and} \qquad A_2 = (1 - \delta_2)[1 - (\delta_1\delta_2)^{1/2}].$$

In the limit then, the updating equations for the mean vector $\mathbf{m}_t = (m_t, b_t)'$ are given by

$$m_t = m_{t-1} + b_{t-1} + A_1 e_t$$

and
$$b_t = b_{t-1} + A_2 e_t$$
where the limiting adaptive coefficients A_1 and A_2 are defined directly in terms of the discount factors. With $0 < \delta_1, \delta_2 < 1$, it follows that $0 < A_1 < 1$ and, for any A_1 in this range, A_2 is restricted to lie between 0 and $A_1[1 - (1 - A_1)^{1/2}]$.

7.3   LINEAR GROWTH MODELS

7.3.1   Introduction

In practice the discount models of Section 7.2.4 are recommended for their simplicity, parsimony and appropriateness. Of great interest and importance, however, are the class of second-order polynomial growth models, or *linear growth models* that have different evolution variance structure. Historically these models have been widely used by practitioners for modelling linear trends with easily interpretable parameters and stochastic components (Harrison, 1965, 1967; Godolphin and Harrison, 1975). As we shall show below, constant, linear growth models are such that the limiting adaptive coefficients in the vector **A** provide all the practically useful values, and so, in a very real practical sense, other second-order models are essentially redundant. A further reason for closely examining these models is for communication with users of other point forecasting metods for linear trends, and comparison with such techniques.

Definition 7.4. *Define the model M by the quadruple*
$$\{\mathbf{E}_2, \mathbf{L}_2, V_t, \mathbf{W}_t\} \tag{7.5a}$$
where, for some scalar variances W_{t1} and W_{t2},
$$\mathbf{W}_t = \begin{pmatrix} W_{t1} + W_{t2} & W_{t2} \\ W_{t2} & W_{t2} \end{pmatrix}. \tag{7.5b}$$
*Then any second-order polynomial DLM equivalent to M is defined as a **Linear Growth Model**. The class of such models is obtained as the variances W_{t1} and W_{t2} vary over all non-negative values.*

This definition is obviously just the special case of Definition 7.2 in which $n = 2$.[†] The standard equations for (7.5) are simply those

---

[†]Note that the form of the variance in (7.5b) is the simplest and preferred form for linear growth models although others are possible. For example, a model equivalent to M exists in which $\mathbf{W}_t = \text{diag}(W_{t1}, W_{t2})$ although this is not pursued here.

in (7.3a,b and c). However, we can now express the evolution errors as

$$\boldsymbol{\omega}_t = \mathbf{L}_2 \boldsymbol{\delta\theta}_t$$

where

$$\boldsymbol{\delta\theta}_t = \begin{pmatrix} \delta\theta_{t1} \\ \delta\theta_{t2} \end{pmatrix} = \begin{pmatrix} \delta\mu_t \\ \delta\beta_t \end{pmatrix},$$

and

$$\boldsymbol{\delta\theta}_t \sim \mathrm{N}[\,\mathbf{0},\ \mathrm{diag}(W_{t1}, W_{t2})\,].$$

It follows that $\omega_{t1} = \delta\mu_t + \delta\beta_t$ and $\omega_{t2} = \delta\beta_t$. Thus we have an alternative representation of the system equations, the usual form $\boldsymbol{\theta}_t = \mathbf{L}_2 \boldsymbol{\theta}_{t-1} + \boldsymbol{\omega}_t$ becomes

$$\boldsymbol{\theta}_t = \mathbf{L}_2(\boldsymbol{\theta}_{t-1} + \boldsymbol{\delta\theta}_t)$$

In this form, β_t has the interpretation of *incremental growth* in the level of the series over the time interval from $t-1$ to t, evolving during that interval according to the addition of the stochastic element $\delta\beta_t$. μ_t is the level itself at time t, evolving systematically via the addition of the growth β_t and undergoing a further stochastic shift via the addition of $\delta\mu_t$. In terms of model equations, this implies the more familiar versions (eg. Harrison and Stevens, 1976)

$$
\begin{array}{lll}
\text{Observation equation:} & Y_t = \mu_t + \nu_t, & \\
\text{System equation:} & \mu_t = \mu_{t-1} + \beta_t + \delta\mu_t, & \\
& \beta_t = \beta_{t-1} + \delta\beta_t, &
\end{array}
$$

with the zero-mean, evolution errors $\delta\mu_t$ and $\delta\beta_t$ uncorrelated.

7.3.2   Constant linear growth models

Suppose that the evolution variances are constant with $W_{t1} = W_1$ and $W_{t2} = W_2$. From Theorem 7.1, we have equations determining the relationships amongst the limiting values of the components of the model analysis. In this special case of the linear growth model, these results simplify as follows.

Theorem 7.2. *In the constant, linear growth model we have limiting behaviour defined by the limiting values*

$$\mathbf{C} = \begin{pmatrix} A_1 & A_2 \\ A_2 & A_2(A_1 - A_2)/(1 - A_2) \end{pmatrix}, \quad \mathbf{R} = \begin{pmatrix} A_1 & A_2 \\ A_2 & A_1 A_2 \end{pmatrix} Q,$$

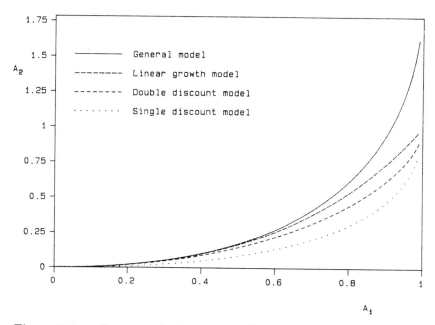

Figure 7.1a.   Ranges of adaptive coefficients.

with the feasible region for the adaptive vector given by

$$0 < A_1 < 1, \quad and \quad 0 < A_2 < A_1^2/(2 - A_1) < A_1.$$

Proof: **R**, **C** and the limits on A_1 follow directly from Theorem 7.1 by substituting the particular form of the **W** matrix. Specifically, we replace W_1 with $W_1 + W_2$ and W_3 with W_2. To determine the bounds on A_2, note that, since $R_2 = A_1 A_2 Q > 0$, then $A_2 > 0$. Also, from (iii) of Theorem 7.1, we obtain $A_1^2 - A_2(1 - A_1) > 0$ or $A_2 < A_1^2/(2 - A_1) < A_1$, as required.

◇

Figure 7.1a provides a graph of the boundary line $A_2 = A_1^2/(2 - A_1)$ in the $A_2 : A_1$ plane. This appears as a dashed line. The region below this line and such that $0 < A_1 < 1$ is the feasible region for the adaptive coefficients given above. The full line in the figure is the corresponding boundary for the general second-order polynomial model provided in Theorem 7.1, defined by $A_2 = 4 - 2A_1 - 4(1 -$

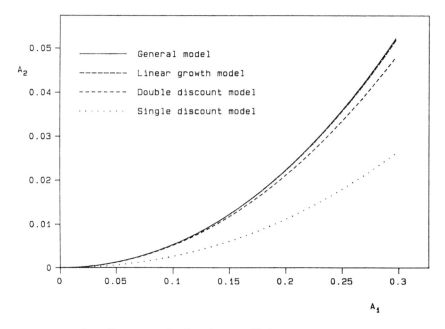

Figure 7.1b.   Ranges of adaptive coefficients.

$A_1)^{1/2}$. Clearly the regions differ appreciably only for larger values
of A_1. Such values are extremely unusual in practice; remember that
these are limiting rates of adaptation in a closed model. For example,
in stock control and monitoring applications, A_1 will rarely exceed
0.3, implying $A_2 < 0.053$. These values are very widely applicable
upper bounds on the adaptive coefficients and so, for most practical
purposes, the linear growth model is an adequate subclass of all
second-order models. Practitioners gain little by entertaining models
outside this class. Figure 7.1b is a close up of the graph over the
range $0 < A_1 < 0.3$. It is clear from this figure that the feasible
regions essentially coincide for this range of values. Also graphed in
Figures 7.1a and 7.1b are the possible values of adaptive coefficients
in the single discount model of Definition 7.3. In this model, $\mathbf{W}_t$ is
defined via a single quantity and the limiting adaptive coefficients are
related via $A_2 = 2 - A_1 - 2(1 - A_1)^{1/2}$. Finally, for further theoretical
comparisons, the double discount approach of Section 7.2.5 leads to
the ranges $0 < A_1 < 1$ and $0 < A_2 < A_1[1 - (1 - A_1)^{1/2}]$. The
boundary line $A_2 = A_1[1 - (1 - A_1)^{1/2}]$ also appears in Figures 7.1a
and 7.1b. The ranges of limiting adaptive coefficients in this double
discount model are little different to those in the linear growth class.

7.3.3 Limiting predictors in the constant model

Assuming a closed, constant linear growth model, the limiting form
of the updating given in Section 7.2.2 implies that

$$m_t = m_{t-1} + b_{t-1} + A_1 e_t,$$
$$b_t = b_{t-1} + A_2 e_t.$$

Furthermore, again from that section, we have a limiting representation of the observations in terms of the second difference equation

$$Y_t - 2Y_{t-1} + Y_{t-2} = e_t + \beta_1 e_{t-1} + \beta_2 e_{t-2}, \qquad (7.6)$$

where $\beta_1 = \lim_{t\to\infty} \beta_{t1} = -(2 - A_1 - A_2)$ and $\beta_2 = \lim_{t\to\infty} \beta_{t2} = 1 - A_1$. This can be written in terms of the backshift operator B as

$$(1 - B)^2 Y_t = (1 + \beta_1 B + \beta_2 B^2) e_t.$$

A number of popular, non-Bayesian methods of point prediction
employ equations of the form (7.6) to predict Y_t from past values.
Some of the most widely used such techniques are discussed for comparison.

(1) **Holt's linear growth method** (Holt, 1957).
    Holt's method produces point forecasts using a linear forecast function as in (7.1) with $a_{t0} = m_t$ and $a_{t1} = b_t$ sequentially updated according to

$$m_t = A Y_t + (1 - A)(m_{t-1} + b_{t-1}),$$
$$b_t = D(m_t - m_{t-1}) + (1 - D)b_{t-1},$$

where $0 < A, D < 1$. After rearrangement, and using $e_t = Y_t - (m_{t-1} - b_{t-1})$, we have

$$m_t = m_{t-1} + b_{t-1} + A e_t,$$
$$b_t = b_{t-1} + AD e_t.$$

Thus, if $A_1 = A$ and $A_2 = AD$, Holt's method is just that
given as a *limiting* case of the constant, linear growth DLM,
provided that $D < A/(2 - A)$.

(2) **Box and Jenkins' ARIMA(0,2,2) method** (Box and Jenkins, 1976).

The ARIMA model of Box and Jenkins assumes the time series to follow the process model

$$Y_t - 2Y_{t-1} + Y_{t-2} = e_t + \beta_1 e_{t-1} + \beta_2 e_{t-2}$$

where the error terms e_t are zero-mean, uncorrelated errors. The forecasts are generated from this model on the basis that it holds as a constant relationship, with constant error variances for the e_t, for all time. The forecast function for this ARIMA model is defined by

$$f_t(0) = Y_t - \beta_2 e_t,$$
$$f_t(1) = 2Y_t - Y_{t-1} + \beta_1 e_t + \beta_2 e_{t-1},$$

and, for $k \geq 2$,

$$f_t(k) = 2f_t(k-1) - f_t(k-2)$$
$$= f_t(0) + [f_t(1) - f_t(0)]k.$$

In view of (7.6), this point predictor clearly coincides with that obtained in the limit from the DLM, provided that $0 < \beta_2 < 1$ and $2\beta_2 + \beta_1 > 0$.

(3) **Exponentially weighted regression** (Brown, 1962).

The exponentially weighted regression, or EWR, technique developed by Brown is basically designed for the estimation and extrapolation of straight lines. The method assumes an essentially infinite history for the series of observations $Y_t, Y_{t-1}, \ldots$, and derives the pair m_t and b_t as those values of μ and β, respectively, minimising the *discounted* sum of squares

$$S_t(\mu, \beta) = \sum_{r=0}^{\infty} \delta^r (Y_{t-r} - \mu - r\beta)^2,$$

for some discount factor δ, $(0 < \delta < 1)$. One interpretation of this method is that of discounted, maximum likelihood estimation of the straight line regression coefficients. It is easily shown that the minimising values satisfy the recurrence equations

$$m_t = m_{t-1} + b_{t-1} + (1 - \delta^2)e_t,$$
$$b_t = b_{t-1} + (1 - \delta)^2 e_t,$$

where $e_t = Y_t - m_{t-1} - b_{t-1}$. Thus the EWR recurrence relationships, limiting in the sense that an infinite history of the series is assumed available, are identical to the limiting forms from the DLM in the case $A_1 = 1-\delta^2$ and $A_2 = (1-\delta)^2$. In particular, the constant, linear growth model in which

$$W_1 = 2(1 - \delta)^2/\delta \quad \text{and} \quad W_2 = (1 - \delta)^4/\delta^2$$

leads to such values of the limiting adaptive coefficients. This model also has $Q = 1/\delta^2$ and

$$C = \begin{pmatrix} 1 - \delta^2 & (1 - \delta)^2 \\ (1 - \delta)^2 & 2(1 - \delta)^3/\delta \end{pmatrix}.$$

Note also that the single discount DLM of Section 7.2.4 provides the same limiting adaptive coefficients. This stresses and reaffirms our earlier remarks about the usefulness and appropriateness of discount models in practice. The relationships between these models can be seen in the following table of discount factors and associated limiting values of some of the model components.

δ	A_1	A_2	$1/W_1$	W_1/W_2	Q
0.95	0.10	0.003	200	800	1.11
0.90	0.19	0.010	45	180	1.23
0.85	0.28	0.023	19	76	1.38
0.80	0.36	0.040	10	40	1.56
0.70	0.51	0.090	4	16	2.04
0.50	0.75	0.250	1	4	4.00

(4) A fuller range of limiting adaptive coefficients, and hence limiting forecast functions, can be obtained through the use of double discounting as explained in Section 7.2.5, if desired. The ranges of limiting adaptive coefficients in that double discount model are little different to those in the linear growth class.

7.3.4 Discussion

Some further discussion of the limiting predictors described above is
given here. This discussion is rather technical and may be omitted
on first reading. However, for practitioners familiar with the non-
Bayesian methods discussed above, this section will provide further
insight into the nature of Bayesian models and more sharply identify
the points of difference.

The limiting updating equations for the constant DLM provide
each of the limiting predictors but with particular restrictions on
the values of the limiting adaptive coefficients. Brown's values $A_1 =
1 - \delta^2$ and $A_2 = (1 - \delta)^2$ are constrained within the feasible region
for the linear growth DLM (see Figures 7.1). However, Holt's region is defined by $0 < A_1, A_2 < 1$, and that of the ARIMA model
is $0 < A_2 < 2$, $0 < A_1 + A_2 < 4$. These both allow values outside the feasible region determined in Theorem 7.1. Now it might
be suggested that these regions should be contained within that for
second-order polynomial models since the latter class of models contains all those having a linear forecast function. The fact that this
is not the case derives from a rather subtle and hidden point. The
reason is that the predictors using $\mathbf{A}$ outside the feasible region from
the DLM are unknowingly superimposing a moving average process,
of order not more than 2, on the linear model. This leads to poly-
nomial/moving average models which, in DLM terms, are obtained
from the superposition of the polynomial model as one component
with another component having at most 2 *zero* eigenvalues. The set
of canonical models has the form

$$\left\{ \begin{pmatrix} \mathbf{E}_2 \\ \mathbf{E}_2 \end{pmatrix}, \begin{pmatrix} \mathbf{J}_2(1) & \mathbf{0} \\ \mathbf{0} & \mathbf{J}_2(0) \end{pmatrix}, V, \mathbf{W} \right\},$$

and will produce feasible regions $0 < A_1 < 2$, $0 < A_1 + A_2 < 4$
depending on the structure assigned to $\mathbf{W}$. The resulting forecast
function is not, however, a polynomial for all time, having irregular
starting values like the ARIMA predictor, $f_t(0) \neq m_t$ and $f_t(1) \neq
m_t + b_t$, in general. Clearly, though, $f_t(k) = m_t + kb_t$ for $k \geq 2$.

This subtlety applies to all DLMs. Referring, for example, to the
first-order polynomial model of Chapter 2, the constant DLM has a
limiting representation of the form

$$Y_t - Y_{t-1} = e_t + (1 - A)e_{t-1}$$

with $0 < A < 1$, whereas the Box and Jenkins ARIMA (0,1,1) predictor, having essentially the same form, allows $0 < A < 2$. This latter region is precisely that valid for the constant DLM

$$\left\{ \begin{pmatrix} 1 \\ 1 \end{pmatrix}, \begin{pmatrix} 1 & 0 \\ 0 & 0 \end{pmatrix}, V, \mathbf{W} \right\},$$

in which $\mathbf{A}$ may exceed unity if and only if $W_3 \neq 0$.

More generally, consider any TSDLM $\{\mathbf{F}, \mathbf{G}, V, \mathbf{W}\}$ with $\mathbf{G}$ of full rank n. We know from Theorem 5.1 of Chapter 5 that the updating equations have stable, limiting forms and, from Theorem 5.2 of Chapter 5, that

$$\lim_{t \to \infty} \left\{ \prod_{r=1}^{n} (1 - \lambda_r B) Y_t - \prod_{r=1}^{n} (1 - \rho_r B) e_t \right\} = 0,$$

where B is the backshift operator, $\lambda_1, \ldots, \lambda_n$ are the n eigenvalues of $\mathbf{G}$, and $\rho_1, \ldots, \rho_n$ are simply linear functions of the elements of the limiting adaptive vector $\mathbf{A}$. The feasible region for these ρ_r coefficients is a subset of the region

$$\{\rho_1, \ldots, \rho_n : |\rho_r| < 1, (r = 1, \ldots, n)\}.$$

This region can be enlarged, however, by superimposing a zero eigenvalue component of dimension n on the model, leading to the $2n$ dimensional DLM

$$\left\{ \begin{pmatrix} \mathbf{F} \\ \mathbf{E}_n \end{pmatrix}, \begin{pmatrix} \mathbf{G} & \mathbf{0} \\ \mathbf{0} & \mathbf{J}_n(0) \end{pmatrix}, V, \mathbf{W}^* \right\}.$$

The above limiting representation of the observation series holds for this extended model and the ρ_r coefficients may take any values in the above region, depending on the structure of the evolution variance matrix $\mathbf{W}^*$.

Clearly this model extension is unnecessary in practice, but it serves here to identify peculiarities of the classical techniques, indicating the need to restrict the range of allowable values for adaptive coefficients further than previously suggested by advocates of such methods. The zero eigenvalue component is superfluous, its addition violates parsimony and clouds interpretability.

7.4 THIRD-ORDER POLYNOMIAL MODELS

7.4.1 Introduction

By definition, a third-order polynomial DLM is any observable, TS-DLM having a forecast function of the quadratic form

$$f_t(k) = a_{t0} + a_{t1}k + a_{t2}k^2,$$

for all $t \geq 0$ and $k \geq 0$. An alternative representation is given by

$$f_t(k) = f_t(0) + [f_t(1) - f_t(0)]k + [f_t(2) - 2f_t(1) + f_t(0)]k(k-1)/2,$$

where the coefficients of the quadratic forecast function are identified in terms of initial values at times $t = 0$ and $t = 1$. These models are rarely used in practice since, for many short term forecasting and micro-forecasting applications, local trends are adequately described using first- or second-order polynomials. When dealing with macro or aggregate data, however, and when forecasts are required for longer lead times, the random variation measured by observational variances V_t is often small relative to movement in trend. In such circumstances, third-order polynomial descriptions may be needed. An example of such a case is given in the application to longer term growth forecasting in Harrison, Gazard and Leonard (1977).

The general, canonical, third-order polynomial model is defined by $\{\mathbf{E}_3, \mathbf{J}_3(1), . , . \}$. As above, however, the more interpretable alternative form $\{\mathbf{E}_3, \mathbf{L}_3, . , . \}$ is considered. Further, as with second-order models, it is a case that a subclass of third-order models, namely those termed **quadratic growth** models, are essentially sufficient in terms of the possible values of limiting adaptive coefficients. Thus we restrict attention to a brief description of the theoretical structure of models in this subclass. The evolution variance sequence in quadratic growth models has a particular, structured form, derived as the case $n = 3$ in Definition 7.2. In practice, the more parsimonious models based on single discount factors wil typically be used without real loss of flexibility. Although apparently rather different in nature, these discount models are intimately related to the canonical quadratic growth models and, in constant DLMs, the rapidly approached limiting behaviour of the two are equivalent. This is directly analogous to the situation with first- or second-order models, and the analogy carries over to polynomials of higher order.

7.4.2 Quadratic growth models

Definition 7.5. *Define the model M by*

$$\{\mathbf{E}_3, \mathbf{L}_3, V_t, \mathbf{W}_t\}$$

where $\mathbf{W}_t$ has the form

$$\mathbf{W}_t = \mathbf{L}_3 \operatorname{diag}(W_{t1}, W_{t2}, W_{t3})\mathbf{L}_3',$$

*for some variances W_{t1}, W_{t2} and W_{t3} for all t. Then any third-order polynomial DLM equivalent to M is called a **Quadratic Growth Model**. The class of all such models is obtained as the defining variances in $\mathbf{W}_t$ vary.*

Writing

$$\boldsymbol{\theta}_t = \begin{pmatrix} \theta_{t1} \\ \theta_{t2} \\ \theta_{t3} \end{pmatrix} = \begin{pmatrix} \mu_t \\ \beta_t \\ \gamma_t \end{pmatrix},$$

the model M may be written as

$$Y_t = \mu_t + \nu_t,$$
$$\mu_t = \mu_{t-1} + \beta_t + \delta\mu_t,$$
$$\beta_t = \beta_{t-1} + \gamma_t + \delta\beta_t,$$
$$\gamma_t = \gamma_{t-1} + \delta\gamma_t$$

where $\nu_t \sim N[0, V_t]$ and

$$\boldsymbol{\delta\theta}_t = (\delta\mu_t, \delta\beta_t, \delta\gamma_t)' \sim N[\, \mathbf{0}, \operatorname{diag}(W_{t1}, W_{t2}, W_{t3})\,].$$

The system equations may be written as

$$\boldsymbol{\theta}_t = \mathbf{L}_3(\boldsymbol{\theta}_{t-1} + \boldsymbol{\delta\theta}_t).$$

The quantities μ_t, β_t and γ_t represent level, incremental growth and change in that incremental growth at time t, respectively. In a continuous time analogue, β_t would represent the first derivative with respect to time, or gradient, of the expected level of the series at time t, and γ_t the second derivative of the expected level. The components of the evolution error $\boldsymbol{\delta\theta}_t$ represent the corresponding stochastic changes in the state vector components. Given the posterior mean

$$E[\boldsymbol{\theta}_t | D_t] = \mathbf{m}_t = (m_t, b_t, g_t)'$$

at time t, the forecast function is clearly identified as

$$f_t(k) = m_{t-1} + k b_{t-1} + k(k-1)g_{t-1}/2,$$

for all $k \geq 0$.

Finally, the updating equations lead to a limiting representation of the observation series just as in the second-order model. With adaptive coefficient vector $\mathbf{A}_t = (A_{t1}, A_{t2}, A_{t3})'$ at time t, we have the third-order difference equation

$$(1-B^3)Y_t = Y_t - 3Y_{t-1} + 3Y_{t-2} - Y_{t-3} = e_t + \beta_{t1}e_{t-1} + \beta_{t2}e_{t-2} + \beta_{t3}e_{t-3},$$

where the coefficients on the left-hand side are given by

$$\beta_{t1} = -(3 - A_{t-1,1} - A_{t-1,2}),$$
$$\beta_{t2} = 3 - 2A_{t-2,1} - A_{t-2,2} + A_{t-2,3},$$
$$\beta_{t3} = -(1 - A_{t-3,1}).$$

7.4.3 Constant, quadratic growth model

Theorem 7.3. *In the constant, quadratic growth model*

$$\{\mathbf{E}_3, \mathbf{L}_3, 1, \mathbf{L}_3 \mathbf{W} \mathbf{L}_3'\},$$

with $\mathbf{W} = \mathrm{diag}(W_1, W_2, W_3)$, *the components* $\mathbf{A}_t$, $\mathbf{C}_t$, $\mathbf{R}_t$ *and* Q_t *have limiting values* $\mathbf{A}$, $\mathbf{C}$, $\mathbf{R}$ *and* Q *defined, respectively, by the equations*

$$Q = 1/(1 - A_1),$$
$$W_3 = A_3^2 Q,$$
$$W_2 = [A_1^2 - A_3(2A_1 + A_2)]Q,$$
$$W_1 = (A_1^2 + A_1 A_2 - 2A_2 + A_3)Q,$$

and

$$\mathbf{R} = \begin{pmatrix} A_1 & A_2 & A_3 \\ A_2 & A_1 A_2 - A_3(1 - 2A_1 - A_2) & A_3(A_1 + A_2) \\ A_3 & A_3(A_1 + A_2) & A_2 A_3 \end{pmatrix} Q,$$

with $\mathbf{C} = \mathbf{R} - \mathbf{A}\mathbf{A}'Q$.

In addition, the feasible region for **A** *is defined by*

$$\{\mathbf{A}: \quad 0 < A_3, A_1 < 1, 0 < A_2 < 2, 0 < A_2^2 - A_3(2A_1 + A_2),$$
$$0 < A_1^2 + A_1 A_2 - 2A_2 + A_3\}.$$

Proof: The limits exist by Theorem 5.1. The proof of the limiting relationships follows the style of Theorem 7.1, and in fact the technique is general and may be used to derive analogous limiting relationships in more general, higher order polynomial DLMs. The proof is only sketched, the details being left to the reader.

From the defining equations $\mathbf{R} = \mathbf{L}_3(\mathbf{C} + \mathbf{W})\mathbf{L}_3'$ and $\mathbf{C} = \mathbf{R} - \mathbf{A}\mathbf{A}'Q$ it follows that

$$\mathbf{L}_3^{-1}\mathbf{R}(\mathbf{L}_3')^{-1} = \mathbf{R} - \mathbf{A}\mathbf{A}'Q + \mathbf{W}.$$

The required representation of the components as functions of **A** and Q can now be deduced by matching elements in the above matrix identity, noting the special diagonal form of **W** and the identity

$$\mathbf{L}_3^{-1} = \begin{pmatrix} 1 & -1 & 1 \\ 0 & 1 & -1 \\ 0 & 0 & 1 \end{pmatrix}.$$

Finally, the feasible region for limiting adaptive coefficients can be deduced using the facts that $0 < W_1, W_2, W_3 < Q$, $1 < Q$ and the positive definiteness of **R**.

$\diamond$

We note the following.

(1) If we define $\beta_1 = \lim_{t\to\infty} \beta_{t1} = -(3 - A_1 - A_2)$, $\beta_2 = \lim_{t\to\infty} \beta_{t2} = 3 - 2A_1 - A_2 + A_3$ and $\beta_3 = \lim_{t\to\infty} \beta_{t3} = -(1 - A_1)$, then the limiting representation of the observation series is simply

$$(1 - B)^3 Y_t = (1 + \beta_1 B + \beta_2 B^2 + \beta_3 B^3)e_t.$$

The predictors of Box and Jenkins' ARIMA (0,3,3) models have just this form, but with coefficients β_j, $(j = 1, 2, 3)$, that may take values in $(-1, 1)$. The DLM limiting predictors thus correspond to a subset of the ARIMA predictors. As discussed in Section 7.3.4, however, the full set of ARIMA predictors are obtained by extending the DLM to have a further component with three zero eigenvalues, although this is of little practical value.

(2) Exponentially weighted regression (or discounted likelihood) techniques provide predictors of similar form. In particular, the values m_t, b_t and g_t minimising the discounted sum of squares

$$S(\mu, \beta, \gamma) = \sum_{r=0}^{\infty} \delta^r [Y_{t-r} - \mu - \beta r - \gamma r(r-1)/2]^2$$

with respect to the parameters μ, β and γ, are given by the DLM updating equations with $A_{t1} = A_1 = 1 - \delta^3$, $A_{t2} = A_2 = 2 - 3\delta + \delta^3$ and $A_{t3} = A_3 = (1 - \delta)^3$. In this case, the observation series has the special form

$$(1 - B)^3 Y_t = (1 - \delta B)^3 e_t.$$

Note that this is also the limiting form given in a discount DLM with the evolution variance sequence structured using a single discount factor δ applied to the whole quadratic trend component.

(3) The above results generalise to higher order models. Some discussion of the limiting representation of the observation series in higher order models is given in Godolphin and Harrison (1975). The following result is given there. In a constant, n^{th}-order polynomial growth DLM $\{\mathbf{E}_n, \mathbf{L}_n(1), 1, \mathbf{W}\}$, with $\mathbf{W} = \mathbf{L}_n \text{diag}(W_1, \ldots, W_n)\mathbf{L}'_n$, suppose that the evolution variances are given by $W_r = \binom{n}{r} c^r$, $(r = 1, \ldots, n)$, where $c = (1 - \delta)^2 / \delta$ for some discount factor δ. Then the limiting representation of the observation series (a special case of Theorem 5.2 of Chapter 5) is

$$(1 - B)^n Y_t = (1 - \delta B)^n e_t,$$

and is just that obtained using the EWR technique; equivalently, it is that of a single discount model in which $\mathbf{R}_t = \mathbf{C}_t / \delta$. The limiting value of the adaptive vector $\mathbf{A} = (A_1, \ldots, A_n)'$ is given by

$$A_1 = 1 - \delta^n,$$

$$A_{r+1} = \binom{n}{r}(1 - \delta)^r - A_r, \qquad (r = 2, \ldots, n-1),$$

$$A_n = (1 - \delta)^n.$$

The limiting one-step forecast variance is $Q = 1/\delta^n$.

7.5   EXERCISES

(1) Consider the constant model $\{\mathbf{E}_n, \mathbf{L}_n, V, V\mathbf{W}\}$. Apply Theorem 5.2 of Chapter 5 to obtain the limiting representation of the observation series in terms of forecast errors. Specifically, prove that as $t \to \infty$,

$$(1 - B)^n Y_t \approx e_t + \sum_{i=1}^{n} \beta_i e_{t-i},$$

for some, real quantities β_i, $(i = 1, \dots, n)$. Note that this is a consequence of the model assumptions *irrespective* of whether or not the data actually follow the assumed polynomial model.

(2) Suppose that Y_t actually follows the polynomial model in Exercise 1. Show directly that

$$(1 - B)^n Y_t = a_t + \sum_{i=1}^{n} \gamma_i a_{t-i},$$

for some, real quantities γ_i, $(i = 1, \dots, n)$, with a_t being a sequence of zero-mean, uncorrelated random quantities.

(3) Verify that the polynomial growth model $\{\mathbf{E}_n, \mathbf{L}_n, . , . \}$ is observable and similar to the canonical model $\{\mathbf{E}_n, \mathbf{J}_n(1), . , . \}$.

(4) Show directly that the forecast function of the model $\{\mathbf{E}_n, \mathbf{L}_n, . , . \}$ is a polynomial of order $n - 1$.

(5) Consider DLMs M and M_1 characterised by quadruples

$$\begin{aligned} M: & \quad \{\mathbf{E}_n, \mathbf{J}_n(1), V_t, \mathbf{W}_t\}, \\ M_1: & \quad \{\mathbf{E}_n, \mathbf{L}_n, V_t, \mathbf{W}_{1t}\}. \end{aligned}$$

(a) Calculate the observability matrices $\mathbf{T}$ and $\mathbf{T}_1$ for M and M_1 respectively.

(b) Show that the similarity matrix $\mathbf{H} = \mathbf{T}^{-1}\mathbf{T}_1$ is given by

$$\mathbf{H} = \begin{pmatrix} 1 & 0 & 0 & 0 & \cdots & 0 & 0 \\ 0 & 1 & 1 & 0 & \cdots & 0 & 0 \\ 0 & 0 & 1 & 1 & \cdots & 0 & 0 \\ \vdots & \vdots & & & \ddots & \vdots & \vdots \\ 0 & 0 & 0 & 0 & \cdots & 1 & 1 \\ 0 & 0 & 0 & 0 & \cdots & 0 & 1 \end{pmatrix}.$$

(c) Interpret the meaning of the state parameters in each model.

(6) Consider the single discount, second-order polynomial model of Definition 7.2.

  (a) Use the identity $\mathbf{C}_t^{-1} = \mathbf{R}_t^{-1} + \mathbf{E}_2 \mathbf{E}_2'$ to directly verify that $\mathbf{C}_t$ converges to the limiting value stated in Section 7.2.4.

  (b) Confirm the limiting values of $\mathbf{A}_t$, $\mathbf{R}_t$ and Q_t.

  (c) Deduce that, in this (rapidly approached) limiting form, the updating equations are just those of a constant DLM in which

$$\mathbf{W} = \begin{pmatrix} 1 - \delta^2 & (1 - \delta)^2 \\ (1 - \delta)^2 & W \end{pmatrix}$$

for *any* variance W.

(7) Verify the limiting value of $\mathbf{C}$ given in the double discount, second-order polynomial model in Section 7.2.5.

(8) In the constant, linear growth model of Section 7.3.2 with observational variance known, calculate, for any $k > 1$, the forecast distributions for

  (a) $(Y_{t+k} | D_t)$, and

  (b) $(X_t(k) | D_t)$ where $X_t(k) = \sum_{r=1}^{k} Y_{t+r}$.

(9) Using the limiting form of the updating equations in the constant linear growth model of Section 7.3, and the definitions of β_1 and β_2 in terms of A_1 and A_2, verify that the limiting forecast function can be written as given in part (2) of Section 7.3.3, namely

$$f_t(0) = Y_t - \beta_2 e_t,$$
$$f_t(1) = 2Y_t - Y_{t-1} + \beta_1 e_t + \beta_2 e_{t-1},$$

with, for $k \geq 2$,

$$f_t(k) = 2f_t(k-1) - f_t(k-2) = f_t(0) + [f_t(1) - f_t(0)]k.$$

(10) Verify that the recurrence equations given in part (3) of Section 7.3.3 do indeed define those values of μ and β that uniquely minimise the discounted sum of squares $S_t(\mu, \beta)$.

CHAPTER 8

SEASONAL MODELS

8.1 INTRODUCTION

Cyclical or periodic behaviour is evident in many time series associated with economic, commercial, physical and biological systems. Annual seasonal cycles, for example, are well accepted as providing the basis of the agricultural calendar. Each year the earth revolves about the sun, the relative positions of the two bodies determining the earth's climatic conditions at any time. This natural cycle is essentially uncontrollable and has to be accepted. It is important to recognise the induced seasonality in, for example, the demand for agricultural goods and many other products, and to include it as a factor in any forecasting model. Such seasonal patterns may have enormous implications for stock control and production planning in agricultural business; it is not uncommon, for example, to encounter demand patterns exhibiting seasonal peak to trough ratios of order of 20 to 1. Various other annual cycles, such as demand for fireworks, Valentine cards, Easter eggs and so forth, are even more pronounced. Although many such annual patterns arise from the natural solar cycle, not all do, nor is the effect always controllable. One example concerns the demand for an anaesthetic whose annual cycle initially surprised the marketing department of the manufacturing company. Upon closer investigation it was found that most of the seasonality was quarterly, coinciding with the quarterly accounting periods of the hospitals to whom most of the anaesthetic was sold. Similar seasonal patterns of demand and sales arise partially as responses to advertising effort and expenditure, campaigns being repeated along similar lines from year to year. In addition to identifying and anticipating cycles of this sort, decision makers may wish to exert control in attempts to alter and, particularly, reduce seasonal fluctuations. In commercial environments this is often desirable to reduce inefficiencies in the use of manufacturing plant and warehouse facilities. One particular example concerns the increase in milk prices that is designed to encourage producers to keep more cows over winter when feed costs are high, thus dampening seasonal fluctuations in milk supplies.

The annual cycle, though of primary importance, is one amongst many of varying periods. The day/night, lunar and other planetary cycles are well-known. So too are natural biorythms in animals and oscillatory phenomena in the physical sciences. Perhaps the most debatable forms of cyclic behaviour arise from system response mechanisms. Examples include the cobweb, pig and beef cycles, and predator-prey interactions. In commercial and economic activities it is important to be aware of the various business cycles, such as those christened Kitchin, Jugler and Kondratieff (van Duijn, 1983). These are not inevitable cycles but they do reflect responses of systems involving decision makers to the imbalances in things such as supply and demand. Often such responses reflect a lack of understanding on the part of decision makers as to the possible consequences of their actions over time. An illuminating example from agricultural, taken from Harrison and Quinn (1978), concerns beef markets facing a supply shortage. Such a shortage often prompts governmental agencies to initiate incentive schemes for beef farmers, resulting in the farmers retaining heifers for breeding and culling fewer calves. Although this action to produce more beef has started, maturing and gestation times lead to a delay of about three years before results are seen. This basic delay is of little interest to the market in the short term, however, and beef prices rise sharply since the breeder's action leads to an even more serious shortage. This signal that action to restore the supply balance is underway is now drastically misinterpreted by breeders and decision makers as indicating that the shortage was originally under-estimated, leading to further expansion of breeding stock. The consequent further drop in supply leads farmers to act similarly the following year. This has disasterous consequences the following year since then the initial expansive action reaches maturity, flooding the beef market and leading to a collapse in prices. As a side effect the increased size of herds over the period has led to soaring feed and housing costs which destroy the breeder's profit margins. The final response is now massive slaughter of cows and calves at birth, sowing the seeds of the next shortage so that the cycle repeats. This example is not hypothetical. In Britain in the mid-1970's, for example, almost 25% of Friesan calves were slaughtered at birth, and farmers in Australia were paid to shoot cows, some farmers going out of business altogether as a result.

The key point of this example is that it specifically illustrates the common, but little understood, phenomenon of a delayed feedback response and cyclical behaviour that may be the result of decision makers' actions. The keys to control lie in understanding the nature of the system, recognising the great time delays between action and effect, and the fact that, once an imbalance occurs, the appropriate corrective action will typically cause an immediate worsening of the situation before balance is restored. It is also important to be aware of the consequences for related activities, such as milk, feed, pigs, leather goods, and so on, in the above example. The modern tendency to divide control responsibilities for these areas as though they were independent only exaggerates crises such as beef shortages and mountains, with the continued reactions of decision makers only perpetuating the swings. The main message for time series forecasting is cyclical patterns should not necessarily be automatically modelled as inevitable seasonality. It is important to consider the background, and, in particular, to decide whether there may be a useful description of an underlying mechanism in terms of transfer response models.

Having identified these issues, we return in this chapter to consider descriptions of observed cyclical behaviour purely in terms of superficial seasonal factors. It is the case in practice that such simple representational models, either alone or combined with trend and/ or regression components, often prove adequate for assessing current and historical seasonal patterns and analysing changes over time, and for short-term forecasting. We use the term seasonality as a label for any observed behaviour of a cyclical or periodic nature, whether or not this corresponds to well-defined and accepted seasons, as, for example, with annual patterns. To begin, we describe the basic structure of linear models for purely deterministic functions exhibiting pure seasonality, setting out the requisite notation and terminology. This leads to suitable DLMs via, as usual, the consideration of cyclical forecast functions. We detail two important classes of models. The first uses seasonal factors and is termed the **form-free** approach since the form of the seasonal patterns is essentially unrestricted. The second approach uses a functional representation of the seasonal factors in terms of trigonometric terms and is termed the **form** approach. Both approaches are useful in practice.

8.2 SEASONAL FACTOR REPRESENTATION OF CYCLICAL FUNCTIONS

Let $g(t)$ be any real-valued function defined on the non-negative integers $t = 0, 1, \ldots$, where t is a time index. Note that we specifically take the function to be defined from time zero in order to conform with the usage of cyclical forecast functions in DLMs, as detailed below.

Definition 8.1.

(1) $g(t)$ is **cyclical** or **periodic** if, for some integer $p \geq 1$, $g(t + np) = g(t)$ for all integers $t \geq 0$ and $n \geq 0$.

(2) Unless otherwise stated, the smallest integer p such that this is true is called the **period** of $g(.)$.

(3) $g(.)$ exhibits a single full **cycle** in any time interval containing p consecutive time points, such as $(t, t + p - 1)$, for any $t > 0$.

(4) The **seasonal factors** of $g(.)$ are the p values taken in any full cycle

$$\psi_j = g(j), \qquad\qquad (j = 0, \ldots, p - 1).$$

Notice that, for $t > 0$, $g(t) = g(j)$ where j is the remainder after division of t by p, denoted by $j = p|t$.

(5) The **seasonal factor vector** at time t is simply the vector of seasonal factors permuted so that the first element is that for the time t, namely

$$\boldsymbol{\psi}_t = (\psi_j, \psi_{j+1}, \ldots, \psi_{p-1}, \psi_0, \ldots, \psi_{j-1})',$$

when the current seasonal factor is ψ_j. In particular, for any integers n and $k = np$, $\boldsymbol{\psi}_k = (\psi_0, \ldots, \psi_{p-1})'$.

(6) In any cycle, the time point corresponding to the relevant seasonal factor ψ_j is given a label $M(j)$. This label then defines the timing within each cycle, as, for example, with months within years where $M(0)$ may be January, $M(1)$ February, and so forth. Clearly the labels are cyclic with period p, and $j = p|t$ if, and only if, $M(j) = t$.

The p seasonal factors describe a general pattern of period p; they may take arbitrary real values. For this reason, the seasonal pattern is termed **form-free**.

Definition 8.2. *The $p \times 1$ vector $\mathbf{E}_p$ and $p \times p$ **permutation** matrix $\mathbf{P}$ are defined as*

$$\mathbf{E}_p = (1, 0, \ldots, 0)',$$

and

$$\mathbf{P} = \begin{pmatrix} 0 & 1 & 0 & \cdots & 0 \\ 0 & 0 & 1 & \cdots & 0 \\ \vdots & \vdots & \vdots & \ddots & \vdots \\ 0 & 0 & 0 & \cdots & 1 \\ 1 & 0 & 0 & \cdots & 0 \end{pmatrix} = \begin{pmatrix} \mathbf{0} & \mathbf{I} \\ 1 & \mathbf{0}' \end{pmatrix}.$$

It is easy to see that $\mathbf{P}$ is p-cyclic, satisfying the equation

$$\mathbf{P}^{k+np} = \mathbf{P}^k, \qquad\qquad (k = 1, \ldots, p),$$

for any $n \geq 0$. In particular, $\mathbf{P}^{np} = \mathbf{I}$ for any such n.

At any time t, the current value of $g(.)$ is just ψ_j given by

$$g(t) = \mathbf{E}_p' \psi_t \tag{8.1a}$$

where $j = p|t$. Using the permutation matrix, it is also clear that the seasonal factors rotate according to

$$\psi_t = \mathbf{P}\psi_{t-1} \tag{8.1b}$$

for all $t \geq 0$. This relationship provides the initial step in constructing a DLM for pure seasonality. Suppose the desired forecast function is cyclical in the sense that $f_t(k) = g(t+k)$. Equations (8.1) imply that the forecast function has the form of that in a Time Series DLM with regression vector $\mathbf{F} = \mathbf{E}_p$ and system matrix $\mathbf{G} = \mathbf{P}$. It can be verified that $\mathbf{P}$ has p distinct eigenvalues given by the p roots of unity, $\exp(2\pi i j/p)$ for $j = 1, \ldots, p$. It follows from Section 5.3 that the model

$$\{\mathbf{E}_p, \mathbf{P}, ., .\},$$

and any observable, similar model, will produce the desired forecast function.

8.3   FORM-FREE SEASONAL FACTOR DLMS

8.3.1   General models

Definition 8.3. *The canonical, form-free seasonal factor DLM having forecast function of period $p \geq 1$ is defined as*

$$\{\mathbf{E}_p, \mathbf{P}, V_t, \mathbf{W}_t\}$$

for any appropriate variances V_t and $\mathbf{W}_t$.

The canonical model can be written as

$$
\begin{aligned}
\text{Observation equation:} \quad & Y_t = \mathbf{E}_p' \boldsymbol{\psi}_t + \nu_t \\
\text{System equation:} \quad & \boldsymbol{\psi}_t = \mathbf{P} \boldsymbol{\psi}_{t-1} + \boldsymbol{\omega}_t
\end{aligned}
\qquad (8.2)
$$

where the state parameter vector $\boldsymbol{\psi}_t$ relates to the seasonal factors as in the last section. We note the following points.

(1) Let $\boldsymbol{\psi}_t' = (\psi_{t0}, \psi_{t1}, \dots, \psi_{t,p-1})$ with posterior mean given by $\mathrm{E}[\boldsymbol{\psi}_t \mid D_t] = \mathbf{m}_t$ as usual, and $\mathbf{m}_t' = (m_{t0}, \dots, m_{t,p-1})$. The forecast function for $k \geq 0$ is simply

$$f_t(k) = \mathbf{E}_p' \mathbf{P}^k \mathbf{m}_t = m_{tj}, \qquad \text{where} \quad j = p|k.$$

(2) The model is observable since, for $k = 0, \dots, p - 1$, the row vector $\mathbf{E}_p' \mathbf{P}^k$ has unity as the $(k + 1)^{st}$ element, being zero elsewhere. It follows that the observability matrix is just the identity, $\mathbf{T} = \mathbf{I}$.

(3) Any similar model will be called a form-free seasonal factor DLM.

8.3.2   Closed, constant models

The structure of the form-free model is easily appreciated in the special case of a closed, constant model $\{\mathbf{E}_p, \mathbf{P}, V, V\mathbf{W}\}$. The equations (8.2) reduce to the components

$$
\begin{aligned}
Y_t &= \psi_{t0} + \nu_t, \\
\psi_{tr} &= \psi_{t-1,r+1} + \omega_{tr}, \qquad (r = 0, \dots, p - 2), \\
\psi_{t,p-1} &= \psi_{t-1,0} + \omega_{t,p-1},
\end{aligned}
$$

where $\nu_t \sim \mathrm{N}[0, V]$ and $\boldsymbol{\omega}_t = (\omega_{t0}, \dots, \omega_{t,p-1})' \sim \mathrm{N}[0, V\mathbf{W}]$ with the usual independence assumptions. Consider the current seasonal level ψ_{t0} and suppose that, without loss of generality, the current time point is labelled M(0). Having observed Y_t, no further observations are made directly on the seasonal factor for times labelled M(0) until p observations hence, at time $t + p$. Over that full period, the factors change stochastically via the addition of the p evolution errors. Any information gained about this particular seasonal factor is due entirely to the correlation structure in $\mathbf{W}$. Clearly, therefore, the form of this matrix is of crucial importance. Generally we specify it according to the discount principle as in Section 6.3, developed further in this particular model below. Initially, however, we explore the theoretical structure of the model a little further with a particularly simple diagonal form for the evolution variance matrix. Throughout note that V is supposed known. If not, then the usual learning procedure applies without altering the basic components of the following discussion.

EXAMPLE 8.1: Special case of $\mathbf{W} = W\mathbf{I}$. With the individual errors ω_{tr} assumed uncorrelated for each time t, the model reduces to a collection of p simple, first-order polynomial models $\{1, 1, V, pVW\}$. For clarity, suppose that $p = 12$ and our data is monthly over the year, M(0) being January, and so on. Then each year in January, one observation is made on the current January level ψ_{t0}, the level then evolving over the next full year by the addition of 12 uncorrelated evolution error terms, each being distributed as N$[0, VW]$. The net result is that

$$\psi_{t+p,0} = \psi_{t0} + \omega_t^*$$

with $\omega_t^* \sim$ N$[0, 12VW]$. It is clear that the only link between the seasonal factors is that deriving from the initial prior covariance terms at time 0. The effect of this initial prior decays with time and so, for simplicity, assume that $(\psi_0 \mid D_0) \sim$ N$[\mathbf{m}_0, V\mathbf{C}_0]$ where $\mathbf{m}_0 = m_0\mathbf{1}$ and $\mathbf{C}_0 = C_0\mathbf{I}$ for some scalars m_0 and C_0. The following results may now be simply derived by applying the updating recurrences and limiting results from the closed, constant, first-order polynomial model of Section 2.4.

(1) For each t, $(\psi_t \mid D_t) \sim$ N$[\mathbf{m}_t, V\mathbf{C}_t]$ with mean vector $\mathbf{m}_t = (m_{t0}, \ldots, m_{t,p-1})'$ and variance matrix $\mathbf{C}_t = \text{diag}(C_{t0}, \ldots, C_{t,p-1})$.

(2) Suppose that $t = np$ for some integer $n \geq 1$ so that n full periods have passed. Then the current time label is M(0) and the updating equations for the corresponding seasonal factor

ψ_{t0} are

$$m_{t0} = m_{t-p+1,0} + A_t e_t,$$
$$C_{t0} = A_t,$$

where $e_t = Y_t - m_{t-p+1,0}$ and $A_t = R_t/(R_t + 1)$ with $R_t = C_{t-p+1,0} + pW$. Similar comments apply to the next time intervals $M(1), M(2), \ldots$, with the subscript 0 updated to $1, 2, \ldots$, respectively.

(3) The limiting behaviour follows by applying the results of Section 2.4 to each of the p models $\{1, 1, V, pVW\}$ here. Thus

$$\lim_{t \to \infty} A_t = \lim_{t \to \infty} C_{t0} = A = \frac{pW}{2}\left[(1 + 4/pW)^{1/2} - 1\right],$$

and

$$\lim_{t \to \infty} R_t = R = A + pW = A/(1 - A).$$

It then follows that, for each $r = 0, \ldots, p - 1$, $\lim_{t \to \infty} C_{tr} = A + rW$, and this implies $(\psi_{tr} \mid D_t) \sim N[m_{tr}, V(A + rW)]$ as $t \to \infty$.

(4) The limiting analysis is equivalent to that obtained by applying the discount approach with a single discount factor $\delta = 1 - A$, leading to a model $\{\mathbf{E}_p, \mathbf{P}, V, V\mathbf{W}_t\}$ with

$$\mathbf{W}_t = (\mathbf{P}\mathbf{C}_{t-1}\mathbf{P}')(1 - \delta)/\delta.$$

Following comment (4) above, it may also be shown that the limiting forecast function is equivalent to that derived using exponentially weighted regression techniques. This suggests a rephrasing of the form-free model and the use of more than one discount factor to structure the evolution variance matrix. Underlying this suggestion is the idea that we generally view a seasonal pattern as relatively stable over time relative to an underlying, deseasonalised level. Harrison (1965) discusses this idea. Thus the seasonal factors may be decomposed into an underlying level, plus seasonal deviations from this level. This allows us the flexibility to model changes in two components separately using the ideas of component, or block, models described in Chapter 6. We now describe this decomposition.

8.4 FORM-FREE SEASONAL EFFECTS DLMS

8.4.1 Introduction and definition

In decomposing a set of seasonal factors into non-seasonal level plus
seasonal deviation from that level, we refer to the seasonal deviations
as *seasonal effects*. Although specifically concerned with seasonality,
many of the points here carry over to more general effects models.
Practitioners familiar with standard statistical models will appreciate the idea of descriptions in terms of an overall mean for observations plus treatment, block and other effects. The effects for any
treatment or block will be subject to an identifiability constraint that
is imposed in one of several forms, either by aliasing one of the effects
or by constraining an average of the effects. Perhaps the commonest
such constraint is zero-sum constraint. The analogy for seasonality
is that the seasonal deviations from the underlying level sum to zero
in any full period. This constraint is used here although others are
possible and may typically be obtained by linear transformation of
the seasonal factors. In generalised linear modelling using GLIM, for
example, the effect at one chosen level, the first level by default, is
constrained to be zero (Baker and Nelder, 1978).

Initially we suppose that the underlying level of the series is zero
so that the seasonal factors already sum to zero. By superposition,
this special case then provides the constrained, seasonal effects component for a series with non-zero level.

Definition 8.4. *Any DLM* $\{\mathbf{E}_p, \mathbf{P}, V_t, \mathbf{W}_t\}$ *whose state vector* $\boldsymbol{\phi}_t$
satisfies the zero-sum constraint $\mathbf{1}'\boldsymbol{\phi}_t = 0$, *for all t, is called a* **form-
free seasonal effects** *DLM. The seasonal effects* ϕ_{tj} *represent
seasonal deviations from the mean 0 of the series, and are simply
constrained seasonal factors.*

In terms of equations, such a model has the form given in (8.2),
with the parameter notation changed from $\boldsymbol{\psi}$ to $\boldsymbol{\phi}$, of course, and
the addition of the constraint

$$\mathbf{1}'\boldsymbol{\phi}_t = 0. \tag{8.3}$$

8.4.2 Imposing constraints

The constraint (8.3) leads to requirements on the model and prior
distributions that have the following forms.

   (1) **Initial prior.** The constraint applies at time $t = 0$ and so
      must be satisfied by the initial prior $(\boldsymbol{\phi}_0 \mid D_0) \sim \mathrm{N}[\mathbf{m}_0, \mathbf{C}_0]$.

Since $(\mathbf{1}'\boldsymbol{\phi}_0 \mid D_0) \sim \mathrm{N}[\mathbf{1}'\mathbf{m}_0, \mathbf{1}'\mathbf{C}_0\mathbf{1}]$ then, necessarily, we require

$$\mathbf{1}'\mathbf{m}_0 = 0$$
$$\mathbf{C}_0\mathbf{1} = 0, \tag{8.4}$$

in order that (8.3) be satisfied with probability one. Thus the initial prior means must sum to zero, as must the elements of each row (and each column) of the initial prior variance matrix.

(2) **Evolution variances.** Since, for all t, $\boldsymbol{\omega}_t$ is an increment to the seasonal effects vector, then clearly it too should satisfy the zero-sum constraint. Since the error already has zero mean, then it will satisfy the constraint with probability one if, in addition,

$$\mathbf{W}_t\mathbf{1} = 0. \tag{8.5}$$

If, more generally, the evolution error has a non-zero mean, then of course the constraint must apply to the mean.

Theorem 8.1. *In the form-free seasonal effects model subject to the constraints $\mathbf{1}'\mathbf{m}_0 = 0$, $\mathbf{C}_0\mathbf{1} = \mathbf{0}$ and $\mathbf{W}_t\mathbf{1} = \mathbf{0}$ for all t, the posterior distribution for the state vector at time t, $(\boldsymbol{\phi}_t \mid D_t) \sim \mathrm{N}[\mathbf{m}_t, \mathbf{C}_t]$ is such that*

$$\mathbf{1}'\mathbf{m}_t = 0 \qquad and \qquad \mathbf{C}_t\mathbf{1} = \mathbf{0}.$$

Thus $\mathbf{1}'\boldsymbol{\phi}_t = 0$ with probability one for all t.

Proof: The proof is by induction. If the constraints apply to the posterior at time $t-1$ we have $\mathbf{1}'\mathbf{m}_{t-1} = 0$ and $\mathbf{C}_{t-1}\mathbf{1} = \mathbf{0}$. Proceeding to the prior at time t we have, as usual, $(\boldsymbol{\phi}_t \mid D_t) \sim \mathrm{N}[\mathbf{a}_t, \mathbf{R}_t]$ with $\mathbf{a}_t = \mathbf{P}\mathbf{m}_{t-1}$ and $\mathbf{R}_t = \mathbf{P}\mathbf{C}_{t-1}\mathbf{P}' + \mathbf{W}_t$. Now, since $\mathbf{P}\mathbf{1} = \mathbf{1}$, it follows that these prior moments also satisfy the constraints, with $\mathbf{1}'\mathbf{a}_t = 0$ and $\mathbf{R}_t\mathbf{1} = \mathbf{0}$. Updating to the posterior at t gives $\mathbf{m}_t = \mathbf{a}_t + \mathbf{R}_t\mathbf{E}_p Q_t^{-1} e_t$ and $\mathbf{C}_t = \mathbf{R}_t - \mathbf{R}_t\mathbf{E}_p\mathbf{E}_p'\mathbf{R}_t'Q_t^{-1}$, whence, directly, $\mathbf{1}'\mathbf{m}_t = 0$ and $\mathbf{C}_t\mathbf{1} = \mathbf{0}$. Since, by assumption, the constraints apply at $t = 0$, it follows by induction that they apply for all time as required.

$$\diamond$$

The conditions (8.4) and (8.5) therefore provide consistency with the requirement (8.3) for all time as information is sequentially processed. Two problems remain. The first is to ensure that the initial prior satisfies (8.4), the second is to design a suitable sequence of evolution variance matrices satisfying (8.5).

8.4.3 Constrained initial priors

It is common for practitioners to specify initial priors about the seasonal effects individually, providing essentially only the marginal priors $(\phi_{0j} \mid D_0)$ for $j = 0, \ldots, p - 1$. Sometimes it may be possible to produce covariance terms for the full, joint prior by further elicitation. Quite often if not typically, however, this will be difficult and we may be content with the *independently* specified marginals plus the constraint (8.3). In such cases, it is necessary to use this partial information to derive a fully coherent, joint prior satisfying (8.4). This may be done formally within a general framework detailed here. Suppose that we start with an initial prior of the form

$$(\phi_0 \mid D_0) \sim N[\mathbf{m}_0^*, \mathbf{C}_0^*], \tag{8.6}$$

that may or may not satisfy the constraint (8.4). Note that the initial (mis-)specification of independent marginals discussed above would imply that the variance matrix here is diagonal. We proceed as follows.

Theorem 8.2. *Imposing the constraint $\mathbf{1}'\phi_0 = 0$ on the prior in (8.6) leads to the revised, joint prior*

$$(\phi_0 \mid D_0) \sim N[\mathbf{m}_0, \mathbf{C}_0],$$

where

$$\mathbf{m}_0 = \mathbf{m}_0^* - \mathbf{A}(\mathbf{1}'\mathbf{m}_0^*)/(\mathbf{1}'\mathbf{C}_0^*\mathbf{1})$$

and

$$\mathbf{C}_0 = \mathbf{C}_0^* - \mathbf{A}\mathbf{A}'/(\mathbf{1}'\mathbf{C}_0^*\mathbf{1}),$$

with $\mathbf{A} = \mathbf{C}_0^*\mathbf{1}$.

Proof: Under (8.6), the joint distribution of the initial seasonal effects and their total $\theta = \mathbf{1}'\phi_0$ is the singular normal given by

$$\begin{pmatrix} \phi_0 \\ \theta \end{pmatrix} \middle| D_0 \Biggr) \sim N\left[\begin{pmatrix} \mathbf{m}_0^* \\ \mathbf{1}'\mathbf{m}_0^* \end{pmatrix}, \begin{pmatrix} \mathbf{C}_0^* & \mathbf{A} \\ \mathbf{A}' & \mathbf{1}'\mathbf{C}_0^*\mathbf{1} \end{pmatrix} \right].$$

Thus, given θ, the conditional distribution of ϕ_0 may be obtained, using normal theory results (Section 16.2, Chapter 16), as

$$(\phi_0 \mid \theta, D_0) \sim N[\mathbf{m}_0^* + \mathbf{A}(\theta - \mathbf{1}'\mathbf{m}_0^*)/(\mathbf{1}'\mathbf{C}_0^*\mathbf{1}), \ \mathbf{C}_0^* - \mathbf{A}\mathbf{A}'/(\mathbf{1}'\mathbf{C}_0^*\mathbf{1})],$$

for any θ. Conditioning on the value $\theta = 0$ implied by the zero-sum constraint leads to the stated result.

◇

Thus, whatever initial prior is elicited, this Theorem should be applied to ensure compliance with the constraint. Often the prior variance $\mathbf{C}_0^*$ will be diagonal as mentioned earlier, sometimes a multiple of the identity. Notice that no changes are made to the prior if the constraints are already satisfied since then $\mathbf{A} = \mathbf{0}$. Otherwise notice that, in a general sense, the total variation in the constrained prior will always be less than that originally specified in the unconstrained prior, the difference being removed by imposing the deterministic constraint.

EXAMPLE 8.2. As an example, suppose that $p = 4$, as for quarterly data, and that the initial estimates of the four seasonal factors are $\mathbf{m}_0^* = (10,\ 5,\ 0, -7)'$, the factors being uncorrelated with standard deviations of unity. Thus $\mathbf{C}_0^*$ is the identity matrix. Imposing the constraints leads to $\mathbf{m}_0 = (8,\ 3,\ -2,\ -9)'$ and

$$\mathbf{C}_0 = \begin{pmatrix} 0.75 & -0.25 & -0.25 & -0.25 \\ -0.25 & 0.75 & -0.25 & -0.25 \\ -0.25 & -0.25 & 0.75 & -0.25 \\ -0.25 & -0.25 & -0.25 & 0.75 \end{pmatrix}.$$

In practice, it may be desired that the trace of $\mathbf{C}_0$ should equal that of $\mathbf{C}_0^*$, thus retaining the original total amount of variation. This can always be achieved by simply scaling $\mathbf{C}_0$ appropriately. In this case, the original trace is 4, reduced to 3 following the imposition of the constraint. Thus the original trace is retained by multiplying the above matrix by $4/3$.

8.4.4   Constrained evolution variances

There are a variety of structures available for the evolution variance matrices that satisfy the constraint (8.5). One such, used in Harrison and Stevens (1976b), is based on the assumption that $\boldsymbol{\omega}_t = \omega_t \mathbf{a}$ where $\mathbf{a}' = (p - 1, -1, \ldots, -1)$ and $\omega_t \sim \mathrm{N}[0, W_t]$ for some $W_t > 0$.

The (questionable) basis for this structure is that information from the current observation relates mainly to the current seasonal effect, equally affecting the others only via the implied renormalisation in applying the zero-sum constraint. In this case, $\mathbf{W}_t = W_t \mathbf{aa}'$ and (8.5) is satisfied since $\mathbf{1}'\mathbf{a} = 0$. An alternative structure is given by starting with $\mathbf{W}_t = W_t \mathbf{I}$ and then following the argument in Theorem 8.2 to transform to a variance matrix that complies with the zero-sum costraint. It is easily seen that this leads to the form

$$\mathbf{W}_t = W_t(p\mathbf{I} - \mathbf{11}').$$

These two forms, and others, have been fairly widely used in practice historically. They are not, however, recommended for general application, being specifically designed to represent particular forms of change in the seasonal effects over time. Instead we recommend, as usual, the approach using discount concepts. For a single discount factor δ, possibly depending on t, the discount idea of Section 6.3 applied to the entire seasonal effects vector leads to

$$\mathbf{W}_t = \mathbf{P}\mathbf{C}_{t-1}\mathbf{P}'(1 - \delta)/\delta, \qquad (8.7)$$

which always satisfies (8.5) since $\mathbf{P}'\mathbf{1} = \mathbf{1}$ and $\mathbf{C}_{t-1}\mathbf{1} = \mathbf{0}$.

Definition 8.5. *The single discount, form-free seasonal effects DLM is defined by Definition 8.4 with evolution variance sequence defined via (8.7), for some discount factor δ, $(0 < \delta \leq 1)$.*

8.5 TREND/FORM-FREE SEASONAL EFFECTS DLMS

As mentioned earlier, the key reason for considering the constrained effects model in the last section is that it provides a seasonal component describing seasonal deviations from a non-seasonal level or trend, using the principle of superposition. The two most important and widely used such models are defined here.

8.5.1 First-order polynomial/seasonal effects model

Definition 8.6. *Any model $\{\mathbf{F}, \mathbf{G}, ., .\}$, with $\mathbf{F}' = (1, \mathbf{E}'_p)$ and $\mathbf{G} = $ block diag$[1, \mathbf{P}]$, is defined as a* **first-order polynomial trend/ form-free seasonal factor** *model. With state vector $\boldsymbol{\theta}_t = (\mu_t, \boldsymbol{\phi}'_t)'$,* *the model is a* **first-order polynomial trend/form-free seasonal effects** *model if the seasonal component state vector ϕ_t satisfies the zero-sum constraint $\mathbf{1}'\phi_t = 0$ for all t.*

Such models are comprised of the superposition of a first-order polynomial model for the non-seasonal trend with a seasonal component. It is easily seen that the seasonal factor model is unobservable, but is constrained observable when subject to zero-sum constraint on the seasonal factor vector, thus being equivalent to the observable effects model. The forecast function has the form

$$f_t(k) = m_t + h_{tj}$$

where $j = p|k$. Here m_t is the expected value of the non-seasonal (or deseasonalised) level of the series, and h_{tj} is the expected seasonal deviation from this underlying level at time $t + k$, with time label $M(j)$. The model may be written as

$$Y_t = \mu_t + \phi_{t0} + \nu_t,$$
$$\mu_t = \mu_{t-1} + \omega_t,$$
$$\phi_{tr} = \phi_{t-1,r+1} + \omega_{tr},$$
$$(r = 0, \ldots, p - 2),$$
$$\phi_{t,p-1} = \phi_{t-1,0} + \omega_{t,p-1}.$$

When subject to the zero-sum constraint it should be clear that this model is similar to the seasonal factor model of Section 8.3. To see this note that, if $\mathbf{S}$ is the $p \times (p+1)$ matrix $\mathbf{S} = [\mathbf{1}, \mathbf{I}]$, then $\psi_t = \mathbf{S}\phi_t$ is the vector of seasonal factors

$$\psi_{tj} = \mu_t + \phi_{tj}, \qquad\qquad (j = 0, \ldots, p - 1).$$

Given the constraint, it follows that

$$p^{-1} \sum_{j=1}^{p} \psi_{tj} = \mu_t,$$

for all t so that μ_t represents the average of the seasonal levels. In the ψ_t parameterisation we have the factor model (8.2). By construction it follows that this model represents the linear composition of the seasonal effects model with a first-order polynomial. Note also that, starting with the seasonal factors, the transformation to the effects model is simply

$$\begin{pmatrix} \mu_t \\ \phi_t \end{pmatrix} = \mathbf{U}\psi_t,$$

where $\mathbf{U}$ is the $(p+1) \times p$ matrix

$$\mathbf{U} = \begin{pmatrix} \frac{1}{p} & \frac{1}{p} & \cdots & \frac{1}{p} \\ 1 - \frac{1}{p} & -\frac{1}{p} & \cdots & -\frac{1}{p} \\ \vdots & \vdots & \ddots & \vdots \\ -\frac{1}{p} & -\frac{1}{p} & \cdots & 1 - \frac{1}{p} \end{pmatrix}.$$

8.5.2 Second-order polynomial/seasonal effects model

Definition 8.7. *Any model* $\{\mathbf{F}, \mathbf{G}, ., .\}$, *with* $\mathbf{F}' = (\mathbf{E}_2', \mathbf{E}_p')$ *and* $\mathbf{G} = $ *block* $diag[\mathbf{J}_2(1), \mathbf{P}]$, *is defined as a* **second-order polynomial trend/form-free seasonal factor** *model. With state vector*

$$\boldsymbol{\theta}_t = (\mu_t, \beta_t, \phi_t')',$$

the model is a **second-order polynomial trend/form-free seasonal effects** *model if the seasonal component vector satisfies the usual zero-sum constraint* $\mathbf{1}'\phi_t = 0$ *for all* t.

As in the previous section, the effect models are observable, being the constrained observable versions of the unobservable factor models. Also such DLMs are obtained from the superposition of a second-order polynomial model for the non-seasonal trend with a seasonal component. The forecast function in the practically important effects model has the form

$$f_t(k) = (m_t + kb_t) + h_{tj}$$

where $j = p|k$ and

$$\sum_{j=0}^{p-1} h_{tj} = 0$$

for all t.

Further discussion of this class of models, easily the most widely used and useful class of Time Series DLMs, is deferred until applications in later chapters.

8.6 FOURIER FORM REPRESENTATION OF SEASONALITY

8.6.1 Introduction

Alternative approaches to representing cyclical patterns use linear combinations of periodic functions with particular forms. The particular approach we favour and develop here uses possibly the simplest and most natural class of periodic functions, namely trigonometric functions, leading to what we refer to as **Fourier form** representations of seasonality.

The main reasons for the use of form models, rather than the flexible and essentially unrestricted seasonal effects models, are economy and interpretation. In certain applications, observations result from sampling what is essentially a simple waveform. Sine/cosine waves are economic and natural representation of such patterns. Simple phenomena exhibiting such behaviour abound in electrical and electronic systems, geophysical studies, including earthquake tremors and marine depth soundings, many resulting from *pure* seasonal responses to the revolution of the earth about the sun. For example, almost all of the variation in average monthly British temperature about the annual mean may be characterised in terms of a single cosine wave of period 12. If such a representation is deemed acceptable, then it is defined in terms of only two quantities determining the phase and amplitude of the cosine waveform. By comparison, a seasonal effects model requires 11 parameters describing the monthly deviations from the annual mean; with weekly data it would require 51. The economy of form models is immediately apparent in such cases. More generally, a Fourier form model can provide an acceptable representation of an apparently erratic seasonal pattern whilst economising (perhaps not so drastically as in the above, extreme example) on parameters compared to an effects model. This may be of great benefit in stochastic models in particular since it leads to greater precision of inferences about the form of seasonality.

8.6.2 Fourier form representation of cyclical functions

Consider the cyclical function $g(t)$ of Section 8.2 defined via the seasonal factors $\psi_0, \ldots, \psi_{p-1}$. The basic result is that these factors can be written as a linear combination of trigonometric terms. This representation depends on the parity of the period p. Throughout, let $\alpha = 2\pi/p$.

(1) **Fourier representation: p odd**

If, for some integer $q > 0$, $p = 2q - 1$, then there exist p real numbers $a_0, a_1, \ldots, a_{q-1}; b_1, \ldots, b_{q-1}$ such that, for $j = 0, \ldots, p - 1$,

$$\psi_j = a_0 + \sum_{r=1}^{q-1} [a_r \cos(\alpha rj) + b_r \sin(\alpha rj)].$$

(2) **Fourier representation: p even**

If, for some integer $q > 0$, $p = 2q$, then there exist p real numbers $a_0, a_1, \ldots, a_q; b_1, \ldots, b_{q-1}$ such that, for $j = 0, \ldots, p - 1$,

$$\psi_j = a_0 + \sum_{r=1}^{q-1} [a_r \cos(\alpha rj) + b_r \sin(\alpha rj)] + a_q \cos(\pi j).$$

Notice that, defining $b_q = 0$, we can always write

$$\psi_j = a_0 + \sum_{r=1}^{q} [a_r \cos(\alpha rj) + b_r \sin(\alpha rj)]$$

whatever the parity of p, remembering that $a_q = 0$ also when p is odd.

(3) **Basic identities**

In dealing with trigonometric terms the following, basic mathematical identities are useful:

- For integer n and any x, $\cos(x + 2\pi n) = \cos(x)$ and $\sin(x + 2\pi n) = \sin(x)$;

- For each $r = 1, \ldots, q$,

$$\sum_{j=0}^{p-1} \cos(\alpha rj) = \sum_{j=0}^{p-1} \sin(\alpha rj) = 0;$$

- For integers h and k,

$$\sum_{j=0}^{p-1} \cos(\alpha hj)\sin(\alpha kj) = 0$$

$$\sum_{j=0}^{p-1} \cos(\alpha hj)\cos(\alpha kj) = \begin{cases} 0, & h \neq k, \\ p, & h = k = \frac{p}{2}, \\ \frac{p}{2}, & h = k \neq \frac{p}{2}; \end{cases}$$

$$\sum_{j=0}^{p-1} \sin(\alpha hj)\sin(\alpha hj) = \begin{cases} 0, & h \neq k, \\ 0, & h = k = \frac{p}{2}, \\ \frac{p}{2}, & h = k \neq \frac{p}{2}. \end{cases}$$

(4) **Fourier coefficients**

The coefficients a_r and b_r are obtained simply by multiplying the ψ_j by the appropriate cos/sin terms and summing over the index $j = 0,\ldots,p-1$. Then, using the basic identities above, we have:

$$a_0 = p^{-1} \sum_{j=0}^{p-1} \psi_j,$$

$$a_q = p^{-1} \sum_{j=0}^{p-1} \psi_j \cos(\pi j),$$

and, for $r = 1,\ldots,q-1$,

$$a_r = 2p^{-1} \sum_{j=0}^{p-1} \psi_j \cos(\alpha r j),$$

$$b_r = 2p^{-1} \sum_{j=0}^{p-1} \psi_j \sin(\alpha r j).$$

Note that the seasonal effects can be obtained as the seasonal factors minus their mean a_0, via $\phi_j = \psi_j - a_0$, for $j = 0,\ldots,p-1$. Clearly $a_0 = 0$ if the factors are already constrained to have zero-sum.

(5) **Notation and terminology**

The quantities a_r and b_r are called the **Fourier coefficients** of the Fourier form representation of the p seasonal factors. Define the quantities S_{rj}, as functions of the time index $j = 0, \ldots, p - 1$, via

$$S_{rj} = a_r \cos(\alpha r j) + b_r \sin(\alpha r j), \qquad (r = 1, \ldots, q - 1),$$

and

$$S_{qj} = \begin{cases} a_q \cos(\pi j) = (-1)^j, & p = 2q, \\ 0, & p = 2q - 1. \end{cases}$$

The quantity S_{rj} is the value of the rth **harmonic** of the Fourier description of the cyclical function $g(.)$. Notice that

$$S_{rj} = A_r \cos(\alpha r j + \gamma_r)$$

where $A_r = (a_r^2 + b_r^2)^{1/2}$ and $\gamma_r = \arctan(b_r/a_r)$. The harmonic may thus be described in terms of either (a_r, b_r) or (A_r, γ_r). A_r is the **amplitude** of the harmonic, the maximum value taken by the S_{rj} over $j = 0, \ldots, p - 1$. The **phase** γ_r of the harmonic determines the position of the maximum. If $p = 2q$, the qth harmonic is defined by a_q and $b_q = 0$ so that $A_q = |a_q|$ and $\gamma_q = 0$, is called the **Nyquist** harmonic. The **frequency** of the rth harmonic is defined as $\alpha r = 2\pi r / p$, with $\alpha \pi$ being the Nyquist frequency, and the **cycle length** of the harmonic is p/r. In particular, the first harmonic is called the **fundamental** harmonic, having fundamental frequency α and fundamental cycle length p. Notice that the rth harmonic completes exactly r full cycles for each single, complete cycle of the fundamental harmonic.

(6) **Seasonal variation**

Using the basic trigonometric identies it is easily shown that

$$p^{-1} \sum_{j=0}^{p-1} (\psi_j - a_0)^2 = \frac{1}{2} \sum_{r=1}^{q-1} (a_r^2 + b_r^2) + a_q^2.$$

Hence the sum of the squares of the seasonal factors about their arithmetic mean a_0 is simply proportional to

$$\frac{1}{2} \sum_{r=1}^{q-1} A_r^2 + A_q^2$$

with, of course, $A_q = 0$ if $p = 2q$, even. The contribution of the rth harmonic to the overall seasonal variation in $g(.)$ can be assessed with reference to the contribution to the sum of squares here and so is directly related to the amplitude.

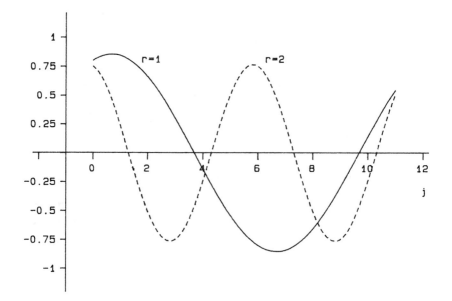

Figure 8.1a.   Harmonics 1 and 2.

EXAMPLE 8.3. As a simple illustration with $p = 12$, let the function $g(.)$ be defined by seasonal factors

$$\psi' = (1.65, 0.83, 0.41, -0.70, -0.47, 0.40, -0.05, -1.51,$$
$$-0.19, -1.02, -0.87, 1.52).$$

The factors are plotted, as vertical lines against $j = 0, \ldots, 11$, in Figure 8.1(d). Figures 8.1(a), (b) and (c) display the corresponding 6 harmonics S_{rj} as functions of j for $0 \leq j \leq 11$. Figure 8.1(d) displays the full Fourier composition of the seasonal factors as a function of t, clearly coinciding with the seasonal factors at the integer values. The mean here is $a_0 = 0$, the factors already constrained to zero-sum as can be easily checked, the other Fourier coefficients, the amplitudes and phases being easily derived, and are displayed in the following table.

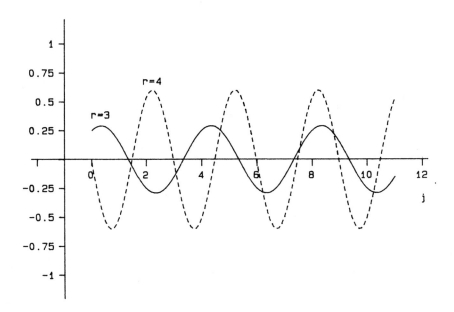

Figure 8.1b. Harmonics 3 and 4.

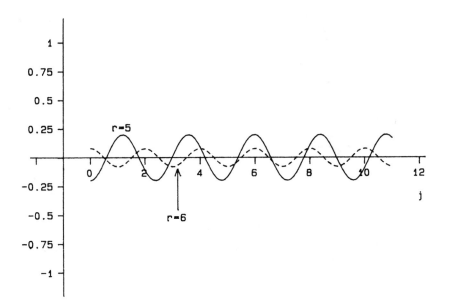

Figure 8.1c. Harmonics 5 and 6.

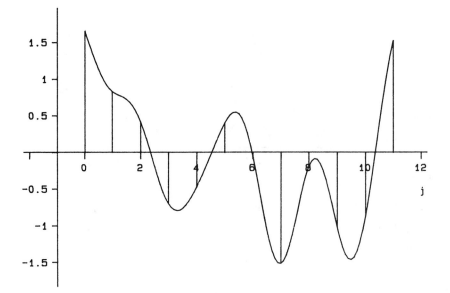

Figure 8.1d.   Seasonal factors: sum of harmonics.

Harmonic r	1	2	3	4	5	6
a_r	0.80	0.75	0.25	−0.03	−0.20	0.08
b_r	0.30	−0.15	0.15	−0.60	0.01	0.00
Amplitude A_r	0.85	0.76	0.29	0.60	0.20	0.08
Phase γ_r	0.36	−0.20	0.55	1.53	−0.04	0.00

In the Fourier form representation we have $g(t)$ expressed as the sum of harmonics so that, for all integers $t > 0$, setting $j = p|t$, we have

$$g(t) = \psi_j = a_0 + \sum_{r=1}^{q} S_{rj}, \qquad (8.8)$$

with the final term given by

$$S_{qj} = \begin{cases} 0, & p = 2q - 1, \text{ odd}; \\ a_q(-1)^j & p = 2q,, \text{ even}. \end{cases} \qquad (8.9)$$

Notice that this representation makes use of the fact the component harmonics are cyclical, with

$$S_{rt} = S_{rj}, \qquad\qquad (r = 1, \ldots, q)$$

where $j = p|t$. This form of $g(t)$ may now be written in a determin-
istic DLM representation, as follows.

Recall from Definition 5.7 of Chapter 5 the **harmonic matrix**

$$\mathbf{J}_2(\lambda, \omega) = \lambda \begin{pmatrix} \cos(\omega) & \sin(\omega) \\ -\sin(\omega) & \cos(\omega) \end{pmatrix}$$

defined for all real ω and $\lambda \neq 0$. This matrix is of full rank, being non-
singular with determinant unity. In addition, simple trigonometric
identities can be used to verify that $\mathbf{J}_2(\lambda, \omega)$ is p-cyclic, with

$$\mathbf{J}_2(\lambda, \omega)^{k+np} = \mathbf{J}_2(\lambda, \omega)^k.$$

This harmonic matrix is used in the DLM construction in the following, particular cases.

Definition 8.8. *For $r = 1, \ldots, q - 1$ define the 2×2 non-singular
matrices $\mathbf{G}_r$ and the 2×1 vectors $\mathbf{a}_r$ by*

$$\mathbf{G}_r = \mathbf{J}_2(1, \omega r) = \begin{pmatrix} \cos(\omega r) & \sin(\omega r) \\ -\sin(\omega r) & \cos(\omega r) \end{pmatrix}$$

and $\mathbf{a}_r = (a_r, b_r)'$, where, as usual, $\omega = 2\pi/p$.

It follows from the p-cyclicity of $\mathbf{J}_2(1, \omega)$ that each $\mathbf{G}_r$ is p-cyclic.
In particular, for any positive integer t, setting $j = p|t$ we have

$$\mathbf{G}_r^t = \begin{pmatrix} \cos(\omega r t) & \sin(\omega r t) \\ -\sin(\omega r t) & \cos(\omega r t) \end{pmatrix} = \begin{pmatrix} \cos(\omega r j) & \sin(\omega r j) \\ -\sin(\omega r j) & \cos(\omega r j) \end{pmatrix} = \mathbf{G}_r^j.$$

In addition, for all $j \geq 0$ we have

$$\mathbf{G}_r^j \mathbf{a}_r = \begin{pmatrix} a_r \cos(\omega r j) + b_r \sin(\omega r j) \\ -a_r \sin(\omega r j) + b_r \cos(\omega r j) \end{pmatrix}, \qquad (r = 1, \ldots, q - 1).$$

The first element of this vector is just the rth harmonic of $g(t)$ at
time $t = j$, namely S_{rj}, so that, using the vector $\mathbf{E}_2 = (1, 0)'$, we
have

$$S_{rj} = \mathbf{E}_2' \mathbf{G}_r^j \mathbf{a}_r, \qquad (r = 1, \ldots, q - 1).$$

In the case of even period, $p = 2q$, the Nyquist harmonic has the
simpler representation in (8.9). From these representations of the

harmonics, it follows that the functions S_{rt} of time t are essentially forecast functions of DLMs of the form

$$\{\mathbf{E}_2, \mathbf{G}_r, ., .\},$$

when $r < q$. The Nyquist component is similar to that of models of the form $\{1, -1, ., .\}$. Since $g(t)$ is expressed as a sum of the harmonics, it follows that $g(t)$ itself may be obtained as the forecast function of a DLM formed from the superposition of any such DLMs for the harmonic components.

8.6.3 Harmonic component DLMs

Definition 8.9. *For any* $\omega \in (0, \pi)$ *let* $c = \cos(\omega)$, $s = \sin(\omega)$ *and*

$$\mathbf{G} = \mathbf{J}_2(1, \omega) = \begin{pmatrix} c & s \\ -s & c \end{pmatrix}.$$

Then any model $\{\mathbf{E}_2, \mathbf{G}, ., .\}$ *is defined as a* **component harmonic DLM** *with frequency* ω. *For Nyquist frequency, the case* $\omega = \pi$, *the models have the form* $\{1, -1, ., .\}$.

The Nyquist component model is easily seen to be observable and requires little discussion, being simply understood. Some comments relevant to other component models as follows.

(1) The observability matrix of any such model is

$$\mathbf{T} = \begin{pmatrix} 1 & 0 \\ c & s \end{pmatrix}$$

which is of full rank. Component harmonic DLMs are thus observable.

(2) With parameterisation $\boldsymbol{\theta}_t = (\theta_{t1}, \theta_{t2})'$ at time t, having posterior mean $\mathrm{E}[\boldsymbol{\theta}_t \mid D_t] = \mathbf{m}_t = (a_t, b_t)'$, the forecast function has the form

$$f_t(k) = \mathbf{E}_2' \mathbf{G}^k \mathbf{m}_t = a_t \cos(\omega k) + b_t \sin(\omega k),$$

a sine/cosine wave of frequency ω, phase $\arctan(b_t/a_t)$ and amplitude $(a_t^2 + b_t^2)^{1/2}$.

(3) As a theoretical side issue, note that the eigenvalues of $\mathbf{G}$ are the distinct, complex conjugate pair $e^{\pm i\omega}$ and the forecast function can be written as

$$f_t(k) = \frac{1}{2}\left[(a_t - ib_t)\,e^{ik\omega} + (a_t + ib_t)\,e^{-ik\omega}\right].$$

This is, of course, real-valued but it is of passing interest to note that this is a special case of a general, complex valued forecast function

$$f_t(k) = d_t e^{ik\omega} + h_t e^{-ik\omega},$$

obtained from canonical, complex DLMs of the form

$$\left\{ \begin{pmatrix} 1 \\ 1 \end{pmatrix}, \begin{pmatrix} e^{ik\omega} & 0 \\ 0 & e^{-ik\omega} \end{pmatrix}, \cdot\,,\cdot\, \right\}.$$

Thus the component harmonic model may itself be regarded as comprising two sub-components. For real valued time series, however, the complex model is of little interest since the parameter vector, $\boldsymbol{\alpha}_t$, say, will be complex valued, requiring additional restrictions to ensure the two elements are conjugate. Thus we use the real canonical form in practice. The two are, of course, similar models, the invertible mapping between them defined via

$$\boldsymbol{\theta}_t = \begin{pmatrix} 1 & 1 \\ i & -i \end{pmatrix} \boldsymbol{\alpha}_t.$$

(4) A second theoretical point of interest concerns the case of a constant model

$$\{\mathbf{E}_2, \mathbf{G}, V, V\mathbf{W}\}.$$

Here it can be shown that the derived series

$$Z_t = Y_t - 2\cos(\omega)Y_{t-1} + Y_{t-2}$$

is a moving average process of order 2.

8.6.4 Full seasonal effects DLMs

As described in Section 8.6.2, any p seasonal factors/effects can be expressed in terms of harmonic components. In this section we specify DLMs having such cyclical forecast functions in terms of the equivalent collection of Fourier component DLMs, combined using the principle of superposition.

Definition 8.10. *Let the* $(p-1)$*-vector* **F** *and* $(p-1) \times (p-1)$ *matrix* **G** *be given by*

$$\mathbf{F}' = \begin{cases} (\mathbf{E}_2', \ldots, \mathbf{E}_2'), & p = 2q-1, \text{ odd;} \\ (\mathbf{E}_2', \ldots, \mathbf{E}_2', 1), & p = 2q, \text{ even} \end{cases}$$

and

$$\mathbf{G} = \begin{cases} block \ diag[\mathbf{G}_1, \ldots, \mathbf{G}_{q-1}], & p = 2q-1, \text{ odd;} \\ block \ diag[\mathbf{G}_1, \ldots, \mathbf{G}_{q-1}, -1], & p = 2q, \text{ even.} \end{cases}$$

Then any model $\{\mathbf{F}, \mathbf{G}, ., .,\}$ *is defined as a* **full seasonal effects Fourier form DLM.**

Any such model is formed from the superposition of the component harmonic DLMs

$$\{\mathbf{E}_2, \ \mathbf{G}_r \ , \ . \ , \ .\}$$

whose structure and features are described in the previous section. We note the following.

(1) The forecast function is just the sum of those of the components, having the form

$$f_t(k) = \sum_{r=1}^{q} S_{rk} = \sum_{r=1}^{q-1} [a_{tr} \cos(\omega r k) + b_{tr} \sin(\omega r k)] + a_{tq}(-1)^k,$$

where a_{tq} is zero for all t if $p = 2q - 1$, and the current posterior mean for the state vector is

$$E[\boldsymbol{\theta}_t \mid D_t] = \mathbf{m}_t = \begin{cases} (a_{t1}, b_{t1}; \ldots; a_{t,q-1}, b_{t,q-1})', & p = 2q-1; \\ (a_{t1}, b_{t1}; \ldots; a_{tq-1}, b_{t,q-1}; a_{t,q})', & p = 2q. \end{cases}$$

Here $\boldsymbol{\theta}_t$ is the vector of Fourier coefficients, the a and b values being their current estimates.

(2) By construction, the Fourier model automatically satisfies the requirement that the seasonal effects sum to zero and so no constraints are required on any of the component distributions.

(3) The observabilty matrix is

$$\mathbf{T} = \begin{pmatrix} \mathbf{F}' \\ \mathbf{F}'\mathbf{G} \\ \vdots \\ \mathbf{F}'\mathbf{G}^{p-2} \end{pmatrix}.$$

This square matrix has $p - 1$ rows $\mathbf{t}'_j$, $(j = 0, 1, \ldots, p - 2)$, determined as follows. Clearly $\mathbf{t}_0 = \mathbf{F}$. For $j = 1, \ldots, p - 2$, let

$$\mathbf{h}'_j = [\cos(\omega j), \sin(\omega j); \cos(2\omega j), \sin(2\omega j);$$
$$\ldots; \cos\{(q - 1)\omega j\}, \sin\{(q - 1)\omega j\}].$$

Then

$$\mathbf{t}'_j = \begin{cases} \mathbf{h}'_j, & p = 2q - 1, \text{ odd}; \\ [\,\mathbf{h}'_j, \cos(q\omega j)\,], & p = 2q, \text{ even}. \end{cases}$$

It can be verified that $\mathbf{T}$ has full rank and so the model is observable. This can be done either by using the trigonometric identities of Section 8.6.2, or by similarity with the complex canonical representation $\{\mathbf{1}, \mathbf{\Lambda}, ., .\}$ with system matrix $\mathbf{\Lambda}$ being diagonal with elements given by the p roots of unity.

(4) A single discount factor may be used to assign the evolution variances, agreeing with the use of a single factor in the form-free model. In some cases, however, it may be desirable that different discount factors should be used for different component harmonics, consistent with a view that some are more durable than others. Let v denote the number of harmonic components

$$v = \begin{cases} q - 1, & p = 2q - 1; \\ q, & p = 2q. \end{cases}$$

Following Section 6.3 of Chapter 6, at time t, the marginal variances of the components of the state vector are given by the relevant diagonal blocks of the full prior variance matrix $\mathbf{P}_t = \mathbf{G}\mathbf{C}_{t-1}\mathbf{G}'$. These are denoted by

$$\mathbf{P}_{1t}, \ldots, \mathbf{P}_{vt},$$

where, of course, the final component is a scalar when $p = 2q$. Following Definition 6.1, the most general framework then supposes that the evolution variances are defined as

$$\mathbf{W}_t = \text{block diag}\,[\mathbf{P}_{1t}(1 - \delta_1)/\delta_1, \ldots, \mathbf{P}_{vt}(1 - \delta_v)/\delta_v]$$

for v discount factors δ_r, $0 < \delta_r \leq 1$, $r = 1, \ldots, v$.

Given the Fourier DLM, the current state of information about
the Fourier coefficients at time t is described by the posterior mean
and variance matrix $\mathbf{m}_t$ and $\mathbf{C}_t$ respectively. (Here we assume a
normal posterior distribution, the case of a Student T being entirely
analogous). To make inferences about the seasonal effects themselves
requires a linear transformation from the Fourier coefficients. Con-
versely, if we start with a model in terms of seasonal effects directly,
inferences can be drawn about the Fourier representation of those
uncertain effects by a second, related linear transformation. This is
of key importance in practice when assessing the importance of the
contribution to an overall seasonal pattern of the various harmonic
components — see also the next section. These transformations are
as follows, the results being easily deduced from the basic trigonometric identities in Section 8.6.2.

- At any time t, the constrained seasonal effects vector $\boldsymbol{\phi}_t$ is
  obtained from the vector $\boldsymbol{\theta}_t$ of Fourier coefficients via the
  linear transformation

$$\boldsymbol{\phi}_t = \mathbf{L}\boldsymbol{\theta}_t \qquad (8.10)$$

where $\mathbf{L}$ is the $p \times (p-1)$ matrix

$$\mathbf{L} = \begin{pmatrix} \mathbf{T} \\ \mathbf{F'G}^{p-1}, \end{pmatrix}$$

with $\mathbf{F}$, $\mathbf{G}$ and $\mathbf{T}$ given above in the specification of the
Fourier model. Thus, given

$$(\boldsymbol{\theta}_t | D_t) \sim \mathrm{N}[\mathbf{m}_t,\ \mathbf{C}_t],$$

then

$$(\boldsymbol{\phi}_t | D_t) \sim \mathrm{N}[\mathbf{Lm}_t,\ \mathbf{LC}_t\mathbf{L'}],$$

with the analogous result (T replacing N) when the posteriors
are Student T rather than normal.

- The reverse transformation is given by

$$\boldsymbol{\theta}_t = \mathbf{H}\boldsymbol{\phi}_t$$

where $\mathbf{H}$ is the $(p-1) \times p$ matrix

$$\mathbf{H} = (\mathbf{L'L})^{-1}\mathbf{L'}.$$

Using the trigonometric identities, it is clear that $\mathbf{L'L}$ is always a diagonal matrix, leading to a simplification of the calculation of $\mathbf{H}$ in practice.

EXAMPLE 8.4. In the case $p = 4$, it follows that

$$\mathbf{F} = \begin{pmatrix} 1 \\ 0 \\ 1 \end{pmatrix}, \qquad \mathbf{G} = \begin{pmatrix} 0 & 1 & 0 \\ -1 & 0 & 0 \\ 0 & 0 & -1 \end{pmatrix}, \qquad \mathbf{T} = \begin{pmatrix} 1 & 0 & 1 \\ 0 & 1 & -1 \\ -1 & 0 & 1 \end{pmatrix}.$$

Thus

$$\mathbf{L} = \begin{pmatrix} 1 & 0 & 1 \\ 0 & 1 & -1 \\ -1 & 0 & 1 \\ 0 & -1 & -1 \end{pmatrix}$$

and

$$\mathbf{H} = \begin{pmatrix} 0.5 & 0 & -0.5 & 0 \\ 0 & 0.5 & 0 & -0.5 \\ 0.25 & -0.25 & 0.25 & -0.25 \end{pmatrix}.$$

8.6.5 Reduced Fourier form models

A modeller sometimes requires a model that describes a seasonal pattern in simpler terms than provided by a full p seasonal effects. The case of annual climatic cycles related to the earth's revolution about the sun are typical. In such cases, an economic representation in terms of only a few harmonics, or even a single component such as the fundamental harmonic with climatological data, may suffice. The construction of such reduced models follows directly from the discussion above of the use of the superposition to provide an overall seasonal pattern, rather than forming the full model using harmonics of all frequencies, the relevant frequencies only are included. For example, with monthly data having an annual cycle, the model

$$\left\{ \begin{pmatrix} \mathbf{E}_2 \\ \mathbf{E}_2 \end{pmatrix}, \begin{pmatrix} \mathbf{G}_1 & \mathbf{0} \\ \mathbf{0} & \mathbf{G}_4 \end{pmatrix}, \dots \right\}$$

provides a Fourier form representation assuming that only the first and fourth harmonics are important. This provides for a full annual cycle of fundamental frequency with a quarterly pattern superimposed. As in the previous section, different discount factors may be used for the two components to reflect the view that one is more durable than the other over time.

If appropriate, a reduced form model is more meaningful and economic than a full model, and will produce better forecasts in terms

of both accuracy and precision. Clearly if harmonic components that
are included in the model truly have little or no effect, then their
inclusion leads to a degrading of model performance via more variable forecasts and correlated forecast errors as the analysis gradually
estimates the associated, near zero coefficients. In a first-order polynomial trend/seasonal effects model, for example, the full Fourier
form representation leads to a simple updating for the current seasonal factor and no others. In contrast, a reduced form model here
would clearly update the entire seasonal pattern, revising forecasts
for the whole period rather than just individual time points. The
message then is that, in addition to considering whether a reduced
form model should be used for reasons related to the application
area, the significance of components in the model should be assessed
over time and possibly removed if deemed neglible. This does not
mean that such components will always be unnecessary, however,
since future changes in seasonal patterns may require that they be
brought back in as components at some future time.

As with full harmonic models, the Fourier coefficients in a reduced
model are related to the constrained seasonal effects by a couple of
linear transformations that allow inferences to be made about effects from coefficients, and vice versa. The orthogonality properties
underlying the trigonometric identities of Section 8.6.2 lead to essentially the same form for these linear transformations whether using
a reduced Fourier model or not. Thus, as in the previous Section,
if $\mathbf{F}$ and $\mathbf{G}$ are the regression vector and evolution matrix of the
(possibly reduced) Fourier model, define

$$\mathbf{L} = \begin{pmatrix} \mathbf{F}' \\ \mathbf{F}'\mathbf{G} \\ \vdots \\ \mathbf{F}'\mathbf{G}^{p-1} \end{pmatrix},$$

and

$$\mathbf{H} = (\mathbf{L}'\mathbf{L}')^{-1}\mathbf{L}'.$$

Note that, writing the dimension of the Fourier model as $n < p$, then
$\mathbf{L}$ is $p \times n$ and $\mathbf{H}$ is $n \times p$. The relationships then are simply

$$\psi_t = \mathbf{L}\theta_t \qquad \text{and} \qquad \phi_t = \mathbf{H}\psi_t,$$

for all t.

EXAMPLE 8.5. In the case $p = 4$, suppose just the first harmonic is used so that $n = 2$,

$$F = \begin{pmatrix} 1 \\ 0 \end{pmatrix}, \qquad G = \begin{pmatrix} 0 & 1 \\ -1 & 0 \end{pmatrix}, \qquad \text{and} \qquad T = \begin{pmatrix} 1 & 0 \\ 0 & 1 \end{pmatrix}.$$

Then

$$L = \begin{pmatrix} 1 & 0 \\ 0 & 1 \\ -1 & 0 \end{pmatrix}$$

and

$$H = \begin{pmatrix} 0.5 & 0 & -0.5 \\ 0 & 0.5 & 0 \end{pmatrix}.$$

The assessment of importance of harmonics is based on the posterior distribution for the harmonic coefficients at any given time. At time t, the current posterior distribution for the coefficient vector $\boldsymbol{\theta}_t$ is given, as usual, by

$$(\boldsymbol{\theta}_t \mid D_t) \sim N[\mathbf{m}_t, \mathbf{C}_t].$$

Note that we are assuming that the observational variance sequence is known, otherwise the normal distribution here would simply be replaced by a multivariate T as usual. The modifications in that case are noted below. Given the normal posterior for the full vector, the marginal posteriors for the coefficients of individual harmonics are, for $r = 1, \ldots, q - 1$, given by the relevant components

$$(\boldsymbol{\theta}_{tr} \mid D_t) \sim N[\mathbf{m}_{tr}, \mathbf{C}_{tr}], \qquad (8.11)$$

where

$$\mathbf{m}_{tr} = (a_{tr}, b_{tr})'$$

comprises the current point estimates of the cosine/sine terms of the rth harmonic. In the case of $p = 2q$, we have, in addition, the Nyquist harmonic with posterior $(\theta_{tq} \mid D_t) \sim N[m_{tq}, C_{tq}]$ with $m_{tq} = a_{tq}$. Notice that a_{tr} is the posterior estimate of the harmonic component of the seasonal pattern at time t, namely S_{tr}. A plot over time t of the estimates a_{tr}, (or the associated filtered values if filtering has been performed), with an indication of the associated uncertainty as measured by the posterior variance, provides a clear and useful visual indication of the importance of the harmonic term to the full seasonal description. More formal assessments may be based on the calculation of HPD (higest posterior density) regions (Section 16.3.5 of Chapter 16) for the individual harmonics. This is done as follows.

The values $\boldsymbol{\theta}_{tr} = \mathbf{0}$ define the hypothesis that the rth harmonic is irrelevant to the seasonal pattern. Referring to (8.11), the probability content of the HPD region defined by $\boldsymbol{\theta}_{tr} = \mathbf{0}$ is given by

$$\Pr\left[\chi_2^2 \le \mathbf{m}'_{tr}\mathbf{C}_{tr}^{-1}\mathbf{m}_{tr}\right],\tag{8.12}$$

where χ_2^2 is a random quantity having a standard chi-square distribution with 2 degrees of freedom. A large probability content indicates little support for the hypothesis so that the harmonic will be retained when this probability is appreciable. The tail-area of the chi-square here may be compared with conventional levels of "significance" for assessing the harmonic. In the case of variance learning, the posterior normal distributions are modified to multivariate T based on n_t degrees of freedom at time t. Now the probability content of the HPD above is based on, not the chi-square, but the F distribution with 2 degrees of freedom in the numerator and n_t in the denominator. Thus (8.12) becomes

$$\Pr\left[F_{2,n_t} \le \frac{1}{2}\mathbf{m}'_{tr}\mathbf{C}_{tr}^{-1}\mathbf{m}_{tr}\right].$$

Notice that, using the updating recurrence equations summarised in Chapter 4 for the case of unknown observational variance, the matrix $\mathbf{C}_t$ already incorporates a scale factor related to the current estimate of observational variance and the associated degrees of freedom n_t. Thus the only correction to the quadratic form in the HPD region probability calculation is to divide by 2. Finally, if $p = 2q$ and a Nyquist harmonic is included in the model, then the relevant probabilities are simply

$$\Pr\left[\chi_1^2 \le m_{tq}^2/C_{tq}\right]$$

and

$$\Pr\left[F_{1,n_t} \le m_{tq}^2/C_{tq}\right]$$

respectively.

EXAMPLE 8.6: Gas consumption data. A simple data series is used to illustrate the above discussion about Fourier decomposition.

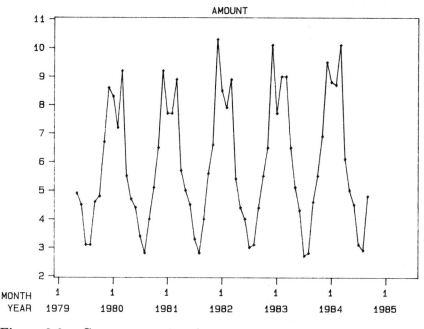

Figure 8.2.   Gas consumption data.

The observations are monthly totals of inland U.K. natural gas consumption over the period May 1979 to September 1984 inclusive, 65 observations in all, displayed in Table 8.1. The units of measurement are millions of tonnes coal equivalent, the data deriving from the Central Statistical Office Monthly Digest.

It is to be expected here that the seasonal pattern will be dominated by the first harmonic, gas usage following the annual temperature cycle. Higher frequency harmonics may enter in to refine the pattern, however, possibly accounting for industrial demand patterns and holiday effects. The data, plotted in Figure 8.2, reflect these suggestions. The simplest model is a first order polynomial,

Table 8.1. UK Gas consumption: Amount in 10^6 tonnes coal equivalent

Year	1	2	3	4	5	6	7	8	9	10	11	12
1979					4.9	4.5	3.1	3.1	4.6	4.8	6.7	8.6
1980	8.3	7.2	9.2	5.5	4.7	4.4	3.4	2.8	4.0	5.1	6.5	9.2
1981	7.7	7.7	8.9	5.7	5.0	4.5	3.3	2.8	4.0	5.6	6.6	10.3
1982	8.5	7.9	8.9	5.4	4.4	4.0	3.0	3.1	4.4	5.5	6.5	10.1
1983	7.7	9.0	9.0	6.5	5.1	4.3	2.7	2.8	4.6	5.5	6.9	9.5
1984	8.8	8.7	10.1	6.1	5.0	4.5	3.1	2.9	4.8			

seasonal effects DLM with 12 parameters. This is used here with a full (6 harmonics) Fourier representation of the 11 seasonal effects.

Consider fitting a static model with $\mathbf{W}_t = \mathbf{0}$ for all t. Here all the level and seasonal parameters are constant, and assessment of the harmonics may be based on the final posterior distributions at $t = 65$. In a reference analysis, using reference initial priors, the F values for the 6 harmonics as at $t = 65$ are given as follows:

Harmonic	1	2	3	4	5	6
F value	837.8	1.18	6.86	68.14	12.13	0.32

In the reference analysis the final degrees of freedom for the Student T posteriors is 53 (65 minus 12), thus the first five F values may be compared to the $F_{2,53}$ distribution, the final value for the Nyquist term to the $F_{1,53}$. On this basis, harmonics 1 and 4 are enormously significant, clearly indicating the dominance of the annual and quarterly cycles. Less significant, but still important, are harmonics 3 and 5. Harmonics 2 and 6 are essentially insignificant. Whilst not a definitive analysis of the data, this static reference analysis serves to identify the key harmonics.

To explore this more fully, and more appropriately, consider a single discount model with discount factor of 0.95. Thus, given the posterior variance matrix $\mathbf{C}_{t-1}$ for level parameter and Fourier coefficients at time $t - 1$, the evolution variance matrix is simply $(0.95^{-1} - 1)\mathbf{G}\mathbf{C}_{t-1}\mathbf{G}'$, and so $\mathbf{R}_t = 1.05\ \mathbf{G}\mathbf{C}_{t-1}\mathbf{G}'$. Thus we have roughly 5% increase in uncertainty about the parameters between observations. Whilst not optimised in terms of discount factors in any sense, this at least allows for some change in parameters. A reference prior based analysis with this model (with, as usual, discount factor of unity during the initial, reference part of the analysis) is partially illustrated in Figures 8.3 to 8.10. In Figure 8.3, the data are plotted with 1-step ahead point forecasts and corresponding 90% probability limits from the one-step forecast Student T distributions. Note that these forecasts appear only after 13 observations have been processed since this is a reference analysis. Clearly the one-step forecasts are fairly good, and the model adapts to the slight changes in pattern from year to year, such change being particularly evident in the December levels of consumption. Figure 8.4 is produced after retrospective smoothing using the backwards filtering algorithms. This displays the retrospective estimates of the seasonal

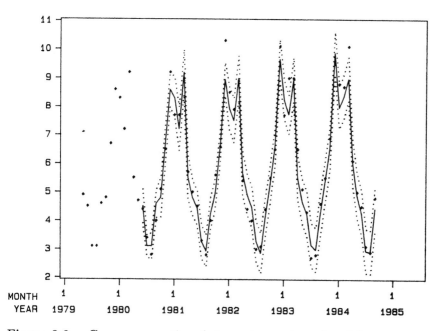

Figure 8.3.   Gas consumption data and one-step ahead forecasts.

effects in each month, the vertical bar in each month representing a
90% posterior probability interval (symmetrically) located about the
posterior mode for the effect in that month. Similar plots appear in
Figures 8.5 to 8.10 for the individual harmonic components of the
seasonal pattern, simply retrospective posterior intervals for the har-
monics $S_{tr} = \mathbf{E}'_2 \boldsymbol{\theta}_{tr}$, $(r = 1,\dots,6)$, with $\boldsymbol{\theta}_{tr}$ from equation (8.11).
The dominance of harmonics 1 and 4 is clear from these plots, as is
the nature of their contribution to the overall pattern.

8.7 EXERCISES

(1) Consider the the permutation matrix $\mathbf{P}$ of Definition 8.2.
   (a) Verify that $\mathbf{P}$ is $p-$cyclic, so that $\mathbf{P}^{k+np} = \mathbf{P}^k$ for all
        integers k and n.
   (b) Verify that the eigenvalues of $\mathbf{P}$ are the p roots of unity,
        $\exp(2\pi ij/p)$, $(j = 1,\dots,p)$.
(2) Verify that $\{\mathbf{E}_p, \mathbf{P}, .,.,\}$ is observable with identity observability
   matrix.

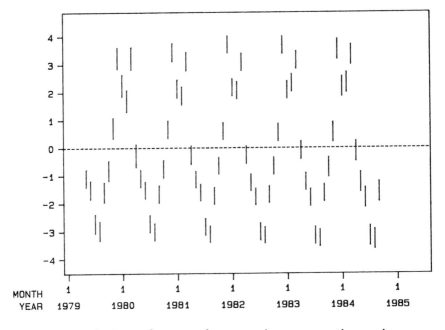

Figure 8.4.   Estimated seasonal pattern in consumption series.

(3) In the first-order polynomial/seasonal effects model, let X_t be
the cumulative sum of the Y values over the coming period,

$$X_t = \sum_{r=0}^{p-1} Y_{t+r}.$$

Verify that X_t does not depend on the seasonal effects over the
period so that all uncertainty in forecasting X_t is due to that
about the future, non-seasonal trend. Comment on possible
benefits of this feature in cases when there is a high degree
of uncertainty about the seasonl pattern. (nb. Of course this
generalises to other models with more complex trends).

(4) For a quarterly series, a seasonal effects component model is
used with $p = 4$. A forecaster unused to thinking about covariances specifies initial prior moments for seasonal factors in
terms of $\mathbf{m}^* = (-100, 100, -300, 400)'$ and $\mathbf{C}^* = \mathrm{diag}(200, 200,$
$200, 400)$. Show that this is an invalid prior for the constrained
seasonal factors, and use Theorem 8.2 to derive a valid prior
subject only to this information.

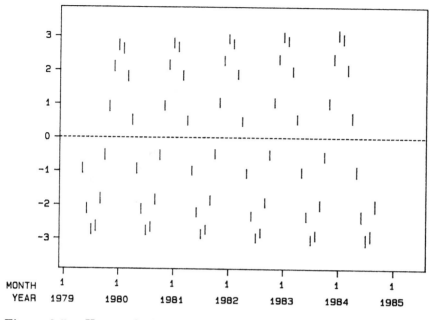

Figure 8.5.   Harmonic 1.

(5) Let $p = 6$ and consider the seasonal factors $\psi = (\psi_0, \ldots, \psi_5)'$ given by

$$\psi' = (0.35, 0.62, 0.04, -0.41, -0.70, 0.10).$$

Calculate the Fourier coefficients a_r and b_r, and the corresponding amplitudes A_r and phases γ_r, ($r = 1, 2, 3$), in the Fourier representation of the cyclic function $g(.)$ given by (8.8) with these values of ψ.

(6) Using the basic, trigonometric identities in Section 8.6.2 (or otherwise), verify that $\mathbf{G}_r$ of Definition 8.8 is p–cyclic.

(7) Consider the constant, single harmonic component model $\{\mathbf{E}_2, \mathbf{G}, V, \mathbf{W}\}$ with system matrix $\mathbf{G} = \mathbf{J}_2(1, \omega)$. Verify result (4) of Section 8.6.3. Specifically, show that $Z_t = Y_t - 2\cos(\omega)Y_{t-1} + Y_{t-2}$ can be written as a linear function of ν_t, ν_{t-1}, ν_{t-2}, ω_t and ω_{t-1}.

(8) Consider the single harmonic component model $\{\mathbf{E}_2, \mathbf{G}, 1, \mathbf{W}_t\}$ with $\mathbf{G} = \mathbf{J}_2(1, \omega)$ and $\mathbf{W}_t$ based on a single discount factor, so that $\mathbf{R}_t = \mathbf{C}_{t-1}/\delta$ for all t.

(a) Verify that $\mathbf{G}$ is non-singular with

$$\mathbf{G}^{-1} = \mathbf{G}'.$$

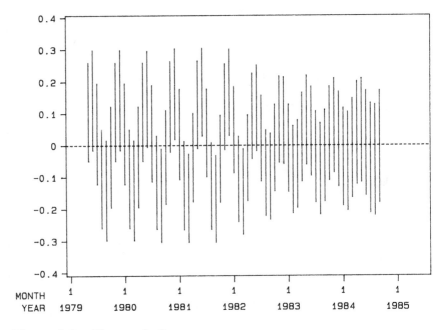

Figure 8.6. Harmonic 2.

Deduce that the variance matrix updating equation can be written in terms of precision matrices $K_t = C_t^{-1}$ as

$$K_t = \delta G K_{t-1} G' + E_2 E_2'.$$

(b) Assume that C_t converges to a limiting value $C = K^{-1}$ so that K satisfies the equation

$$K = \delta G K G' + E_2 E_2'.$$

By matching elements in this equation, deduce that

$$K = \begin{pmatrix} 1 - \delta^2 & (1-\delta)^2 K \\ (1-\delta)^2 K & (1-\delta)L \end{pmatrix}$$

where $K = \cos(\omega)/\sin(\omega)$ and $L = 3 - \delta + (1-\delta)^2/[\delta \sin^2(\omega)]$.

(c) Hence deduce the limiting value of the adaptive vector $A = [1 - \delta^2, (1-\delta)^2 K]'$.

(d) (Harder). Can you prove that the assumed limits exists?

(9) In the Fourier representation of the full seasonal model in Section 8.6.4, prove that the observability matrix is as stated, and verify that the model is observable.

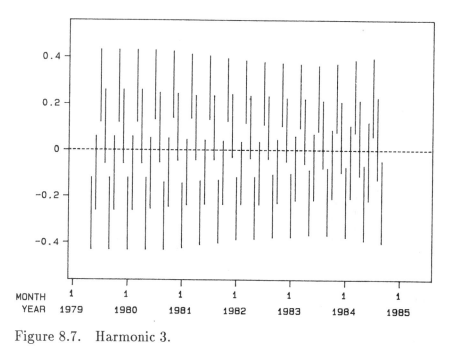

Figure 8.7.   Harmonic 3.

(10) Consider the full seasonal model in Section 8.6.4. Verify that the seasonal factors are given by equation (8.10) as stated. Verify also that, for all p, $\mathbf{L'L}$ is a non-singular diagonal matrix.

(11) In the above exercise, perform the calculations of Example 8.3 in the cases of $p = 5$ and $p = 6$, calculating:

(a) The $p \times (p-1)$ matrix $\mathbf{L}$, in equation (8.10), that provides the linear transformations from Fourier coefficients $\boldsymbol{\theta}_t$ to seasonal factors $\boldsymbol{\phi}_t$.

(b) The $(p-1) \times p$ matrix $\mathbf{H} = (\mathbf{L'L})^{-1}\mathbf{L}$ that provides the transformation from $\boldsymbol{\phi}_t$ to $\boldsymbol{\theta}_t$.

(12) Consider the reduced Fourier models of Section 8.6.5. For $p = 6$, perform the calculations of Example 8.4 in the cases of the two, reduced models given as follows.

(a) The first harmonic only so that $n = 2$, $\mathbf{F} = \mathbf{E}_2$ and $\mathbf{G} = \mathbf{J}_2(1, \omega)$.

(b) The first two harmonics so that $n = 4$, $\mathbf{F'} = (\mathbf{E}_2', \mathbf{E}_2')$ and $\mathbf{G} = \text{block diag}[\mathbf{J}_2(1, \omega), \mathbf{J}_2(1, \omega)]$.

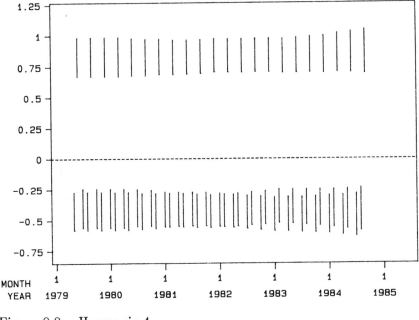

Figure 8.8.   Harmonic 4.

(13) Write a computer program to implement a DLM with second-order polynomial and Fourier form seasonal components.[†] The program should:

  (a) Allow a user to select any number of harmonic components.

  (b) Structure the evolution variance sequence in component discount form as in Definition 6.1 of Chapter 6, with one discount factor δ_T for the trend, and one δ_S for the seasonal component. These discount factors are to be specified by the user.

  (c) Allow for the choice of initialisation via either the reference analysis of Section 4.8, Chapter 4, or through user input priors.

  (d) Produce numerical and graphical summaries of analyses. Useful outputs from the program are graphs of on-line and filtered estimates, with uncertainties indicated by intervals about the estimates, of the time trajectories of model

---

[†]The **BATS** package (West, Harrison and Pole, 1987) implements this, and a wider class of models for micro-computer applications

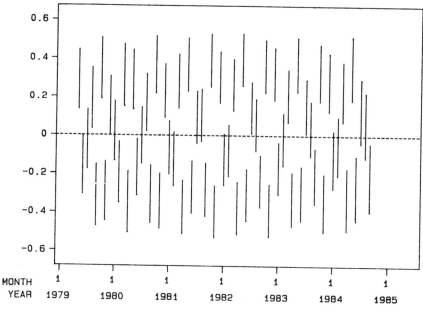

Figure 8.9.   Harmonic 5.

components including (i) the non-seasonal trend, (ii) the seasonal component, (iii) the individual harmonics.

(14) Test the program by analysing the Gas Consumption data of Table 8.1. Gain experience in the use of the model by exploring different analyses of this series. In particular, with the full 6 harmonics, experiment with this data series with a range of different initial priors (including the reference analysis), and different values of the discount factors in the range $0.85 < \delta_T, \delta_S \leq 1$. Note that the analysis in Section 8.6.5 uses a model in which the trend is first-order, rather than second-order polynomial, and that this can be reproduced in the more general model by specifying zero mean and zero variance for the growth parameter.

(15) Reanalyse the Gas Consumption data using a reduced Fourier form model having only two harmonic components, $r = 1$ and $r = 4$, for the fundamental (annual) cycle and the fourth (quarterly) cycle.

(16) Analyse the data below using the same model. The data provide quarterly total sales (in thousands) of one-day old turkey

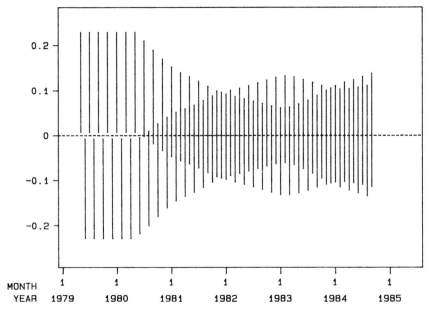

Figure 8.10.   Harmonic 6.

chicks from hatcheries in Eire over a period of years (taken from Ameen and Harrison, 1985a).

Eire turkey chick data

Year	Quarter 1	2	3	4
1974	131.7	322.6	285.6	105.7
1975	80.4	285.1	347.8	68.9
1976	203.3	375.9	415.9	65.8
1977	177.0	438.3	463.2	136.0
1978	192.2	442.8	509.6	201.2
1979	196.0	478.6	688.6	259.8
1980	352.5	508.1	701.5	325.6
1981	305.9	422.2	771.0	329.3
1982	384.0	472.0	852.0	

There is obvious, sustained growth in the series over the years and a marked, annual seasonal pattern. Again explore various models, discount factors etc., comparing them through subjective judgements based on graphical summaries of analysis, and numerical summaries such as the MSE, MAD and LLR measures. Verify that a static model ($\delta_T = \delta_S = 1$) performs poorly relative to dynamic models, and that, in particular, there is a fair degree of change indicated in the seasonal pattern.

(17) Re-analyse the turkey data after transformation of the original Y_t to an alternative scale. Compare graphs of transformed series against time, considering transforms such as $\log(Y_t)$, $Y_t^{1/2}$ and $Y_t^{3/4}$, and try to identify a transformed scale on which the amplitude of the seasonality is most constant (although, as with most real data, there is a fair degree of change in seasonal pattern on essentially all scales).

(18) Consider the regression DLM with $n = 3$, $\mathbf{G}_t = \mathbf{I}$ for all t,

$$\mathbf{F}_1 = \begin{pmatrix} 1 \\ 0 \\ 0 \end{pmatrix}, \qquad \mathbf{F}_2 = \begin{pmatrix} 0 \\ 1 \\ 0 \end{pmatrix}, \qquad \mathbf{F}_3 = \begin{pmatrix} 0 \\ 0 \\ 1 \end{pmatrix}, \qquad \mathbf{F}_4 = \begin{pmatrix} -1 \\ -1 \\ -1 \end{pmatrix}$$

and

$$\mathbf{F}_{t+4} = \mathbf{F}_t, \qquad (t > 4).$$

Show that the forecast function is that of a form-free seasonal pattern of period 4 and thus that the model offers an alternative way of modelling such a seasonal component. Generalise this model to all values of p.

(19) Consider the constant model

$$\left\{ \begin{pmatrix} 1 \\ 1 \end{pmatrix}, \cos(\omega) \begin{pmatrix} 1 & g \\ -g & 1 \end{pmatrix}, V, VW\mathbf{I} \right\},$$

for some variances V and W, and real g, $(g \neq 0)$, with

$$\omega = \arccos[(1 + g^2)^{-1/2}].$$

Show that the forecast function is that of a seasonal model with period $p = 2\pi/\omega$.

CHAPTER 9

REGRESSION, TRANSFER FUNCTION AND NOISE MODELS

9.1 INTRODUCTION

Polynomial and seasonal models provide two basic and very widely used classes of DLMs, both, of course, TSDLMs. A third important class is that based on regression relationships; the three classes combined via the principle of superposition provide for the vast majority of forms of behaviour encountered in practice for which linear models can provide adequate descriptions. Here we describe the theory and analysis of regression DLMs and important related models.

Chapter 3 concerned simple straight line regression DLMs, and, in particular, zero-origin models as a very special, but illuminating, case. In Section 3.2 of that Chapter a general multiple regression DLM was defined, though not analysed. The analysis can now be deduced directly from the general theory of the DLM in Chapters 4,5 and 6, and this is done below. Multiple regression DLMs are simply special cases of the general DLM in which the system evolution matrices G_t are all set to the identity; thus the state vectors of such models change in time according to random walks. More complicated regression type relationships cannot be represented in this way. A key example concerns *transfer response models* with particular functional forms determining the lagged effects over time of regressor variables on a response series. A second, important class provides for correlation structure in the observational error sequence using autoregressions. These classes of models can be incorporated within the DLM framework with minor extensions of the framework as presented in Chapter 4; this is done in this Chapter.

9.2 THE MULTIPLE REGRESSION DLM

9.2.1 Definition

Consider modelling the series Y_t by regressing on a collection of n independent, or regressor, variables labelled as $X_1,\ldots,X_n$. The value of the ith variable X_i at each time t is assumed known, denoted by X_{ti}, $(i = 1, \dots, n; \ t = 1, \dots,)$. Usually a constant term is included in the model, in which case, X_1 is taken as unity, $X_{t1} = 1$ for all t. For $t = 1, \dots$, let the regression vector $\mathbf{F}_t$ be given by $\mathbf{F}'_t = (X_{t1}, \dots, X_{tn})$. Then the multiple regression DLM with regressors $X_1, \dots, X_n$ is defined by the quadruple $\{\mathbf{F}_t, \mathbf{I}, V_t, \mathbf{W}_t\}$, for some observational variances V_t and evolution variance matrices $\mathbf{W}_t$. This is just as specified in Definition 3.1 of Chapter 3.

For each t then, the model equations are:

Observation equation: $Y_t = \mathbf{F}'_t \boldsymbol{\theta}_t + \nu_t,$ $\nu_t \sim \text{N}[0, V_t],$

System equation: $\boldsymbol{\theta}_t = \boldsymbol{\theta}_{t-1} + \boldsymbol{\omega}_t,$ $\boldsymbol{\omega}_t \sim \text{N}[\mathbf{0}, \mathbf{W}_t].$

Write the elements of $\boldsymbol{\theta}_t$ as $\boldsymbol{\theta}'_t = (\theta_{t1}, \dots, \theta_{tn})$. Then the observation equation can be written as

$$Y_t = \mu_t + \nu_t,$$

where the mean response μ_t is given by

$$\mu_t = \mathbf{F}'_t \boldsymbol{\theta}_t = \sum_{i=1}^{n} \theta_{ti} X_{ti}.$$

From this representation it is clear that the model can be viewed as being formed from the superposition of n straight line regressions with zero origin, the simple models of Chapter 3.

Obviously the regression model is a DLM in which the state vector evolves only via the addition of a noise term $\boldsymbol{\omega}_t$, the system matrices being given by $\mathbf{G}_t = \mathbf{I}$, the $n \times n$ identity for all t. The model is thus a dynamic generalisation of standard, static regression models. Setting $\mathbf{W}_t = \mathbf{0}$ for all t provides the specialisation to static regression for then $\boldsymbol{\theta}_t = \boldsymbol{\theta}$ is constant over time, and the model equations reduce to

$$Y_t = \mathbf{F}'_t \boldsymbol{\theta} + \nu_t, \qquad \nu_t \sim \text{N}[0, V_t].$$

Such models, and Bayesian analyses of them, are very well documented and illustrated in, for example, Broemeling (1985), Box and Tiao, (1973), De Groot (1971), Press (1985) and Zellner (1971). Here $\boldsymbol{\theta}$ plays the role of a fixed regression vector and the time ordering of the observations is not so relevant. In the dynamic regression, stochastic variation over time in the regression vector is permitted through the noise terms $\boldsymbol{\omega}_t$, modelling the possibility of change in regression relationships over time. The dynamic model, whilst retaining the basic linear structure, offers a little more flexibility in adapting to observed data in which the relationships between the response series and the regressors cannot be adequately represented by a static linear model. See Chapter 3, notably Section 3.1, for discussion and illustration of the concepts underlying such models, and, for further discussion and illustrations, Ameen and Harrison (1984), and Johnston *et al* (1986).

9.2.2   Common types of regressions

Just as with the simple, static regression, the DLM structure allows for a host of possible forms of relationships through appropriate choice of regression variables, and combinations of them. The basic types are quantitative measurements; indicator variables to group the response data according to an underlying classificatory variable, or factor; and higher-order terms invoving interactions between variables of these types. For examples of each of these, consider the sort of data series analysed in Section 3.3.4 of Chapter 3. Suppose that Y_t represents sales of a product, the data being quarterly figures over several years. Particular examples of the common model forms are as follows.

(1) **Straight line regressions on quantitative variables**
Suppose that $X = X_2$ is a related predictor variable, some form of economic indicator, for example. As with Y, X is a quantitative measurement and often such measurements are viewed as effectively continuous, although this is by no means necessary. A basic, straight line regression model is formed by taking $X_1 = 1$ to define an intercept term, so that $\mu_t = \alpha_t + \beta_t X_t$. Here, of course, $\mathbf{F}'_t = (1, X_t)$ and $\boldsymbol{\theta}'_t = (\alpha_t, \beta_t)$.

(2) **Multiple regressions on quantitative variables**
This may be extended to include other quantitative variables by adding in further terms such as $\gamma_t Z_t$ by superposition, to give a canonical multiple regression on several variables, each contributing a linear term. Here Z_t is the value of a second independent variable, $\mathbf{F}'_t = (1, X_t, Z_t)$ and $\boldsymbol{\theta}'_t = (\alpha_t, \beta_t, \gamma_t)$.

(3) **Lagged variables**
Particular cases of multiple regression of importance in time series forecasting involve the use of regressor variables calculated as *lagged* values of basic independent variables. In the sales forecasting scenario above, it may be thought that the

economic indicator X has predictive power for the sales series into the future as well as just for the current time. A similar example concerns forecasting of sales or demand with the independent variable X relating to advertising and other promotional activities. Here current promotional expenditure can be expected to impact not only on immediate sales, but also on sales further into the future. Hence, in modelling the mean response μ_t at time t, past, or lagged, values of the regressor variable X should be considered in addition to the current value X_t. Generally, suppose that it is felt that appreciable effects of the regressor variable may be sustained up to a maximum lag of k time points for some $k > 1$. The linear regression on lagged values of X then has the general form of

$$\mu_t = \alpha_t + \sum_{i=0}^{k} \beta_{ti} X_{t-i}$$
$$= \alpha_t + \beta_{t0} X_t + \beta_{t1} X_{t-1} + \cdots + \beta_{tk} X_{t-k}.$$

Here $\mathbf{F}'_t = (1, X_t, X_{t-1}, \dots, X_{t-k})$ and $\boldsymbol{\theta}'_t = (\alpha_t, \beta_{t0}, \beta_{t1}, \dots, \beta_{tk})$. Much use of these models has been made, of course, in economic forecasting (eg. Granger and Newbold, 1977; Zellner, 1971). We return to lagged relationships in detail in Section 9.3 below. Also note that *autoregressions* fall in this framework. Here the response series is directly regressed on lagged values of itself, with $X_t = Y_{t-1}$ for all t. Again much further discussion of these models appears below.

(4) **Polynomial surfaces**

Higher-order terms can be used to refine the basic description by defining further variables. For example, the two variable model $\mu_t = \alpha_t + \beta_t X_t + \gamma_t Z_t$ may be refined by including further regressor variables that are quadratic, and higher-order, functions of the two original variables X and Z. Examples include quadratic terms X^2 or Z^2, and cross-product terms such as XZ. These, and higher-order powers and cross-products, can be used to define polynomial regressions on possibly several variables, building up *response surface* descriptions of the regression function (Box and Draper, 1987).

(5) **Classificatory variables**

Classificatory variables, or factors, can be included using dummy X variables to indicate the classifications for each response observation. Seasonality, for example, can be modelled this way as has already been seen in Chapter 8. Specifically in this context, consider the quarterly classification of the sales data and suppose that a simple seasonal factor model is desired, with different levels of sales in each quarter, those levels changing stochastically from year to year. This is just the model of Section 8.1 of Chapter 8, and can be represented in regression DLM form as follows. Let $n = 4$ and define X_i as the indicator variable for the ith quarter of each year. Thus, for $i = 1, \ldots, 4$,

$$
X_{ti} = \begin{cases} 1, & \text{when } Y_t \text{ is in quarter } i \text{ of any year;} \\ 0, & \text{otherwise.} \end{cases}
$$

In this context, denote the elements of the state vector by $\theta_t' = (\phi_{t1}, \ldots, \phi_{t4})$ where ϕ_{ti} is the seasonal factor for quarter i of the year at time t. The mean response $\mu_t = \sum_{i=1}^{4} \phi_{ti} X_{ti}$ at time t is then given simply by $\mu_t = \phi_{ti}$, the relevant seasonal level, when time t corresponds to the ith quarter of the year.

The seasonal effects model, with effects representing seasonal deviations from an underlying, non-seasonal sales level, is also easily, and obviously, representable in regression form. This is necessary with classificatory variables in general if other regressors are to be included by superposition. Set

$$
\alpha_t = \frac{1}{4} \sum_{i=1}^{4} \phi_{ti},
$$

and, for $i = 1, \ldots, 4$,

$$
\theta_{ti} = \phi_{ti} - \alpha_t.
$$

Clearly the θ_{ti} sum to zero at each time t, representing the seasonal effects. Redefine the state vector as $\theta_t' = (\alpha_t, \theta_{t1}, \ldots, \theta_{t4})$ and the regression vector as $F_t' = (1, X_{t1}, \ldots, X_{t4})$ where the X variables are the above indicators of the quarters. Then

$$
\mu_t = F_t' \theta_t = \alpha_t + \sum_{i=1}^{4} \theta_{ti} X_{ti},
$$

subject to the zero-sum restriction on the final four elements of the state vector.

Grouping data according to an underlying factor in this way has wide uses in time series modelling, just as in other areas of statistics. Perhaps the most common use is in designed experiments where the factors relate to treatment groups, block effects and so forth. One other, important example is the use of dummy variables as intervention indicators, the corresponding parameter then representing a shift due to intervention.

(6) **Several factors**

As an example, suppose that the data is also classified according to sales area, there being just two areas for simplicity. The additive model formed by superposition has the form

$$\mu_t = \alpha_t + \sum_{i=1}^{4} \theta_{ti} X_{ti} + \sum_{j=1}^{2} \gamma_{tj} Z_{tj}$$

where, as with seasonality, γ_{t1} and γ_{t2} are the effects added to sales level in the two different sales regions (having zero-sum), with Z_{t1} and Z_{t2} being indicator variables for the two regions. Clearly this could be extended with the addition of further regressors of any types.

(7) **Factor by factor interactions**

Higher order terms involve interactions of two basic types. Firstly, two classificatory variables may interact, producing what is often referred to as a factor by factor interaction. In the above example, this amounts to dealing with the two sales regions separately, having different seasonal factors within each region. One way of modelling this is to add in an *interaction* term of the form

$$\sum_{i=1}^{4} \sum_{j=1}^{2} \beta_{tij} U_{tij}.$$

Here the U_{tij} are dummy, indicator variables with $U_{tij} = 1$ if, and only if, observation Y_t corresponds to quarter i and sales area j. The β_{tij} are the interaction parameters, subject to zero-sum constraints $\sum_{i=1}^{4} \beta_{tij} = 0$ for $j = 1, 2$, and $\sum_{j=1}^{2} \beta_{tij} = 0$ for $i = 1, \ldots, 4$.

(8) **Interaction between factors and quantitative measurements**

The second, and highly important, form of interaction is typified as follows. Consider a straight line regression on the variable X combined by superposition with the seasonal effects model to give a mean response function of the form

$$\mu_t = \alpha_t + \beta_t X_t + \sum_{i=1}^{4} \theta_{ti} X_{ti}.$$

The effects of the variable X and the seasonality are additive, not interacting. Often it may be felt that the effect of X on the response is different in different quarters; more generally, that the regression coefficient of a variable takes different values according to the various levels of a classifying factor. Here the necessary refinement of the model is

$$\mu_t = \alpha_t + \sum_{i=1}^{4} (\theta_{ti} + \beta_{ti} X_t) X_{ti}.$$

In this case the regression vector is $\mathbf{F}'_t = (1; X_{t1}, \ldots, X_{t4}; X_t X_{t1}, \ldots, X_t X_{t4})$, and the parameter vector is $\boldsymbol{\theta}'_t = (\alpha_t; \theta_{t1}, \ldots, \theta_{t4}; \beta_{t1}, \ldots, \beta_{t4})$.

9.2.3 Summary of analysis

The analysis follows from the general theory of DLMs. The results are given in the case of constant, unknown observational variance, $V_t = V = 1/\phi$ for all t, with ϕ unknown, consistent with the summary in Section 4.6 of Chapter 4. Thus $\mathbf{W}_t$ is scaled by the current estimate S_{t-1} of $V = \phi^{-1}$ at each time. Then, as usual, the evolution/updating cycle is based on the following distributions:

$$
\begin{aligned}
(\boldsymbol{\theta}_{t-1} \mid D_{t-1}) &\sim \ \mathrm{T}_{n_{t-1}}[\mathbf{m}_{t-1}, \mathbf{C}_{t-1}], \\
(\boldsymbol{\theta}_t \mid D_{t-1}) &\sim \ \mathrm{T}_{n_{t-1}}[\mathbf{a}_t, \mathbf{R}_t], \\
(\phi \mid D_{t-1}) &\sim \ \mathrm{G}[n_{t-1}/2, d_{t-1}/2] \quad \text{with } S_{t-1} = d_{t-1}/n_{t-1}, \\
(Y_t \mid D_{t-1}) &\sim \ \mathrm{T}_{n_{t-1}}[f_t, Q_t], \\
(\boldsymbol{\theta}_t \mid D_t) &\sim \ \mathrm{T}_{n_t}[\mathbf{m}_t, \mathbf{C}_t], \\
(\phi \mid D_t) &\sim \ \mathrm{G}[n_t/2, d_t/2] \quad \text{with } S_t = d_t/n_t,
\end{aligned}
$$

where $\mathbf{a}_t = \mathbf{m}_{t-1}$, $\mathbf{R}_t = \mathbf{C}_{t-1} + \mathbf{W}_t$, $f_t = \mathbf{F}'_t \mathbf{m}_{t-1}$, $Q_t = \mathbf{F}'_t \mathbf{C}_{t-1} \mathbf{F}_t + S_{t-1}$, and, with $e_t = Y_t - f_t$ and $\mathbf{A}_t = \mathbf{R}_t \mathbf{F}_t / Q_t$, the updating equations $n_t = n_{t-1} + 1$, $d_t = d_{t-1} + S_{t-1} e_t^2 / Q_t$, $\mathbf{m}_t = \mathbf{m}_{t-1} + \mathbf{A}_t e_t$ and $\mathbf{C}_t = [\mathbf{R}_t - \mathbf{A}_t \mathbf{A}'_t Q_t](S_t / S_{t-1})$.

The usual filtering and forecasting equations apply similarly. The former are not reproduced here. For forecasting ahead to time $t + k$ from time t, it follows easily that

$$(\boldsymbol{\theta}_{t+k} \mid D_t) \sim \mathrm{T}_{n_t}[\mathbf{m}_t, \mathbf{R}_t(k)],$$
$$(Y_{t+k} \mid D_t) \sim \mathrm{T}_{n_t}[f_t(k), Q_t(k)]$$

where $\mathbf{R}_t(k) = \mathbf{C}_t + \sum_{r=1}^{k} \mathbf{W}_{t+r}$, $f_t(k) = \mathbf{F}'_{t+k}\mathbf{m}_t$ and $Q_t(k) = S_t + \mathbf{F}'_{t+k}\mathbf{R}_t(k)\mathbf{F}_{t+k}$.

9.2.4 Comments

Various features of the regression model analysis require comment.

(1) Forecasting performance is achieved through the identification of stability in regression relationships. Thus models with small evolution noise terms are to be desired, static regressions in which $\boldsymbol{\theta}_t = \boldsymbol{\theta}$ is constant being ideal so long as they are appropriate. If the time variation in $\boldsymbol{\theta}_t$ is significant, evidenced by large values on the diagonals of the evolution variance matrices $\mathbf{W}_t$, forecasting suffers in two ways. Firstly, forecasting ahead the forecast distributions become very diffuse as the $\mathbf{R}_t(k)$ terms increase due to the addition of further evolution noise variance matrices. Secondly, in updating the weight placed on new data is high so that the posterior distributions adapt markedly from observation to observation. Thus, although very short-term forecasts may be accurate in terms of location and reasonably precise, medium and longer-term forecasts may be poorly located and very diffuse.

As a consequence, effort is needed in practice to identify meaningful and appropriate independent variables to be used as regressors, often after transformation and combination, with a view to identifying regressions whose coefficients are as stable as possible over time. There will usually be a need for some minor stochastic variation to account, at the very least, for changing conditions in the environment of the series, and model misspecification. Thus, in practice, the $\mathbf{W}_t$ matrices determined by the forecaster will be relatively small.

(2) Structuring the $\mathbf{W}_t$ sequence using discount factors can be done in various ways following the development of component models in Section 6.3 of Chapter 6. If the regressors are similar, related variables viewed as modelling an overall effect of

an unobserved, underlying variable, then they should be considered together as one block or component for discounting. For example, the effects of a classifying factor variable, should be viewed as a single component, as is the case in Chapter 8, and elsewhere, with seasonal models. Otherwise, considering regressors as contributing separately to the model implies a need for one discount factor for each regression parameter, the $\mathbf{W}_t$ matrix then being diagonal.

(3) The static regression model obtains when $\mathbf{W}_t = \mathbf{0}$ for all t, implied by unit discount factors for all components. In this case, $\mathbf{R}_t = \mathbf{C}_{t-1}$ for all t, no information decaying on the state vector between observation stages. The update for the posterior variance matrix can be rewritten in terms of the precision, or information matrices $\mathbf{C}_t^{-1}$ and $\mathbf{C}_{t-1}^{-1}$ as

$$\mathbf{C}_t^{-1} = [\mathbf{C}_{t-1}^{-1} + S_{t-1}^{-1}\mathbf{F}_t\mathbf{F}_t'](S_{t-1}/S_t).$$

It follows that, in terms of the scale free matrices $\mathbf{C}_t/S_t$ for all t,

$$S_t\mathbf{C}_t^{-1} = S_{t-1}\mathbf{C}_{t-1}^{-1} + \mathbf{F}_t\mathbf{F}_t'.$$

Repeatedly applying this for times $t-1, t-2, \dots, 1$, leads to

$$S_t\mathbf{C}_t^{-1} = S_0\mathbf{C}_0^{-1} + \sum_{r=1}^{t}\mathbf{F}_r\mathbf{F}_r'.$$

It also follows, in a similar fashion, that

$$\mathbf{m}_t = S_t^{-1}\mathbf{C}_t[S_0\mathbf{C}_0^{-1}\mathbf{m}_0 + \sum_{r=1}^{t}\mathbf{F}_rY_r].$$

These results coincide, naturally, with the standard Bayesian linear regression results in static models (See, for example, De Groot, 1971, Chapter 11; Box and Tiao, 1973, Chapter 2).

(4) The reference analysis of Section 4.8 of Chapter 4 may be applied in cases of little initial prior information. This leads to the dynamic version of standard reference analyses of linear models (see above references). When sufficient data has been processed so that the posterior for $\boldsymbol{\theta}_t$ (and ϕ) is proper, the standard updating equations above apply. The main difference is that the degrees of freedom are not initially updated until sufficient data are available to make the posteriors proper. This results in reduced degrees of freedom n_t thereafter. Section 4.8 provides full details.

The meaning of "sufficient data" in this context brings in directly the notion of collinearity amongst the regressor variables. In the initial reference updating, the posterior distributions become proper at that time t such that the *precision matrix* $\sum_{r=1}^{t} \mathbf{F}_r \mathbf{F}'_r$ first becomes non-singular. At this stage, the standard updating may begin for $\boldsymbol{\theta}_t$, although one further observation is necessary to begin the updating for ϕ. The soonest that this may occur is at $t = n$, the dimension of the parameter vector, when one observation has been observed for each parameter. If missing observations are encountered, then this increases by 1 for each. Otherwise, collinearity amongst the regressor variables can lead to the precision matrix being singular at time n. This is actually rather uncommon in practice, although much more often it is the case that the matrix is close to singularity, having a very small, positive determinant. This indicates strong relationships amongst the regressor variables, the well-known feature of multi-collinearity in regression. Numerical problems can be encountered in inverting the above precision matrix in such cases, so that care is needed. To avoid inverting the matrix, and also to cater for the case of precise singularity, a small number of further observations can be processed retaining the reference analysis updating equations. This means that further terms $\mathbf{F}_{t+1} \mathbf{F}'_{t+1}$, $\mathbf{F}_{t+2} \mathbf{F}'_{t+2}$, and so on, are added to the existing precision matrix until it is better conditioned and does not suffer from numerical instablilities when inverted. Under certain circumstances, however, collinearity amongst regressors may persist over time. In such cases there is a need for action, usually to reduce the number of regressors, removing redundant variables, the model being over-parametrised. The problem of multi-collinearity here is precisely as encountered in standard, static regression.

(5) When dealing with regressors that are observed values of time series themselves the static model concept of *orthogonality* of regressors also applies. The simplest, but important, use of this involves considering independent variables as deviations from some average value. In standard regression, it is common practice to standardise regressors; given a fixed sample of observations, the regressors are standardised primarily by subtracting the arithmetic mean and, secondarily, dividing by the standard deviation. The reason for subtracting the

arithmetic mean is so that the regression effects are clearly separated from the intercept term in the model, being essentially orthogonal to it. Thus, for example, a static straight line model on regressor X, $Y_t = \alpha + \beta X_t$ is rewritten as $Y_t = \alpha^* + \beta X_t^*$ where $X_t^* = X_t - \bar{X}$, with $\bar{X}$ the arithmetic mean of the X_t values in the fixed sample considered, and $\alpha^* = \alpha + \beta \bar{X}$ the new, fixed intercept term. The new regression vectors $\mathbf{F}_t' = (1, X_t^*)$ are such that $\sum \mathbf{F}_t \mathbf{F}_t'$ (the sum being over the fixed sample of observations) is now diagonal and so inverts easily. This *orthogonality*, and its more general versions with several regressors, allows simpler interpretation of the regression.

In dynamic regression the same principles apply, although now the time dependence clouds the issue. If a relatively short series of known number of observations is to be analysed, then the above form of standardisation to zero arithmetic mean for regressors may apply. It is usually to be expected that regression relationships do not change rapidly, and so the features of the static model may be approximately reproduced. Some problems do arise even under such circumstances, however. For a start, $\bar{X}$ may be unknown initially since X values later in the series have yet to be observed; this derives from the time series nature of the problem and applies even if the model is static. Some insight into what form of standardisation may be appropriate can be gained by considering the regression as possibly being derived from a more structured model, one in which Y_t and X_t are initially jointly normally distributed conditional on a collection of time-varying parameters that determine their mean vector and covariance matrix, forming a bivariate time series. Suppose specifically that Y_t has mean α_t and X_t has mean γ_t. It follows that

$$Y_t = \alpha_t + \beta_t(X_t - \gamma_t)$$

where β_t is the regression coefficient from the covariance matrix of Y_t and X_t. In the static model, $\gamma_t = \gamma$ is constant over time. The static model correction is now obvious; the population mean γ is simply estimated by the sample value $\bar{X}$ from the fixed sample of interest. More generally, if γ is assumed constant so that the X values are distributed about a common mean, then a sequentially updated estimate of γ is more appropriate. If, on the other hand, γ_t is possibly time-varying, a *local* mean for the X_t time series is appropriate. Such an estimate may be obtained from a separate time series model for the X_t series. Further development of this is left as an exercise to the reader.

(6) In forecasting ahead the future values of regressors are required. If some or all of the regressors are observed values of related time series, as if often the case in socio-economic modelling, for example, then the required future values may not be available at the time of forecasting. Various possible solutions exist for this problem. One general and theoretically exact method is to construct a joint model for forecasting the X variables as time series along with Y. This introduces the need for multivariate time series modelling and forecasting, a vast topic in its own right and beyond the scope of the present discussion (see Chapter 15 for multivariate DLMs).

Simpler, alternative approaches involve the use of estimated values of the future regressors. These may be simply guessed at or provided by third parties, separate models etc. Consider forecasting Y_{t+k} from time t, with $\mathbf{F}_{t+k}$ uncertain. Typically, information relevant to forecasting $\mathbf{F}_{t+k}$ separately will result in specification of some features of a forecast distribution, assumed to have a density $p(\mathbf{F}_{t+k} \mid D_t)$. Note that, formally, the extra information relevant to forecasting $\mathbf{F}_{t+k}$ should be included in the conditioning here; without loss of generality, assume that this is already incorporated in D_t. Then the step-ahead forecast distribution for Y_{t+k} can be deduced as

$$p(Y_{t+k} \mid D_t) = \int p(Y_{t+k} \mid \mathbf{F}_{t+k}, D_t) p(\mathbf{F}_{t+k} \mid D_t) d\mathbf{F}_{t+k}.$$

The first term in the integrand here is just the standard forecast T distribution from the regression, as specified in Section 9.2.3 above, with $\mathbf{F}_{t+k}$ assumed known and explicitly included in the conditioning. Features of the predictive density will depend of the particular forms of predictions for $\mathbf{F}_{t+k}$. Some generally useful features are available, as follows. Suppose that the forecast mean and variance matrix of $\mathbf{F}_{t+k}$ exist, denoted by $\mathbf{h}_t(k) = \mathrm{E}[\mathbf{F}_{t+k} \mid D_t]$ and $\mathbf{H}_t(k) = \mathrm{V}[\mathbf{F}_{t+k} \mid D_t]$ respectively. Then the forecast mean and variance of Y_{t+k} can be deduced. Simply note that, when the degrees of freedom of the conditional T distribution for Y_{t+k} exceeds unity,

$n_t > 1$, then

$$E[Y_{t+k} \mid D_t] = E\{E[Y_{t+k} \mid \mathbf{F}_{t+k}, D_t] \mid D_t\}$$
$$= E[\mathbf{F}'_{t+k}\mathbf{m}_t \mid D_t] = \mathbf{h}_t(k)'\mathbf{m}_t.$$

Similarly, when $n_t > 2$,

$$V[Y_{t+k} \mid D_t] = E\{V[Y_{t+k} \mid \mathbf{F}_{t+k}, D_t] \mid D_t\}$$
$$+ V\{E[Y_{t+k} \mid \mathbf{F}_{t+k}, D_t] \mid D_t\}$$
$$= E[\frac{n_t}{n_t - 2}Q_t(k) \mid D_t] + V[f_t(k) \mid D_t]$$
$$= \frac{n_t}{n_t - 2}\{S_t + E[\mathbf{F}'_{t+k}\mathbf{R}_t(k)\mathbf{F}_{t+k} \mid D_t]\}$$
$$+ V[\mathbf{F}'_{t+k}\mathbf{m}_t \mid D_t]$$
$$= \frac{n_t}{n_t - 2}[S_t + \mathbf{h}_t(k)'\mathbf{R}_t(k)\mathbf{h}_t(k)$$
$$+ \text{trace}\{\mathbf{R}_t(k)\mathbf{H}_t(k)\}] + \mathbf{m}'_t\mathbf{H}_t(k)\mathbf{m}_t.$$

In this way, uncertainty about the future regressor values are formally incorporated in forecast distributions for the Y series.

(7) Standard modes of inference about regression parameters in static models apply directly in the dynamic case. Consider any $q \leq n$ elements of the state vector $\boldsymbol{\theta}_t$, reordering the elements as necessary, so that the q of interest occupy the first q positions in the vector. Thus $\boldsymbol{\theta}'_t = (\boldsymbol{\theta}'_{t1}, \boldsymbol{\theta}'_{t2})$ where $\boldsymbol{\theta}'_{t1} = (\theta_{t1}, \ldots, \theta_{tq})$ is the subvector of interest. Then, with $\mathbf{m}_t$ and $\mathbf{C}_t$ conformably partitioned, $(\boldsymbol{\theta}_{t1} \mid D_t) \sim T_{n_t}[\mathbf{m}_{t1}, \mathbf{C}_{t1}]$ in an obvious notation. Inferences about $\boldsymbol{\theta}_{t1}$ are based on this marginal posterior distribution. In particular, the contribution of the corresponding regressors to the model may be assessed by considering the support in the posterior for the values $\boldsymbol{\theta}_{t1} = \mathbf{0}$, consistent with no effect of the regressors. The posterior density $p(\boldsymbol{\theta}_{t1} \mid D_t)$ takes values greater than that at $\boldsymbol{\theta}_{t1} = \mathbf{0}$ whenever

$$(\boldsymbol{\theta}_{t1} - \mathbf{m}_{t1})'\mathbf{C}_{t1}^{-1}(\boldsymbol{\theta}_{t1} - \mathbf{m}_{t1}) < \mathbf{m}'_{t1}\mathbf{C}_{t1}^{-1}\mathbf{m}_{t1}.$$

The posterior probability that this occurs is given from the usual F distribution,

$$\text{Pr}[(\boldsymbol{\theta}_{t1} - \mathbf{m}_{t1})'\mathbf{C}_{t1}^{-1}(\boldsymbol{\theta}_{t1} - \mathbf{m}_{t1}) < \mathbf{m}'_{t1}\mathbf{C}_{t1}^{-1}\mathbf{m}_{t1} \mid D_t]$$
$$= \text{Pr}[F_{q,n_t} < q^{-1}\mathbf{m}'_{t1}\mathbf{C}_{t1}^{-1}\mathbf{m}_{t1}],$$

where F_{q,n_t} denotes a random quantity having the standard F distribution with q degrees of freedom in the numerator and n_t in the denominator. Thus the highest posterior density (HPD) based test of the hypothesis that $\boldsymbol{\theta}_{t1} = \mathbf{0}$ is based on the probability level

$$\alpha = \Pr[F_{q,n_t} \geq q^{-1}\mathbf{m}'_{t1}\mathbf{C}^{-1}_{t1}\mathbf{m}_{t1}];$$

a small value of α indicates rejection of the hypothesised value $\boldsymbol{\theta}_{t1} = \mathbf{0}$ as unlikely.

(8) As in the general DLM, the variance matrices $\mathbf{C}_0$ and the sequence $\mathbf{W}_t$ may be structured in order to incorporate modeller's views about relationships amongst the parameters. At an extreme, the parameters may be subject to linear restrictions that relate components or condition some elements to taking known values. This then implies that $\mathbf{C}_0$ is singular with a specific structure determined by the linear constraints, the same form of constraints applying to the evolution variance matrices if the restrictions are to hold to $\boldsymbol{\theta}_t$ over time. More usually, initial views about relationships amongst the parameters will be modelled in terms of stochastic constraints of various kinds, usually leaving the variance matrices non-singular. One important example is the embodiment of beliefs about the likely decay of coefficients of lagged values of variables, related to the use of smoothness prior distributions in lagged regressions and autoregressions (Cleveland, 1974; Leamer, 1972; Young, 1983; Zellner, 1971). Similar structures may be applied to the coefficients of higher-order terms in polynomial regressions (Young, 1977). Other examples include the use of hierarchical models, often based on assumptions of exchangeability amongst subsets of elements of $\boldsymbol{\theta}_t$ (Lindley and Smith, 1972). A typical example concerns symmetry assumptions about the effects of different levels of an underlying factor that groups the data; these may be initially viewed as exchangeable, with changes in the effects over time subject to the same assumption. Though interesting and important in application when appropriate, these topics are not developed further here in a general framework.

9.3 TRANSFER FUNCTIONS OF INDEPENDENT VARIABLES

9.3.1 Form-free transfer functions

Consider regression on current and past values of a single independent variable X, assuming initially that the regression parameters are constant over time. As in Section 9.2.2 above, regression on a fixed and finite number of lagged values falls within the standard regression DLM framework. Generally, if appreciable effects of the regressor variable are expected to be sustained up to a maximum lag of k time points, for some $k > 1$, the linear regression on lagged values determines the contribution to the mean response at time t as

$$\mu_t = \sum_{r=0}^{k} \beta_r X_{t-r} = \beta_0 X_t + \beta_1 X_{t-1} + \cdots + \beta_k X_{t-k}.$$

Here $\mathbf{F}'_t = (X_t, X_{t-1}, \dots, X_{t-k})$ and $\boldsymbol{\theta}'_t = \boldsymbol{\theta}' = (\beta_0, \beta_1, \dots, \beta_k)$. Projecting ahead from the current time t to times $t + r$, for $r \geq 0$, the *effect* of the current level X_t of the regressor variable is then simply the contribution to the mean response, namely $\beta_r X_t$ for $r = 0, 1, \dots, k$, being zero for $r > k$. This defines the **transfer response function** of X,

$$\begin{cases} \beta_r X, & r = 0, 1, \dots, k; \\ 0, & r > k. \end{cases}$$

In words this is just the effect of the current regressor value $X_t = X$ on the mean response at future times r, conditional on $X_{t+1} = \dots = X_{t+r} = 0$. Obviously this model, for large enough k, provides a flexible method of modelling essentially any expected form of transfer response function, the coefficients β_r being arbitrary regression parameters to be specified or estimated. In the more practically suitable dynamic regression, the flexibility increases as stochastic variation in the parameters allows the model to adapt to changing responses, and also to cater for misspecification in the model.

However, whilst the regression structure provides a very general model for lagged responses, there are often good reasons to consider functional relationships amongst the regression coefficients β_r that essentially alter the structure of the model, providing functional forms over time for the lagged effects of X. This leads to the considerations in the following sections.

9.3.2 Functional form transfer functions

One obvious feature of the regression model above is that the effect of X_t on Y_{t+r} is zero when $r > k$. The model is thus inappropriate for cases in which it is felt that lagged effects persist into the future, perhaps decaying smoothly towards zero as time progresses. One simple way of adapting the regression structure to incorporate such features is to consider regression, not on X directly, but on a constructed *effect* variable measuring the combined effect of current and past X values. Some examples provide insight.

EXAMPLE 9.1. Suppose that Y_t represents a monthly consumer series, such as sales or demand for a product, or consumer awareness of the product in the market. Currently, and prior to time $t = 1$, the Y series is supposed to follow a time series model with level parameter $\mu_t = \alpha_t$; this may, for example, be a simple steady model, or include other terms such as trend, seasonality and regressions. In month $t = 1$, the marketing company initiates a promotional campaign for the product involving expenditure on advertising and so forth, the expenditure being measured by a single independent variable $X_t, t = 1, 2, \ldots$. This may be a compound of various factors but is assumed, for simplicity, to measure investment in promoting the product. This is a simple instance of a very common event, and it is generally understood that, with no other inputs, the effect of such promotional expenditure can be expected to be as follows: (a) in month t, the level of the Y series should increase, say in proportion to X_t; (b) without further expenditure at times $t + 1, t + 2, \ldots$, the effect of past expenditure will decay over time, often approximately exponentially; and (c) further expenditure in following months will have the same form of effect. In model terms, the anticipated mean response is given by the original level plus a second term, ξ_t,

$$\mu_t = \alpha_t + \xi_t,$$

where ξ_t is the effect on the current level of the series of current and past expenditure. This effect is modelled as

$$\xi_t = \lambda \xi_{t-1} + \psi X_t,$$

with $\xi_0 = 0$, there being no expenditure prior to $t = 1$. The parameter ψ determines the immediate, *penetration* effect of the monthly advertising, the level being raised initially on average by ψ per unit of expenditure. ψ is a positive quantity whose units depends on those of both the X and Y series. λ represents the *memory* of the market. Extrapolating ahead to time $t + k$, the model implies that

$$\xi_{t+k} = \lambda^k \xi_t + \psi \sum_{r=1}^{k} \lambda^{k-r} X_{t+r}.$$

Thus, if $X_{t+1} = X_{t+2} = \ldots = X_{t+k} = 0$, then

$$\xi_{t+k} = \lambda^k \xi_t.$$

This embodies point (b); with λ a dimensionless quantity in the unit interval, the effect of expenditure up to time t is reduced by a factor λ for each future time point, decaying asymptotically to zero at an exponential rate.

EXAMPLE 9.2. Example 9.1 concerns the exponential decay of effects, where promotional advertising is not anticipated to sustain the sales/demand series at higher levels. Minor modification provides a closely related model for sustained growth or decay. An example concerns increases (or decreases) of sales to a new, higher (or lower) and sustained level following price reductions (or rises) for a product or products in a limited consumer market. Let the initial level α_t again represent previous information about the sales/demand series subject to an original pricing policy. Let X_t now represent the *reduction* in price in month t, either positive, implying a decrease in price, or negative implying an increase. It is to be expected that, in a finite market, sales will tend to increase as the price is reduced, eventually levelling of at some *saturation* level. Similarly, sales tend to decay towards zero as prices increases. This can be modelled via

$$\xi_t = \xi_{t-1} + \theta_t,$$

where

$$\theta_t = \lambda \theta_{t-1} + \psi X_t,$$

with $\xi_0 = \theta_0 = 0$. Suppose, for example, that a single price change is made at time t, with $X_{t+1} = X_{t+2} = \ldots = 0$ and $\xi_{t-1} = \theta_{t-1} = 0$. It follows that the immediate effect on the mean response is simply $\xi_t = \theta_t = \psi X_t$, the positive quantity ψ again measuring the immediate unit response. Projecting to time $t + r$, $(r = 1, \ldots, k)$, under these conditions, we have $\theta_{t+r} = \lambda^r \theta_t$ and thus

$$\xi_{t+k} = \sum_{r=0}^{k} \lambda^r \theta_t = \theta_t (1 - \lambda^k)/(1 - \lambda)$$

if $0 < \lambda < 1$. Thus, if X_t is positive so that the price decreases, then ξ_{t+k} is an increasing function of k, tending to the limit $\psi X_t (1 - \lambda)$ as k increases. Given the initial penetration factor ψ, λ determines both the rate of increase and the eventual saturation level. Similarly, of course, a negative value of X_t consistent with price increase implies a decay in level.

These two examples typify a class of structures for lagged effects of a single independent variable. The general form of such models, detailed here, is apparently an extension of the usual DLM representation, but can easily be rewritten as a standard DLM, as will be seen below. The initial definition is given in terms of the extended representation for interpretability.

Definition 9.1. *Let X_t be the value of an independent, scalar variable X at time t. A general* **transfer function model** *for the effect of X on the response series Y is defined by*

$$Y_t = \mathbf{F}' \boldsymbol{\theta}_t + \nu_t, \tag{9.1a}$$
$$\boldsymbol{\theta}_t = \mathbf{G} \boldsymbol{\theta}_{t-1} + \boldsymbol{\psi}_t X_t + \delta \boldsymbol{\theta}_t, \tag{9.1b}$$
$$\boldsymbol{\psi}_t = \boldsymbol{\psi}_{t-1} + \delta \boldsymbol{\psi}_t, \tag{9.1c}$$

with terms defined as follows. $\boldsymbol{\theta}_t$ is an n-dimensional state vector, $\mathbf{F}$ a constant and known n-vector, $\mathbf{G}$ a constant and known evolution matrix, and ν_t and $\delta \boldsymbol{\theta}_t$ are observation and evolution noise terms. (Note the use of the δ notation for the latter rather than the usual ω_t notation). All these terms are precisely as in the standard DLM, with the usual independence assumptions for the noise terms holding here. The term $\boldsymbol{\psi}_t$ is an n-vector of parameters, evolving via the addition of a noise term $\delta \boldsymbol{\psi}_t$, assumed to be zero-mean normally distributed independently of ν_t (though not necessarily of $\delta \boldsymbol{\theta}_t$).

The state vector $\boldsymbol{\theta}_t$ carries the effect of current and past values of the X series through to Y_t in equation (9.1a); this is formed in (9.1b) as the sum of a linear function of past effects, $\boldsymbol{\theta}_{t-1}$, and the current effect $\boldsymbol{\psi}_t X_t$, plus a noise term.

Suppose that, conditional on past information D_t, the posterior point estimates of the two vectors $\boldsymbol{\theta}_t$ and $\boldsymbol{\psi}_t$ are denoted by

$$\mathbf{m}_t = \mathrm{E}[\boldsymbol{\theta}_t|D_t] \quad \text{and} \quad \mathbf{h}_t = \mathrm{E}[\boldsymbol{\psi}_t|D_t].$$

Extrapolating expectations into the future in (9.1b and c), it follows that

$$\mathrm{E}[\boldsymbol{\theta}_{t+k} \mid D_t] = \mathbf{G}^k \mathbf{m}_t + \sum_{r=1}^{k} \mathbf{G}^{k-r} \mathbf{h}_t X_{t+r}. \tag{9.2}$$

Then, from (9.1a), the forecast function is

$$f_t(k) = \mathrm{E}[Y_{t+k} \mid D_t] = \mathbf{F}'\mathbf{G}^k \mathbf{m}_t + \mathbf{F}' \sum_{r=1}^{k} \mathbf{G}^{k-r} \mathbf{h}_t X_{t+r}. \tag{9.3}$$

Let a_t denote the first term here, $a_t = \mathbf{F}'\mathbf{G}^k \mathbf{m}_t$, summarising the effects of past values of the X series; a_t is known at time t. Also, consider the special case in which, after time $t+1$, there are no input values of the regressor variable, so that

$$X_{t+r} = 0, \quad (r = 2, \dots, k). \tag{9.4}$$

Then (9.3) implies that

$$f_t(k) = a_t + \mathbf{F}'\mathbf{G}^{k-1}\mathbf{h}_t X_{t+1}. \tag{9.5}$$

This can be seen as determining the way in which any particular value of the X series in the next time period is expected to influence the response into the future, the dependence on the step-ahead index k coming, as in TSDLMS, through powers of the system matrix $\mathbf{G}$.

In the special case that $\boldsymbol{\psi}_t = \boldsymbol{\psi}$ is constant over time and known, $\boldsymbol{\psi} = \mathbf{h}_t$ the **transfer response function** of X is given by $f_{t-1}(k+1)$ subject to (9.4) with past effects $a_{t-1} = 0$. Under these circumstances, (9.5) then leads to

$$\mathbf{F}'\mathbf{G}^k \boldsymbol{\psi} X$$

being the expected effect on the response due to $X_t = X$.

EXAMPLE 9.1 (continued). In the exponential decay model as described earlier, we have dimension $n = 1$, $\boldsymbol{\theta}_t = \xi_t$, the effect variable, $\boldsymbol{\psi}_t = \psi$ for all t, $\mathbf{F} = 1$ and $\mathbf{G} = \lambda$, all noise terms assumed zero. Note that $\mathbf{G}$ is just the memory decay term λ, assumed known. The transfer response function of X is simply $\lambda^k \psi X$.

EXAMPLE 9.2 (continued). In the second example of growth or decay to a new level, $n = 2$,

$$\boldsymbol{\theta}_t = \begin{pmatrix} \xi_t \\ \theta_t \end{pmatrix}, \quad \boldsymbol{\psi}_t = \boldsymbol{\psi} = \begin{pmatrix} 0 \\ \psi \end{pmatrix}, \quad \mathbf{F} = \begin{pmatrix} 1 \\ 1 \end{pmatrix}, \quad \mathbf{G} = \begin{pmatrix} 1 & 1 \\ 0 & \lambda \end{pmatrix},$$

with zero noise terms. The transfer response function is simply $\psi(1 - \lambda^{k+1})/(1-\lambda)$.

The general model (9.1) can be rewritten in the standard DLM form as follows. Define the new, $2n$-dimensional state parameters vector $\tilde{\boldsymbol{\theta}}_t$ by catenating $\boldsymbol{\theta}_t$ and $\boldsymbol{\psi}_t$, giving

$$\tilde{\boldsymbol{\theta}}_t' = (\boldsymbol{\theta}_t', \boldsymbol{\psi}_t').$$

Similarly, extend the $\mathbf{F}$ vector by catenating an n-vector of zeros, giving a new vector $\tilde{\mathbf{F}}$ such that

$$\tilde{\mathbf{F}}' = (\mathbf{F}', 0, \dots, 0).$$

For the evolution matrix, define

$$\tilde{\mathbf{G}}_t = \begin{pmatrix} \mathbf{G} & X_t \mathbf{I}_n \\ \mathbf{0} & \mathbf{I}_n \end{pmatrix},$$

where $\mathbf{I}_n$ is the $n \times n$ identity matrix. Finally, let $\boldsymbol{\omega}_t$ be the noise vector defined by

$$\boldsymbol{\omega}_t' = (\delta\boldsymbol{\theta}_t' + X_t\delta\boldsymbol{\psi}_t', \delta\boldsymbol{\psi}_t').$$

Then the model (9.1) can be written as

$$\begin{aligned} Y_t &= \tilde{\mathbf{F}}'\tilde{\boldsymbol{\theta}}_t + \nu_t, \\ \tilde{\boldsymbol{\theta}}_t &= \tilde{\mathbf{G}}_t\tilde{\boldsymbol{\theta}}_{t-1} + \boldsymbol{\omega}_t. \end{aligned} \tag{9.6}$$

Thus the transfer function model (9.1) has standard DLM form (9.6) and the usual analysis applies. Some particular features of the model in this setting require comment.

(a) The model as discussed provides just the transfer function for the variable X. In practice, this will usually be combined by superposition with other components (as in Example 9.1), such as trend, seasonality, regression and maybe even transfer functions of other independent variables.

(b) The unknown parameters in ψ_t play a role similar to the regression parameters in the dynamic regression model of Section 9.2 and is likely to be subject to some variation over time in particular applications. Thus, in some cases, more appropriate versions of the models in the two preceding examples would have ψ time dependent.

(c) Some of the the elements of ψ_t may be fixed and known. This can be modelled by setting to zero the corresponding values of the initial prior variance matrix $\mathbf{C}_0$ for $\tilde{\theta}_0$ and those of the evolution variance matrices $\mathbf{W}_t = V[\omega_t]$. However, it will often be the case that, from the structure of the model, there are zero elements in ψ_t, as is the case with the integrated transfer response function in Example 9.2. Then, for practical application, the general model (9.6) may be reduced in dimension to include just the non-zero elements of ψ_t. This is obviously desirable from a computational viewpoint if n is at all large. In general, suppose that just $p < n$ of the n elements of ψ_t are non-zero. The reader may verify that the model (9.6) may be reduced to one of dimension $n + p$, rather than $2n$, in which $\tilde{\mathbf{F}}' = (\mathbf{F}', 0, \dots, 0)$, having p trailing zero elements, and

$$\tilde{\mathbf{G}}_t = \begin{pmatrix} \mathbf{G} & X_t \Delta \\ \mathbf{0} & \mathbf{I}_p \end{pmatrix},$$

where Δ is an $n \times p$ matrix with just one unit element in each column, all other elements being zero. For instance, the model of Example 9.2 may be written as a 3−dimensional DLM with $\mathbf{F}' = (1, 1, 0)$ and

$$\tilde{\mathbf{G}}_t = \begin{pmatrix} 1 & 1 & 0 \\ 0 & \lambda & X_t \\ 0 & 0 & 1 \end{pmatrix}.$$

Thus, although the general model (9.6) always applies, it is often the case that a reduced form will be utilised.

(d) As specified, the model is developed from (9.1) and this results in a particular, structured form for the evolution noise vector, $\omega_t' = (\delta\theta_t' + X_t\delta\psi_t', \delta\psi_t')$. There are various ways of specifying the evolution variance matrices. The most direct and appropriate is to assume the evolution noise terms $\delta\boldsymbol{\theta}_t$ and $\delta\boldsymbol{\psi}_t$ uncorrelated with variance matrices $V[\delta\boldsymbol{\theta}_t] = \mathbf{U}_t$ and $V[\delta\boldsymbol{\psi}_t] = \mathbf{Z}_t$ respectively. It then follows that

$$\mathbf{W}_t = \begin{pmatrix} \mathbf{U}_t + X_t^2 \mathbf{Z}_t & X_t \mathbf{Z}_t \\ X_t \mathbf{Z}_t & \mathbf{Z}_t \end{pmatrix}.$$

Choice of $\mathbf{U}_t$ and $\mathbf{Z}_t$ is most simply guided by discount factors. The two subvectors $\boldsymbol{\theta}_t$ and $\boldsymbol{\psi}_t$ are naturally separated as distinct components of the model and so the component discounting concept of Section 6.3, Chapter 6 applies.

(e) The specific form of the stochastic structure in the evolution equation is not vital to the model, deriving as it does from the assumptions underlying (9.1). We can simply model the series directly using (9.2) and then impose *any* form on $\mathbf{W}_t$, simplifying the modelling process by choosing simple, discount based forms, for example.

(f) The discussion in Section 9.2.2 about the problems arising in forecasting ahead when future X values are unknown at the time is pertinent here. The only technical differences between the models arise through the appearance of the regressors in the evolution rather than the observation equation.

(g) More general models involve the concept of *stochastic* transfer responses, as the following example illustrates. Suppose a company uses, and re-uses, various alternative styles of advertising films or campaigns in promoting its products. Suppose that advertising effort is characterised by a simple measure X_t, such as advertising expenditure, and a basic transfer response model relates X_t to an output variable Y_t such as consumer demand or estimated awareness of the products. Then, although the model may adequately describe the $X - Y$ relationship over time during any given advertising campaign, it does not capture wider qualitative aspects of differences between films and their effects may differ widely. An appropriate extension of the model to cater for this sort of additional variation is to assume that the transfer response parameters are sampled from a population of such parameters. With reference to the simple decay of effects in Example 9.1, the transfer response function of film i may be taken as $\lambda^k \psi^{(i)} X$. The stochastic (at any given time) nature of the response is modelled by assuming, for example, that $\psi^{(i)} \sim N[\psi, U]$, independently over i. This can obviously be incorporated within the DLM form, as can be verifed by the reader, with extension to time-variation in the overall expected response parameter ψ about which the individual, film specific parameters $\psi^{(i)}$ are distributed. The variance U describes a second level of variation over and above any variation over time in ψ.

(h) $\mathbf{G}$, hence $\tilde{\mathbf{G}}_t$, typically depends on parameters that must be specified for the usual analysis to apply. In the two above examples, the decay parameter λ enters into $\mathbf{G}$. In many applications, it will be possible to specify values in advance; otherwise it may be desired to widen the analysis to allow for uncertainty in some of the elements of $\mathbf{G}$ and to incorporate learning about them. Some preliminary discussion of these sorts of problems appears in the next Section, further development being left to later chapters.

9.3.3 Learning about parameters in G: introductory comments

The final point in the above discussion raises, for the first time, issues of estimation in *non-linear models*, models in which the mean response function of the model has non-linear terms in some parameters. The estimation problems raised typify those of much more general, non-linear models. General concepts and techniques of non-linear estimation appear in later chapters. Here we restrict discussion to some basic, introductory comments in the context of simple transfer function models.

Consider the dynamic version of the model in Example 9.2, a DLM in which the evolution equation is

$$\theta_t = \lambda \theta_{t-1} + \omega_t.$$

For known λ, the usual analysis applies. Otherwise, if it is desired to learn about λ from the data, a much more complicated analysis is implied. With λ constant over time, though unknown, the formal analysis proceeds as follows.

(a) For each value of λ, in this example $0 < \lambda < 1$, specify an initial prior for θ_0 (and the observational variance V if unknown) of standard form. Note that this may now depend on λ and so is denoted $p(\theta_0 \mid \lambda, D_0)$. Also, specify the intial prior distribution, of any desired form, for λ, denoted $p(\lambda \mid D_0)$.

(b) For each value of λ, process the data according to the usual DLM with $\mathbf{G} = \lambda$. At time t, the one-step predictive density

$p(Y_t \mid \lambda, D_{t-1})$ and posteriors $p(\theta_t \mid \lambda, D_{t-1})$ etc. become available, defined by the usual normal or T forms, but with moments depending on the particular value of λ.

(c) Learning about λ proceeds via the sequential updating of the posterior distributions

$$p(\lambda \mid D_t) \propto p(\lambda \mid D_{t-1})p(Y_t \mid \lambda, D_{t-1}),$$

starting from the initial distribution provided in (a); here the observed one-step forecast density from (b) provides the likelihood function for λ. At each t, this posterior for λ is easily calculated as above up to a constant of normalisation, this constant being determined by integrating over the parameter space for λ, in this case the unit integral.

(d) Posterior inferences at time t for θ_t, Y_{t+1}, and other quantities of inference are based on their posterior distributions *marginal* with respect to λ. For example, the posterior for θ_t is defined by

$$p(\theta_t \mid D_t) = \int_0^1 p(\theta_t \mid \lambda, D_t)p(\lambda \mid D_t)d\lambda.$$

The first term in the integrand is the conditional normal or T density from (b), the second the posterior for λ from (c). Point estimates of θ_t and probabilities etc. may be calculated from this posterior via further integrations.

This formally defines the extended analysis. In practice, of course, it is impossible to perform this analysis exactly since it requires an infinite number of DLM analyses to be performed corresponding to the infinite number of values for λ. Thus approximations are used in which the parameter space is discretised, a finite, and often fairly small number of values of λ being considered, resulting in a discrete posterior distribution for each t in (c). Thus the integrals appearing in (d) become summations. This defines what may be called *multi-process models*, comprising a collection of DLMs analysed in parallel and mixed for inferences with respect to the posterior probabilities over values of λ. Chapter 12 is devoted to multi-process modelling. The particular analysis here can be viewed as the use of numerical integration in approximating the various integrals appearing in the formal theory. Uses of these, and other, related and more sophisticated techniques of numerical integration appear in Chapter 13.

9.3.4   Non-linear learning: further comments

Alternative approaches to the analysis of the DLM, extended to include parameters such as λ above are based on analytic approximation such as a *linearisation*. Again in the above simple model, group λ with θ_t in the vector $\theta'_t = (\theta_t, \lambda)'$ as a new model state vector, evolving according to the *non-linear* evolution equation

$$\theta_t = g(\theta_{t-1}) + \omega_t, \tag{9.7}$$

where $g(\theta_{t-1})$ is the non-linear vector function $\lambda(\theta_{t-1}, 1)'$ and $\omega_t = (\omega_t, 0)'$. Major computational problems now arise in the model analysis due to the non-linearity. The linearity basic to the DLM combines with the assumed normality to provide neat, tractable and efficiently updated sufficient summaries for all prior/posterior and predictive distributions of interest. Once this is lost, such distributions, although theoretically easily defined, can only be calculated through the use of numerical integration as described in the previous Section. Approaches using analytic approximations are typically quite easy to develop and apply, and are very commonly used since often more refined approximations are unnecessary. One of the important features of such approaches is that they naturally extend to cover models in which parameters such as λ here are themselves dynamic, varying over time. Various approximations based on linearisation of non-linear fuctions exist, the most obvious, and widely, applied, being based on Taylor series approximations as follows. Assume that, at time $t - 1$, the posterior distribution for $(\theta_{t-1} \mid D_{t-1})$ is adequately approximated by the usual distribution $(\theta_{t-1} \mid D_{t-1}) \sim T_{n_{t-1}}[\mathbf{m}_{t-1}, \mathbf{C}_{t-1}]$. A linear approximation to the non-linear function $g(.)$ in (9.7) based on a Taylor series exapansion about the estimate $\theta_{t-1} = \mathbf{m}_{t-1}$ leads to the *linearised* evolution equation

$$\theta_t \approx g(\mathbf{m}_{t-1}) + \mathbf{G}_t(\theta_{t-1} - \mathbf{m}_{t-1}) + \omega_t = \mathbf{h}_t + \mathbf{G}_t\theta_{t-1} + \omega_t, \tag{9.8}$$

where $\mathbf{h}_t = g(\mathbf{m}_{t-1}) - \mathbf{G}_t\mathbf{m}_{t-1}$ and $\mathbf{G}_t$ is the 2×2 matrix derivative of $g(\theta_{t-1})$ evaluated at $\theta_{t-1} = \mathbf{m}_{t-1}$; in this particular model,

$$\frac{\delta g(\theta_{t-1})}{\delta \theta'_{t-1}} = \begin{pmatrix} \lambda & \theta_{t-1} \\ 0 & 1 \end{pmatrix}.$$

When evaluated at $\theta_{t-1} = \mathbf{m}_{t-1}$, this provides a known matrix $\mathbf{G}_t$ and results in a DLM, albeit with a constant term $\mathbf{h}_t$ added to the evolution equation (a simple extension is discussed at the end of Section 4.3 of Chapter 4). This approximate model can be analysed as usual to give approximate prior, posterior and forecast distributions in standard forms. In particular, it follows from the linearised evolution equation that $(\theta_t \mid D_{t-1}) \sim T_{n_{t-1}}[a_t, R_t]$, where

$$a_t = h_t + G_t m_{t-1} = g(m_{t-1})$$

and

$$R_t = G_t C_{t-1} G_t' + W_t.$$

Note that R_t has the usual form, and the mean a_t is precisely the non-linear function $g(.)$ evaluated at the estimated value of θ_{t-1}. Thus the linearisation technique, whilst leading to a standard analysis, retains the non-linearity in propogating the state vector through time. Note that, as mentioned earlier, this approach clearly extends easily to cover cases in which λ is dynamic, incorporating the relevant terms in the evolution noise. Note finally that the linearised model, though derived as an approximation, may be interpreted as a prefectly valid DLM in its own right without reference to approximation. See Chapter 13 for further details of this, and other, approaches to non-linear models.

9.3.5   Further comments on transfer functions

The regression structure for form-free transfer functions has several atttractions that make it the most widely used approach to modelling lagged effects. One such is the simplicity and familiarity of the regression structure. A second, and probably the most important, attraction is the flexibility of regression models. Whatever the nature of a transfer function, the regression on lagged X values will allow the model to adapt adequately to the observed relationship so long as there is sufficient data and information available to appropriately estimate the parameters. In addition, and in particular with dynamic regression, as time evolves, the estimates will adapt to changes in the series allowing for changes in the response function. A related point is that the coefficients can also rapidly adapt to changes and inaccuracies in the timing of observations that can distort the observed relationship. Johnston and Harrison (1980) provide an interesting application in which form-free transfer functions are used as components of larger forecasting models in consumer sales forecasting.

By contrast, the approach using functional representations of transfer effects is rather less flexible since such models impose a particular form on the transfer function. There is some flexibility to adapt
in parameter estimation, particularly with time varying parameters,
but the imposed form still must be approximately appropriate for the
model to be useful. If the form is basically adequate, then the advantages of the functional form model relative to the form-free, regression model are apparent. Primarily, there will usually be many more
parameters in a regression model in order to adequately represent
a particular form of response. As mentioned earlier, it will usually
be desirable to model anticipated relationships amongst regression
coefficients using structured intial priors and evolution variance matrices, thus implictly recognising the form nature of the response.
A form model, having fewer parameters, provides a more efficient
and parsimonious approach. Finally, as earlier mentioned, regression models directly truncate the effects of lagged X whereas form
models allow for smooth decay over time without truncation.

9.4 NOISE MODELS

9.4.1 Introduction

Consider writing the observation equation of the DLM as $Y_t = \mu_t + \eta_t + \nu_t$, where $\mu_t = \mathbf{F}_t'\boldsymbol{\theta}_t$ is the usual mean response function
based upon a meaningful parameter set $\boldsymbol{\theta}_t$, ν_t is the observational
error, with the usual independence structure, and η_t is a new, unobservable term introduced to describe any persistent residual variation
over time not explained through μ_t and the evolution model for $\boldsymbol{\theta}_t$.
This residual variation, whilst expected to be zero, may be partially
predictable through a model for η_t as a stochastic process over time,
whose values are expected to display some form of dependence. η_t is
thus a correlated noise process, hence the terminology for the models developed here. A general model is simply that of a stochastic
process over time with some joint distribution for η_t, $(t = 1, 2, \dots)$,
through which η_t is related to past values η_{t-1}, η_{t-2}, etc. Popular noise models that form the basis for much of classical time series
analysis are based on the the theory of stationary processes. ARIMA
models, as commonly understood (eg. Box and Jenkins, 1976) are
typical.

A basic principle underlying the development of DLMs is that modelling effort be devoted to explaining variation through identifiable and meaningful model components. Only after as much significant variation as possible is explained in meaningful terms should noise model components be considered. Uninterpreted noise models are essentially confessions of ignorance and imply lack of control. Assuming a single, static noise model for a component of a series may be ill-advised since the peculiarities of past data can then be overly influential on current inferences and forecasts. Also, even if a noise component is generally appropriate, long data series are necessary to establish good estimates of the true character of the process. It is surprising how few people are aware of just how much data is needed to provide precise estimates of parameters in standard, static noise models, such as ARIMA models. And with dynamic noise models, that recognise and allow for time variation in the characteristics of noise processes, precision in estimation is further reduced since past history is discounted.

In Chapters 5 and 6 we saw how all types of ARIMA noise models can be represented as DLMs through their limiting forecast functions. For example, the constant component DLM with $\mathbf{F}' = (\mathbf{1}'_p, \mathbf{E}'_d, \mathbf{E}'_r)$ and $\mathbf{G} = \mathrm{diag}[\lambda_1, \ldots, \lambda_p; \mathbf{J}_d(1); \mathbf{J}_r(0)]$ has limiting forecast function capable of modelling any standard ARIMA(p, d, q) process. Further, dropping the assumed constancy of observation and evolution variances, DLMs generalise the class of ARIMA models. (Incidentally, the qualitative nature of classical model identification procedures is retained in this generalised framework). Here the ARIMA forms are given interpretation through the original, meaningful structure and parametrisation of the DLM. Otherwise, it is worth restating our view that modellers should regard uninterpreted or context unfounded noise models as a last resort in building forecasting systems. Nevertheless, much academic work concerns these models since the associated mathematical structure is tractable, particularly with respect to the assumption of stationarity, and allows the development of a theory of time series. It is the case that a study of this theory gives much insight into the more general area of dynamic modelling and so, with the above comments and cautionary notes in mind, we describe some features of classical noise modelling and illustrate alternative descriptions of noise processes that can be beneficial if wisely used.

9.4.2 Stationarity

Stationarity of the stochastic process, or simply time series, η_t involves assumptions of symmetry over time in the process, that may sometimes be judged appropriate in practice, sometimes not. The definition is conditional on any specified model structure for η_t as given, for example, by a specified observation and evolution equation of a DLM, and conditional on all required model parameters. This state of information is denoted by M (for model).

Definitions 9.2.

  (a) *A time series η_t, $(t = 1, 2, \ldots)$, is said to be* **stationary** *if the joint distribution (conditional on M) of any collection of k values is invariant with respect to arbitrary shifts of the time axis. In terms of joint densities this may be written, for any integers $k \geq 1$ and $s \geq 0$, and any k time points $t_1, \ldots, t_k$, as*

$$p(\eta_{t_1}, \ldots, \eta_{t_k} \mid M) = p(\eta_{s+t_1}, \ldots, \eta_{s+t_k} \mid M).$$

     Such a time series is said to have the property of stationarity.

  (b) *The series is said to be* **weakly stationary**, *alternatively* **second-order stationary**, *if, for all integers $t > 0$ and $s < t$,*

$$E[\eta_t \mid M] = \mu, \qquad \text{constant over time,}$$
$$V[\eta_t \mid M] = W, \qquad \text{constant over time,}$$

    and

$$C[\eta_t, \eta_{t-s} \mid M] = \gamma_s = \rho_s W, \qquad \text{independent of } t,$$

whenever these moments are finite.

Stationarity clearly implies weak stationarity for series whose second-order moments exist, stationarity being a much more far reaching symmetry assumption. A weakly stationary series has a fixed mean μ, variance W and *auto-covariances* γ_s depending only on the lag between values, not on the actual timings. The *auto-correlations* ρ_s, $(s = 0, 1, \ldots)$, thus determine the second-order structure of the series; note $\rho_0 = 1$ and $\gamma_0 = W$. Also, if any collection of any number of values are assumed jointly normally distributed given M, then weak stationarity together with normality implies stationarity. In the normal framework of DLMs, therefore, the two definitions coincide.

Suppose that η_t is a weakly stationary series. A basic result in the theory of stationary processes is that η_t can be *decomposed* into the sum of the mean μ and a linear combination of values in a (possibly infinite) sequence of zero-mean, uncorrelated random quantities. In representing this result mathematically, it is usual to extend the time index t backwards to zero and negative values. This is purely for convenience in mathematical notation and is adopted here for this reason and for consistency with general usage. Introduce a sequence of random quantities ϵ_t, $(t = \ldots, -1, 0, 1, \ldots)$, such that $E[\epsilon_t \mid M] = 0$, $V[\epsilon_t \mid M] = U$ and $C[\epsilon_t, \epsilon_s \mid M] = 0$ for all $t \neq s$, and some variance U. Then the representation of η_t is given by

$$\eta_t = \mu + \sum_{r=0}^{\infty} \pi_r \epsilon_{t-r}, \tag{9.9}$$

for some (possibly infinite) sequence of coefficients $\pi_0, \pi_1, \ldots$, with $\pi_0 = 1$. Note that this formally extends the time series η_t to $t \leq 0$. The representation (9.9) implies some obvious restrictions on the coefficients. Firstly, $V[\eta_t \mid M] = W$ implies that $W = U \sum_{r=0}^{\infty} \pi_r^2$, so that the sum of squared coefficents must converge. Secondly, the auto-covariances γ_s may be written in terms of sums of products of the π_r which must also converge.

There is a huge literature on mathematical and statistical theory of stationary stochastic processes and time series analysis for stationary series. From the time series viewpoint, Box and Jenkins (1976) provide comprehensive coverage of the subject, with many references. Classical time series analysis is dominated by the use of models that can be written in the form (9.9), special cases of which are now introduced. The above reference provides much further and fuller development. Firstly, however, recall the *backshift operator* B such that, for any series X_t and any t, $B^r X_t = X_{t-r}$ for all $r \geq 0$. Write (9.9) as

$$\eta_t = \mu + \sum_{r=0}^{\infty} \pi_r B^r \epsilon_t = \mu + \pi(B)\epsilon_t,$$

where $\pi(.)$ is a polynomial function (of possibly infinite degree), given by $\pi(x) = 1 + \pi_1 x + \pi_2 x^2 + \ldots$, for any x with $|x| < 1$. Formally, this equation may be expressable as $\psi(B)(\eta_t - \mu) = \epsilon_t$, where $\psi(.)$ is another polynomial of the form $\psi(x) = 1 - \psi_1 x - \psi_2 x^2 - \ldots$, satisfying the identity $\pi(x)\psi(x) = 1$ for all x. The coefficients ψ_r

here are determined as functions of the π_r. If such a function $\psi(.)$ exists, then (9.9) has the equivalent representation

$$\eta_t - \mu = \epsilon_t + \sum_{r=1}^{\infty} \psi_r B^r (\eta_t - \mu) = \epsilon_t + \psi_1 (\eta_{t-1} - \mu) + \psi_2 (\eta_{t-2} - \mu) + \ldots .$$

$$(9.10)$$

9.4.3 Autoregressive models

Suppose (9.10) to hold with $\psi_r = 0$ when $r > p$ for some integer $p \geq 1$. Then η_t is said to be an *autoregressive process of order* p, denoted AR(p) for short and written as $\eta_t \sim$ AR(p). Here

$$\eta_t = \mu + \sum_{r=1}^{p} \psi_r (\eta_{t-r} - \mu) + \epsilon_t,$$

the value at time t depending linearly on a finite number of past values. It is an easy consequence (see, for example, Abraham and Ledholter, 1983, Chapter 5; Box and Jenkins, 1976) that the autocorrelations ρ_s may be non-zero for large values of s. The AR(p) process is Markovian in nature, with $p(\eta_t \mid \eta_{t-1}, \ldots, \eta_{t-p}, \eta_{t-p-1}, \ldots, M) = p(\eta_t \mid \eta_{t-1}, \ldots, \eta_{t-p}, M)$. If normality is assumed, then, for $t > p$ and any specified initial information D_0,

$$(\eta_t \mid \eta_{t-1}, \ldots, \eta_{t-p}, D_0) \sim \mathrm{N}[\mu + \sum_{r=1}^{p} \psi_r (\eta_{t-r} - \mu), U].$$

EXAMPLE 9.3: AR(1) model. If $p = 1$ then $\eta_t = \mu + \psi_1 (\eta_{t-1} - \mu) + \epsilon_t$. Here it easily follows, on calculating variances, that $W = \psi_1^2 W + U$. Hence, if $|\psi_1| < 1$, $W = U/(1 - \psi_1^2)$. The condition that the first order autoregression coefficient be less than unity in modulus is called the *stationarity* condition. Further, autocorrelations are given by $\rho_s = \psi_1^s$ so that stationarity is necessary in order that the autocorrelations are valid. Note then that ρ_s decays exponentially to zero in s, although is non-zero for all s. The polynomial $\psi(.)$ here is simply $\psi(x) = 1 - \psi_1 x$, with

$$\pi(x) = 1/\psi(x) = 1 + \psi_1 x + \psi_2 x^2 + \ldots, \qquad (|x| < 1)$$

Thus, in terms of (9.9), $\pi_r = \psi_1^r = \rho_r$ for $r \geq 0$.

Higher order models, $AR(p)$ for $p > 1$, can provide representations of stationary series with essentially any form of correlation structure that is sustained over time but decays eventually (and exponentially) towards zero. In higher-order models too the ψ coefficients are subject to stationarity restrictions that lead to valid autocorrelations.

9.4.4   Moving-average models

Suppose (9.9) to hold with $\pi_r = 0$ for $r > q$, some integer $q \geq 1$. Then η_t is said to be a *moving average process of order* q, denoted $MA(q)$ for short and written as $\eta_t \sim MA(q)$. Thus η_t depends on only a finite number $q + 1$ of the ϵ_t,

$$\eta_t = \mu + \epsilon_t + \pi_1 \epsilon_{t-1} + \ldots + \pi_q \epsilon_{t-q}.$$

It is an easy consequence that η_t and η_{t-s} are unrelated for $s > q$, depending as they do entirely on distinct ϵ terms that are uncorrelated. Thus $\rho_s = 0$ for $s > q$. For q relatively small, therefore, the MA process is useful for modelling *local* dependencies amongst the η_t.

EXAMPLE 9.4: MA(1) model. If $q = 1$ then $\eta_t = \mu + \epsilon_t + \pi_1 \epsilon_{t-1}$, values more than one step apart being uncorrelated. It easily follows that $W = (1 + \pi_1^2)U$ and $\rho_1 = \pi_1/(1 + \pi_1^2)$. Note that $|\rho_1| \leq 0.5$. There is an identifiability problem inherent in this, and all other, MA representations of a stationary process; note that replacing π_1 by π_1^{-1} leads to the same value for ρ_1, so that there are two, essentially equivalent, representations for the MA(1) process with a given value of ρ_1. However, here $\pi(x) = 1 + \pi_1 x$ so that the representation in terms of (9.10) leads to $\eta_t = \mu + \eta_t - \pi_1(\eta_{t-1} - \mu) + \pi_1^2(\eta_{t-2} - \mu) - \ldots$. For $|\pi| < 1$, the powers π^r decrease exponentially to zero, naturally reflecting a rapidly diminishing effect of values in the past on the current value. If, however, $|\pi| > 1$ then this implies an increasing dependence of η_t on values into the distant past, which is obviously embarrassing. For this (and other, similarly ad-hoc) reasons, the MA model is typically subjected to *invertibility* conditions that restrict the values of the π coefficients, just as the stationarity conditions that apply in AR models. Invertibility gives an illusion of uniqueness and serves to identify a single model whose limiting errors converge in probability to the ϵ_t. In this case the restriction is to $|\pi_1| < 1$ (see the earlier references for full discussion of such conditions).

9.4.5 ARMA models

Given the required stationarity conditions, any $AR(p)$ model $\psi(B)(\eta - \mu) = \epsilon_t$ implies the representation (9.9) with coefficients π_r that decay towards zero as r increases. Thus, as an approximation, taking q sufficiently large implies that any AR model can be well described by an $MA(q)$ model, possibly with q large. Similarly, an invertible $MA(q)$ model can be written in autoregressive form with coefficients ψ_r decaying with r, and similar thinking indicates that, for large enough p, this can be well described by an $AR(p)$ model. Thus, with enough coefficients, any stationary process can be well approximated by using either AR or MA models. Combining AR and MA models can, however, lead to adequate representations with many fewer parameters, and the classical modelling of Box and Jenkins (1976) is based on autoregressive-moving average models, ARMA for short. An $ARMA(p, q)$ model for η_t is given by

$$\eta_t = \mu + \sum_{r=1}^{p} \psi_r(\eta_{t-r} - \mu) + \sum_{r=1}^{q} \pi_r \epsilon_{t-r} + \epsilon_t,$$

where, as usual, the ϵ_t are zero-mean, uncorrelated random quantities with constant variance U. The above references fully develop the mathematical and statistical theory of ARMA models. They describe a wide range of possible stationary autocorrelation structures, subject to various restrictions on the ψ and π coefficients.

9.4.6 ARMA models in DLM form

All ARMA noise models can be written in DLM form in a variety of ways and for a variety of purposes. Suppose here that the series η_t is stationary with zero-mean, $\mu = 0$, and is actually observed. In the context of modelling Y_t in Section 9.4.1, this is equivalent to supposing that $Y_t = \eta_t$ for all t.

Suppose first that an $AR(p)$ model is used for η_t. Then the model is essentially a static regression sequentially defined over time as mentioned in Section 9.1. Simply note that $\eta_t = \mathbf{F}'_t \boldsymbol{\theta} + \nu_t$ where $\nu_t = \epsilon_t$, $\mathbf{F}'_t = (\eta_{t-1}, \ldots, \eta_{t-p})$ and $\boldsymbol{\theta}' = (\psi_1, \ldots, \psi_p)$. Thus the sequential analysis of dynamic regression applies to this, particular static model to estimate the elements of $\boldsymbol{\theta}$ and the variance $V = U$ of the independent errors, and to forecast future Y values. Forecasting ahead leads to computational problems in calculating forecast distributions, however, since future values are needed as regressors. Zellner (1971) discusses such problems. See also Broemeling (1985) and Schnatter (1988).

With an MA(q) model for the observed series, there is no standard regression type representation, and hence no neat, standard analysis. Use of an AR approximation of possibly high order is one way of obtaining such an analysis. Alternatively, the model may be written sequentially in DLM form to enable the sequential calculation of a likelihood for the unknown π parameters. There are several, equivalent ways of doing this, but all the representations so obtained easily extend to the ARMA model. For this reason, the general case of an ARMA(p, q) model is given, the MA(q) an obvious special case when $p = 0$. This also provides an alternative to the above representation for pure autoregressive models when $q = 0$.

For the usual ARMA model with $\mu = 0$,

$$\eta_t = \sum_{r=1}^{p} \psi_r \eta_{t-r} + \sum_{r=1}^{q} \pi_r \epsilon_{t-r} + \epsilon_t,$$

define $n = \max(p, q+1)$, and extend the ARMA coefficients to $\psi_r = 0$ for $r > p$, and $\pi_r = 0$ for $r > q$. Introduce the $(n \times 1)$ state vector $\boldsymbol{\theta}_t$ whose first element is the current observation η_t. Further define $\mathbf{F} = \mathbf{E}_n = (1, 0, \ldots, 0)'$,

$$\mathbf{G} = \begin{pmatrix} \psi_1 & 1 & 0 & \cdots & 0 \\ \psi_2 & 0 & 1 & \cdots & 0 \\ \vdots & \vdots & \vdots & \ddots & \vdots \\ \psi_{n-1} & 0 & 0 & \cdots & 1 \\ \psi_n & 0 & 0 & \cdots & 0 \end{pmatrix}$$

and

$$\boldsymbol{\omega}_t = (1, \pi_1, \ldots, \pi_{n-1})' \epsilon_t.$$

With these definitions, it can be verified that the ARMA model may be rewritten as

$$\begin{aligned} \eta_t &= \mathbf{F}' \boldsymbol{\theta}_t, \\ \boldsymbol{\theta}_t &= \mathbf{G} \boldsymbol{\theta}_{t-1} + \boldsymbol{\omega}_t, \end{aligned} \tag{9.11}$$

Thus, conditional on the defining parameters and also on D_{t-1}, the ARMA process has the form of an observational noise-free DLM of dimension n. The evolution noise has a specific structure, giving evolution variance matrix

$$\mathbf{U} = U(1, \pi_1, \ldots, \pi_{n-1})(1, \pi_1, \ldots, \pi_{n-1})'. \tag{9.12}$$

Given an initial prior for $\boldsymbol{\theta}_0$, the standard analysis applies *conditional* on values of the defining parameters ψ_r and π_r being specified. Note that the model is a TSDLM and so the forecast function is given simply by

$$f_t(k) = \mathrm{E}[\eta_{t+k} \mid D_t] = \mathbf{F}'\mathbf{G}^k\mathbf{m}_t, \qquad (k = 1, 2, \dots),$$

where $\mathbf{m}_t = \mathrm{E}[\boldsymbol{\theta}_t \mid D_t]$ as usual. It follows, as with all TSDLMs, that the eigenvalues of $\mathbf{G}$ determine the form of the forecast function and that the model has a canonical Jordan block representation (see Exercise 12). In addition, Theorem 5.1 of Chapter 5 implies that there is a stable limiting form for the updating equations determined by a limiting value for the sequence of posterior variance matrices $\mathbf{C}_t$ of $\boldsymbol{\theta}_t$.

EXAMPLE 9.3 (continued). In the AR(1) case $p = 1$ and $q = 0$ so that $n = 1$, $\mathbf{G} = \psi_1$, and, directly, $\eta_t = \psi_1\eta_{t-1} + \epsilon_t$. Forecasting ahead gives $f_t(k) = \psi_1^k m_t$ where $m_t = \eta_t$. The stationarity condition $|\psi_1| < 1$ implies an exponential decay to zero, the marginal mean of η_{t+k}, as k increases.

EXAMPLE 9.4 (continued). In the MA(1) case $p = 0$ and $q = 1$ so that $n = 2$. Here

$$\mathbf{G} = \begin{pmatrix} 0 & 1 \\ 0 & 0 \end{pmatrix} = \mathbf{J}_2(0)$$

giving the component model $\{\mathbf{E}_2, \mathbf{J}_2(0), ., .\}$. Setting $\boldsymbol{\theta}_t' = (\eta_t, \theta_t)$, the evolution equation gives $\eta_t = \theta_{t-1} + \epsilon_t$ and $\theta_t = \pi_1\epsilon_t$. The standard MA(1) representation follows by substitution, namely $\eta_t = \epsilon_t + \pi_1\epsilon_{t-1}$. Here $\mathbf{G}^k = \mathbf{0}$ for $k > 1$ so that, with $\mathbf{m}_t' = (\eta_t, m_t)$, the forecast function is given by $f_t(1) = f_{t+1} = m_t$ and $f_t(k) = 0$ for $k > 1$.

EXAMPLE 9.5. In an MA(q) model for any q, it similarly follows that the forecast function takes irregular values up to $k = q$, being zero, the marginal mean of η_{t+k}, thereafter. The canonical component model is $\{\mathbf{E}_{q+1}, \mathbf{J}_{q+1}(0), ., .\}$. (Incidentally, it is possible to incorporate the observational noise component ν_t directly into this noise model, if desired.)

Much further development and discussion of similar and related representations of ARMA models (usually referred to as state-space representations) can be found in Abraham and Ledolter (1983), Harvey (1981), Priestley (1980), and Young (1984), for example. Harvey, in particular, discusses initial conditions and limiting forms of the updating equations.

9.4.7   Dynamic noise models as component DLMs

As discussed in Section 9.4.1, noise models, such as ARMA models, have the ability to describe and partially predict residual structure in a series that is not explained by standard components. In classical ARMA modelling, such as exemplified by Box and Jenkins (1976), practical application is approached by first transforming the Y_t series in an attempt to achieve a transformed series η_t that is, at least approximately, zero mean and stationary. In addition to the usual possibilities of instantaneous, non-linear data transformations to correct for variance inhomogeneities, data series are usually subject to *differencing* transforms. The first-order difference of Y_t is simply defined as $(1 - B)Y_t = Y_t - Y_{t-1}$, the second-order $(1 - B)^2 Y_t = Y_t - 2Y_{t-1} + Y_{t-2}$, and higher-order differences similarly defined by $(1 - B)^k Y_t$. In attempting to derive stationarity this way, it is important that the original Y_t series be non-seasonal. If seasonality of period p is evident, then Y_t may be first subjected to seasonal differencing, producing the transformed series $(1 - B^p)Y_t$. Thus the classical strategy is to derive a series of the form $\eta_t = (1 - B)^k (1 - B^p)Y_t$, for some k and p, that may be assumed to be stationary. Once this is approximately obtained, residual autocorrelation structure in the η_t series is modelled within the above ARMA framework. Full discussion of differencing, and the relationship with the assumptions of stationarity of differenced series, can be found in the references in Section 9.4.3.

The DLM approach to modelling the original time series Y_t directly is obviously rather different. In certain, special cases there are similarities between the approaches; these have been made apparent, for example, in deriving restricted ARMA forms of limiting forecast functions in constant models, Theorem 5.2 of Chapter 5, typified by the polynomial growth and seasonal TSDLMs of Chapters 7 and 8 respectively. The concept of a noise model, such as an ARMA process, for residual autocorrelation structure in a DLM provides another point of contact, although the use of differencing and the assumptions of derived stationarity are not necessary nor appropriate in many contexts. In structuring a DLM for the Y_t series, the use of trend, seasonal and regression components provides an interpretable description for the, often highly non-stationary, basic development over time. Given such a model, the application of differencing transformations serves mainly to confuse the interpretation of the model, confound the components, and highlight noise at the expense of meaningful interpretations. In particular, with regression effects the differencing carries over to transform independent variables in the model, obscuring the nature of the regression relationships. This confusion is exacerbated by the appearance of time varying parameters that are so fundamental to short-term forecasting performance. Furthermore, the occurrence of abrupt changes in time series structure, that may be directly modelled through a DLM representation, evidence highly non-stationary behaviour that cannot usually be removed by differencing or other data transformations. Thus noise models are best used directly as "last resort" components in larger DLMs that already describe the non-stationary behaviour of the Y_t series through standard components for trend, seasonality and regressions. ARMA type models of low order, in particular those with $p \leq 2$ and $q \leq 2$, can be employed as descriptions of local residual autocorrelation in the series not already adequately described by the basic DLM. Given a DLM, $\{\mathbf{F}_t, \mathbf{G}_t, V_t, \mathbf{W}_t\}$, superposition of the ARMA process η_t described by (9.11) and (9.12) leads to the model

$$\left\{ \begin{pmatrix} \mathbf{F}_t \\ \mathbf{F} \end{pmatrix}, \begin{pmatrix} \mathbf{G}_t & \mathbf{0} \\ \mathbf{0} & \mathbf{G} \end{pmatrix}, V_t, \begin{pmatrix} \mathbf{W}_t & \mathbf{0} \\ \mathbf{0} & \mathbf{U} \end{pmatrix} \right\}. \tag{9.13}$$

The observation equation has the form introduced in Section 9.1, namely $Y_t = \mu_t + \eta_t + \nu_t$ where μ_t is the standard mean response function from the original DLM and ν_t the usual observational error. Some comments and special cases of interest are as follows.

(1) If $V_t = 0$ so that $\nu_t = 0$ then η_t plays the role of a correlated observational noise process directly. In some applications it may be appropriate to assume that, relative to the noise variance U, additional observational variation is negligible so that this assumption is appropriate. Otherwise, the ν_t term may be important, representing unpredictable, instantaneous variation in Y_t due, for example, to sampling errors, errors in measurement, definition, timing, explained or unexplained outliers, and so forth.

(2) If the original model is null so that $\mu_t = \nu_t = 0$ with probability one, then the simple noise process is observed directly, $Y_t = \eta_t$. Thus the general formulation includes all stationary ARMA models as special cases.

9.4.8 Non-linear learning problems

The model (9.13) is a standard DLM when the parameters ψ_r and π_r of the noise process are specified. Usually, as just mentioned, a rather low order noise model is adequate, representing residual autocorrelations that are appreciable only for small lags. If the DLM defining the μ_t component is adequate, then the noise process will have negligible autocorrelations and be unnecessary. If the DLM is under-adaptive, such as with discount factors that are too small, then the noise process will exhibit locally positive correlations such as modelled by an AR(1) with $\psi_1 > 0$, and usually rather small. With an over-adaptive DLM the correlations are negative and similar comments apply. Sometimes, a forecaster may proceed by specifying a low order model with values of the coefficients assessed directly with these points in mind. However, more generally there is a need to learn about the coefficients. Unfortunately, as with the transfer reponse models in Section 9.3, there is no neat analysis, the parameters combining with the state vector in such a way as to introduce non-linear terms in unknown quantities into the model. Thus, as with those transfer function models, various approximate methods of analysis are necessary.

Theoretically, as in Section 9.3.3 for transfer function models, the extended analysis is well-defined. Group the unknown, but assumedly constant, ARMA parameters in a vector

$$\lambda = (\psi_1, \dots, \psi_p; \pi_1, \dots, \pi_q)'.$$

This is a parameter vector in $p + q$ dimensions. Conditional on any value for λ, the noise process is a standard component of the DLM (9.13) and the Y_t series may be processed according to the usual DLM analysis.

(a) For each value of λ, specify an initial prior for θ_0 (and the observational variance V if unknown) of standard form. Note that this may now depend on λ and so is denoted by $p(\theta_0 \mid \lambda, D_0)$. Also, specify the intial prior distribution, of any desired form, for λ, denoted by $p(\lambda \mid D_0)$.

(b) For each value of λ, process the data according to the usual DLM. At time t, the one-step predictive density $p(Y_t \mid \lambda, D_{t-1})$ and posteriors $p(\boldsymbol{\theta}_t \mid \lambda, D_{t-1})$ etc. become available, defined by the usual normal or T forms, but with moments depending on the particular value of λ.

(c) Learning about λ proceeds via the sequential updating of the posterior distributions

$$p(\lambda \mid D_t) \propto p(\lambda \mid D_{t-1})p(Y_t \mid \lambda, D_{t-1}), \qquad (9.14)$$

starting from the initial distribution provided in (a); here the observed one-step forecast density from (b) provides the likelihood function for λ. At each t, this posterior for λ is easily calculated as above up to a constant of normalisation, this constant being determined by integrating over the parameter space for λ.

(d) Posterior inferences at time t for $\boldsymbol{\theta}_t$, Y_{t+1}, and other quantities of inference are based on their posterior distributions *marginal* with respect to λ. For example, the posterior for $\boldsymbol{\theta}_t$ is defined by

$$p(\boldsymbol{\theta}_t \mid D_t) = \int p(\boldsymbol{\theta}_t \mid \lambda, D_t)p(\lambda \mid D_t)d\lambda. \qquad (9.15)$$

The first term in the integrand is the conditional normal or T density from (b), the second the posterior for λ from (c). Point estimates of $\boldsymbol{\theta}_t$ and probabilities etc. may be calculated from this posterior via further integrations.

The impracticability of the analysis is evident as it was in Section 9.3.3; the best that can be done in practice is to use a discrete approximation, choosing a collection of values Λ for λ and performing a discretised analysis. Thus the posterior for λ in (9.14) becomes a discrete mass function, normalised by summation over Λ. Also the density in (9.15), and all other densities conditional on λ and D_t (for any t) become discrete mixtures. In the case of $\boldsymbol{\theta}_t$ in (9.15), this gives a mixture of normal or T densities

$$p(\boldsymbol{\theta}_t \mid D_t) = \sum_\lambda p(\boldsymbol{\theta}_t \mid \lambda, D_t)p(\lambda \mid D_t)d\lambda.$$

From (9.14) it follows that

$$p(\lambda \mid D_t) \propto p(\lambda \mid D_0) \prod_{r=1}^{t} p(Y_r \mid \lambda, D_{r-1}) = p(\lambda \mid D_0)p(\mathbf{Y}_t \mid \lambda, D_0),$$

where $p(\mathbf{Y}_t \mid \boldsymbol{\lambda}, D_0)$, as a function of $\boldsymbol{\lambda}$, is the likelihood function based on the series of observation up to time t. With a series of fixed length, calculation of this likelihood at a collection of values of $\boldsymbol{\lambda}$ provides a guide as to plausible values in cases when, as is often true in practice, the initial prior is rather diffuse. Calculation of this likelihood over a plausible grid of values is the basis of the analyses of pure ARMA models in Ansley (1979), Box and Jenkins (1976), Gardner, Harvey and Phillips, (1980). Some discussion of these, and related issues for stationary ARMA components, appears also in Harvey (1981).

9.4.9 Dynamic noise model parameters and extended learning problems

Some final comments on learning problems in models with noise components are in order.

(1) As has been stressed throughout, the noise term is a refinement of a structured forecasting system that should usually be based on relatively few parameters. Even in such cases, however, the above analysis can obviously become computationally very demanding. Thus there is a need for the development of computationally efficient approaches to the problem. This is, at time of writing, an area for much further research and development. See Marriot (1987), Schnatter (1988), and Pole and West (1988) for some initial works in this area. This, and other topics such as choice of p and q, initial priors on noise terms, and so forth, are not considered further here.

(2) Analytic approximations to the analysis can be developed using linearisation techniques as for transfer function models in Section 9.3.4. In the basic noise model (9.11), for example, if $\boldsymbol{\theta}_t$ is extended to incorporate both $\boldsymbol{\lambda}$ and ϵ_t, it follows that the extended state vector undergoes an evolution of the form (9.7) for some non-linear function $\mathbf{g}(.)$. The derivation of this is left to the reader. Once in this form, the linearisation as described in Section 9.3.4 may be applied to provide an approximate, linearised DLM for learning about the new state vector, thus about $\boldsymbol{\lambda}$ in addition to $\boldsymbol{\theta}_t$.

(3) In applications in areas of physical and engineering sciences, such as typified by control engineering, there may be underlying process theory that leads to the development of stationary noise models as basic to the application. Here then, constant

parameter ARMA models may be well-founded. By contrast, in commercial and socio-economic applications, stationarity of derived series is often much more of a convenience than an underlying feature. In particular, the assumption that the parameters of an ARMA noise component remain constant over time is one way in which stationarity can be relaxed whilst retaining the basic model form. A simple random walk assumption for the noise model parameters λ provides a generalised model in which the ARMA type autocorrelation structure is essentially retained *locally*, though possibly varying as time progresses. This results in a model of the form (9.11) and (9.12) but in which $\mathbf{G}$ and $\mathbf{U}$ become time dependent since they are functions of λ.

9.5   EXERCISES

(1) Consider the discount regression DLM in which V_t is known for all t and $\mathbf{R}_t = \mathbf{C}_{t-1}/\delta$. Show that the updating equations can be written as

$$\mathbf{m}_t = \mathbf{C}_t(\delta^t \mathbf{C}_0^{-1} \mathbf{m}_0 + \sum_{r=0}^{t-1} \delta^r \mathbf{F}_{t-r} V_{t-r}^{-1} Y_{t-r})$$

and

$$\mathbf{C}_t^{-1} = \delta^t \mathbf{C}_0^{-1} + \sum_{r=0}^{t-1} \delta^r \mathbf{F}_{t-r} \mathbf{F}_{t-r}' V_{t-r}^{-1}.$$

Deduce that, if $\mathbf{C}_0^{-1} \approx \mathbf{0}$, $\mathbf{m}_t$ is approximately given by the exponentially weighted regression (EWR) form

$$\mathbf{m}_t \approx (\sum_{r=0}^{t-1} \delta^r \mathbf{F}_{t-r} \mathbf{F}_{t-r}' V_{t-r}^{-1})^{-1} (\sum_{r=0}^{t-1} \delta^r \mathbf{F}_{t-r} V_{t-r}^{-1} Y_{t-r}).$$

(2) Consider the reference prior analysis of the static model $\{\mathbf{F}_t, \mathbf{I}, V, \mathbf{0}\}$, referring to the general theory of reference analyses in Section 4.8 of Chapter 4. As in that Section, let

$$\mathbf{K}_t = \sum_{r=1}^{t} \mathbf{F}_r \mathbf{F}_r', \qquad \mathbf{k}_t = \sum_{r=1}^{t} \mathbf{F}_r Y_r,$$

and $t = [n]$ be the first time $t \geq n$ such that $\mathbf{K}_t$ is non-singular. Using the results of Theorem 4.6 and Corollary 4.5 of Section 4.8, Chapter 4, verify that the reference analysis leads to

$$\mathbf{C}_{[n]} = \mathbf{K}_{[n]}^{-1} S_{[n]}$$

and

$$\mathbf{m}_{[n]} = \mathbf{K}_{[n]}^{-1} \mathbf{k}_{[n]}.$$

(3) The table below provides data on weekly sales of a product over a number of standardised, four-weekly months, together with the corresponding values of a compound index of market buoyancy and product competitiveness. Take Sales as response, Index as independent variable, and consider fitting dynamic straight line regression models to the series to explain and predict Sales based on Index. Note that several observations are missing, being denoted by asterisks in the table.

Consider single discount models with $\mathbf{W}_t = \mathbf{C}_{t-1}(\delta^{-1} - 1)$ for all t for updating, and assume that the observational variance

Sales and advertising Index series
(Missing data indicated by *)

| Month | SALES Y_t | | | | INDEX X_t | | | |
| | Week | | | | Week | | | |
	1	2	3	4	1	2	3	4
1	102.29	101.18	100.49	100.31	0.00	0.00	0.17	0.26
2	99.39	101.69	99.96	105.25	0.21	0.22	0.23	0.29
3	103.71	99.21	99.82	100.95	0.28	0.05	0.00	0.00
4	101.91	100.46	102.81	101.59	0.13	0.18	0.24	0.05
5	104.93	101.95	101.01	99.03	0.21	0.05	0.00	0.14
6	100.80	101.90	98.35	99.21	0.15	0.03	0.00	0.00
7	99.87	101.30	98.42	98.66	0.15	0.23	0.05	0.14
8	103.16	102.52	103.08	101.37	0.16	0.27	0.24	0.18
9	101.75	100.76	97.21	100.47	0.25	0.21	0.25	0.05
10	99.10	***	***	***	0.00	***	***	***
11	***	***	***	***	***	***	***	***
12	105.72	106.12	103.41	97.75	0.17	0.27	0.06	0.00
13	101.18	101.61	105.78	98.35	0.00	0.22	0.21	0.04
14	104.08	102.02	100.33	101.34	0.17	0.04	0.00	0.13
15	100.37	99.49	103.39	101.32	0.03	0.15	0.24	0.20
16	100.26	102.70	100.40	101.85	0.28	0.20	0.03	0.24
17	103.50	95.41	99.19	103.05	0.06	0.00	0.14	0.03
18	99.79	103.49	102.93	102.63	0.13	0.15	0.27	0.06
19	102.72	101.52	98.55	101.07	0.00	0.00	0.16	0.04
20	99.65	104.59	104.24	100.74	0.16	0.20	0.23	0.05

is constant. Assume the initial information is summarised by $\mathbf{a}_1 = (100, 0)'$, $\mathbf{R}_1 = \text{diag}(25, 10)$, $n_0 = 10$ and $S_0 = 1.5$.

(a) Explore sensitivity to the value of δ by fitting the model to the data for values of δ representing the range $0.75 \leq \delta \leq 1$. Examine sensitivity by plotting the on-line and filtered estimates of the time trajectories of the intercept parameter α_t and the regression parameter β_t.

(b) Assess support from the data for the values of δ by calculating the aggregate LLR measure for each value considered.

(4) Analyse the Sales/Index series above with $\delta = 0.95$, recording the values of the posterior summary quantities $\mathbf{m}_{80}$, $\mathbf{C}_{80}$, S_{80} and n_{80} at the final observation stage $t = 80$, corresponding to week 4 of month 20. Looking ahead, suppose that the step-ahead values of Index over the coming four weeks are calculated as

$$X_{81} = 0.15, \quad X_{82} = 0.22, \quad X_{83} = 0.14 \quad \text{and} \quad X_{84} = 0.03.$$

Taking $\mathbf{W}_{80+k} = \mathbf{W}_{81} = \mathbf{C}_{80}(\delta^{-1} - 1)$, constant for forecasting ahead from time 80 (as usual in discount models), calculate the following forecast distributions.

(a) The marginal forecast distributions $p(Y_{80+k}|D_{80})$ for each of $k = 1, 2, 3$ and 4.

(b) The full, joint forecast distribution for the 4 quantities,

$$p(Y_{81}, \dots, Y_{84}|D_{80}).$$

(c) $p(Z|D_{80})$ where $Z = \sum_{r=81}^{84} Y_r$, the cumulative total Sales over the coming month.

(d) The conditional forecast distribution

$$p(Y_{81}, \dots, Y_{84}|Z, D_{80})$$

for any given value of the total Z.

(5) A forecast system uses the first-order polynomial model $\{1, 1, 100, 5\}$ as a base model for forecasting sales of a group of products. Initially, $(\mu_0|D_0) \sim N[400, 20]$. The effect of a price increase at $t = 1$ is modelled via the addition of a transfer response effect E_t, giving

$$Y_t = \mu_t + E_t + \nu_t, \qquad (t = 1, 2, \dots),$$

where

$$E_t = 0.9 E_{t-1} + \delta E_t, \qquad (t = 2, 3, \ldots),$$

and

$$(\delta E_t | D_0) \sim N[0, 25],$$

independently of $(\mu_1 | D_0)$ and $(\omega_t | D_0)$ for all t. It is also assumed that, initially,

$$(E_1 | D_0) \sim N[-40, 200]$$

independently of $(\mu_1 | D_0)$ and $(\omega_t | D_0)$.

(a) Calculate the forecast distributions $p(Y_t | D_0)$ for each of $t = 1, 2, 3$ and 4.

(b) Given $Y_1 = 377$, calculate $p(\mu_1, E_1 | D_1)$ and revise the forecast distributions for the future, calculating $p(Y_t | D_1)$ for each of $t = 2, 3$ and 4.

(c) Perform similar calculations at $t = 2$ given $Y_2 = 404$. Calculate $p(\mu_2, E_2 | D_2)$ and $p(Y_t | D_2)$ for $t = 3$ and 4.

(6) It is desired to produce a component DLM which models the form response of stock market indicators to stock market news. The general form of response anticipated is that of an increase (or decrease) to a new level after damped, cyclic variation due to under/over reaction of market makers. Discuss the forms of the forecast functions of the two models:

$$\left\{ \begin{pmatrix} 1 \\ 0 \end{pmatrix}, \begin{pmatrix} 1 & 1 \\ 0 & \phi \end{pmatrix}, \ldots \right\}$$

and

$$\left\{ \begin{pmatrix} 1 \\ 0 \end{pmatrix}, \lambda \begin{pmatrix} \cos(\omega) & \sin(\omega) \\ -\sin(\omega) & \cos(\omega) \end{pmatrix}, \ldots \right\},$$

where $0 < \phi$, $\lambda < 1$ and $0 < \omega < 2\pi$ with $\omega \neq \pi$. Verify that a model formed by the superposition of these, or any similar, models has qualitatively the desired form of forecast function.

(7) Identify the form of the forecast function of the model of Definition 9.1 when

$$\mathbf{F} = \begin{pmatrix} 1 \\ 1 \\ 1 \end{pmatrix} \quad \text{and} \quad \mathbf{G} = \begin{pmatrix} 1 & 0 & 0 \\ 0 & \lambda & 0 \\ 0 & 0 & \phi \end{pmatrix},$$

for some λ and ϕ such that $0 < \lambda, \phi < 1$. Describe the form of the implied transfer response function, and explain under what circumstances this form would be appropriate as a description of the response of a sales series to a price increases X_t.

(8) Consider the form response model of Definition 9.1 in which $\mathbf{F} = 1$, $\mathbf{G} = \lambda$, $(0 < \lambda < 1)$, $\boldsymbol{\theta}_t = \theta_t$ is scalar, $\psi_t = \psi$, a scalar constant, and $\boldsymbol{\delta\theta}_t = \boldsymbol{\delta\psi}_t = 0$ for all t. Suppose also that $\nu_t \sim N[0, V]$ with V known, $\theta_0 = 0$ and $(\psi|D_0) \sim N[h_0, H_0]$ with known moments. Finally, suppose that $X_1 = 1$ and $X_t = 0$ for $t > 1$. With $(\psi|D_t) \sim N[h_t, H_t]$ verify that

$$H_t^{-1} = H_0^{-1} + V^{-1}(1 - \lambda^{2t})/(1 - \lambda^2)$$

and

$$h_t = H_t[H_0^{-1}h_0 + V^{-1} \sum_{r=1}^{t} \lambda^{r-1}Y_r.]$$

Find the limiting posterior distribution of $(\psi|D_t)$ as $t \to \infty$.

(9) Consider the first-order polynomial plus AR(1) noise model defined by equations

$$Y_t = \mu_t + \eta_t,$$
$$\mu_t = \mu_{t-1} + \omega_t,$$
$$\eta_t = \psi\eta_{t-1} + \epsilon_t,$$

where $\epsilon_t \sim N[0, U]$ independently of $\omega_t \sim N[0, W]$. Show how the model may be written in DLM form and derive the updating equations for the elements of $\mathbf{m}_t$ and $\mathbf{C}_t$. Identify the form of the forecast function $f_t(k)$.

(10) Consider the first-order polynomial plus MA(1) noise model defined by equations

$$Y_t = \mu_t + \eta_t,$$
$$\mu_t = \mu_{t-1} + \omega_t,$$
$$\eta_t = \pi\epsilon_{t-1} + \epsilon_t,$$

where $\epsilon_t \sim N[0, U]$ independently of $\omega_t \sim N[0, W]$. Show that the model can be written in DLM form. Derive the updating equations for the elements of $\mathbf{m}_t$ and $\mathbf{C}_t$, and identify the form of the forecast function $f_t(k)$.

(11) Suppose that Y_t follows the standard, static AR(1) model $\{Y_{t-1}, 1, V, 0\}$ with autoregressive coefficient ψ.

   (a) Suppose that $(\psi|D_1) \sim N[m_1, C_1]$ for some known m_1 and C_1. Deduce that, for all $t > 1$, $(\psi|D_t) \sim N[m_t, C_t]$ and identify the moments as functions of the past data $Y_t, Y_{t-1}, \ldots, Y_1$.

   (b) Letting $C_1^{-1} \to 0$, verify that

$$C_t^{-1} \to V^{-1} \sum_{r=2}^{t} Y_{r-1}^2.$$

   (c) On the basis of the model assumptions, the series Y_t is stationary with variance $V[Y_t|\psi] = V(1 - \psi^2)^{-1}$ for all t. Deduce that, as $t \to \infty$, $tC_t \to (1 - \psi^2)$ with probability one. Comment on the implications for the precision with with ψ can be estimated in the cases (i) $\psi = 0.9$; and (ii) $\psi = 0.1$.

(12) Consider the ARMA model of (9.11) with $n > p$ and

$$\psi(B) = \prod_{i=1}^{s}(1 - \lambda_i B)^{n_i},$$

where $\sum_{i=1}^{s} n_i = p$, and the λ_i are distinct and non-zero. Obtain $\mathbf{F}$ and $\mathbf{G}$ of the similar canonical DLM, showing that one of the blocks of the system matrix is $\mathbf{J}_{n-p}(0)$.

CHAPTER 10

ILLUSTRATIONS AND EXTENSIONS
OF STANDARD DLMS

10.1 INTRODUCTION

In the preceding Chapters the focus has been on the theoretical structure and analysis of DLMs with little reference to practical aspects of modelling. Here we switch the focus to the latter to consolidate what has been developed in theory, illustrating many basic concepts via analyses of typical datasets. We consider both retrospective analysis of a time series as well as forecasting with an existing model, and describe a variety of modelling activities using the class of models built up from the trend, seasonal and regression components of Chapters 7, 8 and 9. Together these three components provide for the majority of forms of behaviour encountered in commercial and economic areas, and thus this class, of what may be referred to as *standard* models, forms a central core of structures for the time series analyst. In approaching the problem of modelling a new series, the basic trend and seasonal components are a useful first attack. If retrospective analysis is the primary goal, then these simple and purely descriptive models may be adequate in themselves, providing estimates of the trend (or deseasonalised series), seasonal pattern (detrended series) and irregular or random component over time. In addition to retrospective time series decomposition, these models can prove adequate for forecasting in the short term. The inclusion of regression terms is the next step, representing an attempt to move away from simple descriptions via explanatory relationships with other variables. Linking in such variables is also the route to firmer, more credible and reliable short/medium-term forecasting, the key idea being that future changes in a series not catered for within a simpler trend/seasonal description may be adequately predicted (at least qualitatively) by one or a small number of regression variables. Identifying the important variables is, of course and as usual in statistics generally, the major problem.

The next two sections consider several models corresponding to different levels of complexity, inputs based on different states of prior

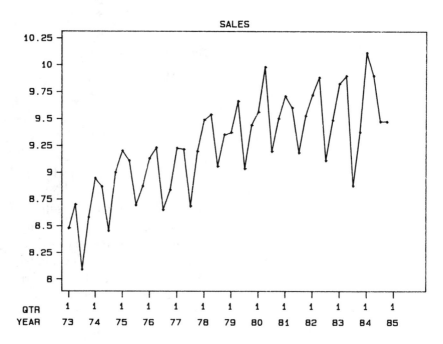

Figure 10.1.   Agricultural sales series.

information, and different objectives in modelling. Initially we concentrate on a reference analysis, as developed in Section 4.8, in order to present the main features of model analyses free from the effects of chosen initial priors. In addition, we assume closed models throughout. That is, once the model is specified, fitting proceeds without further interventions. The final sections of this chapter concern practical issues of error analysis and model assessment, data irregularities, transformations and other modifications and extensions of the basic normal DLM.

10.2   BASIC ANALYSIS WITH A TREND/SEASONAL MODEL

10.2.1   Data and basic model form

Consider the dataset given in Table 10.1 and plotted over time in Figure 10.1. The single time series is of quarterly observations providing a measure, on a standard scale, of the total quarterly sales of a company in agricultural markets over a period of 12 years. Quarters are labelled Qtr running from 1 to 4 within each year, years are labelled Year running from 75 (1975) to 84 (1984). The series, referred to simply as Sales, is evidently highly seasonal, driven by the annual cycle of demand for agricultural supplies following the natural annual pattern of farming activities. This can be clearly seen from the graph, or by examining the figures in the table making comparisons between rows.

Whilst the form of seasonality is stable, the actual quantified pattern does change somewhat from year to year. Underlying the seasonal series is a changing trend that grows at differing rates during different periods of the 12 years recorded.

The most basic, possibly appropriate descriptive model is that comprised of a polynomial trend plus seasonal effects as in Section 8.5 of Chapter 8. Given that any trend term will be quantified only locally, a second-order trend term is chosen. Quarterly changes in level and growth components of such a trend are assumed to model the apparent changes in direction of the trend in the series. This model is very widely used in practice in this sort of context as a first step in retrospective time series analysis. Any movement in trend is ascribed to changes in level and growth parameters of the model rather than being explained by possible regression variables.

In usual DLM notation, the first component of the model is linear trend

$$\{\mathbf{E}_2, \mathbf{J}_2(1), ., .\}.$$

To this is added the seasonal component. Since a full seasonal pattern is specified, this may be in terms of either 4 seasonal factors, constrained to zero sum, or the corresponding 3 Fourier coefficients. We work, as usual, in terms of the latter. The Fourier representation has just 2 harmonics: the first providing the fundamental annual cycle, the second adding in the Nyquist frequency. In DLM notation

Table 10.1. Agricultural sales data

Qtr.	Year											
	73	74	75	76	77	78	79	80	81	82	83	84
1	8.48	8.94	9.20	9.13	9.23	9.49	9.37	9.56	9.71	9.72	9.82	10.11
2	8.70	8.86	9.11	9.23	9.21	9.54	9.66	9.98	9.60	9.88	9.90	9.90
3	8.09	8.45	8.69	8.65	8.68	9.06	9.03	9.19	9.18	9.11	8.87	9.47
4	8.58	9.00	8.87	8.84	9.20	9.35	9.44	9.50	9.53	9.49	9.38	9.47

the Fourier model is, following Section 8.6,

$$\left\{ \begin{pmatrix} 1 \\ 0 \\ 1 \end{pmatrix}, \begin{pmatrix} c & s & 0 \\ -s & c & 0 \\ 0 & 0 & -1 \end{pmatrix}, \ldots \right\},$$

where, with $\omega = \pi/2$, $c = \cos(\omega) = 0$ and $s = \sin(\omega) = 1$. The full, 5 dimensional TSDLM thus has

$$\mathbf{F} = (1, 0; \ 1, 0; 1)'$$

and

$$\mathbf{G} = \begin{pmatrix} 1 & 1 & 0 & 0 & 0 \\ 0 & 1 & 0 & 0 & 0 \\ 0 & 0 & 0 & 1 & 0 \\ 0 & 0 & -1 & 0 & 0 \\ 0 & 0 & 0 & 0 & -1 \end{pmatrix}.$$

The corresponding parameter vector has 5 elements, $\boldsymbol{\theta}_t = (\theta_{t1}, \ldots, \theta_{t5})'$, with the following meaning:

- θ_{t1} is the underlying, deseasonalised level of the series at time t;
- θ_{t2} is the deseasonalised growth between quarters $t-1$ and t;
- θ_{t3} and θ_{t4} are the Fourier coefficients of the first harmonic, the full annual sine/cosine wave, with the contribution of this harmonic to the seasonal factor at time t being, in the notation of Section 8.6, simply $S_{1t} = \theta_{t3}$;
- similarly $S_{2t} = \theta_{t5}$ is the Nyquist contribution to the seasonal factor.

The seasonal factors are obtained from the Fourier coefficients as in Section 8.6.4. As before the factors (or effects since they are constrained to have zero sum) at time t are $\boldsymbol{\phi}_t = (\phi_{t0}, \ldots, \phi_{t3})'$. Thus ϕ_{t0} is the seasonal factor for quarter t. Then, following Example 8.3,

$$\boldsymbol{\phi}_t = \mathbf{L}(\theta_{t3}, \theta_{t4}, \theta_{t5})'$$

where

$$\mathbf{L} = \begin{pmatrix} 1 & 0 & 1 \\ 0 & 1 & -1 \\ -1 & 0 & 1 \\ 0 & -1 & -1 \end{pmatrix}.$$

Finally, we assume that the observational variance V of the model is constant in time and unknown, requiring, as is typical, the modified analysis for on-line variance learning. The model, with 6 parameters (including V), is now essentially specified. In line with the development of reference analyses in Section 4.8, we assume that there is no change in model parameters during the first 6 quarters, there being no information available to inform on any such changes. The reference updating equations apply up to $t = 6$ when the posterior distribution for θ_6 and V given D_6 is fully specified in the usual conjugate normal/inverse gamma form. To proceed from this time point on, the sequence of evolution variance matrices $\mathbf{W}_t$ are required to allow for time varying parameters.

10.2.2 Discount factors for components

The evolution variance matrices are specified, as usual, following the concept of block discounting consistent with the use of component models, as described in Section 6.3. Two discount factors provide the flexibility to allow for a full range of rates of change over time in the trend and seasonal components of the model; these are denoted δ_T (for the trend component), and δ_S (for the seasonal component). The former determines the rate at which the trend parameters θ_{t1} and θ_{t2} are expected to vary between quarters, with $100(\delta_T^{-1} - 1)\%$ of information (as measured by reciprocal variance or precision) about these parameters decaying each quarter, between observations. δ_S plays the same role for the seasonal (Fourier) parameters, (or, equivalently, the seasonal effects).

Thus, at times $t > 6$, the model evolution equation is as follows. With

$$(\theta_{t-1}|D_{t-1}) \sim T_{n_{t-1}}[\mathbf{m}_{t-1}, \mathbf{C}_{t-1}],$$

then

$$(\theta_t|D_{t-1}) \sim T_{n_{t-1}}[\mathbf{a}_t, \mathbf{R}_t],$$

where

$$\mathbf{a}_t = \mathbf{G}\mathbf{m}_{t-1} \quad \text{and} \quad \mathbf{R}_t = \mathbf{P}_t + \mathbf{W}_t$$

with

$$\mathbf{P}_t = \mathbf{G}\mathbf{C}_{t-1}\mathbf{G}',$$

and the evolution variance matrix is defined as

$$\mathbf{W}_t = \text{block diag}\{\ \mathbf{P}_{tT}(\delta_T^{-1} - 1),\ \ \mathbf{P}_{tS}(\delta_S^{-1} - 1)\ \}$$

where $\mathbf{P}_{tT}$ is the upper left 2x2 block of $\mathbf{P}_t$ and $\mathbf{P}_{tS}$ is the lower right 3x3 block.

For the trend discount factor, values in the range 0.8 to 1.0 are typically appropriate, smaller values anticipating greater change. δ_S is typically larger than δ_T to reflect:

(1) relatively less information being obtained about the seasonal parameters than the trend in each quarter, and

(2) the expectation that seasonal patterns will tend to have more sustained, stable forms relative to the potentially more rapidly changing deseasonalised trend.

This second point is consistent with the view that, in using this simple, descriptive TSDLM, we are not attempting to anticipate sustained movement and changes in trend and seasonality, but just to detect and estimate them. Omitted related explanatory variables are more likely to have a marked impact on changes in the underlying trend in the series than on the seasonal pattern, the latter viewed as more durable over time.

Forecasting accuracy comes through the identification of stable structure and relationships and, in the best of all worlds, discount factors of unity would be used giving constant parameters over time. Thus in modelling we may start at this ideal, comparing the predictive performance of such a model with that based on more realistic values, possibly comparing several such models. This sort of activity is described further in Section 10.2.5 below. Here we consider the particular case in which $\delta_t = 0.85$ and $\delta_S = 0.97$.

10.2.3   Model fitting and on-line inferences

From time $t = 6$ onwards then, the usual updating equations apply, one-step ahead forecast and posterior distributions for model parameters being sequentially calculated for times $t = 7,\ldots,48$. Recall that, in the reference analysis, the (posterior) degrees of freedom at time t is $n_t = t-$dimension of the model $= t - 5$. Thus, for $t > 6$, the one-step ahead forecast distribution at $t-1$ has $n_{t-1} = t-6$ degress of freedom, $(Y_t|D_{t-1}) \sim T_{t-6}[f_t, Q_t]$. Figure 10.2 displays information about these sequentially calculated forecasts. Here, superimposed on a scatter plot of the data over time, we have the point forecasts f_t joined up as the full line, and 90% probability limits about the point forecasts as dotted lines. The latter provide the 90% HPD region for Y_t given D_{t-1} from the relevant T predictive distribution, an interval symmetrically located (hence with equal 5% tails) about f_t. Due to

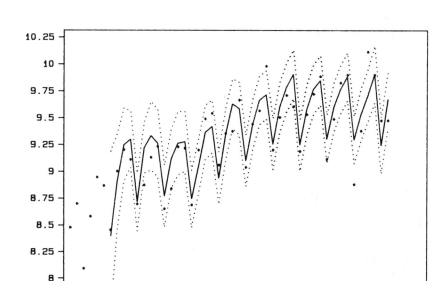

Figure 10.2. Agricultural sales series with one-step ahead
forecasts.

the reference prior used, no forecasts are available initially. Clearly
most of the points lie within the 90% limits; five or six exceptions are
clear. Of course, roughly 5 out of 48 are expected to lie outside the
90% limits. Further insight into the one-step ahead forecasting performance is given by examining Figure 10.3. Here the standardised
forecast errors $e_t/Q_t^{1/2}$ are plotted over time with 90% probability
limits from the one-step ahead forecast distributions. The several
points noted above show up clearly here as leading to large (in absolute value) residuals.

Having reached time $t = 48$ and the end of the data, we can
proceed to examine model inferences by considering the estimated
model components over time. Central to this is the following general
result. Let $\mathbf{x}_t$ be any 5 element vector, and set $x_t = \mathbf{x}_t'\boldsymbol{\theta}_t$. Then,
from the posterior for $\boldsymbol{\theta}_t$ at time t, it follows that

$$(x_t \mid D_t) \sim T_{t-5}[\mathbf{x}_t'\mathbf{m}_t, \mathbf{x}_t'\mathbf{C}_t\mathbf{x}_t].$$

The mode of this T distribution, $\mathbf{x}_t'\mathbf{m}_t$, is referred to as the *on-line*
estimated value of the random quantity x_t. Inferences about model
components are made by appropriately choosing $\mathbf{x}_t$, as follows.

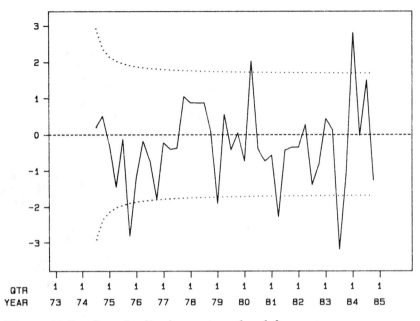

Figure 10.3.   Standardised one-step ahead forecast errors.

(1) DESEASONALISATION

If $\mathbf{x}'_t = (1,0,0,0,0)$ then x_t is the underlying, deseasonalised
level of the series $x_t = \theta_{t1}$. From the sequence of posterior T
distributions, 90% HPD intervals are calculated and plotted
in Figure 10.4 in a similar fashion to the forecast intervals in
Figure 10.2. The local linearity of the trend is evident, as are
changes, both small and more abrupt, in the growth pattern
over time. In particular, there is a marked change in trend
from positive growth during the first two years or so, to little,
even negative, growth during the third and fourth years. This
abrupt change is consistent with the general economic conditions at the time, and could, perhaps, have been accounted
for if not forecast. With respect to the simple, closed model
illustrated here, however, adaptation to the change is apparent over the next couple of observations, although responding
satisfactorily in general to changes that are greater than anticipated requires some form of intervention, as discussed in
later chapters. Taking $\mathbf{x}'_t = (0,1,0,0,0,)$ similarly provides
on-line inference about the annual growth in trend.

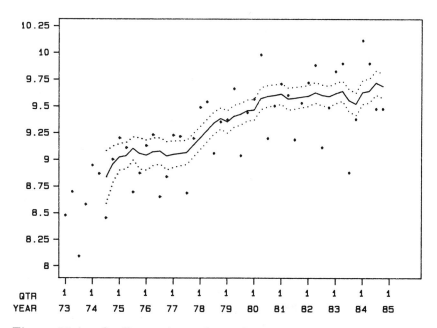

Figure 10.4. On-line estimated trend in sales.

(2) SEASONAL FACTORS/EFFECTS

The seasonal effect at time t is given by $\phi_{t0} = x_t = \mathbf{x}_t' \boldsymbol{\theta}_t$ with $\mathbf{x}_t' = (0, 0, 1, 0, 1)$ to select the two harmonic contributions to the seasonal pattern from the state vector. Thus, as a function over time, this x_t provides the detrended seasonal pattern of the Sales series. Figure 10.5 displays information from the posterior distributions for x_t at each time, analogous to that for the trend although displayed in a different form. At each t on the graph, the vertical bar represents the 90% HPD interval for the seasonal effect, the mid-point of the interval being the on-line estimated value. This is a plot similar to those displayed in Figure 8.4 of Chapter 8. There is evidently some minor variation in the seasonal pattern from year to year, although the basic form of a peak in Spring and Summer and deep trough in Autumn is sustained, consistent with expected variation in demand for agricultural supplies.

Further investigation of the seasonal pattern is possible using the ideas of Section 8.6 to explore harmonics. Taking $\mathbf{x}_t = (0, 0, 1, 0, 0)'$ gives x_t as the contribution of the first harmonic, and $\mathbf{x}_t = (0, 0, 0, 0, 1)'$ leads to the Nyquist harmonic.

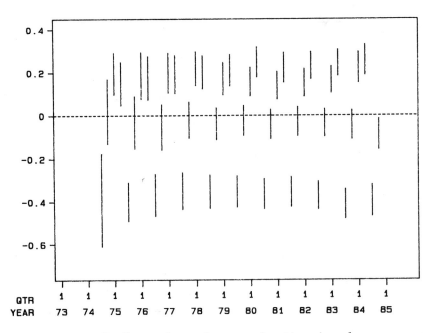

Figure 10.5.   On-line estimated seasonal pattern in sales.

10.2.4   Step ahead forecasting

Consider finally step ahead forecasting of the Sales series from time
4/84, the end of the data series. Without intervention, we simply
forecast by projecting based on the existing model and fit to the
past 12 years of data, as summarised by the posterior distributions
at $t = 48$, using the theory summarised in Section 4.6. The posteriors
at time 48 have defining parameters:

$$\mathbf{m}_{48} = (9.685, \ 0.015, \ -0.017, \ 0.304, \ 0.084)',$$

and

$$\mathbf{C}_{48} = 0.0001 \begin{pmatrix} 45.97 & 3.77 & -2.30 & 1.86 & -1.03 \\ & 0.66 & -0.20 & 0.15 & -0.09 \\ & & 12.87 & -0.10 & -0.13 \\ & & & 13.23 & -0.24 \\ & & & & 6.38 \end{pmatrix},$$

(only the upper part of the symmetric matrix $\mathbf{C}_{48}$ being displayed);

$$n_{48} = 43 \quad \text{and} \quad S_{48} = 0.016.$$

Forecasting ahead requires that we specify, at time $t = 48$, the evolution variance matrices for the future, $\mathbf{W}_{t+r}$ for $r = 1, \ldots, k$, where k is the maximum step ahead of interest. We use the discount model strategy described in Section 6.3 of Chapter 6, using the existing one-step ahead matrix $\mathbf{W}_{t+1}$ in forecasting ahead. Thus, at any time t, forecasting ahead is based on $\mathbf{W}_{t+r} = \mathbf{W}_{t+1}$ for $r = 2, \ldots, k$. Although in updating to future times we use evolution variance matrices defined via discounts based on the information available at those times, in forecasting ahead we use slightly different values given by the currently implied $\mathbf{W}_{t+1}$. As previously discussed in Section 6.3 of Chapter 6, this approach is practically simple and the sequence $\mathbf{W}_{t+r} = \mathbf{W}_{t+1}$ into the future appropriate for forecasting ahead from our current position at time t. Proceeding to time $t + 1$, and observing Y_{t+1}, our view of the future evolution variances for times $t + 2$ onwards is changed slightly, being based on the new, current (at $t + 1$) value. This may be interpreted most satisfactorily as an intervention to change the evolution variance sequence.

Thus, in the example, the evolution variance matrix for forecasting from time 48 is given by $\mathbf{W}_{t+1} = $ block diag$[\mathbf{W}_T, \ \mathbf{W}_S \ ]$ where, by applying the discount concept,

$$\mathbf{W}_T = 0.0001 \begin{pmatrix} 9.56 & 0.78 \\ & 0.12 \end{pmatrix}$$

and

$$\mathbf{W}_S = 0.0001 \begin{pmatrix} 0.41 & 0.003 & 0.008 \\ & 0.40 & -0.004 \\ & & 0.20 \end{pmatrix},$$

(again, of course, each of these is a symmetric matrix), the latter being roughly equal to 0.0001 diag$(0.4, 0.4, 0.2)$.

It is now routine to apply the results of Section 4.6 to project forecast distributions into the future, and this is done up to $k = 24$, giving 6 full years of forecast up to 4/90. The forecast distributions are all Student T with 43 degrees of freedom, therefore roughly normal, with increasing scale parameters as we project further ahead. Figure 10.6 displays the point forecasts $f_{48}(k) = E[Y_{48+k}|D_{48}]$, $(r = k, \ldots, 24)$, joined up as a full line, and 90% prediction intervals symmetrically located about the point forecasts.

10.2.5   Numerical summaries of predictive performance

Finally, we mention basic measures of model predictive performance that allow easy and informal comparison of alternative models. Three

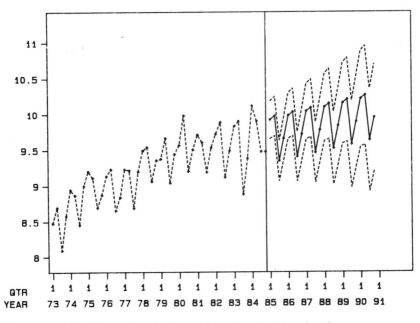

Figure 10.6.   Step-ahead forecasts of agricultural sales.

simple, and well-used, measures of predictive performance, easily calculated from the one-step forecast errors, or *residuals* e_t, and already
used in previous chapters. Each of these is useful only in a relative
sense, providing comparison between different models.

Total absolute deviation is defined as $\sum |e_t|$. With proper initial
priors, the first few errors will reflect the appropriateness of the initial
priors as well as that of the model, and their contribution to the
sum may be important. In reference analyses, however, there are no
defined errors until $t = 6$ (more generally, $n + 1$ in an n-dimensional
model), and so the sum starts at $t = 6$. The *mean absolute deviation*,
or MAD, measure is simply this total divided by the number of
errors in the sum. In this example, the MAD based on the final 43
observations is calculated as 0.128.

Total square error is $\sum e_t^2$, the *mean square error*, or MSE, being
this divided by the number of terms in the sum. In this example, the
MSE based on the final 43 observations is calculated as 0.028. Both
MAD and MSE are simple measures of actual forecasting accuracy.

Model likelihood, a rather more formal measure of goodness of
predictive performance, was introduced in examples in Chapters 2
and 3. West (1986a) gives very general discussion and development

of the ideas in this section, and the notation used there is adopted here. For each observation Y_t, the observed value of the one-step predictive density $p(Y_t|D_{t-1})$ is larger or smaller according to whether the observation accords or disagrees with the forecast distribution. This accounts for raw accuracy of the point forecast f_t, and also for forecast precision, a function of the spread of the distribution. The aggregate product of these densities for observations $Y_t, Y_{t-1}, \ldots, Y_s$, $(s = 1, \ldots, t)$, provides the overall measure

$$p(Y_t|D_{t-1})p(Y_{t-1}|D_{t-2}) \ldots p(Y_s|D_{s-1}) = p(Y_t, Y_{t-1}, \ldots, Y_s|D_{s-1}),$$

just the joint predictive density of those observations from time $s-1$. If proper priors are used, then this can extend back to time $s = 1$ if required. In the reference anlysis, the densities are, of course, only defined for $s > n + 1$ in a n-dimensional model.

This measure, based on however many observations are used, provides a likelihood (in the usual statistical sense of the term) for the parameters, such as discount factors, determining the model. Just as with MAD and MSE, these measures may be used as informal guides to model choice when comparing two, or more, models differing only in values of defining parameters. Suppose we consider two models with the same mathematical structure, model 0 and model 1, differing only, for example, in the values of discount factors. To make clear the dependence of the model on these values, suffix the predictive densities by 0 and 1 respectively. Then the *relative likelihood* of model 0 versus model 1 based on the observation Y_t at time t is just the ratio

$$H_t = p_0(Y_t|D_{t-1})/p_1(Y_t|D_{t-1}).$$

More generally, the observations $Y_t, Y_{t-1}, \ldots, Y_s, (s = 1, \ldots, t)$, provide the the overall likelihood ratio

$$H_t(t - s + 1) = \prod_{r=s}^{t} H_r$$
$$= p_0(Y_t, Y_{t-1}, \ldots, Y_s|D_{s-1})/p_1(Y_t, Y_{t-1}, \ldots, Y_s|D_{s-1}).$$

These likelihood ratios, alternatively called *Bayes' Factors*, or *Weights of evidence* (Jeffreys, 1961; Good, 1985 and references therein; West, 1986a), provide a basic measure of predictive performance of

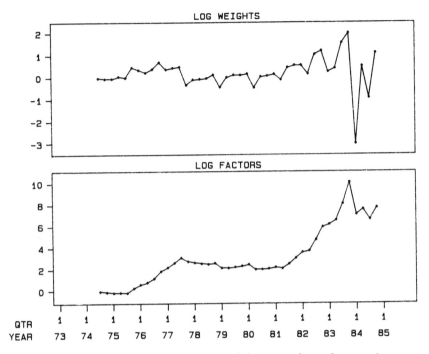

Figure 10.7.   Log weights and Bayes' factors plotted over time.

model 0 relative to model 1. Note that the evidence for or against
the models accumulates multiplicatively as data is processed, via

$$H_t(t - s + 1) = H_t\ H_{t-1}(t - s)$$

for $t > s$, with boundary values $H_t(1) = H_t$ when $s = t$. Many authors, including those referenced above, consider the log-likelihood
ratio, or log Bayes' factor/Weight of evidence, as a natural and interpretable quantity. On the log scale, evidence is additive, with

$$\log[H_t(t - s + 1)] = \log(H_t) + \log[H_{t-1}(t - s)].$$

Following Jeffreys (1961), a log Bayes' factor of 1 (-1) indicates
evidence in favour of model 0   (1), a value of 2 or more $(-2$ or less)
indicating the evidence to be strong. Clearly the value 0 indicates
no evidence either way.

EXAMPLE 10.1. Let model 0 be the model used throughout this
section, having discount factors $\delta_T = 0.85$ and $\delta_S = 0.97$. For comparison, let model 1 be a *static* model, having constant parameters,
defined by unit discount factors $\delta_t = \delta_S = 1$. Figure 10.7 provides

a plot versus time t of the individuals log weights $\log(H_t)$, and a similar plot of the cumulative log Bayes' factors $\log[H_t(t-6)]$. The latter drops below 0 for only a few quarters in year 75, being positive most of the time, essentially increasing up to a final value between 7 and 8, indicating extremely strong evidence in favour of the dynamic model relative to the static. For additional comparison, the static model has MAD=0.158 and MSE=0.039, both well in excess of those in the dynamic model. Similar comparisons with models having different values of discounts indicate support for that used here as acceptable, having MAD and MSE values near the minimum and largest likelihood.

Visual inspection of residual plots, and retrospective residual analyses, are related, important assessment activities. However, whilst providing useful guides to performance, we **do not** suggest that the choice of a model for forecasting be automated by choosing that one (in the example, the values of discount factors) that minimise MAD/MSE and/or maximising the likehood measure. All such activities pay essentially no regard to the longer term predictive ability of models, focussing as they do on how well a model has performed in one-step ahead forecasting in the past. In considering forecasting ahead, a forecaster must be prepared to accept that, often, a model found to be adequate in the past may not perform as well in the future, and that continual model monitoring and assessment of predictive performance is a must. We consider such issues in Chapter 11.

10.3 A TREND/ SEASONAL/ REGRESSION DLM

10.3.1 Data and basic model form

This second illustration again concerns commercial sales data, this time taken from the food industry. The response series of interest, Sales, is the monetary value of monthly total sales, on a standardised, deflated scale, of a widely consumed and well established food product, covering a variety of brands and manufacturers, in UK markets. The data runs over the full 6 years 1976 to 1981 inclusive, thus there are 72 observations, given in Table 10.2. Also given in the table are corresponding values of a second series, Index, that is used as an indicator to partially explain the movement in trend of sales. This is a compound measure constructed by the company concerned, based on market prices, production and distribution costs, and related variables, and is to be used as an independent regressor variable in the model for Sales. The Index variable is standardised by subtracting the arithmetic mean and dividing by the standard deviation, both mean and deviation being calculated from the 72 observations presented. The two data series are graphed over time in Figure 10.8. The response series clearly has a varying trend over the 6 years, is apparently inversely related to Index, and is evidently seasonal with a rather variable seasonal pattern from year to year.

The Sales figures refer, as mentioned above, to a mature, established consumer product. In consequence for short-term forecasting, any sustained growth (positive or negative!) in Sales over time should be attributable to regressor variables relating to pricing policies, costs, advertising and competitor activity, and general market/economic conditions; hence the introduction of the Index series. The model for Sales does not, therefore, include a descriptive growth term, the initial component being a simple steady model, the first-order polynomial component of Chapter 2, $\{1, 1, ., .\}$. The second component is the regression DLM for the effect of the Index series. Note that we include only contemporaneous values of Index, lagged effects being ignored for this illustration, it being assumed

Table 10.2. Sales and Index series

Year	Month											
	1	2	3	4	5	6	7	8	9	10	11	12
	SALES											
1976	9.53	9.25	9.36	9.80	8.82	8.32	7.12	7.10	6.59	6.31	6.56	7.73
1977	9.93	9.75	10.57	10.84	10.77	9.61	8.95	8.89	7.89	7.72	8.37	9.11
1978	10.62	9.84	9.42	10.01	10.46	10.66	11.03	10.54	10.02	9.85	9.24	10.76
1979	10.33	10.62	10.27	10.96	10.93	10.66	10.58	9.66	9.67	10.20	10.53	10.54
1980	9.67	10.40	11.07	11.08	10.27	10.64	11.03	9.63	9.08	8.87	8.65	9.27
1981	10.58	10.52	11.76	11.58	11.31	10.16	10.14	9.81	10.27	8.80	8.62	9.46
	INDEX											
1976	0.54	0.54	1.12	1.32	1.12	1.12	1.12	1.32	1.93	2.13	1.52	1.32
1977	−0.03	0.15	−0.03	−0.22	−0.03	0.15	0.15	0.15	0.73	0.54	0.34	−0.03
1978	−0.22	0.93	1.32	0.93	0.15	0.15	0.34	0.34	−0.41	1.12	0.54	−0.22
1979	0.54	0.54	0.34	−0.03	−0.03	0.15	−0.03	−0.22	−0.03	−0.59	−0.59	−0.78
1980	−1.14	−0.96	−0.96	−0.59	−0.41	−0.03	0.15	0.34	0.54	0.54	0.73	0.54
1981	−1.84	−1.84	−2.70	−2.19	−2.19	−1.84	−0.96	−1.67	−1.67	−0.96	−1.14	−0.96

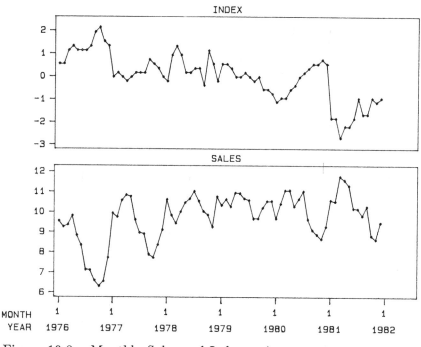

Figure 10.8.   Monthly Sales and Index series.

that the Index has been constructed to account for these. Thus the
explanatory component is the regression with zero origin of Chapter
3, $\{X_t, 1, ., .\}$, where X_t is the value of the Index series at t.
The single parameter of the first-order polynomial component now
provides the (time-varying) intercept for the dynamic regression of
Sales on Index. In models such as these, the possible effects of related, omitted variables are partially accounted for by the purely
random movement allowed in the intercept term and regression parameter.

The third model component is a seasonal term describing monthly
variation within a year free from the effects of regression variables.
The most general model is, of course, a full seasonal effects model
with 12 monthly seasonal effects (with zero sum). As usual, we
adopt the Fourier representation of seasonal effects. To proceed,
we consider hypothetical prior information and its implications for
model specification, forecasting ahead from time 0, and subsequent
data analysis.

10.3.2 Hypothetical initial prior information

Suppose that, at December 1975 where $t = 0$, we are in the position of having to provide step ahead forecasts of Sales based on existing information D_0. Such information includes past data (prior to 1976) on the Sales and Index series, past experience with modelling their relationship and forecasting, and past experience with similar products. In practice, if such a model had been used previously, then there would exist a full posterior distribution for the model parameters at $t = 0$, providing the prior for the new data beginning in January 1976, at $t = 1$. Suppose, however, that past modelling activities have only been partially communicated, and that we adopt initial priors based on the following initial information at $t = 0$. Note that, rather than communicating the posterior for $(\theta_0 \mid D_0)$ as usual, we are specifying information directly for the first time point and will determine the prior for $(\theta_1 \mid D_0)$ directly.

- The parameter $\theta_{1,1}$, the first element of the DLM state vector at $t = 1$, represents the non-seasonal intercept term, the non-seasonal level if the Index value is $X_1 = 0$. Given D_0, the value $m_{1,1} = 9.5$ is taken as the prior mean. Information on the intercept is viewed as rather vague, particularly since the parameter can be expected to change to account for effects of omitted variables, and values as low as 9 or as high as 10 are plausible.

- At $t = 1$, the prior view of the regression effect of Index is that the regression coefficient is negative, and that to achieve an expected change in Sales of 1 unit, the Index would have to change by about 1.5 units. This suggests a value near $-2/3$ for the prior estimate of the regression parameter at $t = 1$, namely $\theta_{1,2}$, the second element of the DLM parameter vector. The value adopted is $m_{1,2} = -0.7$. Uncertainty is such that values as high as -0.5 or as low as -0.9 are plausible.

- Prior information on the seasonal effects $\phi_1 = (\phi_{1,0}, \dots, \phi_{1,11})'$ is provided initially in terms of the prior estimates

$$0.8, \ 1.3, \ 1, \ 1.1, \ 1.2, \ 0, \ -1.2, \ -1, \ -1, \ -1.5, \ -0.5, \ 0,$$

although the uncertainties are fairly large, with values possibly ranging up to plus or minus 0.4 or so from each of these estimates. Note that these estimates do not sum to zero, and will therefore have to be constrained to do so later.

- The observational variance is viewed as constant over time, initially estimated at $S_0 = 0.0225 = 0.15^2$, having fairly low associated degrees of freedom $n_0 = 6$. The 90% equal-tails interval for the standard deviation $V^{1/2}$ is easily calculated as $(0.1, 0.29)$.

This information suffices to determine marginal priors for the intercept parameter, the regression parameter and the seasonal effects separately. There is no information provided to determine prior correlation between the components, so they are taken as uncorrelated. In the usual notation then, marginal priors are as follows.

INTERCEPT:

$(\theta_{1,1} \mid D_0) \sim T_6[9.5, 0.09]$, with 90% HPD interval $(8.92, 10.08)$;

REGRESSION:

$(\theta_{1,2} \mid D_0) \sim T_6[-0.7, 0.01]$, with 90% interval $(-0.89, -0.51)$;

SEASONAL:

$(\boldsymbol{\phi}_1 \mid D_0) \sim T_6$, with mean vector given by the estimates above, and scale matrix $0.04\,\mathbf{I}$, although, since neither conform to the zero-sum constraint, they must be adjusted according to the theory of Section 8.4. Applying Theorem 8.2 leads easily to the constrained mean vector

$$\mathbf{m}_{\phi_1} = (0.783, 1.283, 0.983, 1.083, 1.183, -0.017,$$
$$-1.217, -1.017, -1.017, -1.517, -0.517, -0.017)'$$

(to 3 decimal places). The constrained scale matrix, denoted by $\mathbf{C}_{\phi_1}$, has diagonal elements equal to 0.0367, and off diagonal elements -0.0034 so that the pairwise correlations between the effects are all -0.092. On this basis, 90% prior HPD intervals for the effects are given by the means plus and minus 0.356.

10.3.3 Seasonal component prior

The prior for the seasonal effects suggests a seasonal pattern that grows from 0 in December to a plateau in February to May, decays rapidly thereafter to a trough in October and returns to 0 by the end of the year. If used in the model, this is the pattern that will apply for step ahead forecasts. With such large uncertainty about the effects, as indicated by the 90% interval, it is sensible to proceed with a full seasonal model, having all Fourier harmonics, in order that forthcoming data provide further information about an unrestricted pattern. The main problem with restricting the seasonal pattern lies in the possibility that the prior may be doubtful and that any restrictions imposed on the basis of the prior may be found to be unsatisfactory in the light of further data. For illustration here, however, the seasonal pattern is restricted by a reduced Fourier form representation that is consistent with this initial prior. This is obtained by first considering a full harmonic representation and examining the implications of the prior for the individual harmonic components.

Using the theory of Section 8.6.4, we can deduce by linear tranformation the initial prior for the eleven Fourier coefficients in the full harmonic model. If $\mathbf{H}$ is the transformation matrix of Section 8.6.4, then the coefficients are given by $\mathbf{H}\boldsymbol{\phi}_1$, having a T_6 prior distribution with mean vector $\mathbf{Hm}_{\phi_1}$ and scale matrix $\mathbf{HC}_{\phi_1}\mathbf{H}'$. It is left as an exercise for the reader to verify that this is essentially a diagonal matrix. This follows since harmonics are always orthogonal, so we know in advance that the Fourier coefficients from one harmonic will be uncorrelated with those of any other. Although the two coefficients of any harmonic may be correlated in general, these correlations are negligible in this case. Thus the prior can be summarised in terms of prior means and scale factors for the marginal T_6 distributions of the individual Fourier coefficients. These are given in Table 10.3. Also provided are the F values associated with the hypotheses of zero contribution from the harmonics (Section 8.6.5), although since the

Table 10.3. Initial prior summary for Fourier coefficients

Harmonic	Coefficient	Prior		F value
		Mean	Scale	
1	cos	0.691	0.0067	136.5
	sin	1.159	0.0067	
2	cos	−0.033	0.0067	0.3
	sin	−0.058	0.0067	
3	cos	0.283	0.0067	6.2
	sin	−0.050	0.0067	
4	cos	−0.217	0.0067	5.1
	sin	0.144	0.0067	
5	cos	0.026	0.0067	0.7
	sin	0.091	0.0067	
6	cos	0.033	0.0029	0.3

coefficients are largely uncorrelated, this is redundant information, the assessment may be made from inferences on the T distributed, uncorrelated coefficients directly.

Recall that the first 5 F values refer to the $F_{2,6}$ distribution, the final one for the Nyquist harmonic to the $F_{1,6}$ distribution. Clearly the first harmonic dominates in the prior specification of the seasonal pattern, the third and fourth harmonics being also of interest but the rest are negligible. Accordingly, our earlier comments about the possible benefits of proceeding with the full, flexible seasonal model notwithstanding, the second, fifth and sixth (Nyquist) harmonics are dropped from the model. We assume the corresponding coefficients are zero at time $t = 1$, consistent with the initial prior, and also for all times $t = 1, \ldots, 72$ for the full span of the data to be processed. This restriction can be imposed properly by taking the prior T distribution for the 11 coefficients, and calculating the T conditional distribution of the 6 coefficients for the 3 harmonics retained, given that the other 5 coefficients are zero. However, in view of the fact that the coefficients are essentially uncorrelated and the specified prior rather vague, the marginal distributions already specified are used directly.

10.3.4   Model and initial prior summary

The model is 8 dimensional, given at time t by $\{\mathbf{F}_t, \mathbf{G}, ., , \}$ with

$$\mathbf{F}'_t = (1; \quad X_t; \quad \mathbf{E}'_2; \quad \mathbf{E}'_2; \quad \mathbf{E}'_2),$$

and

$$\mathbf{G} = \text{block diag}[1; \quad 1; \quad \mathbf{G}_1; \quad \mathbf{G}_3; \quad \mathbf{G}_4],$$

where the harmonic terms are defined, as usual, by

$$\mathbf{E}_2 = \begin{pmatrix} 1 \\ 0 \end{pmatrix} \quad \text{and} \quad \mathbf{G}_r = \begin{pmatrix} \cos(\pi r/6) & \sin(\pi r/6) \\ -\sin(\pi r/6) & \cos(\pi r/6) \end{pmatrix},$$

for $r = 1, 3$ and 4.

The initial prior is given by $n_1 = 6$ and $S_1 = 0.15^2$, determining the gamma prior for $(V^{-1} \mid D_0)$, and, for $(\boldsymbol{\theta}_1 \mid D_0) \sim T_6[\mathbf{a}_1, \mathbf{R}_1]$, we have

$$\mathbf{a}'_1 = (9.5; \quad -0.7; \quad 0.691, 1.159; \quad 0.283, -0.050; \quad -0.217, 0.144),$$

and
$$\mathbf{R}_1 = \text{block diag}\,[0.09;\quad 0.01;\quad 0.0067\,\mathbf{I}_6].$$

Finally, note that having converted to a restricted seasonal pattern, the implied seasonal effects are constrained at each time to accord with the combination of just 3 harmonics, and so the initial prior for the effects will be rather different to that for the effects originally unrestricted (apart from zero sum) in Section 10.3.2. The theory of Section 8.6.5 can be used to transform the marginal prior above for the 6 Fourier coefficients to that for the full 12 seasonal effects. It is routine to apply this theory, which leads to the following estimates (the new prior mean for $(\phi_1 \mid D_0)$):

$$0.76,\ 1.36,\ 1.05,\ 0.99,\ 1.18,\ -0.09,\ -1.19,\ -0.89,$$
$$-1.08,\ -1.43,\ -0.71,\ 0.05$$

(to 2 decimal places). These restricted values differ only slightly from the unrestricted values, as is to be expected since the differences lie in the omission of what are insignificant harmonics.

It remains to specify discount factors determining the extent of anticipated time variation in the parameters. This illustration uses, as usual, the block discounting technique of Section 6.3, Chapter 6, requiring one discount factor for each of the intercept, regression and seasonal components. The values used are 0.9, 0.98 and 0.95 respectively. 0.98 for regression reflects the anticipation that the quantified regression relationship will tend to be rather durable, 0.9 for the intercept anticipates greater change in the constant term to cater for effects of independent variables not explicitly recognised and model misspecification generally. These 3 discount factors now complete the model definition, and we can proceed to forecast and process the new data.

10.3.5   'No-data' step ahead forecasting

Initially, step ahead forecast distributions are computed for the three years 1976 to 1978 from time $t = 0$, December 1975. In discount models the evolution variances for forecasting step ahead are taken to be constant, fixed at the currently implied value for forecasting one-step ahead, as described in Section 6.3 of Chapter 6. Thus, at $t = 0$, we require the current evolution variance $\mathbf{W}_1$, using this as the value in forecasting ahead to times $k > 0$. We derive this here after commenting on a feature of general interest and utility.

For general time t, recall the notation $\mathbf{P}_t = \mathbf{G}\mathbf{C}_{t-1}\mathbf{G}'$ and that $\mathbf{R}_t = \mathbf{P}_t + \mathbf{W}_t$. Recall also that, in a component model with r components having discount factors $\delta_1, \ldots, \delta_r$, if the diagonal block variance matrices from $\mathbf{P}_t$ are denoted $\mathbf{P}_{t1}, \ldots, \mathbf{P}_{tr}$, then the evolution variance matrix is defined as $\mathbf{W}_t = $ block diag $[\mathbf{P}_{t1}(\delta_1^{-1} - 1), \ldots, \mathbf{P}_{tr}(\delta_r^{-1} - 1)]$. This defines $\mathbf{W}_t$ in terms of $\mathbf{P}_t$. In the special case that $\mathbf{P}_t$ (equivalently $\mathbf{R}_t$) is block diagonal, the r components being uncorrelated, then it follows immediately that $\mathbf{W}_t$ can be written in terms of $\mathbf{R}_t$ also, viz.

$$\mathbf{W}_t = \text{block diag } [\mathbf{R}_{t1}(1 - \delta_1), \ldots, \mathbf{R}_{tr}(1 - \delta_r)].$$

In our example, we have $r = 3$ components and have specified $\mathbf{R}_1$ above in block diagonal (actually diagonal) form. Thus, given the discount factors for the 3 components at 0.9, 0.98 and 0.95, we have

$$\mathbf{W} = \mathbf{W}_1 = \text{block diag } [0.09(1 - 0.9); \ 0.01(1 - 0.98);$$
$$0.0067(1 - 0.95)\mathbf{I}]$$
$$= \text{block diag } [\ 0.009; \ 0.0002; \ 0.0003 \ \mathbf{I}].$$

Thus, viewed step ahead from $t = 0$, the state parameter vector evolves according to the system equation with evolution noise term having scale matrix $\mathbf{W}$.

Forecast distributions are now routinely calculated from the theory summarised in Section 4.6 of Chapter 4, although, since a regression component is included in the model, values are required for the regression variable, Index, in order that $\mathbf{F}_t$ be available. In this illustration, we are interested in forecasting over the first 36 months. As an hypothetical study, suppose that the first 36 values of the Index series (the values that actually arise in the future) are provided as values to use in forecasting Sales. These X_t values can be viewed either as point estimates/ forecasts of the Index series made prior to 1976, or as values provided for a *What if?* study to assess the consequences for Sales of these particular future possible values of Index. For the purposes of our example, of course, this is ideal since forecast errors will be totally derived from the unpredictable, observational noise, and model misspecification, rather than errors in forecasting the Index series. Finally note that these 3 years of forecasts are *no-data* forecasts, based purely on the model and the initial priors.

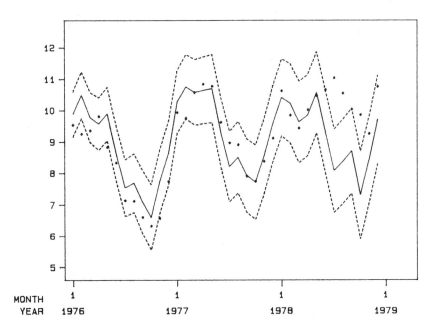

Figure 10.9.   Step-ahead forecasts of Sales made in December
                1975.

Figure 10.9 provides graphical display of the forecast means and
90% symmetric intervals for the outcomes based on the forecast T
distributions. Also plotted are the actual observations that later
become available. If these forecasts had actually been available at
December 1975, then the resulting accuracy would have been fairly
impressive up to the early part of the third year, 1978. However,
although most of the first two and a half years of data lie within
the 90% intervals, there is clear evidence of systematic variation in
the data not captured by the model. Almost all of the point forecast
errors in the first year or so are negative, the model overforecasting to
a small degree. The second year is rather better, but things appear
to breakdown in late 1978 with radical under-forecasting in the last 6
or 7 months. Proceeding to analyse the data in the next Section, we
should therefore expect the model to adapt to the apparent change
in data structure seen here as the observations are processed.

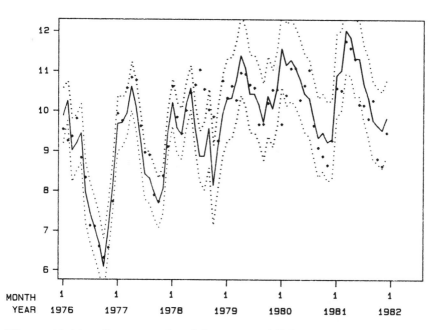

Figure 10.10.   One-step ahead forecasts of Sales.

10.3.6   Data analysis and one-step ahead forecasting of Sales

Figure 10.10 provides a graph of the data with one-step point forecasts and associated 90% intervals as sequentially calculated during the data analysis. During 1976, the model clearly learns and adapts to the data, the forecasts being rather accurate and not suffering the systematic deficiencies of those produced at December 1975. Figure 10.11 displays actual one-step forecast errors, Figure 10.12 gives a similar plot but now of standardised errors with 90% intervals, symmetrically located about 0, from the forecast T distributions. Clearly the model adapts rapidly in the first year, the forecast errors being acceptably small and appearing unrelated. Having adapted to the lower level of the data in the first year (lower, that is, than predicted at time 0), however, there are several consecutive, positive errors in 1977 indicating a higher level than forecast until the model adapts again towards the end of that year. Note that the model has 8 parameters so that there is an inevitable delay of several months before it can completely adapt to change. Subsequent forecasts for late 1977 and early 1978 are acceptable.

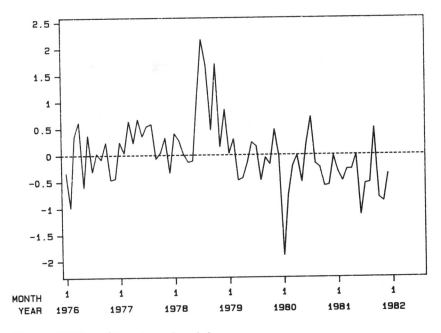

Figure 10.11.   One-step ahead forecast errors.

In mid-1978, forecast performance breaks down dramatically, with one very large error and several, subsequent positive errors. The changing pattern of Sales at this time is not predicted by the model and, again, several observations are required for learning about the change. The nature of the change is made apparent in Section 10.3.7 below. For the moment, note that this sort of event, so commonly encountered with real series, goes outside the model form and so will rarely be adequately handled by a closed model. What is required is some form of intervention, either to allow for more radical change in model parameters than currently modelled through the existing discount factors, or to incorporate some explanation of the change via intervention effects or other independent variables. We proceed, however, with the model as it stands, closed to intervention.

After 1978, the analysis continues with reasonable success during 1979 and 1980, a single, large negative error in January 1980 being essentially an outlier. During late 1980 and 1981, negative errors, (although acceptably small in absolute value) predominate and, in particular, the model has difficulty in forecasting the seasonal troughs, reflecting, perhaps, a change in seasonal pattern greater than anticipated.

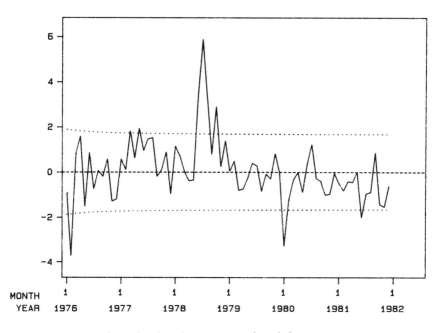

Figure 10.12.   Standardised one-step ahead forecast errors.

One important feature clearly apparent in Figure 10.10 is that
the widths of the one-step forecast 90% intervals increase markedly
towards the end of 1978, being much wider from then on than in
early years. This effect is explained by the presence of the large
forecast errors in late 1978. In the updating for the observational
variance, recall that that the estimate at time t is updated via

$$S_t = S_{t-1}(n_{t-1} + e_t^2/Q_t)/(n_{t-1} + 1).$$

Thus large standardised errors $e_t/Q_t^{1/2}$ lead to an inflation of the
estimate, the posterior distribution for V then favouring larger values
and, consequently, the forecast distribution for the next observation
will be more diffuse. This is an interesting feature; the model/data
mismatch that leads to large errors shows up in the variance estimate,
which can therefore be used as a simple diagnostic, and the model
responds in the way of increased uncertainty about the future. Note,
however, that this is due entirely to the fact that V is unknown, the
effect being diminished for larger degrees of freedom n_{t-1}.

10.3.7 Retrospective time series analysis after filtering

Having processed all the data and examined the one-step forecasting
activity through Figures 10.10 to 10.12, we now move to retrospective
time series analysis more formally by performing the backwards fil-
tering/ smoothing with the model, obtaining posterior distributions
for the model parameters at all times $t = 1, \ldots, 72$ now conditional
on all the data information D_{72}. Of course these are T distributions
with $n_{72} = 78$ degrees of freedom (the initial prior 6 plus 1 for each
observation), and model components may be extracted as in Section
10.2.3 (although there we used the on-line posteriors rather than the
filtered posteriors). In the notation of Section 4.7, we have posteriors

$$(\boldsymbol{\theta}_t \mid D_{72}) \sim \mathrm{T}_{72}[\mathbf{a}_{72}(t - 72), \ \mathbf{R}_{72}(t - 72)]$$

with moments determined by the filtering algorithms of Theorem 4.4
and Corollary 4.3. Thus for any vector $\mathbf{x}_t$, if $x_t = \mathbf{x}_t'\boldsymbol{\theta}_t$, then

$$(x_t \mid D_t) \sim \mathrm{T}_{72}[\mathbf{x}_t'\mathbf{a}_{72}(t - 72), \ \mathbf{x}_t'\mathbf{R}_t(t - 72)\mathbf{x}_t],$$

the mean here providing a *retrospective* estimate of the quantity x_t.
Various choices of $\mathbf{x}_t$ are considered.

(1) SEASONAL PATTERN
   $\mathbf{x}_t' = (0; 0; 1, 0; 1, 0; 1, 0)$ implies that $x_t = \theta_{t3} + \theta_{t5} + \theta_{t7} = \phi_{t0}$,
   the seasonal factor at time t. Figure 10.13 displays 90% in-
   tervals for the effects over time, symmetrically located, as
   usual, about the filtered, posterior mean. Changes in seasonal pattern are evident from this graph. The amplitude of
   seasonality drops noticeably in early 1978, is stable over the
   next 3 years but increases again in 1981. The phase also shifts
   markedly in early 1978 (recall the deterioration in forecasting performance during this period), and continues to shift
   slightly from year to year from then on. Note also the increase in uncertainty about the effects in later years due to
   the inflation in estimation of V earlier described.

(2) NON-SEASONAL TREND
   $\mathbf{x}_t' = (1; X_t; 0, 0; 0, 0; 0, 0)$ implies that $x_t = \theta_{t1} + \theta_{t2}X_t$ is the
   deseasonalised trend at time t, including the regression effect
   of Index. The corresponding posterior mean plus 90% limits
   appear in Figure 10.14.

(3) REGRESSION EFFECT
   A related case is given by $\mathbf{x}_t' = (0; X_t; 0, 0; 0, 0; 0, 0)$, so that

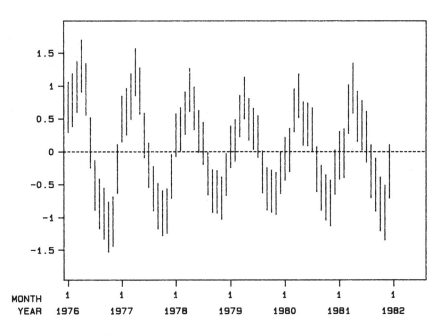

Figure 10.13.   Estimated monthly pattern in Sales.

$x_t = \theta_{t2} X_t$ is the regression effect of Index free from the time-varying intercept term θ_{t1}. The corresponding plot is given in Figure 10.15. This figure is useful in retrospectively assessing the effects of the historic changes in values of the independent variable series. If Index is subject to some form of control by the company (such as via pricing policies), then this graph provides information about the effects of past attempts at controlling the Sales series via changes to Index.

(4) REGRESSION COEFFICIENT

Taking $\mathbf{x}_t' = (0; 1; 0, 0; 0, 0; 0, 0)$ leads to $x_t = \theta_{t2}$, the coefficient in the regression on Index at time t. The corresponding plot is given in Figure 10.16. There is obvious movement in the coefficient, the Sales series being apparently less strongly related to Index in the later years, the coefficient drifting upwards to around 0.4, although the uncertainty about the values is fairly large.

(5) INTERCEPT COEFFICIENT

Figure 10.17 is the display for the intercept term θ_{t1} obtained from (2) above with $X_t = 0$, so that $\mathbf{x}_t' = (1; 0; 0, 0; 0, 0; 0, 0)$.

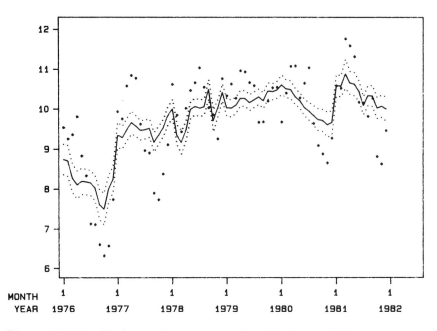

Figure 10.14.   Estimated non-seasonal variation in Sales.

Again there is obvious change here, consistent with the features noted in discussing the model adaptati n to data in
Section 10.3.6. It is of interest to the compa y to attempt,
retrospectively, to explain this movement in the parameter;
if independent variable information exists a. d can be related
to this smooth movement, then future chan ;es may be anticipated and forecasting accuracy improved.

Note that the estimated time evolution of th parameters as described above are a consequence of the model an lysis. Here the data
series are, at times, rather ill-behaved relativ to the model form
specified, and the changes estimated may be artially explained as
the model's reponse to the model:data misr atch. It is inevitable
that, in allowing for change in model compo ents, some change may
be derived from the effects of variables omit ed that could otherwise
have explained some of the observed dev .tions away from model
form. If such explanations could be foun ., then the estimation of
existing components would be protected f m possible biases induced
by their omission. Hence care is needed n interpreting the changes
identified by any given model.

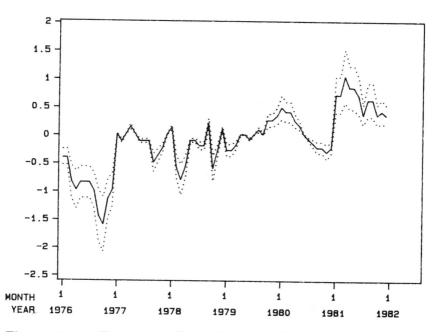

Figure 10.15.   Estimated effect of Index on Sales.

10.3.8   Step ahead forecasting: *What if?* analysis

Finally, at time $t = 72$, December 1982, consider forecasting step
ahead based solely on the model and current posteriors. To forecast
Sales during 1982 to December 1984 requires values for the Index
series as inputs. For illustration of a *What if?* analysis, the values
are taken as

$$X_t = 2, \qquad (t = 73, \ldots, 84),$$
$$X_t = 0, \qquad (t = 85, \ldots, 96),$$
$$X_t = -2, \qquad (t = 97, \ldots, 108).$$

The values of Index over 1976 to the end of 1981 range from a low of
-2.70 to a high of 2.13. Thus the step ahead values used here allow
exploration of the implications of fairly extreme values of Index;
during 1982, Index is unfavourable at 2, and during 1984 favourable
at -2. During 1983, the value of 0 leads to forecasts based only on
the seasonal pattern. Figure 10.18 displays the historical data, and
forecasts over these 3 years based on these *What if?* inputs.

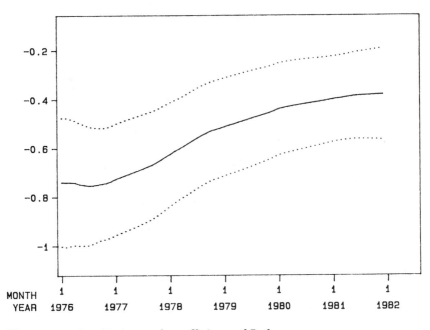

Figure 10.16.   Estimated coefficient of Index.

10.4   ERROR ANALYSIS

10.4.1   General comments

There is a large literature on statistical techniques for error, or residual, analysis in linear models generally and classical time series models in particular. Almost all of the formal testing techniques are non-
Bayesian, though most of the useful approaches are informal and do
not require adherence to any particular inferential framework. Fair
coverage and reference to error analysis in time series can be found
in Box and Jenkins (1976) and Abraham and Ledolter (1985). General Bayesian theory for residual diagnostics appears in Box (1979),
and Smith and Pettit (1985). In our opinion, the most useful error
analyses consist of informal examination of forecast errors, looking at
graphical displays and simple numerical summaries to obtain insight
into possible model inadequacies with a view to refinement. The use
of formal tests of goodness of fit, though possibly of value in certain
contexts, does not address the key issues underlying error analyses.
These can be summarised as:

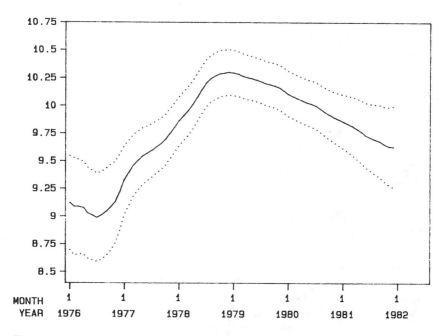

MONTH   1   1   1   1   1   1   1
YEAR   1976   1977   1978   1979   1980   1981   1982

Figure 10.17.   Estimated intercept parameter.

(1) identifying periods of time in the history of the series analysed where model performance deteriorated;

(2) suggesting explanations for this deterioration, or breakdown, in model performance; and

(3) modifying the model for the future in the light of such explanations.

The focus in classical error analyses is usually on *fitted* errors, or residuals, measuring the retrospective departure of the data from the model. How closely the historical data fits the model in retrospect is not really the point in time series forecasting; more incisive are investigations of the *predictive fit* of the model, embodied in the sequence of one-step ahead forecast distributions. The one-step forecast errors e_t are thus the raw material for model assessment. Thus we have presented plots of the e_t, and their standardised analogues, in the illustrations of Sections 10.2 and 10.3. Under the assumptions of the model, the forecast distributions are, as usual, $(Y_t \mid D_{t-1}) \sim \mathrm{T}_{n_{t-1}}[f_t, Q_t]$ so that, as a random quantity prior to observing Y_t, the error e_t has the predictive distribution

$$(e_t \mid D_{t-1}) \sim \mathrm{T}_{n_{t-1}}[0, Q_t].$$

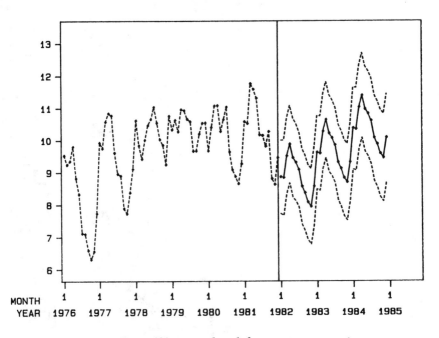

Figure 10.18. 'What-if?' step-ahead forecasts.

Of course, when n_{t-1} is large, or the observational variance sequence is known, these distributions are essentially normal. Also,

$$p(e_t, \dots, e_1 \mid D_0) = \prod_{r=1}^{t} p(e_r \mid D_{r-1}),$$

and so $(e_t \mid D_{t-1})$ is conditionally independent of $(e_s \mid D_{s-1})$ for $s < t$. The observed errors are thus the realisation of what is, under the model assumptions, a sequence of independent random quantities with the above distributions. Observed deviations in the error sequence away from this predicted behaviour is indicative of either irregularitites in the data series, or model inadequacies, or both. The raw errors e_t should be standardised with respect to their scale parameters for examination. Thus consider the sequence of standardised errors

$$u_t = e_t / \sqrt{Q_t},$$

hypothetically (conditionally) independent and distributed as

$$(u_t \mid D_{t-1}) \sim \mathrm{T}_{n_{t-1}}[0, 1], \qquad (t = 1, 2, \dots).$$

The plot over time of the u_t sequence is the most useful starting point in error analysis. Deviations of the error sequence from model predictions show up rather easily in such plots. Common deviations of interest and importance from the model assumptions are as follows.

(a) Individual, extreme errors, possibly due to outlying observations;

(b) Groups, or patches, of a few errors of the same sign suggesting either a local drift away from zero location or the development of positive correlation between the errors;

(c) Patches of negatively related errors, the signs tending to alternate;

(d) Patches of errors large in absolute value, without any clear pattern of relationship.

Though not exclusive, these features are common and account for many types of deviation of the data from the model, and are now examined.

10.4.2 Outliers

A single large error u_t indicates that Y_t is a single observation that deviates markedly from forecast under the model. The notion of extremity with respect to the forecast distribution of u_t involves considering values in the tails of the distribution. The fact that the degrees of freedom may differ for each error is important; Figures 10.3 and 10.12 provide, as a graphical guide, 90% forecast intervals symmetrically located about the mode of zero. Note that these are both equal-tailed and HPD intervals since the T densities are symmetric about zero. Values of u_t outside these intervals occur only 10% of the time, on average. Supplementing these plots with the corresponding 95% intervals would be a useful addition. Such observations occur in every area of application and are commonly referred to as outliers. With socio-economic data series, it is not uncommon to encounter such wild values one observation in 20, and often they are rather more frequent. In the context under study, there may be many possible explanations for outliers. A true outlier provides no information about the observation process, such as the case if the observation has been misrecorded in the data collection process. The same may be true even though the observation is correctly recorded, when there exists auxilliary information as to why the value is wild. For example, in a consumer sales series the announcement of a future price rise can lead to consumers stocking up in advance at the old price, leading to a single, very large observation that is totally correct and acceptable, but which is quite irrelevant to the underlying pattern of sales generally. In ideal circumstances, all outliers should be explained in this way, being of potentially major importance to decision makers. From the point of view of retrospective error analysis, outliers can be identified by their single, large standardised errors. From the point of view of modifying the model for the future in view of these outliers, there is often little to be done other than inform decision makers that outliers can be expected, but not predicted unless suitable auxilliary information arises in advance. Without such information, if outliers are to be anticipated the main way of reflecting their existence is to somehow increase forecast uncertainty; very large values that are rare under the existing model will be less rare under a model generating forecast distributions with heavier tails. Thus a common model modification is to extend the basic, normal DLM to include an observational error distribution that has heavy tails. Some possibilities are developed in Harrison and Stevens (1976), West (1981), and West, Harrison and Migon (1985). Usually the extensions are to heavier-tailed, symmetric error distributions. If large errors of a particular sign predominate, then the suggestion may be that the errors have a skew distribution, possibly indicating the need for a data transformation (see Section 10.6).

A typical example outlier shows up in the January, 1980 error in the example of Section 10.3. Figure 10.12 shows the error in this month to be way outside the 90% interval, and in fact it is ouside the 99% limit. The error corresponds to a much lower sales figure in that month than forecast. Another apparent outlier occurs in February 1976, but this one is explained as due to the inappropriateness of the initial prior estimate of the seasonal factor for that month.

10.4.3   Correlation structure

Continuing with an examination of the errors in Figure 10.12, the very wild observation in the sales series in mid-1978 may be initially thought to be an outlier, but the sequence of positive errors immediately following this one point indicate something more sustained has occurred over a period of a few months. This is consistent with item (b) of Section 10.4.1, the occurrence of a patch of consecutive, positive errors. In this example, there was a marked change in the seasonal pattern of the sales series for the year of 1978 that led to these errors. This is a very common occurrence. If changes occur that are much more marked than allowed for through the existing dynamics of the model, then there will be a subsequent run of errors that are positively correlated. Gradually, the model adapts to the change and, after sufficient observations, the errors are back under control. Note that this feature of autocorrelated errors involves the development, from time to time, of *local* correlation structure in the errors that eventually decays as the model adapts. In a model in which the basic dynamic is too constrained, due to discount factors that are too large, the natural movement in the series will tend to be under-predicted throughout and therefore the positive correlation between consecutive errors will tend to be sustained. This feature appears again during the last few months of the dataset where there is a run of several, negative errors as the model fails to adapt rapidly enough to the lower sales levels in the recession of late 1981.

There are various ways of modifying the model for the future to deal with this. Notably:

(1) Forecasting and updating in the future subject to some form of monitoring of the model to detect the development of correlation between forecast errors. This may simply involve the forecaster in subjective monitoring, calling for intervention when correlation develops. The most useful and usual form of intervention involves corrective action to decrease discount factors, often just temporarily, on some or all of the existing model components. This has the effect of increasing adaptivity to change and the model then responds more rapidly in future.

(2) Explanation of consecutive, positively related deviations from forecast via the inclusion in the model of additional terms. For example, a temporary jump in level of the series can be modelled by including an additional, possibly slowly varying level parameter for a few time periods. Regression type terms are also useful if independent variables can be identifed. Seasonal patterns not modelled will also show up in correlated errors.

(3) The extension of the model to include a noise process term to model the local correlation structure and use past errors to forecast those to come in the short term; see the development of noise models in Section 9.4 of Chapter 9. It should be stressed that such terms be included only to describe local error structure that cannot be otherwise, more constructively, explained via parameter changes or regression type terms. It is sometimes tempting to explain more global movement in

the series by such noise models when in fact they should attributed to changes in trend or other components of the basic DLM. These comments notwithstanding, short-term forecasting accuracy can be improved by incorporating a simple noise model when local correlations develop. Most useful is a the basic AR(1) type model of Section 9.3, with, typically, a coefficient expected to be fairly low, between 0 and 0.4, say.

Consecutive observations exhibiting negative correlation structure also occur, though perhaps less commonly than positive correlation. One obvious explanation for this is the use of a dynamic model that is too adaptive, having discount factors that are too small. The result is that the posterior distributions for model parameters adapt markedly to each observation and thus short term predictive accuracy is lost. Negatively correlated errors result as the model swings from one observation to the next. Again corrections to the model properly require adjustments to discount factors, although simple noise models will improve short-term forecasting accuracy.

10.4.4   Observational variance structure

Patches of errors that are large in absolute value but which do not show any clear structure can arise, and often do, if the assumptions about the observational variances in the normal DLM are inappropriate. Particular reasons for such features are:

(1) The possibility that each observation Y_t is made up as a function of several or many basic quantities, such as a total or average, leading to variances that may differ widely depending on the number of such components of each of the Y_t;

(2) Observational variances that depend on the underlying level of the series, due to original non-normality of the Y_t; and

(3) Possible random fluctuations in the observational variance about the level, representing volatity explained only by the absence of relevant independent variables from the model ie. model misspecification that is evidenced, not by systematic, local correlations in the errors, but rather by random changes in variation.

These, and other, features can be modelled adequately by extending and modifying the basic, normal DLM as in the next section.

10.5 DATA MODIFICATIONS AND IRREGULARITIES

10.5.1 Outliers and missing values

Following the discussion of outliers in Section 10.4.2, consider the treatment of wild observations. For routine updating in the model, the fact that, explainable or otherwise, an observation is quite unrelated to the underlying process means that, for forecasting the future, it provides little or no information and should be discarded. Formal techniques for doing this by extending the model to include outlier generating distributions can be used (Harrison and Stevens, 1976; West, 1981; and West, Harrison and Migon, 1985). Some such models are developed in later chapters. Often, however, simply discarding the observation from the model updating is the best approach. That outlying observations can seriously adversely affect the analysis of a model and degrade predictions is well known and not specific to time series. There is a large literature on outlier handling and modelling techniques. See, for example, Barnett and Lewis (1978), Box (1980), Smith and Pettit (1985), West (1984, 1985a), and references therein. Work specifically concerning outlier problems in time series can be found in Box and Tiao (1968), Fox (1972), Harrison and Stevens (1976), Kleiner, Martin and Thompson (1981), Masreliez and Martin (1977), West (1981, 1986b), and West and Harrison (1986). In the standard updating equations for the DLM recall that the error e_t enters in the equations for the estimates of both $\boldsymbol{\theta}_t$ and V, namely

$$\mathbf{m}_t = \mathbf{a}_t + \mathbf{A}_t e_t,$$

and

$$S_t = S_{t-1}(n_{t-1} + e_t^2/Q_t)/(n_{t-1} + 1).$$

Thus a large, outlying value of $|e_t|$ leads to a major correction to both $\mathbf{a}_t$ and S_{t-1} to obtain the posterior quantities $\mathbf{m}_t$ and S_t. The variance estimate, a quadratic function of the error, can be inflated enormously when n_{t-1} is small. To guard against this, the most appropriate action is to discard the observation from the updating of the model if it has been identified as an outlier. Thus, given this identification (which can sometimes be rather difficult to achieve, but is assumed here), Y_t provides no information about the model, the information set at time t is just $D_t = D_{t-1}$, and the posterior distributions are equal to the priors, viz. $\mathbf{m}_t = \mathbf{a}_t$, $\mathbf{C}_t = \mathbf{R}_t$, $S_t =$

S_{t-1} and $n_t = n_{t-1}$. We stress again that this is appropriate if Y_t has been identified as uninformative about the model. In Chapters 11 and 12 we return to problems of outlier detection and modelling.

An observation that is truly missing, being unrecorded or lost, is treated in just the same way, of course. Other approaches to time series, in particular using ARIMA type models, have enormous difficulties in handling missing values. The sequential Bayesian approach, in marked contrast, trivially accounts for them: no information leads to no changes in distributions for model parameters. In some applications observations associated with discrete, external events may be viewed as missing in order to simply separate the events from the routine model. This can be done, for example, to model the effects of individual, irregular events such as strikes in production plants, responses to budget and tax announcements, irregular holidays, and so forth. This is a simple alternative to modelling the effects of such events with extra parameters to be estimated, appropriate if all that is required is protection of the standard model from the observation.

10.5.2   Irregular timing intervals

Another practically important problem that bedevils classical time series models is the irregular timing of observations. So far we have assumed, as with all classical models, that the observations are equally spaced in time, and that the timing is precise. There will often be timing errors, some random, some systematic, that will be impossible to identify and quantify. Such minor discrepancies are assumed to be accounted for by the error sequences ν_t and ω_t in the model. Of greater significance is the problem of observations arriving at intervals of irregular length. For example, monthly data may be recorded on a particular day towards the end of each month, but the precise day can vary between months. Daily observations may be recorded only to the nearest hour. In these cases, where the data is reported at an aggregate level but may be based on different numbers of time intervals at a lower level of aggregation, it is clear that the timing should, if possible, be accounted for in the model. If the timing is unknown then little can be done. Otherwise, the approach is simple. The answer lies in the fact that there is a basic, original time unit that provides a precise time scale. In the first example above the unit is the day, in the second it is the hour. If we define our model so that the observations are indexed by this basic unit, then the observations are equally spaced, but some, often many, of them are missing. Thus, with monthly data in the first example, if t indexes days and $t = 1$ represents January 1st, then we may receive data, $Y_{29}, Y_{57}, Y_{86}, \ldots$, one in each month, the others, $Y_1, Y_2, \ldots, Y_{28}, Y_{30}, \ldots$, being viewed as missing.

Note that, if the data are aggregates, the Y_t above, for example, purporting to be monthly totals or averages of daily data, then further, obvious modifications are necessary to provide an appropriate model at the daily level.

10.5.3 Discount models with missing data

Some comments are in order concerning the use of discount models with missing data. The inherent, one-step ahead nature of the discount technique, and the too rapid decay of information if it is used to project step ahead, has already been discussed in Section 6.3 of Chapter 6, and Section 10.2.4 above. This feature is apparent when missing values occur. When $\boldsymbol{\theta}_t$ evolves with no incoming data, whether truly missing or due to irregular timing, the evolution error variance matrix should be based on that most recently defined using the standard discount factors. For example, suppose at $t = 4$ that we evolve as usual with $\mathbf{W}_4$ determined by discount factors. If Y_4 and Y_5, say, are missing, then the evolution over the corresponding time intervals is based on $\mathbf{W}_5 = \mathbf{W}_6 = \mathbf{W}_4$, that value most recently used in a time interval when an observation was made. Only when the next observation Y_7 arrives is the discount technique applied again, determining $\mathbf{W}_8$ for evolution to $t = 8$.

10.6 DATA TRANSFORMATIONS

10.6.1 Non-normality of observations

It is rather commonly found that observed data do not adequately conform to the assumptions of normality and constancy of variance of the observational errors. We noted in Section 10.4.4 that the forecast errors can signal apparent changes in variance, consistent with original non-normality of the sampling structure of the series. A typical feature of commercial and socio-economic series of positive data (by far the majority of data in these areas are positive), is an increasing variance with level. Possible, closely related reasons are that the data are aggregated or that a non-normal model is appropriate. As an example, consider sales or demand data for batches of items, the batch sizes being uncorrelated with constant mean f and variance V. For a batch of size n, the number sold has mean nf and variance nV. As a function of varying n (with f and V fixed), the variance of the batch size is then proportional to the mean batch size. With Poisson data, the variance equals the mean. Other non-normal observational distributions for positive data, such as lognormal, gamma or inverse gamma, various compound Poisson forms, and others, have the feature that the variance increases as a function of the mean of the distribution. Empirical or theoretical evidence that such distributions may underlie a series therefore will suggest such a mean/variance relationship. Often a transformation, such as logs, will approximately remove this dependence of varia- tion on level, leading to a transformed series with roughly constant variance and a distribution closer to normality, or at least symme- try. Such data transformations are widely used in statistics, and can provide a satisfactory solution, the data being modelled on the transformed scale. Uncritical use of transformations for these rea- sons of convenience should, however, be guarded against. Often a suitable transformation will be difficult to identify and use of a sim- ple substitute may complicate or obscure patterns and relationships evident on the original data scale. The benefits of the original scale are obvious when we consider the interpretation of a given model, and the communication of its implications to others. In particular, ·the construction of linear models in terms of components for distinct features neatly separates the random error term from the structural component of a series, and operating on a transformed scale will usu- ally mean that the benefits acruing from this component modelling are diminished.

Generally then, it is preferable to model the series on the original data scale. Various techniques are available when we want to retain the original scale but the data are essentially non-normal. The ob- vious approach, in principle, is to use a more appropriate sampling distribution for the raw data, and we do this in Chapter 14. Here we stay with normal models, considering modifications of the basic DLM framework to cater for certain, commonly encountered features of non-normality.

10.6.2 Transformations and variance laws

The above, cautionary notes about transformations notwithstand- ing, consider operating with a DLM for a transformed series. Let $g(.)$ be a known, continuous and monotone function and define the transformed series Z_t by

$$Y_t = g(Z_t), \qquad \text{or} \qquad Z_t = g^{-1}(Y_t), \qquad\qquad t = 1, 2, \ldots,$$

the inverse of g existing by definition. Often, though not always, it is the case that Z_t is real-valued, the transformation g mapping the real line to the, possibly bounded, original range for Y_t. Most commonly the Y_t are positive, and this is assumed throughout this section. The associated observational error distributions are typically skewed to the right, with variances increasing as the location increases, indicating greater random, unpredictable variation in the data at higher levels. Various underlying error distibutions can be supported on theoretical and empirical grounds. Some examples, and general theory, are given in Stevens (1974) and Morris (1983). The primary aims of transformation are to achieve a roughly normal, or, more, realistically, roughly symmetric, distribution for the Z_t with a variance not depending on the location. We concentrate here on the latter feature, the normality being very much of secondary importance. In fact the transformations usually used to achieve constancy of variance tend to lead to rough symmetry too.

For convenience of notation throughout this section, we drop the dependence of the distibutions etc. on the time index t, the discussion being, in any case, rather general. Thus $Y = Y_t$ and $Z = Z_t$ are related via $Y = g(Z)$. Approximate, guiding relationships between the moments of Y and those of Z can be used if we assume, as is usually appropriate, that the function g is at least twice differentiable, having first and second derivatives g' and g'' respectively. If this is so, then simple approximations from Lindley (1965, Part 1, Section 3.4) apply. Suppose that the transformed quantity Z has mean μ and variance V, whether the distribution be approximately normal or otherwise. Neither need, in fact, be known; we simply assume that these are the moments of Z conditional on their values being assumed known, writing $\mu = \mathrm{E}[Z]$ and $V = \mathrm{V}[Z]$. Then, following Lindley as referenced above, we have:

(a) $\mathrm{E}[Y] \approx g(\mu) + 0.5g''(\mu)V$; and
(b) $\mathrm{V}[Y] \approx \{g'(\mu)\}^2 V$.

Thus, approximately, the variance will be constant (as a function of μ) on the transformed scale if, and only if, the original Y has variance $\mathrm{V}[Y] \propto \{g'(\mu)\}^2$. If we ignore the second term in (a), assuming that the first term dominates, then $\mu \approx g^{-1}(\mathrm{E}[Y])$ and the requirement becomes

$$\mathrm{V}[Y] \propto \{g'[g^{-1}(\mathrm{E}[Y])]\}.^2 \qquad\qquad (10.1)$$

This is a particular case of a *variance law* for Y, the variance being functionally related to the mean.

Particular cases of (10.1) are examined in connection with the most widely used class of *power* transformations (Box and Cox, 1964). Here

$$Z = \begin{cases} (Y^\lambda - 1)/\lambda, & \lambda \neq 0; \\ \log(Y), & \lambda = 0, \end{cases}$$

for some *power index* λ. Interesting values of this index, in our context of positive data, lie in the range $0 \leq \lambda < 1$. These provide decreasing transformations that shrink larger values of Y towards zero, reducing the skewness of the sampling distribution. Note that the inverse transformation is just

$$g(Z) = \begin{cases} (\lambda Z + 1)^{1/\lambda}, & \lambda \neq 0; \\ \exp(Z), & \lambda = 0. \end{cases}$$

Using this in (10.1) leads to

$$V[Y] \propto E[Y]^{2(1-\lambda)},$$

and the following special cases:

(a) $\lambda = 0$ leads to $V[Y] \propto E[Y]^2$, with the log transform appropriate. Note that, if in fact Y is lognormally distributed, then the variance is exactly a quadratic function of the mean and the log transform exactly appropriate.

(b) $\lambda = 0.5$ gives $V[Y] \propto E[Y]$ and the square root tranformation is appropriate, consistent with a Poisson distribution, for example.

(c) $\lambda = 0.25$ gives $V[Y] \propto E[Y]^{1.5}$, consistent with compound Poisson like distributions often empirically supported in commercial and economic applications (Stevens, 1974).

10.6.3   Forecasting transformed series

Given a DLM for the transformed Z series, forecasting the original Y series is, in principle, straightforward within the Bayesian framework. We suppose that, from a DLM for the Z series, a forecast distribution (one-step or otherwise) is available in the usual T form, namely

$$Z \sim T_n[f, Q].$$

Note again that we have simplified the notation and dependence on the time series context for clarity and generality. Here n, f and Q are all known at the time of forecasting. Let P_Z denote the cumulative distribution function, or cdf, of this T distribution, so that $P_Z(z) = \Pr[Z \leq z]$ for all real z. Since $g(.)$ is continuous and monotonic, then the corresponding forecast cdf for Y is defined as

$$P_Y(y) = \Pr[Y \leq y] = \Pr[Z \leq g^{-1}(y)] = P_Z(g^{-1}(y)).$$

Point forecasts, uncertainty measures and probabilities can in principle, now be calculated for Y. In practice, these calculations may be difficult to perform without numerical integrations. Practically useful transfromations, such as the power transformations, lead to distributions that are not of standard form so that, unlike the usual T distributions, a little more effort is required to adequately summarise them. Some generally useful features are as follows.

(1) The forecast distribution will usually (though beware, not always) be unimodal, so that point estimates such as means, modes and medians provide reliable guides as to the location of the distribution.

(2) The moments of Y are often obtainable only using numerical integration. Simple quadrature, for example, is relatively straightforward since Y is a scalar. The approximate results used in the last section provide rough values.

(3) The median of Y can always be found easily, being the solution in y to the equation

$$0.5 = P_Y(y) = P_Z(g^{-1}(y)).$$

Thus, since the mode f of P_Z is also the median, then the median of Y is simply $g(f)$.

(4) The result (3) is a special case of the result that percentage points of P_Y are obtained as the transformed values of those of P_Z. Simply replace 0.5 in the equation in (3) by any required probability to see this. Thus, for example, intervals for Z with stated probabilities under the T distribution for Z provide intervals for Y, with the same probabilities, simply by transforming the end-points.

10.6.4   Log transforms and multiplicative models

The most important transformation is the natural log transform. Consider, for illustration, a simple dynamic regression for the transformed series Z_t, given by

$$Z_t = \alpha_t + \beta_t X_t^* + \nu_t, \qquad \nu_t \sim N[0, V],$$

where X_t^* is the observed value of a regressor variable. The level at time t is $\mu_t = \alpha_t + \beta_t X_t^*$, and the transformed series follows a standard DLM with constant, possibly unknown variance V. The original series is given by $Y_t = \exp(Z_t)$, corresponding to the power transformation with index $\lambda = 0$. Some points of relevance to the model for the original series are as follows.

(1) Conditional on the model parameters $\boldsymbol{\theta}_t = (\alpha_t, \beta_t)'$ and V,

$$Y_t = e^{\mu_t + \nu_t} = e^{\mu_t} e^{\nu_t}.$$

Now $\exp(\nu_t)$ is a lognormal random quantity so that Y_t has a lognormal distribution. Properties of lognormal distributions are fully discussed in Aitchison and Brown (1957). In particular, $E[Y_t \mid \boldsymbol{\theta}_t, V] = \exp(\mu_t + V/2)$, the mode of Y_t is $\exp(\mu_t - V/2)$, and the median is $\exp(\mu_t)$. Also, the variance is proportional to the square of the mean, the constant of proportionality being given by $1 - \exp(-V)$.

(2) The model for Y_t is multiplicative, given by

$$Y_t = \gamma_t X_t^{\beta_t} e^{\nu_t},$$

where $\gamma_t = \exp(\alpha_t)$ and $X_t = \exp(X_t^*)$. Often X_t is an original, positive quantity transformed to the log scale along with Y_t.

(3) This form of model, widely used to represent socio-economic relationships, is particularly of use in studying regressions in terms of *rates* of change of the response with the independent variable. An increase of $100\epsilon\%$ in X_t, $(0 \le \epsilon \le 1)$, leads to an expected increase of $100(1 + \epsilon)^{\beta_t}\%$ in Y_t. Thus inferences about β_t from the DLM for Z_t lead directly to inferences about expected *percentage* changes on the original scale.

Similar features arise, of course, in more general DLMs for the log values.

10.7 MODELLING VARIANCE LAWS

10.7.1 Weighted observations

Before considering modelling of variance laws generally, the simple case of weighted observations, involving observational variances known up to a constant scale parameter, is discussed. As a motivating example, consider monthly data Y_t that are arithmetic means, or averages, over daily data in each month t. If the raw daily data are assumed uncorrelated with common, constant variance V, then the variance of Y_t is $V_t = V/n_t$ where n_t is the known number of days in month t, obviously varying from month to month. The observation is said to have be *weighted*, having a known, positive *weight*, or *variance divisor*, $k_t = n_t$. An observation with little or no weight has a large, or infinite variance, one with high weight has a small variance. Unit weight $n_t = 1$ corresponds to the original, unweighted model. This sort of weighting applies generally when the data are aggregates, of some form, of more basic quantities. Note also that the treatment of outliers, missing values and irregularly spaced data of Section 10.5 is a special case, missing values essentially receiving no weight.

The basic DLM requires only a minor modification to allow for weighted observations. The usual theory applies directly to the general model

$$\{\mathbf{F}_t, \mathbf{G}_t, k_t V, \mathbf{W}_t V\},$$

where k_t is, for each t, a known, positive constant, giving Y_t a weight k_t^{-1}. The modification to the updating and forecasting equations is that the scale parameter Q_t now includes the weight. Thus

$$(Y_t \mid D_{t-1}) \sim \mathrm{T}_{n_{t-1}}[f_t, Q_t]$$

where $f_t = \mathbf{F}_t' \mathbf{a}_t$, as usual, but

$$Q_t = k_t S_{t-1} + \mathbf{F}_t' \mathbf{R}_t \mathbf{F}_t.$$

Similarly, future values k_{t+r} appear in the expressions $Q_t(r)$ for the scale parameters of step ahead forecast distributions made at time t for times $t + r$.

10.7.2 Observational variance laws

Suppose we wish to use a DLM for the Y_t series on the original, positive scale, but recognise that the data are non-normal with a variance law identified, at least approximately. Forecasting and updating can usually satisfactorily proceed with a weighted model, the weights provided as rough estimates of the variance law, varying as the level of the data varies. Precise values for the variance multipliers are not essential. What is important is that the variance multipliers change markedly as the level of the series changes markedly, providing an approximate indication of the relative degrees of observational variation at different levels. Denote the variance law by $k(.)$, so that the variance at time t is expected to be approximately given by

$$V_t = k(\mu_t)V,$$

where $\mu_t = \mathbf{F}_t'\boldsymbol{\theta}_t$ is the level of the series. Examples are

    (a) $k(\mu_t) = 1 + b\mu_t^p$, for constants $b > 0$ and $p > 0$; and
    (b) $k(\mu_t) = \mu_t^p$ for $p > 0$.

The latter has been much used in practice, as in Stevens (1974), Harrison and Stevens (1971, 1976b), Smith and West (1983), West, Harrison and Pole (1987), for example. Recall from Section 10.6.2 that a powerlaw variance function of the form in (b) corresponds roughly to a power transformation of Y_t to constant variance with power index $\lambda = 1 - p/2$. Note that, generally, $k(.)$ may depend on t although this is not explicitly considered here. Obviously the model analysis would be lost if variances are allowed to depend on $\boldsymbol{\theta}_t$ in this way, the prior, posterior and forecast distributions no longer being analytically tractable. We stay within the ambit of standard theory, however, if the value of the variance law $k(.)$ is replaced by a known quantity, a variance multiplier. The obvious multiplier is obtained by replacing the conditional mean μ_t in the law by its prior mean, the forecast value f_t, thus using the model $\{\mathbf{F}_t,\ \mathbf{G}_t,\ k_t V,\ \mathbf{W}_t V\}$, where $k_t = k(f_t)$ is known. This simply, and appropriately, accounts for changing variance with *expected* level of the series. Note that, in forecasting ahead from time t, the variance mulipliers used for future observations will now depend upon the time t, being functions of the current (at time t) expected values of the future observations.

10.8   STOCHASTIC CHANGES IN VARIANCE

10.8.1   General considerations

The unknown variance scale parameter V appearing in the observational variance, whether weighted or not, has been assumed throughout to be constant over time. The updating for V based on gamma prior and posterior distributions for the precision parameter $\phi = 1/V$ provides a coherent, effective learning algorithm that eventually leads to convergence. To see this recall the posterior at time t is given by $(\phi \mid D_t) \sim G[n_t/2, d_t/2]$, where the degrees of freedom parameter n_t updates by 1 for each observation. Thus, as $t \to \infty$, $n_t \to \infty$ and the posterior converges about the mode. With the usual point estimate $S_t = d_t/n_t$ of V, the posterior asymptotically degenerates with $|V - S_t| \to 0$ with probability one. Now, although this is an asymptotic result, the posterior can become quite precise rapidly as n_t increases, leading to under-adaptation to new data so far as learning on the variance is concerned. The problem lies with the assumption of constancy of V, that conflicts somewhat with the underlying belief in change over time applied to the parameters in θ_t. Having explored above the possibility of changes in observational variance due to non-normality, we now consider the possibility that, whether using a variance weight or not, the scale parameter V may vary stochastically and unpredictably over time. Some supporting arguments for such variation are as follows.

(1) There may actually be additional, stochastic elements affecting the observational error sequence, that have not been modelled. Some possible sources of extra randomness are inaccuracies in the quoted timing of observations, truncation of readings, changes in data recording and handling procedures, etc. Some of these are present in the application discussed by Smith and West (1983), for example. Some such unexplained errors may be systematic but, if not identified, a general method of allowing for possible extra observational variation that may change in time is to simply suppose that V may change, albeit slowly and steadily.

(2) The variance V_t may change deterministically in a way not modelled, or modelled inappropriately. A variance function may be improperly specified, or omitted, the changes in variance thus not being adequately predicted. If such changes are not too dramatic, then they may be adequately estimated by allowing for stochastic drift in the scale parameter V.

(3) More generally, all features of model misspecification, which will become apparent in the forecast error sequence, can be attributed to changes in V. If V is allowed to vary stochastically, then the estimated trajectory over time of V will provide indications of times of improvement or deterioration in forecast performance; the estimation of V will tend to favour

larger values in the latter case, for example. Thus a simple model for slow, purely random variance changes can be a useful diagnostic tool.

10.8.2 A model for change in V: discounted variance learning

The model for stochastic changes in variation described here is developed in Ameen and Harrison (1985), Harrison and West (1986, 1987), and implemented in West, Harrison and Pole (1987). Consider the general DLM $\{F_t, G_t, k_tV, W_tV\}$, with known weights k_t^{-1}. Assuming the unknown scale V to have been constant up to time t, the usual analysis leads to the posterior

$$(\phi \mid D_{t-1}) \sim G[n_{t-1}/2, d_{t-1}/2].$$

Suppose, however, that we now believe V to be subject to some random disturbance over the time interval $t-1$ to t. The simplest way of modelling steady, stochastic variation is via a random walk, or first-order polynomial model, for V or some function of V. We use such a model here for the precision parameter ϕ rather than V directly, the practical implications of using one rather than the other being of little importance. To reflect the change, subscript ϕ by time. Thus, at $t-1$, the precision ϕ_{t-1} has the posterior

$$(\phi_{t-1} \mid D_{t-1}) \sim G[n_{t-1}/2, d_{t-1}/2]. \tag{10.2}$$

A random walk over the time interval gives the precision at time t as

$$\phi_t = \phi_{t-1} + \psi_t, \tag{10.3}$$

where the random disturbance ψ_t is uncorrelated with $(\phi_{t-1} \mid D_{t-1})$. Let

$$\psi_t \sim [0, U_t],$$

denote a distribution for ψ_t, of form unspecified, with mean 0 and variance U_t. Unpredictable changes in precision suggest the zero-mean for ψ_t, the variance U_t controls the magnitude of changes. Note that, from (10.2),

$$E[\phi_{t-1} \mid D_{t-1}] = n_{t-1}/d_{t-1} = 1/S_{t-1}$$

and

$$V[\phi_{t-1} \mid D_{t-1}] = 2n_{t-1}/d_{t-1}^2 = 2/(n_{t-1}S_{t-1}^2).$$

Thus, evolving to time t via (10.3), the mean $1/S_{t-1}$ is unchanged, but the variance increases to

$$V[\phi_t \mid D_{t-1}] = U_t + 2/(n_{t-1}S_{t-1}^2).$$

As throughout previous chapters, it is practically useful to think of variance increases in a multiplicative sense; thus

$$V[\phi_t \mid D_{t-1}] = 2/(\delta_t n_{t-1}S_{t-1}^2),$$

where δ_t, $(0 < \delta_t \le 1)$, is implicitly defined via

$$U_t = (\delta_t^{-1} - 1)2/(n_{t-1}S_{t-1}^2) = V[\phi_{t-1} \mid D_{t-1}](\delta_t^{-1} - 1). \quad (10.4)$$

Thus the evolution inflates the variance of the precision parameter by dividing it by the factor of δ_t. The analogy with discount factors as used to structure the evolution of θ_t is obvious; δ_t here plays the role of a discount factor and may be used to choose appropriate values of U_t. In particular, a constant discount rate $\delta_t = \delta$ for all t is usually adequate in practice, U_t then following directly from (10.4). Typically only small degrees of stochastic variation in ϕ, hence V, will be desirable, δ taking a value near unity, typically between 0.95 and 0.99. Clearly $\delta = 1$ leads to the original, constant V with $U_t = 0$.

Now to proceed with the analysis at time t, it is desirable that the prior $p(\phi_t \mid D_{t-1})$ be of the usual, conjugate gamma form. No distribution has as yet been specified for the innovation ψ_t; we have only the mean 0 and variance U_t. In fact, specifying a distribution of any given form leads to calculations that are essentially impossible to perform analytically. We therefore assume as an approximation that there exists a distibution for ψ_t, with the specified mean and variance, such that $p(\phi_t \mid D_{t-1})$ is approximately gamma distributed. This gamma prior must, of course, be consistent with the mean and variance $1/S_{t-1}$ and $1/(\delta n_{t-1}S_{t-1}^2)$. It is easily verified that the unique gamma distribution with these moments is defined as

$$(\phi_t \mid D_{t-1}) \sim G[\delta n_{t-1}/2, \delta d_{t-1}/2]. \quad (10.5)$$

This is rather interesting. In evolving via (10.3), the gamma distribution of the precision parameter $\phi = 1/V$ retains the same location $1/S_{t-1}$, the change being to the degrees of freedom which *decreases*, being multiplied by the discount factor δ, this leading also to the same form of change in d_{t-1}. Again the analogy with the standard, first-order polynomial model is evident. The evolution is effected simply by discounting the degrees of freedom a little to refect increased uncertainty about the scale parameter.

This analysis is summarised in Table 10.4. This is analogous to the summary in Section 4.6 of Chapter 4, providing the simple generalisation of the updating and one-step ahead forecasting equations to include discounting of the variance learning procedure.

10.8.3 Limiting behaviour of constant, discounted variance model

The asymptotic behaviour of the discounted variance learning procedure provides insight into the nature of the effect of the discounting. Note from the summary that

$$n_t = 1 + \delta n_{t-1} = \ldots = 1 + \delta + \delta^2 + \ldots + \delta^{t-1} n_1 + \delta^t n_0,$$

and

$$
\begin{aligned}
d_t =& (S_{t-1} e_t^2 / Q_t) + \delta d_{t-1} = \ldots \\
=& (S_{t-1} e_t^2 / Q_t) + \delta(S_{t-2} e_{t-1}^2 / Q_{t-1}) + \delta^2 (S_{t-3} e_{t-2}^2 / Q_{t-2}) \\
& + \ldots + \delta^{t-1}(S_0 e_1^2 / Q_1) + \delta^t d_0.
\end{aligned}
$$

Table 10.4. Summary of updating with variance discounting

Univariate DLM with variance discounting		
Observation:	$Y_t = \mathbf{F}_t' \boldsymbol{\theta}_t + \nu_t$	$\nu_t \sim N[0, k_t/\phi_t]$
System:	$\boldsymbol{\theta}_t = \mathbf{G}_t \boldsymbol{\theta}_{t-1} + \boldsymbol{\omega}_t$	$\boldsymbol{\omega}_t \sim T_{n_{t-1}}[0, \mathbf{W}_t]$
Information:	$(\boldsymbol{\theta}_{t-1} \mid D_{t-1}) \sim T_{n_{t-1}}[\mathbf{m}_{t-1}, \mathbf{C}_{t-1}]$ $(\boldsymbol{\theta}_t \mid D_{t-1}) \sim T_{n_{t-1}}[\mathbf{a}_t, \mathbf{R}_t]$	$\mathbf{a}_t = \mathbf{G}_t \mathbf{m}_{t-1}$ $\mathbf{R}_t = \mathbf{G}_t \mathbf{C}_{t-1} \mathbf{G}_t' + \mathbf{W}_t$
	$(\phi_{t-1} \mid D_{t-1}) \sim G[n_{t-1}/2, d_{t-1}/2]$ $(\phi_t \mid D_{t-1}) \sim G[\delta n_{t-1}/2, \delta d_{t-1}/2]$	$S_{t-1} = d_{t-1}/n_{t-1}$
Forecast:	$(Y_t \mid D_{t-1}) \sim T_{\delta n_{t-1}}[f_t, Q_t]$	$f_t = \mathbf{F}_t' \mathbf{a}_t$ $Q_t = \mathbf{F}_t' \mathbf{R}_t \mathbf{F}_t + k_t S_{t-1}$
Updating Recurrence Relationships		
$(\boldsymbol{\theta}_t \mid D_t) \sim T_{n_t}[\mathbf{m}_t, \mathbf{C}_t]$ $(\phi_t \mid D_t) \sim G[n_t/2, d_t/2]$ with $\mathbf{m}_t = \mathbf{a}_t + \mathbf{A}_t e_t,$ $\mathbf{C}_t = (S_t/S_{t-1})[\mathbf{R}_t - \mathbf{A}_t \mathbf{A}_t' Q_t],$ $n_t = \delta n_{t-1} + 1, \; d_t = \delta d_{t-1} + S_{t-1} e_t^2 / Q_t, \; \text{and} \; S_t = d_t/n_t,$ where $e_t = Y_t - f_t, \; \text{and} \; \mathbf{A}_t = \mathbf{R}_t \mathbf{F}_t / Q_t.$		

As $t \to \infty$, with $0 < \delta < 1$, then

$$n_t \to (1 - \delta)^{-1}.$$

Also $S_t = d_t/n_t$ so that, for large t,

$$S_t \approx (1 - \delta)^{-1} \sum_{r=0}^{t-1} \delta^r (e_{t-r}^2 / Q_{t-r}^*)$$

with $Q_t^* = Q_t/S_{t-1}$ for all t, the scale-free one-step forecast variance at time t. More formally,

$$\lim_{t \to \infty} \{ S_t - (1 - \delta)^{-1} \sum_{r=0}^{t-1} \delta^r (e_{t-r}^2 / Q_{t-r}^*) \} = 0$$

with probability one. Thus n_t converges to the constant, limiting degrees of freedom $(1 - \delta)^{-1}$. $\delta = 0.95$ implies a limit of 20, $\delta = 0.98$ a limit of 50. The usual, static variance model has $\delta = 1$ so that, of course, $n_t \to \infty$. Otherwise, the fact the changes in variance are to be expected implies a limit to the accuracy with which the variance at any time is estimated, this being defined by the limiting degrees of freedom, directly via the discount factor. The point estimate S_t has the limiting form of an exponentially weighted moving average of the standardised forecast errors; e_{t-r}^2 / Q_{t-r}^* asymptotically receives weight $\delta^r/(1-\delta)$, the weights decaying with r and summing to unity. Thus the estimate continues to adapt to new data, whilst further discounting old data, as time progresses.

Note finally that, if it is desired to allow for greater variation in ϕ_t at a given time in response to external infromation about possible marked changes, a smaller, intervention value of the discount factor can replace δ for just that time point. This has the effect of increasing U_t markedly, temporarily, thus considerably increasing uncertainty about ϕ_t through the evolution (10.3).

10.8.4   Filtering with discounted variance

The filtering analyses of Section 4.7, Chapter 4, applied to θ_t may be extended to the now time varying observational precision parameter using linear Bayes' methods. The problem of interest, in the model with (10.3) applying to ϕ_t, is that of updating the on-line posteriors defined by (10.2) in the light of data received afterwards. The problem is of considerable practical interest since it is the revised, filtered distributions that provide the retrospective analysis, identifying periods of stability and points of marked change in the observational variance sequence. The solution must be approximate, however, since the updating is based on the use of approximate gamma priors in (10.5). The approximation uses the generally applicable, linear Bayesian techniques described in Section 4.9 of Chapter 4. The basic ingredients of the derived results are as follows, the notation for filtering being similar to that used in the DLM in Section 4.7. The results are only sketched here, details being left to the reader.

(1) Consider the precision ϕ_{t-k+1}, for integers $(1 \leq k \leq t-1)$, having filtered back from time t to time $t-k+1$. The linear Bayes' approach supplies estimates for the posterior mean and variance of $(\phi_{t-k+1} \mid D_t)$. Now, as above in updating, any given mean (which is positive here, of course) and variance uniquely define a gamma distribution with the same moments. Although it is not necessarily to assume the filtered posteriors to be approximately gamma, adopt the notation consistent with such an assumption. Thus the filtered mean and variance determine quantities $n_t(-k+1)$ and $d_t(-k+1)$ such that the $G[n_t(-k+1)/2, d_t(-k+1)/2]$ distribution has these moments. Define $S_t(-k+1) = d_t(-k+1)/n_t(-k+1)$ to be the corresponding estimate of observational variance, so that $E[\phi_{t-k+1} \mid D_t] = S_t(-k+1)^{-1}$.

(2) From (10.2) and (10.3), the mean and variance matrix of the bivariate distribution of ϕ_{t-k} and ϕ_{t-k+1} given D_{t-k} are:

$$\left(\begin{array}{c|c} \phi_{t-k+1} \\ \phi_{t-k} \end{array} D_{t-k} \right) \sim \left[S_{t-k}^{-1} \begin{pmatrix} 1 \\ 1 \end{pmatrix}, \ \frac{2}{n_{t-k}^2 S_{t-k}} \begin{pmatrix} \delta_{t-k+1}^{-1} & 1 \\ 1 & 1 \end{pmatrix} \right],$$

where δ_{t-k+1} relates to U_{t-k+1} as in (10.4). Note that this is general; any specified U_t sequence determines the δ_t sequence, or vice-versa.

(3) The general result of Theorem 4.9 applies here to give the linear Bayes' estimates of the conditional moments of the distribution $(\phi_{t-k} \mid \phi_{t-k+1}, D_{t-k})$ as follows:

$$E[\phi_{t-k} \mid \phi_{t-k+1}, D_{t-k}] = S_{t-k}^{-1} + \delta_{t-k+1}(\phi_{t-k+1} - S_{t-k}^{-1}),$$

and

$$V[\phi_{t-k} \mid \phi_{t-k+1}, D_{t-k}] = \frac{2}{n_{t-k} S_{t-k}^2}(1 - \delta_{t-k+1}).$$

(4) Using arguments similar to those in Theorem 4.4 for filtering in the standard DLM, the filtered moments for ϕ_{t-k} can be deduced. Details are left to the reader. The mean is

$$S_t(-k)^{-1} = E[\phi_{t-k} \mid D_t] = E\{E[\phi_{t-k} \mid \phi_{t-k+1}, D_{t-k}] \mid D_t\}$$
$$= S_{t-k}^{-1} + \delta_{t-k+1}(S_t(-k+1)^{-1} - S_{t-k}^{-1}).$$

The corresponding filtered variance is

$$E\{V[\phi_{t-k} \mid \phi_{t-k+1}, D_{t-k}] \mid D_t\} + V\{E[\phi_{t-k} \mid \phi_{t-k+1}, D_{t-k}] \mid D_t\},$$

which is given by

$$\frac{2}{n_{t-k}S_{t-k}^2}(1 - \delta_{t-k+1}) + \frac{2}{n_t(-k+1)S_t(-k+1)^2}\delta_{t-k+1}^2.$$

(5) Hence, starting at time t with $S_t(0) = S_t$ and $n_t(0) = n_t$, the one-step back quantities $S_t(-1)$ and $n_t(-1)$ can be deduced, and on backwards over times $k = 2, 3, \ldots, 0$. The corresponding values of $d_t(-k)$ and $n_t(-k)$ can be deduced by the reader. Note that, in the filtering algorithms for the state vector θ_t in Corollaries 4.3 and 4.4 of Chapter 4, the estimated value of the precision $S_t(-k)$ will now replace the value S_t appearing there for that the original case of static observational variance. Also, assuming a gamma approximation for the filtered distribution, the degrees of freedom n_t in those Corollaries will be replaced by the filtered value of the degrees of freedom, $n_t(-k)$.

10.9   EXERCISES

(1) Reanalyse the agricultural sales series of Section 10.2 using the same linear trend/seasonal model. Extend the discussion of Example 10.1 to a fuller exploration of sensitivity of the analysis to variation in the dynamic controlled through the discount factors. Do this as follows.

   (a) For any values $\delta = (\delta_T, \delta_S)'$ of the trend and seasonal discount factors, explicitly recognise the dependence of the analysis on δ by including it in the conditioning of all distributions. Then the aggregate predictive density at time t is $p(Y_t, \ldots, Y_1 \mid \delta, D_0)$, calculated sequentially as in Section 10.2.5. As a function of δ, this defines the likelihood function for δ from the data. Calculate this quantity at $t = 48$ for each pair of values of discount factors with $\delta_T = 0.8, 0.85, \ldots, 1$ and $\delta_S = 0.9, 0.925, \ldots, 1$. What values of δ are supported by the data? In particular, assess the support for the static model defined by $\delta = (1,1)'$ relative to other, dynamic models fitted.

(b) Explore sensitivity of inferences to variation in δ amongst values supported by the data from (a). In particular, do the estimated time trajectories of trend and seasonal components change significantly as δ varies? What about step ahead forecasts over one or two years form the end of the data series?

(c) Can you suggest how you might combine forecasts (and other inferences) made at time t from two (or more) models with different values of δ?

(d) Compare models with different values of δ using MSE and MAD measures of predictive performance as alternatives to the above model likelihood measures. Do the measures agree as to the relative support for different values of δ? If not, describe how and why they differ.

(2) Consider the prior specification for the seasonal factors ϕ_1 of the model for the Sales/Index series in Section 10.3.2, given initially by

$$(\phi_1 | D_0) \sim T_6[\mathbf{m}^*_{\phi_1}, \mathbf{C}^*_{\phi_1}]$$

where

$$\mathbf{m}^*_{\phi_1} = (0.8,\ 1.3,\ 1,\ 1.1,\ 1.2,\ 0,\ -1.2,\ -1,\ -1,\ -1.5,\ -0.5,\ 0)'$$

and

$$\mathbf{C}^*_{\phi_1} = (0.2)^2 \mathbf{I}.$$

(a) Verify that this initial prior does not satisfy the zero-sum constraint on seasonal factors. Verify also that Theorem 8.2 of Chapter 8 leads to the constrained prior as given in Section 10.3.2, namely

$$(\phi_1 | D_0) \sim T_6[\mathbf{m}_{\phi_1}, \mathbf{C}_{\phi_1}]$$

where

$$\mathbf{m}_{\phi_1} = (0.783,\ 1.283,\ 0.983,\ 1.083,\ 1.183,\ -0.017,\ -1.217,$$
$$-1.017,\ -1.017,\ -1.517,\ -0.517,\ -0.017)',$$

and

$$
\mathbf{C}_{\phi_1} = (0.19)^2 \begin{pmatrix}
1 & -0.09 & -0.09 & \cdots & -0.09 \\
-0.09 & 1 & -0.09 & \cdots & -0.09 \\
-0.09 & -0.09 & 1 & \cdots & -0.09 \\
\vdots & \vdots & \vdots & \ddots & \vdots \\
-0.09 & -0.09 & -0.09 & \cdots & -0.09
\end{pmatrix}.
$$

(b) Let **H** be the transformation matrix of Section 8.6.4 used to transform from seasonal effects to Fourier coefficients. Calculate **H** and verify the stated results of Section 10.3.3, Table 10.3, by calculating the moments $\mathbf{Hm}_{\phi_1}$ and $\mathbf{HC}_{\phi_1}\mathbf{H}'$.

(3) Reanalyse the Sales/Index data of Section 10.3 using the same model. Instead of initialising the analysis with the prior used in that Section, perform a reference analysis as described in Section 4.8 of Chapter 4 (beginning with $\mathbf{W}_t = \mathbf{0}$, corresponding to unit discount factors, over the reference period of the analysis). Describe the main differences between this analysis and that discussed in Section 10.3.

(4) Consider the transformation $Y = g(Z)$ of Section 10.6.2 with $E[Z] = \mu$ and $V[Z] = V$. By expanding $g(Z)$ in a Taylor series about $Z = \mu$ and ignoring all terms after the quadratic in $Z - \mu$, verify the approximate values for the mean and variance of Y given by

$$
E[Y] \approx g(\mu) + 0.5g''(\mu)V \quad \text{and} \quad V[Y] \approx \{g'(\mu)\}^2 V.
$$

(5) Suppose that Y_t is the sum of an uncertain number k_t quantities $Y_{tj}, (j = 1, \ldots, k_t)$, where the basic quantities are uncorrelated with common mean β_t and common, known variance V_t. This arises when items are produced and sold in batches of varying sizes.

(a) Show that $E[Y_t|k_t, \beta_t] = \mu_t$ where $\mu_t = k_t\beta_t$, and $V[Y_t|k_t, \beta_t] = k_t V_t$.

(b) Suppose that a forecaster has a prior distribution for the batch size k_t with finite mean and variance, and that k_t is viewed as independent of β_t. Show that, unconditional on k_t, $E[Y_t|\beta_t] = \beta_t E[k_t]$ and $V[Y_t|\beta_t] = V_t E[k_t] + \beta_t^2 V[k_t]$.

(c) In the special case of Poisson batch or lot sizes, $E[k_t] = V[k_t] = a$, for some $a > 0$. Letting $\mu_t = E[Y_t|\beta_t]$ as usual, show that Y_t has a quadratic variance function, $V[Y_t|\beta_t] = V_t^*[1 + b_t\mu_t^2]$ where $V_t^* = aV_t$ and $b_t^{-1} = a^2 V_t$.

(6) Reanalyse the agricultural sales data of Section 10.2. Use the same model but now allow for changes in observational variation about the level through a variance discount factor δ as in Section 10.8. (Note that the reference analysis can be performed as usual assuming all discount factors, including that for the variance, are unity initially over the reference part of the analysis).

(a) Perform an analysis with $\delta = 0.95$. Verify that, relative to the static variance model, variance discounting affects only the uncertainties in forecasting and inferences about time trajectories of model components.

(b) Analyse the data for several values of δ over the range $\delta = 0.90, 0.92, \ldots, 1$, and assess relative support from the data using the likelihood function over δ provided by the aggregate predictive densities from each model.

(7) Consider the two series in the table below. The first series, Sales, is a record of the quarterly retail sales of a confectionary product over a period of eleven years. The product cost has a major influence on Sales, and the second series, Cost, is a compound, quarterly index of cost to the consumer constructed in an attempt to explain nonseasonal changes in Sales.

Consider fitting DLMs with first-order polynomial, regression on Cost and full seasonal components. This data is analysed using a similar model in West, Harrison and Pole (1987b), with extensive summary information from the analysis provided there.

	SALES				COST			
		Quarter				Quarter		
Year	1	2	3	4	1	2	3	4
1975	157	227	240	191	10.6	8.5	6.7	4.1
1976	157	232	254	198	1.9	0.4	1.1	1.9
1977	169	234	241	167	0.2	2.9	4.1	−1.4
1978	163	227	252	185	−4.0	−4.5	−5.3	−8.4
1979	179	261	264	196	−12.8	−13.2	−10.1	−4.6
1980	179	248	256	193	−1.1	−0.1	0.0	−2.5
1981	186	260	270	210	−5.1	−6.4	−8.0	−6.5
1982	171	227	241	170	−3.7	−1.3	6.1	16.5
1983	140	218	208	193	22.9	23.9	18.0	8.3
1984	184	235	245	209	2.9	0.7	−2.4	−7.0
1985	206	260	264	227	−9.8	−10.6	−12.3	−13.2

Based on previous years of data, the forecaster assesses initial prior information as follows.

- The initial, underlying level of the series when Cost is zero is expected to be about 220, with a nominal standard error of 15.
- The regression coefficient of Cost is estimated as -1.5 with a standard error of about 0.7.
- The seasonal factors for the four quarters of the first year are expected to be $-50(25)$, $25(15)$, $50(25)$ and $-25(15)$, the nominal standard errors being given in brackets.
- The trend, regression and seasonal components are assumed, initially, to be uncorrelated.
- The observational variance is estimated as 100, with initial degrees of freedom of 12.

  Analyse the series along the following lines.

(a) Using the information provided for the seasonal factors above, apply Theorem 8.2 of Chapter 8 to derive the appropriate intial prior that satisfies the zero sum constraint.

(b) Identify the full initial prior quantities for the 6 dimensional state vector $\boldsymbol{\theta}_1$ and the observational variance V. Write down the defining quantities $\mathbf{a}_1$, $\mathbf{R}_1$, n_0 and S_0 based on the above, initial information.

(c) Use three discount factors to structure the evolution variance matrices of the model: δ_T for the constant, intercept term, δ_R for the regression coefficient, and δ_S for the seasonal factors. Following West, Harrison and Pole (1987b), consider initially the values $\delta_T = \delta_S = 0.9$ and $\delta_R = 0.98$. Fit the model and perform the retrospective, filtering calculations to obtain filtered estimates of the state vector and all model components over time.

(d) Based on this analysis, verify the findings in the above reference to the effect that the regression parameter on Cost is, in retrospect, rather stable over time.

(e) Produce step ahead forecasts from the end of the data in the fourth quarter of 1985 for the next three years. The estimated values of Cost to be used in forecasting ahead are given by:

Year	COST Quarter			
	1	2	3	4
1986	8.4	10.6	7.2	13.0
1987	−2.9	−0.7	−6.4	−7.0
1988	−14.9	−15.9	−18.0	−22.3

CHAPTER 11

INTERVENTION AND MONITORING

11.1   INTRODUCTION

In Section 2.3.2 of Chapter 2 we introduced simple intervention ideas and considered in detail intervention into a first-order polynomial model. That intervention fed-forward information anticipating a major change in the level of a time series, modelled by altering the prior distribution for the level parameter to accommodate the change. Had the model used in that example been closed to intervention, then the subsequent huge change in the level of the series would have been neither forecast nor adequately estimated afterwards, the model/ data match breaking down entirely. In practice, all models are only components of forecasting systems which include the forecasters as integral components. Interactions between forecasters and models is necessary to adequately cater for events and changes that go beyond the existing model form. This is evident also in the illustrations of standard, closed models in Chapter 10 where deterioration in forecasting performance, though small, is apparent. In this Chapter, we move closer to illustrating forecasting systems rather than simply models, considering ways in which routine interventions can be incorporated into existing DLMs, and examples of why and when such interventions may be necessary to sustain predictive performance. The mode of intervention used in the example of Section 2.3.2 was simply to represent departures from an existing model in terms of major changes in the parameters of the model. This is the most widely used and appropriate method, although others, such as extending the model to include new parameters, are also important. An example dataset highlights the need for intervention.

EXAMPLE 11.1.   The dataset in Table 11.1 perfectly illustrates many of the points to be raised in connection with intervention. This real data series, referred to by the codename CP6, provides monthly total sales, in monetary terms on a standard scale, of tobacco and related products marketed by a major company in the UK. The time of the data runs from January 1955 to December 1959 inclusive. The

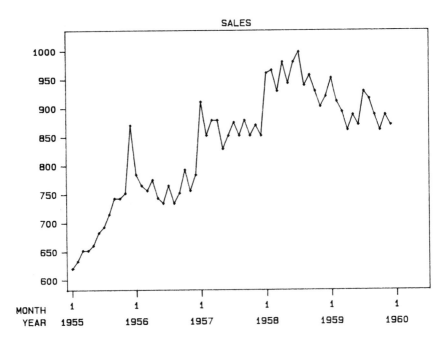

Figure 11.1. CP6 sales series.

series is graphed over time in Figure 11.1. Initially, during 1955, the
market clearly grows at a fast but steady rate, jumps markedly in
December, then falls back to pre-December levels and flattens off
for 1956. There is a major jump in the sales level in early 1957,
and another in early 1958. Throughout the final two years 1958
and 1959, there is a steady decline back to late 1957 levels. An
immediate reaction to this series, which is not atypical of real series
in consumer markets, is that it is impossible to forecast with a simple
time series model. This is correct. However, the sketch in Figure 11.2

Table 11.1. CP6 Sales data

Year	Month											
	1	2	3	4	5	6	7	8	9	10	11	12
1955	620	633	652	652	661	683	693	715	743	743	752	870
1956	784	765	756	775	743	734	765	734	752	793	756	784
1957	911	852	879	879	829	852	875	852	879	852	870	852
1958	961	966	929	980	943	980	998	939	957	929	902	920
1959	952	911	893	861	888	870	929	916	888	861	888	870

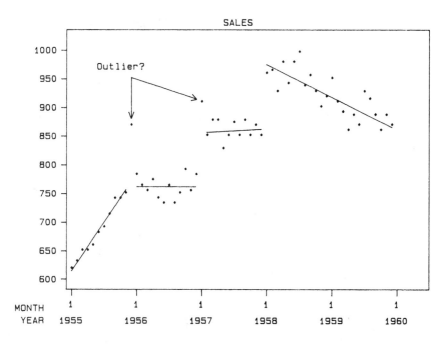

Figure 11.2.   Heuristic model for CP6 sales.

suggests that in fact a simple model may be appropriate for short-term forecasting (up to twelve months ahead, say), *if* that model is open to interventions to explain (and possibly anticipate) some of the discontinuities the series exhibits. From this figure, it is apparent that a simple possible description of the trend in the series is as a sequence of roughly linear segments (ie. *piecewise linear*), with major changes in the quantified linear form at three places. Thus a second-order polynomial model, providing a linear predictor in the short term, may be used so long as the abrupt changes in parameter values are catered for. Other features to note are the possible *outliers* at two points in the series. These observations deviate markedly from the general pattern and should be considered individually, perhaps being omitted from the analysis. This is not to say that such extreme observations should be ignored, of course, since they may be critical commercially. For updating the model however, they are ignored since they convey little or no information relevant to forecasting the underlying trend within the model structure. Finally, note that the random scatter in Sales about the sketched trend is apparently higher at higher levels of the series, suggesting that intervention, or some other appropriate device such as a variance law as in Section 10.7, be required to adjust the observational variance to model greater randomness at such levels.

The sources of information available to a forecaster are not always restricted to historical data and other, related information, but also include information about forthcoming events affecting the environment of the series. In cases when it is perceived that these events may materially affect the development of the series, then action is taken to intervene into the existing model to feed-forward information, allowing the model to anticipate and predict change. For example, a marketing department of a company may know that an export licence for their products has just been granted, that a new type of "special offer" campaign is to be launched, that a patent expires or that, unexpectedly, a competitive product has been banned. In each case, the likely effects on demand, sales and inventory must be assessed in order to anticipate changes and plan accordingly. The decisions resting on forecasts in such cases may be of major importance to the company. Some of the events in the CP6 series may have been anticipated at the time in this way. In forming views of the likely outcomes, the forecasters may have prior experience of related events to base their judgements upon, but, clearly, such views are usually largely subjective. This does, of course, raise questions about possible personal biases and errors in intervention, and thus such facilities require caution in use and should always be subject to calibration and retrospective assessment.

Interventions can be roughly classified as either *feed-forward* or *feed-back*. The former is anticipatory, in nature, as in the above examples. The latter is corrective, responding to events that had not been foreseen or adequately catered for. Corrective actions often arise when it is seen that forecasting performance has deteriorated, thus prompting the forecaster to examine the environment of the series for explanation. In such instances, any such information, which should ideally have been available beforehand, must be used retrospectively in an attempt to adjust the model appropriately to the current, local, conditions. It is evident that a complete forecasting system should be open to both feed-forward and feed-back interventions. In ideal circumstances, forecasting systems should operate according to a principle of *Management by Exception*. That is, based upon experience and analysis, a formal statistical model is adopted and is used routinely to process data and information, providing forecasts and inferences that are used unless exceptional circumstances arise. The exceptions occur in the two ways introduced above; the first relating to the receipt of information providing the basis for feed-forward interventions, the second provided by the detection of deterioration in forecasting performance, usually detected by some form of forecast monitoring activity. Such monitoring may be largely subjective and informal, involving the forecaster in considering the forecast performance from a subjective standpoint, or may take the form of an automatic, statistical error analysis or control scheme that continually monitors the model/ data match and issues signals of breakdown as necessary. Often it is the case that exceptions matter most in forecasting. When circumstances change markedly there are major opportunities for both losses and gains. Thus the more vigilant and successful the forecaster is in anticipating major events, the more effective the decisions. Also, concerning feed-back interventions, the more perceptive and immediate the diagnosis is, the better. Hence it is to be recommended that there should always be a user response to exceptions identified and signalled by a monitoring system. However, in some circumstances there may be no identifiable reason for the exceptions, but the very fact that something has occurred raises uncertainty about the future. Then the forecaster's response will not necessarily identify possible sources of model breakdown, but must communicate the problem to the model by increasing total uncertainty in the model, making it more adaptive to new data so that changes that may have taken place are rapidly identified and estimated.

Many of these ideas are made more concrete in this Chapter where an existing model is assumed to be operating subject to both forms of intervention. Much of the development is based on West and Harrison (1989). We consider throughout a single intervention at time t into the existing model; clearly the concepts and theory apply generally to analyses subject to possibly many interventions. In addition, although we consider interventions at time t that represent changes to the model at time t, it should be clear that feed-forward interventions can also be made for times after t, although this is not specifically described.

For clarity the development is in terms of a model with observational variance sequence known, so that all prior, posterior and predictive distributions are normal. The results apply, of course and with no essential difference, to the case of variance learning when all such distributions are T rather than normal. We comment on minor differences between the cases as necessary.

Thus, for reference, we have the model at the current time t given, as usual, by

$$Y_t = \mathbf{F}'_t \boldsymbol{\theta}_t + \nu_t, \qquad\qquad \nu_t \sim \mathrm{N}[0, V_t], \qquad\qquad (11.1)$$
$$\boldsymbol{\theta}_t = \mathbf{G}_t \boldsymbol{\theta}_{t-1} + \boldsymbol{\omega}_t, \qquad\qquad \boldsymbol{\omega}_t \sim \mathrm{N}[\mathbf{0}, \mathbf{W}_t], \qquad\qquad (11.2)$$

with the usual assumptions of Section 4.2 and 4.3, Chapter 4. The historical information D_{t-1} (including past data and any previous interventions) is summarised in terms of the posterior for $\boldsymbol{\theta}_{t-1}$, namely

$$(\boldsymbol{\theta}_{t-1} \mid D_{t-1}) \sim \mathrm{N}[\mathbf{m}_{t-1}, \ \mathbf{C}_{t-1}], \qquad\qquad (11.3)$$

where the mean and variance matrix are known. Thus the prior for the state vector at the current time is, via the evolution equation (11.2) as usual,

$$(\boldsymbol{\theta}_t \mid D_{t-1}) \sim \mathrm{N}[\mathbf{a}_t, \ \mathbf{R}_t], \qquad\qquad (11.4)$$

with

$$\mathbf{a}_t = \mathbf{G}_t \mathbf{m}_{t-1} \qquad \text{and} \qquad \mathbf{R}_t = \mathbf{G}_t \mathbf{C}_{t-1} \mathbf{G}'_t + \mathbf{W}_t.$$

Suppose now that feed-forward intervention is to be made at the current time into the model (11.1—11.4). This may be a response to additional information that can be viewed as extra observational data; a response to new information about related variables, such as competitor activity in a consumer market, legislation changes, and so forth; forecasts from other individuals or models; control actions, such as changes in advertising campaigns, pricing policy etc; or it may simply reflect a dissatisfaction with the current prior, and the forecasts of the future that it implies, thus being a purely subjective intervention by the forecaster. Whatever the case, we represent the intervention at time t by I_t. This is an information set that identifies time t as a point of intervention and includes all the information used to effect the intervention, such information being made specific below in the various modes of intervention we consider. Note now that, following the intervention, the available information set prior to observing Y_t is $\{I_t, \ D_{t-1}\}$, rather than just D_{t-1}.

The following section details the basic modes of intervention into the existing model at time t, and the theory necessary to combine interventions of essentially any desired form into the existing model, allowing the intervention effects to be used both in forecasting from the current time onwards and in retrospective time series analysis.

11.2 MODES OF FEED-FORWARD INTERVENTION

11.2.1 Ignoring observation Y_t

The first mode of intervention involves simply treating Y_t as an out-lier. Examples include the effects of strikes on sales and inventory levels, pricing changes that lead to forward purchasing of goods, and other interventions in the environment of the time series that may lead to a single observation being quite discrepant and essentially unrelated to the rest of the series. In such cases, although the observation is of critical importance to the company, perhaps it should not be used in updating the model for forecasting the future since it provides no relevant information.

EXAMPLE 11.1 (continued). In the CP6 series, such an event is (retrospectively) apparent at December 1955, where the Sales leap upwards by about 15% for just that month. This is the market response, in terms of immediate purchasing of stocks, to a company announcement of a forthcoming price rise. A second possible outlier is the very high value in January 1957 that presages a change in the overall level of Sales, but to a lower level than the single point in January.

When such information exists so that a discrepant observation is anticipated, one possible, fail-safe reaction is to just omit the observation from the analysis, treating it as if it is a true missing or unrecorded value. Y_t is uninformative about the future and so should be given no weight in updating the model distributions. Thus

$$I_t = \{ \; Y_t \text{ is missing} \; \}$$

so that $D_t = \{I_t, \; D_{t-1}\}$ is effectively equal to D_{t-1} alone. The posterior for the state vector at time t is just the original prior, namely $(\boldsymbol{\theta}_t \mid D_t) \sim N[\mathbf{m}_t, \; \mathbf{C}_t]$ where

$$\mathbf{m}_t = \mathbf{a}_t \qquad \text{and} \qquad \mathbf{C}_t = \mathbf{R}_t.$$

Formally, this can be modelled in the DLM format in a variety of ways, the simplest, and most appropriate, is just to view the observation as having a very large variance V_t; formally, let V_t tend to infinity, or V_t^{-1} tend to 0, in the model equations, so that the observation provides no information for $\boldsymbol{\theta}_t$ (nor for the scale parameter in the case of variance learning). In the updating equations, the one-step ahead forecast variance Q_t tends to infinity with V_t and it follows easily that the posterior for $\boldsymbol{\theta}_t$ is just the prior as required. Formally, then,

$$I_t = \{ \ V_t^{-1} = 0 \ \}$$

in this case.

A modification of this mode of intervention is often desirable, stemming from the uncertainties associated with the sorts of events that suggest Y_t be ignored. Immediately following such events, it may be that the series will develop in a rather different way than currently forecast, exhibiting knock-on, delayed effects of the events. Thus, for example, a huge increase in sales at time t that represents forward buying before a previously announced increase in price at $t+1$ can be expected to be followed by a drop off in sales at times $t+1$, and possibly later times, after the price increase takes place. In order to adapt to the changing pattern after the omitted observation, an additional intervention may be desired to increase uncertainty about components of $\boldsymbol{\theta}_t$. In CP6, for example, the possible outlier in January 1957 is followed by Sales at a higher level than during 1956. This calls for intervention of the second type, considered in the following section.

11.2.2   Additional evolution noise

A common response to changes in conditions potentially affecting the development of the series is increased uncertainty about the future, reflected by increased uncertainties about some or all of the existing model parameters. The withdrawal of a major competing product in a consumer market may be anticipated by feeding forward an estimated increase in the level of sales, but it is surely the case that the uncertainty about the new level will be greater than about the current level, possibly much greater. In other instances, the response to an exception signalled by a forecast monitoring system may be simply to increase prior variances of some or all components of the model. This is an appropriate reflection of the view that, although *something* has changed, it is difficult to attribute the change to a particular component. As a catch-all measure, the entire variance matrix $\mathbf{R}_t$ may be altered to reflect increased uncertainty about all parameters without a change in the prior mean $\mathbf{a}_t$ that would anticipate the direction of change. This does, of course, lead to a loss of information on the entire state vector and is therefore an omnibus technique to be used in cases of complete neutrality as to sources of change. Otherwise it is well to be selective, increasing uncertainties only on those components that are viewed as potentially subject to major change. In the above example, the change in the sales market as it expands to meet the increased demand will evidence itself in a marked increase in sales level, and also possibly in the observational variance about the level, but is unlikely, for example, to seriously impact upon the phases of components of seasonal patterns in sales.

Generally, the model is open to interventions on components of $\boldsymbol{\theta}_t$ that increase uncertainties in this way, simply by adding in further evolution noise terms paralleling those in equation (11.2). As in the sales example, it is common that such interventions include a shift in mean in addition to an inflation in uncertainty. All such interventions can be formally represented in DLM form by extending the model to include a second evolution of the state vector in addition to that in the routine model in (11.2).

Generally, suppose the intervention information to be given by

$$I_t = \{\mathbf{h}_t, \ \mathbf{H}_t\}$$

where $\mathbf{h}_t$ is the mean vector and $\mathbf{H}_t$ the covariance matrix of a random quantity $\boldsymbol{\xi}_t$, with

$$\boldsymbol{\xi}_t \sim \mathrm{N}[\mathbf{h}_t, \ \mathbf{H}_t].$$

Suppose also that the $\boldsymbol{\xi}_t$ is uncorrelated with $(\boldsymbol{\theta}_{t-1} \mid D_{t-1})$ and with $\boldsymbol{\omega}_t$, so that it is also uncorrelated with $(\boldsymbol{\theta}_t \mid D_{t-1})$. The intervention is effected by adding the additional noise term $\boldsymbol{\xi}_t$ to $\boldsymbol{\theta}_t$ after (11.2); equivalently, the post-intervention prior distribution is defined via the extended evolution equation

$$\boldsymbol{\theta}_t = \mathbf{G}_t \boldsymbol{\theta}_{t-1} + \boldsymbol{\omega}_t + \boldsymbol{\xi}_t \tag{11.5}$$

replacing (11.2). Thus

$$(\boldsymbol{\theta}_t \mid I_t, \ D_{t-1}) \sim \mathrm{N}[\mathbf{a}_t^*, \ \mathbf{R}_t^*],$$

where

$$\mathbf{a}_t^* = \mathbf{a}_t + \mathbf{h}_t \quad \text{and} \quad \mathbf{R}_t^* = \mathbf{R}_t + \mathbf{H}_t.$$

This caters for arbitrary shifts in the prior mean vector to the post-intervention value $\mathbf{a}_t^*$. Some elements of $\mathbf{h}_t$ may be zero, not anticipating the direction of changes in the corresponding parameters. In practice these mean changes may be assigned by directly choosing the $\mathbf{h}_t$ vector, the expected increment in $\boldsymbol{\theta}_t$ due to intervention. Alternatively, the adjusted mean $\mathbf{a}_t^*$ can be specified directly and then $\mathbf{h}_t$ deduced as $\mathbf{h}_t = \mathbf{a}_t^* - \mathbf{a}_t$.

This mode of intervention caters for many possible and practically useful increases in variance through $\mathbf{H}_t$. Note that some of the variances on the diagonal of $\mathbf{H}_t$ (and the corresponding covariance terms off-diagonal) may be zero, with the result that the corresponding elements of $\boldsymbol{\theta}_t$ are not subject to change due to intervention. This allows the forecaster the flexibility to protect some components of the model from intervention, when they are viewed as durable and unlikely to be subject to change. There will, of course, be some correlation due to the off-diagonal terms in $\mathbf{R}_t$. Again it is sometimes the case that the intervention variance matrix $\mathbf{H}_t$ will be specified directly as the variance of the change in $\boldsymbol{\theta}_t$ due to intervention, the elements representing the uncertainty as to the nature and extent of the change forecast as $\mathbf{h}_t$. Although obviously very flexible, it is difficult in general to assign appropriate values here, although as a general rule it is desirable to err on the side of caution, with any change forecast via $\mathbf{h}_t$ being hedged with sufficient uncertainty that the model will be adaptive to future data, rapidly identifying and estimating changes. In line with the use of discount factors in structuring evolution variance matrices in standard models, it is appropriate to extend the discount concept to intervention. Thus if $\mathbf{W}_t$ is structured using a standard set of discount factors, a matrix $\mathbf{H}_t$, with the same structure, models an extra change in $\boldsymbol{\theta}_t$ with the same correlation pattern in the evolution noise. The appropriateness of this is apparent when the intervention is incorporated into standard DLM form, as follows.

THEOREM 11.1. *Conditional on information* $\{I_t,\ D_{t-1}\}$, *the DLM (11.1—11.4) holds with the evolution equation (11.2) amended according to (11.5), written now as*

$$\boldsymbol{\theta}_t = \mathbf{G}_t \boldsymbol{\theta}_{t-1} + \boldsymbol{\omega}_t^*,$$

where $\boldsymbol{\omega}_t^* = \boldsymbol{\omega}_t + \boldsymbol{\xi}_t$ *is distributed as*

$$\boldsymbol{\omega}_t^* \sim \mathrm{N}[\mathbf{h}_t,\ \mathbf{W}_t^*],$$

with $\mathbf{W}_t^* = \mathbf{W}_t + \mathbf{H}_t$. *In addition,* $\boldsymbol{\omega}_t$ *is independent of* $\boldsymbol{\theta}_{t-1}$.

Proof: Obvious from (11.5), and left to the reader.

◇

This makes it clear that the intervention can be written in the usual DLM form, with a generalisation to a possibly non-zero mean for the evolution noise vector.

This simple sort of structuring is particularly appropriate when $h_t = 0$, when the addition of ξ_t simply increases uncertainty about θ_t. An automatic intervention technique described and ilustrated in Section 11.5 below uses this approach and more discussion appears there. Similar techniques are used in West and Harrison (1986a, 1989).

EXAMPLE 11.1 (continued). Consider the marked jump in level of CP6 Sales in January 1958. Suppose that this change was anticipated prior to occurrence, being the result of planned market expansion. In line with the previous sections, this change could be modelled through intervention, feeding forward prior information about the change and allowing the level parameter of any model to change. For concreteness, suppose a second-order polynomial model with level and growth parameters μ_t and β_t at time t corresponding to January 1958. Thus $n = 2$ and $\theta'_t = (\mu_t,\ \beta_t)$, $F'_t = E'_2 = (1,0)$ and

$$G_t = G = J_2(1) = \begin{pmatrix} 1 & 1 \\ 0 & 1 \end{pmatrix}.$$

Suppose also that $V_t = 15$, $a'_t = (865, 0)$, and

$$R_t = \begin{pmatrix} 100 & 10 \\ 10 & 2 \end{pmatrix}.$$

Available information is such that a jump of roughly 100 is anticipated in the level of Sales. Possible interventions, in the mode of this section, include:

(1) a simple shift in level, setting $h'_t = (100, 0)$ and $H_t = 0$;
(2) more realistically, a shift in level as above but hedged with uncertainty about the size of the shift via $H_t = \text{diag}(100, 0)$, for example;
(3) as in (2) but with increased uncertainty about the new growth as well as the new level, via $H_t = \text{diag}(100, 25)$, say;
(4) as in (3) but including correlation between the changes in level and growth via, for example,

$$H_t = \begin{pmatrix} 100 & 25 \\ 25 & 25 \end{pmatrix},$$

thus suggesting that larger changes in growth will be associated with larger changes in level.

11.2.3   Arbitrary subjective intervention

The most general mode of intervention into the existing model is to simply change the prior moments of $\boldsymbol{\theta}_t$ to new values anticipating changes in the series. Thus suppose that the intervention information is given by

$$I_t = \{\mathbf{a}_t^*, \; \mathbf{R}_t^*\}$$

where the *post-intervention* values $\mathbf{a}_t^*$ and $\mathbf{R}_t^*$ are given by the fore-caster. Note that this covers the case of the previous section, where, in addition to a possible mean shift, the post-intervention uncertainty always exceeds that pre-intervention. It goes well beyond that special case however. As an extreme example, taking $\mathbf{R}_t^* = \mathbf{0}$ implies that $\boldsymbol{\theta}_t = \mathbf{a}_t^*$ with probability 1, thus intervention informs precisely on the values of the parameters at time t. More practically, it allows for cases in which uncertainty about some of the parameters may decrease.

For forecasting Y_t and further into the future, the post-intervention moments replace those in (11.4), and they are then updated as usual when data is observed. A problem arises, however, when considering filtering and smoothing the series for retrospective analysis. The problem is that the post-intervention prior

$$(\boldsymbol{\theta}_t \mid I_t, \; D_{t-1}) \sim \mathrm{N}[\mathbf{a}_t^*, \; \mathbf{R}_t^*] \tag{11.6}$$

is no longer consistent with the model (11.1—11.4). Filtering requires a joint distribution for $\boldsymbol{\theta}_t$ and $\boldsymbol{\theta}_{t-1}$ conditional on D_{t-1} *and the intervention information* I_t, and the arbitrary changes made on the moments of $\boldsymbol{\theta}_t$ to incorporate intervention do not provide a coherent joint distribution. To do so, we need to be able to express the changes due to intervention in a form consistent with the DLM, in a way similar to that used in the previous section in Theorem 11.1. This can be done in Theorem 11.2 below, based on the following, general result.

Lemma 11.1. *Let $\mathbf{K}_t$ be a an $n-$square, upper triangular, non-singular matrix, and $\mathbf{h}_t$ any $n-$vector, and define*

$$\boldsymbol{\theta}_t^* = \mathbf{K}_t \boldsymbol{\theta}_t + \mathbf{h}_t$$

where $\mathrm{E}[\boldsymbol{\theta}_t] = \mathbf{a}_t$ and $\mathrm{V}[\boldsymbol{\theta}_t] = \mathbf{R}_t$. Then $\boldsymbol{\theta}_t^*$ has moments $\mathbf{a}_t^*$ and $\mathbf{R}_t^*$ if $\mathbf{K}_t$ and $\mathbf{h}_t$ are chosen as follows:

$$\mathbf{K}_t = \mathbf{U}_t \mathbf{Z}_t^{-1}$$
$$\mathbf{h}_t = \mathbf{a}_t^* - \mathbf{K}_t \mathbf{a}_t$$

where $\mathbf{U}_t$ and $\mathbf{Z}_t$ are the unique, upper triangular, non-singular square root matrices of $\mathbf{R}_t^*$ and $\mathbf{R}_t$ respectively, thus $\mathbf{R}_t^* = \mathbf{U}_t\mathbf{U}_t'$ and $\mathbf{R}_t = \mathbf{Z}_t\mathbf{Z}_t'$.

Proof: The matrices $\mathbf{U}_t$ and $\mathbf{Z}_t$ exist and are unique since $\mathbf{R}_t^*$ and $\mathbf{R}_t$ are symmetric, positive definite matrices (see, for example, Graybill, 1969). They define the Cholesky decompositon of these variance matrices, and are easily computed. From the definition of $\boldsymbol{\theta}_t^*$ it follows that

$$\mathbf{a}_t^* = \mathbf{K}_t\mathbf{a}_t + \mathbf{h}_t$$

and so the expression for $\mathbf{h}_t$ is immediate for any given $\mathbf{K}_t$. Secondly,

$$\mathbf{R}_t^* = \mathbf{K}_t\mathbf{R}_t\mathbf{K}_t',$$

thus

$$\mathbf{U}_t\mathbf{U}_t' = (\mathbf{K}_t\mathbf{Z}_t)(\mathbf{K}_t\mathbf{Z}_t)'.$$

Now $\mathbf{K}_t\mathbf{Z}_t$ is a square, non-singular, upper-triangular matrix and, since the matrix $\mathbf{U}_t$ is unique, it follows that $\mathbf{U}_t = \mathbf{K}_t\mathbf{Z}_t$. The expression for $\mathbf{K}_t$ follows since $\mathbf{Z}_t$ is non-singular.

$\diamond$

The Lemma shows how a second evolution of $\boldsymbol{\theta}_t$ to $\boldsymbol{\theta}_t^*$ can be defined to achieve any desired moments in (11.6). It is useful to think of the intervention in these terms, but, for calculations, it is often desirable to incorporate this second intervention into the original DLM, as follows.

THEOREM 11.2. *Suppose that the moments $\mathbf{a}_t^*$ and $\mathbf{R}_t^*$ in (11.6) are specified to incorporate intervention, and define $\mathbf{K}_t$ and $\mathbf{h}_t$ as in Lemma 11.1. Then (11.6) is the prior obtained in the DLM (11.1–11.4) with evolution equation (11.2) amended to*

$$\boldsymbol{\theta}_t = \mathbf{G}_t^*\boldsymbol{\theta}_{t-1} + \boldsymbol{\omega}_t^*, \qquad \boldsymbol{\omega}_t^* \sim \mathrm{N}[\mathbf{h}_t, \mathbf{W}_t^*], \qquad (11.7)$$

where, given D_{t-1} and I_t, $\boldsymbol{\omega}_t^*$ is uncorrelated with $\boldsymbol{\theta}_{t-1}$ and

$$\mathbf{G}_t^* = \mathbf{K}_t\mathbf{G}_t,$$
$$\boldsymbol{\omega}_t^* = \mathbf{K}_t\boldsymbol{\omega}_t + \mathbf{h}_t,$$
$$\mathbf{W}_t^* = \mathbf{K}_t\mathbf{W}_t\mathbf{K}_t'.$$

Proof: An easy deduction from the Lemma, and left to the reader.

$$\diamond$$

Thus any interventions modelled by (11.6) can be formally, and routinely, incorporated into the model by appropriately amending the evolution equation at time t, reverting to the usual equations for future times not subject to intervention. As with the intervention modes in Sections 11.2.1 and 11.2.3, this is important since it means that the usual updating, forecasting, filtering and smoothing algorithms apply directly with the post-intervention model. Note that, in forecasting more than one-step ahead, interventions for future times can be simply incorporated in the same fashion, by appropriately changing the model based on the forecast moments for the $\boldsymbol{\theta}$ vector at those times, pre- and post-intervention.

EXAMPLE 11.1 (continued). Consider again intervention into the CP6 model in January 1958. The four example interventions earlier considered can all be phrased in terms of (11.7). The details are left as exercises for the reader. Additionally, of course, other, arbitrary changes can be accommodated via this mode of intervention. As an example, suppose that $\mathbf{a}_t^{*\prime} = (970, 0)$ and $\mathbf{R}_t^* = \mathrm{diag}(50, 5)$. This represents direct intervention to anticipate a new level of 970, with variance 50 *decreased* from the pre-intervention value of 100, new growth estimated at 0 with increased variance of 5. Additionally, the post-intervention level and growth are uncorrelated.

11.2.4   Inclusion of intervention effects

There is one further mode of intervention to be explored. The preceding modes allow for the information I_t by appropriately amending the model at the time (or in advance if the information is available); in each case the dimension n of the model remains fixed, the intervention providing changes to the model parameters. Sometimes it is of interest to isolate the effects of an intervention, providing extra parameters that define the model changes.

EXAMPLE 11.1 (continued). Consider once more the CP6 intervention. To isolate and estimate the jump in level let γ_t represent the change so that, post-intervention, the new level is $\mu_t^* = \mu_t + \gamma_t$. γ_t is an additional model parameter that can be included in the parameter vector from now on. Thus, at the intervention time t, extend the model to 3 parameters

$$\boldsymbol{\theta}_t^* = (\mu_t^*, \beta_t, \gamma_t)'.$$

The DLM at time t is now subject to an additional evolution after (11.2), namely

$$\begin{pmatrix} \mu_t^* \\ \beta_t \\ \gamma_t \end{pmatrix} = \begin{pmatrix} 1 & 0 \\ 0 & 1 \\ 0 & 0 \end{pmatrix} \begin{pmatrix} \mu_t \\ \beta_t \end{pmatrix} + \begin{pmatrix} 1 \\ 0 \\ 1 \end{pmatrix} \gamma_t,$$

with the distribution assigned to γ_t determining the expected change in level. In vector/ matrix form

$$\boldsymbol{\theta}_t^* = \mathbf{K}_t \boldsymbol{\theta}_t + \boldsymbol{\xi}_t,$$

where

$$\mathbf{K}_t = \begin{pmatrix} 1 & 0 \\ 0 & 1 \\ 0 & 0 \end{pmatrix} \text{ and } \boldsymbol{\xi}_t = \begin{pmatrix} 1 \\ 0 \\ 1 \end{pmatrix} \gamma_t.$$

For example, an anticipated change of 100 with a variance of 50 implies $\gamma_t \sim N[100, 50]$. As a consequence,

$$(\boldsymbol{\theta}_t^* \mid I_t, \ D_{t-1}) \sim N[\mathbf{a}_t^*, \ \mathbf{R}_t^*]$$

where

$$\mathbf{a}_t^* = \mathbf{K}_t \mathbf{a}_t + \mathbf{h}_t,$$
$$\mathbf{R}_t^* = \mathbf{K}_t \mathbf{R}_t \mathbf{K}_t' + \mathbf{H}_t,$$

with $\mathbf{h}_t' = E[\boldsymbol{\xi}_t' \mid I_t, \ D_{t-1}] = 100(1, 0, 1) = (100, 0, 100)$ and

$$\mathbf{H}_t = V[\boldsymbol{\xi}_t \mid I_t, \ D_{t-1}] = 50 \begin{pmatrix} 1 & 0 & 1 \\ 0 & 0 & 0 \\ 1 & 0 & 1 \end{pmatrix}.$$

In addition, the observation equation is altered so that $\mathbf{F}_t' = (1, 0, 0)$ at time t. From t onwards, the model remains 3 dimensional, the $\mathbf{F}_t$

vectors all extended to have a third element of 0, the $\mathbf{G}_t$ matrices extended from $\mathbf{J}_2(1)$ to

$$
\begin{pmatrix}
1 & 1 & 0 \\
0 & 1 & 0 \\
0 & 0 & 0
\end{pmatrix},
$$

and, similarly, the 3x3 evolution variance matrices for times after t having third rows and columns full of zeroes. Thus, as new data is processed, the posterior for γ_t is revised, learning about the change that took place at time t. An important variation on this technique is to allow for such changes in level to occur over two (or more) time periods, with incremental changes γ_t, γ_{t+1}, for example, leading to a gradual step up to a new level at time $t+1$. This requires a model extension to 4 (and possibly more) parameters. (It is possible and sometimes appropriate to relate the parameters in order to restrict the increase in dimension to just one, as is done in modelling advertising campaign effects in Harrison (1988), for example). This technique is particularly useful following intervention *action* in the environment of the series to attempt control and produce changes. Changes in pricing policy or advertising strategy, for example, are effected in an attempt to changes sales levels. Intervention effects such as γ_t (and γ_{t+1}, etc.) then provide measures of just how much change occurred.

This sort of model extension intervention mode can be phrased in DLM terms as in Theorem 11.2, although in this case the matrix $\mathbf{G}_t^* = \mathbf{K}_t \mathbf{G}_t$ will not be square. Extending the model to include an additional k parameters at a particular time t means that $\mathbf{G}_t^*$ will be $(n+k) \times n$ if the parameters are unrelated so as to be separately estimated. The details of such extensions, similar to those in the example, are essentially as in Theorem 11.2 with the additional feature of an increase in model dimension. See also Harrison (1988) for approaches in which the increase in dimensionality is restricted.

11.2.5   Model analysis with intervention

The main reason for expressing all modes of intervention in DLM form is that the existing theory can be applied to formally incorporate interventions of essentially arbitrary forms, with the model quadruple $\{\mathbf{F}_t,\ \mathbf{G}_t,\ V_t,\ \mathbf{W}_t\}$ amended according to the mode of intervention. It then follows that the theory applies, with, for example, the elements $\mathbf{a}_t$ and $\mathbf{R}_t$ replaced throughout by their post-intervention (starred) values. In particular, in filtering/ smoothing for retrospective time series analysis and estimation of the time trajectories of model components and parameters over time, it is vital that these subjective changes at intervention times be formally incorporated in model form so that their implications for times past, as well as for forecasting the future, are properly understood.

There are certainly other modes of intervention and subjective interference with a routine mathematical/ statistical model that may be used in practice. In particular, it is worth pointing out that if, in retrospect, it is seen that an intervention in the past was inappropriate, then in filtering back beyond that point of intervention it may be well to ignore the intervention. Thus, for example, the filtered distribution for time $t-1$ just prior to intervention at t can be taken as it was at the time, simply the posterior given D_{t-1}, and filtering continuing on back in time from this new starting point.

Finally note that the concepts apply directly to models with unknown observational variances, the actual techniques being directly appropriate (although T distributions replace normal distributions, as usual). Also, it should be apparent that the specific normal, linear structure used throughout is quite secondary, the intervention models being obviously appropriate and useful more generally. If we simply drop the normality assumptions and use distributions only partially specified in terms of means and variance matrix, for example, the techniques apply directly. However, no further development is given here of this, or other, intervention ideas. We proceed in the next Section with illustrations of the main modes of intervention detailed above.

11.3   ILLUSTRATIONS

11.3.1   CP6 Sales

An illustrative example is given using the CP6 Sales data. The objective of this example is to demonstrate just how effective simple interventions can be if relevant feed-forward information exists and is appropriate. This example is purely hypothetical, the interventions being used are easily seen to be adequate descriptions of the discontinuities in the series. In practice, of course, the interventions

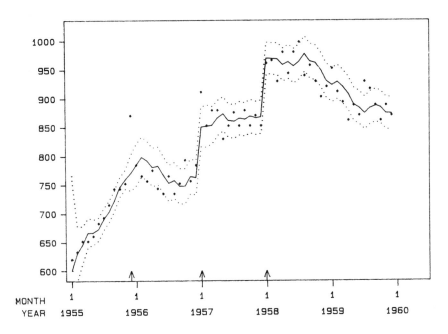

Figure 11.3.   One-step ahead forecasts for CP6 with ideal
              interventions.

are usually made speculatively, and, though often they may be dramatically effective, will usually not be ideal.

The first-order polynomial model introduced in Example 11.1 for
the CP6 Sales data is used here. The basic time variation in monthly
level and growth parameters is determined by a single discount factor
of 0.9; thus, each month, the uncertainty about level and growth
increases by roughly 10%. The observational variance is constant,
set initially at $V_t = V = 64$, with standard deviation of 8 determining
purely random variation about the underlying level.

The features of interest are as follows.

(1) At time $t = 1$, January 1955, the analysis begins with an
    initial prior for the level and growth $\boldsymbol{\theta}_1 = (\mu_1,\ \beta_1)'$ speci-
    fied by prior mean $\mathbf{a}_1 = (600, 10)'$, and variance matrix $\mathbf{R}_1 =$
    diag(10000, 25). The prior standard deviation for the level,
    set at 100, is extremely large, representing a very vague ini-
    tial prior. The one-step forecast distribution for Y_1 thus has
    mean 600, and variance 10064; the 90% forecast interval sym-
    metrically located about the mean appearing in Figure 11.3
    reflects this uncertainty. As data is sequentially processed as

usual, the one month ahead forecasts are calculated and, for each month, the forecast means with 90% forecast intervals appear in the Figure. During the first year of data, Sales closely follow a steep, linear growth and one-step forecasts are accurate.

(2) Intervention is made at December 1955 to cater for an anticipated outlier — forward buying due to a future price rise announcement. Thus, for $t = 12$, the observational variance is infinite, or $V_{12}^{-1} = 0$ as in Section 11.2.1.

(3) In addition, following the price rise, a change in growth is anticipated in the New Year and so, rather than waiting until January to anticipate the change, an additional intervention is made to the model in December. The intervention is neutral, specifying an increase in uncertainty about level and growth with no specific direction of change in mind, thus allowing the data in January and February to inform on the changes. Specifically, as in Section 11.2.2, the prior variance matrix

$$\mathbf{R}_{12} = \begin{pmatrix} 41.87 & 6.11 \\ 6.11 & 1.23 \end{pmatrix}$$

is increased to $\mathbf{R}_{12}^*$ by the addition of

$$\mathbf{H}_{12} = \begin{pmatrix} 100 & 25 \\ 25 & 25 \end{pmatrix}.$$

Thus additional zero-mean, normal changes to level and growth have standard deviations of 10 and 5 respectively, and correlation of 0.5 so that positive growth changes are associated with positive level changes.

(4) A third change is made at $t = 12$, this being to alter the observational variance for the future from $V = 64$ to $V = 225$, the standard deviation increasing from 8 to 15. This reflects the views that, firstly, variation is likely to be more erratic after this first year of fast market growth, and, secondly, that higher variation in Sales is expected at higher levels.

   The effects of these interventions are clear in Figure 11.3, an arrow on the time axis indicating the intervention. After ignoring the December 1955 observation, the forecast interval width increases reflecting greater uncertainty about θ_{12} following intervention, and the increased value of V. Observing Sales sequentially throughout 1956, the model adapts and proceeds adequately to the end of the year. Note also the wider forecast intervals due to the larger value of V.

(5) The second set of interventions takes place for $t = 25$, January 1957, where the current prior for level and growth is given by

$$\mathbf{a}_{25} = \begin{pmatrix} 770.8 \\ 1.47 \end{pmatrix} \quad \text{and} \quad \mathbf{R}_{25} = \begin{pmatrix} 114.4 & 13.2 \\ 13.2 & 2.2 \end{pmatrix}.$$

Anticipating a marked increase in Sales level following a take-over, an estimated change in level of 80 units is fed-forward, with variance of 100. No effects on growth are expected, or allowed, by using an additional evolution term $\boldsymbol{\xi}_{25}$ as in Section 11.2.2, with mean and variance matrix

$$\mathbf{h}_{25} = \begin{pmatrix} 80 \\ 0 \end{pmatrix} \quad \text{and} \quad \mathbf{H}_{25} = \begin{pmatrix} 100 & 0 \\ 0 & 0 \end{pmatrix},$$

respectively.

(6) In addition to this estimated change in level, the January 1957 observation is discarded as an outlier, reflecting a view that the marked change anticipated in the New Year will begin with a maverick value, as the products which are to be discontinued are sold cheaply.

The interventions have a clear effect on short-term forecast accuracy, seen in the Figure. Again for the remainder of 1957 things are stable.

(7) The third and final intervention in this short history of the Sales series comes in the following year, for January 1958. Another jump in level is anticipated, this time of about 100 units. Unlike the previous change, however, there is a feeling of increased certainty about the new level. Also, it is anticipated that growth may change somewhat more markedly than already modelled through the routine discount factor, and so the prior variance of the growth is to be increased. This intervention, then, is of the mode in Section 11.2.3. The prior moments, namely

$$\mathbf{a}_{37} = \begin{pmatrix} 864.5 \\ 0.86 \end{pmatrix} \quad \text{and} \quad \mathbf{R}_{37} = \begin{pmatrix} 91.7 & 9.2 \\ 9.2 & 1.56 \end{pmatrix},$$

are simply altered to

$$\mathbf{a}_{37}^* = \begin{pmatrix} 970 \\ 0 \end{pmatrix} \quad \text{and} \quad \mathbf{R}_{37}^* = \begin{pmatrix} 50 & 0 \\ 0 & 5 \end{pmatrix}.$$

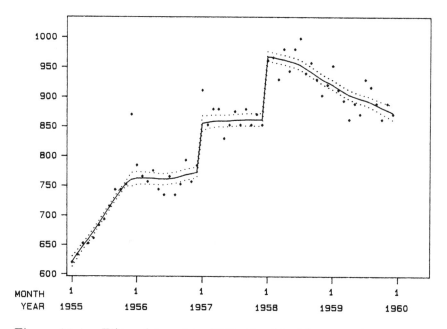

Figure 11.4.   Filtered trend in CP6 after ideal interventions.

The forecasts, in Figure 11.3, adapt accordingly and analysis proceeds as normal until the end of the series. Looked at as a whole, the one-step forecasting performance has been rather good due to the appropriate, though entirely subjective, interventions.

(8) At the end of the analysis, $t = 60$ in December 1959, the backward filtering computations are carried out to provide a retrospective look at the development over time. Details are standard and therefore omitted. Figure 11.4 provides a plot of the retrospectively estimated level of the series, with mean and symmetric 90% interval taken from the filtered posterior distributions $p(\mu_t|D_{60})$, $(t = 1, \ldots, 60)$. Note, of course, that D_{60} incorporates the interventions. The local linearity of the trend is evident, and startlingly so, as is the fairly high precision with which the trend is estimated. Clearly the random noise in the data about the trend is a dominant feature of the series, the underlying trend being smooth and sustained apart from the abrupt changes catered for by intervention.

11.3.2   UK Marriages

The second illustration concerns the data in Table 11.2, providing
quarterly total numbers of marriages registered in the UK during the
years 1965 to 1970 inclusive. The data, taken from the UK Monthly
Digest of Statistics, are given in thousands of registrations.

Before examining the data more closely, consider the pattern that
the marriage figures could be expected to take over a year on general grounds. It might be strongly believed that the majority of
marriages take place in late Spring and Summer in the UK, many
people preferring to get married when there is a reasonable chance
of good weather. Certainly there are short exceptional times, such
as Christmas and Easter, that are particularly popular, but, generally, a strong annual cycle is anticipated, with a high peak in the
third quarter which contains the favoured holiday months of July
and August. In addition, the winter months of November, December, January and February should see the trough in the seasonal
pattern.

Consider now the data in the table, restricting attention to the
first 4 years 1965 to 1968 inclusive. The data for these four years
are plotted as the first part of Figure 11.5. Over these years there is
certainly a strong seasonal pattern evident, though not of the form
anticipated in the previous paragraph. The summer boom in marriages is there in the third quarter, but there is a secondary peak
in the first quarter, the months January, February and March. In
addition, the major trough in numbers comes in the second, Spring/

Table 11.2. Numbers of marriages in UK (in thousands)

Year	Quarter			
	1	2	3	4
1965	111.2	83.5	129.5	97.8
1966	114.7	84.6	131.1	106.3
1967	117.5	80.6	143.3	97.6
1968	131.3	77.5	145.9	108.0
1969	88.1	112.6	152.0	98.5
1970	91.2	117.5	160.0	102.2

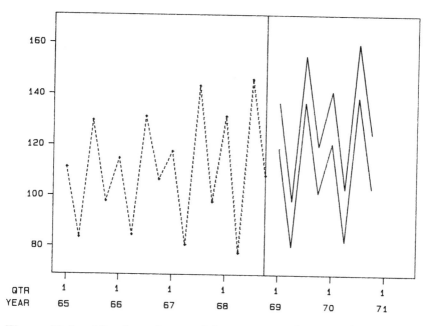

Figure 11.5.   Marriage data and forecasts: no interventions.

early Summer, quarter rather than in the fourth, early Winter quarter. These features are rather surprising though clearly apparent in
each of these 4 years as they were prior to 1965. It can be seen that
the amplitude of the seasonality increases somewhat in mid-1967
and that this increase is sustained during 1968, but the form of the
pattern is still the same: secondary peak in quarter 1, deep trough
in quarter 2, high peak in the Summer, quarter 3, and secondary
trough in quarter 4.

   Before proceeding to explain the rather surprising seasonality here,
consider a simple model for forecasting the series as it stands. Suppose that we use a second-order polynomial for the trend in the non-
seasonal level of marriage, with a full 4 seasonal effects provided, as
usual, by the Fourier representation with 3 Fourier coefficients. Thus
the model has 5 parameters in $\boldsymbol{\theta}_t$; the level and growth and 3 Fourier
coefficients. In addition, assume now that the observational variance
is to be estimated, being assumed constant, $V_t = V$ for all time. The
initial prior information assumed at $t = 1$, the first quarter of 1965,
is as follows.

   • The observational variance has an initial inverse gamma dis-
     tribution with 12 degrees of freedom and estimate of 9. Thus

the prior estimate of standard deviation is 3. Equivalently, V has a scaled, inverse χ^2_{12} distribution.

- The initial posterior for $\boldsymbol{\theta}_0$, the parameters at end of 1964 given historical data and experience, is multivariate T_{12}, the mean and scale matrix taken as

$$\mathbf{m}_0 = (100,\ 1;\ -7.5,\ -7.5,\ 17.5)',$$

and

$$\mathbf{C}_0 = \operatorname{diag}(16,\ 1;\ 4.5,\ 4.5,\ 2.5).$$

Thus the level in late 1964 is estimated as 100, with scale 16, thus having variance (from the T_{12} distribution) of $16 \times 12/10 = 19.2$. Similarly the quarterly growth in level at the time initially is estimated as 1 with scale of 1. Transforming from Fourier coefficients to seasonal factors (Section 8.6.4, Chapter 8), the prior for the 3 Fourier coefficients, with mean $(-7.5,\ -7.5,\ 17.5)'$ and scale matrix $\operatorname{diag}(4.5, 4.5, 2.5)$, is seen to be consistent with initial estimates of seasonal effects given by $(10, -25, 25, -10)'$, these being based on pre-1965 data and experience and clearly anticipating the seasonal pattern over 1965-1968.

- Two discount factors, one for the linear trend and one for the seasonal pattern, are required to complete the model definition. Both components are fairly stable over time and so the discount factors are chosen to be fairly high, both set at 0.95. Concerning the seasonal component, this should allow for adequate adaptation to the observed increases in amplitude of the seasonal pattern at higher levels of the data.

With this model, the sequential, one-step ahead forecasting and updating analysis proceeds and is illustrated in the first part of Figure 11.7. Up to the end of 1968, the graph gives one-step ahead forecast means and symmetric 90% intervals about the means, with the data superimposed. The stong seasonal pattern comes through in the forecasts, and it is clear that forecast accuracy is reasonably good. There is a deterioration in accuracy during late 1967/ early 1968 when the amplitude of the seasonal swings in the data increases, but the model adapts to this in late 1968, the final 2 observations in that year being well inside the forecast intervals.

Up to this point, everything is essentially routine, the data being forecast in the short-term with a simple, standard model closed to interventions. At the end of 1968, time $t = 16$, the prior distribution for the next quarter, $t = 17$ is summarised by

$$(\boldsymbol{\theta}_{17} \mid D_{16}) \sim T_{28}[\mathbf{a}_{17}, \ \mathbf{R}_{17}],$$

where

$$\mathbf{a}_{17} = (117.1, \ 0.84; \ -8.35, \ -9.81, \ 19.03)', \qquad (11.8)$$

and

$$\mathbf{R}_{17} \approx \text{block diag} \left\{ \begin{pmatrix} 5.71 & 0.56 \\ 0.56 & 0.07 \end{pmatrix}; \ 2.79, \ 2.66, \ 1.35 \right\}. \qquad (11.9)$$

The first 2×2 matrix here refers to the level and growth elements, the final 3 to the (essentially uncorrelated) Fourier coefficients. For the observational variance, V has a scaled, inverse χ^2_{28} distribution with point estimate $S_{16} = 16.16$. The latter part of Figure 11.5 provides a graphical display of the implications for the future. Forecasting from the final quarter of 1968 with no changes to the model, the intervals displayed are, as usual, 90% forecast intervals symetrically located about the step ahead forecast means.

Consider an explanation for the seasonal pattern and why, in late 1968, this would have changed our view of the future of the marriage series, prompting an intervention at the time. The explanation lies in the UK income tax laws. Income is only taxed above certain threshold levels, the non-taxable portion of income being referred to as a tax-free allowance, and every employed person has a basic allowance. On getting married, one of the marriage partners is elegible for a higher tax-free allowance, referred to as the married person's allowance (although during the time span of this data it was called a married *man's* allowance). Now, in the good old days of the 1960's, the law was such that a couple could claim this extra allowance for the entire tax year during which they were married, the tax years running from April to March inclusive. Thus, for example, a wedding in late March of 1967 led to an entitlement to reclaim some portion of tax paid during the previous 12 months when both partners were single. Delaying this wedding for a week or two to early April would mean that this extra income would be lost since a new tax year has begun. The seasonal pattern is now explained; many marriages were obviously held in the first quarter of each year to maximise financial benefit, and the second quarter saw a huge slump in numbers since the resulting tax benefits were minimal.

In forecasting ahead from late 1968 over the next 2 years, the latter part of Figure 11.5 is perfectly acceptable on the basis of the historical information as it stands. However, in early 1968 there was an announcement that the tax laws were being revised. The change was simply that, beginning in 1969, this entitlement to reclaim tax paid during the current tax year was abolished; from then on, there would be no tax incentive for couples to avoid marrying in the second quarter. Knowing this and understanding the effect that the old law had on marriages, a change in the forecasting model is called for. On the grounds of a simple cycle following the UK seasons as earlier discussed, it is assumed in late 1968 that the secondary peak in the first quarter will disappear, with marriages transferring to the more attractive second and third quarters. The levels in the second quarter will increase markedly from the totally artificial trough, those in the third quarter rather less so.

As an illustration of intervention to incorporate this view, suppose that, in predicting from the final quarter of 1968, the expected seasonal effects are taken as $(-16, 0, 40, -24)'$. Thus the anticipation is that numbers of marriages grow consistently throughout the calendar year to a high peak during the summer, then crash down to a low trough during the final quarter of the year. Transforming to Fourier coefficients (Section 8.6.4, Chapter 8), it is seen that these expected seasonal effects are consistent with expected Fourier coefficients of about $(-28, 12, 12)'$, although there is clearly a fair degree of uncertainty about just how the pattern will change. What is fairly acceptable, however, is the view that the tax changes are unlikely to affect the non-seasonal level of marriages, nor the rate of growth of this level. In the mode and notation of Section 11.2.3, an arbitrary change is made to the current prior moments a_{17} and R_{17} to accommodate this view. The revised values in (11.6) are taken as

$$a_{17}^* = (117.1, 0.84; -28, 12, 12)',$$

and

$$R_{17}^* = \text{block diag} \left\{ \begin{pmatrix} 5.71 & 0.56 \\ 0.56 & 0.07 \end{pmatrix}; 8, 8, 4 \right\}.$$

These replace the values in (11.8) and (11.9). Note that the level and growth components are unchanged, and they are taken as uncorrelated with the seasonal parameters. Figure 11.6 displays the post-intervention step ahead forecasts made in late 1968 for the next 2 years, the display being analogous to that in Figure 11.5 made pre-intervention. The forecast intervals are slightly wider reflecting the

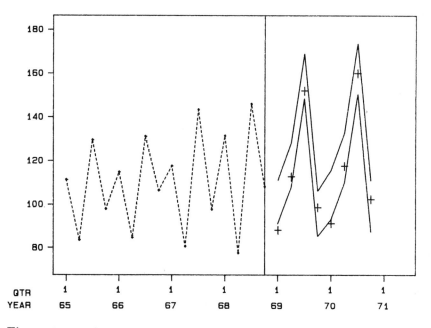

Figure 11.6.   Marriage forecasts with intervention at Qtr.1/Year
69.

increase in uncertainty about the seasonal pattern. The crosses on
the graph are the actual data for those 2 years. Note that the pattern did in fact change markedly in line with the intervention based
forecasts, although several of the actual values are rather low relative to forecast. This suggests, in retrospect, that the non-seasonal
level was very slightly over estimated at the end of 1968. Moving
through these 2 years 1969 and 1970, the data is sequentially processed and the model adapts as usual. The latter part of Figure 11.7
now completes the display of the one-step forecasting activity, and
the adequacy of the model is apparent.

The main point here to be restressed is that an understanding of
the mechanisms influencing, and possibly driving, systems is the key
to effective interventions necessary to adapt to changing conditions.

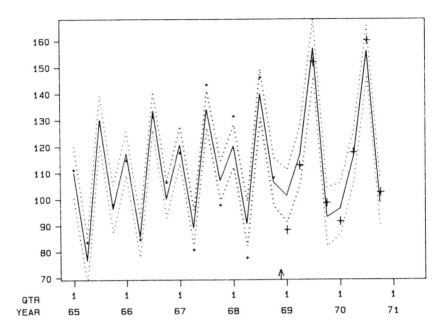

Figure 11.7.   One-step ahead Marriage forecasts with intervention.

11.4   MODEL MONITORING

11.4.1   Bayes' factors for model assessment

The focus is now switched to problems of feed-back intervention,
considering models operating subject to continual monitoring to detect deteriorations in predictive performance that are consistent with
some form of model breakdown (eg. changes in parameters, etc). We
discuss the use of automatic methods of sequentially monitoring the
forecasting activity to detect breakdowns, the assessment of model
performance being based on purely statistical measures of accuracy.
At this most basic level, the problem of model assessment is simply
one of examining the extent to which the observed values of the time
series are consistent with forecasts based on the model. In the DLM
framework, the focus is on consistency of each observation with the
corresponding one-step ahead forecast distribution.   Equivalently,
the assessment can be made on the basis of standardised, one-step
ahead forecast errors, measuring the extent to which they deviate
from model hypothesis of standard, normal or T distributed, uncorrelated quantities. There are many statistical techniques with which

to examine such questions, some specifically designed with time series data in mind, others of use more widely. Central to all such techniques is the notion of assessing model performance relative to that obtained through using one or more alternative models. With the objectives of detecting changes in parameters, outlying observations, and so forth, in mind, these alternatives should be designed to allow for the forms of behaviour in the series that are consistent with such changes, wild observation etc. There are clearly many possible forms that these alternatives can take, the specific forms chosen to be based on the specific forms of departure from the existing, routine model that are anticipated. We consider some possible, generally useful forms below, and describe how they may be used in automatic model monitoring and assessment. Initially, however, the basic concepts of model assessment and sequential monitoring are detailed in a general setting. Much of the basic material here is developed from West (1986a). The basic mathematical ingredients of sequential model monitoring have already been introduced in informal settings in previous chapters, notably Section 10.2.5 of Chapter 10, the keystone being provided by Bayes' factors, defined formally below.

Consider any two models with the same mathematical structure, differing only through the values of defining parameters, for example, in the values of discount factors. Denote any model by the symbol M, and let these two models be denoted by M_0 and M_1. M_0 will have a special status, being the routine or standard model that is used subject to continual assessment. M_1, (and possibly further models M_2, etc. at a later stage), is an alternative that is introduced to provide assessment of M_0 by comparison. At time t, each model provides a predictive distribution for Y_t given D_{t-1} as usual. Formally including the specification of the models in the conditioning, we write these densities as

$$p(Y_t|D_{t-1}, M_i), \qquad (i = 1, 2)$$

with D_{t-1} being the historical information that is common to the two models at time t. The inclusion of M_i differentiates between the two models. For simplicity of notation in this section, however, we temporarily discard this general notation, making clear the dependence of the distributions on the model using the subscript 0 or 1. Thus the predictive densities at time t are here denoted by

$$p_i(Y_t|D_{t-1}) = p(Y_t|D_{t-1}, M_i), \qquad (i = 1, 2)$$

respectively.

Definitions 11.1.

(i) *The* **Bayes' Factor** *for M_0 versus M_1 based on the observed value of Y_t is defined as*

$$H_t = p_0(Y_t|D_{t-1})/p_1(Y_t|D_{t-1}).$$

(ii) *For integers $k = 1, \ldots, t$, the Bayes' Factor for M_0 versus M_1 based on the sequence of k consecutive observations $Y_t, Y_{t-1}, \ldots, Y_{t-k+1}$ is defined as*

$$H_t(k) = \prod_{r=t-k+1}^{t} H_r$$
$$= p_0(Y_t, Y_{t-1}, \ldots, Y_{t-k+1}|D_{t-k})/p_1(Y_t, Y_{t-1}, \ldots, Y_{t-k+1}|D_{t-k}).$$

These Bayes' factors, or *Weights of evidence* (Jeffreys, 1961; Good, 1985 and references therein; West, 1986a), provide the basic measures of predictive performance of M_0 relative to M_1. For each k, $H_t(k)$ measures the evidence provided by the most recent (up to and including time t), k consecutive observations. Some basic features of Bayes' factors are noted.

(1) Setting $k = 1$ in (ii) leads to the special case (i): $H_t(1) = H_t$, for all t.
(2) Taking $k = t$, the Bayes' factor based on all the data is $H_t(t)$.
(3) The Bayes' factors for M_1 versus M_0 are the reciprocals of those for M_0 versus M_1, $H_t(k)^{-1}$.
(4) Evidence for or against the model M_0 accumulates multiplicatively as data is processed. Specifically, for each $t > 1$,

$$H_t(k) = H_t \, H_{t-1}(k-1), \qquad\qquad (k = 2, \ldots, t).$$

(5) On the log scale, evidence is additive, with

$$\log[H_t(k)] = \log(H_t) + \log[H_{t-1}(k-1)], \qquad\qquad (k = 2, \ldots, t).$$

(6) Following Jeffreys' (1961), a log Bayes' factor of 1 (-1) indicates evidence in favour of model 0 (1), a value of 2 or more $(-2$ or less) indicating the evidence to be strong. Clearly the value 0 indicates no evidence either way.

The definition of Bayes' factors is general, obviously applying outside the confines of normal DLMs. In considering specifically the possibilities that, in model M_0, Y_t may be a wild, outlying observation or that the defining parameters in M_0 may have changed at (or before) time t, the alternative M_1 should provide for the associated forms of departure of the observations from prediction under M_0. Some possible forms for $p_1(Y_t \mid D_{t-1})$ are discussed in Section 11.4.3. Notice, however, that it is not actually necessary to construct a fully specified alternative model for the data, all that is needed is a suitable sequence of alternative, one-step forecast densities that provide the denominators of the Bayes' factors. Once appropriate densities are defined, they may be used without further consideration of particular forms of departure from M_0 that may lead to such a density. This is important in practice since the derived Bayes' factor is easily computed and forms the basis for sequential monitoring even though a formal model M_1 has not been constructed.

11.4.2 Cumulative Bayes' factors

In considering how the Bayes' factors may be applied in monitoring M_0, assume for the moment that appropriate densities have been defined, thus the *sequence* $p_1(Y_t|D_{t-1})$, $t = 1, \ldots$, is available to base the Bayes' factors upon. Assume here that the alternatives are appropriate for the deviations from M_0 and forms of departure of the data from M_0 that are of interest.

The overall Bayes' factor $H_t(t)$ is a basic tool in overall model assessment. In the monitoring context, however, the focus is on *local* model performance and here the individual measures H_t and the cumulative measures $H_t(k)$ for $k < t$ are key. As an illustration of this, suppose that $t = 6$ and that $H_r = 2$ for $r < 6$. Thus each of the first five observations are well in accord with the standard model M_0, their individual Bayes' factors all being equal to 2. Consequently, $H_5(5) = 32$, representing the cumulative evidence for M_0 relative to M_1 from the first five observations. Suppose that Y_6 is very unlikely under M_0, out in the tails of the forecast distribution. Then even though it may be extremely discrepant, the cumulative Bayes' factor $H_6(6)$ may still exceed 1, and even be much larger than 1, thus indicating no evidence against M_0. Clearly we would require $H_6 \leq 1/32$, an extremely small Bayes' factor, to even begin to doubt M_0. The problem is that the evidence in favour of the standard model from earlier observations *masks* that against it at

time 6 in the overall measure. Hence the need to consider the Bayes'
factors from individual observations as they arise. In addition, it
is important to look back over groups of recent observations when
considering the possibility of small or gradual change in a time series, consistent with a shift in the Bayes' factors from favouring M_0
to favouring M_1. In such cases, where observations gradually drift
away from forecasts, the individual Bayes' factors, although small,
may not be small enough that they signal the changes individually,
needing to be cumulated to build up evidence against M_0. For each
k, $H_t(k)$ assesses the fit of the most recent k observations. A single
small value of $H_t(1) = H_t$ provides a warning of a possible outlier or
the onset of change in the series at time t. A small $H_t(k)$ for $k > 1$
is indicative of possible changes having taken place (at least) k steps
back in the past. To focus on the most likely point of change, we
can identify the most discrepant group or recent, consecutive observations by minimising the Bayes' factors $H_t(k)$ with respect to k.
This may be done simply, sequentially as the standard model M_0 is
updated, using the following result (West, 1986a).

THEOREM 11.3. *With H_t and $H_t(k)$ as in Definition 11.1, let*

$$L_t = \min_{1 \le k \le t} H_t(k)$$

*with $L_1 = H_1$. Then the quantities L_t are updated sequentially over
time by*

$$L_t = H_t \min\{1, \ L_{t-1}\}$$

*for $t = 2, 3, \ldots$. The minimum at time t is taken at $k = l_t$, with
$L_t = H_t(l_t)$, where the integers l_t are sequentially updated via*

$$l_t = \begin{cases} 1 + l_{t-1}, & \text{if } L_{t-1} < 1, \\ 1, & \text{if } L_{t-1} \ge 1. \end{cases}$$

Proof: Since $H_t(1) = H_t$ and, for $2 \le k \le t$, $H_t(k) = H_t H_{t-1}(k - 1)$, then

$$\begin{aligned} L_t &= \min\{H_t, \ \min_{2 \le k \le t} H_t H_{t-1}(k - 1)\} \\ &= H_t \ \min\{1, \ \min_{2 \le k \le t} H_{t-1}(k - 1)\} \\ &= H_t \ \min\{1, \ \min_{1 \le j \le t-1} H_{t-1}(j)\} \\ &= H_t \ \min\{1, \ L_{t-1}\}, \end{aligned}$$

as stated. Note that $L_t = H_t$ if and only if $l_t = 1$, otherwise $L_t = H_t L_{t-1}$ and $l_t = 1 + l_{t-1}$, providing the stated results.

◇

The focus on the local behaviour of the series is evident in L_t. If, at time $t - 1$, the evidence favours the standard model M_0 so that $L_{t-1} \geq 1$, then $L_t = H_t$ and decisions about possible inadequacies of the model are based on Y_t alone. If H_t is very small, then Y_t is a possible outlier or may indicate the onset of change. If the evidence is against M_0 before time t with $L_{t-1} < 1$, then evidence is cumulated via $L_t = H_t L_{t-1}$, and l_t increases by 1. l_t is termed the *run-length* at time t, counting the number of recent, consecutive observations that contribute to the minimum Bayes' factor. The local focus is geared to detecting slow, gradual change. To see this note that a relatively slow or gradual change beginning at time t leads to a sequence of consecutive values H_{t+1}, $H_{t+2}, \ldots$, that are small though not exceedingly so. Hence, starting from $L_t = 1$, subsequent values $L_{t+1}, \ldots$ will drop rapidly as the evidence against M_0 is built up. The fact that M_0 was adequate prior to time t does not now mask this local breakdown since that evidence, being of historical relevance only, has been discarded in moving at time t to $L_t = 1$.

The sequence $\{L_t\}$ provides a sequential monitor or tracking of the predictive performance of M_0 relative to M_1. The simplest mode of operation involves monitoring the sequence until evidence of inadequacy of M_0, as measured by a small value of L_t at the time, is sufficiently great to warrant intervention. This simple detection of model breakdown is the primary goal here; once detected, the forecaster may intervene in feed-back mode, attempting to correct the model retrospectively. This use of the Bayes' factors has much in common with standard tracking signals based on sequential probability ratio tests (SPRTs) which has a long history, going back to the early work of Wald in the 1940s (Page, 1954; Barnard, 1959; Berger, 1985, Chapter 7). Specifically, let τ be a prespecified threshold for Bayes factors, defining the lower limit on acceptability of L_t. τ lies between 0 and 1, values between 0.1 and 0.2 being most appropriate. If L_t exceeds τ then M_0 operates as usual. Even though the evidence may be against M_0 in the sense that $L_t < 1$, it is not viewed as strong evidence unless $L_t < \tau$. From Theorem 11.3, if in fact $\tau < L_t < 1$, then $L_{t+1} < L_t$ if and only if $H_{t+1} < 1$. If, however, L_t falls below the threshold, breakdown in predictive performance of M_0 is indicated, and we have the following considerations.

If $L_t < \tau$, then:

- If $l_t = 1$ then $L_t = H_t$ and the single observation Y_t has led to the monitor signal. There are several possible reasons for such an extreme or discrepant observations. Y_t may be an outlier under M_0, in which case it may be rejected and M_0 used as usual for the next observation stage. Alternatively, Y_t may represent a major departure from M_0 at time t, such as a level change, for example, that M_1 is designed to cater for. This inability to distinguish between an outlying observation and true model changes based on a single, discrepant observation dogs any automatic monitoring scheme. The need for some form of intervention is paramount.
- If $l_t > 1$ then there are several observations contributing individual Bayes' factors to L_t, suggesting departure of the series from M_0 at some time past. The suggested time of onset of departure is l_t steps back at time $t - l_t + 1$.

It is obviously possible to extend the approach to consider two, or more, alternative forecast densities, leading to a collection of monitoring signals. For example, with two alternatives M_1 and M_2, the application of Theorem 11.3 leads to sequentially updated quantities $L_{i,t}$ for $i = 1$ and 2 respectively. Then the standard analysis will proceed as usual unless either of the minimum Bayes' factors $L_{1,t}$ and $L_{2,t}$ falls below prespecified thresholds τ_1 and τ_2.

11.4.3   Specific alternatives for the DLM

Within the DLM framework, the predictive distributions are, of course, normal or T depending on whether the observational variances are known or unknown. We consider the normal case for illustration, the following, fundamental, framework being used throughout this section. Suppose that the basic, routine model M_0 to be assessed is a standard normal DLM producing one-step ahead forecast distributions $(Y_t \mid D_{t-1}) \sim N[f_t, Q_t]$ as usual. Assessing consistency of the observed values Y_t with these distributions is essentially equivalent to assessing consistency of the standardised forecast errors $e_t/Q_t^{1/2} = (Y_t - f_t)/Q_t^{1/2}$ with their forecast distributions, standard normal, and the hypothesis that they be uncorrelated. Thus, for discussion in this section, Bayes' factors are based on the predictive densities of the forecast errors, simply linear functions of the original observations. In the context of sequential monitoring, the focus lies, not on the historical peformance of the model, but on the local performance; it is the extent to which the current and most recent observations accord with the model that determines whether or not some form of intervention is desirable. To start then, consider only the single observation, equivalently, the single forecast error, and, without loss of generality, take $f_t = 0$ and $Q_t = 1$ so that, under M_0, the forecast distribution for $e_t = Y_t$ is simply

$$(e_t \mid D_{t-1}) \sim N[0,1],$$

and so

$$p_0(e_t \mid D_{t-1}) = (2\pi)^{-1/2} \exp\{-0.5e_t^2\}.$$

Within the normal model, there are various possible alternatives M_1 that provide for the types of departure form M_0 encountered in practice. Key examples are as follows.

EXAMPLE 11.2. An obvious alternative is the level change model M_1 in which e_t has a non-zero mean h with

$$p_1(e_t \mid D_{t-1}) = (2\pi)^{-1/2} \exp\{-0.5(e_t - h)^2\}.$$

For any fixed shift h, the Bayes' factor at time t is

$$H_t = p_0(e_t \mid D_{t-1})/p_1(e_t \mid D_{t-1}) = \exp\{0.5(h^2 - 2he_t)\}.$$

Ranges of appropriate values of h may be considered by reference to values of the Bayes' factor at various, interesting values of the error e_t. As an illustration, Suppose that the error is positive and consider the point at which $H_t = 1$ so that there is no evidence from e_t alone to discriminate between M_0 and M_1. At this point, $\log(H_t) = 0$ and so, since h is non-zero, $h = 2e_t$. To be indifferent between the models on the basis of an error $e_t = 1.5$, for example, (at roughly the upper 90% point of the forecast distribution), suggests that $h = 3$. Similarly, suppose that a threshold $H_t = \tau$, $(0 < \tau \ll 1)$, is specified, below which the evidence is accepted as a strong indication that e_t is inconsistent with M_0. A threshold of -2 for the log-Bayes' factor implies $\tau = e^{-2} \approx 0.135$, for example. Fixing e_t at an acceptably extreme value when $H_t = \tau$ leads to a quadratic equation for h, namely $h^2 - 2he_t - 2\log(\tau) = 0$. Thus $e_t = 2.5$, (roughly the upper 99% point of the forecast distribution), and $\log(\tau) = -2$ implies $h = 1$ or $h = 4$. Thus values of h between 3 and 4 lead to indifference between M_0 and M_1 when e_t is near 1.5, and fairly strong evidence $(\tau = e^{-2})$ against M_0 for values as high as $e_t = 2.5$, Note that, if e_t is negative, the sign of h must be reversed.

EXAMPLE 11.3. Consider the previous example, and suppose that (as will typically be the case in practice) it is desired to cater for the possibilities of change in either direction. Then two alternatives are needed. Treating the cases symmetrically, suppose that the first, denoted M_1, has the form in Example 11.2 with mean shift $h > 0$, and the second, M_2, has the same form though with mean shift $-h$. The mode of operation of the sequential monitor now involves applying Theorem 11.3 to each alternative and leads to minimum Bayes' factors $L_{i,t}$, $(i = 1, 2)$. Then the standard analysis will proceed as usual unless either of the minimum Bayes' factors $L_{1,t}$ and $L_{2,t}$ falls below prespecified thresholds. Consistent with the symmetric treatment of the alternatives, suppose that these thresholds are both equal to some value τ. It follows from Example 11.2 that, with a common run-length l_t,

$$\log(L_{1,t}) = 0.5h^2 l_t - hE_t$$

where

$$E_t = \sum_{r=0}^{l_t - 1} e_{t-r}$$

is the sum of the most recent l_t errors. Similarly,

$$\log(L_{2,t}) = 0.5h^2 l_t + hE_t.$$

Hence

$$L_{1,t} \geq 1 \quad \text{if, and only if,} \quad E_t \leq \quad 0.5hl_t,$$
$$L_{2,t} \geq 1 \quad \text{if, and only if,} \quad E_t \geq -0.5hl_t.$$

In either case, $|E_t| \leq 0.5hl_t$, consistent with the standard model and monitoring is reinitialised with $l_t = 0$ and $L_{1,t} = L_{2,t} = 1$ before proceeding to time $t + 1$. The monitor will signal in favour of one of the alternatives if either Bayes' factor drops below the prespecified threshold τ,

$$L_{1,t} \leq \tau \quad \text{if, and only if,} \quad E_t \geq \quad 0.5hl_t - \frac{\log(\tau)}{h},$$

$$L_{2,t} \leq \tau \quad \text{if, and only if,} \quad E_t \leq -0.5hl_t + \frac{\log(\tau)}{h}.$$

Thus the double monitoring is based on E_t alone, the operation here being equivalent to standard *backward cusum* techniques (Harrison and Davies, 1964). E_t is a cusum, meaning cumulative sum, of the most recent l_t errors. The standard model is accepted as satisfactory at time t so long as E_t satisfies

$$0.5hl_t - \frac{\log(\tau)}{h} \geq E_t \geq -0.5hl_t + \frac{\log(\tau)}{h}.$$

If E_t is so large or so small as to lie outside one or other of these bounds, then the monitor signals in favour of the corresponding alternative model, and intervention is needed. If, on the other hand, E_t lies within the bounds and, in fact, $|E_t| < 0.5hl_t$, then monitoring is reinitialised with $l_t = 0$ and $L_{1,t} = L_{2,t} = 1$ before proceeding to time $t + 1$. Otherwise monitoring proceeds as usual. The bounds for signalling model breakdown or reinitialising the monitor define what are sometimes referred to as *moving V-masks* for the backward cusum (Harrison and Davies, 1964).

EXAMPLE 11.4. A useful single, alternative is the scale shift model M_1 in which e_t has standard deviation k rather than unity, with

$$p_1(e_t \mid D_{t-1}) = (2\pi k^2)^{-1/2} \exp\{-0.5(e_t/k)^2\}.$$

The Bayes' factor at time t is then

$$H_t = k \exp\{-0.5e_t^2(1 - k^{-2})\}.$$

Modelling a scale inflation with $k > 1$ provides a widely useful alternative. It is immediately clear that it is appropriate for changes in observational variance, or *volatility*, in the series. It is also a rather robust alternative for more general changes, being designed as it is to cater for observations that are extreme relative to M_0. The use of such models has a long history in Bayesian statistics, particularly in modelling outliers (Box and Tiao, 1968; and, more recently, Smith and Pettit, 1985). The effects of different values of k can be assessed as in the previous example, by considering indifference and extreme points, this being left to exercises for the reader. With a view to catering for large errors, we draw on experience with outlier models, as just referenced, anticipating that, above a certain level, the particular value of k chosen is largely irrelevant. The key point is that the alternative provides a larger variance than M_0, thus large errors will tend to be more consistent with M_1 no matter just how large the variance inflation is. Some indication of the variation with

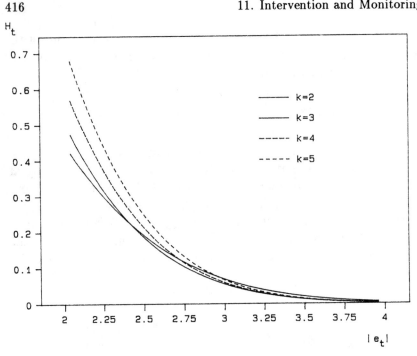

Figure 11.8.   Bayes' factors H_t in normal scale inflation model.

respect to k appears in Figure 11.8. The curves plotted here pro-
vide Bayes' factor as a function of $|e_t|$ over the range 2—4, for $k = $
2, 3, 4 and 5. This range of values of $|e_t|$ focuses attention on the
region in which we would doubt the model M_0. Obviously, evidence
in favour of M_0 decreases as $|e_t|$ increases. Of major interest is the
fact that, when the error becomes extreme relative to M_0, the choice
of k matters little, the Bayes' factors all being close to 0.2 when the
error is 2.5, and below 0.1 when the error is at 3. A scale inflation
of $k = 3$ or $k = 4$ with a threshold of $\tau \approx 0.15$ provides a useful,
general alternative.

In considering specifically the possibilities that, in model M_0, e_t
may be a wild, outlying observation or that the defining parameters
in M_0 may have changed at (or before) time t, then either of these
alternatives may be suitable. The level shift model allows for a jump
in the series in a particular direction, determind by the sign of h,
the scale shift model allows for changes in either direction, and also
for changes in variation. In this sense, the latter alternative may
be viewed as a general alternative, catering for a variety of types of
changes in addition to outlying observations.

11.5 FEED-BACK INTERVENTION

11.5.1 Automatic exception detection and diagnosis

Signalling model breakdown using simple monitoring techniques is the first step in automatic exception handling. Most appropriately, such signals should prompt a user response, suggesting subjective, feed-back interventions to correct M_0 and adapt to new conditions. In automatic mode, the monitoring of model performance to *detect* deterioration in predictions needs to be supplemented with techniques for *diagnosis* of the problem and, subsequently, with *adaptation* to control and correct for the problem. When changes in parameters values are the primary causes of breakdown of M_0, adaptation is required rapidly to improve future predictions. The possibility of outlying observations confuses the diagnosis issue; on the basis of a single, discrepant observation and no further information, parametric change is indistinguishable from a wild data point. Generally, if an outlier can be identified as such, the policy of omitting the observation from the analysis, treating it as a missing value, is a simple and sound one. The updating of distributions in M_0 will then be unaffected by the wild value, and possible causes and consequences can be explored separately. The following, logical scheme provides a guide to the use of the Bayes' factors in detecting and diagnosing model breakdown in such circumstances.

At time t, whatever has occurred previously, proceed with the monitor as follows.

(A) Calculate the single Bayes' factor H_t. If $H_t \geq \tau$, then Y_t is viewed as consistent with M_0; proceed to (B) to assess the possibilities of model failure (ie. marked changes in parameters) prior to time t. If, on the other hand, $H_t < \tau$, then Y_t is a *potential* outlier and should be omitted from the analysis, being treated as a missing value (from the point of view of updating and revising M_0). However, the possibility that Y_t presages a changes in model parameters must be catered for after rejecting the observation, thus the need for intervention is signalled and we proceed to (C).

(B) Calculate the cumulative Bayes' factor L_t and the corresponding run-length l_t to assess the possibility of changes prior to time t. If $L_t \geq \tau$, then M_0 is satisfactory and so proceed to (D) to perform standard updates etc. Otherwise, $L_t < \tau$ indicates change that should be signalled, requiring intervention;

proceed to (C). Note that the sensitivity of this technique to slow changes can be increased by signalling a possible breakdown of M_0 if either $L_t < \tau$ or $l_t > 3$ or 4, say. The rationale here is that several (3 or 4) recent observations may provide evidence very marginally favouring M_1 over M_0, but this may be so small that L_t, whilst being less than 1, still exceeds τ.

(C) Issue signal of possible changes consistent with deterioration of predictions from M_0 and call for feed-back interventions to adapt the model for the future. Following such interventions, update the time index to $t+1$ for the next observation stage, and proceed to (A), reinitialising monitoring by setting $l_t = 0$ and $L_t = 1$.

(D) Perform usual analysis and updating with M_0, proceeding to (A) at time $t+1$.

With this scheme in mind, it remains to specify the forms of intervention at points of possible changes that are detected. The full range of user interventions of Sections 11.2 are, of course, available. Automatic alternatives for routine use are described in the next section.

11.5.2   Automatic adaptation in cases of parametric change

It is a general principle that the onset of change brings with it increased uncertainties that, if appropriately incorporated in the model, lead naturally to more rapid adaptation in the future. Thus models can be made self-correcting if uncertainties about parameters can be markedly increased at points of suspected change. This is quite consistent with the use of increased variances on model parameters in the various forms of subjective intervention in Section 11.2, and the mode of intervention described there suggests a general, automatic procedure for adapting to changes once detected. The basic idea is simple; on detecting change, the prior variances of model parameters can be increased by the addition of further evolution noise terms to allow further data to more heavily influence the updating to posterior distributions. It is possible to use this automatic mode of intervention retrospectively, feeding in greater uncertainty about parameters at the most likely point of change l_t observations past, although, for forecasting, it currently matters little that the change was not catered for at the time. What is important is a response now to adapt for the future.

With this in mind, consider again the scheme (A) to (D) above, and suppose that we are at (C), having identified that changes beyond those allowed for in M_0 may have occurred. An intervention of the form (11.5), additional evolution noise at the current time, permits greater changes in θ_t and, if the extra evolution variance $\mathbf{H}_t$ is appropriately large, leads to automatic adaptation of M_0 to the changes. Additionally, unless particular parameters are identified by the user as subject to changes in preferred directions, then the additional evolution noise should not anticipate directions of change, therefore having zero mean. Thus the feed-back interventions called for at (C) may be effected simply, automatically and routinely via the representation from Theorem 11.1: $\theta_t = \mathbf{G}_t\theta_{t-1} + \omega_t + \xi_t$ where $\omega_t \sim N[\mathbf{0}, \mathbf{W}_t]$, as usual, being independent of the automatic intervention noise term $\xi_t \sim N[\mathbf{0}, \mathbf{H}_t]$. The scheme (A) to (D) with this form of automatic intervention is represented diagramatically in the flow-chart of Figure 11.9.

The additional variance $\mathbf{H}_t$ must be specified. It is important that $\mathbf{H}_t$ provides increased uncertainty for those parameters most subject to abrupt change, and obviously the user has much scope here to model differing degrees of durability of parameters, just as in designing the standard evolution matrix $\mathbf{W}_t$. It is possible, as an extreme example, to take some of the elements of $\mathbf{H}_t$ to be zero, indicating that no additional changes are allowed in the corresponding parameters of θ_t. As a generally useful approach, the magnitude and structure of $\mathbf{H}_t$ may be modelled on that of the standard evolution variance matrix $\mathbf{W}_t$. In particular, if all parameters are subject to possible changes, an appropriate setting, neutral as to source and magnitude of changes, is to simply take

$$\mathbf{H}_t = (c - 1)\mathbf{W}_t$$

where $c > 1$. This has the effect of an additional noise term with the same covariance structure as the usual ω_t, but with an inflation of the overall variation in θ_t by a factor of c.

A generalisation of this simple technique has been used extensively in connection with the use of discount factors to routinely structure $\mathbf{W}_t$ (West and Harrison, 1986a; Harrison and West, 1986, 1987; West, Harrison and Pole, 1987). Discount factors for components of θ_t provide suitable structure and magnitudes for the standard evolution noise variance matrix $\mathbf{W}_t$, controlling the nature and extent of the basic dynamic. At points of abrupt change that go beyond this

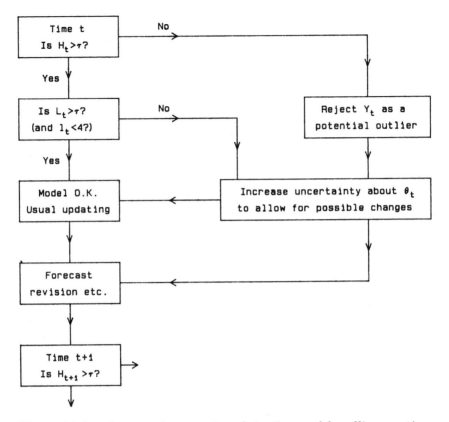

Figure 11.9.   Automatic exception detection and handling routine.

basic dynamic, a more heavily discounted version of the matrix $\mathbf{W}_t$ is an obvious way of simply and automatically extending the model. Thus $\mathbf{H}_t$ may be specified so that $\mathbf{W}_t + \mathbf{H}_t$ is just of that form — based on the discount concept that applies to $\mathbf{W}_t$, but with smaller discount factors than standard. The following section illustrates the use of this in automatic adaptation to abrupt changes.

11.5.3   Industrial sales illustration

Illustration is provided in an analysis combining the use of the above scheme with simple feed-forward intervention based on externally available information.   The series of interest is displayed in Figure 11.10, the data given in Table 11.3. The analysis reported was performed using the **BATS** package (West, Harrison and Pole,

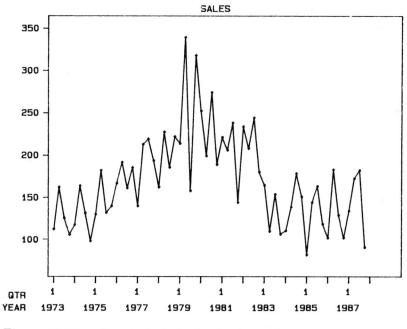

Figure 11.10. Quarterly industrial sales series.

Table 11.3. Quarterly industrial sales data

Year	Quarter			
	1	2	3	4
1973	112.08	162.08	125.42	105.83
1974	117.50	163.75	131.25	97.92
1975	130.00	182.08	131.67	139.58
1976	166.67	191.67	160.83	185.42
1977	139.58	212.88	219.21	193.33
1978	162.00	227.50	185.42	221.92
1979	213.79	339.08	157.58	317.79
1980	251.83	199.08	274.13	188.79
1981	221.17	205.92	238.17	143.79
1982	233.75	207.92	244.17	179.58
1983	164.04	109.58	153.67	106.25
1984	110.42	138.75	178.33	150.83
1985	81.67	144.17	163.33	118.33
1986	101.67	182.92	128.75	101.67
1987	134.17	172.50	182.08	90.42

1987), as were most others throughout the book. The data are quarterly sales figures over a period of years, representing total sales of a chemical product in international industrial markets. Seasonality is evident, though the extent and durability over time of the seasonal component is unclear from the graph, the non-seasonal trend also changing markedly from time to time.

Consider the position of a forecaster at $t = 1$, corresponding to the first quarter of 1973, using a basic first-order polynomial component DLM for the trend in sales, plus a full seasonal effects component. For all t, $\boldsymbol{\theta}_t$ has 6 elements comprising the level growth, and 4 seasonal factors constrained to have zero sum. Suppose the initial prior for $\boldsymbol{\theta}_1$ is specified as

$$(\boldsymbol{\theta}_1|D_0) \sim \mathrm{T}_{20}[\mathbf{a}_1, \mathbf{R}_1]$$

and, for the observational precision,

$$(\phi|D_0) \sim \mathrm{G}[20, 4500],$$

where
$$\mathbf{a}_1 = (130, 0;\ 0, 0, 0, 0)'$$

and

$$\mathbf{R}_1 = 225 \begin{pmatrix} 225 & 0 & 0 & 0 & 0 & 0 \\ 0 & 100 & 0 & 0 & 0 & 0 \\ 0 & 0 & 300 & -100 & -100 & -100 \\ 0 & 0 & -100 & 300 & -100 & -100 \\ 0 & 0 & -100 & -100 & 300 & -100 \\ 0 & 0 & -100 & -100 & -100 & 300 \end{pmatrix}.$$

Thus the initial estimate of $V = 1/\phi$ is $S_0 = 4500/20 = 225$, an estimated observational standard deviation of 15, with 20 degrees of freedom. The expected level of the series is 130, thus the prior on the variance represents a reasonably strong belief that the random variation will have standard deviation of about 10 or 11% of the expected level. $\mathbf{R}_1$ indicates a rather uncertain view about the components initially, particularly in the seasonal component where no pattern is anticipated, with large variances so that the model will rapidly adapt to the data over the first couple of years. The model specification is completed through the values of discount factors. The component discount approach provides $\mathbf{W}_t$ in block diagonal form (like $\mathbf{R}_1$) for all t through the use of two discount factors, one for trend and one for seasonality. Trend and seasonality are viewed to be of similar durability, and this is reflected in the choice of discount factors of 0.95 for each component. In addition, some minor variation over time in V is modelled through the use of an observational variance discount factor of 0.99 (Section 10.8 of Chapter 10).

The model is subject to monitoring and automatic adaptation at points of change identified. At such points, the above, standard values of the discount factors are dropped to lower values consistent with more marked change in the parameters. Note that this automatic shift to smaller, exceptional discount factors applies only at points identified as change points through the routine monitoring scheme of the previous Section. Otherwise the standard values apply. Thus, at points of possible change, the use of smaller discounts simulates a user intervention to increase uncertainty about θ_t and V above and beyond that already modelled through the standard discounts. In the analysis reported here, these exceptional values are 0.1 for the trend and seasonal components, and 0.9 for the observational variance. Concerning the trend and seasonal components, the standard discount of 0.95 implies an increase of roughly 2.5% in the standard deviations of the parameters between observations in the standard analysis. At exceptions, the drop to 0.1 implies roughly a three-fold increase, allowing much more marked adaptation to forthcoming data. Monitoring is based on the use of the single, robust, scale-inflation alternative in Example 11.4. Thus predictive performance of the model under standard conditions is compared through the Bayes' factors with an alternative whose one-step ahead forecast distributions have the same location but increased spread, the forecast standard deviations under the alternative inflated by a factor $k > 1$. The value used for k here is 2.5, the threshold level τ taken as 0.2. Now the one-step forecast distributions are Student T, but the discussion of the case of normal distributions in the examples of the previous Section provides a guide to the implications of these chosen values (and the comments about insensitivity to them apply). With a normal forecast distribution the Bayes' factor H_t based on a single observation will be unity for a standardised forecast error of roughly ± 1.5. For errors larger in absolute value the evidence weighs against the standard model, reaching the threshold $\tau = 0.2$ at standardised errors of roughly ± 2.5. Following the discussion about monitor sensitivity in cases of less marked change, the decision is taken to signal for automatic intervention based on the size of the minimum Bayes' factor L_t and the run-length l_t combined, whenever $L_t < \tau$ or $l_t \geq 3$.

The data is sequentially observed and processed subject to the monitoring and adaptation scheme thus defined. The background to the data series puts the data in context and provides information relevant to modelling the development over time. In the early years of the data, representing the early life of the product in newly created markets, the product sales grow noticeably, the company manufacturing the product essentially dominating the market. In early 1979, however, information becomes available to the effect that a second major company is to launch a directly competitive product later that year. In addition to the direct effect on sales that this is expected to have over coming years, it is anticipated that it will also encourage other competing companies into the market. Thus change in trend, and possibly seasonality, are expected. For the analysis here, however, such information is not fed-forward, all interventions to adapt to observed changes being left to the automatic scheme. One feed-forward intervention is made, however, on the basis of company actions taken immediately on receiving the information about the competitor. In an attempt to promote the existing product to consumers and to combat the initial marketing of the competitor, the company launches a marketing drive in early 1979 that includes price reductions throughout that year. The market response to this is anticipated to lead to a major boost in sales but there is great uncertainty about the response, the effects of the promotional activitities possibly continuing to the end of the year. In order that the underlying market trend, including competitor effects, be estimated as accurately as possible free from the effects of this promotion, some form of intervention is necessary. The simplest such intervention is to assume that the effects of the promotions lead to observations that are essentially uninformative about the underlying market trend and seasonal pattern, and the decision is taken simply to omit the observations in the last three quarters of 1979 when the promotions are expected to significantly impact on sales. Thus these three observations are effectively ignored by the analysis, treated as missing values.

As the analysis proceeds from $t = 0$, updating as usual, the values of L_t and l_t are calculated from the simple recursions in Theorem 12.3. The logged values of the former are plotted against t in Figure 11.11. There are six points at which the monitor signals.

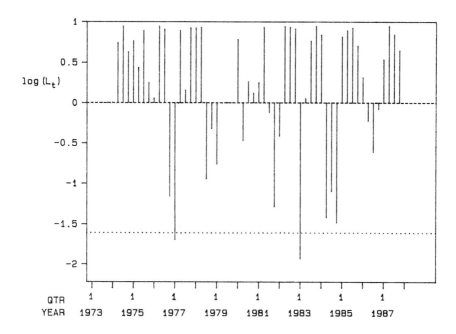

Figure 11.11.   Tracking signal for industrial sales model.

(1) The first point at which evidence weighs against the standard model is at $t = 16$, where $L_{16} = H_{16}$ is less than unity for the first time. The monitor does not signal, however, since it remains above the threshold and $l_{16} = 1$. At $t = 17$, corresponding to the fourth quarter of 1977, $H_{17} < 1$ too, and $L_{17} = H_{16}H_{17} < \tau$, signalling breakdown based on two observations, $l_{17} = 2$.

(2) After automatic intervention using the exceptional discount factors at $t = 17$, $L_t > 1$, so that $l_t = 1$, until $t = 23$. At $t = 25$, the first quarter of 1979, L_{25} exceeds τ but $l_{25} = 3$ and the monitor signals. Three consecutive observations provide evidence against the model which, though not so extreme as to cross the chosen threshold, is viewed as requiring intervention. Note that this use of the run-length in addition to the Bayes' factor may be alternatively modelled by directly linking the threshold to the run-length.

(3) A similar exception occurs at $t = 37$, the first quarter of 1982, again based on the run-length.

(4) At $t = 41$, the single observation Y_{41} is discrepant enough so that $H_{41} < \tau$ and the monitor signals with $l_{41} = 1$.

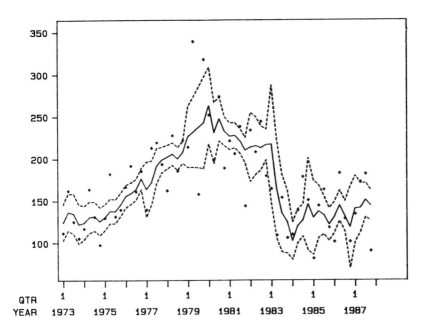

Figure 11.12.   On-line estimated trend in industrial sales.

(5) At times $t = 46, 47$ and 48 L_t is low, close to the threshold though not dropping below it. There is clear evidence here accumulating against the model and the monitor signals when $l_{48} = 3$, in the fourth quarter of 1984.

(6) Finally, the run-length reaches 3 again at $t = 56$, the fourth quarter of 1986, and an exception is signalled. In contrast to the previous exception, note that the evidence here is only marginally against the standard model, certainly compared to that at the preceeding exception.

These events are interpreted and discussed with reference to the graphs in Figures 11.12 to 11.17. Figure 11.12 displays the usual on-line estimate, with two standard deviation interval, of the non-seasonal trend component at each time t. The adequacy of the locally linear trend description is apparent, with marked changes in the trend at three points of intervention. The trend apparently grows roughly linearly up to the start of the 1980s, plateaus off for a few years, drops markedly and rapidly during 1983 from levels near 200 to between 100 and 150, remaining at these, lower levels thereafter. Uncertainty about this on-line estimated trend is fairly high, particularly after the crash to lower levels in 1983. At the intervention times,

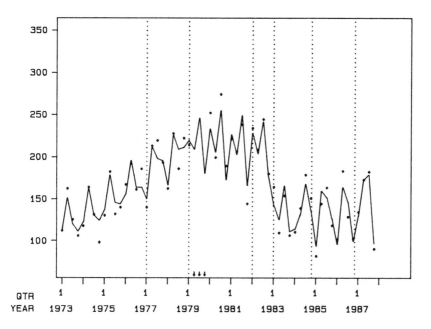

Figure 11.13.   Retrospective fitted values of industrial sales.

the interval about the estimated trend increases in width due to the intervention, allowing the model to adapt to any changes. In 1979, the intervention in the first quarter has this effect, and is followed by three quarters in which, due to the feed-forward intervention to isolate the effects of the company promotions, the observations are ignored. As a consequence, the uncertainty about the trend, and the other model components, increases throughout 1979, no information being obtained.

As always, the on-line estimates of model components can appear rather erratic, responding as they do to data as it is processed. The interventions increase this response and so on-line trajectories provide only a first, tentative indication of the pattern of behaviour over time. More appropiate for retrospective analysis are the smoothed and filtered distributions for model components calculated using the filtering algorithms backwards over time from the end of the series. Figure 11.13 provides a plot of the data (with the three observations omitted from analysis removed and indicated on the time axis), together with a plot of the filtered estimates of the mean response function. The latter, appearing as a line through the data over time, provide smoothed or retrospectively fitted values of the sales figures.

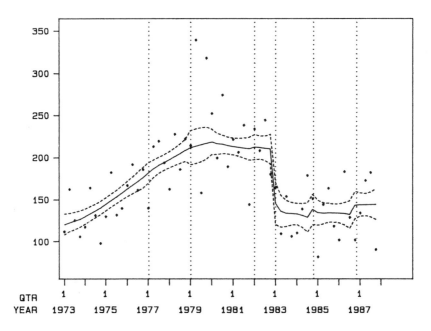

Figure 11.14.   Filtered trend in industrial sales.

The points of intervention are indicated by vertical dotted lines. Note that uncertainty about the smoothed values is not indicated on the graph so as not to obscure the main features noted here. Uncertainties are indicated, in terms of 2 standard deviation intervals, in the corresponding plots for the trend and seasonal components appearing in

Figures 11.14 and 11.15. Additionally, the seasonal form is decomposed into the two harmonic components and filtered estimates of each of these, with intervals, appear in Figures 11.16 and 11.17.

Figure 11.14, the smoothed version of Figure 11.12, indicates that the non-seasonal trend in sales grows steadily until 1979, then flattens off at around 220, decaying very slightly, until the end of 1982. At this point there is a huge drop in level to around 130, growth therafter being small though increasing towards the end of the series in late 1986. Consider now the effects of the automatic interventions that resulted form the monitor signals. At each of these six times, the exceptional discount factors used allow for abrupt change in all model components. The trend, however, changes markedly only at the beginning of 1983. Clearly the monitor signal here was vital in terms

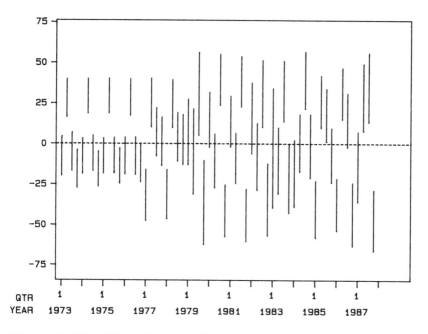

Figure 11.15.   Filtered seasonal component.

of rapidly adapting to this change. The earlier signals do not appear to have led to the trend adapting, so that they are attributable
to some other feature of model:data mismatch. The intervention in
early 1979, followed by further increased uncertainty in the model
due to the three omitted observations in that year, inflates uncertainty sufficiently that the levelling off in trend is identified and appropriately estimated without further interventions. This behaviour
may, in retrospect, be attributed to the competitor incursion into
the market. The huge drop in sales level in early 1983 bears study,
possibly relating to several factors. Company decisions to withdraw
from part of the market is one possibility which, if true, could and
should have led to further feed-forward intervention to anticipate
the change. Further, major and sustained incursion by competitors
is another. The international industrial recession is a third, important consideration, which could easily explain the crash in sales level
through the withdrawal of demand from one or a small number of
consumer companies who had hitherto comprised a major part of the
market. Whatever reasons are considered retrspectively, the model,
operating without feed-forward information, identifies and appropriately estimates the crash. The final two interventions in 1984 and

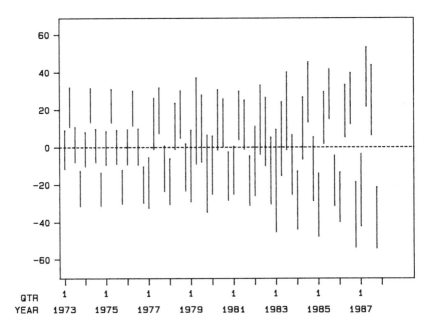

Figure 11.16. Filtered first harmonic.

1986 appear to be attributable, at least in part, to minor, though important, increases in non-seasonal trend.

The form of the seasonal pattern and its changes over time are evident from the fitted sales values in Figure 11.13 and the seasonal component in Figure 11.15, together with the harmonics in Figures 11.16 and 11.17. From Figure 11.13, the seasonal form appears stable up to the first intervention point in early 1977. Here, and again at the second intervention in early 1979, the form of seasonality apparently changes. In particular, the contribution of the second harmonic is more marked after each intervention, very noticeably so after that in 1979. After the drop in level in 1983, seasonality again changes noticeably, the second harmonic becoming again less important whilst the amplitude of the first is much increased.

In retrospect, the three observations omitted from the analysis can now be clearly explained as the response to promotion. The first of these, in the second quarter of 1979, is a huge peak in sales resulting from the price cutting promotional activity. In the third quarter, sales drop off, partially due to seasonal demand but probably also due to stocks having been depleted in the previous quarter. Sales

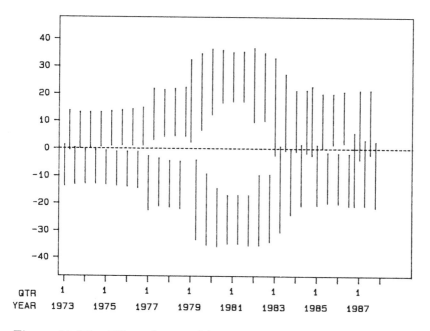

Figure 11.17. Filtered second harmonic.

bounce back to high levels in the final quarter as supplies come through again.

Figure 11.18 displays step ahead forecast sales, with 2 standard deviation intervals, made for each quarter in 1988 and 1989 from the end of the series in the fourth quarter of 1987. Here the effects of the later interventions show up in the forecasts, particularly through the final shift to higher non-seasonal level in sales.

11.6 EXERCISES

(1) Verify the general results in Theorem 11.1 and Theorem 11.2.

(2) In the first-order polynomial model $\{1, 1, V_t, W_t\}$ with state parameter $\mu_t = \theta_t$, suppose that $(\mu_{t-1}|D_{t-1}) \sim N[m_{t-1}, C_{t-1}]$ as usual. Intervention is to be performed to achieve the prior $(\mu_t|D_{t-1}) \sim N[m_{t-1}, R_t^*]$ where $R_t^* = C_{t-1} + W_t + H_t = R_t + H_t$ for some $H_t > 0$.

This can obviously be done directly through Theorem 11.1 by adding a further evolution noise term $\xi_t \sim N[0, H_t]$ to the existing evolution equation. Alternatively, the more general

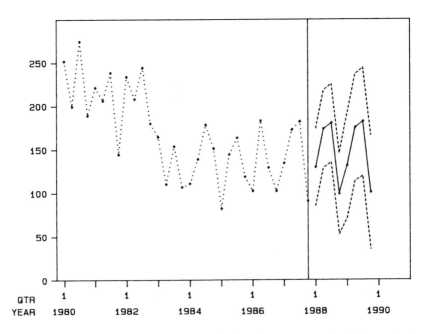

Figure 11.18.   Forecast industrial sales for 1988 and 1989 made in
                 late 1987.

approach through Theorem 11.2 can be used. Show that this
approach leads to the revised evolution equation at time t given
by

$$\mu_t = G_t^* \mu_{t-1} + \omega_t^*, \qquad \omega_t^* \sim N[h_t, W_t^*],$$

where

$$G_t^* = (1 + H_t/R_t)^{1/2},$$
$$h_t = (1 - G_t^*)m_{t-1},$$

and

$$W_t^* = (1 + H_t/R_t)W_t.$$

(3) Suppose that $\mathbf{R}$ is a 2×2 variance matrix,

$$\mathbf{R} = \begin{pmatrix} R_1 & R_3 \\ R_3 & R_2 \end{pmatrix}.$$

Let $\mathbf{Z}$ be the unique, upper triangular matrix

$$\mathbf{Z} = \begin{pmatrix} Z_1 & Z_3 \\ 0 & Z_2 \end{pmatrix}$$

satisfying $\mathbf{R} = \mathbf{Z}\mathbf{Z}'$. Prove that $\mathbf{Z}$ is given by

$$Z_1 = (R_1 - R_3^2/R_2)^{1/2},$$
$$Z_2 = R_2^{1/2}$$

and

$$Z_3 = R_3/R_2^{1/2}.$$

(4) Consider Example 11.1 in Section 11.2.3, supposing that, at $t = 37$ (January 1958), we have

$$\mathbf{a}_{37} = \begin{pmatrix} 865 \\ 0 \end{pmatrix} \quad \text{and} \quad \mathbf{R}_{37} = \begin{pmatrix} 100 & 10 \\ 10 & 2 \end{pmatrix}.$$

A forecaster considers five possible interventions, leading to the five possible post-intervention moment pairs $\mathbf{a}_{37}^*$ and $\mathbf{R}_{37}^*$ in (a)—(e) below. In each case, use the result of the previous exercise to calculate the quantities $\mathbf{G}_t^*$, $\mathbf{h}_t$ and $\mathbf{W}_t^*$ in the amended evolution equation as in Theorem 11.2.

(a)
$$\begin{pmatrix} 970 \\ 0 \end{pmatrix} \quad \text{and} \quad \begin{pmatrix} 100 & 10 \\ 10 & 2 \end{pmatrix}.$$

(b)
$$\begin{pmatrix} 970 \\ 0 \end{pmatrix} \quad \text{and} \quad \begin{pmatrix} 100 & 0 \\ 0 & 0 \end{pmatrix}.$$

(c)
$$\begin{pmatrix} 970 \\ 0 \end{pmatrix} \quad \text{and} \quad \begin{pmatrix} 100 & 0 \\ 0 & 25 \end{pmatrix}.$$

(d)
$$\begin{pmatrix} 970 \\ 0 \end{pmatrix} \quad \text{and} \quad \begin{pmatrix} 100 & 25 \\ 25 & 25 \end{pmatrix}.$$

(e)
$$\begin{pmatrix} 970 \\ 0 \end{pmatrix} \quad \text{and} \quad \begin{pmatrix} 50 & 5 \\ 5 & 2 \end{pmatrix}.$$

(5) A quarterly seasonal effects component of a DLM at time t has the usual normal prior distribution $(\boldsymbol{\theta}_t|D_{t-1}) \sim \mathrm{N}[\mathbf{a}_t, \mathbf{R}_t]$ with

$$\mathbf{a}_t = \begin{pmatrix} 25 \\ 131 \\ -90 \\ -66 \end{pmatrix} \quad \text{and} \quad \mathbf{R}_t = \begin{pmatrix} 12 & -4 & -4 & -4 \\ -4 & 12 & -4 & -4 \\ -4 & -4 & 12 & -4 \\ -4 & -4 & -4 & 12 \end{pmatrix}.$$

(a) Verify that this prior is consistent with the zero-sum constraint $\mathbf{1}'\boldsymbol{\theta}_t = 0$.

(b) A forecaster considers it likely that the peak level in the
second quarter can be expected to increase from 131 to
about 165 before the next observation, but that there is
additional uncertainty about this change. This is modelled
by an additional evolution noise term and the moments are
directly revised to

$$\mathbf{a}_t^* = \begin{pmatrix} 25 \\ 165 \\ -90 \\ -66 \end{pmatrix} \quad \text{and} \quad \mathbf{R}_t^* = \begin{pmatrix} 28 & -4 & -4 & -4 \\ -4 & 12 & -4 & -4 \\ -4 & -4 & 12 & -4 \\ -4 & -4 & -4 & 12 \end{pmatrix}.$$

Show that this revised prior does not now satisfy the zero-
sum constraint, and apply Theorem 8.2 of Chapter 8 to
appropriately correct it.

(6) In monitoring based on the scale inflation alternative of Example 11.4, define E_t to be the sum of squares of the most recent
l_t forecast errors at time t,

$$E_t = \sum_{r=0}^{l_t-1} e_{t-r}^2.$$

(a) Let x be any real quantity such that $0 < x \leq 1$. Prove
that $\log(L_t/x) \leq 0$ if, and only if,

$$E_t \geq al_t + b(x)$$

where

$$a = 2k^2\log(k)/(k^2 - 1) \quad \text{and} \quad b(x) = -2k^2x/(k^2 - 1).$$

Verify that a and b are both positive.

(b) Using (a), prove that $L_t \leq 1$, so that evidence weighs
against the standard model, when $l_t^{-1}E_t \geq a$. Comment
on this result.

(c) Show also that the monitor signals model failure when
$E_t \geq al_t + b(\tau)$.

(7) In Exercise 6, suppose that the data actually follow the standard
model being monitored so that the errors e_t are independent,
standard normally distributed random quantities.

(a) Deduce that $E_t \sim \chi^2_{l_t}$, a chi-square random quantity with l_t degrees of freedom.

(b) With $k = 2.5$ and $\tau = 0.2$, calculate the probability that $L_t \leq 1$ for each of $l_t = 1, 2, 3$ and 4. Comment on meaning of these probabilities.

(c) Recalculate the above probabilities for each combinations of pairs of values of k and τ given by $k = 2$ or 3 and $\tau = 0.1, 0.2$ and 0.3. Comment on the sensitivity of the derived probabilities as functions of k and τ and the implications for the monitor in practice.

(8) In Exercises 6 and 7, suppose now that, in fact, $e_t \sim N[0, k^2]$ for all t.

(a) Deduce the distribution of E_t.

(b) Perform the probability calculations in (e) above and comment on the meaning of these probabilities.

(9) Following Example 11.3, describe how the scale inflation alternative can be extended to two alternatives, with scale factors $k > 1$ and k^{-1} respectively, to monitor for the possibilities of either increased or decreased variation.

(10) Write a computer program to implement the monitoring and automatic adaptation routine described in Section 11.5, based on the scale inflation alternative model of Example 11.4 for the monitor. This should apply to any component DLM using component discounting, the automatic adaptation being modelled through the use of exceptional discount factors as in Sections 11.5.2 and 11.5.3. Test the program by performing the analysis of the industrial sales data in 11.5.3, reproducing the results there.

(11) Analyse the CP6 sales data in Table 11.1 and Figure 11.1 using a first-order polynomial model with a single discount factor $\delta = 0.9$ for the trend component, and a variance discount factor $\beta = 0.95$. Apply the automatic monitoring as programmed in the previous example with exceptional trend and variance discount factors of 0.05 and 0.5 respectively.

(12) Reanalyse the CP6 sales data as in the previous example. Instead of discounting the observational variance to account for changes, explore models that, rather more appropriately, use variance laws as in Section 10.7.2 of Chapter 10. In particular, consider models in which the observational variance at time t is estimated by the weighted variance $V_t = k(f_t)V$, where V is unknown as usual and $k(f) = f^p$ for some $p > 1$. Explore the

relative predictive fit of models with values of p in the range $1 \leq p \leq 2$.

(13) Reanalyse the confectionary sales and cost series from Exercise 7 of Section 10.9, Chapter 10 subject to monitoring and adaptation using the program from Exercise 7 above (This data is analysed using a similar model in West, Harrison and Pole (1987), with extensive summary information from the analysis provided there).

CHAPTER 12

MULTI-PROCESS MODELS

12.1 INTRODUCTION

Discussion of interventionist ideas and monitoring techniques is an implicit acknowledgement of the principle that, although an assumed DLM form may be accepted as appropriate for a series, the global behaviour of the series may only be adequately mirrored by allowing for changes from time to time in model parameters and defining structural features. Intervention allows for parametric changes, and goes further by permitting structural changes to be made to the defining quadruple $\{\mathbf{F}, \mathbf{G}, V, \mathbf{W}\}_t$. In using automatic monitoring techniques, as in Section 11.4 of Chapter 11, the construction of specific alternative models rather explicity recognises the global inadequacy of any single DLM, introducing possible explanations of the inadequacies. The simple use of such alternatives to provide comparison with a standard, chosen model is taken much further in this chapter. We formalise the notion of explicit alternatives by considering classes of DLMs, the combination of models across a class providing an overall super-model for the series. This idea was originally developed in Harrison and Stevens (1971), and taken further in Harrison and Stevens (1976a). We refer to such combinations of basic DLMs as *multi-process* models; any single DLM defines a process model, the combination of several defines a multi-process model. Loosely speaking, the combining is effected using discrete probability mixtures of DLMs, and so multi-process models may be alternatively referred to simply as mixture models. Following Harrison and Stevens (1976a), we distinguish two classes of multi-process models, Class I and Class II, that are fundamentally different in structure and serve rather different purposes in practice. The two classes are formally defined and developed throughout the chapter after first providing a general introduction.

Generically, we have a DLM defined by the usual quadruple at each time t, here denoted by

$$M_t \quad : \quad \{\mathbf{F}, \mathbf{G}, V, \mathbf{W}\}_t,$$

conditional on initial information D_0. Suppose that any defining parameters of the model that are possibly subject to uncertainty are denoted by $\boldsymbol{\alpha}$. Examples include discount factors, transformation parameters for independent variables, eigenvalues of $\mathbf{G}$ in the case when $\mathbf{G}_t = \mathbf{G}$, constant for all time, and so forth. Previously, all such quantities, assumed known, had been incorporated in the initial information set D_0 and therefore not made explicit. Now we are considering the possibility that some of these quantities are uncertain, and so we explicity include them in the conditioning of all distributions in the model analysis, reserving the symbol D_0 for all other, known and certain quantities. Represent the dependence of the model on these uncertain quantities by writing

$$M_t \;=\; M_t(\boldsymbol{\alpha}), \qquad\qquad (t = 1, 2, \dots).$$

Note also that the initial prior in the model may also depend on $\boldsymbol{\alpha}$ although this possibility is not specifically considered here. For any given value of $\boldsymbol{\alpha}$, $M_t(\boldsymbol{\alpha})$ is a standard DLM for each time t. It is the possibility that we do not precisely know, or are not prepared to assume that we know, the value of $\boldsymbol{\alpha}$ that leads us to consider multi-process models. Let $\mathcal{A}$ denote the set of possible values for $\boldsymbol{\alpha}$, whether it be uncountably infinite, with $\boldsymbol{\alpha}$ taking continuous values, discrete and finite, or even degenerate at a single value. The class of DLMs at time t is given by

$$\{\, M_t(\boldsymbol{\alpha}) \;:\; \boldsymbol{\alpha} \in \mathcal{A} \,\}. \qquad\qquad (12.1)$$

The two, distinct possibilities for consideration are as follows.

(I) For some $\boldsymbol{\alpha}_0 \in \mathcal{A}$, $M_t(\boldsymbol{\alpha}_0)$ holds for all t.   (12.2a)

(II) For some sequence of values $\boldsymbol{\alpha}_t \in \mathcal{A}$, $(t = 1, 2, \dots ,)$

$M_t(\boldsymbol{\alpha}_t)$ holds at time t.   (12.2b)

In (I), a single DLM is viewed as appropriate for all time, but there is uncertainty as to the "true" value of the defining parameter vector $\boldsymbol{\alpha} = \boldsymbol{\alpha}_0$. In (II), by contrast, and usually more realistically, there is no single DLM accepted as adequate for all time. The possibility that different models are appropriate at different times is explicitly recognised and modelled through different defining parameters $\boldsymbol{\alpha}_t$.

EXAMPLE 12.1. Suppose that $\boldsymbol{\alpha}$ is a set of discount factors used to determine the evolution variance matrices $\mathbf{W}_t$. Under (I), a fixed set of discount factors is assumed as appropriate, although just which set of values is uncertain. Sometimes this assumption, leading to a stable and sustained degree of variation in model parameters $\boldsymbol{\theta}_t$ over time, is tenable. This might be the case, for example, with time series observed in physical or electronic systems with stable driving mechanisms. More often in socio-economic areas, this assumption will be plausible only over rather short ranges of time, these ranges being interspersed by marked discontinuities in the series. To cater adequately for such discontinuities within an existing DLM, the parameters $\boldsymbol{\theta}_t$ must change appropriately and this may often be effected by altering discount factors temporarily. The interventionist ideas of Chapter 11 are geared to this sort of change, and (II) above formalises the notion in terms of alternative models.

Other examples include unknown power transformation parameters, and, analogously, variance power law indices (see Section 10.6). Whichever form of model uncertainty, (12.2a) or (12.2b), is assumed to be appropriate, the basic analysis of multi-process models rests upon manipulation of collections of models within a discrete mixture framework. The basics are developed in the next section in connection with the first approach, (I) of (12.2a).

12.2 MULTI-PROCESS MODELS: CLASS I

12.2.1 General framework

The basic theory underlying mixture models is developed in the context of the introduction under assumption (I) of (12.2a). Much of this follows Harrison and Stevens (1976a), and general background material can be found in Titterington, Smith and Makov (1985).

Thus Y_t follows a DLM $M_t(\boldsymbol{\alpha})$, the precise value $\boldsymbol{\alpha} = \boldsymbol{\alpha}_0$ being uncertain. The following components of analysis are basic.

(1) Given any particular value $\boldsymbol{\alpha} \in \mathcal{A}$, the DLM $M_t(\boldsymbol{\alpha})$ may be analysed as usual, producing sequences of prior, posterior and forecast distributions that are sequentially updated over time as data is processed. The means, variances and other features of these distributions all depend, usually in complicated ways, on the specific value $\boldsymbol{\alpha}$ under consideration. We make this dependence explicit in the conditioning of distributions

and densities. At time t, let $\mathbf{X}_t$ be any vector of random quantities of interest; for example, the full $\boldsymbol{\theta}_t$ state vector, a future observation Y_{t+k}, $(k > 0)$, etc. Inference about $\mathbf{X}_t$ in $M_t(\boldsymbol{\alpha})$ is based on the density

$$p(\mathbf{X}_t \mid \boldsymbol{\alpha}, D_t). \tag{12.3}$$

Many such densities may exist, one for each $\boldsymbol{\alpha} \in \mathcal{A}$.

(2) Starting with an initial prior density $p(\boldsymbol{\alpha} \mid D_0)$ for $\boldsymbol{\alpha}$, information is sequentially processed to provide inferences about $\boldsymbol{\alpha}$ via the posterior $p(\boldsymbol{\alpha} \mid D_t)$ at time t. This is sequentially updated, as usual, using Bayes' theorem,

$$p(\boldsymbol{\alpha} \mid D_t) \propto p(\boldsymbol{\alpha} \mid D_{t-1})p(Y_t \mid \boldsymbol{\alpha}, D_{t-1}). \tag{12.4}$$

Here $p(Y_t \mid \boldsymbol{\alpha}, D_{t-1})$ is the usual, one-step ahead predictive density from $M_t(\boldsymbol{\alpha})$, simply (12.3) with $\mathbf{X}_t = Y_t$. The posterior $p(\boldsymbol{\alpha} \mid D_t)$ informs about $\boldsymbol{\alpha}$, identifying interesting values and indicating the relative support, from the data and initial information, for the individual DLMs $M_t(\boldsymbol{\alpha})$, $\quad(\boldsymbol{\alpha} \in \mathcal{A})$.

(3) To make inferences about $\mathbf{X}_t$ without reference to any particular value of $\boldsymbol{\alpha}$, the required unconditional density is

$$p(\mathbf{X}_t \mid D_t) = \int_{\mathcal{A}} p(\mathbf{X}_t \mid \boldsymbol{\alpha}, D_t)p(\boldsymbol{\alpha} \mid D_t)d\boldsymbol{\alpha}, \tag{12.5}$$

or, simply, the expectation of (12.3) with respect to $\boldsymbol{\alpha}$ having density (12.4). If, for example, $\mathbf{X}_t = Y_{t+1}$, then (12.5) is the one-step forecast density for Y_{t+1} at time t.

This is, in principle, how we handle uncertain parameters $\boldsymbol{\alpha}$. In practice, all is not so straightforward since, generally, we do not have a tractable, easily calculated and manageable density in (12.4). The calculations implicity required are as follows.

(i) For each $\mathbf{X}_t$ of interest, calculate (12.3) for all possible values of $\boldsymbol{\alpha}$.

(ii) For each $\boldsymbol{\alpha}$, calculate (12.4), and then integrate over $\mathcal{A}$ to normalise the posterior density $p(\boldsymbol{\alpha} \mid D_t)$.

(iii) For each $\mathbf{X}_t$ of interest, perform the integration in (12.5), and the subsequent integrations required to find posteriors moments and probabilities etc. for $\boldsymbol{\alpha}$.

Unfortunately, we will typically not now have a neat set of sequential updating equations for these calculations. The parameters in α will enter into the likelihood $p(Y_t \mid \alpha, D_{t-1})$ in (12.4) in such a complicated way that there will be no conjugate form for the prior for α, no simple summary of D_{t-1} for α in terms of a fixed and small number of sufficient statistics. Thus all the required operations in (i) to (iii) must be done numerically. Generally this will be a daunting task if the parameter space A is at all large, and impossible to carry out if, in particular and often the case, α is continuous. The computations are really only feasible when the parameter space is discrete and relatively small. Two cases arise, namely when either (a) α truly takes only a small number of discrete values, and (b) when a discrete set of values are chosen as representative of a large and possibly continuous "true" space A. Under (b), large spaces A can often be adequately approximated for some purposes by a fairly small discrete set that somehow span the larger space, leading to the consideration of a small number of distinct DLMs. We now consider the use of such collections of models, whether they be derived under (a) or (b).

12.2.2 Definitions and basic probability results

Definition 12.1. *Suppose that, for all t, $M_t(\alpha)$ holds for some $\alpha \in A$, the parameter space being the finite, discrete set $A = \{\alpha_1, \ldots, \alpha_k\}$ for some integer $k \geq 1$. Then the series Y_t is said to follow a* **Multi-process, Class I** *model.*

The general theory outlined in the previous section applies here with appropriate specialisation the discretisation of A. Thus the densities for α in (12.4) and (12.5) are mass functions. For convenience, the notation is simplified in the discrete case as follows.

Definition 12.2.

(i) $p_t(j)$ is the posterior mass at $\alpha = \alpha_j$ in (12.4), namely

$$p_t(j) = p(\alpha_j \mid D_t) = \Pr[\alpha = \alpha_j \mid D_t], \qquad (j = 1, \ldots, k)$$

for all t, with specified initial prior probabilities $p_0(j)$.

(ii) $l_t(j)$ is the value of the density $p(Y_t \mid \alpha_j, D_{t-1})$, providing the likelihood function for α in (12.4) as j varies,

$$l_t(j) = p(Y_t \mid \alpha_j, D_{t-1}), \qquad (j = 1, \ldots, k)$$

for each t.

The discrete versions of (12.4) and (12.5) are as follows.

- *The posterior masses are updated via $p_t(j) \propto p_{t-1}(j)l_t(j)$, or*

$$p_t(j) = c_t p_{t-1}(j)l_t(j) \tag{12.6}$$

where $c_t^{-1} = \sum_{j=1}^{k} p_{t-1}(j)l_t(j)$. Note that c_t^{-1} is just the observed value of the unconditional predictive density for Y_t, namely

$$c_t^{-1} = \sum_{j=1}^{k} p(Y_t \mid \boldsymbol{\alpha}_j, D_{t-1})p_{t-1}(j).$$

- *Marginal posterior densities are now simply*

$$p(\mathbf{X}_t \mid D_t) = \sum_{j=1}^{k} p(\mathbf{X}_t \mid \boldsymbol{\alpha}_j, D_t)p_t(j). \tag{12.7}$$

From (12.7) it follows that all posterior distributions for linear functions of $\boldsymbol{\theta}_t$, and predictive distributions for future observations, are *discrete probability mixtures* of the standard T or normal distributions. Some basic features of such mixtures, fundamental to their use in inference, are as follows.

(a) Probabilities are calculated as discrete mixtures. Generally, for any set of values of interest $\mathbf{X}_t \in \mathcal{X}$,

$$\Pr[\mathbf{X}_t \in \mathcal{X} \mid D_t] = \sum_{j=1}^{k} \Pr[\mathbf{X}_t \in \mathcal{X} \mid \boldsymbol{\alpha}_j, D_t]p_t(j),$$

the conditional probabilities $\Pr[\mathbf{X}_t \in \mathcal{X} \mid \boldsymbol{\alpha}_j, D_t]$ being calculated from the relevant T or normal distribution as usual.

(b) Moments are similar mixtures, thus

$$E[\mathbf{X}_t \mid D_t] = \sum_{j=1}^{k} E[\mathbf{X}_t \mid \boldsymbol{\alpha}_j, D_t]p_t(j),$$

with the conditional moments calculated from the relevant T or normal.

(c) Particular results of use derived from (b) are as follows. Let $\mathbf{h}_t = E[\mathbf{X}_t \mid D_t]$ and $\mathbf{H}_t = V[\mathbf{X}_t \mid D_t]$. For $j = 1, \dots, k$, let $\mathbf{h}_t(j) = E[\mathbf{X}_t \mid \boldsymbol{\alpha}_j, D_t]$, and $\mathbf{H}_t(j) = V[\mathbf{X}_t \mid \boldsymbol{\alpha}_j, D_t]$. Then we have:

- Posterior means:

$$\mathbf{h}_t = \sum_{j=1}^{k} \mathbf{h}_t(j) p_t(j),$$

- Posterior variances:

$$\mathbf{H}_t = \sum_{j=1}^{k} \{\mathbf{H}_t(j) + [\mathbf{h}_t(j) - \mathbf{h}_t][\mathbf{h}_t(j) - \mathbf{h}_t]'\} p_t(j).$$

In previous chapters we have used Bayes' factors in model comparison. The formalism of alternatives within the multi-process framework here brings Bayes' factors into the picture as follows. Consider any two elements of $\mathcal{A}$, namely $\boldsymbol{\alpha}_i$ and $\boldsymbol{\alpha}_j$ for some i and j. From (12.6) it follows that

$$\frac{p_t(i)}{p_t(j)} = \frac{p_{t-1}(i)}{p_{t-1}(j)} \frac{l_t(i)}{l_t(j)} = \frac{p_{t-1}(i)}{p_{t-1}(j)} H_t(\boldsymbol{\alpha}_i, \boldsymbol{\alpha}_j),$$

where, in an extension of the earlier notation for Bayes' factors,

$$H_t(\boldsymbol{\alpha}_i, \boldsymbol{\alpha}_j) = \frac{l_t(i)}{l_t(j)} = \frac{p(Y_t \mid \boldsymbol{\alpha}_i, D_t)}{p(Y_t \mid \boldsymbol{\alpha}_j, D_t)},$$

just the Bayes' factor for $\boldsymbol{\alpha} = \boldsymbol{\alpha}_i$ relative to $\boldsymbol{\alpha} = \boldsymbol{\alpha}_j$ based on the single observation Y_t. Thus, at each observation stage, the ratio of posterior probabilities for any two values of $\boldsymbol{\alpha}$, hence any two of the DLMs in the multi-process mixture, is modified through multiplication by the corresponding Bayes' factor. As information builds up in favour of one value, the Bayes' factors for that value relative to each of the others will increase, resulting in an increased posterior probability on the corresponding DLM.

12.2.3   Discussion and illustration

In application forecasts and decisions will usually be based on information relating to the entire multi-process model, inferences from individual DLMs $M_t(\boldsymbol{\alpha}_j)$ in the mixture being combined, as detailed above, in proportion to their current posterior model probabilities $p_t(j)$. It is worth reiterating that a single DLM representation of a

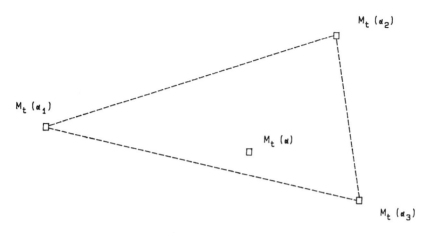

Figure 12.1.   Mixtures of three distinct models.

series is often a rather hopeful and idealistic assumption, although it can be useful if subject to careful user management and monitoring. These comments notwithstanding, the mixture modelling procedure provided by a multi-process, class I model may be used in a manner similar to classical model identification and parameter estimation techniques to identify a single DLM, or a restricted class of them, for future use. The posterior probabilities across $\mathcal{A}$ identifying interesting, supported values of α, those with low weight contributiing little to the multi-process. It is well to be aware that these probabilities can change markedly over time, different DLMs best describing the data series at different times reflecting the global inadequaciy of any single DLM. Under very general conditions, however, the posterior probabilities will converge to zero on all but one value in $\mathcal{A}$, the probability on that value tending to unity. This convergence may take a very large number of observations, but, if suitable in the context of the application, can serve to identify a single value for α and hence a single DLM. The use of a representative set of values $\mathcal{A}$ hopefully suitably spanning a larger, continuous space is the usual setup, and here convergence to a particular value identifies that model closest to the data even though no single model actually generates the series. This basic principle of mixture modelling is schematically illustrated in Figure 12.1, where four DLMs are symbolically displayed. The generic model $M_t(\alpha)$ is seen to lie in the convex hull of the mixture components $M_t(\alpha_j)$, $(j = 1, 2, 3)$, being identified as one possible

DLM represented as a mixture of the three components with respect to particular values of the posterior model probabilities.

If, as is usually the case in practice, no single DLM actually generates the series, the mixture approach allows the probabilities to vary as the data suggests thus adapting to process change.[†]

A rather different context concerns cases when α is simply an index for a class of distinct, and possibly structurally different, DLMs. An example concerns the launch of a new product in a consumer market subject to seasonal fluctuations in demand, the product being one of several or many such similar products. Initially there is no sales data to base forecasts and decisions upon, the only relevant information being subjective, derived from market experience with similar products. In the market of interest, suppose that there are a small number of rather distinct patterns of demand across items, reflected in differences in seasonal patterns. Here α simply indicates which of these structurally different patterns of demand the sales of the new product will follow, it being believed that one and only one such pattern will hold for each product line. The particular DLMs may thus differ generally in structure and dimensions of $\mathbf{F}_t$, $\mathbf{G}_t$. After processing a relatively small amount of data, say several months, major decisions are to be made concerning purchasing of raw materials and stock levels for the new product. At this stage then, it is desirable to choose a particular, single DLM to base such decisions upon, the mixture only being used initially due to the uncertainty about the likely seasonal form.

EXAMPLE 12.2. Reconsider the exchange rate ($ USA/£ UK) index data analysed in Section 2.6 of Chapter 2. The analyses there concerned a first-order polynomial model for the series, various models differing in the value of the defining discount factor being explored. Here that study is extended, the model and initial priors are as used in Section 2.6, to which the reader may refer. The single discount factor defining the evolution variance sequence is δ, viewed here as uncertain so that $\alpha = \delta$. Now δ is a continuous quantity, taking values in the range 0—1, typically between 0.7 and 1. With

---

[†]Sometimes it is appropriate to view a series as adequately described by a DLM that *slowly* varies within the convex hull. Then model probabilities based on past data are less appropriate as time evolves, and it is desirable to intervene between observations to allow for this. One approach is to discount the effect of past data with a discount factor $0 < \beta \leq 1$, adjusting (12.6) to $p_t(j) \propto \{p_{t-1}(j)\}^\beta l_t(j)$. Details are left to the reader.

$k = 4$, the four distinct values $\mathcal{A} = \{1.0, \ 0.9, \ 0.8, \ 0.7\}$ are taken as representing the a priori plausible range of values, hopefully spanning the range of values appropriate for the series. For example, a mixture of models 1 and 2, with discount factors of 1.0 and 0.9 respectively, can be viewed as providing the flexibility to approximate the forms of behaviour of models with discount factors between 0.9 and 1. Initial probabilities are assigned as $p_0(j) = 0.25$, $\qquad (j = 1, \ldots, 4)$, not favouring any particular value in $\mathcal{A}$. The level parameter of the series is $\mu_t = \theta_t$. Conditional on any particular value $\alpha = \alpha_j \in \mathcal{A}$, we have

$$(Y_t \mid \alpha_j, D_{t-1}) \sim \mathrm{T}_{n_{t-1}}[f_t(j), Q_t(j)], \qquad (12.8)$$

and

$$(\mu_t \mid \alpha_j, D_t) \sim \mathrm{T}_{n_t}[m_t(j), C_t(j)], \qquad (12.9)$$

for each $t = 1, 2, \ldots$, where the quantities $f_t(j)$, $Q_t(j)$, etc., are sequentially calculated as usual. Now the dependence on δ is made explicit through the arguments j. Recall that Figure 2.7 of Chapter 2 displays a plot of the data with point forecasts from the models with discount factors of 1.0 and 0.8, models corresponding to $j = 1$ and $j = 3$. In this example, the initial degrees of freedom parameter is $n_0 = 1$ as in Section 2.6, so that $n_t = t + 1$ for each value of j, not differing between the DLMs. Thus in (12.6) we have $p_t(j) \propto p_{t-1} l_t(j)$ with

$$l_t(j) = \frac{\Gamma[(n_{t-1} + 1)/2]}{\Gamma[n_{t-1}/2]\sqrt{n_{t-1}\pi Q_t(j)}} \left\{ 1 + \frac{(Y_t - f_t(j))^2}{n_{t-1}Q_t(j)} \right\}^{-(n_{t-1}+1)/2}$$

Figure 12.2 displays a plot over time of the four model probabilities $p_t(j), (j = 1, \ldots, 4; t = 1, 2, \ldots,)$, from this multi-process model. In examining the time variation in these probabilities the reader should refer also to the plot of the data in Figure 2.7, Chapter 2. Overall, $p_t(2)$ increases up to high values around 0.6, $p_t(3)$ increases and levels off near 0.4, the other two probabilities becoming essentially negliglible. This indicates the suitability of discount values between 0.8 and 0.9, the larger probabilities $p_t(2)$ indicating more support for values nearer 0.9. Certainly model 1, with discount factor of 1, has essentially no support from the data. The extremely adaptive model 4, with discount factor 0.7, has slightly higher probabilities but they are still essentially negligible. In addition to these general conclusions, note the occurrence of one or two rather marked, though transient, changes in the general trend in the probabilities. In early

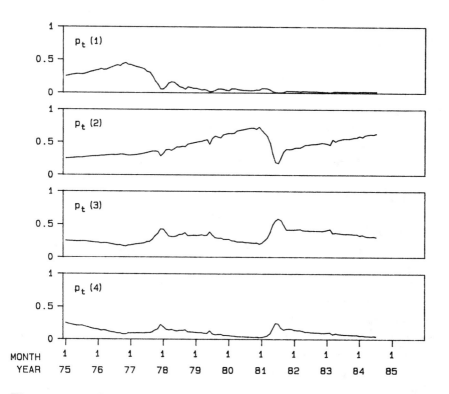

Figure 12.2.   Posterior model probabilities in USA/UK index
example.

1981, in particular, the posterior probabilities on the more adaptive
models 3 and, to a lesser extent, 4, increase at the expense of those on
the generally favoured, though less adaptive, model 2. This continues
for a few months when, in late 1981, $p_t(2)$ begins to grow again to
its previous high levels. The reason for this can be clearly seen in
Figure 2.7. During early 1981, the level of the series drops markedly
over a period of months, clearly deviating from the steady behaviour
anticipated under a first-order polynomial model with a relatively
high discount factor. Models 3 and 4, having lower discount factors
than model 2, respond more rapidly to this drift in level with a
consequent improvement in forecasting performance, as judged by
Bayes' factors, relative to models 1 and 2. Later in the year, model
2 has had time to adapt to the data and the subsequent, steady
behaviour of the series is again more consistent with this model.

This example also serves to illustrate some features of inference with discrete mixtures of distributions that users of multi-process models need to be aware of. In each of the DLMs $M_t(\alpha_j)$, the prior, posterior and forecast distributions are all standard, unimodal distributions, of well known and understood forms, whether they be normal or T (or inverse χ^2 for the observational variance). Inferences use point estimates/ forecasts which are posteriors means or modes, uncertainty measures such as standard deviations, and probabilities that are easily calculated from the standardised forms of the normal or T distributions. If one of these models has a very high posterior probability at the time of interest, then it may be used alone for inference. Otherwise, the formal procedure is to use the full, unconditional mixture provided by (12.7), and the features of such mixtures can be far from standard (Titterington, Smith and Makov, 1985). In particular, simple point forecasts such as means or modes of mixtures can mislead if they are used without further investigation of the shape of the distribution and, possibly, calculation of supporting probabilities, hpd regions, etc. Consider, for example, inferences made at $t = 34$, corresponding to October 1977, about the level of the series $(\mu_{34} \mid D_{34})$ and the next observation $(Y_{35} \mid D_{34})$. The relevant, conditional posterior and forecast distributions, equations (12.8) and (12.9) with $t = 34$ and $n_t = 35$, are summarised in Table 12.1.

Consider first inference about μ_{34}. The individual posterior densities (12.9) summarised in the table are graphed in Figure 12.3. They are clearly rather different, model 1, in particular, favouring negative values of μ_{34} whilst the others favour positive values. More adaptive models with smaller discount factors have more diffuse posteriors. The two central densities, with discounts of 0.9 and 0.8, have probabilities 0.36 and 0.32 respectively, dominating the mixture as is clear from the overall mixture density graphed in the figure. The

Table 12.1. Summary of multi-process model at $t = 34$.

j	$\alpha_j = \delta$	$p_{34}(j)$	$m_{34}(j)$	$\sqrt{C_{34}(j)}$	$f_{35}(j)$	$\sqrt{Q_{35}(j)}$
				μ_{34}	Y_{35}	
1	1.0	0.18	−0.008	0.005	−0.008	0.027
2	0.9	0.36	0.002	0.008	0.002	0.026
3	0.8	0.32	0.009	0.011	0.009	0.027
4	0.7	0.14	0.013	0.013	0.013	0.028

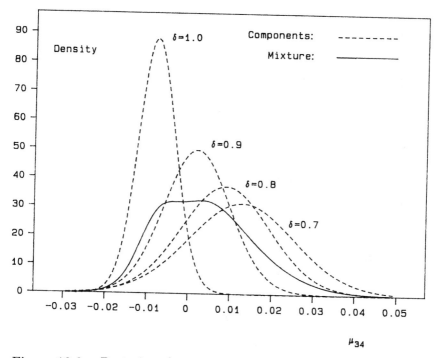

Figure 12.3.   Posteriors for μ_{34} at $t = 34$.

mixture is skewed to the right and very flat in the central region
between roughly -0.007 and 0.006. The mean, calculated from the
formula in Section 12.2.2, is 0.004. In fact the density is unimodal
with the mode near 0.005. If the probability on model 1, actually
0.18, had been slightly higher then the mixture would have more
density on negative values and could have become bimodal. Generally, with a mixture of k symmetric and unimodel distributions, (as
is the case here with 4 T distributions), there is the possibility of
multimodality with up to k modes. Note finally that the mixture is
rather spread, reflecting appreciable posterior probability on each of
the rather distinct components.

Consider the problem of forecasting $(Y_{35} \mid D_{34})$ based on the mixture of components (12.8) at $t = 35$. The component one-step ahead
T densities are graphed in Figure 12.4, together with the mixture.
By contrast with Figure 12.3, the components here are very similar. Although they have the same means as the posteriors for μ_{34},
the scale parameters are very much larger — note the differences between the figures in the scales on the axes. The reason for this is that

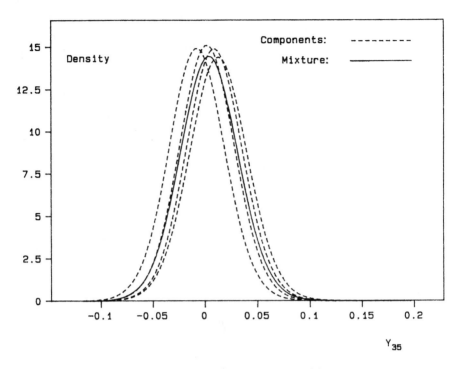

Figure 12.4.   Forecast densities for Y_{35} at $t = 34$.

in each of the DLMs, the uncertainty in the Y_t series is dominated
by observational variation about the level, the differences between
models evident in the posteriors in Figure 12.3 being totally masked,
and quite unimportant, for forecasting Y_{35}. As a consequence, the
overall mixture density is very similar to each of the components.

Further illustration of inference using mixtures, and much more
discussion, appears in the sections below concerning Multi-process,
Class II models.

12.2.4   Model identification via mixtures

In the previous section we referred to the possible use of multi-
process class I models as an automatic model identification technique. Some further comments about this aspect of mixture modelling are given here, specifically concerning the feature of convergence of the posterior model probabilities as t increases. The basic
asymptotic theory for posterior distributions is rather general and so
we revert to the framework of Section 12.2, embedding the discrete

set of values of α in a general, continuous parameter space $\mathcal{A}$ for the purposes of this section. We do not not provide formal, mathematical theory relating to convergence, preferring only to discuss concepts and motivate the key results. Theoretical details of convergence posterior distributions, albeit in a much less general framework, can be found in Berk (1970). Related material on asymptotic normality of posterior distributions, in a rather general framework, appears in Heyde and Johnstone (1979), and Sweeting and Adekola (1987). Related results specific to multi-process, normal DLMs, though less general, appear in Anderson and Moore (1979, Section 10.1).

In the general setting with α continuous, suppose that $p(\alpha \mid D_0) > 0$ for all $\alpha \in \mathcal{A}$. For any two, distinct values α_1 and α_2, the Bayes' factor measuring the evidence from Y_t in favour of $\alpha = \alpha_1$ relative to $\alpha = \alpha_2$ is

$$H_t(\alpha_1, \alpha_2) = p(Y_t \mid \alpha_1, D_{t-1})/p(Y_t \mid \alpha_2, D_{t-1}).$$

The focus below is on the log-Bayes' factors $\log[H_t(\alpha_1, \alpha_2)]$, assumed to be finite for all t and all α_1, $\alpha_2 \in \mathcal{A}$. The corresponding overall log-Bayes' factor, or log-likelihood ratio, from the first t observations $Y_t, \ldots, Y_1$ is simply

$$\log[\prod_{r=1}^{t} H_r(\alpha_1, \alpha_2)] = \sum_{r=1}^{t} \log[H_r(\alpha_1, \alpha_2)].$$

Denote the average value of the individual log-factors by

$$L_t(\alpha_1, \alpha_2) = t^{-1} \sum_{r=1}^{t} \log[H_r(\alpha_1, \alpha_2)].$$

Now, for $j = 1, 2$, we have

$$p(\alpha_j | D_t) = p_t(j) \propto p_0(j) \prod_{r=1}^{t} p(Y_r | \alpha_j, D_{r-1}),$$

and so

$$p(\alpha_1 | D_t)/p(\alpha_2 | D_t) = [p_0(1)/p_0(2)] \prod_{r=1}^{t} H_r(\alpha_1, \alpha_2)$$

for all t. It follows that

$$\log[p(\alpha_1|D_t)/p(\alpha_2|D_t)] = \log[p_0(1)/p_0(2)] + t\, L_t(\alpha_1, \alpha_2).$$

As t increases, the effect of the initial prior on the posterior for α becomes negligible in the sense that, as $t \to \infty$,

$$t^{-1}\log[p(\alpha_1 \mid D_t)/p(\alpha_2 \mid D_t)] - L_t(\alpha_1, \alpha_2) \to 0.$$

Thus the limiting behaviour of the posterior is defined by that of the functions L_t. Two assumptions provide the required structure for convergence.

(A) Assume that, as t increases, $L_t(\alpha_1, \alpha_2)$ converges, in an appropriate probabilistic sense, to a finite limiting value for all α_1 and α_2. Denote this value by $L(\alpha_1, \alpha_2)$, so that

$$L(\alpha_1, \alpha_2) = \lim_{t \to \infty} L_t(\alpha_1, \alpha_2).$$

(B) Assume that there exists a unique value $\alpha_0 \in \mathcal{A}$ such that

$$L(\alpha_0, \alpha) > 0$$

for all $\alpha \neq \alpha_0$.

These assumptions require some comment. Concerning (A), the mode of convergence, whether it be with probability one, in probability etc., is of little importance. When the underlying mechanism generating the data can be assumed to give rise to a joint distribution for the data $Y_t, \ldots, Y_1$, for all t, then (A) can be made more concrete. Note that this does not assume that this true underlying distribution is known, nor that it bears any relation to the class of models used. An interesting example is, however, the hypothetical case when the data truly follow a DLM within the class, identified by a particular, unknown value of α, although this is not necessary. Whatever this true distribution may be, (B) applies (with probability one) if, for all t, $E\{\log[H_t(\alpha_1, \alpha_2)] \mid D_0\}$ is finite, where the expectation is with respect to the full t random quantities $Y_t, \ldots, Y_1$ having this true, joint distribution. If this is the case, then

$$L(\alpha_1, \alpha_2) = \lim_{t \to \infty} t^{-1} \sum_{r=1}^{t} E\{\log[H_r(\alpha_1, \alpha_2)] \mid D_0\}.$$

Assumptions (A) and (B) are both valid in cases when the posterior for α is asymptotically normal. This will usually be the case. Formal regularity conditions required, and proofs of the asymptotic normality, can be found Heyde and Johnstone (1979), and Sweeting and Adekole (1987). In such cases, for large values of t the posterior density $p(\alpha \mid D_t)$ concentrates increasingly about a unique posterior mode $\hat{\alpha}_t$. Assuming the posterior to be twice differentiable, then, for α near $\hat{\alpha}_t$,

$$\log[p(\alpha \mid D_t)] \approx -\,0.5t(\alpha - \hat{\alpha}_t)'\mathbf{I}_t(\hat{\alpha}_t)(\alpha - \hat{\alpha}_t)$$
$$+ \text{ constants not involving } \alpha$$
$$+ \text{ other, negligible terms,}$$

where

$$\mathbf{I}_t(\alpha) = t^{-1}\frac{-\delta^2}{\delta\alpha\,\delta\alpha'}\log[p(\alpha \mid D_t)]$$

so that $\mathbf{I}_t(\hat{\alpha}_t)$ is a symmetric, positive definite matrix. Thus, for large t,

$$L_t(\alpha_1, \alpha_2) \approx t^{-1}\log[p(\alpha_1 \mid D_t)/p(\alpha_2 \mid D_t)]$$
$$\approx (\alpha_2 - \alpha_1)'\mathbf{I}_t(\hat{\alpha}_t)(\bar{\alpha} - \hat{\alpha}_t),$$

where $\bar{\alpha} = (\alpha_1 + \alpha_2)/2$. Under general conditions (as given in the above references), the posterior eventually becomes degenerate at a particular value $\alpha = \alpha_0$, with $\lim_{t\to\infty} \hat{\alpha}_t = \alpha_0$ and a positive definite limit $\mathbf{I}_0 = \lim_{t\to\infty} \mathbf{I}_t(\hat{\alpha}_t)$. Thus

$$L(\alpha_1, \alpha_2) = \lim_{t\to\infty} L_t(\alpha_1, \alpha_2) = (\alpha_2 - \alpha_1)'\mathbf{I}_0(\bar{\alpha} - \alpha_0).$$

In this case then, Assumptions (A) and (B) clearly hold, the limiting posterior distribution identifying the value α_0 in the formal sense that $\lim_{t\to\infty} \Pr[\|\alpha - \alpha_0\| > \epsilon \mid D_t] = 0$ for all $\epsilon > 0$.

Under assumptions (A) and (B), then, it follows, for all $\alpha_1, \alpha_2 \in \mathcal{A}$, that

$$\lim_{t\to\infty} t^{-1}\log[p(\alpha_1 \mid D_t)/p(\alpha_2 \mid D_t)] = L(\alpha_1, \alpha_2).$$

Thus

$$\lim_{t\to\infty}[p(\alpha_1 \mid D_t)/p(\alpha_2 \mid D_t)] = \begin{cases} \infty, & \text{if } L(\alpha_1, \alpha_2) > 0; \\ 0, & \text{if } L(\alpha_1, \alpha_2) < 0. \end{cases}$$

This applies in the general case of continuous $\mathcal{A}$, and also, as a corollary, when working with finite, discrete set $\{\alpha_1, \ldots, \alpha_k\}$ in a multi-process model. Note that we can *always* introduce a larger, continuous space in which to embed a discrete set $\mathcal{A}$, and so the following results apply to multi-process, class I models.

THEOREM 12.1. *In the multi-process, class I model of Defini-tions 12.1 and 12.2, suppose that* $A = \{\alpha_1, \ldots, \alpha_k\}$ *is embedded in a continuous space such that assumptions (A) and (B) above hold, with* α_0 *defined as in (B) (although unknown and not necessarily in* A). *Suppose also that for a single element* $\alpha_i \in A$,

$$L(\alpha_0, \alpha_i) = \min_{\alpha_j \in A} L(\alpha_0, \alpha_j).$$

Then the posterior model probabilities converge as

$$\lim_{t \to \infty} p_t(i) = 1,$$

and

$$\lim_{t \to \infty} p_t(j) = 0, \quad (j \neq i).$$

The numbers $L(\alpha_0, \alpha_i)$ are all non-negative, being positive if, as is usual, $\alpha_0 \notin A$. These numbers are the limiting versions of the average Bayes' factors, or log-likelihood ratios, comparing the values α_j with the "optimal" value α_0. As a corollary to the theorem, suppose that there exists an unknown, underlying joint distribution for the data $Y_t, \ldots, Y_1$, for all t, as previously discussed. Now, for all j,

$$L(\alpha_0, \alpha_j) = \lim_{t \to \infty} t^{-1} \sum_{r=1}^{t} \mathrm{E}\{\log[H_r(\alpha_0, \alpha_j)] \mid D_0\}$$

$$= \lim_{t \to \infty} t^{-1} \sum_{r=1}^{t} \mathrm{E}\{\log[p(Y_r \mid \alpha_0, D_{r-1})] \mid D_0\}$$

$$- \lim_{t \to \infty} t^{-1} \sum_{r=1}^{t} \mathrm{E}\{\log[p(Y_r \mid \alpha_j, D_{r-1})] \mid D_0\}.$$

The first term here does not involve j and so can be ignored if the second term is finite. We thus have the following result.

Corollary 12.1. *In the framework of Theorem 12.1, suppose that, for all* $\alpha_j \in A$, *the quantity*

$$L^*(\alpha_j) = \lim_{t \to \infty} t^{-1} \sum_{r=1}^{t} \mathrm{E}\{\log[p(Y_r \mid \alpha_j, D_{r-1})] \mid D_0\},$$

exists, the expectation being with respect to the true distribution of the random quantities $(Y_t, \dots, Y_1 \mid D_0)$. *Then the posterior probability on model* i *converges to unity, where*

$$L^*(\alpha_i) = \max_{\alpha_j \in \mathcal{A}} L^*(\alpha_j).$$

The quantity $L^*(\alpha)$, for any α of interest, is the limiting average log-likelihood of the model $M_t(j)$ with respect to the true distribution of the series. This is related to the limiting average **Kullback-Leibler Divergence**, otherwise directed divergence, measuring the discrepancy between the model and the data. A basic reference is Kullback and Leibler (1951), although such *entropy* based measures have been used extremely widely in statistics and other areas for many years and under many names. Kullback (1983) gives further details. The measure $L^*(\alpha)$ is the limiting average of the individual quantities $\mathrm{E}\{\log[p(Y_t \mid \alpha, D_{t-1})] \mid D_0\}$.

Some further comments are in order.

(1) These results apply when there is a single value i, $(1 \le i \le k)$ such that α_i *uniquely* maximises $L^*(\alpha)$ over $\alpha \in \mathcal{A}$ — equivalently, minimises $L(\alpha_0, \alpha)$ as in the statement of Theorem 12.1. In some (rare) cases $\mathcal{A}$ may contain two, or more, values of α for which $L^*(\alpha)$ attains the maximum over $\mathcal{A}$. The above discussion then applies with minor modification to show that the posterior model probabilities converge to zero on all but these maximising values of α, being asymptotically uniform across these values.

(2) Assume a model of the form $M_t(\alpha_0)$ actually generates the series. If $\alpha_0 \in \{\alpha_1, \dots, \alpha_k\}$ then the limiting, preferred value is $\alpha_i = \alpha_0$, the true value. If, on the otherhand, $\alpha_0 \notin \{\alpha_1, \dots, \alpha_k\}$, then α_i is closest to α_0 in the sense of divergence, which usually corresponds to closeness in terms of Euclidean distance.

(3) If $M_j(t)$ is a Time Series DLM (TSDLM) for each j, Theorem 5.1 of Chapter 5 implies that $p(Y_t \mid \alpha, D_{t-1})$ has a stable, limiting form. In particular, as $t \to \infty$,

$$(Y_t \mid \alpha_j, D_{t-1}) \sim \mathrm{N}[f_t(j), Q(j)],$$

for some constant variances $Q(j)$. Thus, for large t,

$$\mathrm{E}\{\log[p(Y_t \mid \alpha_j, D_{t-1}] \mid D_0\} = -0.5\log(2\pi Q(j))$$
$$-0.5\mathrm{E}[(Y_t - f_t(j))^2 \mid D_0]/Q(j)$$
$$= -0.5\log(2\pi Q(j)) - 0.5V_0(j)/Q(j),$$

say, where $V_0(j) = \mathrm{E}[(Y_t - f_t(j))^2 \mid D_0]$ is assumed to exist. Then, for all j,

$$L^*(\alpha_j) = -0.5\log(2\pi Q(j)) - 0.5V_0(j)/Q(j).$$

EXAMPLE 12.3. In the first-order polynomial model of Example 12.2 where $\alpha = \delta$, the discount factor, suppose the observational variance to be known and equal to V. Suppose that the data are generated by such a model with true discount factor of δ_0. For any $\delta_j \in \{\delta_1, \ldots, \delta_k\}$, recall from Section 2.4 that, asymptotically, we have $(Y_t \mid \delta_j, D_{t-1}) \sim \mathrm{N}[f_t(j), V/\delta_j]$, and the errors $Y_t - f_t(j)$ are zero mean even though the model is incorrect, having true variances

$$V_0(j) = \mathrm{V}[Y_t - f_t(j) \mid D_0] = [1 + (\delta_j - \delta_0)^2/(1 - \delta_j)^2]V/\delta_0.$$

Thus, from note (2) above, the limiting, preferred discount factor δ_i is that value maximising

$$\log(\delta_j/\delta_0) + (\delta_0/\delta_j)[1 + (\delta_j - \delta_0)^2/(1 - \delta_j)^2].$$

It is left as an exercise for the reader to explore this as a function of δ_j.

12.3   MULTI-PROCESS MODELS: CLASS II

12.3.1   Introduction and definition

The reader will by now be conversant with the basic, underlying principle of Bayesian forecasting, that at any time, historical information relevant to forecasting the future is sufficiently sumarised in terms of posterior distributions for model parameters. Given such summaries, a model used in the past may be altered or discarded at will in modelling for the future, whether such modifications be based on formal interventions or otherwise. Thus the notion that there may be uncertainty as to which of a possible class of DLMs is most appropriate at any time, irrespective of what has happened in the past, is natural and acceptable as a norm. This notion is formalised via the multi-process, class II models (Harrison and Stevens, 1976a), that embody the assumption (12.2b).

Definition 12.3. *Suppose that, at each time t, α takes a value in the discrete set $\mathcal{A} = \{\alpha_1, \ldots, \alpha_k\}$, the values possibly differing over time. Then, irrespective of the mechanism by which the values of α are chosen, the Y_t series is said to follow a* **Multi-process, Class II Model.**

This situation, in which no single DLM is assumed to adequately describe the series but any one of a discrete collection may obtain at each observation stage, rather often describes the real situation in practice. The variety of possible multi-process, class II models is clearly enormous. We motivate the development of an important special class, that has been used rather widely with success, in the following example.

EXAMPLE 12.3. As in the multi-process, class I case in Example 12.2, let $\alpha = \delta$, the discount factor of a first-order polynomial model. Take the parameter space to contain just two values, $\mathcal{A} = \{0.9, 0.1\}$. Under multi-process class II assumptions, δ takes either the value 0.9 or the value 0.1 at each time t. If 0.9, the dynamic movement in the level of the series is steady, information decaying at a typical rate of about 10% between observations. It is to be expected that this will be the case most of the time. If 0.1, the movement allowed in the level is much greater, the change in level having a standard deviation that is 9 times that obtained with the discount of 0.9. This value is appropriate, usually rather infrequently, as an alternative to permit more marked, abrupt changes in level. Compare the use of similar, alternative discounts in automatic interventions in Section 11.5 of Chapter 11.

In order to utilise the multi-process idea, it remains to specify the mechanisms by which a particular value of α is chosen at each time. There are many possibilities, including, for example, subjective intervention by the forecaster to determine the values. We restrict attention, however, to probabilistic mechanisms that provide multi-processes based on discrete probability mixtures of DLMs.

Definition 12.4. *Suppose that, in the multi-process, class II framework of Definition 12.3, the value $\alpha = \alpha_j$ at time t, defining the model $M_t(j)$, is selected with known probability*

$$\pi_t(j) = Pr[M_t(j) \mid D_{t-1}].$$

Then the series Y_t *follows a* **Multi-Process, Class II Mixture Model**.

Here $\pi_t(j) = \Pr[\alpha = \alpha_j$ at time $t \mid D_{t-1}]$ may, in general, depend upon the history of the process. Some practically important possibilities follow, in order of increasing complexity of the resulting analyses.

(1) Fixed model selection probabilities

$$\pi_t(j) = \pi(j) = \Pr[M_t(j) \mid D_0]$$

for all t. Thus, prior to observing Y_t, $M_t(j)$ has prior probability $\pi(j)$ of obtaining independently of what has previously occurred.

(2) First-order Markov probabilities in which the model obtaining at time t depends on which of the models obtained at time $t-1$, but not on what happened prior to $t-1$. Here we have fixed, and known, transition probabilities

$$\pi(j \mid i) = \Pr[M_t(j) \mid M_{t-1}(i), D_0],$$

for all i and j, $(i = 1, \dots, k; j = 1, \dots, k)$, and all t. Given these probabilities, this model supposes that

$$\Pr[M_t(j) \mid M_{t-1}(i), D_{t-1}] = \pi(j \mid i).$$

Thus the chance of any model obtaining at any time depends on which model obtained at the immediately preceding time, but not on any other features of the history of the series. The *marginal* probabilities of models at time t are calculated via

$$\pi_t(j) = \Pr[M_t(j) \mid D_{t-1}]$$

$$= \sum_{i=1}^{k} \Pr[M_t(j) \mid M_{t-1}(i), D_{t-1}]\Pr[M_{t-1}(i) \mid D_{t-1}]$$

$$= \sum_{i=1}^{k} \pi(j \mid i)p_{t-1}(i),$$

where $p_{t-1}(i)$ is the posterior probability, at time $t-1$, of model $M_{t-1}(i)$.

(3) Higher-order Markov probabilities which extend the dependence to models at times $t-2, t-3, \dots$, etc.

EXAMPLE 12.3 (continued). In the discount factor example, fixed selection probabilities imply the same chance of the lower discount factor $\delta = 0.1$ applying at each time. Viewing $\delta = 0.1$ as consistent with the possibility of marked changes, or jumps, in the level of the series, this models a series subject to the possibility of jumps at random times. If $\pi(2) = \Pr[\delta = 0.1] = 0.05$, for example, then jumps may occur roughly 5% of the time. A first-order Markov model as in (2) refines this random jump mechanism according to expected behaviour of the series. For example, if $\delta = 0.1$ at time $t - 1$ it may be felt that $\delta = 0.1$ at time t is rather less likely than otherwise, consistent with the view that jumps are unlikely to occur in runs. This can be modelled by taking $\pi(1 \mid 2) > \pi(2 \mid 2)$. The reverse would be the case if jumps were expected to occur in groups or runs.

Theoretical developments of the basic probability results for Markov models are, in principle, straightforward, though obviously rather more complicated than those with fixed selection probabilities. We concentrate exclusively on the latter for the remainder of the Chapter. Details of analyses for Markov models are therefore left to the interested reader. Although working with mixture models as in Definition 12.4, the mixture terminology is dropped for simplicity; we refer to the models simply as multi-process, class II models, the mixture feature being apparent and understood.

12.3.2 Fixed selection probability models: Structure of analysis

The analysis of multi-process, class I models in Section 12.2 introduced the use of discrete mixtures of standard DLMs. These are also central to the analysis of class II models. We simplify the model notation here and throughout the rest of the Chapter. The possible DLMs, hitherto indexed by parameters α, are now distinguished only by integers indices, setting $\alpha_j = j$, $(j = 1, \ldots, k)$. Thus the model index set is now simply $\mathcal{A} = \{1, \ldots, k\}$, and we refer to $M_t(j)$ as model j at time t. Some further notation serves to simplify the presentation.

Definition 12.5. *For each t and integer h, $(0 \leq h < t)$, define the probabilities*

$$p_t(j_t, j_{t-1}, \ldots, j_{t-h}) = \Pr[M_t(j_t), M_{t-1}(j_{t-1}), \ldots, M_{t-h}(j_{t-h}) \mid D_t].$$

Thus

$$p_t(j_t) = \Pr[M_t(j_t) \mid D_t],$$

consistent with previous usage,

$$p_t(j_t, j_{t-1}) = \Pr[M_t(j_t), M_{t-1}(j_{t-1}) \mid D_t],$$

and so on.

The analysis of class II models is far more complicated than that of class I. The nature of the complications can be appreciated by considering the position at time $t = 1$, assuming that $(\boldsymbol{\theta}_0 \mid D_0)$ has a standard, normal or T, distribution.

At $t = 1$, there are k possible DLMs $M_1(j_1)$ with prior probabilities $\pi(j_1)$, $(j_1 = 1, \dots, k)$. Within each DLM, analysis proceeds in the usual way, providing posterior distributions for the state vector $\boldsymbol{\theta}_1$ which, although of the usual normal or T forms, generally differ across models. Thus, in DLM j_1, the state vector has posterior $p(\boldsymbol{\theta}_1 \mid M_1(j_1), D_1)$, the DLM having posterior probability

$$p_1(j_1) = \Pr[M_1(j_1) \mid D_1] \propto p(Y_1 \mid M_1(j_1), D_0)\pi(j_1).$$

Unconditionally, inferences about the state vector are based on the discrete mixture

$$p(\boldsymbol{\theta}_1 \mid D_1) = \sum_{j_1=1}^{k} p(\boldsymbol{\theta}_1 \mid M_1(j_1), D_0)p_1(j_1).$$

Proceeding to time $t = 2$, any of the k possible DLMs $M_2(j_2)$, $(j_2 = 1, \dots, k)$, may be selected, again with probabilities $\pi(j_2)$. Now it is only possible to retain the components of standard DLM analyses if, in addition, the particular models possible at $t = 1$ are considered. Thus, conditional on both $M_2(j_2)$ and $M_1(j_1)$ applying, for some j_2 and j_1, the posterior for $\boldsymbol{\theta}_2$ given D_2 follows from the usual DLM analysis, depending, of course, on j_2 and j_1, denoted by $p(\boldsymbol{\theta}_2 \mid M_2(j_2), M_1(j_1), D_2)$. Unconditionally, the posterior for inference is

$$p(\boldsymbol{\theta}_2 \mid D_2) = \sum_{j_2=1}^{k} \sum_{j_1=1}^{k} p(\boldsymbol{\theta}_2 \mid M_2(j_2), M_1(j_1), D_2)p_2(j_2, j_1).$$

Again this is a discrete mixture of standard posteriors. However, whereas at $t = 1$ the mixture contained k components, one for each possible model at time $t = 1$, this contains k^2 components, one for each combination of models possible at time $t = 1$ and $t = 2$. Another way of looking at the mixture is to write

$$p(\boldsymbol{\theta}_2 \mid M_2(j_2), D_2) = \sum_{j_1=1}^{k} p(\boldsymbol{\theta}_2 \mid M_2(j_2), M_1(j_1), D_2)\Pr[M_1(j_1) \mid D_2].$$

Thus, conditional on model j_2 at time $t = 2$, the posterior is a mixture of k standard forms, depending on which of the model obtained at time $t = 1$. Then, unconditionally,

$$p(\boldsymbol{\theta}_2 \mid D_2) = \sum_{j_2=1}^{k} p(\boldsymbol{\theta}_2 \mid M_2(j_2), D_2)p_2(j_2).$$

This development continues as time progresses. Then, at time t, the posterior density can be written hierarchically as

$$p(\boldsymbol{\theta}_t \mid D_t) = \sum_{j_t=1}^{k} p(\boldsymbol{\theta}_t \mid M_t(j_t), D_t)p_t(j_t), \qquad (12.10a)$$

or, equivalently, as

$$\sum_{j_t=1}^{k} \sum_{j_{t-1}=1}^{k} p(\boldsymbol{\theta}_t \mid M_t(j_t), M_{t-1}(j_{t-1}), D_t)p_t(j_t, j_{t-1}), \qquad (12.10b)$$

and so on, ... , down to the final stage

$$\sum_{j_t=1}^{k} \sum_{j_{t-1}=1}^{k} \cdots \sum_{j_1=1}^{k} p(\boldsymbol{\theta}_t \mid M_t(j_t), M_{t-1}(j_{t-1}), \ldots, M_1(j_1), D_t)$$

$$\times p_t(j_t, j_{t-1}, \ldots, j_1).$$

$$(12.10c)$$

Only at the final stage of elaboration in (12.10c), where the sequence of models obtaining at each of the times $1, \ldots, t$ are assumed, are the posteriors

$$p(\boldsymbol{\theta}_t \mid M_t(j_t), M_{t-1}(j_{t-1}), \ldots, M_1(j_1), D_t)$$

of standard DLM form. Thus, to obtain the marginal posterior as a mixture, it is necessary to consider all k^t possible combinations that may apply. Within any particular combination $M_t(j_t), M_{t-1}(j_{t-1})$, $\ldots, M_1(j_1)$, the usual DLM analysis applies. The posterior probabilities on each combination, namely $p_t(j_t, j_{t-1}, \ldots, j_1)$, weight the components in the overall mixture.

At each level of elaboration of the mixtures in equations (12.10), the number of components of the mixture corresponds to the number of steps back in time that are being explicitly considered. In (12.10a) there are k components corresponding to k possible models at time t. In (12.10b) there are k^2 possible combinations of models at times t and $t-1$, and so on down to the full k^t possibilities in (12.10c). At the intermediate levels (12.10a,b), etc., the conditional posteriors in the mixture are themselves discrete mixtures of more basic components.

Obvious problems arise with such an analysis, largely relating to the computational demands made by this explosion of the number of mixture components as time progresses. In each possible combination of models, the posterior for the state vector, predictive distributions for future observations, etc., are summarised in terms of a fixed number of means, variances and so forth. After t observations there are then k^t collections of such quantities requiring calculation and storage. If k is relatively small and the analysis is to be performed only on a very short series, then the computations may be feasible with sufficient computing capacity. Schervish and Tsay (1988) provide several illustrations of analysis of series of up to t around 200 or so observations in models with $k = 4$. In practice, however, the possibly immense computational demands of the analysis will often be prohibitive. Fortunately, it is usually possible to avoid the explosion of the size of mixture models by exploiting further features of the particular class II structure used, and a full analysis with all the computational problems will often not be necessary. On general grounds, mixtures of posterior distributions with many components tend to suffer major redundancies in the sense that components can be grouped together with others that are similar in location and spread. Thus, for all practical purposes, it may be possible to reduce the number of components by approximations, leading to a smaller, manageable mixture. This reduction can be quite dramatic in terms of numbers of discarded components. Some general discussion of this follows.

12.3.3 Approximation of mixtures: General discussion

It is basic to dynamic modelling that, as time progresses, what occurred in the past becomes less and less relevant to inference made
for the future. This applies to mixtures, the possible models obtaining in the past losing relevance to inferences made at the current
time t as t increases. It is therefore to be expected that the full,
conditional posterior $p(\boldsymbol{\theta}_t \mid M_t(j_t), M_{t-1}(j_{t-1}), \dots, M_1(j_1), D_t)$ will
depend negligibly on $M_1(j_1), M_2(j_2)$, etc. when t is large. Depend-
ing on the dimension n of the state vector and the complexity of the
multi-process model, it is thus to be expected that, for some fixed
integer $h \geq 1$, the full posterior will depend essentially upon those
models applying only up to h steps back in time, viz.

$$p(\boldsymbol{\theta}_t \mid M_t(j_t), M_{t-1}(j_{t-1}), \dots, M_1(j_1), D_t)$$
$$\approx p(\boldsymbol{\theta}_t \mid M_t(j_t), M_{t-1}(j_{t-1}), \dots, M_{t-h}(j_{t-h}), D_t).$$

$$(12.11)$$

If this is assumed, the number of components of the mixture posterior
at any time will not exceed k^{h+1}. As a consequence, it is only necessary to consider models up to h steps back in time when performing
the analysis. Thus the full mixture (12.10c) will be approximately

$$\sum_{j_t=1}^{k} \sum_{j_{t-1}=1}^{k} \cdots \sum_{j_{t-h}=1}^{k} p(\boldsymbol{\theta}_t \mid M_t(j_t), M_{t-1}(j_{t-1}), \dots, M_{t-h}(j_{t-h}), D_t)$$
$$\times p_t(j_t, j_{t-1}, \dots, j_{t-h}).$$

$$(12.12)$$

This is a mixture whose components are all still of standard DLM
form, but now containing a fixed number k^{h+1} components rather
than the original, increasing number k^t. The posterior model probabilities weighting the posteriors in this mixture are calculated as
follows. Firstly, and as usual, by Bayes' Theorem,

$$p_t(j_t, j_{t-1}, \dots, j_{t-h}) \propto$$
$$\Pr[M_t(j_t), \dots, M_{t-h}(j_{t-h}) \mid D_{t-1}]$$
$$\times p(Y_t \mid M_t(j_t), \dots, M_{t-h}(j_{t-h}), D_{t-1}).$$

$$(12.13)$$

The second term in (12.13) is given by

$$p(Y_t|M_t(j_t), \ldots, M_{t-h}(j_{t-h}), D_{t-1}) =$$

$$\sum_{j_{t-h-1}=1}^{k} p(Y_t|M_t(j_t), \ldots, M_{t-h}(j_{t-h}), M_{t-h-1}(j_{t-h-1}), D_{t-1})$$

$$\times \Pr[M_{t-h-1}(j_{t-h-1})|D_{t-1}],$$

an average of the standard, normal or T one-step predictive densities for Y_t given the combination of models in the conditionings. The averaging is with respect to models $h+1$ steps back, those at time $t-h-1$. The probabilities weighting these terms are available from the identity

$$\Pr[M_{t-h-1}(j_{t-h-1})|D_{t-1}]$$

$$= \sum_{j_{t-1}=1}^{k} \cdots \sum_{j_{t-h}=1}^{k} p_{t-1}(j_{t-1}, \ldots, j_{t-h}, j_{t-h-1}).$$

The first term in (12.13) is similarly calculated via

$$\Pr[M_t(j_t), \ldots, M_{t-h}(j_{t-h}) \mid D_{t-1}]$$

$$= \Pr[M_t(j_t) \mid M_{t-1}(j_{t-1}), \ldots, M_{t-h}(j_{t-h}), D_{t-1}]$$

$$\times p_{t-1}(j_{t-1}, \ldots, j_{t-h})$$

$$= \pi(j_t)p_{t-1}(j_{t-1}, \ldots, j_{t-h})$$

$$= \pi(j_t) \sum_{j_{t-h-1}=1}^{k} p_{t-1}(j_{t-1}, \ldots, j_{t-h-1}). \qquad (12.14)$$

This is directly available since the summands here are just the k^{h+1} posterior model probabilities at time $t-1$.

Further approximations to the mixture (12.12) can often be made to reduce the number of components. Three key considerations are as follows.

(A) Ignore components that have very small posterior probabilities.

(B) Combine components that are roughly equal into a single component, also combining the probabilities.

(C) Replace the contribution of a collection of components by a component that somehow represents their contribution.

These points apply generally to the use of mixtures, not only to the time series context. To provide insight, consider the following example where we drop the notation specific to the time series context for simplification.

EXAMPLE 12.4. Suppose $\theta_t = \theta$ has density

$$p(\theta) = \sum_{j=1}^{4} p_j(\theta)p(j)$$

where, for $j = 1,\ldots,4$, $p_j(.)$ is a T density with mode, scale and degrees of freedom possibly depending on the index j. Thus, in model j,

$$\theta \sim T_{n(j)}[m(j), C(j)].$$

Note that this can be viewed as a very special case of (12.12). The approach (A) to approximating the mixture would apply if, for example, $p(4) = 0.005$, the fourth component receiving only 0.5% of the probability. In such a case,

$$p(\theta) \approx \sum_{j=1}^{3} p_j(\theta)p^*(j)$$

where $p^*(j) = p(j)/(1-p(4))$, $(j = 1, 2, 3)$, in order that $\sum_{j=1}^{3} p^*(j) = 1$. Case (B) would apply in the ideal, and extreme, case $p_3(\theta) = p_4(\theta)$ for all θ. Then

$$p(\theta) = p_1(\theta)p(1) + p_2(\theta)p(2) + p_3(\theta)[p(3) + p(4)],$$

the final two, equal, components being combined to leave a mixture of only three components. More realistically, if the two densities are very similar rather than exactly equal, then the same form of approximation is appropriate. Suppose, for example, $n(3) = n(4)$, $C(3) = C(4)$ and $m(3) = m(4) + \epsilon$ where ϵ is small relative to the scale of the distributions, so that the densities look similar (as do some of the T densities in Figure 12.4, for example). Then the contribution $p_3(\theta)p(3) + p_4(\theta)p(4)$ to the overall density may be approximated by a single component $p^*(\theta)[p(3) + p(4)]$ where $p^*(.)$ is a T density with the same degrees of freedom and scale as the two component densities, and mode $am(3) + (1 - a)m(4)$ where $a = p(3)/[p(3) + p(4)]$. This is in the spirit of approximations developed in Harrison and Stevens (1971 and 1976a), and those termed *quasi-Bayes* in Titterington, Smith and Makov (1985). One interpretation is that $p^*(.)$ is a form of average of the two densities it replaces. Using the third approach (C), we can often reduce the size

of a mixture by approximation even though the components removed
are not, apparently, very similar. The basic technique is to replace
a mixture of components of a given functional form with a single
density of the same form. With T distributions here, and generally
in DLM models, the combination in a mixture may be unimodal
and roughly symmetric. Suppose $p(\theta)$ is the density in Figure 12.4,
with components also graphed there. The mixture is unimodal and
apparently very close to symmetry, suggesting that it can be well
approximated by a single T density. This is the case here, and quite
often in practice. More generally, some subset of the components of
a mixture may be approximated in this way, and substituted with
some combined probability. It remains, of course, to specify how the
approximating density is chosen, this being the subject of the next
section. Sometimes, however, it will not be appropriate. Suppose,
for example, that the mixture is as displayed in Figure 12.3. The
mixture is far from symmetric and cannot be well approximated by
a single T density. Recall that mixtures can become multi-modal,
clearly pin-pointing the dangers of uncritical use of unimodal approximations.

12.3.4   Approximation of mixtures: Specific results

The mixture approximating, or collapsing, techniques derived in this
section are fundamental to the application of multi-process, class II
models (Harrison and Stevens, 1971, 1976a and b; Smith and West,
1983). What is needed is a method by which an approximating
density can be chosen to represent a mixture in cases when such an
approximation is desirable and sensible. We begin rather generally,
again ignoring the time series context and notation since the results
here are not specific to that context, assuming that the density of
the random vector $\boldsymbol{\theta}$ is the mixture

$$p(\boldsymbol{\theta}) = \sum_{j=1}^{k} p_j(\boldsymbol{\theta})p(j). \qquad (12.15)$$

Here the component densities may generally take any forms, although often they will have the same functional form, such as normal
or T, differing only through defining parameters such as means, variances, etc. The probabilities $p(j)$ are known. This density is to be
approximated by a density $p^*(\boldsymbol{\theta})$ of specified functional form, the parameters defining the approximation to be chosen. Thus a mixture

of normal densities may be approximated by a normal with mean and variance matrix to be chosen in some optimal way. The notion that $p^*(.)$ should be *close* to $p(.)$ brings in the concept of a measure of how close, and the need for a *distance* measure between densities, or distributions. There are many such measures, some leading to similar or equivalent results, and we focus on just one.

Viewing $p(.)$ as the true density of θ to be approximated by $p^*(.)$, consider the quantity

$$-E\{\log[p^*(\theta)]\} = -\int \log[p^*(\theta)]p(\theta)d\theta. \qquad (12.16)$$

For any approximating distribution with density $p^*(.)$, this, *entropy* related quantity is a natural measure of the closeness of approximation to the true distribution. Similar measures abound in Bayesian, and non-Bayesian, statistics. Recall, for example, the appearance of this sort of measure in the convergence results in multi-processs, class I models of Section 12.2.4. Choosing the approximating density to achieve a small value of (12.16) is clearly equivalent to attempting to minimise the quantity $K(p^*)$ defined by

$$K(p^*) = E\{\log\left[\frac{p(\theta)}{p^*(\theta)}\right]\} = \int \log\left[\frac{p(\theta)}{p^*(\theta)}\right]p(\theta)d\theta. \qquad (12.17)$$

This is the Kullback-Leibler directed divergence between the approximating distribution whose density is $p^*(.)$ and the true distribution with density $p(.)$. The divergence is defined for continuous and discrete distributions alike, although we restrict attention here to continuous distributions and assume that $p(.)$ and $p^*(.)$ have the same support. Thus $p(\theta) > 0$ if, and only if, $p^*(\theta) > 0$. Although not a true distance measure, the properties of the divergence are appropriate to the problem of density approximation. Two key properties are as that

- $K(p^*) \geq 0$ for all densities $p^*(.)$, and
- $K(p^*) = 0$ if, and only if, $p^* = p$, the true density, almost everywhere.

From now on, the divergence is used to measure closeness of approximations to mixtures, its use being illustrated in examples pertinent to the multi-process context. Related discussion can be found in Titterington, Makov and Smith (1985). Some of the results quoted, and used in exercises, are to be found in Quintana (1987), and related material appears in Amaral and Dunsmore (1980).

EXAMPLE 12.5. Suppose $\boldsymbol{\theta} = \theta$ is scalar whose true distribution has finite mean and variance. Suppose also that the approximation $p^*(.)$ is a normal density with mean m and variance C to be determined. It can be easily shown that

$$2K(p^*) = constant + \log(C) + V[\theta]/C + (m - E[\theta])^2/C.$$

It follows directly that the optimal normal approximation to any distribution with finite mean and variance just matches the moments, setting $m = E[\theta]$ and $C = V[\theta]$.

EXAMPLE 12.6. The multivariate generalisation of Example 12.5 assumes that the vector $\boldsymbol{\theta}$ has true distribution with finite mean vector and variance matrix, and that the approximating distribution is multivariate normal, $N[\mathbf{m}, \mathbf{C}]$. Then, as a function of $\mathbf{m}$ and $\mathbf{C}$,

$$2K(p^*) = constant + \log(|\mathbf{C}|) + E\{(\boldsymbol{\theta} - \mathbf{m})'\mathbf{C}^{-1}(\boldsymbol{\theta} - \mathbf{m})\}$$
$$= constant + \log(|\mathbf{C}|) + trace(\mathbf{C}^{-1}V[\boldsymbol{\theta}])$$
$$+ (E[\boldsymbol{\theta}] - \mathbf{m})'\mathbf{C}^{-1}(E[\boldsymbol{\theta}] - \mathbf{m}).$$

It is left as an exercise to the reader to verify that this is minimised by taking $\mathbf{m} = E[\boldsymbol{\theta}]$ and $\mathbf{C} = V[\boldsymbol{\theta}]$. In the case when $p(.)$ is a mixture as in (12.15), suppose that the mixture components $p_j(.)$ have means $\mathbf{m}(j)$ and variance matrices $\mathbf{C}(j)$. Then the optimal values for the approximating moments are the true moments, namely

$$\mathbf{m} = \sum_{j=1}^{k} \mathbf{m}(j)p(j)$$

and

$$\mathbf{C} = \sum_{j=1}^{k} \{\mathbf{C}(j) + (\mathbf{m} - \mathbf{m}(j))(\mathbf{m} - \mathbf{m}(j))'\}p(j).$$

In the multi-process context various mixture collapsing procedures may be considered. The basic approach for the posteriors for the state vector $\boldsymbol{\theta}_t$ uses the normal mixture technique of Example 12.6. However, this needs a little refinement to cover the cases when the observational variance sequence is unknown and estimated, leading to mixtures of T posteriors for the state vector and Gamma mixtures for the reciprocal variance, or precision, parameters. The following examples provide the relevant theory for these cases.

EXAMPLE 12.7. Suppose $\theta = \phi$, a scalar, having a true distribution with finite mean. Suppose also that the approximating density $p^*(\phi)$ is Gamma, $G[n/2, d/2]$ for some degrees of freedom parameter $n > 0$ and $d > 0$ to be identified.

We consider first a special case of relevance to multi-process, in which the degrees of freedom parameter n is fixed in advance, the optimisation problem therefore being restricted to the choice of d. It is easily shown that, as a function of d alone, $2K(p^*) = dE[\phi] - n \log(d)$ which is minimised by setting $n/d = E[\phi]$. This simply equates the mean of the Gamma distribution to the true mean and provides $d = n/E[\phi]$.

If n is to be considered too, the problem is rather more complex. Here, as a function of n and d,

$$2K(p^*) = constant + dE[\phi] - n \log(d/2) - nE[\log(\phi)] + 2 \log[\Gamma(n/2)].$$

It follows by differentiation that the minimising values of n and d satisfy

(a) $E[\phi] = n/d$, and

(b) $E[\log(\phi)] = \gamma(n/2) + \log(2E[\phi]/n)$.

Here $\gamma(.)$ is the digamma function defined, for all $x > 0$, by $\gamma(x) = \Gamma'(x)/\Gamma(x)$. Let $S = d/n$. Then, from (a), $S^{-1} = E[\phi]$, the approximating Gamma distribution having the true mean. The value of n can be found from (b), although the unique solution can be found generally only via numerical methods. Given S from (a), (b) leads to $E[\log(\phi)] = \gamma(n/2) - \log(nS/2)$, an implicit equation in n that may be solved numerically. If it is clear that n is relatively large, then the approximation $\gamma(x) \approx \log(x) - 1/(2x)$, for x large, may be applied, leading from (b), to $n^{-1} = \log(E[\phi]) - E[\log(\phi)]$. This approximation is certainly adequate for $n > 20$.

Consider the special case that $p(.)$ is a discrete mixture of Gamma distributions, the component $p_j(.)$ being $G[n(j)/2, d(j)/2]$ with means $S(j)^{-1} = n(j)/d(j)$. Now the required quantities $E[\phi]$ and $E[\log(\phi)]$ in (a) and (b) are given by

$$E[\phi] = \sum_{j=1}^{k} S(j)^{-1} p(j)$$

$$E[\log(\phi)] = \sum_{j=1}^{k} \{\gamma(n(j)/2) - \log(d(j)/2)\} p(j).$$

EXAMPLE 12.8. Specifically in connection with the multi-process DLM, suppose that the q-vector $\boldsymbol{\theta}$ and the scalar ϕ have a joint distribution that is a mixture of normal/gamma forms. The mixture has k components, the jth component being defined as follows.

(1) Given ϕ, $(\boldsymbol{\theta} \mid \phi) \sim N[\mathbf{m}(j), \mathbf{C}(j)/\{S(j)\phi\}]$ for some mean vector $\mathbf{m}(j)$, variance matrix $\mathbf{C}(j)$, and estimate $S(j) > 0$ of ϕ^{-1}.

(2) For ϕ marginally, $\phi \sim G[n(j)/2, d(j)/2]$ for some degrees of freedom $n(j) > 0$, and parameter $d(j) > 0$, having mean $E[\phi] = S(j)^{-1} = n(j)/d(j)$.

(3) Putting (a) and (b) together and integrating out ϕ gives the marginal, multivariate T distribution for $\boldsymbol{\theta}$ in model j, $\boldsymbol{\theta} \sim T_{n(j)}[\mathbf{m}(j), \mathbf{C}(j)]$.

Now, suppose that the mixture is to be approximated by $p^*(\boldsymbol{\theta}, \phi)$, a single, normal/gamma distribution defined by parameters $\mathbf{m}$, $\mathbf{C}$, n and d, giving $(\boldsymbol{\theta} \mid \phi) \sim N[\mathbf{m}, \mathbf{C}/\{S\phi\}]$ where $S = d/n$, $\phi \sim G[n/2, d/2]$ and $\boldsymbol{\theta} \sim T_n[\mathbf{m}, \mathbf{C}]$. Calculation of the Kullback-Liebler divergence requires integration of $\log[p^*(\boldsymbol{\theta}, \phi)]$ with respect to the joint, mixture distribution of $\boldsymbol{\theta}$ and ϕ. We can write

$$K(p^*) = constant - E\{\log[p^*(\boldsymbol{\theta}, \phi)]\}$$
$$= constant - E\{\log[p^*(\boldsymbol{\theta} \mid \phi)] + \log[p^*(\phi)]\},$$

specifying the joint density in terms of the conditional/marginal pair in (1) and (2). It is left as an exercise to the reader to verify that, on substituting the densities from (1) and (2), this results in

$$2K(p^*) = constant$$
$$- n\log(n/2) + 2\log[\Gamma(n/2)] - (n + q - 2)E[\log(\phi)]$$
$$- n\log(S) + nSE[\phi] + \log(|S^{-1}\mathbf{C}|)$$
$$+ S \sum_{j=1}^{k} S(j)^{-1}\{\text{trace}(\mathbf{C}^{-1}\mathbf{C}(j))$$
$$+ (\mathbf{m} - \mathbf{m}(j))\mathbf{C}^{-1}(\mathbf{m} - \mathbf{m}(j))'\}p(j).$$

It follows that minimisation with respect to $\mathbf{m}$, $\mathbf{C}$, S and n is achieved as follows.

(a) Minimisation with respect to $\mathbf{m}$ leads, as in Example 12.6, to

$$\mathbf{m} = \left\{\sum_{j=1}^{k} S(j)^{-1}p(j)\right\}^{-1} \sum_{j=1}^{k} S(j)^{-1}\mathbf{m}(j)p(j).$$

(b) Minimisation with respect to **C** leads, again as in Example 12.6, to

$$\mathbf{C} = S \sum_{j=1}^{k} S(j)^{-1}\{\mathbf{C}(j) + (\mathbf{m} - \mathbf{m}(j))(\mathbf{m} - \mathbf{m}(j))'\}p(j).$$

(c) Minimisation with respect to S leads, as in Example 12.7(a), to

$$S^{-1} = \mathrm{E}[\phi] = \sum_{j=1}^{k} S(j)^{-1}p(j).$$

This also implies that, in (a) and (b),

$$\mathbf{m} = \sum_{j=1}^{k} \mathbf{m}(j)p^*(j),$$

and

$$\mathbf{C} = \sum_{j=1}^{k}\{\mathbf{C}(j) + (\mathbf{m} - \mathbf{m}(j))(\mathbf{m} - \mathbf{m}(j))'\}p^*(j),$$

where the revised weights $p^*(j) = p(j)S/S(j)$, $(j = 1, \dots, k)$, sum to unity.

(d) Minimisation with respect to n leads, as in Example 12.7(b), to

$$\mathrm{E}[\log(\phi)] = \gamma(n/2) - \log(nS/2).$$

With S obtained from (c), this again must generally be solved numerically for n. Since $p(\phi)$ is a mixture of gamma densities, then $\mathrm{E}[\log(\phi)]$ may be calculated as in Example 12.7.

Two special cases for consideration are as follows.

(i) Results (a) and (b) clearly specialise to the corresponding versions in Example 12.6 if it is assumed that $\phi = S^{-1}$ is known. Formally, let each $n(j)$ tend to infinity and $S(j) = S$.

(ii) As in Example 12.7, consider the case that the $n(j)$ are equal. The above results apply, of course, to define the optimal, approximating gamma distribution, the resulting value of n being typically different to that common to the mixture components. It may be viewed as appropriate to fix the degrees of freedom n at the common value, in which case the optimisation problem is restricted to the choice of S, defined in (c), part (d) being irrelevant.

12.4   CLASS II MIXTURE MODELS ILLUSTRATED

12.4.1   Second-order polynomial models with exceptions

As an illustration we consider a widely used class II model that was introduced in Harrison and Stevens (1971, 1976a and b). See also Green and Harrison (1973), Smith and West (1983), and Ameen and Harrison (1985b) for extensions and applications. The model is designed for series that behave generally according to a second-order polynomial, or *locally linear* model, but which are subject to exceptional events including outliers, and changes in level or growth. These are just the sorts of commonly occurring exceptions discussed in Chapter 11, and which were handled using subjective intervention and simple, automatic exception detection and adaptation methods. The multi-process models developed here can be seen as a formal, model based technique for monitoring and adaptation. Clearly the multi-process technique may also be applied to other, more complex models.

To begin, consider the usual, second-order polynomial model for the Y_t series, assuming the observational variance sequence known. We use the linear growth form of Section 7.3, Chapter 7, with level and growth parameters μ_t and β_t comprising the state vector $\boldsymbol{\theta}_t = (\mu_t, \beta_t)'$ at time t. Stochastic changes are given in 'delta' notation so that the model is

$$Y_t = \mu_t + \nu_t, \qquad \nu_t \sim \mathrm{N}[0, V_t],$$
$$\mu_t = \mu_{t-1} + \beta_t + \delta\mu_t,$$
$$\beta_t = \beta_{t-1} + \delta\beta_t.$$

where $\delta\mu_t$ and $\delta\beta_t$ are the changes in level and growth, having a joint normal distribution with zero-mean vector and variance matrix to be specified. In matrix notation, the evolution equation is

$$\boldsymbol{\theta}_t = \mathbf{G}\boldsymbol{\theta}_{t-1} + \boldsymbol{\omega}_t$$

where

$$\mathbf{G} = \begin{pmatrix} 1 & 1 \\ 0 & 1 \end{pmatrix},$$

and

$$\boldsymbol{\omega}_t = \begin{pmatrix} \delta\mu_t + \delta\beta_t \\ \delta\beta_t \end{pmatrix} = \mathbf{G}\begin{pmatrix} \delta\mu_t \\ \delta\beta_t \end{pmatrix}.$$

Thus the evolution variance matrix $\mathbf{W}_t$ is given by

$$\mathbf{W}_t = \mathbf{G}V[(\delta\mu_t, \delta\beta_t)']\mathbf{G}'. \qquad (12.18)$$

In addition, the usual independence assumptions hold. As a specific example, suppose that $V_t = V = 1$, $V[\delta\mu_t] = 0.1$, $V[\delta\beta_t] = 0.01$ and $C[\delta\mu_t, \delta\beta_t] = 0$ for all t. Then

$$\mathbf{W}_t = \mathbf{W} = \begin{pmatrix} 0.1 & 0.01 \\ 0.01 & 0.01 \end{pmatrix}.$$

The model is constant, having an evolution variance matrix $\mathbf{W}$ of a linear growth form, and the constant defining quadruple is

$$\left\{ \begin{pmatrix} 1 \\ 0 \end{pmatrix}, \begin{pmatrix} 1 & 1 \\ 0 & 1 \end{pmatrix}, 1, \begin{pmatrix} 0.1 & 0.01 \\ 0.01 & 0.01 \end{pmatrix} \right\}. \qquad (12.19)$$

Note that the model is standardised to have unit observational variance for convenience here, so that $\mathbf{W}_t$ is scale-free.

Consider modelling outliers and changes in trend in a series thought to behave generally according to this model. An outlying observation Y_t may be modelled via a large (positive or negative) observational error ν_t. Under existing model assumptions, $\Pr[|\nu_t| > 3] < 0.01$ so that values larger than 3 in absolute value are exceptional. Larger values can be generated by replacing $V = 1$ in the model with a larger variance. A single, extreme value $\nu_t = 5$, for example, is an exception in the model as it stands but perfectly consistent with a model having a larger observational variance, say $V_t = 100$, for example, at time t alone. If outliers of this sort of magnitude are expected to occur some (small) percentage of the time, then the alternative DLM

$$\left\{ \begin{pmatrix} 1 \\ 0 \end{pmatrix}, \begin{pmatrix} 1 & 1 \\ 0 & 1 \end{pmatrix}, 100, \begin{pmatrix} 0.1 & 0.01 \\ 0.01 & 0.01 \end{pmatrix} \right\} \qquad (12.20)$$

will adequately model them, whilst (12.19) remains appropriate for uncontaminated observations. This sort of outlier model, commonly referred to as a scale shift or scale inflation model, has been widely used in modelling outliers in Bayesian analyses, and its use is not restricted to time series (Box and Tiao, 1968; Smith and Pettit, 1985). The larger observational variance leads to larger observational errors than in the standard model (12.19), but remains neutral as to the sign, the errors still having zero mean.

Similarly, changes in level μ_t much greater than predicted by (12.19) can be accommodated by replacing $\mathbf{W}$ with an alternative in which the variance of $\delta\mu_t$ is inflated. An inflation factor of 100, as used in (12.20) for the outlier model, for example, leads to

$$\mathbf{W} = \begin{pmatrix} 10 & 0.01 \\ 0.01 & 0.01 \end{pmatrix} \tag{12.21}$$

for just the times of change. With this particular matrix, the increment $\delta\mu_t$ to the level parameter at time t can be far greater in absolute value than under (12.19), leading to the possibility of marked, abrupt changes or jumps in level. Note again that the direction of change is not anticipated in this model.

Finally, the same idea applies to modelling jumps in growth of the series. Taking $V[\delta\beta_t] = 1$, for example, to replace the original variance of 0.01, leads to

$$\mathbf{W} = \begin{pmatrix} 1.1 & 1 \\ 1 & 1 \end{pmatrix} \tag{12.22}$$

for the times of possible abrupt changes in growth.

In line with the discussion of Sections 12.3.1 and 12.3.2, Chapter 12, consider collections of DLMs each of which may apply at any given time, the selection being according to fixed model probabilities. Suppose, for example, that

- the standard DLM (12.19) applies with probability 0.85;
- the outlier generating DLM (12.20) applies with probability 0.07;
- the level change DLM (12.21) applies with probability 0.05;
- the growth change DLM (12.22) applies with probability 0.03.

With these probabilities, the series is expected to accord to the standard DLM (12.19) about 85% of the time. The chance of an outlier at any time is 7%, thus one outlying observation in 20 is to be expected. Abrupt changes in the trend of the series are viewed as likely to occur about 8% of the time, with level changes more likely at 5% then growth changes at 3%. This sort of setup is appropriate for many real series, the precise values of the probabilities chosen here are also representative of the forms of behaviour evident in many macro commercial and economic series.

This is an example of a mixture of DLMs with $k = 4$ possible models applying at any time. Some important variations in this particular context are as follows.

(i) Reduction to 3 DLMs by combining level and growth changes into a single model. For example, taking (12.19) with $\mathbf{W}$ replaced by

$$\mathbf{W} = \begin{pmatrix} 11 & 1 \\ 1 & 1 \end{pmatrix}$$

would give an appropriate trend change model, allowing for changes in either or both components μ_t and β_t of the trend. This reduces the number of models, and is sensible if it is not of interest to distinguish between changes in trend and changes in growth. Also, it can often be difficult to distinguish between the two, particularly when growth changes are small, in which case the combination of the two into a single, overall model loses little in practice.

(ii) It may be desired to use discount factors to assign evolution variances $\mathbf{W}_t$ rather than using a constant matrix. If this is so, then exceptionally small discount factors apply to model abrupt changes applies just as in Section 11.5 of Chapter 11. However, the specific structure of the linear growth model here cannot be obtained by using the standard discounting technique. Approaches using discount factors can take the following forms.

Firstly, and generally often appropriately, the quantities $\delta\mu_t$ and $\delta\beta_t$ may be taken as uncorrelated, with variances determined as fractions of the posterior variances of μ_{t-1} and β_{t-1}. Thus, given the usual posterior

$$(\boldsymbol{\theta}_{t-1} \mid D_{t-1}) \sim N[\mathbf{m}_{t-1}, \mathbf{C}_{t-1}]$$

at $t - 1$, write

$$\mathbf{C}_{t-1} = \begin{pmatrix} C_{t-1,\mu} & C_{t-1,\mu\beta} \\ C_{t-1,\mu\beta} & C_{t-1,\beta} \end{pmatrix}.$$

Separate discount factors for level and growth, denoted by δ_μ and δ_β respectively, now define $V[\delta\mu_t] = C_{t-1,\mu}(\delta_\mu^{-1} - 1)$ and $V[\delta\beta_t] = C_{t-1,\beta}(\delta_\beta^{-1} - 1)$. The linear growth structure is retained with evolution variance matrix

$$\mathbf{W}_t = \begin{pmatrix} C_{t-1,\mu}(\delta_\mu^{-1} - 1) + C_{t-1,\beta}(\delta_\beta^{-1} - 1) & C_{t-1,\beta}(\delta_\beta^{-1} - 1) \\ C_{t-1,\beta}(\delta_\beta^{-1} - 1) & C_{t-1,\beta}(\delta_\beta^{-1} - 1) \end{pmatrix}.$$

$$(12.23)$$

The possibilities of abrupt changes are modelled using smaller, exceptional discount factors δ_μ^* and δ_β^*, simply inflating the elements of $\mathbf{W}_t$ appropriately.

(iii) All alternative discounting methods lead to evolution variance matrices that do not have the linear growth structure (Section 7.3). However, they are still perfectly valid second-order polynomial models and the differences in practical application typically small. In line with the usual component modelling ideas applied to discounting, the key alternative approach is appropriate if the level and growth are viewed as a single trend component, the two parameters changing together at the same rate. Here a single discount factor δ controls the time variation. With

$$V[\boldsymbol{\theta}_t \mid D_{t-1}] = \mathbf{P}_t = \mathbf{G}\mathbf{C}_{t-1}\mathbf{G}',$$

this gives

$$\mathbf{W}_t = \mathbf{P}_t(\delta^{-1} - 1).$$

Again abrupt changes in trend are modelled by altering δ to a smaller, exceptional value δ^* to define a trend change DLM.

We proceed to analyse and apply the multi-process with 4 possible DLMs. By way of general notation,

$$\mathbf{F} = \mathbf{E}_2 = \begin{pmatrix} 1 \\ 0 \end{pmatrix} \quad \text{and} \quad \mathbf{G} = \mathbf{J}_2(1) = \begin{pmatrix} 1 & 1 \\ 0 & 1 \end{pmatrix}.$$

Suppose initially, for notational simplicty, that the observational variance sequence V_t is known (the extension to the case of an unknown, constant variance will be summarised below). The basic DLM quadruple is then

$$\{\mathbf{F}, \ \mathbf{G}, \ V_t, \ \mathbf{W}_t\}.$$

The special cases for the 4 possible models in the multi-process are as follows.

(1) **Standard DLM:** $\{\mathbf{F}, \ \mathbf{G}, \ V_t V(1), \ \mathbf{W}_t(1)\}$, where $V(1) = 1$ and $\mathbf{W}_t(1) = \mathbf{W}_t$ a standard evolution variance matrix.

(2) **Outlier DLM:** $\{\mathbf{F}, \ \mathbf{G}, \ V_t V(2), \ \mathbf{W}_t(2)\}$, where $V(2) > 1$ is an inflated variance consistent with the occurrence of observations that would be extreme in the standard DLM (a), and $\mathbf{W}_t(2) = \mathbf{W}_t(1) = \mathbf{W}_t$.

(3) **Level change DLM:** $\{\mathbf{F}, \mathbf{G}, V_t V(3), \mathbf{W}_t(3)\}$, where $V(3) = 1$ and $\mathbf{W}_t(3)$ is an evolution variance matrix consistent with level changes.

(4) **Growth change DLM:** $\{\mathbf{F}, \mathbf{G}, V_t V(4), \mathbf{W}_t(4)\}$, where $V(4) = 1$ and $\mathbf{W}_t(4)$ is an evolution variance matrix consistent with growth changes.

Formally, we have models $M_t(j)$ indexed by $\alpha_j = j$, $(j = 1, \dots, 4)$, the generic parameter α taking values in the index set $\mathcal{A}$.

12.4.2 Model analysis (V_t known)

In providing for the possibility of various exceptions, it is clear that after the occurrence of any such event, further observations will be required before the nature of the event can be identivfied. Thus an outlier is indistinguishable from the onset of change in either level or growth, or both, until the next observation is available. Hence, in approximating mixtures by collapsing over possible models in the past (Section 12.3.3), it will be usual that $h = 1$ at least in (12.11). The original material in Harrison and Stevens (1971, 1976a) and Smith and West (1983) use $h = 1$ and indeed this will often be adequate. In some applications $h = 2$ may be necessary, retaining information relevant to all possible models up to two steps back in time. This may be desirable, for example, if exceptions are rather frequent, possibly occurring consecutively over time. It will be very rare, however, that $h > 2$ is necessary. For illustration here we follow the above referenced authors in using $h = 1$.

We have models

$$M_t(j_t) : \qquad \{\mathbf{F}, \mathbf{G}, V_t V(j_t), \mathbf{W}_t(j_t)\}, \qquad (j_t = 1, \dots, 4).$$

For each j_t, it is assumed that $M_t(j_t)$ applies at time t with known probability $\pi(j_t) = \Pr[M_t(j_t) \mid D_{t-1}] = \Pr[M_t(j_t) \mid D_0]$. Thus, at time t, the model is defined by observation and evolution equations

$$(Y_t \mid \boldsymbol{\theta}_t, M_t(j_t)) \sim \mathrm{N}[\mathbf{F}'\boldsymbol{\theta}_t, V_t V(j_t)], \qquad (12.24)$$

and

$$(\boldsymbol{\theta}_t \mid \boldsymbol{\theta}_{t-1}, M_t(j_t)) \sim \mathrm{N}[\mathbf{G}\boldsymbol{\theta}_{t-1}, \mathbf{W}_t(j_t)], \qquad (12.25)$$

with probability $\pi(j_t)$, conditionally independently of the history of the series D_{t-1}.

Assume also that, at $t = 0$, the initial prior for the state vector is the usual normal form

$$(\boldsymbol{\theta}_0 \mid D_0) \sim N[\mathbf{m}_0, \mathbf{C}_0],$$

irrespective of possible models obtaining at any time, where $\mathbf{m}_0$ and $\mathbf{C}_0$ are known and fixed at $t = 0$. Given this setup, the development of Section 12.3.3 applies in this special case of $h = 1$, the position at any time $t - 1$ being summarised as follows.

Firstly, historical information D_{t-1} is summarised in terms of a 4-component mixture posterior distribution for the state vector $\boldsymbol{\theta}_{t-1}$, the mixture being with respect to the four possible models obtaining at time $t - 1$. Within each component, the posterior distributions have the usual conjugate normal forms. Thus

(a) For $j_{t-1} = 1, \dots, 4$, model $M_{t-1}(j_{t-1})$ applied at time $t - 1$ with posterior probability $p_{t-1}(j_{t-1})$. These probabilities are now known and fixed.

(b) Given $M_{t-1}(j_{t-1})$,

$$(\boldsymbol{\theta}_{t-1} \mid M_{t-1}(j_{t-1}), D_{t-1}) \sim N[\mathbf{m}_{t-1}(j_{t-1}), \mathbf{C}_{t-1}(j_{t-1})]. \quad (12.26)$$

Note that, generally, the quantities defining these distributions depend on the model applying at $t - 1$, hence the index j_{t-1}.

Evolving to time t, statements about $\boldsymbol{\theta}_t$ and Y_t depend on the combinations of possible models applying at both $t - 1$ and t.

(c) Thus, from (12.25) and (12.26), for each j_{t-1} and j_t we have

$$(\boldsymbol{\theta}_t \mid M_t(j_t), M_{t-1}(j_{t-1}), D_{t-1}) \sim N[\mathbf{a}_t(j_{t-1}), \mathbf{R}_t(j_t, j_{t-1})]$$

where $\mathbf{a}_t(j_{t-1}) = \mathbf{G}\mathbf{m}_{t-1}(j_{t-1})$, and $\mathbf{R}_t(j_t, j_{t-1}) = \mathbf{G}\mathbf{C}_{t-1}(j_{t-1})\mathbf{G}' + \mathbf{W}_t(j_t)$. Note that the conditional mean $E[\boldsymbol{\theta}_t \mid M_t(j_t), M_{t-1}(j_{t-1}), D_{t-1}] = \mathbf{a}_t(j_{t-1})$ does not differ across the $M_t(j_t)$, depending only upon possible models applying at $t - 1$ since $\mathbf{G}$ is common to these models.

(d) Similarly, the one-step ahead forecast distribution is given, for each possible combination of models, by the usual form

$$(Y_t \mid M_t(j_t), M_{t-1}(j_{t-1}), D_{t-1}) \sim N[f_t(j_{t-1}), Q_t(j_t, j_{t-1})], \quad (12.27)$$

where

$$f_t(j_{t-1}) = \mathbf{F}'\mathbf{a}_t(j_{t-1}),$$

also common across the models at time t, and

$$Q_t(j_t, j_{t-1}) = \mathbf{F}'\mathbf{R}_t(j_t, j_{t-1})\mathbf{F} + V_t V(j_t).$$

Note again that, due to the particular structure of the multi-process, the means of these sixteen possible forecast distributions take only four distinct values,

$$E[Y_t \mid M_t(j_t), M_{t-1}(j_{t-1}), D_{t-1}] = f_t(j_{t-1})$$

for each of the four values j_t. This follows since the $M_t(j_t)$ differ only through scale parameters and not in location. Note, however, that this feature is specific to the particular model here and may be present or absent in other models. Calculation of the forecast distribution unconditional on possible models then simply involves the mixing of these standard normal components with respect to the relevant probabilities, calculated as follows.

(e) For each j_t and j_{t-1},

$$\begin{aligned} &\Pr[M_t(j_t), M_{t-1}(j_{t-1}) \mid D_{t-1}] \\ &= \Pr[M_t(j_t) \mid M_{t-1}(j_{t-1}), D_{t-1}] \Pr[M_{t-1}(j_{t-1}) \mid D_{t-1}]. \end{aligned}$$

Now, by assumption, models apply at t with constant probabilities $\pi(j_t)$ irrespective of what happened previously, so that

$$\Pr[M_t(j_t) \mid M_{t-1}(j_{t-1}), D_{t-1}] = \pi(j_t);$$

additionally,

$$\Pr[M_{t-1}(j_{t-1}) \mid D_{t-1}] = p_t(j_{t-1}),$$

and then

$$\Pr[M_t(j_t), M_{t-1}(j_{t-1}) \mid D_{t-1}] = \pi(j_t) p_{t-1}(j_{t-1}).$$

This is just a special case of the general formula (12.14) with $h = 1$.

(f) The marginal predictive density for Y_t is now the mixture of the $4^2 = 16$ components (12.27) with respect to these probabilities, given by

$$\begin{aligned} p(Y_t &\mid D_{t-1}) \\ &= \sum_{j_t=1}^{4} \sum_{j_{t-1}=1}^{4} p(Y_t \mid M_t(j_t), M_{t-1}(j_{t-1}), D_{t-1}) \pi(j_t) p_{t-1}(j_{t-1}). \end{aligned}$$

$$(12.28)$$

Now consider updating the prior distributions in (c) to posteriors when Y_t is observed. Given j_{t-1} and j_t, the standard updating equations apply within each of the sixteen combinations, the posterior means, variances, etc. obviously varying across combinations.

(g) Thus,

$$(\boldsymbol{\theta}_t \mid M_t(j_t), M_{t-1}(j_{t-1}), D_t) \sim \mathrm{N}[\mathbf{m}_t(j_t, j_{t-1}), \mathbf{C}_t(j_t, j_{t-1})] \quad (12.29)$$

where:

$$\mathbf{m}_t(j_t, j_{t-1}) = \mathbf{a}_t(j_{t-1}) + \mathbf{A}_t(j_t, j_{t-1})e_t(j_{t-1}),$$
$$\mathbf{C}_t(j_t, j_{t-1}) = \mathbf{R}_t(j_t, j_{t-1}) - \mathbf{A}_t(j_t, j_{t-1})\mathbf{A}_t(j_t, j_{t-1})'Q_t(j_t, j_{t-1}),$$
$$e_t(j_{t-1}) = Y_t - f_t(j_{t-1})$$

and

$$\mathbf{A}_t(j_t, j_{t-1}) = \mathbf{R}_t(j_t, j_{t-1})\mathbf{F}/Q_t(j_t, j_{t-1}).$$

Posterior probabilities across the sixteen possible models derive directly from the general formula (12.13) in the case $h = 1$.

(h) Thus

$$p_t(j_t, j_{t-1}) = \Pr[M_t(j_t), M_{t-1}(j_{t-1}) \mid D_t]$$
$$\propto \pi(j_t)p_{t-1}(j_{t-1})p(Y_t \mid M_t(j_t), M_{t-1}(j_{t-1}), D_{t-1}).$$

The second term here is the observed value of the predictive density (12.27), providing the model likelihood, and so these probabilities are easily calculated. These are simply given by

$$p_t(j_t, j_{t-1}) = \frac{c_t \pi(j_t)p_{t-1}(j_{t-1})}{Q_t(j_t, j_{t-1})^{1/2}} \exp\{-0.5e_t(j_t, j_{t-1})^2/Q_t(j_t, j_{t-1})\},$$
$$(12.30)$$

where c_t is a constant of normalisation such that $\sum_{j_t=1}^{4} \sum_{j_{t-1}=1}^{4} p_t(j_t, j_{t-1}) = 1$.

Inferences about $\boldsymbol{\theta}_t$ are based on the unconditional, sixteen component mixtures that average (12.29) with respect to the posterior model probabilities (12.30).

(i) Thus

$$p(\boldsymbol{\theta}_t \mid D_t) = \sum_{j_t=1}^{4} \sum_{j_{t-1}}^{4} p(\boldsymbol{\theta}_t \mid M_t(j_t), M_{t-1}(j_{t-1}), D_t)p_t(j_t, j_{t-1}).$$
$$(12.31)$$

with components given by the normal distributions in (12.29).

These calculations essentially complete the evolution and updating steps at time t. To proceed to time $t + 1$, however, we need to remove the dependence of the joint posterior (12.31) on possible models obtaining at time $t - 1$. If we evolve (12.31) to time $t + 1$ directly, the mixture will expand to $4^3 = 64$ components for $\boldsymbol{\theta}_{t+1}$, depending on all possible combinations of $M_{t+1}(j_{t+1})$, $M_t(j_t)$ and $M_{t-1}(j_{t-1})$. However, the principle of approximating such mixtures by assuming that the effects of different models at $t - 1$ are negligible for time $t + 1$ applies. Thus, in moving to $t + 1$ the sixteen component mixture (12.31) will be reduced, or collapsed, over possible models at $t - 1$. The method of approximation of Example 12.6 applies here. For each $j_t = 1, \dots, 4$, it follows that

$$\bullet \quad p_t(j_t) = \Pr[M_t(j_t) \mid D_t] = \sum_{j_{t-1}=1}^{4} p_t(j_t, j_{t-1}),$$

$$\bullet \quad \Pr[M_{t-1}(j_{t-1}) \mid D_t] = \sum_{j_t=1}^{4} p_t(j_t, j_{t-1}), \qquad (12.32)$$

$$\bullet \quad \Pr[M_{t-1}(j_{t-1}) \mid M_t(j_t), D_t] = p_t(j_t, j_{t-1})/p_t(j_t).$$

The first equation here gives current model probabilities at time t. The second gives posterior probabilities over the possible models one-step back in time, at $t - 1$. These, one-step back or *smoothed* model probabilities are of great use in retrospective assessment of which models were likely at the previous time point. The third equation is of direct interest here in collapsing the mixture (12.31) with respect to time $t - 1$; it gives the posterior (given D_t) probabilities of the various models at time $t - 1$ *conditional* on possible models at time t. To see how these probabilities, and those in (a), feature in the posterior (12.31) note that this distribution can be rewritten as

$$p(\boldsymbol{\theta}_t \mid D_t) = \sum_{j_t=1}^{4} p(\boldsymbol{\theta}_t \mid M_t(j_t), D_t) p_t(j_t), \qquad (12.33)$$

the first terms of the summands being given by

$$p(\boldsymbol{\theta}_t \mid M_t(j_t), D_t)$$
$$= \sum_{j_{t-1}=1}^{4} p(\boldsymbol{\theta}_t \mid M_t(j_t), M_{t-1}(j_{t-1}), D_t) p_t(j_t, j_{t-1})/p_t(j_t). \qquad (12.34)$$

In (12.33), the posterior is represented as a four component mixture, the components being conditional only on models at time t and being calculated as four component mixtures themselves in (12.34). Only in the latter mixture are the component densities of standard normal form. Now (12.33) is the exact posterior, the dependence on possible models at time t being explicit, whilst that on models at $t-1$ is implicit through (12.34). Thus, in moving to time $t+1$, the posterior will have the required form (12.26) if each of the components (12.34) is replaced by a normal distribution. For each j_t, the mixture in (12.34) has precisely the form of that in Example 12.6, and may be collapsed to a single approximating normal (optimal in the sense of minimising the Kullback-Leibler divergence) using the results of that example. For each j_t, define the appropriate mean vectors $\mathbf{m}_t(j_t)$ by

$$\mathbf{m}_t(j_t) = \sum_{j_{t-1}=1}^{4} \mathbf{m}_t(j_t, j_{t-1}) p_t(j_t, j_{t-1})/p_t(j_t);$$

the corresponding variance matrices $\mathbf{C}_t(j_t)$ are given by

$$\sum_{j_{t-1}=1}^{4} \{\mathbf{C}_t(j_t, j_{t-1}) + (\mathbf{m}_t(j_t) - \mathbf{m}_t(j_t, j_{t-1}))(\mathbf{m}_t(j_t) - \mathbf{m}_t(j_t, j_{t-1}))'\}$$
$$\times\ p_t(j_t, j_{t-1})/p_t(j_t).$$

Then (12.34) is approximated by the single normal posterior having the same mean and variance matrix, namely

$$(\boldsymbol{\theta}_t \mid M_t(j_t), D_t) \sim \mathrm{N}[\mathbf{m}_t(j_t), \mathbf{C}_t(j_t)]. \tag{12.35}$$

The distributions (12.35) replace the components (12.34) of the mixture (12.33), collapsing from sixteen to four standard normal components. In doing so, we complete the cycle of evolution/ updating/ collapsing; the resulting four component mixture is analogous to the starting, four component mixture defined by components (12.26) with the time index updated from $t-1$ to t.

12.4.3   Summary of full model analysis

The above analysis extends to include learning about a known and constant observational variance scale parameter $V_t = V$ for all t. The conditionally conjugate normal/gamma analysis applies as usual within any collection of DLMs applying at all times. Differences arise only through the requirement that mixtures be approximated by collapsing with respect to models h steps back before evolving to the next time point. Since we now have an extended, conditional normal/gamma posterior at each time, this collapsing is based on that in Example 12.8. Full details of the sequential analysis are summarised here.

At $t = 0$, the initial prior for the state vector and observational scale has the usual conjugate form irrespective of possible models obtaining at any time. With precision parameter $\phi = V^{-1}$, we have

$$(\boldsymbol{\theta}_0 \mid V, D_0) \sim \mathrm{N}[\mathbf{m}_0, \mathbf{C}_0 V/S_0],$$
$$(\phi \mid D_0) \sim \mathrm{G}[n_0/2, d_0/2],$$

where $\mathbf{m}_0$, $\mathbf{C}_0$, n_0 and d_0 are known and fixed at $t = 0$. The initial point estimate S_0 of V is given by $S_0 = d_0/n_0$, and the prior for $\boldsymbol{\theta}_0$ marginally with respect to V is just $(\boldsymbol{\theta}_0 \mid D_0) \sim \mathrm{T}_{n_0}[\mathbf{m}_0, \mathbf{C}_0]$.

At times $t-1$ and t, the components of analysis are as follows. Historical information D_{t-1} is summarised in terms of a 4-component mixture posterior distribution for the state vector $\boldsymbol{\theta}_{t-1}$ and the variance scale parameter V, the mixture being with respect to the four possible models obtaining at time $t - 1$. Within each component, the posterior distributions have the usual conjugate normal/gamma forms. Thus

(i) For $j_{t-1} = 1, \ldots, 4$, model $M_{t-1}(j_{t-1})$ applied at time $t - 1$ with posterior probability $p_{t-1}(j_{t-1})$. These probabilities are now known and fixed.

(ii) Given $M_{t-1}(j_{t-1})$, $\boldsymbol{\theta}_{t-1}$ and ϕ have a joint normal/gamma posterior with marginals

$$(\boldsymbol{\theta}_{t-1} \mid M_{t-1}(j_{t-1}), D_{t-1}) \sim \mathrm{T}_{n_{t-1}}[\mathbf{m}_{t-1}(j_{t-1}), \mathbf{C}_{t-1}(j_{t-1})],$$
$$(\phi \mid M_{t-1}(j_{t-1}), D_{t-1}) \sim \mathrm{G}[n_{t-1}/2, d_{t-1}(j_{t-1})/2],$$

$$(12.36)$$

where $S_{t-1}(j_{t-1}) = d_{t-1}(j_{t-1})/n_{t-1}$ is the estimate of $V = \phi^{-1}$ in model $M_{t-1}(j_{t-1})$. Note that, generally, the quantities defining these distributions depend on the model applying at $t-1$, hence the index j_{t-1}. The exception here is the degrees of freedom parameter n_{t-1} common to each of the four possible models.

Evolving to time t, statements about $\boldsymbol{\theta}_t$ and Y_t depend on the combinations of possible models applying at both $t-1$ and t.

(iii) The prior for $\boldsymbol{\theta}_t$ and ϕ is normal/gamma with marginals, for each j_{t-1} and j_t, given by

$$(\boldsymbol{\theta}_t \mid M_t(j_t), M_{t-1}(j_{t-1}), D_{t-1}) \sim T_{n_{t-1}}[\mathbf{a}_t(j_{t-1}), \mathbf{R}_t(j_t, j_{t-1})],$$
$$(\phi \mid M_t(j_t), M_{t-1}(j_{t-1}), D_{t-1}) \sim G[n_{t-1}/2, d_{t-1}(j_{t-1})/2],$$

where $\mathbf{a}_t(j_{t-1}) = \mathbf{G}\mathbf{m}_{t-1}(j_{t-1})$, and $\mathbf{R}_t(j_t, j_{t-1}) = \mathbf{G}\mathbf{C}_{t-1}(j_{t-1})\mathbf{G}' + \mathbf{W}_t(j_t)$. Again it should be noted that $E[\boldsymbol{\theta}_t \mid M_t(j_t), M_{t-1}(j_{t-1}), D_{t-1}] = \mathbf{a}_t(j_{t-1})$ does not differ across the $M_t(j_t)$, depending only upon possible models applying at $t-1$ since $\mathbf{G}$ is common to these models. Also, $p(\phi \mid M_t(j_t), M_{t-1}(j_{t-1}), D_{t-1}) = p(\phi \mid M_{t-1}(j_{t-1}), D_{t-1})$.

(iv) Forecasting one-step ahead,

$$(Y_t \mid M_t(j_t), M_{t-1}(j_{t-1}), D_{t-1}) \sim T_{n_{t-1}}[f_t(j_{t-1}), Q_t(j_t, j_{t-1})],$$

$$(12.37)$$

where

$$f_t(j_{t-1}) = \mathbf{F}'\mathbf{a}_t(j_{t-1}),$$

also common across the models at time t, and

$$Q_t(j_t, j_{t-1}) = \mathbf{F}'\mathbf{R}_t(j_t, j_{t-1})\mathbf{F} + S_{t-1}(j_{t-1})V(j_t).$$

As in the previous Section with V known, the particular structure of the multi-process leads to the modes (and means when $n_{t-1} > 1$) of these sixteen possible forecast distributions taking only four distinct values,

$$E[Y_t \mid M_t(j_t), M_{t-1}(j_{t-1}), D_{t-1}] = f_t(j_{t-1})$$

for each of the four values j_t. Calculation of the forecast distribution unconditional on possible models then simply involves the mixing of these standard T components with respect to the relevant probabilities calculated as in part (e) of the previous Section.

(v) Forecasting Y_t is based on the mixture

$$p(Y_t \mid D_{t-1}) =$$

$$\sum_{j_t=1}^{4} \sum_{j_{t-1}=1}^{4} p(Y_t \mid M_t(j_t), M_{t-1}(j_{t-1}), D_{t-1})\pi(j_t)p_{t-1}(j_{t-1}).$$

$$(12.38)$$

Updating proceeds as usual given j_{t-1} and j_t.

(vi) $p(\boldsymbol{\theta}_t, \phi | M_t(j_t), M_{t-1}(j_{t-1}), D_t)$ is normal/gamma posterior having marginals

$$(\boldsymbol{\theta}_t \mid M_t(j_t), M_{t-1}(j_{t-1}), D_t) \sim T_{n_t}[\mathbf{m}_t(j_t, j_{t-1}), \mathbf{C}_t(j_t, j_{t-1})],$$
$$(\phi \mid M_t(j_t), M_{t-1}(j_{t-1}), D_t) \sim G[n_t/2, d_t(j_t, j_{t-1})/2],$$

$$(12.39)$$

where:

$$\mathbf{m}_t(j_t, j_{t-1}) = \mathbf{a}_t(j_{t-1}) + \mathbf{A}_t(j_t, j_{t-1})e_t(j_{t-1}),$$
$$\mathbf{C}_t(j_t, j_{t-1}) = [S_t(j_t, j_{t-1})/S_{t-1}(j_{t-1})]$$
$$\times [\mathbf{R}_t(j_t, j_{t-1}) - \mathbf{A}_t(j_t, j_{t-1})\mathbf{A}_t(j_t, j_{t-1})'Q_t(j_t, j_{t-1})],$$
$$e_t(j_{t-1}) = Y_t - f_t(j_{t-1}),$$
$$\mathbf{A}_t(j_t, j_{t-1}) = \mathbf{R}_t(j_t, j_{t-1})\mathbf{F}/Q_t(j_t, j_{t-1}),$$
$$n_t = n_{t-1} + 1,$$
$$d_t = d_{t-1}(j_{t-1}) + S_{t-1}(j_{t-1})e_t(j_t, j_{t-1})^2/Q_t(j_t, j_{t-1}),$$

and

$$S_t(j_t, j_{t-1}) = d_t(j_t, j_{t-1})/n_t.$$

Note that n_t is common across models. Posterior model probabilities are given, following part (h) of the previous Section, from equation (12.30) with the normal one-step forecast density replaced by the corresponding T form here.

(vi) Note that, since the degrees of freedom of the T distributions (12.39) are all n_{t-1}, then the probabilities are simply given by

$$p_t(j_t, j_{t-1}) = \frac{c_t\pi(j_t)p_{t-1}(j_{t-1})}{Q_t(j_t, j_{t-1})^{1/2}\{n_{t-1} + e_t(j_t, j_{t-1})^2/Q_t(j_t, j_{t-1})\}^{n_t/2}},$$

$$(12.40)$$

where c_t is a constant of normalisation such that $\sum_{j_t=1}^4 \sum_{j_{t-1}=1}^4 p_t(j_t, j_{t-1}) = 1.$

Inferences about $\boldsymbol{\theta}_t$ and V are based on the unconditional, sixteen component mixtures that average (12.39) with respect to the posterior model probabilities (12.40).

(vii) The marginal for $\boldsymbol{\theta}_t$ is

$$p(\boldsymbol{\theta}_t \mid D_t) = \sum_{j_t=1}^4 \sum_{j_{t-1}} p(\boldsymbol{\theta}_t \mid M_t(j_t), M_{t-1}(j_{t-1}), D_t)p_t(j_t, j_{t-1}).$$

$$(12.41)$$

with components given by the T distributions in (12.39). Similarly the posterior for $\phi = 1/V$ is a mixture of sixteen gamma distributions.

In evolving to $t+1$ the sixteen component mixture (12.41) is collapsed over possible models at $t-1$. The method of approximation of Example 12.8 applies here. For each $j_t = 1, \ldots, 4$, it follows, as in the previous Section, that the collapsed posterior for $\boldsymbol{\theta}_t$ and V is defined by

$$p(\boldsymbol{\theta}_t, \phi \mid D_t) = \sum_{j_t=1}^{4} p(\boldsymbol{\theta}_t, \phi \mid M_t(j_t), M_{t-1}(j_{t-1}), D_t) p_t(j_t), \quad (12.42)$$

the component densities in this sum being approximated as follows (cf. Example 12.8).

For each j_t, define the variance estimates $S_t(j_t)$ by

$$S_t(j_t)^{-1} = \sum_{j_{t-1}=1}^{4} S_t(j_t, j_{t-1})^{-1} p_t(j_t, j_{t-1})/p_t(j_t),$$

with

$$d_t(j_t) = n_t S_t(j_t).$$

Define the weights

$$p_t^*(j_t) = S_t(j_t) S_t(j_t, j_{t-1})^{-1} p_t(j_t, j_{t-1})/p_t(j_t),$$

noting that they sum to unity, viz. $\sum_{j_t=1}^{4} p_t^*(j_t) = 1$. Further define the mean vectors $\mathbf{m}_t(j_t)$ by

$$\mathbf{m}_t(j_t) = \sum_{j_{t-1}=1}^{4} \mathbf{m}_t(j_t, j_{t-1}) p_t^*(j_{t-1}),$$

and the variance matrices $\mathbf{C}_t(j_t)$ by the formulae

$$\sum_{j_{t-1}=1}^{4} \{ \mathbf{C}_t(j_t, j_{t-1})$$
$$+ (\mathbf{m}_t(j_t) - \mathbf{m}_t(j_t, j_{t-1}))(\mathbf{m}_t(j_t) - \mathbf{m}_t(j_t, j_{t-1}))' \} p_t^*(j_{t-1}).$$

Then, for each j_t, the joint, four component posteriors $p(\boldsymbol{\theta}_t, \phi | M_t(j_t)$, $M_{t-1}(j_{t-1}), D_t)$ are approximated by single normal/gamma posteriors having marginals

$$(\boldsymbol{\theta}_t \mid M_t(j_t), D_t) \sim \mathrm{T}_{n_t}[\mathbf{m}_t(j_t), \mathbf{C}_t(j_t)],$$
$$(\phi \mid M_t(j_t), D_t) \sim \mathrm{G}[n_t/2, d_t(j_t)/2].$$

These approximate the components in the mixture (12.42), thus collapsing from sixteen to four standard normal/gamma components. In doing so, we complete the cycle of evolution/updating/collapsing; the resulting four component mixture is analogous to the starting, four component mixture defined by components (12.36) with the time index updated from $t - 1$ to t.

12.4.4 Illustration: CP6 series revisited

The model analysis is illustrated using the CP6 series. The basic linear growth form described in Section 12.4.1 is used, the linear growth evolution variance sequence being defined by separate level and growth discount factors δ_μ and δ_β as in (12.23). The four component models at time t have defining quantities as follows.

(1) **Standard model:** $V(1) = 1$ and $\mathbf{W}_t(1) = \mathbf{W}_t$ given by (12.23) with $\delta_\mu = \delta_\beta = 0.9$, having model probability $\pi(1) = 0.85$;

(2) **Outlier model:** $V(2) = 100$ and $\mathbf{W}_t(2) = \mathbf{W}_t$ having model probability $\pi(2) = 0.07$;

(3) **Level change model:** $V(3) = 1$ and $\mathbf{W}_t(3)$ given by (12.23) with $\delta_\mu = 0.01$ and $\delta_\beta = 0.9$, having model probability $\pi(3) = 0.05$; and

(4) **Growth change model:** $V(4) = 1$ and $\mathbf{W}_t(4)$ given by (12.23) with $\delta_\mu = 0.9$ and $\delta_\beta = 0.01$, having model probability $\pi(4) = 0.03$.

Initial priors are defined by $\mathbf{m}_0 = (600, 10)'$, $\mathbf{C}_0 = \mathrm{diag}(10000, 25)$, $d_0 = 1440$ and $n_0 = 10$. It should be remarked that these values are chosen to be in line with previous analyses of the series (Section 11.3.1). The values of discount factors etc. are not optimised in any sense. Also it is clear that the variance scale factor V apparently changes at later stages of the data, inflating with the level of the series. Although this was catered for and modelled in earlier analyses V is assumed constant, though uncertain, here for clarity. Figures 12.5—12.9 illustrate selected features of the multi-process

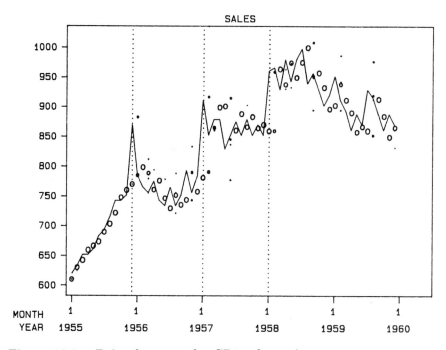

Figure 12.5.   Point forecasts for CP6 sales series.

analysis, these features being common to the use of the models generally and chosen in an attempt to identify key points. Before discussing the analysis, recall that, from earlier analyses, the notable
features of the CP6 series are: (i) an outlier at $t = 12$, followed by a
switch from positive to negative growth; (ii) a jump to higher levels
between $t = 24$ and $t = 26$, with a possible outlier at $t = 25$; (iii) a
further jump at $t = 37$; and (iv) higher random variation at higher
levels of the data in later stages, with one or two events possibly
classifiable as level/growth changes, though obscured by the greater
random variation.

An overall impression of the analysis can be obtained from Figure 12.5. Here the data are plotted and joined over time by the full
line (unlike most previous plots where the forecasts etc. are joined,
the data plotted as separate symbols; this is a temporary change of
convention for clarity in this example). Superimposed on the data
are one-step ahead point forecasts from the distributions (12.38) for
all t. Each forecast distribution has sixteen component T distributions whose modes are taken as individual point forecasts, although,

as noted following equation (12.37), the particular structure of the model means that there are only four distinct such point forecasts at each time. This follows from (12.37) where it is clear that the mode in the component conditional on $(M_t(j_t), M_{t-1}(j_{t-1}), D_{t-1})$ is simply $f_t(j_{t-1})$, for all 4 values of j_t. Thus $f_t(1)$, for example, is the point forecast for Y_t in each of the four models that include $M_{t-1}(1)$, and so applies with probability $p_{t-1}(1)$. Generally, the forecast $f_t(j_{t-1})$ applies with probability $p_{t-1}(j_{t-1})$. Now, rather than plotting all four modes at each t, only those modes that apply with reasonably large probabilities are drawn. Specifically, $f_t(j_{t-1})$ is plotted only if $p_{t-1}(j_{t-1}) > 0.05$. In addition, all forecasts plotted appear as circles whose radii are proportional to the corresponding probabilities; thus more likely point forecasts are graphed as larger circles. This serves to give a relatively simple visual summary of overall forecasting performance (although without indications of uncertainty), having the following features.

- In stable periods where the basic, linear growth form is adequate, the standard model (whose prior probability at any time is 0.85) clearly dominates. This results in a preponderance of large circles denoting the point forecasts $f_t(1)$ at each time, the mode of Y_t conditional on the standard model applying at the previous timepoint.
- At times of instability, the radii of these larger circles decreases, reflecting lower probability, and up to three further circles appear denoting those remaining forecasts with probabilities in excess of 0.05. These tend to be graphed as circles with very small radii, reflecting low probability, sometimes just appearing as points. This occurs at those times t when the previous observation Y_{t-1} (and sometimes the previous one or two observations) are poorly forecast under the standard model. In such cases, the probabilities $p_{t-1}(j_{t-1})$ spread out over the four models reflecting uncertainty as to the behaviour of the series at time $t-1$. This uncertainty feeds directly through to the forecast distribution for Y_t as illustrated on the graph. Some of the occurrences are now described.

Consider the position at November 1955, time 11. Up until this time, the series is stable and is well modelled by the standard DLM; hence the single point forecasts in Figure 12.5. This is clearly evident in Figure 12.9a. Here the four posterior model probabilities $p_t(j_t)$ at each time t are plotted as vertical bars. Up to $t = 11$

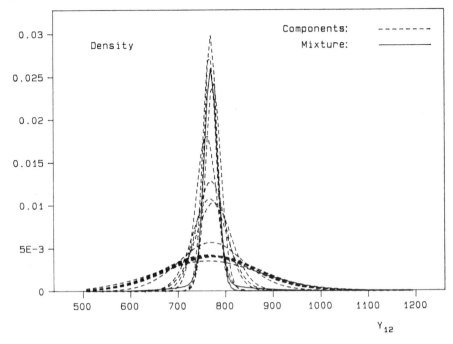

Figure 12.6a.   Forecasts for Y_{12} at $t = 11$.

the standard model has posterior probability near 1 at each time. Figures 12.6a and 12.7a reflect existing information given D_{11}. Figure 12.7a displays the posterior density of the current level parameter, $p(\mu_{11} \mid D_{11})$, the mixture of components (12.36) with $t - 1 = 11$. The four components (12.36) are also plotted on the graph. The components are all located between roughly 750 and 760, and are similarly spread over 730 to 780. That based on $M_{11}(1)$ is the most peaked and has corresponding posterior probability $p_{11}(1)$ very close to 1. This reflects the previous stability of the series and consistency with the standard DLM. As a result, the mixture of these four components is essentially equal to the first, standard component. This is a typical picture in stable periods. Similar comments apply to the one-step ahead forecast distribution $p(Y_{12} \mid D_{11})$ (from (12.28) with t=12), graphed in Figure 12.6a. This is a sixteen component mixture, the components also appearing on the graph. It is difficult to distinguish the overall mixture since it corresponds closely to the highly peaked components in the centre. The only additional feature of note is that several of the components are much more diffuse than the majority here, being those that condition on the outlier model at time 12. The observational variance inflation factor of 100 in the

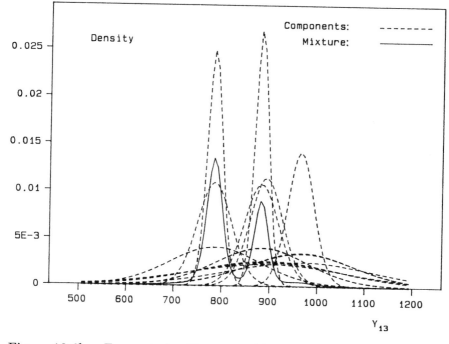

Figure 12.6b. Forecasts for Y_{13} at $t = 12$.

definition of $M_t(2)$ produces this spread. In combining the components however, these outlier components (along with those for level
and growth changes) have small probability and so contribute little
to the mixture. Again this is a typical picture in stable periods.

Proceed now to December 1955, time 12. $Y_{12} = 870$ is a very
wild observation relative to the standard forecast distribution, and
most of the probability under the mixture density in Figure 12.6a is
concentrated between 700 and 820. The outlier components, however, give appreciable probability to values larger than 870. Hence,
in updating to posterior model probabilities given D_{12}, those four
components that include $M_{12}(2)$ will receive much increased weights.
This is clear from Figure 12.9a; the outlier and level change models
share most of the posterior probability at $t = 12$, the former being
the more heavily weighted due to the initial balance of prior probabilities that slightly favour the outlier model. Figure 12.7b plots the
posterior density $p(\mu_{12} \mid D_{12})$ together with the four components;
this is the analogue at time 12 of Figure 12.7a at time 11. Here,
in contrast to Figure 12.7a, the components are rather disparate.
The peaked component located near 770 corresponds to the outlier
model for Y_{12}, being the density $p(\mu_{12} \mid M_{12}(2), D_{12})$. In this case,

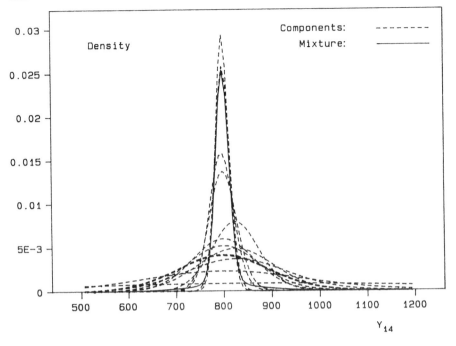

Figure 12.6c.   Forecasts for Y_{14} at $t = 13$.

the observation has effectively been ignored as an outlier, the inference being that the level remains between about 740 and 800. The
two peaked components located in the region of $Y_{12} = 870$ come
from the level and growth change models, the former providing the
more peaked posterior. If the observation is due to marked change
in level and/or growth parameters, then the inference is that the
current level actually lies in the region of 840 to 900. The fourth,
more diffuse component located near 840 is $p(\mu_{12} \mid M_{12}(1), D_{12})$, the
posterior from the standard model at time 12. If Y_{12} is a reliable
observation and no level or growth change has occurred, then clearly
the posterior for the level is a compromise between the prior, located
near 770, and the likelihood from Y_{12}, located at 870. The extra
spread in this posterior arises as follows. Conditional on $M_{12}(1)$, the
extreme observation leads to very large forecast errors resulting in inflated estimates of V in the corresponding components; thus, for each
$j_{11} = 1, \dots, 4$ the variance estimates $S_{12}(1, j_{11})$ are all inflated, and,
consequently, $S_{12}(1)$ is much larger than $S_{12}(j_{12})$ for $j_{12} > 1$. As a result, the four posterior T distributions $p(\mu_{12} \mid M_{12}(1), M_{11}(j_{11}), D_{12})$

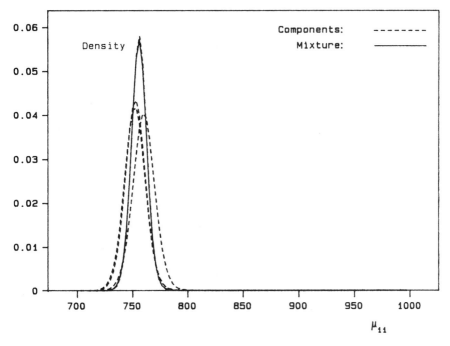

Figure 12.7a.   Posteriors for μ_{11} at $t = 11$.

are all rather diffuse and hence so is $p(\mu_{12} \mid M_{12}(1), D_{12})$, the plotted density. As it happens, this, obviously inappropriate (in retrospect) density receives essentially no posterior probability as can be
seen from Figure 12.9a, contributing negligibly to the overall mixture
posterior plotted in Figure 12.7b. The posterior is clearly bimodal.
This represents the ambiguity as to whether Y_{12} is an outlier or indicates the onset of a change in level or growth. If the former is true,
then the more peaked mode near 770 is the correct location, otherwise the second mode near 870 is correct. Until further information
is processed, there is a complete split between the two inferences.
Forecasting ahead to $t = 13$, January 1956, the bimodality carries
over to $p(Y_{13} \mid D_{12})$, graphed, along with the sixteen components,
in Figure 12.6b. The appearance of two distinct point forecasts with
appreciable probability in Figure 12.5 also evidences the bimodality.
An additional feature to note here concerns the components located
around 960 to 980. These correspond to the four model combinations
that include $M_{12}(4)$, the growth change model at time 12, although
they receive little weight in the overall mixture.

Moving on now to observe $Y_{13} = 784$ it becomes clear that Y_{12} was
in fact an outlier. Figure 12.9b presents posterior model probabilities

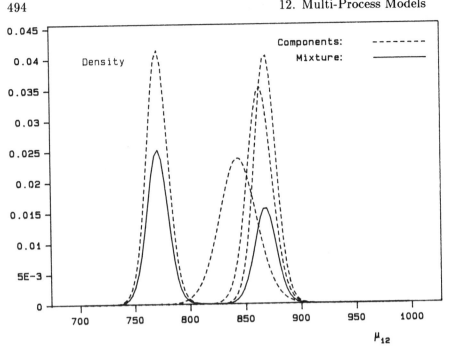

Figure 12.7b.   Posteriors for μ_{12} at $t = 12$.

analogous to those in Figure 12.9a, although these are the *one-step back*, or *smoothed* probabilities referring to models at the previous time point, calculated as in (12.32). For each t, the vertical bars in the figure are the retrospective probabilites $\Pr[M_{t-1}(j_{t-1}) \mid D_t]$ for each $j_{t-1} = 1,\dots,4$. These probabilities, calculated in (12.34), are very useful for retrospective assessment and diagnosis of model occurrence at any time given one further, confirmatory observation. Thus, for example, observing Y_{13} clarifies the position at time 12, with $\Pr[M_{12}(2) \mid D_{13}] \approx 1$. The outlier is clearly identified and, in the updating, has therefore been essentially ignored. After updating, $p(\mu_{13} \mid D_{13})$ appears in Figure 12.7c, along with its four components. The corresponding one-step ahead forecast densities appear in Figure 12.6c. From these graphs it is clear that things have reverted to stability with unimodal distributions once the outlier has been identified and accommodated.

In summary, the posteriors for model parameters tend to comprise components with similar locations in stable periods. These components separate out at the onset of an event, with posterior probabilities reflecting ambiguity as to whether the event relates to an outlier or a change-point. A further observation, or sometimes

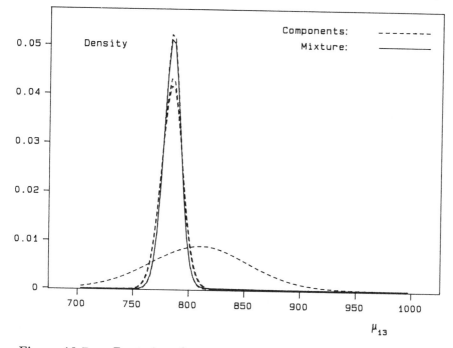

Figure 12.7c.   Posteriors for μ_{13} at $t = 13$.

several further observations, will then serve to identify the event and the posteriors reflect this as the series reverts to stability and consistency with the standard DLM.

   To further illustrate the analysis, consider the level change at $t = 37$, January 1958. Figures 12.8a and 12.8b present $p(\mu_t \mid D_t)$ for $t = 37$ and 38, along with the four components of each $p(\mu_t \mid M_t(j_t), D_t)$. Here the initial effect of the extreme observation Y_{37} leads again to bimodality in the posterior for μ_{37} at the time, Figure 12.8a. The observation is either an outlier or indicates a change-point, but the model cannot as yet distinguish the possibilities. Observing Y_{38} apparently confirms a level change, from around 860 to near 950, and the posterior for the level at time 38 reflects this. Again, finally, current and one-step back posterior model probabilities indicate the switching between models over time and the diagnosis of events.

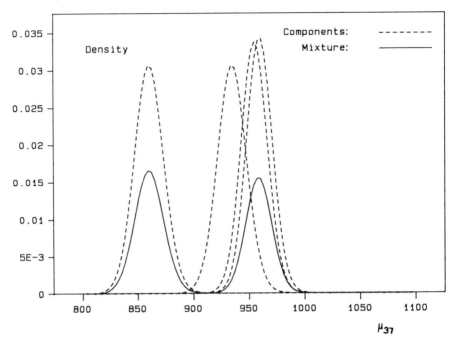

Figure 12.8a.   Posteriors for μ_{37} at $t = 37$.

12.4.5   Other applications

Several notable applications of this, and related multi-process models
exist in commercial and economic fields, where models have been developed and implemented primarily for short term forecasting (Harrison and Stevens, 1971, 1976a, b), Johnston and Harrison (1980).
The adaptablility of the multi-process approach to abrupt changes
in trend can be of crucial benefit to decision makers in such areas.

An interesting application in medicine is documented in West
(1982), Smith and West (1983), Smith *et al* (1983), and Trimble *et al*
(1983). That application concerns a problem in clinical monitoring
typical of many situations (not restricted to medicine) in which time
series observations relate to the state of a monitored subject, and
the detection and interpretation of abrupt changes in the pattern of
the data are of paramount importance. Often series are noisy and
difficult to interpret using simple methods, and the changes of interest are obscured by noise inherent in the measurement process, and
outliers in the data. In addition, series may be subject to abrupt
changes of several types, with only a subset corresponding to the

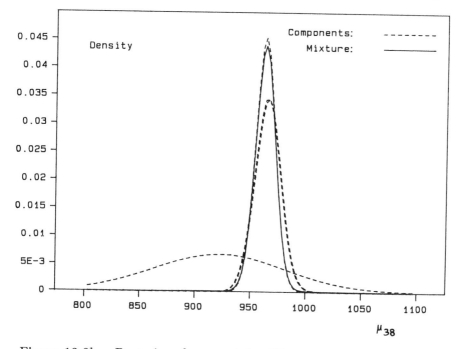

Figure 12.8b.   Posteriors for μ_{38} at $t = 38$.

changes of interest. It is therefore important to be able to distinguish, or diagnose, types of change as well as simply detect change of any kind.

The referenced medical application concerns the monitoring of the progress of kidney function in individual patients who had recently received a transplant. The level of renal functioning is indicated by the rate at which chemical substances are cleared from the blood, and this can be inferred indirectly from measurements on blood and/or urine levels of such substances. This process involves the use of well determined physiological relationships, but is subject to various sources of error thus producing noise corrupted measurements of filtration rates. Following transformations and corrections made for both physiological and statistical reasons, this process results in a response series that is inversely related to kidney well-being, and that can be expected to behave essentially according to a linear growth model in periods of consistent kidney function. In particular, (a) if the kidney functions at a stable, roughly constant rate, the response series should exhibit no growth; (b) if kidney function is improving with roughly constant growth between equally spaced observations, the response series should decay roughly linearly; and (c) the reverse

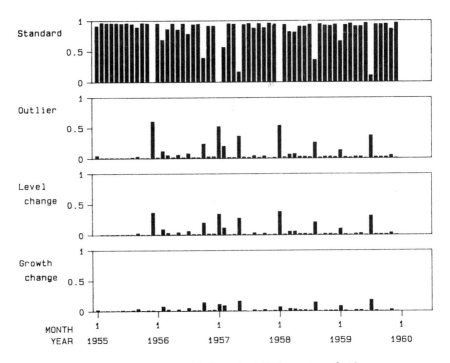

Figure 12.9a.   Posterior model probabilities at each time.

is the case if kidney function deteriorates at a roughly constant rate. Each of these cases is encountered in practice. Stable function (a) may occur at different times during post-operative patient care, with a successful transplant eventually leading to stable function at a high level, corresponding to a low level in the response series. Case (b) is anticipated immediately post-transplant as the transplanted organ is accepted by the patient's body and begins to function, eventually reaching normal levels of chemical clearance. Case (c) is expected if, at any time after the transplant, kidney function deteriorates. This may be attributable to rejection of the transplant and is obviously of paramount importance. One of the main objectives in monitoring the data series is thus to detect an abrupt change from either (a) or (b) to (c), consistent, in model terms, with an abrupt change in growth in the series from non-positive to positive. The appropriateness of the multi-process model is apparent, as is the emphasis on the use of the posterior model probabilities to detect and diagnose change when it occurs. A high posterior probability for the growth change model is used to signal the possibility of such a change. However, since only

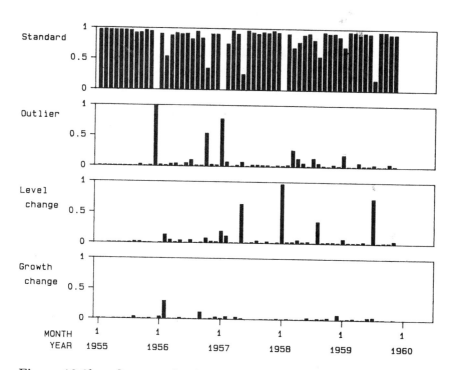

Figure 12.9b.   One-step back smoothed model probabilities at each
time.

changes from non-positive to postitive growth are consistent with
rejection, then only a subset of all possible growth changes give cause
for concern. For example, an abrupt change from positive to negative
growth is consistent with a marked and sustained improvement in
renal function. These particular features of this application raise
questions that are easily answered using the posterior distributions
for growth parameters; a change at time t of the sort consistent with
rejection is indicated if $\beta_{t-1} \leq 0$ and $\beta_t > 0$, which can be readily
assessed using the relevant posterior $p(\beta_t \mid D_t)$ and $p(\beta_{t-1} \mid D_t)$
(nb. the latter is a filtered distribution whose calculation has not
hitherto been discussed in the multi-process context; see Section
12.4.6 below).

The renal monitoring application has various other interesting features from a modelling viewpoint. Concerning the use of the four
component multi-process mixture, the response series are certainly
subject to outlying observations a small fraction (about 5%) of the

time due to mistakes in data transcription, equipment/operator mal-
function and blood/urine sample contamination. The level change
model is also very necessary, being consistent with marked increases
in level of kidney function when the patient is subject to dialysis
treatment to assist renal function. This is common and frequent in
early stages of post-operative care. Further, minor features of the
application include timing errors and irregularly spaced observations
and the use of quadratic powerlaw functions for the observational
variance. Fuller details of these, and other, features of the application are found in the above references.

12.4.6   Step ahead forecasting and filtering

Further important components of the analysis of multi-process mixture models are step ahead forecast distributions and retrospective
filtered distributions for models parameters in the past. Some features of these distributions are detailed here, fuller development being left to the interested reader. The development here is in terms
of the four component, linear growth model analysed above providing a concrete setting for the discussion of features that apply more
generally.

All step ahead forecast distributions are, of course, mixtures. Consider the position at time t with the posterior distributions summarised in (12.42). The one-step ahead forecast distribution for Y_{t+1}
is then simply the analogue of that in (12.38) with time index increased from t to $t+1$, namely the 4^2 component mixture

$$p(Y_{t+1} \mid D_t) = \sum_{j_{t+1}=1}^{4} \sum_{j_t=1}^{4} p(Y_{t+1} \mid M_{t+1}(j_{t+1}), M_t(j_t)) \pi(j_{t+1}) p_t(j_t).$$

The component densities of the mixture are the standard normal or
T forecast densities derived under the standard DLM determined by
$M_{t+1}(j_{t+1})$ and $M_t(j_t)$ jointly holding. Consider now forecasting two
steps ahead for Y_{t+2}. Given any combination of models at times t,
$t+1$ and $t+2$ it follows that the two-step ahead forecast distribution
is a standard form with density

$$p(Y_{t+2} \mid M_{t+2}(j_{t+2}), M_{t+1}(j_{t+1}), M_t(j_t), D_t)$$

for each j_t, j_{t+1} and j_{t+2}, each running from 1 to 4. The overall distribution is a mixture of these 4^3 components with mixing probabilities

$$\Pr[M_{t+2}(j_{t+2}), M_{t+1}(j_{t+1}), M_t(j_t) \mid D_t]$$
$$= \Pr[M_{t+2}(j_{t+2}) \mid M_{t+1}(j_{t+1}), M_t(j_t), D_t]$$
$$\times \Pr[M_{t+1}(j_{t+1}) \mid M_t(j_t), D_t]\Pr[M_t(j_t) \mid D_t]$$
$$= \pi(j_{t+2})\pi(j_{t+1})p_t(j_t).$$

Specifically,

$$p(Y_{t+2} \mid D_t)$$
$$= \sum_{j_{t+2}=1}^{4} \sum_{j_{t+1}=1}^{4} \sum_{j_t=1}^{4} p(Y_{t+2} \mid M_{t+2}(j_{t+2}), M_{t+1}(j_{t+1}), M_t(j_t)D_t)$$
$$\times \pi(j_{t+2})\pi(j_{t+1})p_t(j_t).$$

Forecasting further ahead obviously increases the number of components in the mixture to account for all possible models obtaining between time t and the forecast time point. The general details are left to the reader.

In specific applications there may be features of the model that can be exploited to simplify the problem of summarising these forecast distributions. In the outlier/level and growth change linear growth model here, for example, the structure of the multi-process at each time is such that the $M_t(j)$ differ only through evolution and observational variances, for all t and all j. This implies that, whilst the components of all mixture forecast distributions may have widely differing variances, the differences in point forecasts, the modes or means of the components, derive only from the differences in the means of θ_t across the models $M_t(j_t)$. $E[Y_{t+1} \mid M_{t+1}(j_{t+1}), M_t(j_t), D_t]$ does not depend on j_{t+1}; nor $E[Y_{t+2} \mid M_{t+2}, (j_{t+2}), M_{t+1}(j_{t+1}), M_t(j_t), D_t]$ on either of j_{t+2} or j_{t+1}, and so forth. Hence there are only four distinct point forecasts for each step ahead, clearly reducing calculations required. Further simplifications usually derive from consideration of the comments on approximation of mixtures in Section 12.3.3. For example, if one of the $p_t(j_t)$ is close to 1, then the corresponding components of forecast distributions dominate the mixtures and the others may be ignored.

Consider now the calculation of filtered distributions for retrospective analysis. As with forecast distributions, it should be immediately clear that filtered distributions are mixtures of standard components. The calculations for one-step back from time t are as follows. Conditional on $M_t(j_t)$ and $M_{t-1}(j_{t-1})$, the one-step back filtered distribution for the state vector $\boldsymbol{\theta}_{t-1}$ has the standard form derived in Section 4.7, Chapter 4, with density

$$p(\boldsymbol{\theta}_{t-1} \mid M_t(j_t), M_{t-1}(j_{t-1}), D_t).$$

There are 4^2 such component densities in the overall mixture filtered distribution

$$p(\boldsymbol{\theta}_{t-1} \mid D_t) = \sum_{j_t=1}^{4} \sum_{j_{t-1}=1}^{4} p(\boldsymbol{\theta}_{t-1} \mid M_t(j_t), M_{t-1}(j_{t-1}), D_t) p_t(j_t, j_{t-1}),$$

where the mixing probabilities

$$p_t(j_t, j_{t-1}) = \Pr[M_t(j_t), M_{t-1}(j_{t-1}) \mid D_t]$$

are as previously calculated in (12.40). Often filtering one-step back suffices, recent data providing little additional information about state vectors further back in time. Two, and more, step back distributions are calculated as multi-component mixtures in a similar fashion, the details again being left to the interested reader.

12.4.7 Further comments and extensions

There are other features of the analysis of even this relatively simple multi-process mixture model that may assume importance in application. Some particular features and extensions of the basic model are noted here. Again these points apply to general models although they are highlighted with reference to the particular model used here for the purposes of exposition.

Consider first the structuring of the multi-process. Here the four components of the model at each time differ only through the evolution and observational variances. These are designed to model the possibilities of evolution and observational errors that are significantly larger in absolute value than usual, whilst the retention of zero-means implies that the signs of the errors remain unanticipated. The sizes of the variances in the outlier and parameters changes states are determined in advance. The particular values are not critical, all that is necessary is that they provide significantly larger values than the standard model. The robustness to these values is similar to that in more usual mixture models for outliers, and the experience of others working with such models may be drawn upon in assigning values (Box and Tiao, 1968; Box, 1980; Smith and Pettit, 1985). See also the related discussion in Section 11.4.3 of Chapter 11. The use of discount factors to specify the evolution variances in the level and growth change states is a point of difference, although the basic ideas are similar. It is, of course, possible to use alternative methods, such as choosing the evolution variance matrices in the change models as multiples of those in the standard, which would be more in line with earlier uses of multi-processes (Harrison and Stevens, 1971, 1976a, b; Smith and West, 1983). However, the self-consistency and simplicity of the discount approach makes it the more attractive, and the end results would be similar. Note finally that the design of the multi-process structure is similar in principle whatever approach to discounting is taken; recall the various alternatives described in Section 12.4.1.

Other multi-processes might have component DLMs differing in one or more of the other components $\mathbf{F}_t$ and $\mathbf{G}_t$. This also allows for components having different numbers of parameters in the state vector, a possibility not hitherto considered although related ideas were discussed in connection with subjective intervention in Section 11.2.4 of Chapter 11.

The remaining inputs to the model required from the user are the prior model probabilities $\pi(j)$. These must be assessed and specified in advance, reflecting the forecaster's opinion as to the likely rate of occurrence of outliers, level changes and growth changes. Again experiences with outlier modelling more widely, as referenced above, for example, guide the choice of values of the outlier model probability. This will typically be near 0.05, rarely very much smaller but usually less than 0.1. Very precise values for this, and the change model probabilities, are unneccessary since the occurrence of a forecast error deemed extreme under the standard model will lead to posterior model probabilities that heavily favour the outlier and change models so long as their prior probabilities are not negligible. The discussion in Section 11.4.3, Chapter 11, of cross-over points between the predictive density under the standard model and those in the outlier/change models is again relevant here. It is also usual, though not a panacea, that outliers occur more frequently than changes so that the prior probabilities will often marginally favour the former. This is certainly true of data arising in the majority of commercial applications, though it may not be the case in other areas of application. Thus, when *something* happens at a given time t, the posterior probabilities $p_t(j_t)$ will tend to favour the outlier model. A further observation will identify what has happened as illustrated in Figures 12.9a and b in the CP6 example.

Often previous data on the series, or on similar, related series, will be available to guide the choice of prior probabilities. Also, although here assumed constant over time, there are obvious ways in which the probabilities may be updated as data is processed to reflect a changing view as to the frequency of occurrence of events. This may be based solely on the data, may use externally available information, or be based purely on subjective opinion. Going further with this notion, the model probabilities may even be linked to independent variables to allow for change with external circumstances, time, and so forth. The extension to incorporate Markov transitions between models discussed in Section 12.3.1 is a simple case of this; others might incorporate independent variable information that attempts to predict change through a separate model for the transition probabilities. In this way it is possible, for example, to produce models related to catastrophe theoretic ideas (Harrison and Smith, 1979), and the threshold switching models of classical time series (Tong and Lim, 1980; Tong, 1983).

The modelling of change points in dynamic models using multi-processes complements more conventional Bayesian approaches to change-point analysis in regression and related models. Key works in change-point modelling have been done by Smith (1975, 1980), Smith and Cook (1980). See also Pole and Smith (1985). Various related approaches and many references are found in Broemeling and Tsurumi (1988).

One important extensions of the basic multi-process model used here is to include a component model that is designed to allow for marked change in the observational variance V, hitherto assumed constant. In many areas of application, time series are subject to changes in volatility that can be represented by abrupt increases or decreases in V. Share indices, exchange rates and other financial time series are prime examples. In addition, inappropriate data transformation or variance functions lead to observational variances that change with level so that if the level shifts abruptly, then so will the variance. In Section 10.8 of Chapter 10 variance discounting was introduced to cater for minor, sustained, stochastic changes in V. The effect produced was simply to discount the degrees of freedom parameter n_t between observations, reducing it by multiplying by a discount factor less than, but close to, unity. This extends directly to model abrupt changes by including a component in the multi-process that is the same as the standard model except for the inclusion of a small variance discount factor. Details are straightforward and left to the reader; the main, technical point of difference is that, in this variance change state, the degrees of freedom parameter will be less than that in other states so that in collapsing mixture posterior distributions as in Section 12.4.2, the general results of Example 12.8 apply.

Concerning the collapsing of mixtures of posteriors more widely, the approximation here based on ignoring dependence on models more than $h = 1$ step back in time is adequate in the context of this particular model but may need relaxing in others. In particular, when working with higher dimensional models with possibly more components, it may take up to 3 or more observations after an exceptional event to determine the nature of the event and appropriately estimate the state parameters. In such case, $h > 1$ is needed and the number of components in all mixture distributions may increase dramatically. Schervish and Tsay (1988) provide illustrations and discussions of several analyses in which h is effectively allowed to increase with the number of observations analyses. Quite often, of course, there will be little to gain in doing this, as is the case when the series is essentially well-behaved, not deviating markedly from the standard DLM description. In such circumstances the technical and computational complexity of the multi-process model is highly redundant.

This potential redundancy of the multi-process approach can be obviated if computational restrictions make it infeasible. One possible extension of the approach is developed in Ameen and Harrison (1985b) for just this reason. The underlying idea is that, rather often, a series is stable over periods of observations, conforming to the standard DLM a good deal of the time — in fact approximately $100\pi(1)\%$ of the time. Thus, over long stretches of the series, the multi-process is redundant, posterior model probabilities being close to unity on the standard model. Only at times of exceptional events, and over one or a few observations stages following exceptions, is the full complexity of the multi-process really necessary. This is clearly illustrated with the CP6 analysis where the posterior model probabilities in Figures 12.9a and b clearly indicate that the mixture is only really necessary a small proportion of the time. The approach developed in this reference recognises this, and develops along the following lines:

(a) Model the series with the standard DLM subject to a simple
monitoring scheme such as the sequential Bayesian tests or
cusums described in Section 11.4 of Chapter 11. So long as
the monitoring indicates the data to be consistent with the
standard model, proceed with the usual DLM analysis. In
parallel, use alternative DLMs that provide for the anticipated
forms of departure from the standard. Although this is similar
to a multi-process, class I approach, the parallel DLMs only
come into play at stage (b) and are simply updated to provide
initialisation at that stage.

(b) When the monitor signals deterioration in forecast performance of the standard DLM, begin a period of operation with
a multi-process model, class II model, introducing the parallel
DLMs from (a).

(c) Continue this until the the exceptional event that triggered
the monitor signal has been identified, and the multi-process
analysis settles down to conformity with the standard DLM.
This occurs when the posterior model probability on the standard model is close to unity.

(d) At this stage, revert to operation with only the standard DLM
taken from the relevant component of the multi-process. Return to (a) and continue.

Under such a scheme, the use of mixtures is restricted to periods
of uncertainty about the series where they are needed to identify
exceptional events. In stable periods, the standard DLM analysis is
used, the computational economy being evident. Additionally, the
problems of inference and prediction are simplified since, for much
of the time, a single DLM is used rather than a mixture of possibly
many.

12.5 EXERCISES

(1) Refine the multi-process analysis of Example 12.2 by considering a finer grid of values for the discount factor δ. Specifically,
take
$$\mathcal{A} = \{1.0, 0.975, 0.95, \dots, 0.725, 0.7\},$$

with initial probabilities $p_0(j) = 1/13$, $(j = 1, \dots, 13)$. Plot
the posterior probabilities $p_t(j)$ versus j at each of $t = 34$ at
$t = 116$, and comment on the support provided by the data for
discount factors over this chosen range.

(2) In the framework of Example 12.3, verify that the posterior distribution for δ concentrates at that value δ_i such that $|\delta_i - \delta_0| = \min_j |\delta_j - \delta_0|$.

(3) Consider the multi-process, class I model defined for $t = 1$ by

$$M(j): \qquad \{1, 1, 100, W(j)\}, \qquad\qquad (j = 1, 2),$$

where $W(1) = 0$ and $W(2) = 50$. Initially, $p_0(j) = \Pr[M(j)|D_0] = 0.5$ and

$$(\mu_0|D_0, M(j)) \sim N[100, 100]$$

for each model, $j = 1, 2$.

(a) With $p_1(j) = \Pr[M(j)|D_1]$, show that $p_1(1) < p_1(2)$ if, and only if, $|e_1| > 14.94$ where $e_1 = Y_1 - 100$, the forecast error common to the two models.

(b) Deduce that

$$\Pr[p_1(1) \geq p_1(2)|M(1), D_0] = 0.74$$

and

$$\Pr[p_1(1) < p_1(2)|M(2), D_0] = 0.34.$$

(c) Comment on the results in (b)

(4) Consider a multi-process model for Y_1 given by

$$M(j): \qquad \{1, 1, V(j), W\}, \qquad\qquad (j = 1, 2),$$

where $V(1) = 1$ and $V(2) = 33$, and $p_0(1) = \Pr[M(1)|D_0] = 0.9$. Suppose also that $(\mu_1|D_0) \sim N[0, 3]$.

(a) Show that $p(Y_1|D_0)$ is a mixture of two normals,

$$(Y_1|D_0) \sim \begin{cases} N[0, 4], & \text{with probability } 0.9; \\ N[0, 36], & \text{with probability } 0.1. \end{cases}$$

(b) Show that $p(\mu_1|D_1)$ is also a mixture of two normals,

$$(\mu_1|D_1) \sim \begin{cases} N[\frac{3Y_1}{4}, \frac{3}{4}], & \text{with probability } p_1(1) = q(Y_1); \\ N[\frac{Y_1}{12}, \frac{11}{4}], & \text{with probability } 1 - q(Y_1), \end{cases}$$

where

$$q(Y_1)^{-1} = 1 + 0.037 e^{Y_1^2/9}.$$

(c) Deduce that

$$E[\mu_1|D_1] = [1 + 8q(Y_1)]Y_1/12.$$

(d) Write a computer program to graph each of $q(Y_1)$, $E[\mu_1|D_1]$ and $V[\mu_1|D_1]$ as functions of Y_1 over the interval $-5 \le Y_1 \le 5$. Comment on the form of this functions as Y_1 varies.

(e) Write a computer program to graph the posterior density $p(\mu_1|D_1)$ as a function of μ_1 for any given value of Y_1. Comment on the form of the posterior for the three possible values $Y_1 = 0$, $Y_1 = 4$ and $Y_1 = 8$.

(5) A forecast system uses the first-order polynomial model $\{1, 1, V_t, W_t\}$ as a base model for forecasting sales of a group of products. Intervention information is available to incorporate into the model analysis. Initially, $(\mu_0|D_0) \sim N[400, 20]$. Produce forecast distributions $p(Y_t|D_0, I)$ for each of $t = 1, 2, 3$ and 4 given the following intervention information I available at $t = 0$.

(a) $V_1 = 100$ and $\omega_1 \sim N[100, W_1]$ with $W_1 = 400$. This, non-zero mean evolution error describes the likely impact on sales level of the receipt of a new export licence in the first time period. Verify that $(Y_1|D_0, I) \sim N[500, 520]$.

(b) $\nu_2 \sim N[500, V_2]$ with $V_2 = 200$, describing the anticipated effect of an additional spot order at $t = 2$. Also $W_2 = 5$. Verify that $(Y_2|D_0, I) \sim N[1000, 625]$.

(c) The expected effect of a price increase at $t = 3$ is modelled via a transfer response function, with

$$Y_t = \mu_t + E_t + \nu_t, \qquad\qquad (t = 3, 4, \ldots),$$

where $V_t = 100$, $W_t = 5$ and

$$E_t = 0.8E_{t-1}, \qquad\qquad (t = 4, 5, \ldots),$$

with

$$(E_3|D_0, I) \sim N[-40, 200],$$

independently of $(\mu_0|D_0)$.
Verify that $(Y_3|D_0, I) \sim N[460, 730]$.

(d) At $t = 4$, there is a prior probability of 0.4 that a further export licence will be granted. If so, then $\omega_4 \sim N[200, 600]$, otherwise $\omega_4 \sim N[0,5]$ as usual. Show that $p(Y_4|D_0, I)$ is a mixture of two normals,

$$(Y_4|D_0, I) \sim \begin{cases} N[668, 1258], & \text{with probability } 0.4; \\ N[468, 663], & \text{with probability } 0.6. \end{cases}$$

Plot the corresponding forecast density over the range $400 < Y_4 < 750$. Calculate the forecast mean $E[Y_4|D_0, I]$ and variance $V[Y_4|D_0, I]$. From (c), calculate the joint distribution of $(Y_3, E_3|D_0, I)$ and deduce the posterior $(E_3|Y_3, D_0, I)$. Verify that $(E_3|Y_3 = 387, D_0, I) \sim N[-60, 145]$.

(f) Calculate the joint posterior $p(\mu_4, E_4|Y_4, D_0, I)$ as a function of Y_4, and deduce the marginal posterior for μ_4.

(6) Write a computer program to implement the multi-process, class II model used in Section 12.4. Verify the program by reproducing the analysis of the CP6 Sales series in 12.4.3.

CHAPTER 13

NON-LINEAR DYNAMIC MODELS

13.1 INTRODUCTION

In previous chapters we have encountered several models which depend on parameters that introduce parameter non-linearities into otherwise standard DLMs. Although the full class of DLMs provides an enormous variety of useful models, it is the case that, sometimes, elaborations to include models with *unknown* parameters result in such non-linearities, thus requiring extensions of the usual linear model analysis. Some typical, and important, examples are as follows.

EXAMPLE 13.1. In many commercial series in which seasonality is a major factor, an apparent feature of the form of seasonal patterns is that amplitudes of seasonal components appear to increase markedly at higher levels of the series. This occurs rather commonly in practice with positive data series. With a dynamic linear model for seasonality, it is certainly possible to adapt to changes in seasonal parameters by allowing for major variation through the evolution errors. However, when the changes are very marked and have a systematic form related to the level of the series, they are more appropriately modelled in an alternative way. Let α_t represent the underlying, non-seasonal level of the series at time t, that may include trend and regression terms. The usual, linear model for seasonality of period p is defined via the mean response $\mu_t = \alpha_t + \phi_t$ where ϕ_t is the seasonal effect at time t. Now the observed inflation in seasonal deviations from level as α_t increases can be directly modelled via a mean response that is non-linear in α_t and ϕ_t, the simplest such model being multiplicative of the form

$$\mu_t = \alpha_t(1 + \phi_t).$$

More elaborate non-linearities are also possible, but this simple, multiplicative model alone is of major use in practice, and is referred to as a *multiplicative seasonal effects model*. Two possible models for Y_t use this multiplicative form.

(a) A DLM obtained via log-transformation. With $Y_t > 0$ (as is essentially always the case when multiplicative seasonality is apparent), assume that

$$Y_t = \alpha_t(1 + \phi_t)e^{\nu_t}$$

where ν_t is normally distributed observational noise. Then Y_t follows a a lognormal model (Section 10.6 of Chapter 10) and

$$\log(Y_t) = \alpha_t^* + \phi_t^* + \nu_t$$

where $\alpha_t^* = \log(\alpha_t)$ and $\phi_t^* = \log(1 + \phi_t)$. Thus the series may be modelled using a standard seasonal DLM after log transformation, α_t^* and ϕ_t^* both being linear functions of the state vector. Note also that this is consistent with an observational variance increasing with level on the original scale (Section 10.7, Chapter 10).

(b) For reasons of interpretation of model parameters (and others, see Section 10.6, Chapter 10) it is often desirable to avoid transformation, suggesting a model of the form

$$Y_t = \alpha_t(1 + \phi_t) + \nu_t.$$

Note that this representation may also require the use of a variance power law to model increased observational variation with level.

Model (b) is considered here. The state vector $\boldsymbol{\theta}_t' = (\boldsymbol{\theta}_{t1}', \boldsymbol{\theta}_{t2}')$ comprises trend and/or regression parameters $\boldsymbol{\theta}_{t1}$ such that $\alpha_t = \mathbf{F}_{t1}'\boldsymbol{\theta}_{t1}$, and seasonal parameters $\boldsymbol{\theta}_{t2}$ such that $\phi_t = \mathbf{F}_{t2}'\boldsymbol{\theta}_{t2}$, for some known vectors $\mathbf{F}_{t1}$ and $\mathbf{F}_{t2}$. Thus

$$Y_t = \mathbf{F}_{t1}'\boldsymbol{\theta}_{t2}[1 + \mathbf{F}_{t2}'\boldsymbol{\theta}_{t2}],$$

a bilinear function of the state vector. In summary, the model here involves parameter non-linearities in the observational equation simply through the mean response function, and the parameters involved are standard time-varying elements of the state vector.

See also Abraham and Ledholter (1985, Chapter 4), Harrison (1965), Harrison and Stevens (1971) and Gilchrist (1976, Chapter 8) for further discussion of multiplicative seasonality. In West, Harrison and Pole (1987), multiplicative seasonal models are implemented as components of trend/ seasonal/ regression DLMs, the analysis being based on the use of linearisation as describe below in Section 13.2.

EXAMPLE 13.2. Recall the simple transfer response model in Example 9.1, Chapter 9, in which a DLM is defined by the equations $Y_t = \mu + \nu_t$ and $\mu_t = \lambda\mu_{t-1} + \psi_t X_t$, X_t being the observed value of a regressor variable at t. With λ assumed known, this is a DLM. Otherwise, as introduced in Section 9.3.3, the evolution of the state parameter μ_t involves a non-linear term, multiplicative in λ and μ_{t-1}, and the DLM analysis is lost. The same applies to the general transfer function models of Definition 9.1, or the alternative representation in equation (9.6). In these models, the evolution equation for the state vector at time t involves bilinear terms in state vector parameters from time $t-1$ and other, constant but unknown, parameters from $\mathbf{G}$.

EXAMPLE 13.3. Long term growth towards an asymptote may be modelled with a variety of parametric forms, often called growth or trend curves (Harrison and Pearce, 1972; Gilchrist, 1976, Chapter 9). Most such curves are basically non-linear in their defining parameters. A model similar to those in Example 13.2 is obtained when modelling long term growth using Gompertz growth curves, as follows. Suppose that $\mu_t = \alpha - \beta\exp(-\gamma t)$ for some parameters α, β and γ, with $\gamma > 0$. Then $\mu_0 = \alpha - \beta$, $\lim_{t\to\infty} \mu_t = \alpha$ and μ_t grows exponentially over time from $\alpha - \beta$ to β. The form

$$e^{\mu_t} = \exp[\alpha - \beta e^{-\gamma t}], \qquad (t > 0),$$

is known as a Gompertz curve, and often provides an appropriate qualitative description of growth to an asymptote. If such growth is assumed for the series Y_t, then transforming to logs suggest the model $\log(Y_t) = \mu_t + \nu_t$. Now, let $\alpha_t = \alpha$ and $\beta_t = -\beta\exp(-\gamma t)$, so that $\beta_t = \lambda\beta_{t-1}$ where $\lambda = \exp(-\gamma)$ and $0 < \lambda < 1$. Then $\log(Y_t)$ follows a DLM $\{\mathbf{F}, \mathbf{G}, V_t, \mathbf{0}\}$ where

$$\mathbf{F} = \mathbf{E}_2 = \begin{pmatrix} 1 \\ 0 \end{pmatrix} \qquad \text{and} \qquad \mathbf{G} = \begin{pmatrix} 1 & \lambda \\ 0 & \lambda \end{pmatrix}.$$

For fixed λ, this is an evolution noise-free DLM (with a second-order polynomial as the special case when $\lambda = 1$). Such a model with a non-zero evolution variance matrix $\mathbf{W}_t$ provides the basic Gompertz form with parameters that vary stochastically in time, allowing for random deviation in the growth away from an exact Gompertz curve. In practice, it is usually desirable to learn about λ, so that the evolution equation becomes bilinear in unknown parameters.

EXAMPLE 13.4. The DLM representations of ARMA processes, and their extensions to non-stationary noise models based on ARMA processes with time-varying coefficients, have a structure similar to the models above when these parameters are uncertain. See Section 9.4 of Chapter 9. Again the non-linearities involve bilinear terms in evolution and/or observation equations of what would be DLMs if the parameters were known.

EXAMPLE 13.5. A wide range of problems with rather different structure is typified by Example 12.1 of Chapter 12. There a discount factor was viewed as uncertain; more generally, the evolution variance matrices $\mathbf{W}_t$ may depend on constant but uncertain parameters to be estimated.

Whatever the particular structure of parameter non-linearities is in any application, the basic DLM analysis must be extended to cater for it. We have come a long way with linear models, the associated normal theory being analytically tractable and satisfyingly complete. When extending the framework to allow non-linearities, formal theory defines the analysis as usual but, essentially without exception, the calculation of the required components of analysis becomes burdensome. Implementation of the formally well-defined analysis requires the use of numerical integration to arbitrarily well approximate mathematically defined integrals, and so is in fact impossible. However, a variety of approximation techniques exist and, in practice, much can be done using such techniques. We begin by discussing the simplest, and currently by far the most widely used, types of approximation in the next Section. Following this, more sophisticated (though not necessarily more useful in practice) techniques based on numerical integration are developed.

13.2 LINEARISATION AND RELATED TECHNIQUES

Many models with parameter non-linearities may be written in the following form.

$$
\begin{aligned}
Y_t &= F_t(\boldsymbol{\theta}_t) + \nu_t, \\
\boldsymbol{\theta}_t &= \mathbf{g}_t(\boldsymbol{\theta}_{t-1}) + \boldsymbol{\omega}_t,
\end{aligned}
\tag{13.1}
$$

where $F_t(.)$ is a known, non-linear regression function mapping the $n-$vector $\boldsymbol{\theta}_t$ to the real line, $\mathbf{g}_t(.)$ a known, non-linear vector evolution function, and ν_t and ω_t are error terms subject to the usual assumptions. Various *linearisation* techniques have been developed for such models, all being based in essence on the use of linear approximations to non-linearities. The most straightforward, and easily interpreted, approach is that based on the use of first order Taylor series approximations to the non-linear regression and evolution functions in (13.1). This requires the assumptions that both $f_t(.)$ and $\mathbf{g}_t(.)$ be (at least once-) differentiable functions of their vector arguments.

Suppose, as usual, that $\nu_t \sim N[0, V]$ for some constant but unknown variance V. Assume also that, at time $t-1$, historical information about the state vector $\boldsymbol{\theta}_{t-1}$ and V is (approximately) summarised in terms of standard posterior distributions:

$$(\boldsymbol{\theta}_{t-1} \mid V, D_{t-1}) \sim N[\mathbf{m}_{t-1}, \mathbf{C}_{t-1} V / S_{t-1}],$$

and, with $\phi = 1/V$,

$$(\phi \mid D_{t-1}) \sim G[n_{t-1}/2, d_{t-1}/2],$$

with estimate of V given by $S_{t-1} = d_{t-1}/n_{t-1}$. Also,

$$(\boldsymbol{\omega}_t \mid V, D_{t-1}) \sim N[\, \mathbf{0}, \mathbf{W}_t V / S_{t-1}],$$

with $\mathbf{W}_t$ known. Then

$$(\boldsymbol{\theta}_{t-1} \mid D_{t-1}) \sim T_{n_{t-1}}[\mathbf{m}_{t-1}, \mathbf{C}_{t-1}],$$

being uncorrelated with

$$(\boldsymbol{\omega}_t \mid D_{t-1}) \sim T_{n_{t-1}}[\, \mathbf{0}, \mathbf{W}_t].$$

$\mathbf{m}_{t-1}$ provides the estimate of $\boldsymbol{\theta}_{t-1}$ about which a Taylor series expansion of the evolution function gives

$$\mathbf{g}_t(\boldsymbol{\theta}_t) = \mathbf{g}_t(\mathbf{m}_{t-1}) + \mathbf{G}_t(\boldsymbol{\theta}_{t-1} - \mathbf{m}_{t-1})$$
$$+ \text{ quadratic and higher order terms}$$
$$\text{in elements of } (\boldsymbol{\theta}_{t-1} - \mathbf{m}_{t-1}),$$

where $\mathbf{G}_t$ is the $n \times n$ matrix derivative of the evolution function evaluated at the estimate $\mathbf{m}_{t-1}$, namely

$$\mathbf{G}_t = \left[\frac{\delta \mathbf{g}_t(\boldsymbol{\theta}_{t-1})}{\delta \boldsymbol{\theta}'_{t-1}} \right]_{\boldsymbol{\theta}_{t-1} = \mathbf{m}_{t-1}};$$

obviously $\mathbf{G}_t$ is known. Assuming that terms other than the linear term are negligible, the evolution equation becomes

$$\boldsymbol{\theta}_t \approx \mathbf{g}_t(\mathbf{m}_{t-1}) + \mathbf{G}_t(\boldsymbol{\theta}_{t-1} - \mathbf{m}_{t-1}) + \boldsymbol{\omega}_t = \mathbf{h}_t + \mathbf{G}_t\boldsymbol{\theta}_{t-1} + \boldsymbol{\omega}_t, \quad (13.2)$$

where $\mathbf{h}_t = \mathbf{g}_t(\mathbf{m}_{t-1}) - \mathbf{G}_t\mathbf{m}_{t-1}$ is known. (13.2) is a *linearised* version of the evolution equation, linearised about the expected value of $\boldsymbol{\theta}_{t-1}$.

Assuming (13.2) as an adequate approximation to the model, it follows immediately that the usual DLM evolution applies, with the minor extension to include an additional, known term $\mathbf{h}_t$ in the evolution equation. Thus the prior for $\boldsymbol{\theta}_t$ is determined by

$$(\boldsymbol{\theta}_t \mid V, D_{t-1}) \sim \mathrm{N}[\mathbf{a}_t, \mathbf{R}_t V / S_{t-1}]$$

so that

$$(\boldsymbol{\theta}_t \mid D_{t-1}) \sim \mathrm{T}_{n_{t-1}}[\mathbf{a}_t, \mathbf{R}_t], \quad (13.3)$$

with defining quantities

$$\begin{aligned} \mathbf{a}_t &= \mathbf{h}_t + \mathbf{G}_t\mathbf{m}_{t-1} = \mathbf{g}_t(\mathbf{m}_{t-1}), \\ \mathbf{R}_t &= \mathbf{G}_t\mathbf{C}_{t-1}\mathbf{G}'_t + \mathbf{W}_t. \end{aligned} \quad (13.4)$$

Proceeding to the observation equation, similar ideas apply. Now the non-linear regression function is linearised about the expected value $\mathbf{a}_t$ for $\boldsymbol{\theta}_t$, leading to

$$\begin{aligned} F_t(\boldsymbol{\theta}_t) = F_t(\mathbf{a}_t) &+ \mathbf{F}'_t(\boldsymbol{\theta}_t - \mathbf{a}_t) \\ &+ \text{quadratic and higher order terms} \\ &\quad \text{in elements of} \quad (\boldsymbol{\theta}_t - \mathbf{a}_t), \end{aligned}$$

where $\mathbf{F}_t$ is the n−vector derivative of $F_t(.)$ evaluated at the prior mean $\mathbf{a}_t$, namely

$$\mathbf{F}_t = \left[\frac{\delta F_t(\boldsymbol{\theta}_t)}{\delta \boldsymbol{\theta}_t} \right]_{\boldsymbol{\theta}_t = \mathbf{a}_t},$$

which is known. Assuming the linear term dominates the expansion leads to the linearised observation equation

$$Y_t = f_t + \mathbf{F}'_t(\boldsymbol{\theta}_t - \mathbf{a}_t) + \nu_t = (f_t - \mathbf{F}'_t\mathbf{a}_t) + \mathbf{F}'_t\boldsymbol{\theta}_t + \nu_t, \quad (13.5)$$

where $f_t = F_t(\mathbf{a}_t)$. Combining this with (13.3) and (13.4) leads to a DLM with, in addition to the extra term in the evolution equation, a similar term $f_t - \mathbf{F}'_t \mathbf{a}_t$ in the observation equation. From these equations it follows that, for forecasting one-step ahead,

$$(Y_t \mid D_{t-1}) \sim \mathrm{T}_{n_{t-1}}[f_t, Q_t]$$

where $Q_t = \mathbf{F}'_t \mathbf{R}_t \mathbf{F}_t + S_{t-1}$. Also, once Y_t is observed giving forecast error $e_t = Y_t - f_t$, the standard updating equations for $\boldsymbol{\theta}_t$ and V apply directly. Thus the linearised model is a DLM. Some comments on this analysis are in order.

(1) Firstly, the one-step ahead point forecast $f_t = F_t[\mathbf{g}_t(\mathbf{m}_{t-1})]$ retains precisely the form of the non-linear mean response from the model (13.1) with $\boldsymbol{\theta}_{t-1}$ assumed known and equal to its estimate $\mathbf{m}_{t-1}$. Hence predictions accord with the model; if, for example, either regression or evolution functions impose bounds on the mean response, the predictor accords with these bounds.

(2) Consider forecasting ahead to time $t+k$ from time t. Applying linearisation to the evolution equations successively over time, it follows that

$$(\boldsymbol{\theta}_{t+k}|D_t) \sim \mathrm{T}_{n_t}[\mathbf{a}_t(k), \mathbf{R}_t(k)]$$

with moments successively defined as follows. Setting $\mathbf{a}_t(0) = \mathbf{m}_t$ and $\mathbf{R}_t(0) = \mathbf{C}_t$, we have, for $k = 1, 2, \ldots,$

$$\mathbf{a}_t(k) = \mathbf{g}_{t+k}(\mathbf{a}_t(k-1)),$$
$$\mathbf{R}_t(k) = \mathbf{G}_t(k)\mathbf{R}_t(k-1)\mathbf{G}_t(k)' + \mathbf{W}_{t+k},$$

where

$$\mathbf{G}_t(k) = \left[\frac{\delta \mathbf{g}_{t+k}(\boldsymbol{\theta}_{t+k-1})}{\delta \boldsymbol{\theta}'_{t+k-1}}\right]_{\boldsymbol{\theta}_{t+k-1}=\mathbf{a}_t(k-1)}$$

Then,

$$(Y_{t+k} \mid D_t) \sim \mathrm{T}_{n_t}[f_t(k), Q_t(k)]$$

where

$$f_t(k) = F_{t+k}(\mathbf{a}_t(k)),$$
$$Q_t(k) = \mathbf{F}_t(k)'\mathbf{R}_t(k)\mathbf{F}_t(k) + S_t,$$

with

$$\mathbf{F}_t(k) = \left[\frac{\delta F_{t+k}(\boldsymbol{\theta}_{t+k})}{\delta \boldsymbol{\theta}_{t+k}}\right]_{\boldsymbol{\theta}_{t+k} = \mathbf{a}_t(k)}.$$

The forecast function $f_t(k)$ can be written as

$$f_t(k) = F_{t+k}[\mathbf{g}_{t+k}\{\mathbf{g}_{t+k-1}(\ldots \mathbf{g}_{t+1}(\mathbf{m}_t)\ldots)\}],$$

retaining the required non-linear form into the future.

(3) In truncating Taylor series expansions, the implicit assumption is that the higher order terms are negligible relative to the first. To be specific, consider the linear approximation to the regression function about $\boldsymbol{\theta}_t = \mathbf{a}_t$. When linearising smooth functions, local linearity is evident so that, in a region near the centre $\mathbf{a}_t$ of linearisation, the approximation will tend to be adequate. Far away from this region, it may be that the approximation error increases, often dramatically. However, if most of the probability under $p(\boldsymbol{\theta}_t \mid D_{t-1})$ is concentrated tightly about the estimate $\mathbf{a}_t$, then there is high probability that the approximation is adequate.

(4) The above comments notwithstanding, it should be noted that the linearised model as defined in equations (13.2) and (13.5) defines a valid DLM without reference to its use as an approximation to (13.1). Guided by (13.1), the linearised model and DLM analysis may be accepted for forecasting, producing forecast functions with the desired, non-linear features, its appropriateness being judged on the usual basis of forecast accuracy. With this view, the extent to which the DLM approximates (13.1) is irrelevant; note, however, that the resulting posterior distributions cannot be considered as valid for inference about the state vector in (13.1) without such considerations.

(5) As in non-linear modelling generally, non-linear transformations of original parameters in $\boldsymbol{\theta}_t$ may lead to evolution or regression functions that are much closer to linearity in regions of interest than with the original metric.

(6) Further on the topic of approximation, the evolution (and observation) errors can be viewed as accounting for the neglected terms in the Taylor series expansions.

(7) Some or all of the elements of $\boldsymbol{\theta}_t$ may be constant over time. Thus models such as in Examples 13.2 and 13.3 above lie

within this class; the constant parameters are simply incorporated into an extended state vector.

(8) If either $F_t(.)$ or $g_t(.)$ are linear functions, then the corresponding linearised equations are exact.

(9) Note that the model (13.2) is easily modified to incorporate a stochastic drift in θ_{t-1} *before* applying the non-linear evolution function in (15.36), rather than by adding ω_t *after* transformation. The alternative is simply

$$\theta_t = g_t(\theta_{t-1} + \delta_t),$$

for some zero-mean error δ_t with a known variance matrix U_t. Then a_t and R_t are as defined in (13.4) with

$$W_t = G_t U_t G_t'.$$

EXAMPLE 13.1 (continued). Models in which the non-linearities are bilinear abound. Examples 13.1 to 13.4 inclusive are of this form. In Example 13.1, the multiplicative seasonal model, we have state vector $\theta_t' = (\theta_{t1}', \theta_{t2}')$, a linear evolution equation with $g_t(\theta_{t-1}) = G_t\theta_{t-1}$ for some known evolution matrix G_t, and a bilinear regression function

$$F_t(\theta_t) = (F_{t1}'\theta_{t1})[1 + (F_{t1}'\theta_{t2})].$$

Thus the model simplifies, linearisation is only used in the observation equation. With $a_t' = (a_{t1}', a_{t2}')$, it follows that the prior estimate for the non-seasonal trend component α_t is $f_{t1} = F_{t1}'a_{t1}$, whilst that for the seasonal effect ϕ_t is $f_{t2} = F_{t2}'a_{t2}$. Hence the one-step ahead point forecast is $f_t = f_{t1}(1 + f_{t2})$, the product of the estimated components. This model, and form of approximation, is the basis for many practical schemes of analysis of time series exhibiting the forms of behaviour consistent with multiplicative seasonality (Harrison, 1965; West, Harrison and Pole, 1987).

This approach, along with many variants, has been widely used in various fields of application, particularly in non-linear state space models used in communications and control engineering. Here, and elsewhere, non-Bayesian techniques using linearisation combined with classical least squares theory led, in the 1960's and 1970's, to a plethora of related techniques under such names as extended Kalman filtering, generalised Kalman filtering, non-linear filtering, and so forth. The books of Anderson and Moore (1979), Jazwinski (1970) and Sage and Melsa (1971) provide good discussion and references.

One obvious variant on the approach is to extend the Taylor series expansion to include higher order terms in the relevant state vectors. Considering the regression function for example, the linear approximation can be refined by including the second order term to give

$$F_t(\boldsymbol{\theta}_t) = F_t(\mathbf{a}_t) + \mathbf{F}_t'(\boldsymbol{\theta}_t - \mathbf{a}_t) + \frac{1}{2}(\boldsymbol{\theta}_t - \mathbf{a}_t)'\mathbf{H}_t(\boldsymbol{\theta}_t - \mathbf{a}_t)$$

$$+ \text{ cubic and higher order terms in elements of } (\boldsymbol{\theta}_t - \mathbf{a}_t),$$

where

$$\mathbf{H}_t = \left[\frac{\delta \mathbf{F}_t(\boldsymbol{\theta}_t)}{\delta \boldsymbol{\theta}_t'}\right]_{\boldsymbol{\theta}_t = \mathbf{a}_t} = \left[\frac{\delta^2 \mathbf{F}_t(\boldsymbol{\theta}_t)}{\delta \boldsymbol{\theta}_t \delta \boldsymbol{\theta}_t'}\right]_{\boldsymbol{\theta}_t = \mathbf{a}_t}$$

With this quadratic approximation to $F_t(\boldsymbol{\theta}_t)$, the distribution implied for Y_t is no longer T or normal. However, based on the assumed T or normal prior for $\boldsymbol{\theta}_t$, moments can be calculated. Assuming $n_{t-1} > 2$, of course, it follows directly that

$$f_t = \mathrm{E}[Y_t \mid D_{t-1}] = F_t(\mathbf{a}_t) + \frac{1}{2} \text{ trace } (\mathbf{H}_t\mathbf{R}_t).$$

Note the additional term here that accounts for some of the uncertainty in $\boldsymbol{\theta}_t$, introducing the *curvature* matrix $\mathbf{H}_t$ at the point $\boldsymbol{\theta}_t = \mathbf{a}_t$. Similarly, if $n_{t-1} > 4$ the variance of Y_t may be calculated. Notice that this involves calculation of mixed, fourth order moments of the multivariate T distribution. For updating once Y_t is observed, however, this higher order approximation leads to a difficult analysis.

One other variant of the basic linearisation technique specific to the case of models with bilinearities is suggested, and implemented in a case study, in Migon (1984) and Migon and Harrison (1985). Suppose that $\mathbf{g}_t(\boldsymbol{\theta}_{t-1})$ in (13.1) is a bilinear function of $\boldsymbol{\theta}_{t-1}$. Then given the conditional normal distribution for $\boldsymbol{\theta}_{t-1}$, the exact mean and variance of $\mathbf{g}_t(\boldsymbol{\theta}_{t-1})$ can be calculated using standard normal theory. Thus the exact prior mean and variance matrix of $\boldsymbol{\theta}_t$ can be calculated. Assuming approximate normality for $\boldsymbol{\theta}_t$, then a linear or bilinear regression function $\mathbf{F}_t(\boldsymbol{\theta}_t)$ implies, by a similar argument, the joint first and second order moments of Y_t and $\boldsymbol{\theta}_t$. Thus, again assuming approximate normality of all components, the forecast and posterior distributions are deduced. Given the bilinearity, the approximations here involve the assumptions of approximate normality of products of normal components. In practice, this approach

is typically rather similar to direct linearisation up to second order although remains to be further investigated from a theoretical
standpoint.

13.3 CONSTANT PARAMETER NON-LINEARITIES: MULTI-PROCESS MODELS

A second class of models may be analysed rather more formally
within the framework of multi-process, class I models of Chapter
12. These are models whose non-linearities are due to the appearance of constant, but unknown, parameters α in one or more of
the components. Examples 13.2 to 13.4 are special cases. Thus α
could comprise elements of a constant G matrix, transformation parameters, ARMA parameters, discount factors, and so forth. The
parameter space for α, denoted by $\mathcal{A}$, is usually continuous (at least
in part), and often bounded. It is assumed that α is constant over
time but uncertain. In such cases, the series follows a standard DLM
conditional on any chosen value of $\alpha \in \mathcal{A}$. Generally, therefore, write
the defining quadruple as

$$\{\mathbf{F}_t(\alpha),\ \mathbf{G}_t(\alpha),\ V_t(\alpha),\ \mathbf{W}_t(\alpha)\}. \tag{13.6}$$

The formal theory for analysing such models is developed in Section
12.2 of Chapter 12, to which the reader is referred. It is made clear
there that the theoretically well-specified analysis is practically infeasible due to the abundance of integrals that cannot be analytically
calculated. Multi-process, class I models provide a method of numerical integration that allows the approximation of the formal analysis,
discussion of a particular example appearing in Section 12.2.3. A
summary of the relevant multi-process theory is given here.

Let $\mathbf{X}_t$ be any vector of random quantities of interest at time t.
Thus $\mathbf{X}_t$ may be a function of $\boldsymbol{\theta}_t$ or past values of the state vector, Y_t or future observations, the unknown observational variance,
and so forth. Conditional on any value of α, the required posterior
distribution for inference about $\mathbf{X}_t$ is derived from the conditional
DLM (13.6) in standard, analytically manageable form. Denote the
corresponding density by $p(\mathbf{X}_t \mid \alpha, D_t)$ as usual. Also, D_t informs
on the parameters α in terms of a posterior density $p(\alpha \mid D_t)$. Formally, the marginal posterior for $\mathbf{X}_t$ is the object of interest, given
by

$$p(\mathbf{X}_t \mid D_t) = \int_{\mathcal{A}} p(\mathbf{X}_t \mid \alpha, D_t) p(\alpha \mid D_t) d\alpha. \tag{13.7}$$

It is at this point that the analysis becomes difficult, many such integrals of this form are required for posterior inference and prediction, but most will be impossible to evaluate analytically. Numerical integration is called for. The multi-process, class I framework provides approximations to these integrals based on the use of a fixed and finite *grid* of points for the parameters, $\{\alpha_1, \dots, \alpha_k\}$, as a discrete approximation, in some sense, to the full parameter space $\mathcal{A}$. The analysis is thus based on the use of a finite collection of DLMs, each corresponding to a different choice of the, possibly vector-valued, parameter α, these DLMs being analysed in parallel. It is supposed that the collection of k values chosen for α in some sense adequately represents the, possibly much larger or continuous true parameter space $\mathcal{A}$. Large spaces $\mathcal{A}$ can often be adequately approximated for some purposes by a fairly small discrete set that somehow span the larger space, leading to the consideration of a small number of distinct DLMs. When the dimension of α is small, then k can often be chosen large enough so that the points fairly well cover the parameters space. Otherwise, the notion that they be chosen to appropriately span $\mathcal{A}$, representing different regions that may lead to rather different conditional DLM analyses, underlies the use of multi-processes. Obviously, if $\mathcal{A}$ is discrete to begin with, then this approach can be exact if the finite collection of k points coincides with $\mathcal{A}$.

With this in mind, α is a discretised random quantity whose posterior at time t is a mass function rather than a density, with weights

$$p_t(j) = p(\alpha_j \mid D_t) = \Pr[\alpha = \alpha_j \mid D_t], \qquad (j = 1, \dots, k)$$

for all t. The integral in (13.7) is replaced by the discretised form

$$p(\mathbf{X}_t \mid D_t) = \sum_{j=1}^{k} p(\mathbf{X}_t \mid \alpha, D_t) p_t(j). \qquad (13.8)$$

As time progresses and data are obtained, the conditional DLM analyses proceed in parallel. The additional learning about α proceeds via the updating of the posterior masses defined by

$$p_t(j) \propto p_{t-1}(j) p(Y_t \mid \alpha_j, D_{t-1}),$$

where $p(Y_t \mid \alpha_j, D_{t-1})$ is the standard, normal or T, one-step forecast density at time t. Thus, with normalising constant c_t defined by $c_t^{-1} = \sum_{j=1}^{k} p_{t-1}(j)p(Y_t \mid \alpha_j, D_{t-1})$, the posterior probabilities are $p_t(j) = c_t p_{t-1}(j)p(Y_t \mid \alpha_j, D_{t-1})$. It follows from (13.8) that all posterior distributions for linear functions of θ_t, and predictive distributions for future observations, are *discrete probability mixtures* of the standard T or normal distributions. Refer to Sections 12.2 and 12.3 of Chapter 12 for further discussion.

The general approach using mixtures of standard models is rather well known and quite widely used. Apart from the usages discussed in Chapter 12, the main applications of mixtures have been to just this problem of parameter non-linearities. In the control literature, such models are referred to under various names. Analogues of mixtures of normal DLMs (each with known variances) were used by Sorenson and Alspach (1971) and Alspach and Sorenson (1972) under the name of *Gaussian sums*. Anderson and Moore (1979, Chapter 9) discuss this and provide various related references. See also Chapter 10 of Anderson and Moore for further reference to the use of mixtures under the heading of *parallel processing*.

13.4 CONSTANT PARAMETER NON-LINEARITIES: EFFICIENT NUMERICAL INTEGRATION[†]

13.4.1 The need for efficient techniques of integration

The multi-process approach to numerical integration described above and in Chapter 12 attempts to approximate integrals by discretising them into summations, the chosen grid of values for α being fixed for all time. Generally, the quality of approximation increases with k, the size of the grid; a finer grid improves the representation of the space $\mathcal{A}$ and so the posterior probabilities $p_t(j)$ can more accurately follow the variation in the implictly defined "true" posterior for α. However, with more than very few parameters α, a reasonably fine grid typically involves the use of many points and the resulting computational demands can be enormous. Thus the use of more refined techniques of numerical approximation are called for.

Such numerical problems are a central feature of the use of Bayesian methods in all areas of application. At the time of writing, efficient techniques of numerical approximation of posterior distributions is the focus of much research and development work worldwide. One

---

[†]This Section, and the remainder of the Chapter, concerns more advanced material that, at the time of writing, is recent research and the subject of continuing development work. This can be omitted without loss in a first reading.

reasonably well developed approach to efficient integration is that using Gaussian quadrature, based on a general strategy initiated by Naylor and Smith (1982), and described by Smith *et al* (1985, 1987). This is described here in general terms, and the underlying theory and techniques extended and developed for application to models as in (13.6).

The approach to numerical integration uses *Gaussian quadrature*. There are other approaches that have been developed (and, at time of writing, are being developed) for Bayesian analyses, that are beyond the scope of our discussion. The interested reader should refer to Shaw (1988), Smith *et al* (1987), and Van Dijk, Hop and Louter (1987), for example, for further reading.

13.4.2   Gaussian quadrature in Bayesian analyses

In a general setting, consider a parameter vector $\alpha \in \mathcal{A}$, assuming that the object of interest is the posterior density $p(\alpha)$ from an analysis. The strategy for numerical approximation of the posterior described in the above references is based on the assumption that the posterior (possibly following appropriate transformation of α) is well-behaved in the sense that it can be adequately approximated by a normal density multiplied by a polynomial function of α. The normal component is usually a basic approximation to the posterior that is refined by multiplication by a further function, the latter being fairly well approximated by a polynomial. Many analyses satisfy this assumption, which is taken as read here. A concomitant assumption is that the mean vector and variance matrix of α are finite.

Consider initially $\alpha = \alpha$, a scalar. Suppose that $p(\alpha)$ satisfies the above assumption so that, for some a and A,

$$p(\alpha) = h(\alpha)(2\pi A)^{-1/2}\exp\{-(\alpha - a)^2/(2A)\} \qquad (13.9)$$

where $h(.)$ is a suitably well-behaved function. Note that the behaviour of $h(.)$ far out into the tails of the basic $N[a, A]$ component is largely irrelevant since the normal component decays rapidly to zero there, so that the regularity of $h(.)$ is of interest only over the central range. Gaussian (or Gauss-Hermite) quadrature is designed to approximate integrals of the form

$$I(q) = \int_{\mathcal{A}} q(\alpha)p(\alpha)d\alpha \qquad (13.10)$$

for functions of interest $q(.)$. Approximation is via sums of the form

$$I(q) \approx \sum_{i=1}^{n} q(\alpha_i) w_i p(\alpha_i) \qquad (13.11)$$

where the number n controls the accuracy of the approximation. The integral is effectively "discretised" at *grid points* $\alpha_1, \ldots, \alpha_n$, the density $p(\alpha_i)$ at these points being multiplied in the summation by positive *weights* $w_i, \ldots, w_n$. The Gauss-Hermite quadrature formula defines weights and grid points via

$$w_i = \frac{2^{n-1} n! \pi^{1/2} \exp(t_i^2)}{n^2 [H_{n-1}(t_i)]^2} \quad \text{and} \quad \alpha_i = a + (2A)^{1/2} t_i. \quad (13.12)$$

Here the t_i are the zeroes of the Hermite polynomial $H_n(t)$. For fuller and further discussion, see Davis and Rabinowitz (1967), and, for tables of the t_i and w_i for $1 \leq n \leq 20$, Salzer, Zucker and Capuano (1952).

The closeness of approximations to integrals obviously improves with n, and depends on the regularity of $h(.)$ and $q(.)$. In particular, if the product $q(\alpha)h(\alpha)$ is well approximated (over the central range of the normal density in (13.9)) by a polynomial of degree not exceeding $2n - 1$, then the approximation will tend to be good. If this product is exactly such a polynomial, then the approximating sum (13.11) is exactly equal to the required integral (13.10).

A key feature of (13.12) is that the weights and grid points depend only on the mean a and variance A of the basic normal component, applying to a variety of functions $q(.)$. Thus, *for the purpose of approximating integrals alone*, (13.11) essentially defines a discrete approximation to the posterior; the grid point α_i have approximate probability masses $w_i p(\alpha_i)$ and the expectation of any function $q(\alpha)$ is the usual sum. Given a and A, therefore, the discrete approximation can be determined and expectations of functions $q(\alpha)$ of interest deduced. If, as is often the case, the posterior is known only up to an unknown, positive constant of normalisation c, so that, for any $\alpha \in \mathcal{A}$, it is possible to evaluate only $l(\alpha) = c \, p(\alpha)$, then the normalisation constant is defined by $c = \int_{\mathcal{A}} l(\alpha) d\alpha$ and can be approximately evaluated using (13.11) with $p(.)$ replaced by $l(.)$ and $q(\alpha) = 1$. Then $p(\alpha) = l(\alpha)/c$ can be approximately calculated. The above theory is the basis of the following strategy for integration, discussed in the above references.

(1) Choose initial estimates of the posterior mean and variance of α; call these a_1 and A_1. These may be based, for example,

on a standard normal approximation to the posterior about the posterior mode.

(2) For some chosen n, use (13.12) to calculate grid points and weights based on the estimated moments $a = a_1$ and $A = A_1$.

(3) Use (13.11) with $q(\alpha) = \alpha$ to calculate a revised estimate a_2 of the posterior mean, and then with $q(\alpha) = (\alpha - a_2)^2$, to calculate a revised estimate A_2 of the posterior variance.

(4) Return to (2) and repeat this procedure sequentially revising the grid points, weights and estimated posterior moments. Stop when the changes in revised moments are acceptably small.

The earlier references apply this sort of procedure in a variety of applications, and for n relatively small, even in single figures, the approximations for common functions $q(.)$ of interest can be excellent. Variations allow n to change at each iteration. An important point is that the basic form (13.9) may be more adequate as an approximation to the posterior if the parametrisation α is appropriately chosen (Smith *et al*, 1985, 1987). Convergence, in well-behaved problems, is usually rapid; see the above references for much further discussion of this and of problems of non-convergence. Having achieved convergence, the final step has identified a final set of grid points at which the posterior is evaluated. This can form the basis of estimates of the posterior density over the range of α generally; some form of interpolation method can be used to estimate the posterior at values of α between and outside the grid points. The most commonly used such technique is cubic spline interpolation on the log-density scale (Shaw, 1987; Smith *et al*, 1985, 1987).

Return now to the more general, multiparameter case in which α is a $d-$vector for some integer d, $\mathcal{A}$ being a subspace of $d-$dimensional Euclidean space. Now $p(\alpha)$ is a density in d dimensions, typically exhibiting dependence amongst the component parameters. Obviously the quadrature theory for scalar quantities may be applied if the elements are (at least approximately) independent, the above development applying separately to the elements of α in turn. This feature is exploited by the above authors to provide the multi-parameter version of the above integration scheme as follows. If the parameters are essentially uncorrelated, then a Cartesian product of independent grids in each of the d dimensions forms the basis of discrete approximations to integrals. With n_i points in the ith dimension, this leads, in an obvious notation, to approximations of the form

$$\int \cdots \int q(\boldsymbol{\alpha})p(\boldsymbol{\alpha})d\boldsymbol{\alpha}$$

$$\approx \sum_{i_1=1}^{n_1} \cdots \sum_{i_d=1}^{n_d} q(\alpha_{1,i_1}, \dots, \alpha_{d,i_d}) w_{1,i_1} \cdots w_{d,i_d} p(\alpha_{1,i_1}, \dots, \alpha_{d,i_d}).$$

$$(13.13)$$

The implicit masses here are

$$w_{1,i_1} \cdots w_{d,i_d} p(\alpha_{1,i_1}, \dots, \alpha_{d,i_d}) \qquad (13.14)$$

at each of the $n_1 \times \dots \times n_d$ grid points

$$(\alpha_{1,i_1}, \dots, \alpha_{d,i_d}) \in \mathcal{A}. \qquad (13.15)$$

The point that the elements of $\boldsymbol{\alpha}$ be uncorrelated here is crucial, providing an extra ingredient in the iterative integration strategy in the multi-parameter case.

(1) Choose initial estimates of the posterior mean and variance matrix of $\boldsymbol{\alpha}$, based, for example, on a normal approximation to the posterior about the posterior mode.

(2) Based on the estimated variance matrix, linearly transform $\boldsymbol{\alpha}$ to an uncorrelated vector of parameters $\boldsymbol{\alpha}^*$. Since $p(\boldsymbol{\alpha})$ can be calculated (at least up to a normalising constant), then so can $p(\boldsymbol{\alpha}^*)$.

(3) Construct the discrete approximation (13.13) using weights and grid points (13.14, 13.15) in the $\boldsymbol{\alpha}^*$ parametrisation. Use this to calculate new estimates of the mean and variance of $\boldsymbol{\alpha}^*$.

(4) Proceed to (2) and iterate, determining a new linear transformation to orthogonality at each stage, followed by the use of quadrature in each transformed dimension separately to identify grid points and associated weights. Stop when the changes in transformation matrix and estimated posterior moments are small.

The efficacy of this scheme has been amply demonstrated in the earlier references. Estimation of the posterior density for the original $\boldsymbol{\alpha}$ vector is difficult; see Shaw (1987) for discussion of spline techniques applied to the final discrete approximation to the log-density on the final transformed scale. Full discussion of this topic is well beyond our current scope, although some illustration of its use is given below.

13.4.3   Efficient integration for non-linear parameters

The use of quadrature techniques in dynamic models is introduced in Pole (1988a) and Pole and West (1988, 1989b). These early works in this area are very exploratory, concerning only the simplest of models, this area being open for much further research and development work at the time of writing. The extent of the success of efficient integration techniques in other areas of Bayesian analysis strongly indicates the potential in dynamic modelling. However, the sequential analysis of dynamic models introduces problems in numerical analyses that are not encountered in more standard applications. This section describes the use of efficent integration and some of these problems encountered in models with constant parameter non-linearities, restricting attention to the case of a single, non-linear parameter.

Consider the general model (13.6) at time t, supposing that $\boldsymbol{\alpha} = \alpha$ is a scalar. In Section 13.3, the multi-process model assumes a fixed grid of k values for α that are chosen to represent the full parameter space $\mathcal{A}$. As data is processed, the posterior weights over the grid are updated providing learning about α. Obvious questions now arise about the accuracy and efficiency of this approach to learning about α. Firstly, concerning accuracy, there is an implict assumption that the grid will adequately represent regions of the parameter space deemed plausible *a priori* and supported by the data. However, as time progresses, it may be the case that regions not adequately covered by grid points are supported by the data, in which case the extent to which the discrete, approximate posterior for α appropriately reflects the information available will be questionable. What is needed in such cases is a change in grid values and, sometimes, an increase in the number of grid values. The second, related point concerns efficiency and is pertinent in the opposite case when the posterior masses concentrate more and more around one or a small number of grid points. Here grid values with little or no posterior probability may be dropped from the grid and more added in the high probability region. Thus there is a general need for the grid to evolve over time, ie. for a *dynamic grid* underlying a multi-process model; an approach using the quadrature technique of the previous

Section has such features. The analysis proceeds as follows between times $t - 1$ and t.

In the class of models (13.6) with $\alpha = \alpha$ scalar, suppose, for convenience in exposition, that the observational variance sequence is known for all α. Consider the position posterior to time $t - 1$. For any α, the posterior for the state vector at time $t - 1$ summarises the history of the analysis; here the posterior is normal

$$(\boldsymbol{\theta}_{t-1}|\alpha, D_{t-1}) \sim \mathrm{N}[\mathbf{m}_{t-1}(\alpha), \mathbf{C}_{t-1}(\alpha)]. \qquad (13.16)$$

Suppose also that the following quantities are available:

- (an approximation to) the full posterior density

$$p(\alpha|D_{t-1}), \qquad (\alpha \in \mathcal{A}); \qquad (13.17)$$

- the associated (approximations to the) posterior mean and variance of $(\alpha|D_{t-1})$, denoted by a_{t-1} and A_{t-1}.

Inferences about $\boldsymbol{\theta}_{t-1}$ (and other quantities of interest) are based on the unconditional posterior

$$p(\boldsymbol{\theta}_{t-1}|D_{t-1}) = \int_{\mathcal{A}} p(\boldsymbol{\theta}_{t-1}|\alpha, D_{t-1})p(\alpha|D_{t-1})d\alpha. \qquad (13.18)$$

Of course, this, and related integrals for moments of $\boldsymbol{\theta}_{t-1}$ and so forth, requires that (a) the conditional moments of $(\boldsymbol{\theta}_{t-1}|\alpha, D_{t-1})$ in (13.16) be available for *all* values of $\alpha \in \mathcal{A}$, and (b) that the integration be performed, typically requiring numerical approximation. Whilst (b) can now be entertained, (a) is impossible generally. The conditional density (13.16), summarised by its moments, can only be evaluated for a finite number of values of α. The previous Section suggests that such a set of values be determined using quadrature.

Suppose then, that Gaussian quadrature has been applied to the posterior (13.17). Based on the posterior mean and variance of α and a chosen grid size k_{t-1}, the quadrature scheme leads to a grid of values

$$\mathcal{A}_{t-1} = \{\alpha_{t-1,1}, \dots, \alpha_{t-1,k_{t-1}}\}, \qquad (13.19)$$

with associated quadrature weights $w_{t-1,1}, \dots, w_{t-1,k_{t-1}}$, such that integrals defining expectations of functions of α are optimally estimated by summations. In particular, (13.18) is approximated as

$$p(\boldsymbol{\theta}_{t-1}|D_{t-1}) \approx \sum_{j=1}^{k_{t-1}} p(\boldsymbol{\theta}_{t-1}|\alpha_{t-1,j}, D_{t-1})w_{t-1,j}p(\alpha_{t-1,j}|D_{t-1}).$$

$$(13.20)$$

Similarly the posterior mean of $\boldsymbol{\theta}_{t-1}$ is estimated by

$$E[\boldsymbol{\theta}_{t-1}|D_{t-1}] \approx \sum_{j=1}^{k_{t-1}} \mathbf{m}_{t-1}(\alpha_{t-1,j})w_{t-1,j}p(\alpha_{t-1,j}|D_{t-1}).$$

Thus the moments in (13.16) are required only for those values of α in the grid $\mathcal{A}_{t-1}$. The particular points in this grid at time $t-1$ will depend generally on D_{t-1}, and the accuracy of approximation upon the size of the grid. Notice that, in terms of inference for $\boldsymbol{\theta}_{t-1}$ and related quantities, equations such as (13.20) resemble the marginal posterior distributions derived in standard multi-process modelling as described in Section 13.3. With reference to equation (13.8) in particular, it is as if the posterior for α at time $t-1$ were approximated by a discrete mass function placing masses $p_{t-1}(j) = w_{t-1,j}p(\alpha_{t-1,j}|D_{t-1})$ at the grid values $\alpha_{t-1,j}$. This is essentially the case so far as the calculation of expectations of functions of α is concerned, although inference about α itself is based on the full density (13.17).

In summary, the assumed components summarising the history of the analysis and providing the basis of posterior inferences at time $t-1$ are as follows.

For α we have

   (i) the density function (13.17), and
   (ii) the associated mean and variance a_{t-1} and A_{t-1}.

For given k_{t-1} these moments define the grid (13.19) and associated weights. Then, for $\boldsymbol{\theta}_{t-1}$ we have

   (iii) the defining moments $\mathbf{m}_{t-1}(\alpha)$ and $\mathbf{C}_{t-1}(\alpha)$ for values of $\alpha \in \mathcal{A}_{t-1}$.

For forecasting ahead from time $t-1$, the defining quadruple (13.6) is also obviously required only for those values of $\alpha \in \mathcal{A}_{t-1}$. The one-step ahead forecast density $p(Y_t|D_{t-1}) = \int p(Y_t|\alpha, D_{t-1})p(\alpha|D_{t-1})d\alpha$ is, for each Y_t, an integral with respect to α of standard normal forms. For each α, $(Y_t|\alpha, D_{t-1}) \sim N[f_t(\alpha), Q_t(\alpha)]$ with moments defined by the usual equations for the particular value of α. The quadrature approximation is then the mixture of normal components

$$p(Y_t|D_{t-1}) = \sum_{j=1}^{k_{t-1}} p(Y_t|\alpha_{t-1,j}, D_{t-1})w_{t-1,j}p(\alpha_{t-1,j}|D_{t-1}).$$

Consider now updating when Y_t is observed. The steps in the analysis are as follows.

(A) As in multi-process models, the k_{t-1} DLMs corresponding to values of $\alpha \in \mathcal{A}_{t-1}$ are updated in parallel. This leads to standard posteriors

$$(\boldsymbol{\theta}_t | \alpha_{t-1,j}, D_t) \sim N[\mathbf{m}_t(\alpha_{t-1,j}), \mathbf{C}_t(\alpha_{t-1,j})]$$

for $j = 1, \ldots, k_{t-1}$. Obviously, if the observational variance is unknown, the usual extension to normal/gamma distributions for the state vector and the unknown variance parameter is made conditional on each value of α. The above normal posterior is then replaced by the relevant T distribution.

(B) Bayes' Theorem for α leads to $p(\alpha | D_t) \propto p(Y_t | \alpha, D_{t-1}) \cdot p(\alpha | D_{t-1})$ for all α. Now, whilst the prior density is assumed known from (13.17) for all α, the likelihood component (just the observed value of the one-step ahead forecast density) is only known for $\alpha \in \mathcal{A}_{t-1}$. Fuller information about the posterior is reconstructed in the next two steps.

(C) Bayes' Theorem above provides the values $p(\alpha_{t-1,j} | D_t)$ for $j = 1, \ldots, k_{t-1}$, the ordinates of the posterior density over the existing grid. Using the existing weights $w_{t-1,j}$ on this grid, the posterior is approximately normalised and a first approximation to the posterior mean and variance of α calculated by quadrature. For normalisation, $p(\alpha | D_t) \approx c \, p(Y_t | \alpha, D_{t-1}) \cdot p(\alpha | D_{t-1})$ where

$$c^{-1} = \sum_{j=1}^{k_{t-1}} p(Y_t | \alpha_{t-1,j}) w_{t-1,j} p(\alpha_{t-1,j} | D_{t-1}),$$

the mean and variance being estimated by similar summations. Having calculated these initial estimates, the iterative quadrature strategy of Section 13.4.2 is used. At each stage, the current estimates of posterior mean and variance for α are used to calculate a revised set of grid points with associated weights. Note that the number of grid points is to be chosen, it may differ from k_{t-1} if required and also change between iterations. Using the revised grid at each stage, revised estimates of the normalising constant c, posterior mean and variance are computed and the cycle continues. Iterations halt when these estimates do not change markedly. At this point, the posterior density is evaluated (including normalising constant) at points on a new grid of chosen size k_t points,

$$\mathcal{A}_t = \{\alpha_{t,1}, \dots, \alpha_{t,k_t}\}.$$

The associated quadrature weights $w_{t,1}, \dots, w_{t,k_t}$ are available, as are estimates of the posterior mean a_t and variance A_t of α.

(D) To reconstruct the full posterior density function $p(\alpha|D_t)$, some form of interpolation is needed based on the k_t evaluations over $\mathcal{A}_t$. Cubic spline interpolation (Shaw, 1987) is used in Pole (1988a), and Pole and West (1988, 1989b), splines being fitted to the density on the log scale. This ensures positive density everywhere after interpolation, and linear extrapolation outside the grid leads to exponential decay to zero in the tails of the posterior. Fuller details can be found in these references. This follows the use of splines mentioned earlier in Smith *et al* (1987). In more general models with α vector valued, the basic development is similar but the problems of density reconstruction much more difficult. Following Shaw (1987), Pole and West (1988, 1989b) use tensor product splines. Returning to the model here in which α is scalar, the net result is that the posterior density $p(\alpha|D_t)$ may be (approximately) evaluated at any value of α, forming the basis of posterior inference for α. Thus far, then, the updating analysis is completed as far as α is concerned.

(E) In addition, the analysis has identified a new grid $\mathcal{A}_t$ of k_t values of α which, with the associated weights, provides the basis for approximating expectations of functions of α needed for inference about $\boldsymbol{\theta}_t$ etc. At this point, it becomes clear that further work is required to obtain such inferences. The problem lies in the fact that the conditional posteriors for $\boldsymbol{\theta}_t$ in (A) are based on values of α over the original grid $\mathcal{A}_{t-1}$; what is required are posteriors based on the new grid $\mathcal{A}_t$. Formally, the unconditional posterior for $\boldsymbol{\theta}_t$ is defined by the quadrature formula

$$p(\boldsymbol{\theta}_t|D_t) = \sum_{j=1}^{k_t} p(\boldsymbol{\theta}_t|\alpha_{t,j}, D_t) w_{t,j} p(\alpha_{t,j}|D_t),$$

so that the standard form, conditional posteriors $p(\boldsymbol{\theta}_t|\alpha_{t,j}, D_t)$ are required. Again the answer to the problem of their calculation lies in the use of some form of interpolation. The approach in Pole and West (1988, 1989b) uses direct linear interpolation. For each value $\alpha_{t,j} \in \mathcal{A}_t$, the required mean vector $\mathbf{m}_t(\alpha_{t,j})$ is calculated by linear interpolation between appropriate values in the set $\mathbf{m}_t(\alpha_{t-1,j})$ from (A). The posterior variance matrix is similarly, linearly interpolated. See the above references for further details. Of course, linear interpolation is not possible outside the range of the old grid. To avoid this problem, the referenced works ensure that the two grids always have extreme or boundary points in common so that no interpolation is necessary at the extremes. This is always possible since the choice of the number of grid points is under control and extra points at the extremes can be added as desired. Once this interpolation is complete, the updating is complete with interpolated posteriors over the new grid

$$(\boldsymbol{\theta}_t|\alpha_{t,j}, D_t) \sim N[\mathbf{m}_t(\alpha_{t,j}), \mathbf{C}_t(\alpha_{t,j})], \qquad (j = 1, \dots, k_t).$$

At this point, the components assumed available at time $t-1$ have all been updated to time t, and so the full updating analysis ends.

13.4.4 An example and further comments

The analysis in the previous Section is generally applicable to models with constant parameter non-linearities. There are many technical difficulties with the approach, however, that, at the time of writing, are the subject of further investigations. Some of the most difficult problems lie in the reconstruction of continuous posterior densities from discrete approximations via interpolation, such problems arising in all applications of efficient numerical integration not just in the dynamic modelling context. Some preliminary studies of the approach in simple models appear in Pole (1988a), and Pole and West (1988). The former paper is concerned with transfer responses using the model of Example 13.2 (and Example 9.1 of Chapter 9). There $\alpha = \lambda$, a scalar parameter determining the decay rate for the effect of past values of a regressor variable. Different examples are considered in the second paper; one of these examples is briefly reviewed here, and some of the technical issues arising are discussed.

An example concerns a simple MA(1) noise model as discussed in Example 9.4 of Section 9.4.4, Chapter 9. Consider a TSDLM with level μ and MA(1) noise components, the latter determined by the MA parameter $\pi_1 = \alpha$, so that

$$Y_t = \mu + \epsilon_t + \alpha\epsilon_{t-1},$$

with $\epsilon_t \sim N[0, V]$ independently over time. With μ and α unknown (though here assumed constant over time) the standard DLM representation (cf. equation (9.11) of Chapter 9) is defined by the (observation noise-free) 3$-$dimensional DLM with quadruple

$$\left\{ \begin{pmatrix} 1 \\ 1 \\ 0 \end{pmatrix}, \begin{pmatrix} 1 & 0 & 0 \\ 0 & 0 & 1 \\ 0 & 0 & 0 \end{pmatrix}, 0, V \begin{pmatrix} 0 & 0 & 0 \\ 0 & 1 & \alpha \\ 0 & \alpha & \alpha^2 \end{pmatrix} \right\}.$$

In this special case there is a rather simpler, alternative but equivalent representation in terms of a 2$-$dimensional model with non-zero mean for the evolution noise. Setting $\theta_t = \alpha\epsilon_{t-1}$ and $\boldsymbol{\theta}_t = (\mu, \theta_t)'$, it follows that

$$Y_t = \mathbf{F}'\boldsymbol{\theta}_t + \epsilon_t \qquad \text{and} \qquad \boldsymbol{\theta}_t = \mathbf{G}(\alpha)\boldsymbol{\theta}_{t-1} + \mathbf{h}_t(\alpha),$$

where

$$\mathbf{F} = \begin{pmatrix} 1 \\ 1 \end{pmatrix}, \quad \mathbf{G}(\alpha) = \begin{pmatrix} 1 & 0 \\ -\alpha & -\alpha \end{pmatrix}, \quad \text{and} \quad \mathbf{h}_t(\alpha) = \begin{pmatrix} 0 \\ \alpha Y_{t-1} \end{pmatrix}.$$

This simpler representation is used here (nb. Note that, since Y_{t-1} appears in the evolution eqaution of this alternative representation, it cannot be used in the event of missing data. The original representation does not suffer this problem). This model is applied to the series of 100 observations given in Fuller (1976, p.355) that are generated from the model with $\mu = 0$ and $\alpha = 0.7$. The data appear in Table 13.1.

The analyses reported includes V as an unknown, observational variance parameter, so that the discussion of the previous Section is appropriately extended. At time $t - 1$, the posterior distributions are summarised by

$$(\boldsymbol{\theta}_{t-1}|\alpha, D_{t-1}) \sim T_{n_{t-1}}[\mathbf{m}_{t-1}(\alpha), \mathbf{C}_{t-1}(\alpha)]$$

and, with $\phi = 1/V$,

$$(\phi|\alpha, D_{t-1}) \sim G[n_{t-1}/2, d_{t-1}(\alpha)/2],$$

with estimate $S_{t-1}(\alpha) = d_{t-1}(\alpha)/n_{t-1}$ of V, for any value of α. Evolving to t,

$$(\boldsymbol{\theta}_t|\alpha, D_{t-1}) \sim T_{n_{t-1}}[\mathbf{a}_{t-1}(\alpha), \mathbf{R}_{t-1}(\alpha)]$$

with

$$\mathbf{a}_t(\alpha) = \mathbf{G}(\alpha)\mathbf{m}_{t-1}(\alpha) + \mathbf{h}_t(\alpha)$$

and

$$\mathbf{R}_t(\alpha) = \mathbf{G}(\alpha)\mathbf{C}_{t-1}(\alpha)\mathbf{G}(\alpha)'.$$

In addition, the one-step ahead forecast distribution is the usual T form

$$(Y_t|\alpha, D_{t-1}) \sim T_{n_{t-1}}[f_t(\alpha), Q_t(\alpha)]$$

with $f_t(\alpha) = \mathbf{F}'\mathbf{a}_t(\alpha)$ and $Q_t(\alpha) = \mathbf{F}'\mathbf{R}_t(\alpha)\mathbf{F} + S_{t-1}(\alpha)$.

Two sets of example analyses are reported. In all analyses, the initial prior for $\boldsymbol{\theta}_1$ and ϕ is taken as independent of α, defined by $\mathbf{a}_1 = (0,0)'$, $\mathbf{R}_1 = \mathbf{I}$, $n_0 = 1$ and $d_0 = 1$ so that $S_0 = 1$ also. All that remains for the analysis to proceed directly is to specify the intial grid of values of α, namely $\mathcal{A}_0$, and the number of grid points k_t for each time t.

EXAMPLE 13.6. Recall from Example 9.4 the usual invertibility constraint $-1 < \alpha < 1$ on the MA component. Classical time series analysis using stationarity and invertibility assumptions for derived

Table 13.1. MA(1) data from Fuller (1976)

MA(1) series with $\alpha = 0.7$ (read across rows)									
1.432	−0.343	−1.759	−2.537	−0.295	0.689	−0.633	−0.662	−0.229	−0.851
−3.361	−0.912	1.594	1.618	−1.260	0.288	0.858	−1.752	−0.960	1.738
−1.008	−1.589	0.289	−0.580	1.213	1.176	0.846	0.079	0.815	2.566
1.675	0.933	0.284	0.568	0.515	−0.436	0.567	1.040	0.064	−1.051
−1.845	0.281	−0.136	−0.992	0.321	2.621	2.804	2.174	1.897	−0.781
−1.311	−0.105	0.313	−0.890	−1.778	−0.202	0.450	−0.127	−0.463	0.344
−1.412	−1.525	−0.017	−0.525	−2.689	−0.211	2.145	0.787	−0.452	1.267
2.316	0.258	−1.645	−1.552	−0.213	2.607	1.572	−0.261	−0.686	−2.079
−2.569	−0.524	0.044	−0.088	−1.333	−1.977	0.120	1.558	0.904	−1.437
0.427	0.061	0.120	1.460	−0.493	−0.888	−0.530	−2.757	−1.452	0.158

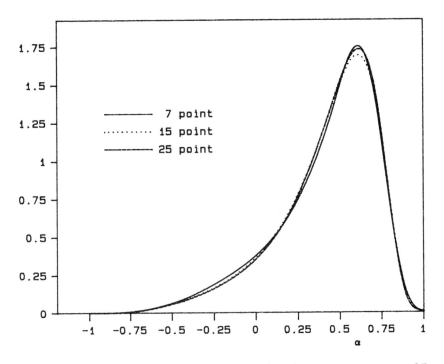

Figure 13.1a.   Posteriors for constrained MA parameter at $t = 25$.

series estimate α subject to this constraint. Fuller (1976, pages 356
and 357) approximately evaluates the constrained maximum likelihood estimate as 0.732 with a nominal standard error of 0.069. To
impose this constraint, the numerical analysis, quadrature and spline
interpolation calculations are performed on the transformed param-
eter $\beta = \log[(1 - \alpha)/(\alpha + 1)]$. With β real valued, it follows that
$|\alpha| < 1$. Thus the calculations are performed over grids of values of
the transformed parameter, the posterior for α at any time being obtained simply by transforming back. We choose a constant grid size
for all time, $k_t = k$, and compare results based on various grid sizes k
between $k = 4$ and $k = 25$. In each of these analyses, the initial grid
points for α are taken as equally spaced over $(-1, 1)$, with uniform
weights k^{-1} at each point. In all respects other than the grid size,
the analyses are similar. Figure 13.1a graphs the spline-interpolated
posterior densities for α at $t = 25$ from three analyses with grid sizes
$k = 7, 15$ and 25, respectively. Obviously the results are very similar, as are those from other analyses with grid sizes varying between
4 and 25. Similar graphs appear in Figure 13.1b at time $t = 100$,

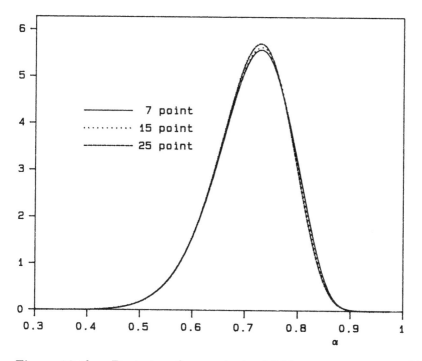

Figure 13.1b.   Posteriors for constrained MA parameter at $t = 100$.

having processed all the data. These graphs demonstrate a notable robustness to grid size, and indicate the potential efficiency of analyses based on grids as small as $k = 4$ (although the analysis is, of course, restricted to a single parameter).

Accuracy of these numerical approximations can be assessed by comparison with fixed grid analysis using the multi-process, class I approach of Section 13.3. Such an analysis is reported, based on $k = 100$ fixed points equally spaced over the interval $-1 < \alpha < 1$, with the initial prior weights for α uniform over this, large grid. Figures 13.2a and 13.2b graph the resulting posterior weights for α over this grid at $t = 25$ and $t = 100$, respectively. Comparison of these graphs with those in Figures 13.1a and 13.1b confirm that there is are no differences of any practical relevance, confirming the usefulness and practicability of the efficient, dynamic grid based technique. A key reason for this is that the posterior distributions for the transformed parameter β are well-behaved, smooth and unimodal, and capable of being well-approximated by normal

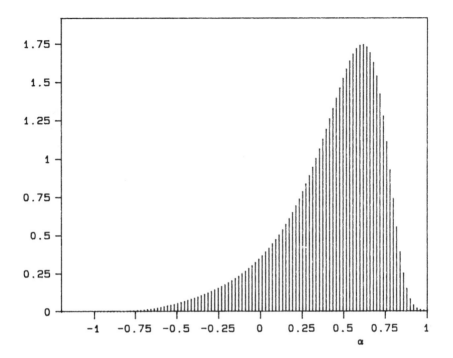

Figure 13.2a.   Posterior from constrained fixed grid analysis at
                $t = 25$.

times polynomial functions as required in the quadrature approach. Under less congenial circumstances, as illustrated in the following example, results may be less satisfactory. For final comparison here, Table 13.2 displays the final posterior means $E[\alpha|D_{100}]$ and standard deviations $V[\alpha|D_{100}]^{1/2}$ from various analyses. Recall that the data actually come from this model with $\alpha = 0.7$, and that the

Table 13.2. Final estimates of α in constrained analysis

Grid	Mean	SD
4	0.722	0.068
7	0.711	0.073
15	0.710	0.073
25	0.709	0.073
100 (fixed)	0.708	0.072

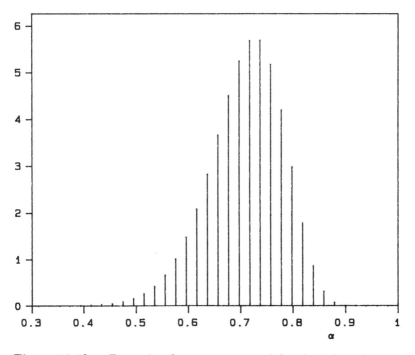

Figure 13.2b.   Posterior from constrained fixed grid analysis at
                $t = 100$.

constrained maximum likelihood estimate is 0.732 with approximate
standard deviation 0.069.

EXAMPLE 13.7. A further set of analyses, in which α is not subject
to the invertibility constraint, are reported. These illustrate some
important features and problems of the numerical integration strategy not evident in the previous, very well-behaved example. Ignoring
the constraint (ie. not restricting the MA component of the model to
invertibility), the analyses here are again based on grid sizes constant
over time with initial priors weights for α uniform over the grids in
each case. Figure 13.3 graphs posteriors for α at time and $t = 100$
from four models with grid sizes $k = 7, 15, 16$ and 25. These are simply the unconstrained versions of Figures 13.1b from the previous
analysis.

It is clear that, relative to those of the constrained analysis in Figures 13.1a and 13.1b, the posteriors here are rather more sensitive
to choice of grid size. For each choice of grid (and other choices

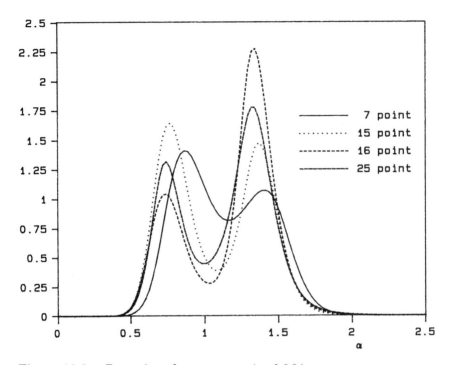

Figure 13.3.   Posteriors for unconstrained MA parameter at
                $t = 100$.

up to $k = 25$), the final posterior is clearly bimodal with modes located near 0.7 and 1.4. The fact that the reconstructed posteriors
differ markedly with grid size is due to the use of spline interpolation to smooth the discrete approximations derived from the use of
quadrature. Quadrature itself leads, in fact, to discrete approximations that are much closer than these figures suggest. However, the
irregular behaviour of the density, sharply increasing to the left of
each mode and then decreasing to right, leads to the fitted splines
"over-shooting" in order to accommodate the gradients around the
modes. This problem is a general feature of the spline interpolation technique that is easy to identify in this, a single parameter
model, but which may be difficult to identify in higher dimensions
suggesting potential for misleading results. In spite of this technical
difficulty with spline reconstructions, the bimodality of the posterior
at $t = 100$ is clearly identified even with a relatively small number of
grid points. For comparison, the "exact" posterior based on a multi-
process mixture model with a fixed grid of 200 points is presented in

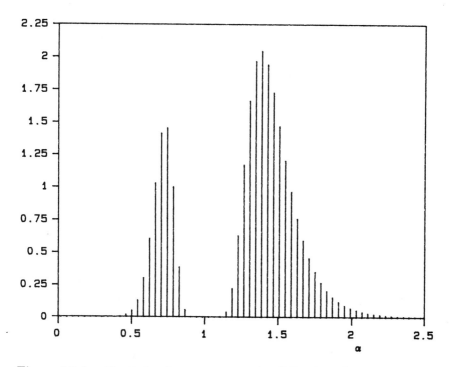

Figure 13.4.   Posterior from unconstrained fixed grid analysis at
           $t = 100$.

Figure 13.4. Here the 200 fixed grid points, defining a multi-process, class I model, are equally spaced over the interval $-4 < \alpha < 4$, and initially equally weighted. The posterior at $t = 100$ is highly bimodal, and it is now clear that the dynamic grid models cannot be expected to adequately pick up the dramatic variation in the density using only moderate grid sizes. Further discussion, and suggestions for refinements of the technique, are given in Pole and West (1988, 1989b).

Finally, a comment about the bimodality in the posterior. This arises in this analysis where α is unconstrained as a result of the equivalence, in terms of correlation structure, of the MA process with parameter α with that having parameter α^{-1}. The data here actually come from such a process with $\alpha = 0.7$. Each of the posteriors in the figures has one mode located around this value , the other being located around $1.4 \approx 0.7^{-1}$. Though a side issue here, it is worth commenting that the "exact", unconstrained posterior in Figure 13.4 highly favours the non-invertible region $\alpha > 1$ that is given probability zero throughout in the constrained analysis of the previous example. This feature is encountered with more general MA models, and derives from the fact that the innovations variance in any invertible MA representation is larger than that in the equivalent non-invertible representation.

13.5 EFFICIENT INTEGRATION FOR GENERAL DYNAMIC MODELS

The ideas underlying the use of quadrature techniques for efficient integration have been extended in Pole and West (1988, 1989b) to apply to much more general models than those above, restricted as they are to constant parameter non-linearities. In fact the strategy presented in this reference is essentially as general as possible within the context of dynamic models having Markovian evolution of state vectors. However, as with the above, more restricted framework, this strategy is recent at time of writing and so discussion here is restricted to an outline of the class of models and the underlying principles.

The primary need is for analyses of models in which parameters that introduce non-linearities are time dependent. If this is the case, then there is no way in which the above ideas of standard analyses conditional on non-linear parameters applies. To be specific, consider the model in (13.1), namely:

Observation equation:	$Y_t = F_t(\boldsymbol{\theta}_t) + \nu_t,$	(13.21)
Evolution equation:	$\boldsymbol{\theta}_t = \mathbf{g}_t(\boldsymbol{\theta}_{t-1}) + \boldsymbol{\omega}_t,$	(13.22)

where $F_t(.)$ is a known, non-linear regression function mapping the n-vector $\boldsymbol{\theta}_t$ to the real line, $\mathbf{g}_t(.)$ a known, non-linear vector evolution function, and ν_t and $\boldsymbol{\omega}_t$ are error terms subject to the usual zero-mean and normality assumptions with known variances V_t and $\mathbf{W}_t$. Suppose also that an initial prior density $p(\boldsymbol{\theta}_0|D_0)$ is specified as usual, though not necessarily in standard, normal form.
Note that:

(a) The assumption of known variances is purely for convenience in notation and exposition here. If the observational variance is unknown, for example, then the vector $\boldsymbol{\theta}_t$ can be extended to include it as a component to be estimated along with the original elements of the state vector. In addition, the errors ν_t and $\boldsymbol{\omega}_t$ may, of course, have known, non-zero means.

(b) These equations define a rather general normal model. However, the normality of the errors is not a crucial assumption. Since the analysis is to be via numerical integration, there is no need to restrict to normality any longer. For example, the errors may have heavy-tailed distributions, such as T distributions, in order to permit larger changes in the state vector and outlying observations from time to time (West, 1981; Pole and West, 1988)

(c) In fact the model need not be defined directly in terms of these two equations. A rather more general framework will specify the model via observational and evolution densities $p(Y_t|\boldsymbol{\theta}_t, D_{t-1})$ and $p(\boldsymbol{\theta}_t|\boldsymbol{\theta}_{t-1}, D_{t-1})$ respectively. This is very general. The observational distribution may be derived from an equation such as (13.21), but allows for essentially any continuous or discrete distribution that may depend on the state vector in complicated, non-linear ways. Similarly the evolution density is general. This too may take the usual DLM form, but the framework allows for any state evolution density, normal or otherwise, with any desired dependence on the state vector at $t - 1$. The standard DLM is obviously a special case, as is the DLM with unknown observational variance since the variance parameter can be incorporated into the state vector as a constant component. In fact all models considered in this book can be written in the form of these two densities, hence, in principle, all models can be analysed within the framework here. In practice, this is currently not the case; the technical and computational difficulties of analysis have yet to be thoroughly investigated even in the simplest special cases.

Conceptually, the analysis of this model involves the following three steps at each time t, familiar from the usual normal DLM.

(1) Assuming knowledge of the posterior distribution for the state vector at time $t - 1$, defined via a density function

$$p(\boldsymbol{\theta}_{t-1}|D_{t-1}), \tag{13.23}$$

compute the implied prior density for $\boldsymbol{\theta}_t$, namely

$$p(\boldsymbol{\theta}_t|D_{t-1}) = \int p(\boldsymbol{\theta}_t|\boldsymbol{\theta}_{t-1}, D_{t-1})p(\boldsymbol{\theta}_{t-1}|D_{t-1})d\boldsymbol{\theta}_{t-1}. \tag{13.24}$$

for all values of $\boldsymbol{\theta}_t$. Typically this implies the need for many numerical integrations over the $n-$dimensional state parameter space, one for each value of $\boldsymbol{\theta}_t$ required.

(2) Use this to produce forecast distributions one, and more steps ahead. The one-step ahead density requires the further integrations implicit in the definition

$$p(Y_t|D_{t-1}) = \int p(Y_t|\boldsymbol{\theta}_t, D_{t-1})p(\boldsymbol{\theta}_t|D_{t-1})d\boldsymbol{\theta}_t. \qquad (13.25)$$

Again the implication is that numerical integrations are needed here.

(3) Update the prior (13.24) to the posterior for $\boldsymbol{\theta}_t$ when Y_t is observed to complete the cycle at time t. As usual,

$$p(\boldsymbol{\theta}_t|D_t) \propto p(\boldsymbol{\theta}_t|D_{t-1})p(Y_t|\boldsymbol{\theta}_t, D_{t-1}), \qquad (13.26)$$

for all $\boldsymbol{\theta}_t$, the constant of proportionality being the observed value of (13.25). Summarisation of (13.26) involves the calculation, in particular, of moments and posterior probabilities that require further integrations with respect to elements of $\boldsymbol{\theta}_t$. Subsidiary activities include filtering and smoothing which are not discussed here — see Pole and West (1988).

Kitagawa (1987) describes implementation of the cycle defined in these three steps using the direct approach through numerical integration based on a fixed grid of values of $\boldsymbol{\theta}_t$. This approach requires that the grid of values, fixed for all time, remain appropriate for all time. We have commented earlier on the need for such grids to evolve as $\boldsymbol{\theta}_t$ evolves. Also, the very heavy computational demands of fixed grid methods with very large numbers of grid points in models of even very moderate size can be prohibitive. These points mitigate against any fixed grid based method being of general applicability, and thus we explore the extension of efficient integration techniques to this context. The ideas underlying the numerical integration in Section 13.4 can be considered to apply to each of the steps summarised in equations (13.23)—(13.26), as follows.

Step 1:
Assume that $p(\boldsymbol{\theta}_{t-1}|D_{t-1})$ has been the subject of Gaussian quadrature, being evaluated on a grid of k_{t-1} values $\mathcal{A}_{t-1} = \{\boldsymbol{\theta}_{t-1,1}, \dots, \boldsymbol{\theta}_{t-1,k_{t-1}}\}$ with associated quadrature weights $w_{t-1,1}, \dots, w_{t-1,k_{t-1}}$. From the evolution equation, this provides an estimate of the prior mean $\mathbf{a}_t = \mathrm{E}[\boldsymbol{\theta}_t|D_{t-1}]$ via the usual quadrature formula

$$\mathbf{a}_t \approx \sum_{j=1}^{k_{t-1}} \mathbf{g}_t(\boldsymbol{\theta}_{t-1,j})w_{t-1,j}p(\boldsymbol{\theta}_{t-1,j}|D_{t-1}).$$

Given this estimate of $\mathbf{a}_t$, the prior variance matrix $\mathbf{R}_t = V[\boldsymbol{\theta}_t|D_{t-1}]$ is estimated similarly by

$$\mathbf{R}_t \approx \sum_{j=1}^{k_{t-1}} [\mathbf{g}_t(\boldsymbol{\theta}_{t-1,j}) - \mathbf{a}_t][\mathbf{g}_t(\boldsymbol{\theta}_{t-1,j}) - \mathbf{a}_t]' w_{t-1,j} p(\boldsymbol{\theta}_{t-1,j}|D_{t-1}).$$

These initial estimates form the starting point for quadrature applied to the prior (13.24), using the general theory outlined and referenced in Section 13.3. The iterative strategy is applied to $\boldsymbol{\theta}_t$, the prior density at any point being evaluated via

$$p(\boldsymbol{\theta}_t|D_{t-1}) \approx \sum_{j=1}^{k_{t-1}} p(\boldsymbol{\theta}_t|\boldsymbol{\theta}_{t-1,j}, D_{t-1}) w_{t-1,j} p(\boldsymbol{\theta}_{t-1,j}|D_{t-1}). \quad (13.27)$$

Once the iterative strategy converges, the result is final estimates of $\mathbf{a}_t$ and $\mathbf{R}_t$, and a *prior* quadrature grid of some k_t points $\mathcal{A}_t = \{\boldsymbol{\theta}_{t,1}, \ldots, \boldsymbol{\theta}_{t,k_t}\}$ with associated quadrature weights $w_{t,1}, \ldots, w_{t,k_t}$. Note that there is no need for interpolation in this approach, the density actually being evaluated (approximately) at all points of interest.

Step 2:
Forecasting ahead simply repeats this basic procedure to produce approximations to distributions for future state vectors and hence for future observations. Restricting attention to one-step ahead, for any value Y_t the one-step density (13.25) is estimated via

$$p(Y_t|D_{t-1}) = \sum_{j=1}^{k_t} p(Y_t|\boldsymbol{\theta}_{t,j}, D_{t-1}) w_{t,j} p(\boldsymbol{\theta}_{t,j}|D_{t-1}).$$

Moments and predictive probabilites may be similarly evaluated.
Step 3:
Updating is directly achieved again using quadrature. The posterior density is evaluated at all points in $\mathcal{A}_t$ via (13.26), which then leads to initial estimates of posterior moments via the usual quadrature formula. These form the basis of a new set of grid points and weights, the density is evaluated at these points, and the process iterated to convergence. The result is a revision of the prior grid $\mathcal{A}_t$ to a final, posterior grid with associated final weights. Thus the analysis at time t is complete, the position now being just at at the start of Step 1 with $t - 1$ increased to t.

The key problem underlying this sort of analysis, and the basic way in which this application of efficient integration differs from more usual analyses, is that the posterior density for θ_t at any time is used as an input into the model for future times. Pole and West (1988) discuss some of the problems that arise with its application, and illustrate them in DLMs in which the evolution and observation errors terms have non-normal distributions, in particular, using heavy-tailed, T distributions for these error terms rather than normal distributions as in Masreliez (1975), Masreliez and Martin (1977), Kitagawa (1988), and West (1981). The interested reader should consult the referenced paper by Pole and West for further details. APL software for the implementation of analyses of this and the previous Section is described in Pole (1988b).

CHAPTER 14

EXPONENTIAL FAMILY DYNAMIC MODELS

14.1 INTRODUCTION

Having explored extensions of normal DLMs to incorporate parameter non-linearities, we now turn to a related area of models with non-normal components. The primary development here concerns data series with non-normal observational distributions. DLMs have been extended and generalised to various non-normal problems, the largest class being that based on the use of *exponential family models* for observational distributions. This Chapter is devoted to these models, the primary references being Migon (1984), Migon and Harrison (1985), and West and Harrison (1986a), West, Harrison and Migon (1985).

A primary need for extension of the class of normal DLMs is to allow for data distributions that are not likely to be adequately modelled using normality, even after transformation. With continuous data plausibly having skewed distributions such as exponential, Weibull or gamma, it is often the case that a data transformation can be used to achieve approximate symmetry on the transformed scale, and normal models can then be quite useful. However, the meaning and interpretation of model parameters is usually obscured through transformation, the original data scale being natural and interpretable, and it may often be viewed as desirable to develop a model directly for the data on the original scale. See also the related comments in Section 10.6 of Chapter 10. With discrete data in the form of counts, transformations are often not sensible and approximation using continuous normal distributions may be radically inappropriate. Discrete data often arise in the form of counts from finite or conceptually infinite distributions, and event indicators. Sample surveys, for example, typically result in data that are essentially binomial-like in character. Poisson, and compound Poisson data arise in demand and inventory processes through "random" arrival of orders and supplies. At the extreme, binary series arise as indicators of the occurrence, or non-occurrence, of sequences of events, such as rise/fall in a financial index, daily rainfall indicators, and so forth. Thus we need to consider non-normal observational distributions.

Major progress in non-normal modelling, and hence in practical statistical methodology, has flowed from the development of the theoretical framework of Generalised Linear Models (GLMs), beginning with Nelder and Wedderburn (1972). The ensuing development of the interactive GLIM computer package (Baker and Nelder, 1985) presaged an explosion in the application and development of formal, parametric statistical models. GLIM provides regression techniques in the context of non-normal models with regression effects on a non-linear scale, one of the most important modelling tools for applied statisticians and researchers. See McCullagh and Nelder (1983) for coverage of theoretical and applied aspects of these, and related models, and West (1985a) for Bayesian modelling within this class. The theoretical framework of GLMs provides a starting point for non-normal, dynamic models. We now describe such models, beginning with an exposition of the basic structure of exponential family analyses.

14.2 EXPONENTIAL FAMILY MODELS

14.2.1 Observational distributions in the EF

Consider the time series of scalar observations Y_t, $(t = 1, 2, \ldots)$, continuous or discrete random quantities taking values in the sample space $\mathcal{Y}$. If Y_t is assumed to have a sampling distribution in the *exponential family*, then the density (if discrete, the probability mass function) of Y_t may be described as follows. For some defining quantities η_t and V_t, and three known functions $y_t(Y_t)$, $a(\eta_t)$ and $b(Y_t, V_t)$, the density is

$$p(Y_t | \eta_t, V_t) = \exp\{V_t^{-1}[y_t(Y_t)\eta_t - a(\eta_t)]\}b(Y_t, V_t), \qquad (Y_t \in \mathcal{Y}).$$
$$(14.1)$$

The primary properties of the distribution may be found in the references above. These properties, some terminology and further points of note, are summarised as follows.

(1) η_t is the *natural parameter* of the distribution, a continuous quantity.

(2) $V_t > 0$ is a *scale parameter*; the *precision parameter* of the distribution is defined as $\phi_t = V_t^{-1}$.

(3) As a function of the natural parameter for fixed Y_t, (14.1), viewed as a likelihood for η_t, depends on Y_t through the transformed value $y_t(Y_t)$.

(4) The function $a(\eta_t)$ is assumed twice differentiable in η_t. It follows that

$$\mu_t = E[y_t(Y_t)|\eta_t, V_t] = \frac{da(\eta_t)}{d\eta_t} = \dot{a}(\eta_t).$$

(5) It follows that

$$V[y_t(Y_t)|\eta_t, V_t] = V_t \ddot{a}(\eta_t).$$

(6) Rather often (in the key practical cases in particular) $y_t(.)$ is the identity function. In such cases,

$$p(Y_t|\eta_t, V_t) = \exp\{V_t^{-1}[Y_t\eta_t - a(\eta_t)]\}b(Y_t, V_t), \qquad (Y_t \in \mathcal{Y}). \ (14.2)$$

Also

$$E[Y_t|\eta_t, V_t] = \mu_t = \dot{a}(\eta_t), \qquad (14.3)$$
$$V[Y_t|\eta_t, V_t] = V_t \ddot{a}(\eta_t). \qquad (14.4)$$

Here, in an obvious terminology, $\dot{a}(.)$ is the *mean function* and $\ddot{a}(.)$ the *variance function* of the distribution. Thus the precision parameter $\phi_t = V_t^{-1}$ appears as a divisor of the variance function to provide the variance in (14.4). $a(.)$ is assumed convex so that $\dot{a}(.)$ is a monotonically increasing function and η_t and μ_t are related via the one-one transformation (14.3); the inverse is simply $\eta_t = \dot{a}^{-1}(\mu_t)$.

EXAMPLE 14.1. The usual normal model $(Y_t|\mu_t, V_t) \sim N[\mu_t, V_t]$ is a key special case. Here $y_t(Y_t) = Y_t$, $a(\eta_t) = 0.5\eta_t^2$ so that $\mu_t = \eta_t$, and $b(Y_t, V_t) = (2\pi V_t)^{-1/2}\exp(-0.5V_t^{-1}Y_t^2)$.

EXAMPLE 14.2. Consider the binomial model where Y_t is the number of successes in $n_t > 0$ Bernoulli trials with success probability π_t. Here $\mathcal{Y}$ is the positive integers, and the mass function is

$$p(Y_t|\mu_t, n_t) = \begin{cases} \binom{n_t}{Y_t}\mu_t^{Y_t}(1 - \mu_t)^{n_t - Y_t}, & (Y_t = 0, 1, \dots, n_t), \\ 0, & \text{otherwise.} \end{cases}$$

This is a special case of (14.2) with defining quantities $y_t(Y_t) = Y_t/n_t$, $\eta_t = \log[\mu_t/(1 - \mu_t)]$ (the *log-odds*, or *logistic* transform of the probability μ_t), $V_t^{-1} = \phi_t = n_t$, $a(\eta_t) = \log[1 + \exp(\eta_t)]$, and $b(Y_t, V_t) = \binom{n_t}{Y_t}$.

Several other important distributions, including Poisson and gamma, are also special cases. It is left to the exercises for the reader to verify that, in each such case, the density or mass function can be written in exponential family form. Thus the development for the general model (14.1) or (14.2) applies to a wide and useful variety of distributions.

14.2.2 Conjugate analyses

The development will, with no loss of generality, be in terms of (14.2). In addition, the scale parameter V_t is assumed known for all t. So in modelling the observation, the only unknown quantity in the sampling density (14.2) is the natural parameter η_t, or, equivalently, the conditional mean μ_t of Y_t. At time $t - 1$, historical information relevant to forecasting Y_t is denoted, as usual, by D_{t-1}. Note that the density for Y_t may depend in some way on D_{t-1} (in particular, through the assumed value of V_t), so that (14.2) provides $p(Y_t|\eta_t, V_t, D_{t-1})$. For notational convenience, and since V_t is assumed known, the explicit dependence in the conditioning is dropped from here on. Thus the sampling density (14.1) or (14.2) is denoted simply by $p(Y_t|\eta_t)$, $(Y_t \in \mathcal{Y})$, the dependence on D_{t-1} being understood.

Now the only uncertainty about the distribution for Y_t given the past history D_{t-1} is due to the uncertainty about η_t. This will be summarised in terms of a prior (to time t) for η_t, the density denoted by $p(\eta_t|D_{t-1})$ as usual. It then follows, as usual, that the one-step ahead forecast distribution is defined via

$$p(Y_t|D_{t-1}) = \int p(Y_t|\eta_t)p(\eta_t|D_{t-1})d\eta_t, \qquad (14.5)$$

the integration being over the full parameter space for η_t, a subset of the real line. Similarly, once Y_t is observed, the prior is updated to a posterior for η_t by Bayes' Theorem,

$$p(\eta_t|D_t) \propto p(\eta_t|D_{t-1})p(Y_t|\eta_t). \qquad (14.6)$$

The calculations in (14.5) and (14.6), so familiar and easily performed in the cases of normal models, are analytically tractable in this exponential family framework when the prior belongs to the *conjugate family*. The use of exponential models with conjugate priors is well known in Bayesian statistics generally; see Aitchison and

Dunsmore (1976, Chapter 2) for a thorough discussion and technical details.

With reference to (14.2), a prior density from the conjugate family has the form

$$p(\eta_t|D_{t-1}) = c(r_t, s_t)\exp[r_t\eta_t - s_t a(\eta_t)], \qquad (14.7)$$

for some defining quantities r_t and s_t (known functions of D_{t-1}, and a known function $c(.\,,.)$ that provides the normalising constant $c(r_t, s_t)$. Some comments and properties are as follows.

(1) Given the quantities r_t and s_t, the conjugate prior is completely specified. Here $s_t > 0$ and, defining $x_t = r_t/s_t$, (14.7) can be written as

$$p(\eta_t|D_{t-1}) \propto \exp\{s_t[x_t\eta_t - a(\eta_t)]\}.$$

As a function of η_t, the prior thus has the same form as the likelihood function (14.2) when Y_t is fixed; s_t is analogous to ϕ_t and x_t is to Y_t.

(2) All such conjugate priors are unimodal (in fact, *strongly* unimodal) by virtue of the fact that $a(.)$ is a convex function. The prior mode for η_t is defined via $x_t = \dot{a}(\eta_t)$, and is therefore just $\dot{a}^{-1}(x_t)$. Thus x_t defines the *location* of the prior.

(3) s_t is the *precision parameter* of the prior; larger values of $s_t > 0$ imply a prior increasingly concentrated about the mode.

(4) The roles of x_t and s_t as defining location and precision are further evident in cases when (as is often true) the prior density and its first derivative decay to zero in the tails. Under such conditions, it is easily verified that $E[\mu_t|D_{t-1}] = x_t$. It follows that the one-step ahead forecast mean of Y_t is given by $E[Y_t|D_{t-1}] = x_t$. In addition, $V[\mu_t|D_{t-1}] = E[\ddot{a}(\eta_t)|D_{t-1}]/s_t$. See West (1985a, 1986a) for further details.

Assuming the values r_t and s_t are specified, it easily follows (and the verification is left to the reader) that the predictive density (14.5) and the posterior (14.6) are determined by:

$$p(Y_t|D_{t-1}) = \frac{c(r_t, s_t)b(Y_t, V_t)}{c(r_t + \phi_t Y_t, s_t + \phi_t)}, \qquad (14.8)$$

and

$$p(\eta_t|D_t) = c(r_t + \phi_t Y_t, s_t + \phi_t)\exp[(r_t + \phi_t Y_t)\eta_t - (s_t + \phi_t)a(\eta_t)]. \qquad (14.9)$$

The predictive distribution may now be easily calculated through (14.8). The posterior (14.9) has the same form as the conjugate prior, with defining quantities r_t and s_t updated to $r_t + \phi_t Y_t$ and $s_t + \phi_t$ respectively. The posterior precision parameter is thus the sum of the prior precision s_t and the precision parameter $\phi_t = V_t^{-1}$ of the sampling model for Y_t. Hence precision increases with data since $\phi_t > 0$. The prior location parameter $x_t = r_t/s_t$ is updated to

$$\frac{r_t + \phi_t Y_t}{s_t + \phi_t} = \frac{s_t x_t + \phi_t Y_t}{s_t + \phi_t} = (1 - \alpha_t)x_t + \alpha_t Y_t,$$

where $\alpha_t = \phi_t/(s_t + \phi_t)$ so that $0 < \alpha_t < 1$. Thus the posterior location parameter is a convex linear combination of the prior location parameter x_t and the observation Y_t. The weight α_t on the observation in the combination is an increasing function of the relative precision ϕ_t/s_t.

Full discussion of conjugate analyses can be found in Aitchison and Dunsmore (1976, Chapter 2), De Groot (1970, Chapter 9). Further details of the properties of conjugate priors appear in West (1985a, 1986a). The normal and binomial cases of Examples 14.1 and 14.2 are illustrated here, being special cases of great practical interest. Futher special cases are left to the exercises for the reader. See Aitchison and Dunsmore for a full list and a complete summary of prior, posterior and forecast distributions.

EXAMPLE 14.1 (continued). In the normal model the conjugate prior is also normal, $(\eta_t|D_{t-1}) \sim N[x_t, s_t^{-1}]$. The forecast density in (14.8) is normal, $(Y_t|D_{t-1}) \sim N[f_t, Q_t]$ with $f_t = x_t$ and $Q_t = V_t + s_t^{-1}$. So, of course, is the posterior (14.9). The posterior mean of $\eta_t = \mu_t$ may be written in the familiar form $x_t + \alpha_t(Y_t - x_t)$, the prior mean plus a weighted correction proportional to the forecast error $Y_t - x_t$. The posterior variance is $(s_t + \phi_t)^{-1}$ which may also be written in a more familiar form $s_t^{-1} - \alpha_t^2 Q_t$.

EXAMPLE 14.2 (continued). In the binomial model the conjugate prior is beta for the probability parameter $\mu_t = [1 + \exp(-\eta_t)]^{-1}$, namely $(\mu_t|D_{t-1}) \sim \text{Beta}[r_t, s_t]$. This has density

$$p(\mu_t|D_{t-1}) \propto \mu_t^{r_t-1}(1 - \mu_t)^{s_t-1}, \qquad (0 \le \mu_t \le 1),$$

with normalising constant

$$c(r_t, s_t) = \frac{\Gamma(r_t + s_t)}{\Gamma(r_t)\Gamma(s_t)}.$$

Here both r_t and s_t must be positive. Transforming to the natural parameter $\eta_t = \log[\mu_t/(1 - \mu_t)]$ provides (14.7) although it is not necessary to work on the η_t scale, and much more convenient to work directly with the conjugate beta on the μ_t scale. The forecast distribution (14.8) is the beta-binomial, and the posterior (14.9) beta, $(\mu_t|D_{t-1}) \sim \text{Beta}[r_t + Y_t, s_t + n_t - Y_t]$.

14.3 DYNAMIC GENERALISED LINEAR MODELS

14.3.1 Dynamic regression framework

The class of generalised linear models (Nelder and Wedderburn, 1972; Baker and Nelder, 1983; West, 1985a) assumes data conditionally independently drawn from distributions with common exponential family form, but with parameters η_t and V_t possibly differing. Regression effects of independent variables are modelled by relating the natural parameters for each observation to a linear function of regression variables. In the time series context, the use of time-varying regression type models is appropriate, applying to define the following *dynamic generalised linear model*.

Definition 14.1. *Define the following quantities at time t.*

- $\boldsymbol{\theta}_t$ *is a $n-$dimensional state vector at time t;*
- $\mathbf{F}_t$ *a known, $n-$dimensional regression vector;*
- $\mathbf{G}_t$ *a known, $n \times n$ evolution matrix;*
- $\boldsymbol{\omega}_t$ *a $n-$vector of evolution errors having zero mean and known variance matrix $\mathbf{W}_t$, denoted by*

$$\boldsymbol{\omega}_t \sim [\, \mathbf{0}, \mathbf{W}_t];$$

- $\lambda = \mathbf{F}_t'\boldsymbol{\theta}_t$, *a linear function of the state vector parameters;*
- $g(\eta_t)$ *a known, continuous and monotonic function mapping η_t to the real line.*

Then the dynamic generalised linear model (DGLM) for the series Y_t, $(t = 1, 2, \ldots,)$ is defined by the following components.

Observation model: $\quad p(Y_t|\eta_t) \quad$ *as in (14.2),* $\quad g(\eta_t) = \lambda_t = \mathbf{F}_t'\boldsymbol{\theta}_t,$

$$(14.10)$$

Evolution equation: $\quad \boldsymbol{\theta}_t = \mathbf{G}_t\boldsymbol{\theta}_{t-1} + \boldsymbol{\omega}_t, \quad \boldsymbol{\omega}_t \sim [\, \mathbf{0}, \mathbf{W}_t].$

$$(14.11)$$

This definition is an obvious extension of the standard DLM in the observation model. The sampling distribution is now possibly non-normal, and the linear regression λ_t affects the observational distribution through a possibly non-linear function of the mean response function $\mu_t = E[Y_t|\eta_t] = \dot{a}(\eta_t)$. The additional component $g(.)$ provides a transformation from the natural parameter space to that of the linear regression. In practice, it is very common that this will be the identity map. This provides, for example, logistic regression in binomial models and log-linear models when the sampling distribution is Poisson. See McCullagh and Nelder (1983) for much further discussion. The evolution equation for the state vector is precisely as in the normal DLM, although normality is not assumed, only the first and second order moments of the evolution errors being so far assumed. As in the DLM, the zero mean assumption for ω_t may be relaxed to include a known, non-zero mean without any essential change to the structure of the model. These evolution errors are assumed uncorrelated over time, as usual, and, conditional on η_t, Y_t is independent of ω_t. The standard, static GLM is a special case in which $\theta_t = \theta$ for all time, given by taking $\mathbf{G}_t$ as the identity and $\mathbf{W}_t$ as the zero matrix.

This class of models provides a generalisation of the DLM to non-normal error models for the time series, and of the GLM to time varying parameters. However, having left the standard, normal/linear framework, analysis becomes difficult, the nice theory of normal models being lost. Thus their is a need for some form of approximation in analysis. The possibilities, as in the non-linear models of Chapter 13, include, broadly, (a) analytic approximations, such as the use of transformations of the data to approximate normality if sensible, and normal approximations for prior/posterior distributions; and (b) numerical approximations, using some form of numerical integration technique. An approach of type (a) is described here, that attempts to retain the essential non-normal features of the observational model and forecast distributions by exploiting the conjugate analysis of Section 14.2.2. This analysis follows West, Harrison and Migon (1985). Other possible approaches, such as the use of normal approximations of various forms, are certainly possible, though, at the time of writing, remain largely unexplored and so are not discussed here.

14.3.2   The DLM revisited

The components of the analysis are motivated by first considering a reformulation of the standard, exact analysis in the special case of the normal DLM.

In the special case that $p(Y_t|\eta_t)$ is the normal density, $(Y_t|\eta_t) \sim N[\mu_t, V_t]$ with $\mu_t = \eta_t$, the DLM is obtained by taking $g(\eta_t) = \eta_t$ so that $\mu_t = \eta_t = \lambda_t = \mathbf{F}_t'\boldsymbol{\theta}_t$. We work in terms of the μ_t notation. In the usual sequential analysis, the model at time t is completed by the posterior for the state vector at $t-1$, namely

$$(\boldsymbol{\theta}_{t-1}|D_{t-1}) \sim N[\mathbf{m}_{t-1}, \mathbf{C}_{t-1}], \qquad (14.12)$$

where the moments are known functions of the past information D_{t-1}. Also, the evolution errors are normal and so, given the assumption of no correlation between $\boldsymbol{\omega}_t$ and $\boldsymbol{\theta}_{t-1}$ in (14.12), the prior for $\boldsymbol{\theta}_t$ is deduced as

$$(\boldsymbol{\theta}_t|D_{t-1}) \sim N[\mathbf{a}_t, \mathbf{R}_t] \qquad (14.13)$$

where $\mathbf{a}_t = \mathbf{G}_t\mathbf{m}_{t-1}$ and $\mathbf{R}_t = \mathbf{G}_t\mathbf{C}_{t-1}\mathbf{G}_t' + \mathbf{W}_t$. So much is standard theory. From (14.13) and the normal sampling model, the one-step ahead forecast distribution for Y_t and the posterior for $\boldsymbol{\theta}_t$ are obtained via the usual equations.

Consider, however, the following alternative derivations of these key components of the analysis.

Step 1: Prior for μ_t.

$\mu_t = \lambda_t = \mathbf{F}_t'\boldsymbol{\theta}_t$ is a linear function of the state vector. Hence, under the prior (14.13), μ_t and $\boldsymbol{\theta}_t$ have a joint (singular) normal prior distribution

$$\left(\begin{matrix}\mu_t \\ \boldsymbol{\theta}_t\end{matrix} \middle| D_{t-1}\right) \sim N\left[\begin{pmatrix}f_t \\ \mathbf{a}_t\end{pmatrix}, \begin{pmatrix}q_t & \mathbf{F}_t'\mathbf{R}_t \\ \mathbf{R}_t\mathbf{F}_t & \mathbf{R}_t\end{pmatrix}\right], \qquad (14.14)$$

where

$$f_t = \mathbf{F}_t'\mathbf{a}_t \qquad \text{and} \qquad q_t = \mathbf{F}_t'\mathbf{R}_t\mathbf{F}_t. \qquad (14.15)$$

Step 2: One-step ahead forecasting.

The sampling distribution of Y_t depends on $\boldsymbol{\theta}_t$ only via the single quantity μ_t, and thus the historical information relevant to forecasting Y_t is completely summarised in the marginal prior $(\mu_t|D_{t-1}) \sim$

$N[f_t, q_t]$. The one-step ahead forecast distribution is given in Example 14.1 as $(Y_t|D_{t-1}) \sim N[f_t, Q_t]$ where $Q_t = q_t + V_t$. The reader will verify that this is the usual result.

Step 3: Updating for μ_t.

Observing Y_t, Example 14.2 provides the posterior for μ_t as normal,

$$(\mu_t|D_t) \sim N[f_t^*, q_t^*]$$

where

$$f_t^* = f_t + (q_t/Q_t)e_t \qquad \text{and} \qquad q_t^* = q_t - q_t^2/Q_t, \qquad (14.16)$$

with $e_t = Y_t - f_t$, the forecast error.

Step 4: Conditional structure for $(\theta_t|\mu_t, D_{t-1})$.

The objective of the updating is to calculate the posterior for θ_t. This can be derived from the *joint* posterior for μ_t and θ_t, which, by Bayes' Theorem, is

$$\begin{aligned}
p(\mu_t, \theta_t|D_t) &\propto p(\mu_t, \theta_t|D_{t-1})\, p(Y_t|\mu_t) \\
&\propto [p(\theta_t|\mu_t, D_{t-1})p(\mu_t|D_{t-1})]\, p(Y_t|\mu_t) \\
&\propto p(\theta_t|\mu_t, D_{t-1})\, [p(\mu_t|D_{t-1})p(Y_t|\mu_t)] \\
&\propto p(\theta_t|\mu_t, D_{t-1})\, p(\mu_t|D_t).
\end{aligned}$$

Hence, given μ_t, θ_t is conditionally independent of Y_t and it follows that

$$p(\theta_t|D_t) = \int p(\theta_t|\mu_t, D_{t-1})p(\mu_t|D_t)d\mu_t. \qquad (14.17)$$

The key point is that information about the state vector from Y_t is channelled through the posterior for μ_t. The first component in the integrand is, using standard normal theory in (14.14), just the conditional normal distribution

$$(\theta_t|\mu_t, D_{t-1}) \sim N[a_t + R_t F_t(\mu_t - f_t)/q_t, R_t - R_t F_t F_t' R_t/q_t], \quad (14.18)$$

for all μ_t. The second is that derived in Step 3 above.

Step 5: Updating for θ_t.

Now, since all components are normal, the required posterior (14.17) is normal and is thus defined by the mean and variance matrix. By analogy with (14.17), we can express these as follows. Firstly the mean is given by

$$
\begin{aligned}
\mathbf{m}_t &= \mathrm{E}[\boldsymbol{\theta}_t | D_t] \\
&= \mathrm{E}[\mathrm{E}\{\boldsymbol{\theta}_t | \mu_t, D_{t-1}\} | D_t] \\
&= \mathrm{E}[\mathbf{a}_t + \mathbf{R}_t \mathbf{F}_t (\mu_t - f_t)/q_t | D_t] \\
&= \mathbf{a}_t + \mathbf{R}_t \mathbf{F}_t (\mathrm{E}[\mu_t | D_t] - f_t)/q_t \\
&= \mathbf{a}_t + \mathbf{R}_t \mathbf{F}_t (f_t^* - f_t)/q_t.
\end{aligned}
\tag{14.19}
$$

Similarly,

$$
\begin{aligned}
\mathbf{C}_t &= \mathrm{V}[\boldsymbol{\theta}_t | D_t] \\
&= \mathrm{V}[\mathrm{E}\{\boldsymbol{\theta}_t | \mu_t, D_{t-1}\} | D_t] + \mathrm{E}[\mathrm{V}\{\boldsymbol{\theta}_t | \mu_t, D_{t-1}\} | D_t] \\
&= \mathrm{V}[\mathbf{a}_t + \mathbf{R}_t \mathbf{F}_t (\mu_t - f_t)/q_t | D_t] + \mathrm{E}[\mathbf{R}_t - \mathbf{R}_t \mathbf{F}_t \mathbf{F}_t' \mathbf{R}_t/q_t | D_t] \\
&= \mathbf{R}_t \mathbf{F}_t \mathbf{F}_t' \mathbf{R}_t \mathrm{V}[\mu_t | D_t]/q_t^2 + \mathbf{R}_t - \mathbf{R}_t \mathbf{F}_t \mathbf{F}_t' \mathbf{R}_t/q_t \\
&= \mathbf{R}_t - \mathbf{R}_t \mathbf{F}_t \mathbf{F}_t' \mathbf{R}_t (1 - q_t^*/q_t)/q_t.
\end{aligned}
\tag{14.20}
$$

Substituting the values of f_t^* and q_t^* from (14.15) leads to the usual expressions $\mathbf{m}_t = \mathbf{a}_t + \mathbf{A}_t e_t$ and $\mathbf{C}_t = \mathbf{R}_t - \mathbf{A}_t \mathbf{A}_t' Q_t$ where $\mathbf{A}_t = \mathbf{R}_t \mathbf{F}_t / Q_t$ is the usual adaptive vector.

Steps 1 to 5 provide an alternative proof of the usual forecasting and updating equations at time t, this route being rather circuitous relative to the usual, direct calculations. However, in non-normal models, an approximate analysis along these lines leads to a parallel analysis for DGLMs.

14.3.3 DGLM updating

Return now to the DGLM of Definition 14.1. Given the non-normality of the observational model and, in general, the non-linearity of the observation mean μ_t as a function of $\boldsymbol{\theta}_t$, there is no general, exact analysis. The approach in West, Harrison and Migon (1985) develops as follows, paralleling the steps above in the DLM.

Definition 14.1 provides the basic observation and evolution model at time t. To complete the model specification for time t, we need to fully define two more component distributions: (a) that of the evolution error $\boldsymbol{\omega}_t$, as yet only specified in terms of mean and variance matrix; and (b) $p(\boldsymbol{\theta}_{t-1} | D_{t-1})$ that sufficiently summarises the historical information and analysis prior to time t. In the DLM, of course, these are both normal. Whatever forms they may take in the DGLM, prior and posterior distributions for the state vector at any time will not now be normal. Without considerable development of numerical integration based methodology for the DGLM, there is no way in which, in any generality, such distributions can be adequately calculated and summarised. West, Harrison and Migon (1985) proceed to develop an approximate analysis based on assuming that these required distributions are only partially specified in terms of their first and second order moments.

Assumption: In the DGLM of Definition 14.1, suppose that the distribution of ω_t is unspecified apart from the moments in (14.11). In addition, suppose that the posterior mean and variance of $(\theta_{t-1}| D_{t-1})$ are as in the DLM case (14.12) but without the normality of the distribution.

Thus the model definition is completed at time t via

$$(\theta_{t-1}|D_{t-1}) \sim [\mathbf{m}_{t-1}, \mathbf{C}_{t-1}], \qquad (14.21)$$

the full distributional form being unspecified. It follows that, from (14.11), the prior moments of θ_t are

$$(\theta_t|D_{t-1}) \sim [\mathbf{a}_t, \mathbf{R}_t] \qquad (14.22)$$

where $\mathbf{a}_t = \mathbf{G}_t\mathbf{m}_{t-1}$ and $\mathbf{R}_t = \mathbf{G}_t\mathbf{C}_{t-1}\mathbf{G}'_t + \mathbf{W}_t$. This parallels the normal theory; the price payed for the greater generality in not assuming normality is that full distributional information is lost, the prior being only partially specified. The parallel with normal theory is now developed in each of the five steps of the previous Section. Now the notation is a little more complex, and worth reiterating. As usual, μ_t is the observation mean, related to the natural parameter η_t via $\mu_t = \dot{a}(\eta_t)$. Similarly, η_t relates to the linear function $\lambda_t = \mathbf{F}'_t\theta_t$ of state parameters via $\lambda_t = g(\eta_t)$. Since $\dot{a}(.)$ and $g(.)$ are bijective, we can (and do) work in terms of μ_t, η_t or λ_t interchangeably.

Step 1: Prior for λ_t.

$\lambda_t = \mathbf{F}'_t\theta_t$ is a linear function of the state vector. Hence, under (14.22), λ_t and θ_t have a joint prior distributions that is only partially specified in terms of moments

$$\begin{pmatrix} \lambda_t \\ \theta_t \end{pmatrix} \Big| D_{t-1} \sim \left[\begin{pmatrix} f_t \\ \mathbf{a}_t \end{pmatrix}, \begin{pmatrix} q_t & \mathbf{F}'_t\mathbf{R}_t \\ \mathbf{R}_t\mathbf{F}_t & \mathbf{R}_t \end{pmatrix} \right], \qquad (14.23)$$

where

$$f_t = \mathbf{F}_t'\mathbf{a}_t \quad \text{and} \quad q_t = \mathbf{F}_t'\mathbf{R}_t\mathbf{F}_t. \tag{14.24}$$

These moments coincide precisely, of course, with those of the special case of normality in (14.14) and (14.15).

Step 2: One-step ahead forecasting.

As in the DLM, the sampling distribution of Y_t depends on $\boldsymbol{\theta}_t$ only via $\eta_t = g^{-1}(\lambda_t)$, and thus the historical information relevant to forecasting Y_t is completely summarised in the marginal prior for $(\eta_t|D_{t-1})$. However, this is now only partially specified through the mean and variance of $\lambda_t = g(\eta)$ from (14.23),

$$(\lambda_t|D_{t-1}) \sim [f_t, q_t]. \tag{14.25}$$

In order to calculate the forecast distribution (and to update to the posterior for η_t), further assumptions about the form of the prior for η_t are necessary. At this point it should be stressed that the form of the prior is arbitrary; apart from (14.25), no further restrictions have been made on the prior. Thus there is no prior form to be calculated or approximated in any sense, the forecaster may choose any desired form consistent with the mean and variance. The prior may be assumed approximately normal, for example, or take any other convenient form. In view of the conjugate analysis described in Section 14.2.2, it is clear that the most convenient form is that of the conjugate family, and thus a conjugate prior is supposed. This requires, of course, that such a prior can be found consistent with the mean and variance of λ_t; happily, priors in the conjugate family typically permit a full range of unrelated values for the mean and variance, so that this poses no problem.

Thus, given (14.25), assume that the prior for η_t has the conjugate form in (14.7), namely

$$p(\eta_t|D_{t-1}) = c(r_t, s_t)\exp[r_t\eta_t - s_t a(\eta_t)]. \tag{14.26}$$

The defining parameters r_t and s_t are chosen to be consistent with the moments for λ_t in (14.25), thus implicitly satisfying the equations

$$E[g(\eta_t)|D_{t-1}] = f_t \quad \text{and} \quad V[g(\eta_t)|D_{t-1}] = q_t. \tag{14.27}$$

The one-step ahead forecast distribution now follows from the density in (14.8). For all $Y_t \in \mathcal{Y}$,

$$p(Y_t|D_{t-1}) = \frac{c(r_t, s_t)b(Y_t, V_t)}{c(r_t + \phi_t Y_t, s_t + \phi_t)}. \tag{14.28}$$

Step 3: Updating for η_t.

Observing Y_t, (14.9) provides the posterior for η_t in the conjugate form

$$p(\eta_t|D_t) = c(r_t + \phi_t Y_t, s_t + \phi_t)\exp[(r_t + \phi_t Y_t)\eta_t - (s_t + \phi_t)a(\eta_t)].$$

By analogy with the prior, denote the posterior mean and variance of $\lambda_t = g(\eta_t)$ by

$$f_t^* = \mathrm{E}[g(\eta_t)|D_t] \qquad \text{and} \qquad q_t^* = \mathrm{V}[g(\eta_t)|D_t]. \tag{14.29}$$

Step 4a: Conditional structure for $(\boldsymbol{\theta}_t|\lambda_t, D_{t-1})$.

As in the normal model, the objective of the updating is to calculate the posterior for $\boldsymbol{\theta}_t$. Again following the normal theory, these can be derived from the *joint* posterior for λ_t and $\boldsymbol{\theta}_t$. The joint density is, by Bayes' Theorem,

$$
\begin{aligned}
p(\lambda_t, \boldsymbol{\theta}_t|D_t) &\propto p(\lambda_t, \boldsymbol{\theta}_t|D_{t-1})\, p(Y_t|\lambda_t) \\
&\propto [p(\boldsymbol{\theta}_t|\lambda_t, D_{t-1})p(\lambda_t|D_{t-1})]\, p(Y_t|\lambda_t) \\
&\propto p(\boldsymbol{\theta}_t|\lambda_t, D_{t-1})\, [p(\lambda_t|D_{t-1})p(Y_t|\lambda_t)] \\
&\propto p(\boldsymbol{\theta}_t|\lambda_t, D_{t-1})\, p(\lambda_t|D_t).
\end{aligned}
$$

Hence, given λ_t and D_{t-1}, $\boldsymbol{\theta}_{t-1}$ is conditionally independent of Y_t and it follows that

$$p(\boldsymbol{\theta}_t|D_t) = \int p(\boldsymbol{\theta}_t|\lambda_t, D_{t-1})p(\lambda_t|D_t)d\lambda_t. \tag{14.30}$$

As in the DLM, information about the state vector from Y_t is channelled through the posterior for λ_t. The second component in the integrand $p(\lambda_t|D_t)$ may be obtained directly from the conjugate form posterior for η_t in (14.26). The first component, defining the conditional prior for $\boldsymbol{\theta}_t$ given λ_t, is, of course, not fully specified. Note, however, that, to complete the updating cycle, we only need to calculate the posterior mean and variance matrix of $\boldsymbol{\theta}_t$, the full posterior remaining unspecified and indeterminate. From (14.30) and in parallel with (14.18) in the normal model, the key ingredients in these calculations are the *prior* mean and variance matrix of $(\boldsymbol{\theta}_t|\lambda_t, D_{t-1})$. Unfortunately, due to the incomplete specification of the joint prior, these conditional moments are unknown, non-linear and *indeterminate* functions of λ_t. They cannot be calculated without imposing further structure. However, given the partial moments specification in (14.23), they can be *estimated* using standard Bayesian techniques, as follows.

Step 4b: Linear Bayes' estimation of moments of $(\boldsymbol{\theta}_t|\lambda_t, D_{t-1})$.

From (14.23) we have the joint prior mean and variance matrix of $\boldsymbol{\theta}_t$ and λ_t, and are interested in the conditional moments $E[\boldsymbol{\theta}_t|, \lambda_t, D_{t-1}]$ and $V[\boldsymbol{\theta}_t|\lambda_t, D_{t-1}]$. Under our assumptions, these functions of λ_t cannot be calculated, but they can be estimated using methods of Linear Bayesian Estimation (LBE), as in Section 4.9 of Chapter 4. Recall that LBE provides a Bayesian decision theoretic approach to estimation of unknown, non-linear functions by linear "regressions". Section 4.9.2 of Chapter describes the general theory of LBE, which is applied here. Theorem 4.9 of that Section applies here directly as follows. Within the class of *linear* functions of λ_t, and subject only to the prior information in (14.23), the LBE optimal estimate of $E[\boldsymbol{\theta}_t|, \lambda_t, D_{t-1}]$ is given by

$$\hat{E}[\boldsymbol{\theta}_t|, \lambda_t, D_{t-1}] = \mathbf{a}_t + \mathbf{R}_t \mathbf{F}_t(\lambda_t - f_t)/q_t, \qquad (14.31)$$

for all λ_t. The associated estimate of $V[\boldsymbol{\theta}_t|\lambda_t, D_{t-1}]$ is

$$\hat{V}[\boldsymbol{\theta}_t|, \lambda_t, D_{t-1}] = \mathbf{R}_t - \mathbf{R}_t \mathbf{F}_t \mathbf{F}'_t \mathbf{R}_t/q_t, \qquad (14.32)$$

for all λ_t. These estimates of conditional moments are adopted. They obviously coincide with the exact values in the normal case.

Step 5: Updating for $\boldsymbol{\theta}_t$.

From (14.30), it follows that (again as in the normal case)

$$E[\boldsymbol{\theta}_t|D_t] = E[E\{\boldsymbol{\theta}_t|\lambda_t, D_{t-1}\}|D_t],$$

and

$$V[\boldsymbol{\theta}_t|D_t] = V[E\{\boldsymbol{\theta}_t|\lambda_t, D_{t-1}\}|D_t] + E[V\{\boldsymbol{\theta}_t|\lambda_t, D_{t-1}\}|D_t].$$

These may be estimated by substituting the LBE estimates for the arguments of the expectations. This leads to the posterior moment

$$(\boldsymbol{\theta}_t|D_t) \sim [\mathbf{m}_t, \mathbf{C}_t],$$

the posterior moments defined as follows. Firstly,

$$
\begin{aligned}
\mathbf{m}_t &= \mathrm{E}[\hat{\mathrm{E}}\{\boldsymbol{\theta}_t|\lambda_t, D_{t-1}\}|D_t] \\
&= \mathrm{E}[\mathbf{a}_t + \mathbf{R}_t\mathbf{F}_t(\lambda_t - f_t)/q_t|D_t] \\
&= \mathbf{a}_t + \mathbf{R}_t\mathbf{F}_t(\mathrm{E}[\lambda_t|D_t] - f_t)/q_t \\
&= \mathbf{a}_t + \mathbf{R}_t\mathbf{F}_t(f_t^* - f_t)/q_t.
\end{aligned}
\tag{14.33}
$$

Similarly,

$$
\begin{aligned}
\mathbf{C}_t &= \mathrm{V}[\hat{\mathrm{E}}\{\boldsymbol{\theta}_t|\lambda_t, D_{t-1}\}|D_t] + \mathrm{E}[\hat{\mathrm{V}}\{\boldsymbol{\theta}_t|\lambda_t, D_{t-1}\}|D_t] \\
&= \mathrm{V}[\mathbf{a}_t + \mathbf{R}_t\mathbf{F}_t(\lambda_t - f_t)/q_t|D_t] + \mathrm{E}[\mathbf{R}_t - \mathbf{R}_t\mathbf{F}_t\mathbf{F}_t'\mathbf{R}_t/q_t|D_t] \\
&= \mathbf{R}_t\mathbf{F}_t\mathbf{F}_t'\mathbf{R}_t\mathrm{V}[\lambda_t|D_t]/q_t^2 + \mathbf{R}_t - \mathbf{R}_t\mathbf{F}_t\mathbf{F}_t'\mathbf{R}_t/q_t \\
&= \mathbf{R}_t - \mathbf{R}_t\mathbf{F}_t\mathbf{F}_t'\mathbf{R}_t(1 - q_t^*/q_t)/q_t.
\end{aligned}
\tag{14.34}
$$

Substituting the values of f_t^* and q_t^* from (14.29) completes the updating. Note that these equations are just as in the normal case, equations (14.19) and (14.20), the differences lying in the ways that f_t^* and q_t^* are calculated in the conjugate updating component in Step 3. In particular, in non-normal models the posterior variance matrix $\mathbf{C}_t$ will depend on the data Y_t through q_t^*. This is not the case in the DLM where $\mathbf{C}_t$ is data independent.

EXAMPLE 14.3. Consider the binomial framework of Example 14.2, Y_t being the number of "successes" in a total of n_t trials. Note that this includes the important special case of binary time series, when $n_t = 1$ for all t. Recall that $\mu_t = \mathrm{E}[Y_t/n_t|\eta_t]$ is the binomial probability, $(0 \le \mu_t \le 1)$, and $\eta_t = \log[\mu_t/(1 - \mu_t)]$, is the log-odds or logistic transform of μ_t. Perhaps the most common regression structure with binomial data is *logistic* regression (McCullagh and Nelder, 1983, Chapter 4). This corresponds to $g(.)$ being the identity function so that $\eta_t = \lambda_t = \mathbf{F}_t'\boldsymbol{\theta}_t$ is given by

$$
\eta_t = \lambda_t = \log\left[\frac{\mu_t}{1 - \mu_t}\right], \qquad (0 \le \mu_t \le 1).
$$

Here the conjugate prior (14.26) is beta on the μ_t scale, $(\mu_t|D_{t-1}) \sim$ Beta$[r_t, s_t]$ having density

$$
p(\mu_t|D_{t-1}) = c(r_t, s_t)\mu_t^{r_t-1}(1 - \mu_t)^{s_t-1}, \qquad (0 \le \mu_t \le 1).
$$

The normalising constant is $c(r_t, s_t) = \Gamma(r_t + s_t)/[\Gamma(r_t)\Gamma(s_t)]$. Given the prior moments f_t and q_t for η_t, this prior is specified by calculating the corresponding values of r_t and s_t. For any positive quantity x, let $\gamma(x) = \dot{\Gamma}(x)/\Gamma(x)$ denote the digamma function at x. It can be verified that, under this prior,

$$f_t = E[\eta_t|D_{t-1}] = \gamma(r_t) - \gamma(s_t)$$

and

$$q_t = V[\eta_t|D_{t-1}] = \dot{\gamma}(r_t) + \dot{\gamma}(s_t).$$

Given f_t and q_t, these equations implicitly define the beta parameters r_t and s_t, and can be solved numerically by an iterative method.[†] For moderate and large values of r_t and s_t (eg. both exceeding 3) the prior for μ_t is reasonably concentrated away from the extremes of 0 and 1 and the digamma function may be adequately approximated using

$$\gamma(x) \approx \log(x) + \frac{1}{2x}.$$

The corresponding approximation for the derivative function is

$$\dot{\gamma}(x) \approx \frac{1}{x}\left(1 - \frac{1}{2x}\right).$$

For even larger values of x, $\gamma(x) \approx \log(x)$ and $\dot{\gamma}(x) \approx x^{-1}$. Thus, for large r_t and s_t, the mean and variance of η_t are approximately given by

$$f_t \approx \log(r_t/s_t) \qquad \text{and} \qquad q_t \approx \frac{1}{r_t} + \frac{1}{s_t}.$$

These may be inverted directly to give

$$r_t = q_t^{-1}[1 + \exp(f_t)] \qquad \text{and} \qquad s_t = q_t^{-1}[1 + \exp(-f_t)].$$

In fact these approximations may be used with satisfactory results even when f_t and q_t are consistent with rather small values of r_t and s_t. West, Harrison and Migon (1985) provide discussion and examples of this.

[†]See Abramowitz and Stegun, 1965, for full discussion of the digamma function. The recursions $\gamma(x) = \gamma(x + 1) - x^{-1}$ and $\dot{\gamma}(x) = \dot{\gamma}(x + 1) + x^{-2}$ are useful for evaluations. They also provide useful tables of numerical values and various analytic approximations.

Given r_t and s_t, the one-step ahead beta-binomial distribution for Y_t is calculable. Updating, $(\mu_t|D_t) \sim \text{Beta}[r_t + Y_t, s_t + n_t - Y_t]$ so that

$$f_t^* = \text{E}[\eta_t|D_t] = \gamma(r_t + Y_t) - \gamma(s_t + n_t - Y_t)$$

and

$$q_t^* = \text{V}[\eta_t|D_t] = \dot{\gamma}(r_t + Y_t) + \dot{\gamma}(s_t + n_t - Y_t).$$

These are easily calculated from the digamma function, and updating proceeds via (14.33) and (14.34).

EXAMPLE 14.4. A binomial *linear* regression structure is given by taking $\mu_t = \lambda_t$ for all t, so that $g^{-1}(.)$ is the logistic or log-odds transform. Thus the expected level of Y_t is just as in the DLM, $\mu_t = \mathbf{F}_t'\boldsymbol{\theta}_t$, but variation about this level is binomial rather than normal. With $(\mu_t|D_{t-1}) \sim \text{Beta}[r_t, s_t]$, $\mu_t = \lambda_t$ implies directly that

$$f_t = \text{E}[\mu_t|D_{t-1}] = \frac{r_t}{(r_t + s_t)} \quad \text{and} \quad q_t = \text{V}[\mu_t|D_{t-1}] = \frac{f_t(1 - f_t)}{(r_t + s_t + 1)}.$$

These equations invert to define the beta prior via

$$r_t = f_t[q_t^{-1}f_t(1 - f_t) - 1] \quad \text{and} \quad s_t = (1 - f_t)[q_t^{-1}f_t(1 - f_t) - 1].$$

After updating,

$$f_t^* = \text{E}[\mu_t|D_t] = \frac{r_t + Y_t}{(r_t + s_t + n_t)}$$

and

$$q_t^* = \text{V}[\mu_t|D_{t-1}] = \frac{f_t^*(1 - f_t^*)}{(r_t + s_t + n_t + 1)}.$$

14.3.4 Step ahead forecasting and filtering

Forecasting ahead to time $t + k$ from time t Steps 1 and 2 of the one-step analysis extend directly. At time t, the posterior moments of $\boldsymbol{\theta}_t$ are available,

$$(\boldsymbol{\theta}_t|D_t) \sim [\mathbf{m}_t, \mathbf{C}_t].$$

From the evolution equation (14.11) applied at times $t+1, \ldots, t+k$, it follows that

$$(\boldsymbol{\theta}_{t+k}|D_t) \sim [\mathbf{a}_t(k), \mathbf{R}_t(k)],$$

with moments defined sequentially into the future via, for $k = 1, 2, \ldots$,

$$\mathbf{a}_t(k) = \mathbf{G}_{t+k}\mathbf{a}_t(k-1)$$

and

$$\mathbf{R}_t(k) = \mathbf{G}_{t+k}\mathbf{R}_t(k-1)\mathbf{G}'_{t+k} + \mathbf{W}_{t+k},$$

with $\mathbf{a}_t(0) = \mathbf{m}_t$ and $\mathbf{R}_t(0) = \mathbf{C}_t$. These equations are, of course, just as in the DLM (see Theorem 4.2 of Chapter 4). Now, $\lambda_{t+k} = \mathbf{F}'_{t+k}\boldsymbol{\theta}_{t+k}$ has moments

$$(\lambda_{t+k}|D_t) \sim [f_t(k), q_t(k)]$$

where

$$f_t(k) = \mathbf{F}'_{t+k}\mathbf{a}_t(k) \qquad \text{and} \qquad q_t(k) = \mathbf{F}'_{t+k}\mathbf{R}_t(k)\mathbf{F}_{t+k}.$$

The conjugate prior for $\eta_{t+k} = g^{-1}(\lambda_{t+k})$ consistent with these moments is now determined; denote the corresponding defining parameters by $r_t(k)$ and $s_t(k)$ so that, by analogy with (14.28), the required forecast density is just

$$p(Y_{t+k}|D_t) = \frac{c(r_t(k), s_t(k))b(Y_{t+k}, V_{t+k})}{c(r_t(k) + \phi_{t+k}Y_{t+k}, s_t(k) + \phi_{t+k})}.$$

Note that, if using discount factors to construct $\mathbf{W}_t$ at each time t, then, as usual, $\mathbf{W}_{t+k}$ is the above equations should be replaced by $\mathbf{W}_t(k) = \mathbf{W}_t$ for extrapolating.

Filtering backwards over time from time t, the ideas underlying the use of Linear Bayes' methods in updating may be applied to derive optimal (in the LBE sense) estimates of the posterior moments of $\boldsymbol{\theta}_{t-k}$ given D_t, for all t and $k > 0$. It is clear from the results in Section 4.9 of Chapter 4 that the Linear Bayes' estimates of filtered moments are derived just as in the case of normality. Thus the estimated moments of the filtered distributions are given by Theorem 4.4 of Chapter 4.

14.3.5 Linearisation in the DGLM

Some practical models will introduce parameter nonlinearities into either of the model equations (14.10) or (14.11). Within the approximate analysis here, such nonlinearities can be treated as in Section

13.2; ie. using some form of linearisation technique. Such an approach underlies the applications in Migon and Harrison (1985), and West and Harrison (1986a). To parallel the use of linearisation applied to the normal, non-linear model (13.1) of Chapter 13, consider the following non-normal, non-linear verison of the DGLM of Definition 14.1. Suppose that the sampling model is just as described in (14.10), but now that

- for some known, non-linear regression function $F_t(.)$,

$$\lambda_t = F_t(\boldsymbol{\theta}_t); \tag{14.35}$$

- for some known, non-linear evolution function $\mathbf{g}_t(.)$,

$$\boldsymbol{\theta}_t = \mathbf{g}_t(\boldsymbol{\theta}_{t-1}) + \boldsymbol{\omega}_t. \tag{14.36}$$

Based on the use of linearisation in Section 13.2 of Chapter 13, the above DGLM analysis can be applied to this non-linear model at time t as follows.

(1) Linearise the evolution equation to give

$$\boldsymbol{\theta}_t \approx \mathbf{h}_t + \mathbf{G}_t \boldsymbol{\theta}_{t-1} + \boldsymbol{\omega}_t,$$

where $\mathbf{G}_t$ is the known evolution matrix

$$\mathbf{G}_t = \left[\frac{\delta \mathbf{g}_t(\boldsymbol{\theta}_{t-1})}{\delta \boldsymbol{\theta}'_{t-1}} \right]_{\boldsymbol{\theta}_{t-1}=\mathbf{m}_{t-1}},$$

and

$$\mathbf{h}_t = \mathbf{g}_t(\mathbf{m}_{t-1}) - \mathbf{G}_t \mathbf{m}_{t-1}$$

is also known. Then the usual DLM evolution applies, with the minor extension to include $\mathbf{h}_t$ in the evolution equation. The prior moments for $\boldsymbol{\theta}_t$ are

$$\begin{aligned}
\mathbf{a}_t &= \mathbf{h}_t + \mathbf{G}_t \mathbf{m}_{t-1} = \mathbf{g}_t(\mathbf{m}_{t-1}), \\
\mathbf{R}_t &= \mathbf{G}_t \mathbf{C}_{t-1} \mathbf{G}'_t + \mathbf{W}_t.
\end{aligned}$$

(2) Proceeding to the observation equation, the non-linear regression function is linearised to give

$$\lambda_t = f_t + \mathbf{F}'_t(\boldsymbol{\theta}_t - \mathbf{a}_t)$$

where

$$f_t = F_t(\mathbf{a}_t)$$

and $\mathbf{F}_t$ is the known regression vector

$$\mathbf{F}_t = \left[\frac{\delta F_t(\boldsymbol{\theta}_t)}{\delta \boldsymbol{\theta}_t} \right]_{\boldsymbol{\theta}_t = \mathbf{a}_t}$$

Assuming an adequate approximation, the prior moments of λ_t and $\boldsymbol{\theta}_t$ in (14.23) are thus given by a modification of the usual equations.

$\mathbf{R}_t$ and q_t are as usual, with regression vector $\mathbf{F}_t$ and evolution matrix $\mathbf{G}_t$ derived here via linearisation of possibly non-linear functions. The means $\mathbf{a}_t$ and f_t retain the fundamental, non-linear characteristics of (14.35) and (14.36), given by

$$\mathbf{a}_t = \mathbf{g}_t(\mathbf{m}_{t-1})$$

and

$$f_t = F_t(\mathbf{a}_t).$$

As noted in comment (9) in Section 13.2 of Chapter 13, the model is easily modified to incorporate a stochastic drift in $\boldsymbol{\theta}_{t-1}$ *before* applying the non-linear evolution function in (14.36), rather than by adding $\boldsymbol{\omega}_t$ *after* transformation. The alternative to (14.36) is simply

$$\boldsymbol{\theta}_t = \mathbf{g}_t(\boldsymbol{\theta}_{t-1} + \boldsymbol{\delta}_t), \tag{14.37}$$

for some zero-mean error $\boldsymbol{\delta}_t$, with a known variance matrix $\mathbf{U}_t$. Then $\mathbf{a}_t$ and $\mathbf{R}_t$ are as defined above with

$$\mathbf{W}_t = \mathbf{G}_t \mathbf{U}_t \mathbf{G}_t'. \tag{14.38}$$

This method is applied in the case study reported in the following Section. Note that the comments in Section 13.2 concerning the use of this, and alternative, techniques of linearisation in normal models apply equally here in the DGLM. Also, step ahead forecast and filtered moments and distributions may be derived in an obvious manner, as the interested reader may easily verify.

14.4   CASE STUDY IN ADVERTISING AWARENESS

14.4.1   Background

An illustration of the above type of analysis is described here. Extensive application of a particular class of non-normal, non-linear dynamic models has been made in the assessment of the impact on consumer awareness of the advertising of various products. Some of this work is reported in Migon and Harrison (1985), Migon (1984), and, for less technical discussion, Colman and Brown (1983). Related models are discussed in West, Harrison and Migon (1985), West and Harrison (1986a), and implemented in West and Harrison (1986b).

Analysts of consumer markets define and attempt to measure many variables in studies of the effectiveness of advertising. The *awareness* in a consumer population of a particular advertisement is one such quantity, the subject of the above referenced studies. Attempts to measure consumer awareness may take one of several forms, though a common approach in the UK is to simply question sampled members of the TV viewing population as to whether or not they have seen TV commercials for any of a list of branded items during a recent, fixed time interval such as a week. The proportion of positive respondents for any particular brand then provides a measure of population awareness for the advertisement of that brand. In assessing advertising effectiveness, such awareness measures can be taken as defining the response variable to be related to independent variables that describe the nature of the advertising campaign being assessed. The extent of television advertising of a product, or a group of products, of a company is often measured in standardised units known as TVR's, for television ratings. TVRs are based on several factors, including the length of the TV commercial in the TV areas in which the sampled awareness is measured. Broadbent (1979) discusses the construction of TVR measurement. Here the key point is that they provide standardised measurements of the expenditure of the company on TV commercials. The main issues in the area now concern the assessment of the relationship between TVRs and awareness, and the comparision of this relationship at different times and across different advertising campaigns.

The basic information available in this study is as follows. Suppose that weekly surveys sample a given number of individuals from the population of TV viewers, counting the number "aware" of the current or recent TV commercial for a particular branded item marketed by a company. Suppose that the number of sampled individuals in week t is n_t, and that Y_t of them respond positively. Finally, let X_t

denote the weekly TVR measurement for week t.[†] The problem now concerns the modelling of the effect of X_s, $(s = t, t-1, \ldots,)$ on Y_t.

Migon and Harrison (1985) discuss elements of the modelling problem and previous modelling efforts, notably that of Broadbent (1979), and develop some initial models based essentially on extensions of normal DLMs. Practically more appropriate and useful models are then developed by moving to non-normal and non-linear structures that are consistent with the nature of the problem, and more directly geared to answering the major questions of interest and importance to the company. Several key features to be modelled are as follows.

(1) The data collection mechanism leads to a binomial-like sampling structure for the observations; the data count the number of positive responses out of a total number sampled. Thus normal, constant variance models are generally inappropriate (except as approximations).

(2) The level of awareness in the population is expected to be bounded below by a minimum, usually non-zero lower threshold value. This base level of awareness represents a (small) proportion of the population who would tend to respond positively in the sample survey even in the absence of any historical advertising of the product (due to as the effects of advertisements for other, related products, misunderstanding and misrepresentation).

(3) Awareness is similarly expected to be bounded above, the maximum population proportion aware being less than some upper threshold level which will typically be less than unity.

(4) Given no advertising into the future, the effect of past advertising on awareness is expected to decay towards the lower threshold.

(5) The effect of TVR on awareness is expected to be non-linear, exhibiting a *diminishing returns* form. That is, at higher values of X_t, the marginal effect of additional TVR is likely to be smaller than at lower values of X_t.

(6) The thresholds and rates of change with current and past TVR will tend to be relatively stable over time for a given advertising campaign, but can be expected to change to some small degree due to factors not included in the model (as is

---

[†]In fact, due to uncertainties about the precise timing of surveys, the TVR measurements used in the referenced studies are usually formed from a weighted average of that in week t and week $t-1$.

usual in regression models in time series; see Chapters 3 and 9). In addition, such characteristics can be expected to change markedly in response to a major change in the advertising campaign.

Points (1) to (6) are incorporated in a model described in the next Section, along with several other features of the practical data collection and measurement scheme that bear comment. The primary objectives in modelling are as follows.

(a) To incorporate these features and provide monitoring of the awareness/TVR relationship through sequential estimation of model parameters relating to these features.

(b) To allow the incorporation of subjective information. Initially this is desired to produce forecasts before data is collected. Initial priors for model parameters may be based, for example, on experience with other, similar branded items in the past. Such inputs are also desired when external information is available. In particular, information about changing campaigns should be fed-forward in the form of increased uncertainty about (some) model parameters to allow for immediate adaptation to observed change.

(c) To cater for weeks with no sample, ie. missing data.

(d) To produce forecasts of future awareness levels based on projected TVR values, ie. so-called *What-if?* analyses.

(e) To produce filtered estimates of past TVR effects: *What-Happened?* analyses.

Clearly the sequential analysis within a dynamic model provides for each of these goals once an appropriate model is constructed.

14.4.2 The non-linear binomial model

The model is defined as follows, following points (1) to (6) itemised in the previous Section.

(1) Sampling model
At time t, the number Y_t of positive respondents out of the known total sample size n_t is assumed binomially distributed. Thus, if μ_t represents the population proportion aware in week t, $(0 \le \mu_t \le 1)$, the sampling model is defined by

$$p(Y_t|\mu_t) = \binom{n_t}{Y_t} \mu_t^{Y_t}(1 - \mu_t)^{n_t - Y_t}, \qquad (Y_t = 0, 1, \ldots, n_t).$$

Let E_t denote the *effect* of current and past weekly TVR on the current expected awareness μ_t. Then, following points (2) to (6), introduce the following components.

(2) Lower threshold for awareness

The minimum level of awareness expected at time t is denoted by the lower threshold parameter α_t. Thus $\mu_t \geq \alpha_t \geq 0$ for all time.

(3) Upper threshold for awareness

The maximum level of awareness expected at time t is denoted by the upper threshold parameter β_t. Thus $\mu_t \leq \beta_t \leq 1$ for all time.

(4) TVR effect: decay of awareness

Suppose that $X_t = 0$. Then the effect on awareness level is expected to decay between times $t - 1$ and t by a factor of $100(1 - \rho_t)\%$, moving from E_{t-1} to $E_t = \rho_t E_{t-1}$, for some ρ_t between 0 and 1, typically closer to 1. ρ_t measures the persistence or memory effect of the advertising campaign between weeks $t - 1$ and t, and is typically expected to be rather stable, even constant over time. Larger values are obviously associated with more effective (in terms of retention of the effect) advertising. If $\rho_t = \rho$ for all time, then a period of s weeks with no TVR implies an exponential decay from E_{t-1} to $\rho^s E_{t-1}$. The *half-life* σ associated with any particular value of ρ is the number of weeks taken for the effect to decay to exactly half any initial value with no further input TVR. Thus $\rho^\sigma = 0.5$ so that

$$\sigma = -\frac{\log(2)}{\log(\rho)},$$

and, obviously, $\sigma > 0$. With time-varying ρ_t, the associated half-life parameter is denoted by σ_t. Clearly $\rho_t = 2^{-1/\sigma_t}$.

(5) TVR effect: diminishing returns

At time t with current TVR input X_t, the input of X_t is expected to instantaneously increase the advertising effect by a fraction of the remaining available awareness. To model diminishing returns at higher values of TVR, this fraction should decay with increasing X_t. Specifically, suppose that, for some $\kappa_t > 0$, this fraction is determined by the penetration function $\exp(-\kappa_t X_t)$, decaying exponentially as TVR increases. Then, before including the instantaneous effect of X_t, the expected level of awareness is the base level α_t from (2) plus the effect E_{t-1} from the previous week discounted by the memory decay parameter ρ_t from (3), giving $\alpha_t + \rho_t E_{t-1}$. With upper threshold level β_t from (3), the remaining possible awareness is $\beta_t - (\alpha_t + \rho_t E_{t-1})$ so that the new expected effect at time t is given by

$$E_t = \rho_t E_{t-1} + [1 - \exp(-\kappa_t X_t)][\beta_t - (\alpha + \rho_t E_{t-1})]$$
$$= (\beta_t - \alpha_t) - (\beta_t - \alpha_t - \rho_t E_{t-1})\exp(-\kappa_t X_t).$$
$$(14.39)$$

The expected level is then

$$\mu_t = \alpha_t + E_t. \qquad (14.40)$$

As with ρ_t, the penetration parameter κ_t is a measures advertising effectiveness, here in terms of the immediate impact of expected consumer awareness of weekly TVR. Also, by analogy with σ_t, the *half-penetration* parameter $\xi_t > 0$ is defined as the TVR needed to raise the effect (from any initial level) by exactly half the remaining possible. Thus $1 - \exp(-\kappa_t \xi_t) = 0.5$ so that $\xi_t = \log(2)/\kappa_t$ and $\kappa_t = \log(2)/\xi_t$.

(6) **Time varying parameters**
Define the 5−dimensional state vector $\boldsymbol{\theta}_t$ by

$$\boldsymbol{\theta}'_t = (\alpha_t, \beta_t, \rho_t, \kappa_t, E_t) \qquad (14.40)$$

for all t. All five parameters are viewed as possibly subject to minor stochastic variation over time, due to factors not included in the model. Some pertinent comments are as follows.

(i) Changes in the composition of the sampled population can be expected to lead to changes in the awareness/TVR relationship.

(ii) In a stable population, the maturity of the product is a guide to stability of the model. Early in the life of a product, base level awareness tends to rise to a stable level, with an associated rise in the upper threshold for awareness. With a mature product, threshold levels will tend to be stable.

(iii) The memory decay rate ρ_t will tend to be stable, even constant, over time and across similar products and styles of advertisement.

(iv) The penetration parameter κ_t will also tend to be stable over time, though differing more widely across advertising campaigns.

(v) A point related to (i) concerns the possibility of additional, week to week variation in the data due to surveys being made in different regions of the country. Such regional variation is best modelled, not by changes in model parameters, but as unpredictable, "extra-binomial" variation.

(vi) A degree of stochastic change in all parameters can be viewed as a means of accounting for model mis-specification generally (as in all dynamic models), in particular in the definition of X_t and the chosen functional forms of the model above.

14.4.3 Formulation as a dynamic model

The model formulation is as follows. The observation model is binomial with probability parameter $\mu_t = \lambda_t$. With $\boldsymbol{\theta}_t$ given by (14.40) we have

$$\mu_t = \alpha_t + E_t = \mathbf{F}'\boldsymbol{\theta}_t$$

where $\mathbf{F}' = (1,0,0,0,1)$ for all t. For the evolution, suppose that $\boldsymbol{\theta}_{t-1}$ changes by the addition of a zero-mean evolution error $\boldsymbol{\delta}_t$ between $t-1$ and t. The only element of the state vector that is then expected to change systematically is the cumulative TVR effect variable E_t, evolving according to (14.39). Thus the evolution is a special case of the general, non-linear model (14.37),

$$\boldsymbol{\theta}_t = \mathbf{g}_t(\boldsymbol{\theta}_{t-1} + \boldsymbol{\delta}_t),$$

where

- $\boldsymbol{\delta}_t$ is an evolution error, with

$$\boldsymbol{\delta}_t \sim [\, \mathbf{0}, \mathbf{U}_t\,]$$

for some known variance matrix $\mathbf{U}_t$;
- For any 5–vector $\mathbf{z} = (z_1, z_2, z_3, z_4, z_5)'$,

$$\mathbf{g}_t(\mathbf{z})' = [z_1, z_2, z_3, z_4, (z_2 - z_1) - (z_2 - z_1 - z_3 z_5)\exp(-z_4 X_t)].$$

The linearisation of this evolution equation follows Section 14.3.5. The derivative matrix of $\mathbf{g}_t(.)$ is given by

$$\mathbf{G}_t(\mathbf{z}) = \frac{\delta \mathbf{g}_t(\mathbf{z})}{\delta \mathbf{z}'} = \begin{pmatrix} 1 & 0 & 0 & 0 & e^{-z_4 X_t} - 1 \\ 0 & 1 & 0 & 0 & 1 - e^{-z_4 X_t} \\ 0 & 0 & 1 & 0 & z_5 e^{-z_4 X_t} \\ 0 & 0 & 0 & 1 & (z_2 - z_1 - z_3 z_5) X_t e^{-z_4 X_t} \\ 0 & 0 & 0 & 0 & z_3 e^{-z_4 X_t} \end{pmatrix}.$$

Then the linearised evolution equation at time t is specified as in the previous Section. With

$$(\boldsymbol{\theta}_{t-1}|D_{t-1}) \sim [\mathbf{m}_{t-1}, \mathbf{C}_{t-1}],$$

we have

$$(\boldsymbol{\theta}_t|D_{t-1}) \sim [\mathbf{a}_t, \mathbf{R}_t]$$

with

$$\mathbf{a}_t = \mathbf{g}_t(\mathbf{m}_{t-1}), \quad \mathbf{R}_t = \mathbf{G}_t\mathbf{C}_{t-1}\mathbf{G}'_t + \mathbf{W}_t \quad \text{and} \quad \mathbf{W}_t = \mathbf{G}_t\mathbf{U}_t\mathbf{G}'_t,$$

where

$$\mathbf{G}_t = \mathbf{G}_t(\mathbf{m}_{t-1}).$$

For calculation, note that $\mathbf{R}_t = \mathbf{G}_t(\mathbf{C}_{t-1} + \mathbf{U}_t)\mathbf{G}'_t$.

Some points of detail are as follows.

(i) The error term $\boldsymbol{\delta}_t$ controls the extent and nature of stochastic change in the state vector through the variance matrix $\mathbf{U}_t$. Following the earlier comments, it is clear that, with a stable product, such changes are likely to be relatively small. Supposing this to be defined via discount factors, all such factors should be close to unity. In Migon and Harrison (1985), various possible cases are considered. In some of these, subsets of $\boldsymbol{\theta}_t$ are assumed constant over time, so that the corresponding rows and columns of $\mathbf{U}_t$ are zero, and the comments in Section 14.4.2 about the relative stability of the elements indicate that, generally, there is a need to construct $\mathbf{U}_t$ from components with different discount factors. For the illustration here, however, a single discount factor is applied to the five element vector as a single block, permitting a very small amount of variation over time in all parameters (so long as the initial prior does not constrain any to be fixed for all time; a point reconsidered below).

(ii) As always with discount models, in forecasting ahead a fixed evolution variance matrix is used. Thus, with $\mathbf{W}_t$ is defined at time $t - 1$ via discount factors based on the current $\mathbf{C}_{t-1}$, forecasting ahead from time t to time $t+k$ we use $\mathbf{W}_{t+k} = \mathbf{W}_t$ for extrapolation.

(iii) There may be a need to consider complications arising due to the bounded ranges of the state parameters; all are positive and some lie in the unit interval. The model assumes, at any time, only prior means, variances and covariances for these quantities. If the model is generally appropriate, then it will be found that the data processing naturally restricts the means, as point estimates, to the relevant parameter ranges. Also, if initial priors are reasonably precise within the ranges, then the restrictions should not cause problems. However, with less precise priors, posterior estimates may stray outside the bounds if correction is not made. One possibility, used in West and Harrison (1986b), is simply to constrain the individual prior and posterior means to lie within the bounds, adjusting them to a point near but inside the boundary if they stray outside. An alternative approach, not explored further here, is to reparametrise the model in terms of unbounded parameters. If this is done, the evolution equation can be re-expressed in terms the new parameters, and linearised in that parametrisation.

(iv) Migon and Harrison (1985) note the possibility of extra-binomial variation due to surveys taken in different regions from week to week (point 6(vi) of Section 14.4.2). This can easily be incorporated within the dynamic model by taking $\mu_t = \mathbf{F}'\boldsymbol{\theta}_t + \delta\mu_t$ where $\delta\mu_t$ is an additional zero-mean error. With $\delta\mu_t = 0$, the original model is obtained. Otherwise, the variance of this error term introduces extra variation in the binomial level that caters for such region to region changes (and also for other sources of extra-binomial variation). This sort of consideration is often important in practice; the inclusion of such a term protects the estimation of the state parameters from possible corruption due to purely random, extraneous variation in the data not adequately modelled through the basic binomial distribution. See Migon and Harrison (1985) for further comments, and Migon (1985) for discussion of similar points in different models.

14.4.4   Initial priors and predictions

Consider application to the monitoring of advertising effectiveness of a forthcoming TV commercial campaign for a single product, to begin at $t = 1$. Suppose that the product has been previously advertised and that the forecaster concerned has some information from the previous campaigns (as well as from experience with other products). In fact the product here is confectionary, the data to be analysed below taken from the studies of Migon and Harrison (1985). The levels of TVR in the forthcoming campaign are expected to be consistent with past advertising, lying (on a standardised scale) in the range 0 − 10 each week. In the past, similar campaigns have been associated with awareness levels that rarely exceeded 0.5. In forming an initial view about the effectiveness of a forthcoming advertising campaign, this background information D_0 is summarised as follows to provide the initial prior for forecasting future awareness.

(1) The base level for awareness is currently thought to be almost certainly less that 20%, and probably close to 0.1. The initial prior for α_0 has mean 0.1 and standard deviation 0.025.

(2) In the past, observed awareness levels have never exceeded 0.5 so that there has been little opportunity to obtain precise information about the upper threshold. It is the case, of course, that if levels remain low in the future, then the model analysis should not be unduly affected by uncertainty about the upper threshold. Based on general experience with consumer awareness, the upper threshold is expected to lie between 0.75 and 0.95, the uncertainty being relatively high. The initial prior is specified in terms of the mean and variance for the *available* awareness $\gamma_0 = \beta_0 - \alpha_0$ rather then for β_0 directly; the mean is 0.75 and standard deviation 0.2. Assuming, in addition, that γ_0 and α_0 are uncorrelated, this implies that β_0 has mean 0.85 and standard deviation approximately 0.202, and that the correlation between α_0 and β_0 is 0.124.

(3) The memory decay rate parameter ρ_t is viewed as stable over time and across products, with value near 0.9 for weekly awareness decay (anticipating approximately 10% decay of the advertising effect each week). The initial prior for ρ_0 has mean 0.9 and standard deviation 0.01. In terms of the memory half-life σ_0, a value of 0.9 for ρ_0 corresponds to a half-life of 6.6 weeks, 0.87 to a half-life of 5 weeks, and 0.93 to a half-life of 9.6 weeks.

(4) The penetrative effect of the advertising on awareness is expected to be fairly low, consistent with previous advertising. The previously low TVR have never achieved marked penetration, so the previous maximum value with TVR near 10 units per week is considered to be well below the half-penetration level. Thus ξ_0 is expected to be much greater

than 10, the forecaster assessing the most likely value as about 35 units. Converting to κ_0, this provides the initial mean as $\log(2)/35 \approx 0.02$. The prior standard deviation of κ_0 is chosen as 0.015. Note that κ_0 values of 0.01 and 0.04 correspond approximately to ξ_0 at 69 and 17 respectively.

(5) The previous campaign has only very recently terminated, and had been running at reasonably high TVR levels. Thus the retained effect E_0 of past advertising is considered to be fairly high although there is a reasonable degree of uncertainty. The prior mean is taken as 0.3, the standard deviation is 0.1.

(6) For convenience and in the absence of further information, the covariances that remain to be specified are set to zero. Thus

$$
\mathbf{m}_0 = \begin{pmatrix} 0.10 \\ 0.85 \\ 0.90 \\ 0.02 \\ 0.30 \end{pmatrix} \quad \text{and } \mathbf{C}_0 = 0.0001 \begin{pmatrix} 6.25 & 6.25 & 0 & 0 & 0 \\ 6.25 & 406.25 & 0 & 0 & 0 \\ 0 & 0 & 1 & 0 & 0 \\ 0 & 0 & 0 & 2.25 & 0 \\ 0 & 0 & 0 & 0 & 100 \end{pmatrix}.
$$

$$(14.41)$$

(7) Week to week variation in model parameters is expected to be fairly low and modelled (in a convenient and parsimonious manner) through the use of a single discount factor of 0.97. Thus the variance matrix of δ_t in (14.37) is defined as $\mathbf{U}_t = (0.97^{-1} - 1)\mathbf{C}_{t-1} \approx 0.03\mathbf{C}_{t-1}$.

The initial prior, based on previous experience with this, and similar, campaigns, represents a reasonable degree of precision about the model parameters. It is therefore directly useful in forecasting the forthcoming campaign without further data, such forecasting providing an assessment of model implications, and the effects of the specified priors. Note that the models allow the forecaster the facility to impose constraints on parameters by assuming the relevant prior variances and covariances to be zero. The applied models in Migon and Harrison (1985), and West and Harrison (1986b), make some use of such constraints. In initial forecasting, this allows an assessment of the implications of prior plus model assumptions for

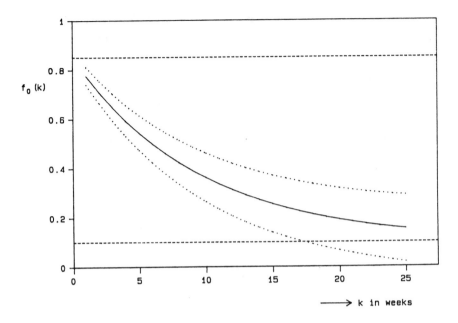

Figure 14.1.   Initial forecasts of expected memory decay.

each of the component parameters, free from the effects of the uncertainties about the constrained components. This is done here,
temporarily constraining E_0.

First, consider the implied decay of awareness into the future in
the absence of any advertising, $X_k = 0$ for $k = 1, \ldots, 25$. Take the
prior in (14.41) but modified so that E_0 has prior mean 0.75 and zero
variance. Thus, initially, the effect of past advertising is assumed to
be saturated at the maximum level. Looking ahead, awareness is
expected to decay towards the base level, initially expected to be
0.1 but with some uncertainty. For forecasting ahead, the theory
in Sections 14.3.4 and 14.3.5 applies with $t = 0$; for $k = 1, \ldots, 25$,
$\mu_k = \lambda_k \sim [f_0(k), q_0(k)]$. The conjugate distribution consistent with
these moments is $(\mu_k | D_0) \sim \text{Beta}[r_0(k), s_0(k)]$ with parameters de-
fined as in Example 14.4. Figure 14.1 provides a plot of $f_0(k)$ against
k, clearly illustrating the expected exponential decay in the effect of
awarness in the absence of further advertising. To give an indication
of uncertainty, the dotted lines provide 1 standard deviation limits
above and below the forecast mean, $f_0(k) \pm \sqrt{q_0}(k)$. This increases
in k due to the change over time expected in θ_k. From the graph,
awareness is forecast as decaying to roughly half the initial value

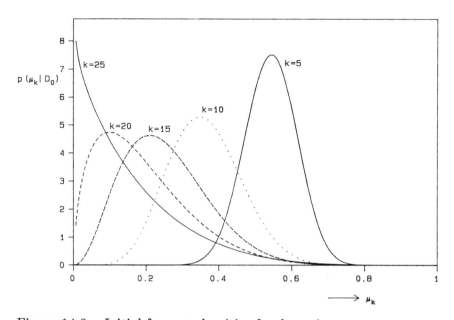

Figure 14.2. Initial forecast densities for decay in awareness.

(whatever this initial value may be) after roughly 7 weeks. To examine the uncertainty in forecasting more closely, Figure 14.2 plots the beta distributions of $(\mu_k|D_0)$ at values $k = 5, 10, 15, 20$ and 25. Note the highly skewed forms as the expected level moves towards the extreme at 0 and the variance increases.

We can similarly explore the implications for expected penetration due to various initial levels of TVR. Take the prior (14.41) modified to have zero mean and variance for E_0. Thus there is no retained effect of past advertising. For any given TVR X_1 in the first week of the campaign, awareness is expected to increase towards the threshold level with expected value 0.85. The one-step ahead forecast distribution for μ_1 has moments $f_0(1)$ and $q_0(1)$ that depend on X_1. Figure 14.3 plots $f_0(1)$ with one standard deviation limits for $X_1 = 1, \ldots, 25$. The increase in uncertainty about the level as X_1 increases is marked, as is the fact that penetration is assumed to be rather low even for TVR values as high as 25 — recall that the prior estimate of the penetration parameter is consistent with an estimated half-penetration of $X_1 \approx 35$. Since the forthcoming campaign will have TVR values less that 10 units per week, it is apparent that penetration is expected to be rather low. Figure 14.4 provides the beta densities of $(\mu_1|D_0)$ for $X_1 = 5, 10, 25$.

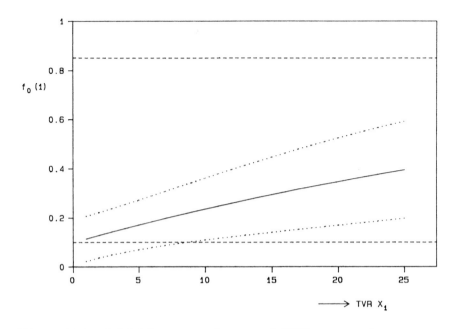

Figure 14.3. Initial forecasts of expected TVR penetration.

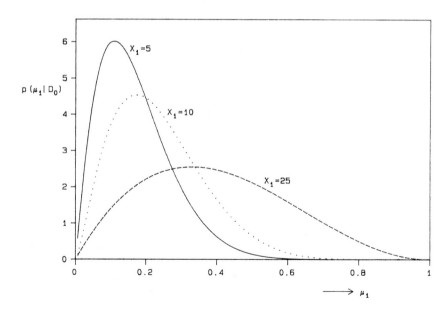

Figure 14.4. Initial forecast densities for TVR penetration.

14.4.5 Data analysis

The campaign data now becomes available, week by week. As the data are observed, any available external information will also be incorporated in the model via subjective intervention as in Chapter 11, and typically applied models will also be subject to continuous monitoring. Suppose here that intervention and monitoring are not considered, and that the weekly observations are simply processed as normal via the analysis of Section 14.3.3 and 14.3.4.

The sample surveys in this dataset count the number of positive respondents out of a nominal total of $n_t = 66$ for all t. The full set of observations becoming available during the first 75 weeks of the campaign are given in Table 14.1. The awareness measurements are given in terms of proportions out of a nominal 66 (rounded to 2 decimal places). These are also plotted in the upper frame of Figure 14.5, the TVR measurements appearing in the lower frame of that Figure. Note that three observations during the campaign are missing, corresponding to weeks in which no sample was taken. Of course these are routinely handled by the sequential analysis, the posterior moments for state vectors following a missing observation being just the prior moments. Note also that advertising over the 75 weeks comes in essentially two bursts. The first, very short spell during weeks 3 to 7 is followed by no TVR for a period of weeks, the second, long spell beginning in week 16 and ending in week 48. The

Table 14.1. Awareness response proportions and TVR

Weekly TVR measurements (to be read across rows)														
0.05	0.00	0.20	7.80	6.10	5.15	1.10	0.00	0.00	0.00	0.00	0.00	0.00	0.00	0.00
1.50	4.60	3.70	1.45	1.20	2.00	3.40	4.40	3.80	3.90	5.00	0.10	0.60	3.85	3.50
3.15	3.30	0.35	0.00	2.80	2.90	3.40	2.20	0.50	0.00	0.00	0.10	0.85	4.65	5.10
5.50	2.30	4.60	0.00	0.00	0.00	0.00	0.00	0.00	0.00	0.00	0.00	0.00	0.00	0.00
0.00	0.00	0.00	0.00	0.00	0.00	0.00	0.00	0.00	0.00	0.00	0.00	0.00	0.00	0.00

Awareness response proportions (read across rows, * denotes missing values)														
0.40	0.41	0.31	0.40	0.45	0.44	0.39	0.50	0.32	0.42	0.33	0.24	0.25	0.32	0.28
0.25	0.36	0.38	0.36	0.29	0.43	0.34	0.42	0.50	0.43	0.43	0.52	0.45	0.30	0.55
0.33	0.32	0.39	0.32	0.30	0.44	0.27	0.44	0.30	0.32	0.30	*	*	*	0.33
0.48	0.40	0.44	0.40	0.34	0.37	0.37	0.23	0.30	0.21	0.23	0.22	0.25	0.23	0.14
0.21	0.16	0.19	0.07	0.26	0.16	0.21	0.07	0.22	0.10	0.15	0.15	0.22	0.11	0.14

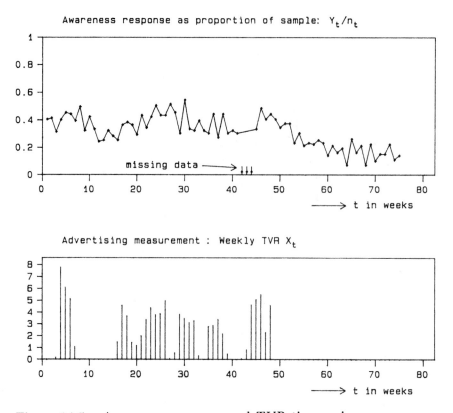

Figure 14.5. Awareness response and TVR time series.

decay of awareness in the final period of no advertising is clear from
the graph.

Based on the prior moments in (14.41), the sequential one-step
ahead forecasting and updating equations apply directly. Figure 14.6
provides the data with a line plot of the one-step point forecasts f_t,
$(t = 1, \ldots, 75)$, as they are calculated over time. These are simply
the one-step forecast means since, under the binomial model for Y_t,
we have $E[Y_t/n_t|\mu_t] = \mu_t$ and so

$$E[Y_t/n_t|D_{t-1}] = E[\mu_t|D_{t-1}] = f_t.$$

Note the response to renewed advertising after the periods of no
TVR, and the smooth decay of forecast awareness during these periods. For clarity, forecast uncertainties are not indicated on the
graph (see below). However, under the binomial model $V[Y_t/n_t|\mu_t] =$

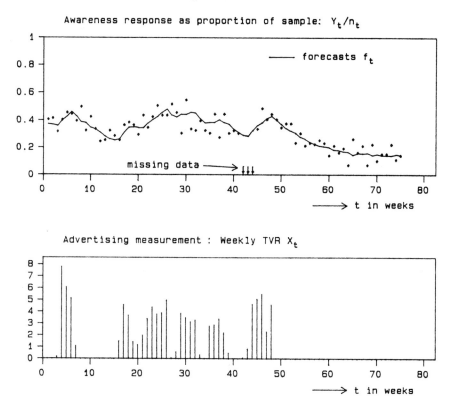

Figure 14.6. Awareness data and one-step ahead point forecasts.

$\mu_t(1 - \mu_t)/n_t$ so that

$$
\begin{aligned}
V[Y_t/n_t|D_{t-1}] &= E[V\{Y_t/n_t|\mu_t\}|D_{t-1}] + V[E\{Y_t/n_t|\mu_t\}|D_{t-1}] \\
&= E[\mu_t(1 - \mu_t)/n_t|D_{t-1}] + V[\mu_t|D_{t-1}] \\
&= f_t(1 - f_t)/n_t + q_t(1 - 1/n_t).
\end{aligned}
$$

Thus the variation in the one-step forecast (beta/binomial) distribution naturally exceeds that of a standard binomial distribution with μ_t estimated by the value f_t. This variance decreases as μ_t moves towards 0 or 1. At the lowest values of f_t in the graph, $f_t \approx 0.1$ corresponds to $f_t(1 - f_t)/66 \approx 0.035^2$, so that the forecast standard deviation for all t exceeds 0.035. With this in mind, the variation in the observed proportions about their forecast means appears to be

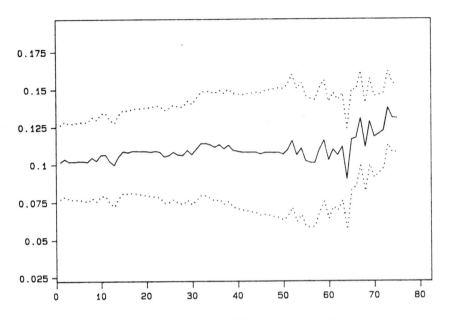

Figure 14.7a.   On-line trajectory of lower threshold parameter α_t.

reasonably attributable to unpredictable, though expected, sampling variation.[†]

Figures 14.7a to 14.10b provide graphs of the on-line and retro-spectively filtered, or smoothed, trajectories of the four state pa-rameters α_t, β_t, ρ_t, κ_t. The on-line trajectories are, as usual, just the posterior means from $\mathbf{m}_t$ over time, together with one standard deviation limits using the relevant posterior variances from $\mathbf{C}_t$. The smoothed trajectories are similar but now the estimates are based on all the data, derived through the standard filtering algorithms (Section 4.9 of Chapter 4; see comments in Section 14.3.4 above). The following features are apparent.

(a) The on-line trajectory of α_t in Figure 14.7a is stable, the esti-mate remaining near the prior mean of 0.1, until the last 10 or 15 observations. Until this time, little information is obtained from the data about the lower threshold since since awareness does not decay sufficiently. At the end, the TVR is zero for

---

[†]One point of detail concerns the last few observations which, at low awareness levels, are somewhat more erratic than perhaps expected. Here there is a degree of evidence for extra-binomial variation at low levels, that could be incoporated in the model for forecasting ahead as previously mentioned.

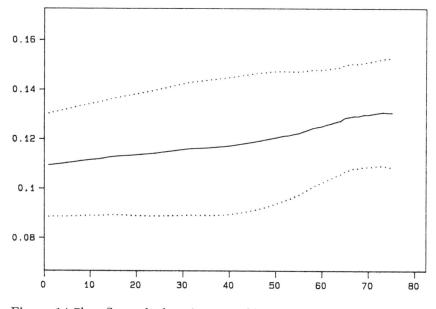

Figure 14.7b.   Smoothed trajectory of lower threshold parameter α_t.

a long period and awareness decays, the data thus informing about the lower threshold. The graph then indicates a slight increase to values nearer 0.12 at the end. The smoothed trajectory in Figure 14.7b, re-estimating the threshold at each time based on all the data, confirms this.

(b) The initial variance for β_0 is rather large, resulting in the possibility of marked adaptation to the data in learning about the upper threshold. This is apparent in Figure 14.8a. The smoothed version in Figure 14.8b confirms that the upper thesbold is fairly stable over time, though estimated as rather lower than the initial prior, around 0.8 at the end of the series. Note, however, the wide intervals about the estimates; there is really very little information in the data about the upper threshold levels.

(c) Figures 14.9a and 14.9b illustrate the stability of the decay parameter ρ_t at around 0.9.

(d) There is a fair degree of movement in the on-line trajectory for the penetration parameter κ_t in Figure 14.10a. In part this is due to conflict between the information in the data and that in the initial prior. This is confirmed in the smoothed version,

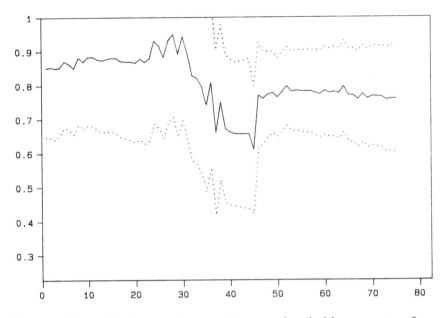

Figure 14.8a.   On-line trajectory of upper threshold parameter β_t.

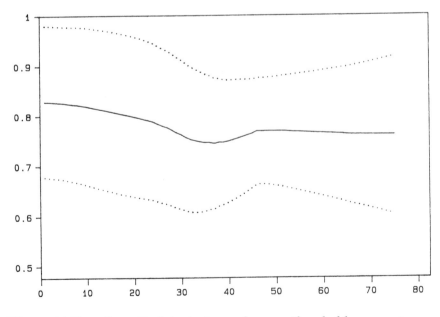

Figure 14.8b.   Smoothed trajectory of upper threshold parameter
β_t.

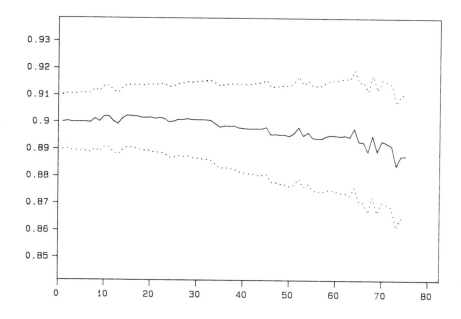

Figure 14.9a.   On-line trajectory of memory decay parameter ρ_t.

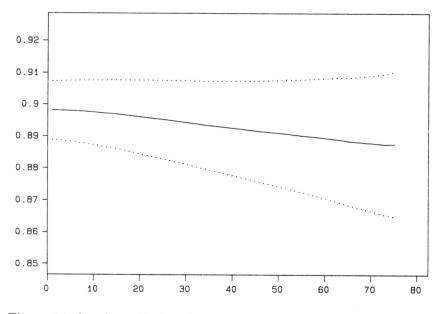

Figure 14.9b.   Smoothed trajectory of memory decay parameter
ρ_t.

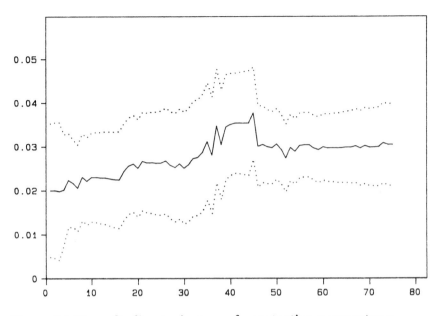

Figure 14.10a.   On-line trajectory of penetration parameter κ_t.

Figure 14.10b, which indicates that κ_t is really fairly stable, taking values nearer 0.03 than the initial mean 0.02. In terms of the half-penetration effect, 0.02 corresponds to TVR levels of 35 whereas the more appropriate value of 0.03 corresponds to TVR at around 24.

These points are further illustrated in Figures 14.10 and 14.11. These are the posterior versions of Figures 14.1 and 14.2, constructed in precisely the same way but now at $t = 75$ rather than at $t = 0$. Hence they represent forecasts based on the final posterior moments $\mathbf{m}_{75}$ and $\mathbf{C}_{75}$ at $t = 75$, modified as follows.

(e) E_{75} is constrained to the expected maximum value, namely, $E[\beta_{75} - \alpha_{75}|D_{75}]$. Figure 14.11 then provides point forecasts $f_{75}(k)$, with limits $f_{75}(k) \pm \sqrt{q_{75}(k)}$, for the next 25 weeks assuming no further TVR, $X_{75+k} = 0$ for $k = 1,\ldots,75$. The decay of awareness forecast is similar in expectation to that initially (Figure 14.1), although the forecast distributions are more precise due to the processing of the 75 observations.

(f) Constraining E_{75} to be zero, Figure 14.12 provides one-step point forecasts $f_{75}(1)$ with standard deviation limits. The expected penetration is rather larger than that initially in

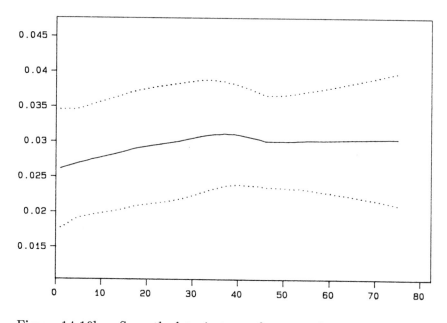

Figure 14.10b.   Smoothed trajectory of penetration parameter κ_t.

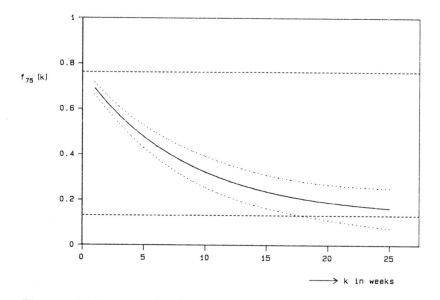

Figure 14.11.   Posterior forecasts of expected memory decay.

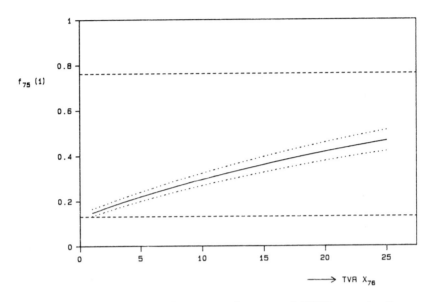

Figure 14.12.   Posterior forecasts of expected TVR penetration.

Figure 14.2, consistent with the posterior view expressed in
(d) above, that κ_{75} is very likely to be larger than the initial
estimate of κ_0.

Finally, model validation *What-if?* forecasts are represented in
Figure 14.13. At $t = 75$, consider forecasting ahead to times $75 + k$,
$(k = 1,\ldots,75)$ based on hypothesised values of TVR over the next
75 weeks. In particular, suppose that the company considers repeating the advertising campaign, reproducing the original TVR series
precisely. Starting with the full posterior moments $\mathbf{m}_{75}$ and $\mathbf{C}_{75}$,
the expected value of E_{75} is altered to 0.3, consistent with initial
expectations. Figure 14.13 then provides step ahead forecasts forecasts of awareness proportions over the coming weeks. The forecasts
are given in terms of means and standard deviation limits from the
beta/binomial predictive distribution. These moments are defined,
for all k and with $t = 75$, by

$$E[Y_{t+k}/n_{t+k}|D_t] = f_t(k),$$

and

$$V[Y_{t+k}/n_{t+k}|D_t] = f_t(k)(1 - f_t(k))/n_{t+k} + q_t(k)(1 - 1/n_{t+k}).$$

Given the past stability of the model parameters, these predictions

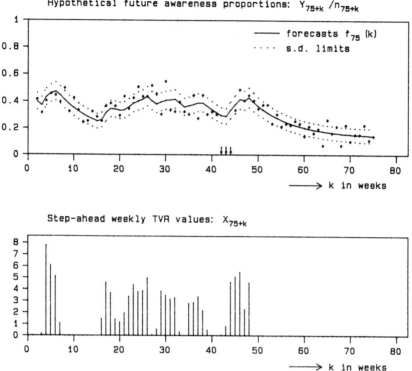

Figure 14.13. 'What-if?' step-ahead forecast awareness.

of course resemble those in the past. The historical data are also plotted in this figure for comparison. This sort of exercise can be viewed as a form of model validation, examining the question of just how well the model performs in forecasting a further data series that happens to coincide precisely with the past data. Any systematic discrepancies will show up most sharply here, although in this case the forecasts are adequate.

14.5 FURTHER COMMENTS AND EXTENSIONS

14.5.1 Monitoring, change-points and outliers

The panoply of techniques for subjective intervention of Chapter 11 apply to non-normal models directly. Interventions may be made on the prior and posterior moments of the state vector to incorporate external information routinely. Thus, for example, abrupt changes in parameters can be modelled through additional evolution error terms, as in the normal DLM.

Concerning the automatic detection of outliers and monitoring for change-points, West (1986a) develops a framework that provides non-normal analogues of the feedback intervention schemes in Section 11.4 and 11.5 of Chapter 11. Recall from that Chapter that the central component of any model testing scheme is the use of one or more alternative models to provide relative assessment of the predictive ability of the original model. Here consider the use of a single alternative model in the framework of Section 11.5, to provide an automatic monitor on the predictive performance of the basic model. As in Section 11.4, comparison between the original model, denoted M_0, and any single alternative M_1 involves the data at time t through the Bayes' factor $H_t = p_0(Y_t|D_{t-1})/p_1(Y_t|D_{t-1})$, where the subscripts denote the relevant model. Sequential model monitoring is based on the sequence of Bayes' factors, and the cumulative versions in Section 11.4.2, and all that is required to extend to the non-normal case is a suitable class of alternative predictive densities $p_1(Y_t|D_{t-1})$. The ideas in Section 11.4.3, in particular Example 11.3, suggest the following alternatives, based on West (1986a). At time t in M_0, the predictive distribution is defined by (14.28) with defining quantities r_t and s_t given in (14.26) as functions of f_t and q_t in (14.23). In particular, the precision of the forecast distribution is a decreasing function of q_t, equivalently an increasing function of the prior precision parameter s_t. In assessing the predictive density $p_0(Y_t|D_{t-1})$ with a view to identifying observation that are poorly predicted, a suitable alternative will give greater probability to regions not heavily favoured whilst being more diffuse in regions of high density. The idea, as in Example 11.3, is that a *flatter* version of the standard forecast density is appropriate. A class of such alternatives is easily constructed by adapting the standard model to have smaller prior precision parameter. In modelling terms, the standard equation $\lambda_t = \mathbf{F}_t'\boldsymbol{\theta}_t$ is revised to

$$\lambda_t = \mathbf{F}_t'\boldsymbol{\theta}_t + \delta\lambda_t,$$

where $\delta\lambda_t$ is an additional zero-mean error term with known variance δq_t, say. With this defining the alternative model at time t, the analogues of equations (14.23) to (14.28) define the components of the alternative model analysis, the only difference being that q_t is replaced by $q_t + \delta q_t$ throughout. Thus the resulting conjugate prior distribution under M_1 is more diffuse than under M_0, and this carries over to the predictive density $p_1(Y_t|D_{t-1})$. Full details about this construction, in particular concerning ways of appropriately structuring and quantifying of δq_t, are given in West (1986a). See also West and Harrison (1986a) for illustration of the use of such alternatives in automatic outlier detection and change-point estimation.

West (1986b) shows how the use of the above ideas provides non-normal extension of the multi-process models for change-point estimation and outlier accommodation. It is clear that, so far as modelling change-points is concerned, the multi-process framework extends directly, abrupt changes in θ_t being ascribed to evolution errors with appropriately large variances. One difference in the non-normal case is, of course, that the predictive densities forming the components of mixtures in a multi-process model are not normal or T, but this is a technicality, the calculations proceed just as in Chapter 12. The main point of difference requiring thought concerns multi-process models with an outlier modelling component. West (1986b) uses the ideas above here, modelling outliers with a component having a diffuse predictive distribution.

Alternative approaches to monitoring and model assessment are found in West, Harrison and Migon (1985), based on developments in West (1985a, 1986a). These have more in common with standard statistical approaches to outlier detection and accommodation, and influence assessment, but are not discussed further here.

14.5.2 Applications, extensions and related approaches

Several further illustrations and applications of the DGLM in forecasting and smoothing time series appear in West, Harrison and Migon (1985), and West and Harrison (1986a). Various Poisson and binomial models are covered. An application of exponential models in reliability growth analysis and forecasting is reported in Mazzuchi and Soyer (1987). Also, related models are considered from different viewpoints by Azzalini (1983), Smith (1979), Souza (1981), and Smith and Miller (1986).

The special case of binary time series is discussed in West and Mortera (1987). The observational model here is a special case of the binomial model in Example 14.2, in which $n_t = 1$ for all t, the observations being simple event indictors, 0 or 1. A variety of applications are discussed in this paper, and related topics in the area of subjective probability assessment, calibration and combination in West (1985b and 1985c). For example, simple Markov models can be written as special static cases of the DGLM for binary data. In the framework of Example 14.2 with $n_t = 1$, suppose the simplest case of a time-homogenous, one-step Markov process. Thus the "success" probability μ_t at time t actually depends on D_{t-1} but only through the previous observation Y_{t-1}. This may be written in many forms, the most obvious being

$$\mu_t = \alpha Y_{t-1} + \beta(1 - Y_{t-1}),$$

where α and β are the probabilities *conditional* on $Y_{t-1} = 1$ and $Y_{t-1} = 0$ respectively. If $\boldsymbol{\theta} = (\alpha, \beta)'$ and $\mathbf{F}_t = (Y_{t-1}, 1 - Y_{t-1})'$, then the model is a special, static case of the DGLM. Various similar models appear in the above references. Easy extensions cover Markov depedencies to order higher than the first, dynamic versions in which α and β vary stochastically over time to provide non-homogenous processes, and similar models but with μ_t undegoing a non-linear transformation, such as logistic, before being related to the state vector.

A major extension and application of the DGLM approach has been made in Gamerman (1985, 1987a) to models for the analysis of survival data. Discussion of this important area is far beyond the scope of this book, although the underlying modelling ideas are essentially those of dynamic modelling of time series. Interested readers should consult the above references for full theoretical details and illustrations. Applications in medicine and economics, along with further theoretical and practical details, can be found in Gamerman (1987b), Gamerman and West (1987a and b), the computer package of Gamerman, West and Pole (1987), and West (1987).

14.6   EXERCISES

(1) Suppose a Poisson sampling model, $(Y_t | \mu_t) \sim P[\mu_t]$ where $\mu_t > 0$ is the Poisson mean, the density being given by

$$p(Y_t | \mu_t) = \mu_t^{Y_t} e^{-\mu_t} / Y_t!, \qquad (Y_t = 0, 1, \dots ,).$$

(a) Verify that this is the density of an exponential family distribution, identifying the natural parameter η_t and the defining functions $y_t(.)$, $a(.)$ and $b(. , .)$ in (14.1).

(b) Verify that μ_t is both the mean and the variance of the distribution.

(c) Identify the conjugate prior family defined through (14.7). Verify that the conjugate distributions are such that μ_t is gamma distributed.

(d) For given prior quantities r_t and s_t in (14.7), calculate the one-step forecast density $p(Y_t|D_{t-1})$.

(2) Suppose Y_t is gamma distributed, $(Y_t|\mu_t, n_t) \sim G[n_t, \mu_t]$ for some n_t and μ_t, both positive quantities, with n_t known. The density is

$$p(Y_t|\mu_t, n_t) = \mu_t^{n_t} Y_t^{n_t-1} e^{-\mu_t Y_t}/\Gamma(n_t), \qquad (Y_t > 0).$$

(a) Verify that this is the density of an exponential family distribution, identifying the natural parameter η_t and the defining functions $y_t(.)$, $a(.)$ and $b(. , .)$ in (14.1).

(b) Calculate the mean and the variance of the distribution as functions of n_t and μ_t, and verify that the distribution has a quadratic variance function, $V[Y_t|\mu_t, n_t] \propto E[Y_t|\mu_t, n_t]^2$.

(c) Identify the conjugate prior family defined through (14.7). Verify that the conjugate distributions are such that μ_t is also gamma distributed, and identify the prior mean and variance as functions of r_t and s_t.

(d) Calculate the one-step forecast density $p(Y_t|D_{t-1})$.

(3) Verify the form of the predictive and posterior densities given in (14.8) and (14.9).

(4) Verify the results used in Example 15.3 that, in a model in which $(\mu_t|D_{t-1}) \sim \text{Beta}[r_t, s_t]$ and $\eta_t = \log[\mu_t/(1-\mu_t)]$, then

$$E[\eta_t|D_{t-1}] = \gamma(r_t) - \gamma(s_t)$$

and

$$V[\eta_t|D_{t-1}] = \dot\gamma(r_t) - \dot\gamma(s_t).$$

(5) Consider Poisson or gamma models above in which the conjugate priors are gamma, $(\mu_t|D_{t-1}) \sim G[r_t, s_t]$. Suppose as usual that $(\lambda_t|D_{t-1}) \sim [f_t, q_t]$.

(a) In a linear regression model, $E[Y_t|\mu_t] = \lambda_t$, show that the conjugate gamma prior is defined via $r_t = f_t^2/q_t$ and $s_t = f_t/q_t$.

(b) In a log-linear regression model, $E[Y_t|\mu_t] = \log(\lambda_t)$, show that the conjugate gamma prior has r_t and s_t implictly defined by $f_t = \gamma(r_t) - \log(s_t)$ and $q_t = \dot\gamma(s_t)$.

(6) Verify that, given the moments (14.16), the equations (14.19) and (14.20) reduce to the usual DLM updating recurrences.

(7) Verify that equations (14.19) and (14.20) reduce to the standard DLM updating equations.

CHAPTER 15

MULTIVARIATE MODELLING AND FORECASTING

15.1 INTRODUCTION

In this final chapter we return to the linear/normal framework to explore models for multivariate time series. Univariate DLMs can be extended in an obvious way to multivariate problems simply by taking the observations at each time as vectors rather than scalars. In fact such models have already been defined in Definition 4.1 of Chapter 4. This means that the observational errors become vectors too, the regression vectors become matrices, but the remaining model components are similar to the univariate case. The basic theory of the univariate DLM as developed in Chapter 4 extends directly to such models. The observational errors are now vectors, so that model specification requires observational variance *matrices* defining the joint stochastic structure of the observations conditional on state parameters. In DLMs having vector observations with observational errors following a multivariate normal distribution, the univariate theory extends easily only when it is assumed that the observational error variance matrices are known for all time. However, as soon as uncertainties about observational variance matrices are admitted, the tractability of analysis is lost. In general, there is no neat, conjugate analysis of multivariate DLMs whose observational errors are multivariate normal with unknown (constant or otherwise) variance matrices. The theory for models with known variance matrices is developed in Section 15.2.

Concerning unknown covariance structures, the picture is fortunately not as gloomy as the above comments may seem to apply. In fact there is a wide class of models which are amenable to fully conjugate analyses, and these are described in Section 15.4. These models are extensions of the basic multivariate DLM in which the state parameters are naturally involved through a state *matrix*, rather than the usual vector, and the analysis most easily developed in terms of *matrix normal*, rather than multivariate normal, distribution theory. The relevant theory is developed below, as is the associated theory for learning about variance matrices within a matrix-normal framework. Some applications illustrate the scope for these models in assessing *cross-sectional* structure of several, possibly many, similar time series.

Section 15.3 is concerned with practical issues arising in multivariate forecasting of hierarchically related time series. Here we identify and discuss problems that arise when forecasting at different levels of aggregation within such hierarchies. Related issues also arise concerning the combination of forecasts made at different levels of aggregation and also possibly by different forecasters or models.

15.2   THE GENERAL MULTIVARIATE DLM

15.2.1   General framework

The general, multivariate normal DLM for a vector time series of observations $\mathbf{Y}_t$ is given in Definition 4.1 of Chapter 4. Suppose, for $t = 1, \ldots$, that $\mathbf{Y}_t$ is a (column) vector of r observations on the series following a multivariate DLM as in Definition 4.1. The model is defined via a quadruple

$$\{\mathbf{F}, \mathbf{G}, \mathbf{V}, \mathbf{W}\}_t = \{\mathbf{F}_t, \mathbf{G}_t, \mathbf{V}_t, \mathbf{W}_t\}$$

for each time t, where:

(a) $\mathbf{F}_t$ is a known $(n \times r)$ dynamic regression matrix;
(b) $\mathbf{G}_t$ is a known $(n \times n)$ state evolution matrix;
(c) $\mathbf{V}_t$ is a known $(r \times r)$ observational variance matrix;
(d) $\mathbf{W}_t$ is a known $(n \times n)$ evolution variance matrix.

The corresponding model equations are:

$$\begin{aligned}
\mathbf{Y}_t &= \mathbf{F}_t'\boldsymbol{\theta}_t + \boldsymbol{\nu}_t, & \boldsymbol{\nu}_t &\sim \mathrm{N}[\mathbf{0}, \mathbf{V}_t], \\
\boldsymbol{\theta}_t &= \mathbf{G}_t\boldsymbol{\theta}_{t-1} + \boldsymbol{\omega}_t, & \boldsymbol{\omega}_t &\sim \mathrm{N}[\mathbf{0}, \mathbf{W}_t],
\end{aligned} \tag{15.1}$$

where the errors sequences $\boldsymbol{\nu}_t$ and $\boldsymbol{\omega}_t$ are independent and mutually independent. As usual in univariate DLMs, $\boldsymbol{\theta}_t$ is the $n-$dimensional state vector. With all components of the defining quadruple known, the following results are immediate extensions of the standard updating, forecasting and filtering results in the univariate DLM.

15.2.2   Updating, forecasting and filtering

Suppose the model to be closed to inputs of external information so that, given initial prior information D_0 at $t = 0$, the information set available at any time t is simply $D_t = \{\mathbf{Y}_t, D_{t-1}\}$. Suppose also that the initial prior at $t = 0$ is the usual multivariate normal,

$$(\boldsymbol{\theta}_0|D_0) \sim \text{N}[\mathbf{m}_0, \mathbf{C}_0], \tag{15.2}$$

for some known moments $\mathbf{m}_0$ and $\mathbf{C}_0$. The following results parallel those in the univariate DLM.

Theorem 15.1. *One-step forecast and posterior distributions in the model just defined are given, for each t, as follows.*

(a) *Posterior at $t - 1$:*
For some mean $\mathbf{m}_{t-1}$ and variance matrix $\mathbf{C}_{t-1}$,

$$(\boldsymbol{\theta}_{t-1} \mid D_{t-1}) \sim \text{N}[\mathbf{m}_{t-1}, \mathbf{C}_{t-1}].$$

(b) *Prior at t:*
$$(\boldsymbol{\theta}_t \mid D_{t-1}) \sim \text{N}[\mathbf{a}_t, \mathbf{R}_t],$$

where

$$\mathbf{a}_t = \mathbf{G}_t \mathbf{m}_{t-1} \quad and \quad \mathbf{R}_t = \mathbf{G}_t \mathbf{C}_{t-1} \mathbf{G}_t' + \mathbf{W}_t.$$

(c) *One-step forecast:*

$$(\mathbf{Y}_t \mid D_{t-1}) \sim \text{N}[\mathbf{f}_t, \mathbf{Q}_t],$$

where
$$\mathbf{f}_t = \mathbf{F}_t' \mathbf{a}_t \quad and \quad \mathbf{Q}_t = \mathbf{F}_t' \mathbf{R}_t \mathbf{F}_t + \mathbf{V}_t.$$

(d) *Posterior at t:*
$$(\boldsymbol{\theta}_t \mid D_t) \sim \text{N}[\mathbf{m}_t, \mathbf{C}_t],$$

with

$$\mathbf{m}_t = \mathbf{a}_t + \mathbf{A}_t \mathbf{e}_t \quad and \quad \mathbf{C}_t = \mathbf{R}_t - \mathbf{A}_t \mathbf{Q}_t \mathbf{A}_t',$$

where
$$\mathbf{A}_t = \mathbf{R}_t \mathbf{F}_t \mathbf{Q}_t^{-1} \quad and \quad \mathbf{e}_t = \mathbf{Y}_t - \mathbf{f}_t.$$

Proof: Proof is by induction, completely paralleling that for the univariate DLM in Theorem 4.1 of Chapter 4. Suppose (a) to hold. Then (b) follows directly as in the univariate case, the evolution equation being no different here. Using (b) together with the evolution equation implies that $\mathbf{Y}_t$ and $\boldsymbol{\theta}_t$ are jointly normally distributed conditional on D_{t-1}, with covariance matrix

$$
\begin{aligned}
C[\mathbf{Y}_t, \boldsymbol{\theta}_t | D_{t-1}] &= C[\mathbf{F}'_t\boldsymbol{\theta}_t + \boldsymbol{\nu}_t, \boldsymbol{\theta}_t \mid D_{t-1}] \\
&= \mathbf{F}'_t V[\boldsymbol{\theta}_t \mid D_{t-1}] \\
&= \mathbf{F}'_t \mathbf{R}_t = \mathbf{A}'_t \mathbf{Q}_t,
\end{aligned}
$$

where $\mathbf{A}_t = \mathbf{R}_t\mathbf{F}_t\mathbf{Q}_t^{-1}$. The mean $\mathbf{f}_t$ and variance $\mathbf{Q}_t$ of $\mathbf{Y}_t$ follow easily from the observation equation and (b), establishing (c). It is also clear that

$$
\left(\left. \begin{matrix} Y_t \\ \boldsymbol{\theta}_t \end{matrix} \right| D_{t-1} \right) \sim N\left[\begin{pmatrix} \mathbf{f}_t \\ \mathbf{a}_t \end{pmatrix}, \begin{pmatrix} \mathbf{Q}_t & \mathbf{Q}_t\mathbf{A}'_t \\ \mathbf{A}_t\mathbf{Q}_t & \mathbf{R}_t \end{pmatrix} \right].
$$

Hence, using normal theory from Section 16.2 of Chapter 16, the conditional distribution of $\boldsymbol{\theta}_t$ given $D_t = \{Y_t, D_{t-1}\}$ is directly obtained. The $n \times r$ matrix of regression coefficients in the regression of $\boldsymbol{\theta}_t$ on $\mathbf{Y}_t$ is just $\mathbf{A}_t$, and (d) follows.

$$\diamond$$

$\mathbf{e}_t$ is the $r-$vector of one-step forecast errors and $\mathbf{A}_t$ the $n \times r$ matrix of adaptive coefficients. Clearly the standard results for the univariate case are given when $r = 1$. Note also that the model definition may be marginally extended to incorporate known, non-zero means for the observational or evolution noise terms. Thus, if $E[\boldsymbol{\nu}_t]$ and/or $E[\boldsymbol{\omega}_t]$ are known and non-zero, the above results apply with the modifications that $\mathbf{a}_t = \mathbf{G}_t\mathbf{m}_{t-1} + E[\boldsymbol{\omega}_t]$ and $\mathbf{f}_t = \mathbf{F}'_t\mathbf{a}_t + E[\boldsymbol{\nu}_t]$.

Theorem 15.2. *For each time t and $k \geq 0$, the $k-$step ahead distributions for $\boldsymbol{\theta}_{t+k}$ and $\mathbf{Y}_{t+k}$ given D_t are given by:*

(a) *State distribution:* $(\boldsymbol{\theta}_{t+k} \mid D_t) \sim N[\mathbf{a}_t(k), \mathbf{R}_t(k)],$

(b) *Forecast distribution :* $(\mathbf{Y}_{t+k} \mid D_t) \sim N[\mathbf{f}_t(k), \mathbf{Q}_t(k)],$

with moments defined recursively by:

$$
\mathbf{f}_t(k) = \mathbf{F}'_t\mathbf{a}_t(k) \quad and \quad \mathbf{Q}_t(k) = \mathbf{F}'_t\mathbf{R}_t(k)\mathbf{F}_t + \mathbf{V}_{t+k},
$$

where

$$\mathbf{a}_t(k) = \mathbf{G}_{t+k}\mathbf{a}_t(k-1) \quad and \quad \mathbf{R}_t(k) = \mathbf{G}_{t+k}\mathbf{R}_t(k-1)\mathbf{G}'_{t+k} + \mathbf{W}_{t+k},$$

with starting values $\mathbf{a}_t(0) = \mathbf{m}_t$, *and* $\mathbf{R}_t(0) = \mathbf{C}_t$.

Proof: The state forecast distributions in (a) are directly deduced from Theorem 4.2 since the evolution into the future is exactly as in the univariate DLM. The forecast distribution is deduced using the observation equation at time $t + k$, the details are left as an exercise for the reader.

$\diamond$

Theorem 15.3. *The filtered distributions* $p(\boldsymbol{\theta}_{t-k} \mid D_t)$, $(k > 1)$, *are defined as in the univariate DLM in Theorem 4.4.*

Proof: Also left to the reader.

$\diamond$

15.2.3   Comments

The above results may be applied in any context where the defining components are known. In particular, they are based on the availabilty of known or estimated values of the observational variance matrices $\mathbf{V}_t$ for all t. This is clearly a limiting assumptions in practice, such variance matrices will often be uncertain, at least in part. In general, there is no neat, conjugate analysis available to enable sequential learning about unknown variance matrices in the model of (15.1).

In principle the Bayesian analysis with unknown variance matrix is formally well-defined in cases when $\mathbf{V}_t = \boldsymbol{\Sigma}$, constant for all time t. The unknown parameters in $\boldsymbol{\Sigma}$ introduce complications that can, in principle, be catered for using some form of approximate analysis. Analytic approximations are developed in West (1982, Chapter 4), and Barbosa and Harrison (1989). Numerical approximation techniques include multi-process, class I models, and more efficient numerical approaches as described in Chapter 13. In these a prior for $\boldsymbol{\Sigma}$ is specified over a discrete set of values at $t = 0$, this prior (and possibly the set of values) being sequentially updated over time to give an approximate posterior at each time t. Though well-defined in principle, this approach has, at the time of writing, been little developed. Some features to be aware of in developing an analysis along thses lines are as follows.

(a) The variance matrix has $r(r+1)/2$ free parameters so that the dimension of the problem grows rapidly with the dimension r of the time series observation vector. Numerical approximations involving grids of points in the parameter space are therefore likely to be highly computationally demanding unless r is fairly small.

(b) There are restrictions on the parameter space in order that the variance matrix be non-negative definite. Unless structure is imposed on Σ, it can be rather difficult to identify the relevant subspace in $r(r+1)/2$ dimensions over which to determine a prior distribution for Σ. Even then, the problems of assessing prior distributions for variance matrices are hard (Dickey, Dawid and Kadane, 1986).

(c) In some applications it may be suitable to impose structure on the elements of Σ, relating them to a small number of unknown quantities. This structures the covariances across the elements of $\mathbf{Y}_t$ and also reduces computational burdens. For example, an assumption of equi-correlated observational errors implies that, for some variance σ^2 and correlation ρ,

$$\Sigma = \sigma^2 \begin{pmatrix} 1 & \rho & \rho & \cdots & \rho \\ \rho & 1 & \rho & \cdots & \rho \\ \vdots & \vdots & \vdots & \ddots & \vdots \\ \rho & \rho & \rho & \cdots & 1 \end{pmatrix}.$$

Here the dimension of the parameter space for Σ is effectively reduced to 2, that of σ and ρ.

(d) Much can be done in practice using off-line estimated values for Σ, and, more generally, a possibly time-varying varying variance matrix $\mathbf{V}_t$, possibly updated sequentially over time externally and rather less than formally. Substituting estimated values will always lead to the uncertainties in forecasting and posterior inferences being understated, however, and this must be taken into account in decisions based on such inferences. Harrison, Leonard and Gazard (1977) describe an application to hierarchical forecasting. The particular context is in forecasting industrial demand/sales series that are hierarchically disaggregated into sub-series classified by geographic region, although the approach is very widely applicable. One component of the models developed in this reference involves a vector time series that essentially form compositions of a whole. Thus a *multinomial* type of covariance structure is postulated, the DLM observation equation forming a normal approximation to the multinomial. In more detail, write the elements of $\mathbf{Y}_t$ as $\mathbf{Y}_t = (Y_{t1}, \ldots, Y_{tr})'$ supposing that $Y_{tj} > 0$ for all t and j, being compositions of the total

$$N_t = \mathbf{1}'\mathbf{Y}_t = \sum_{j=1}^{r} Y_{tj},$$

where $\mathbf{1} = (1, \ldots, 1)'$. The mean response is given by

$$\mu_t = \mathbf{F}_t'\boldsymbol{\theta}_t = (\mu_{t1}, \ldots, \mu_{tr})',$$

and, with an underlying multinomial model,

$$\mu_{tj} = N_t p_{tj}, \qquad (j = 1, \ldots, r),$$

where the proportions p_{tj} sum to unity for all t. Also, the multinomial structure implies that, given N_t,

$$V[Y_{tj}|\boldsymbol{\theta}_t] = N_t p_{tj}(1 - p_{tj}), \qquad (j = 1, \ldots, r),$$

and

$$C[Y_{tj}, Y_{ti}|\boldsymbol{\theta}_t] = -N_t p_{tj} p_{ti}, \qquad (i, j = 1, \ldots, r; \ i \neq j).$$

Thus the observational variance matrix $\mathbf{V}_t$ is time-varying and depends on $\boldsymbol{\theta}_t$ (similar in general terms to univariate models with variance laws; see Section 10.7 of Chapter 10), given by

$$\mathbf{V}_t = N_t[\mathrm{diag}(\mathbf{p}_t) - \mathbf{p}_t \mathbf{p}_t'],$$

where $\mathbf{p}_t = (p_{t1}, \ldots, p_{tr})'$. The form of approximation in Harrison, Leonard and Gazard (1977) is to simply approximate this matrix by substituting current estimates of the elements of $\mathbf{p}_t$. For example, given D_{t-1} the estimate of $\boldsymbol{\theta}_t$ is the prior mean $\mathbf{a}_t$, so that $\mathbf{p}_t$ is estimated by $\mathbf{a}_t/N_t$ for known N_t. The latter may be estimated externally from a different model for the total or by combining such an estimate with the forecast total $\mathbf{1}'\mathbf{f}_t$ where $\mathbf{f}_t$ is the one-step forecast mean for $\mathbf{Y}_t$ — see relevant comments on top-up and bottom down forecasting and forecast combination in the next Section. For further discussion and details, see Harrison, Leonard and Gazard (1977).

15.3 AGGREGATE FORECASTING

15.3.1 Bottom-up and top-down forecasting

> *"Just as processes cannot be predicted upward from a lower*
> *level, they can never be completely analysed downward into*
> *their components. To analyse means to isolate parts from*
> *the whole, and the functioning of a part in isolation is not*
> *the same as its functioning in the whole."*
> ... Arthur Koestler, "The Yogi and the Commissar"

Koestler's words accord entirely with the fundamental tenet of fore-casting several or many series in a system or hierarchy, namely that, in contributing to the movement and variation in series, different factors dominate at different levels of aggregation. Unfortunately, this is an often neglected fact. It is still very common practice, particularly in commerical forecasting, for forecasts at an aggregate level to be made by aggregating the forecasts of the constituents. In manufacturing industries, for example, stock control systems often produce forecasts and control on individual, component items of a product or process with little or no control at the product level. As a result it is commonly found that stocks build up in slump peri-ods and that shortages are encountered in boom periods. Planning at the macro level based on economic and market factors is rarely adequately used to feed-down information relevant to control at the micro level of the individual products. These factors, which domi-nate variations at the macro level, often have relatively little *apparent* effect at the disaggregate level and so are ignored.

Now, whilst this neglect of factors that are apparently insignifi-cant at the component level is perfectly sensible when forecasting individual components, the mistake is often made of combining such individual forecasts to produce overall forecasts of the aggregate. Following this "bottom-up" strategy can lead to disastrous results since moving to the aggregate level means that the previously unim-portant factors are now critical. Poor forecasts often result, and the resulting inefficiency and consequent losses can be great.

A simple example illustrates this. Here, in this and the next Sec-tion, we drop the time subscript t for simplicity in notation and since it is not central to the discussion.

EXAMPLE 15.1. The following simple illustration appears in Green and Harrison (1972). Consider a company selling 1000 individual products. In one month, let Y_i be the sales of the ith product. Suppose that Y_i may be expressed as

$$Y_i = f_i + X + \epsilon_i, \qquad (i = 1, \dots, 1000),$$

where f_i is the expected value of Y_i, assumed known, X and ϵ_i are zero-mean, uncorrelated random quantities with variances

$$V[X] = 1 \quad \text{and} \quad V[\epsilon_i] = 99.$$

Assume that ϵ_i represents variation affecting only the ith product, so that ϵ_i and ϵ_j are uncorrelated for $i \neq j$. X is a factor common to all products, related, for example, to changes in consumer disposable income through taxation changes, interest rates, a seasonal effect of a business cycle, etc.

It is clear that, for each individual product, $V[Y_i] = 100$ and that only 1% of this variance is contributed by the common factor X. Variation in individual sales is dominated, to the tune of 99%, by individual, uncorrelated factors. Even if X were precisely known, little improvement in forecasting accuracy would result, the forecast variance reducing by only 1%. Hence it is justifiable, from the point of view of forecasting the individuals Y_i, to omit consideration of X and simply adopt the model $Y_i = f_i + \epsilon_i^*$, where the ϵ_i^* are zero-mean, uncorrelated with common variance 100.

Consider now total sales

$$T = \sum_{i=1}^{1000} Y_i.$$

It is commonly assumed that X, having essentially no effect on any of the individual products, is of little or no importance when forecasting the total T. This is generally not the case, suprising things happen when aggregating or disaggregating processes. To see this, write

$$f = \sum_{i=1}^{1000} f_i \quad \text{and} \quad \epsilon = \sum_{i=1}^{1000} \epsilon_i.$$

Then

$$T = \sum_{i=1}^{1000} (f_i + X + \epsilon_i) = f + 1000X + \epsilon.$$

Now, $E[T] = f$ so that the point forecasts of individual sales are simply summed to give the point forecast of the total, but

$$V[T] = V[1000X] + V[\epsilon] = 10^6 + 1000 \times 99 = 1,099,000.$$

X, unimportant at the disaggregated level, is now of paramount importance. The term $1000X$ accounts for over 91% of the variance of the total T, the aggregate of the individual ϵ_i terms contributing the remaining 9%. At this level, precise knowledge of X reduces the forecast variance dramatically, leading to

$$E[T|X] = f + 1000X \qquad \text{and} \qquad V[T|X] = 99,000.$$

The key points arising are summarised as follows.

(a) Suppose that X remains unknown but the fact that X is common to the 1000 individual series is recognised. Then the total as forecast from the model for the individuals has mean f and variance 1,099,000. Although, relative to knowing X, the point forecast has a bias of $1000X$ units, the high degree of uncertainty in the forecast distribution is appropriately recognised. This is so since the proper, joint distribution for the Y_i has been used, with the relevant positive correlation amongst the Y_i induced by X multiplies in aggregating. Formally, writing

$$\mathbf{Y} = (Y_1, \dots, Y_{1000})',$$

$$\mathbf{f} = (f_1, \dots, f_{1000})'$$

and

$$\mathbf{Q} = \begin{pmatrix} 100 & 1 & 1 & \cdots & 1 \\ 1 & 100 & 1 & \cdots & 1 \\ 1 & 1 & 100 & \cdots & 1 \\ \vdots & \vdots & \vdots & \ddots & \vdots \\ 1 & 1 & 1 & \cdots & 100 \end{pmatrix},$$

we have $E[\mathbf{Y}] = \mathbf{f}$ and $V[\mathbf{Y}] = \mathbf{Q}$ with $C[Y_i, Y_j] = 1$. In aggregating, $V[T] = \mathbf{1}'\mathbf{Q}\mathbf{1}$ is appropriately large due to the large number of positive covariance terms.

(b) A forecaster who unwittingly follows the argument that X is irrelevant will end up with the representation

$$T = f + \sum_{i=1}^{1000} \epsilon_i^*,$$

so that

$$E[T] = f \quad \text{and} \quad V[T] = 100,000,$$

dramatically under-estimating the uncertainty about T.

(c) It is very worthwhile devoting effort to learning further about X, possibly using a macro model, since this corrects the bias and dramatically reduces the uncertainty about T.

The related procedure of decomposing a point forecast into parts, often referred to as "top-down" forecasting, must also be carefully considered since it can lead to poor individual predictions. Typically, such procedures convert a total forecast to a set of individuals using a set of forecast proportions. Thus, in the previous example, a forecast mean f and variance V for T will provide a forecast mean $p_i f$ and variance $p_i^2 V$ for Y_i where p_i is the assumed or estimated proportion that Y_i contributes to total sales. This will sometimes be refined to include uncertainties about the proportions. If the proportions are indeed well determined and stable, then such a strategy can perform well even though the proportions are themselves forecasts. However, it is clear that if major events occur on individual products, the directly disaggregated forecasts, not accounting for such events, may be dismal. Related discussion of these ideas is found in Green and Harrison (1972), and Harrison (1985c).

15.3.2   Combination of forecasts

If consistent and efficient forecasts are to be obtained in a hierarchy at all levels, the previous Section underpins the need for a framework in which forecasting at any level is primarily based on consideration of issues relevant at that level. It is also clear, however, that information relevant at any level will tend to be of relevance, though typically of much reduced importance, at other levels. Hence there is a need to consider how forecasts made at different levels, by different forecasters based on different, though possibly related information sets, may be appropriately combined.

Combination of forecast information more generally is a many faceted subject about which much has been written. Many authors have considered the problems of combining point forecasts of a single random quantity, or random vector, such forecasts being obtained from several forecasters or models. Common approaches have been to form some kind of weighted average of point forecasts, attempting to correct for biases and information overlap amongst the forecasters. Key references include Bates and Granger (1969), Dickinson (1975), Bunn (1975), Bordley (1982), and Granger and Ramanathan (1984). Such simple rules of forecast combination can sometimes be justified within conceptually sound frameworks that have been developed for more general consideration of forecasts made in terms of full or partially specified probability forecast distributions. Such approaches are developed in Lindley, Tversky and Brown (1979), Lindley (1983, 1985, and, with review and references, 1988), Harrison (1985c), Morris (1983), West (1985b, 1988, 1989a,b), Winkler (1981), and by other authors. Importantly, such approaches allow consideration of problems of interdependencies amongst forecasters and models, coherence, calibration and time variation in such characteristics.

Forecast combination, and, more generally, the synthesis of inferences from different, possibly subjective, sources, is a wide ranging subject, a full discussion being beyond the scope of this book. Attention is restricted to combination of forecasts within the above aggregation framework, with forecasts of linear combinations of a random vector being made in terms of forecast means and variances. Normality is assumed throughout although this is not central to the development.

Consider a forecaster interested in the vector of observations $\mathbf{Y} = (Y_1,\dots,Y_n).'$ It is instructive to bear in mind the archetypical example in which all observations are on sales of items produced by a company. The vector $\mathbf{Y}$ may then represent sales across a collection of, possibly a large number n of individual products or product lines, sales disaggregated according to market sector or geographical region, and so forth. Interest lies in forecasting the individual sales Y_i, $(i = 1,\dots,n)$, total sales $T = \sum_{i=1}^{n} Y_i$, and also in subtotals of subsets of the elements of $\mathbf{Y}$. Assume that the forecaster has a current, joint forecast distribution, assumed normal though this is a side detail, given by

$$\mathbf{Y} \sim \mathrm{N}[\mathbf{f}, \mathbf{Q}],$$

with

$$\mathbf{f} = \begin{pmatrix} f_1 \\ \vdots \\ f_n \end{pmatrix} \quad \text{and} \quad \mathbf{Q} = \begin{pmatrix} q_{1,1} & q_{1,2} & q_{1,3} & \cdots & q_{1,n} \\ q_{1,2} & q_{2,2} & q_{2,3} & \cdots & q_{2,n} \\ q_{1,3} & q_{2,3} & q_{3,3} & \cdots & q_{3,n} \\ \vdots & \vdots & \vdots & \ddots & \vdots \\ q_{1,n} & q_{2,n} & q_{3,n} & \cdots & q_{n,n} \end{pmatrix}.$$

Several considerations arise in aggregate forecasting, typified by the following questions that the forecaster may wish to answer.

(A) Based only on the forecast information embodied in $p(\mathbf{Y})$, what is the forecast of the total T? More generally, what is the forecast of any linear function $\mathbf{X} = \mathbf{LY}$ where $\mathbf{L}$ is a matrix of dimension $k \times n$ with k an integer, $1 \le k < n$. $\mathbf{X}$, for example, may represent sales in k product groups or market sectors.

(B) Suppose that, in a *What-if?* analysis, the total T is assumed fixed. What is the revised distribution for $\mathbf{Y}$? More generally, what is the forecast distribution $p(\mathbf{Y}|\mathbf{X})$ where $\mathbf{X} = \mathbf{LY}$ with $\mathbf{L}$ as in (A)? This sort of analysis is often required in considering forecasts of $\mathbf{Y}$ subject to various aggregates being constrained to nominal, "target" values.

(C) An additional forecast, in terms of a mean and variance for T, is provided from some other source, such as a macro model for the total sales involving factors not considered in forecasting the individual Y_i. What is the revised forecast for T?

(D) Under the circumstances in (C), what is the implied, revised forecast for the vector of individuals $\mathbf{Y}$?

(E) More generally, an additional forecast of a vector of subaggregates $\mathbf{X} = \mathbf{LY}$ is obtained, with $\mathbf{L}$ as in (A). What is the revised forecast for $\mathbf{Y}$?

These points are considered in turn under the assumptions that all forecast distributions are assumed normal.

(A). Forecasting linear functions of Y

This is straightforward, $T \sim N[f, Q]$, with

$$f = \mathbf{1}'\mathbf{f} = \sum_{i=1}^{n} f_i$$

and

$$Q = \mathbf{1}'\mathbf{Q}\mathbf{1} = \sum_{i=1}^{n} \sum_{j=1}^{n} q_{ij}.$$

Note that the comments in Example 15.1 apply here; sensible results are obtained if it can be assumed that the appropriate correlation structure amongst the individual Y_i is embodied in $\mathbf{Q}$. More generally, it is clear that $\mathbf{X} = \mathbf{LY}$ may be forecast using

$$\mathbf{X} \sim N[\mathbf{Lf}, \mathbf{LQL'}],$$

for any $k \times n$ matrix $\mathbf{L}$.

(B). Conditioning on subaggregates

For any $k \times n$ matrix $\mathbf{L}$ with $\mathbf{X} = \mathbf{LY}$, we have the joint (singular) distribution

$$\begin{pmatrix} \mathbf{Y} \\ \mathbf{X} \end{pmatrix} \sim \mathrm{N} \left[\begin{pmatrix} \mathbf{f} \\ \mathbf{Lf} \end{pmatrix}, \begin{pmatrix} \mathbf{Q} & \mathbf{QL'} \\ \mathbf{LQ} & \mathbf{LQL'} \end{pmatrix} \right].$$

Then, using standard multivariate normal theory from Section 16.2, Chapter 16, the conditional distribution for $\mathbf{Y}$ when $\mathbf{X}$ is known is

$$(\mathbf{Y}|\mathbf{X}) \sim \mathrm{N}[\mathbf{f} + \mathbf{A}(\mathbf{X} - \mathbf{Lf}), \mathbf{Q} - \mathbf{ALQLA'}],$$

where

$$\mathbf{A} = \mathbf{QL'}(\mathbf{LQL'})^{-1}.$$

Consider the special case $k = 1$ and $\mathbf{L} = \mathbf{1'}$ so that $\mathbf{X} = T$, the total. Write

$$\mathbf{q} = \mathbf{Q1} = (q_1, \dots, q_n)'$$

where

$$q_i = \sum_{j=1}^{n} q_{ij}, \qquad (i = 1, \dots, n).$$

We then have

$$(\mathbf{Y}|T) \sim \mathrm{N}[\mathbf{f} + \mathbf{A}(T - f), \mathbf{Q} - \mathbf{AA'}Q],$$

with $\mathbf{A} = \mathbf{q}/Q$. For the individual Y_i, the marginal forecasts conditional on the total are

$$(Y_i|T) \sim \mathrm{N}[f_i + (q_i/Q)(T - f), q_{ii} - q_i^2/Q].$$

It may be remarked that this is just the approach used to constrain seasonal factors to zero-sum, as in Theorem 8.2 of Chapter 8, and is obviously applicable generally when precise, linear constraints are to be imposed.

(C). Revising $p(T)$ based on an additional forecast

Suppose an additional forecast of T is made in terms of a forecast mean m and variance M. Consider how this should be used to revise the prior forecast distribution for the total, namely $T \sim N[f, Q]$ from (A). Let $H = \{m, M\}$ denote the information provided. Prior to learning the values of m and M, the information H is a random vector, and the objective is to calculate the posterior distribution for T when H is known, identifying the density $p(T|H)$. Formally, via Bayes' Theorem,

$$p(T|H) \propto p(T)p(H|T) \propto p(T)p(m, M|T).$$

The problem is thus to identify an appropriate *model* for the distribution of m and M conditional on T. This special case is an example of a wider class of problems in which T is a very general random quantity and H may represent a variety of information sets. The general approach here is based on the foundational works of Lindley (1983, 1985 and 1988). Obviously there are many possible models for $p(m, M|T)$, the choice of an appropriate form depending on the features and circumstances of the application. Some important possibilities are as follows.

(1) Various models are developed in Lindley (1983, 1988). Harrison (1985c) develops similar ideas, both authors discussing the following, important special case. Write

$$p(m, M|T) = p(m|M, T)p(M|T),$$

and suppose that these two densities are given as follows.

Firstly, the conditional density of the forecast mean m when both M and T are known, is normal

$$(m|M, T) \sim N[T, M].$$

Thus the point forecast m is assumed to be unbiased and distributed about the correct value T with variance equal to that stated. Secondly, the component, $p(M|T)$, is assumed not to depend on T, so contributes nothing to the likelihood for T. Full discussion of such assumptions is given by Lindley (1988, Section 17).

This results in

$$p(H|T) = p(m, M|T) \propto p(m|M, T),$$

as a function of T, and the revised, posterior forecast density for for the total is given through Bayes' Theorem by

$$p(T|H) \propto p(T)p(H|T) \propto p(T)p(m|M, T).$$

From the normal prior $p(T)$ and likelihood $p(m|M,T)$ here, standard normal theory leads to a normal posterior

$$(T|H) \sim N[f + \rho(m - f), (1 - \rho^2)Q - \rho^2 M],$$

where $\rho = Q/(Q+M)$. The posterior variance here may be alternatively written as ρM; in terms of precision, $V[T|H]^{-1} = Q^{-1} + M^{-1}$.

(2) The above approach is a special case of more general models in Lindley (1983, 1988), an important class of such allowing for anticipated biases in the forecast data m and M. Such biases can be modelled by writing, for some, known quantities a, b and $c > 0$,

$$(m|M,T) \sim N[a + bT, cM].$$

This generalises (1) to allow for systematic biases a and/or b in the point forecast m, and also, importantly, to allow for over- or under- optimism in the forecast variance M through the scale factor c. With this replacing the original assumption in (1) above, (but retaining the assumption that $p(M|T)$ does not involve T), Bayes' Theorem easily leads to the revised forecast distribution

$$(T|H) \sim N[f + \rho(m - a - bf), (1 - \rho^2 b^2)Q - \rho^2 cM],$$

where $\rho = bQ/(b^2Q + cM)$. The posterior variance here may be alternatively written as $\rho(cM/b)$; in terms of precision, $V[T|H]^{-1} = Q^{-1} + b^2(cM)^{-1}$. Obviously the result in (1) is obtained when the forecast information is assumed unbiased so that $a = 0$ and $b = c = 1$.

Further extensions of this model are also considered by Lindley. One such allows for the possibilities of uncertainties about some, or all, of the quantities a, b and c, for example. In a time series context where $\mathbf{Y} = \mathbf{Y}_t$, $(t = 1, 2, \ldots,)$, and independent forecast information $H_t = \{m_t, M_t\}$ for the total T_t is sequentially obtained, these biases can be estimated, and also allowed to vary, over time.

(3) In the above approaches, the additional forecast information is viewed essentially as data informing about T, whether corrected for anticipated biases as in (2) or not, such data to be combined with the information summarised in the original

forecast distribution in more or less standard ways. In some circumstances, this direct combination may be viewed as inappropriate. One of its consequences, for example, concerns the resulting forecast precision. In (1), ρ decreases towards unity as the stated variance M of the additional forecast decreases, with $p(T|H)$ concentrating about m. Often this will lead to spuriously precise inferences that may also be seriously biased. Approach (2) can adequately cater for spurious precision and biases through the use of, possibly uncertain, bias parameters a, b and c. Obviously some form of correction is desirable if the forecast information H is derived from external sources that may suffer from biases and over-, or under-, optimism.

This problem is clearly evident when dealing with additional forecasts provided from other forecasters, models or agencies. In such cases, the concepts of forecasting *expertise* and *calibration* must be considered. West (1989a and b) directly addresses these issues in a very general framework, and develops models for the synthesis of various types of forecast information deriving from several different, possibly subjective, sources. Within the specific, normal and linear framework here, and with a single additional forecast, this approach leads to results similar to those of (1) and (2), though with important differences. Refer to the source of the forecast information as an *agent*, whether this be an individual, a group of individuals or an alternative, possibly related, forecasting model. The essence of the approach is to put a numerical limit on the extent to which the agent's forecast information affects the revised distribution for T. Consider the extreme case in which the forecast provided is perfectly precise, given by taking $M = 0$ so that the information is $H_0 = \{X, 0\}$, the agent's model indicating that T is takes the forecast value X. Starting with $T \sim N[f, Q]$, the question now concerns just how reliable this "perfect" forecast is. One special case of the models in West (1989b) leads to $p(T|H_0) = p(T|X)$ with

$$(T|X) \sim N[f + \rho(X - f), (1 - \rho^2)Q]$$

for some quantity ρ, $(0 \le \rho \le 1)$. The quantity ρ represents a measure of assessed expertise of the agent in forecasting. A value near unity is consistent with the view that the agent is rather good in forecasting, $\rho = 1$ leading to direct acceptance of the stated value X. For $\rho < 1$, there is an implied limit on the posterior precision, unlike the approaches in (1) and (2). Hence this case of hypothetical, perfect forecast information H_0 serves to provide the expertise measure ρ. The effect of the received information $H = \{m, M\}$ is modelled by substituting the moments m and M as those of X, taking $(X|H) \sim N[m, M]$. It should be noted that bias corrections, as in (2), may be used to appropriately correct for miscalibration in forecasts made by other individuals or models, writing $(X|H) \sim N[a + bm, cM]$, for example. For clarity suppose here that such biases are not deemed necessary, the agent forecast being assumed calibrated. Then, putting together the components

$$(T|X) \sim N[f + \rho(X - f), (1 - \rho^2)Q]$$

and

$$(X|H) \sim N[m, M],$$

it follows that the revised forecast for T is given by the posterior

$$(T|H) \sim N[f + \rho(m - f), (1 - \rho^2)Q + \rho^2 M].$$

The posterior mean here is similar in form to that of the earlier approaches, although now the extent ρ of the correction made to the point forecast m directly relates to the assessed expertise of the agent. The posterior variance, unlike the earlier approaches, will exceed the prior variance Q if $M > Q$, never being less than the value $(1 - \rho^2)Q$ obtained when $M = 0$, providing a rather more conservative sythesis of the two information sources.

Many variations of this model, and other, more general models, appear in West (1989a and b). Forms of distribution other than normal lead to results with practically important differences, as do variations in which ρ is uncertain. Details are beyond the scope of the current discussion, however, and the interested reader is referred to the above sources.

(D). Revising $p(Y)$ using an additional forecast of T

Whatever approach is used in (C) above, the result is a revised forecast distribution $(T|H) \sim N[f^*, Q^*]$ for some moments f^* and Q^*.

Given that the additional information H is relevant only to forecasting the total T, it follows that

$$p(\mathbf{Y}|T, H) = p(\mathbf{Y}|T),$$

ie. given T, $\mathbf{Y}$ is conditionally independent of H. This distribution is given in (B) by

$$(\mathbf{Y}|T) \sim \mathrm{N}[\mathbf{f} + \mathbf{q}(T - f)/Q, \mathbf{Q} - \mathbf{q}\mathbf{q}'/Q],$$

with $\mathbf{q} = \mathbf{Q}\mathbf{1}$.

Hence the implications of H for forecasting $\mathbf{Y}$ are derived through the posterior

$$p(\mathbf{Y}|H) = \int p(\mathbf{Y}|T, H)p(T|H)dT = \int p(\mathbf{Y}|T)p(T|H)dT.$$

It easily follows (the proof left as an exercise for the reader) that

$$(\mathbf{Y}|H) \sim \mathrm{N}[\mathbf{f} + \mathbf{q}(f^* - f)/Q, \mathbf{Q} - \mathbf{q}\mathbf{q}'(Q - Q^*)/Q^2].$$

(E).   Additional forecasts of aggregates

Models in (D) generalise directly to the case of independent forecasts of subaggregates, and linear combinations generally. Let $\mathbf{X} = \mathbf{L}\mathbf{Y}$ as above, so that, from (A),

$$\mathbf{X} \sim \mathrm{N}[\mathbf{s}, \mathbf{S}],$$

with moments $\mathbf{s} = \mathbf{L}\mathbf{Q}$ and $\mathbf{S} = \mathbf{L}\mathbf{Q}\mathbf{L}'$. Independent forecast information is obtained providing a forecast mean $\mathbf{m}$ and variance matrix $\mathbf{M}$. The models of (D), in the special case that $\mathbf{L} = \mathbf{1}'$, extend directly. Under model (1) of part (C), the additional forecast information is taken at face value. The direct generalisation of the results there has $p(H|T)$ is defined via the two components $p(\mathbf{m}|\mathbf{M}, \mathbf{X})$ and $p(\mathbf{M}|\mathbf{X})$, given as follows. Firstly,

$$(\mathbf{m}|\mathbf{M}, \mathbf{X}) \sim \mathrm{N}[\mathbf{X}, \mathbf{M}],$$

and, secondly, $p(\mathbf{M}|\mathbf{X})$ does not depend on $\mathbf{X}$. Under these assumptions, Bayes' Theorem leads to the posterior

$$(\mathbf{X}|H) \sim \mathrm{N}[\mathbf{s}^*, \mathbf{S}^*]$$

where the revised moments are given by

$$\mathbf{s}^* = \mathbf{s} + \mathbf{B}(\mathbf{m} - \mathbf{s})$$

and

$$\mathbf{S}^* = \mathbf{S} - \mathbf{B}\mathbf{S}\mathbf{B}',$$

where

$$\mathbf{B} = \mathbf{S}(\mathbf{S} + \mathbf{M})^{-1}.$$

Biases can be modelled, generalising (2) of part (C), the details being left as an exercise for the reader. Similarly, using approaches in West (1989a and b), additional information from an agent may be combined through models generalising that described in part (3) of (C) above.

Given any revised moments $\mathbf{s}^*$ and $\mathbf{S}^*$, and again assuming that the additional information H is relevant only to forecasting $\mathbf{X}$, it follows that

$$p(\mathbf{Y}|\mathbf{X}, H) = p(\mathbf{Y}|\mathbf{X}),$$

ie. given $\mathbf{X}$, $\mathbf{Y}$ is conditionally independent of H. Also, from (B),

$$(\mathbf{Y}|\mathbf{X}) \sim N[\mathbf{f} + \mathbf{A}(\mathbf{X} - \mathbf{s}), \mathbf{Q} - \mathbf{A}\mathbf{S}\mathbf{A}'],$$

where $\mathbf{A} = \mathbf{Q}\mathbf{L}'\mathbf{S}^{-1}$.

Hence $p(\mathbf{Y})$ is revised in the light of H to

$$p(\mathbf{Y}|H) = \int p(\mathbf{Y}|\mathbf{X}, H)p(\mathbf{X}|H)d\mathbf{X} = \int p(\mathbf{Y}|\mathbf{X})p(\mathbf{X}|H)d\mathbf{X},$$

which (as can be verified by the reader) is given by

$$(\mathbf{Y}|H) \sim N[\mathbf{f}^*, \mathbf{Q}^*],$$

with moments given by

$$\mathbf{f}^* = \mathbf{f} + \mathbf{A}(\mathbf{s}^* - \mathbf{s})$$

and

$$\mathbf{Q} - \mathbf{A}(\mathbf{S} - \mathbf{S}^*)\mathbf{A}'.$$

Finally, the revised forecasts for any other aggregates $\mathbf{KY}$, say, can be deduced from part (A).

15.4 MATRIX NORMAL DLMS

15.4.1 Introduction and general framework

A general framework for multivariate time series analysis when the covariance structure across series is unknown is presented in Quintana (1985, 1987), and developed and applied in Quintana and West (1986, 1988). The resulting models are extensions of the basic DLM which allow fully conjugate, closed form analyses of covariance structure when it may be assumed that the scalar component time series follow univariate DLMs with common $\mathbf{F}_t$ and $\mathbf{G}_t$. Thus the models are appropriate in applications when several similar series are to be analysed. Such a setup is common in many application areas. In economic and financial modelling, such series arise as measurements of similar financial indicators, shares prices or exchange rates, and compositional series such as energy consumption by energy source. In medical monitoring, collections of similar time series are often recorded, such as with measurements on each of several, related biochemical indicators in post-operative patient care.

The framework developed in the above references is as follows. Suppose that we have q univariate series Y_{tj} with, for each $j = 1,\dots,q$, Y_{tj} following a standard, univariate DLM with defining quadruple

$$\{\mathbf{F}_t, \mathbf{G}_t, V_t\sigma_j^2, \mathbf{W}_t\sigma_j^2\}.$$

The model is n−dimensional, and the defining quantities of the quadruple are assumed known apart from the scale factors σ_j^2, ($j = 1,\dots,q$). In terms of observation and evolution equation, the univariate series Y_{tj} is given by

Observation Equation: $\quad Y_{tj} = \mathbf{F}_t'\boldsymbol{\theta}_{tj} + \nu_{tj}, \qquad \nu_{tj} \sim \mathrm{N}[0, V_t\sigma_j^2],$

$$(15.3\mathrm{a})$$

Evolution Equation: $\quad \boldsymbol{\theta}_{tj} = \mathbf{G}_t\boldsymbol{\theta}_{t-1,j} + \boldsymbol{\omega}_{tj}, \quad \boldsymbol{\omega}_{tj} \sim \mathrm{N}[\mathbf{0}, \mathbf{W}_t\sigma_j^2],$

$$(15.3\mathrm{b})$$

Note the key feature of the model here: the defining components $\mathbf{F}_t$, $\mathbf{G}_t$ and $\mathbf{W}_t$ are the same for each of the q series. In addition, the model permits a common, known, observational scale factor V_t across all q series, to allow, for example, for common measurement scales, common sampling variation, common occurrence of missing values, outliers and so forth. Otherwise, the series have individual state vectors $\boldsymbol{\theta}_{tj}$ that are typically different. They also possibly vary through the scales of measurement, defined via individual variances σ_j^2, which are assumed uncertain. Note also that, as usual, σ_j^2 appears as a multiplier of the known evolution variance matrix $\mathbf{W}_t$. Of course the model equations above are defined conditional upon these variances. Finally, the usual conditional independence assumptions are made: given all defining parameters, we assume, for all j, that the errors ν_{tj} are independent over time, the evolution errors $\boldsymbol{\omega}_{tj}$ are independent over time, and that the two sequences are mutually independent.

The joint, *cross-sectional* structure across series at time t comes in via the covariances between the observational errors of each of the q series, and also between evolution errors. Introduce a $q \times q$ covariance matrix $\boldsymbol{\Sigma}$ given by

$$\boldsymbol{\Sigma} = \begin{pmatrix} \sigma_1^2 & \sigma_{1,2} & \sigma_{1,3} & \cdots & \sigma_{1,q} \\ \sigma_{1,2} & \sigma_2^2 & \sigma_{2,3} & \cdots & \sigma_{2,q} \\ \vdots & \vdots & \vdots & \ddots & \vdots \\ \sigma_{1,q} & \sigma_{2,q} & \sigma_{3,q} & \cdots & \sigma_q^2 \end{pmatrix}$$

where, for all i and j, $(i = 1, \ldots, q; \ j = 1, \ldots, q; \ i \neq j)$, σ_{ij} determines the covariance between series Y_{ti} and Y_{tj}. Then the model equations (15.3a and b) are supplemented by the cross-sectional assumptions that, conditional on $\boldsymbol{\Sigma}$,

$$C[\nu_{ti}, \nu_{tj}] = V_t \sigma_{ij}, \tag{15.4a}$$

$$C[\boldsymbol{\omega}_{ti}, \boldsymbol{\omega}_{tj}] = \mathbf{W}_t \sigma_{ij}, \tag{15.4b}$$

for $i \neq j$.

The individual series Y_{tj}, $(j = 1, \ldots, q)$, follow the same *form* of DLM, the model parameters $\boldsymbol{\theta}_{tj}$ being different across series. Correlation structure induced by $\boldsymbol{\Sigma}$ affects both the observational errors through (15.4a) and the evolution errors $\boldsymbol{\omega}_{tj}$ through (15.4b). Thus if, for example, σ_{ij} is large and positive, series i and j will tend to follow similar patterns of behaviour in both the underlying movements in their defining state parameters and in the purely random, observational variation about their levels. Of course the scales σ_i and σ_j may differ.

The model equations may be written in matrix notation. For notation, define the following quantities for each t:

- $\mathbf{Y}_t = (Y_{t1}, \ldots, Y_{tq})'$, the q-vector of observations at time t;

- $\nu_t = (\nu_{t1}, \ldots, \nu_{tq})'$, the q-vector of observational errors at time t;
- $\Theta_t = [\theta_{t1}, \ldots, \theta_{tq}]$, the $n \times q$ matrix whose columns are the state vectors of the individual DLMs (15.3a);
- $\Omega_t = [\omega_{t1}, \ldots, \omega_{tq}]$, the $n \times q$ matrix whose columns are the evolution errors of the individual DLMs in (15.3b).

With these definitions, (15.3a and b) are re-expressible as

$$Y_t' = F_t'\Theta_t + \nu_t',$$
$$\Theta_t = G_t\Theta_{t-1} + \Omega_t. \qquad (15.5)$$

Note here that the observation is a *row* vector and the state parameters are in the form of a *matrix*. The fact that F_t and G_t are common to each of the q univariate DLMs is fundamental to this new representation. To proceed we need to identify the distributions of the observational error vector ν_t and the evolution error *matrix* Ω_t, all, of course, conditional on Σ (as well as V_t and W_t for all t). The former is obviously multivariate normal,

$$\nu_t \sim N[\mathbf{0}, V_t\Sigma]$$

independently over time, where Σ defines the cross-sectional covariance structure for the multivariate model. The latter is a *matrix-variate normal distribution* (Dawid, 1981; see also Press, 1985), described as follows.

15.4.2 The matrix normal distribution for Ω_t

Clearly any collection of the qn elements of Ω_t are multivariate normally distributed, as is the distribution of any linear function of the elements. Dawid (1981) describes the class of matrix-variate (or just matrix) normal distributions that provides a concise mathematical representation and notation for such matrices of jointly normal quantities. Full theoretical details of the structure and properties of such distributions, and their uses in Bayesian analyses, are given in Dawid's paper to which the interested reader may refer. See also Quintana (1987, Chapter 3). Some of these features are described here.

The random matrix Ω_t has a matrix normal distribution with **mean matrix 0, left variance matrix** W_t and **right variance matrix** Σ. The density function is given by

$$p(\Omega_t) = k(W_t, \Sigma)\exp\{-\frac{1}{2}\mathrm{trace}[\Omega_t'W_t^{-1}\Omega_t\Sigma^{-1}]\}$$

where
$$k(\mathbf{W}_t, \boldsymbol{\Sigma}) = (2\pi)^{-qn/2} |\mathbf{W}_t|^{-q/2} |\boldsymbol{\Sigma}|^{-n/2}.$$

Some important properties are that (i) all marginal and conditional distributions of elements of $\boldsymbol{\Omega}_t$, and linear functions of them, are (uni-, multi-, or matrix-) variate normal; (ii) the definition of the distribution remains valid when either or both of the variance matrices is non-negative definite; (iii) the distribution is non-singular if, and only if, each variance matrix is positive definite; and (iv) if either of $\mathbf{W}_t$ and $\boldsymbol{\Sigma}$ are identically zero matrices, then $\boldsymbol{\Omega}_t$ is zero with probability one.

Concerning the model (15.5), note that $q = 1$ leads to a standard, univariate DLM with unknown observational scale factor $\boldsymbol{\Sigma} = \sigma_1^2$. Also, for any q, if $\boldsymbol{\Sigma}$ is diagonal then the q series Y_{tj} are unrelated. The distribution for $\boldsymbol{\Omega}_t$ is denoted, in line with the notation in the above references, by

$$\boldsymbol{\Omega}_t \sim \mathrm{N}[\mathbf{0}, \mathbf{W}_t, \boldsymbol{\Sigma}]. \qquad (15.6)$$

A simple extension of (15.5) to include a known mean vector for each of the evolution error vectors poses no problem. Suppose that $\mathrm{E}[\boldsymbol{\omega}_{tj}] = \mathbf{h}_{tj}$ is known at time t for each j, and let $\mathbf{H}_t$ be the $n \times q$ mean matrix $\mathbf{H}_t = [\mathbf{h}_{t1}, \dots, \mathbf{h}_{tq}]$. Then $\boldsymbol{\Omega}_t - \mathbf{H}_t$ has the distribution (15.6); equivalently, $\boldsymbol{\Omega}_t$ has a matrix normal distribution with the same variance matrices, but now with mean matrix $\mathbf{H}_t$, the notation being $\boldsymbol{\Omega}_t \sim \mathrm{N}[\mathbf{H}_t, \mathbf{W}_t, \boldsymbol{\Sigma}]$.

15.4.3   The matrix normal/inverse Wishart distribution

Having introduced the matrix normal distribution for the matrix of evolution errors, it is no surprise that the same form of distribution provides the basis of prior and posterior distributions for the state matrices $\boldsymbol{\Theta}_t$ for all t. This is directly parallel to the use of multivariate normals in standard DLMs when the state parameters form a vector. Analysis conditional on $\boldsymbol{\Sigma}$ is completely within the class of matrix normal distributions, but, in line with the focus of this Chapter, the results are developed below in the more general framework in which $\boldsymbol{\Sigma}$ is unknown. Thus we consider the class of **matrix normal/inverse Wishart** distributions suitable for learning jointly about $\boldsymbol{\Theta}_t$ and $\boldsymbol{\Sigma}$. Before proceeding to the model analysis in the next Section, the structure of this class of distributions is summarised here (free from notational complications due to the t subscripts).

(A) The inverse Wishart distribution

Consider a $q \times q$, positive definite random matrix Σ. Following Dawid (1981), Σ has an **inverse (or inverted) Wishart** distribution if, and only if, the density of the distribution of Σ is given (up to a constant of normalisation) by

$$p(\Sigma) \propto |\Sigma|^{-(q+n/2)} \exp[-\frac{1}{2}\text{trace}(n\mathbf{S}\Sigma^{-1})],$$

where $n > 0$ is a known, scalar *degrees of freedom* parameter, and $\mathbf{S}$ a known, $q \times q$ positive definite matrix. Full discussion and properties appear in Box and Tiao (1973, Section 8.5), and Press (1985, Chapter 5), some features of interest being as follows.

- $\Phi = \Sigma^{-1}$ has a Wishart distribution with n degrees of freedom and mean $E[\Phi] = \mathbf{S}^{-1}$. Thus $\mathbf{S}$ is an estimate of Σ, the harmonic mean under the inverse Wishart distribution. If $\mathbf{S}$ has elements

$$\mathbf{S} = \begin{pmatrix} S_1 & S_{1,2} & S_{1,3} & \cdots & S_{1,q} \\ S_{1,2} & S_2 & S_{2,3} & \cdots & S_{2,q} \\ \vdots & \vdots & \vdots & \ddots & \vdots \\ S_{1,q} & S_{2,q} & S_{3,q} & \cdots & S_q \end{pmatrix},$$

then S_j is an estimate of the variance σ_j^2, and S_{jk} an estimate of the covariance σ_{jk}, for all j and k, $(j \neq k)$.

- If $q = 1$ so that Σ is scalar, then so are $\Phi = \phi$ and $\mathbf{S} = S$. Now ϕ has the usual gamma distribution, $\phi \sim G[n/2, nS/2]$ with mean $E[\phi] = 1/S$.
- For $n > 2$, $E[\Sigma] = \mathbf{S}n/(n-2)$.
- The marginal distributions of the variances σ_j^2 on the diagonal off Σ are inverse gamma; with precisions $\phi_j = \sigma_j^{-2}$, $\phi_j \sim G[n/2, nS_j/2]$.
- As $n \to \infty$, the distribution concentrates around $\mathbf{S}$, ultimately degenerating there.

By way of notation here, the distribution is denoted by

$$\Sigma \sim W_n^{-1}[\mathbf{S}]. \tag{15.7}$$

(B) The matrix normal/inverse Wishart distribution

Suppose that (15.7) holds, and introduce a further random matrix

Θ of dimension $p \times q$. Suppose that, conditional on Σ, Θ follows a matrix-normal distribution with $p \times q$ mean matrix $\mathbf{m}$, $p \times p$ left variance matrix $\mathbf{C}$, and right variance matrix Σ. The defining quantities $\mathbf{m}$ and $\mathbf{C}$ are assumed known. Thus, following (15.6),

$$(\Theta|\Sigma) \sim N[\mathbf{m}, \mathbf{C}, \Sigma]. \tag{15.8}$$

Equations (15.7) and (15.8) define a joint distribution for Θ and Σ that is referred to as a **matrix normal/inverse Wishart** distribution. The special case of scalar Σ, when $q = 1$, corresponds to the usual normal/inverse gamma distribution (Section 16.3, Chapter 16) used in variance learning in DLMs throughout the earlier chapters. Dawid (1981) details this class of distributions, deriving many important results. By way of notation, if (15.7) and (15.8) hold then the joint distribution is denoted by

$$(\Theta, \Sigma) \sim NW_n^{-1}[\mathbf{m}, \mathbf{C}, \mathbf{S}]. \tag{15.9}$$

Equation (15.9) is now understood to imply both (15.7) and (15.8). In addition, the marginal distribution of Θ takes the following form.

(C) The matrix T distribution

Under (15.9) the marginal distribution of the matrix Θ is a matrix-variate analogue of the multivariate T distribution (Dawid, 1981). This is completely analogous to the matrix normal, the differences lying in the tail weight of the marginal distributions of elements of Θ. As with the matrix normal, the component columns of Θ themselves follow p–dimensional multivariate T distributions with n degrees of freedom. Write

$$\Theta = [\boldsymbol{\theta}_1, \dots, \boldsymbol{\theta}_q]$$

and

$$\mathbf{m} = [\mathbf{m}_1, \dots, \mathbf{m}_q].$$

It follows (Dawid, 1981) that

$$\boldsymbol{\theta}_j \sim T_n[\mathbf{m}_j, \mathbf{C}S_j], \qquad (j = 1, \dots, q).$$

If $n > 1$, $E[\boldsymbol{\theta}_j] = \mathbf{m}_j$. If $n > 2$, $V[\boldsymbol{\theta}_j] = \mathbf{C}S_j n/(n-2)$ and the covariance structure between the columns is given by

$$C[\boldsymbol{\theta}_j, \boldsymbol{\theta}_k] = \mathbf{C}S_{jk} n/(n-2), \qquad (j, k = 1, \dots, p; \; j \neq k).$$

As with the matrix normal notation in (15.8), the matrix T distribution of Θ is denoted by

$$\Theta \sim T_n[\mathbf{m}, \mathbf{C}, \mathbf{S}]. \tag{15.10}$$

15.4.4 Updating and forecasting equations

The model (15.5) is amenable to a conjugate, sequential analysis that generalises the standard analysis for univariate series with variance learning. The analysis is based on the use of matrix normal/inverse Wishart prior and posterior distributions for the $n \times q$ state matrix Θ_t and the $q \times q$ variance matrix Σ at all times t. The theory is essentially based on the Bayesian theory in Dawid (1981). Full details are given in Quintana (1985, 1987), the results being stated here without proof and illustrated in following sections.

Suppose $\mathbf{Y}_t$ follows the model defined in equations (15.3—15.6), summarised as

$$\begin{aligned}
\mathbf{Y}'_t &= \mathbf{F}'_t \Theta_t + \boldsymbol{\nu}'_t, & \boldsymbol{\nu}_t &\sim N[\mathbf{0}, V_t \Sigma], \\
\Theta_t &= \mathbf{G}_t \Theta_{t-1} + \boldsymbol{\Omega}_t, & \boldsymbol{\Omega}_t &\sim N[\mathbf{0}, \mathbf{W}_t, \Sigma].
\end{aligned} \tag{15.11}$$

Here $\boldsymbol{\nu}_t$ are independent over time, $\boldsymbol{\Omega}_t$ are independent over time, and the two series are mutually independent. Suppose also that the initial prior for Θ_0 and Σ is matrix normal/inverse Wishart, as in (15.9), namely

$$(\Theta_0, \Sigma | D_0) \sim NW_{n_0}^{-1}[\mathbf{m}_0, \mathbf{C}_0, \mathbf{S}_0], \tag{15.12}$$

for some known defining parameters $\mathbf{m}_0$, $\mathbf{C}_0$, $\mathbf{S}_0$ and n_0. Then, for all times $t > 1$, the following results apply.

Theorem 15.4. *One-step forecast and posterior distributions in the model (15.11) and (15.12) are given, for each t, as follows.*

(a) Posteriors at $t - 1$:
For some $\mathbf{m}_{t-1}$, $\mathbf{C}_{t-1}$, $\mathbf{S}_{t-1}$ and n_{t-1},

$$(\Theta_{t-1}, \Sigma | D_{t-1}) \sim NW_{n_{t-1}}^{-1}[\mathbf{m}_{t-1}, \mathbf{C}_{t-1}, \mathbf{S}_{t-1}].$$

(b) Priors at t:

$$(\Theta_t, \Sigma | D_{t-1}) \sim NW_{n_{t-1}}^{-1}[\mathbf{a}_t, \mathbf{R}_t, \mathbf{S}_{t-1}],$$

where

$$\mathbf{a}_t = \mathbf{G}_t\mathbf{m}_{t-1} \qquad and \qquad \mathbf{R}_t = \mathbf{G}_t\mathbf{C}_{t-1}\mathbf{G}_t' + \mathbf{W}_t.$$

(c) One-step forecast:

$$(\mathbf{Y}_t|\boldsymbol{\Sigma}, D_{t-1}) \sim N[\mathbf{f}_t, Q_t\boldsymbol{\Sigma}]$$

with marginal

$$(\mathbf{Y}_t|D_{t-1}) \sim T_{n_{t-1}}[\mathbf{f}_t, Q_t\mathbf{S}_{t-1}],$$

where

$$\mathbf{f}_t' = \mathbf{F}_t'\mathbf{a}_t \qquad and \qquad Q_t = V_t + \mathbf{F}_t'\mathbf{R}_t\mathbf{F}_t.$$

(d) Posteriors at t:

$$(\boldsymbol{\Theta}_t, \boldsymbol{\Sigma}|D_{t-1}) \sim NW_{n_t}^{-1}[\mathbf{m}_t, \mathbf{C}_t, \mathbf{S}_t],$$

with

$$\mathbf{m}_t = \mathbf{a}_t + \mathbf{A}_t\mathbf{e}_t' \qquad and \qquad \mathbf{C}_t = \mathbf{R}_t - \mathbf{A}_t\mathbf{A}_t'Q_t,$$

$$n_t = n_{t-1} + 1 \qquad and \qquad \mathbf{S}_t = n_t^{-1}[n_{t-1}\mathbf{S}_{t-1} + \mathbf{e}_t\mathbf{e}_t'/Q_t],$$

where

$$\mathbf{A}_t = \mathbf{R}_t\mathbf{F}_t/Q_t \qquad and \qquad \mathbf{e}_t = \mathbf{Y}_t - \mathbf{f}_t.$$

Proof: Omitted, an exercise for the reader. Full details appear in Quintana (1985, 1987).

$\diamond$

As mentioned, this is a direct extension of the univariate theory. Theorem 4.4 of Chapter 4 is the special case $q = 1$ (and, with no real loss of generality, $V_t = 1$).[†] Here, generally, $\boldsymbol{\Theta}_t$ is a matrix, the prior and posterior distributions in (a), (b) and (d) being matrix normal. Thus the prior and posterior means $\mathbf{a}_t$ and $\mathbf{m}_t$ are both $n \times q$ matrices, their columns providing the means of the q state vectors of the individual DLMs (15.3a and b). Notice the key features that,

---

[†]Note that results are given here in the original form with the variance matrices $\mathbf{C}_t$, $\mathbf{W}_t$ etc. all being subject to multiplication by scale factors from $\boldsymbol{\Sigma}$, as in the univariate case in Theorem 4.4. The alternative, and preferred, representation summarised in the table in Section 4.6 of Chapter 4, and used throughout the book, has no direct analogue here in the multivariate framework.

since $\mathbf{F}_t$, $\mathbf{G}_t$ and $\mathbf{W}_t$ are common to these q DLMs, the elements $\mathbf{R}_t$, $\mathbf{C}_t$, $\mathbf{A}_t$ and Q_t are common too. Thus, although there are q series analysed, the calculation of these elements need only be done once. Their dimensions are as in the univariate case, $\mathbf{R}_t$ and $\mathbf{C}_t$ are both $n \times n$, the common adaptive vector $\mathbf{A}_t$ is $n \times 1$, and the common variance Q_t is a scalar.

The analysis essentially duplicates that for the individual univariate DLMs, this being seen in detail by decomposing the vector/matrix results as follows.

(a) For each of the q models, the posterior in (a) of the Theorem has following marginals:

$$(\boldsymbol{\theta}_{t-1,j}|\sigma_j^2, D_{t-1}) \sim \mathrm{N}[\mathbf{m}_{t-1,j}, \mathbf{C}_{t-1}\sigma_j^2],$$

and

$$(\sigma_j^{-2}|D_{t-1}) \sim \mathrm{G}[n_{t-1}/2, d_{t-1,j}/2],$$

where $d_{t-1,j} = n_{t-1}S_{t-1,j}$. Unconditional on the variance σ_j^2,

$$(\boldsymbol{\theta}_{t-1,j}|D_{t-1}) \sim \mathrm{T}_{n_{t-1}}[\mathbf{m}_{t-1,j}, \mathbf{C}_{t-1}S_{t-1,j}].$$

Thus, for each j, the prior is the standard normal/gamma. Note again that the scale-free variance matrix $\mathbf{C}_{t-1}$ is common across the q series.

(b) Evolving to time t, similar comments apply to the prior distribution in (b). Writing $\mathbf{a}_t$ in terms of its columns

$$\mathbf{a}_t = [\mathbf{a}_{t1}, \ldots, \mathbf{a}_{tq}],$$

it is clear that

$$\mathbf{a}_{tj} = \mathbf{G}_t \mathbf{m}_{t-1,j}, \qquad (j = 1, \ldots, q).$$

Thus, for each of the q univariate DLMs, the evolution is standard, $(\boldsymbol{\theta}_{tj}|D_{t-1}) \sim \mathrm{T}_{n_{t-1}}[\mathbf{a}_{tj}, \mathbf{R}_t S_{t-1,j}]$, where $\mathbf{a}_{tj}$ and $\mathbf{R}_t$ are calculated as usual. Again note $\mathbf{R}_t$ is common to the q series and so need only be calculated once.

(c) In the one-step ahead forecasting result (c), the q−vector of forecast means $\mathbf{f}_t = (f_{t1}, \ldots, f_{tq})'$ has elements $f_{tj} = \mathbf{F}_t' \mathbf{a}_{tj}$, $(j = 1, \ldots, q)$ which are as usual for the univariate series. So is the one-step ahead, scale-free variance Q_t. For the jth series, the forecast distribution is Student T,

$$(Y_{tj}|D_{t-1}) \sim \mathrm{T}_{n_{t-1}}[f_{tj}, Q_t S_{t-1,j}]$$

as usual.

(d) In updating in part (d) of the Theorem, the equation for the common variance matrix C_t is as usual, applying to each of the q series. The mean matrix update can be written column by column, as

$$\mathbf{m}_{tj} = \mathbf{a}_{tj} + \mathbf{A}_t e_{tj}, \qquad (j = 1, \dots, q),$$

where, for each j, $e_{tj} = Y_{tj} - f_{tj}$ is the usual forecast error and $\mathbf{A}_t$ the usual, and common, adaptive vector. For Σ, n_t increases by one degree of freedom, being common to each series. The update for the estimate S_t can be decomposed into elements as follows. For the variances on the diagonal,

$$S_{tj} = n_t^{-1}(n_{t-1}S_{t-1,j} + e_{tj}^2/Q_t).$$

With $d_{tj} = n_t S_{tj}$, it follows that $d_{tj} = d_{t-1,j} + e_{tj}^2/Q_t$ and $(\sigma_j^{-2}|D_t) \sim G[n_t/2, d_{tj}/2]$. This is again exactly the standard updating, with a common value of n_t across the q series.

Thus, although we are analysing several, related series together, there is no effect on the posterior and forecast distributions generated. The univariate theory applies separately to each of the q series, although the calculations are reduced since some of the components are common. The difference lies in the fact that the covariance structure across series is also identified through part (d) of the Theorem. Here the full posterior of Σ is given by

$$(\Sigma|D_t) \sim W_{n_t}^{-1}[S_t],$$

having mean $S_t n_t/(n_t - 2)$ if $n_t > 2$, and harmonic mean S_t. The covariance terms are updated via

$$S_{tjk} = n_t^{-1}(n_{t-1}S_{t-1,jk} + e_{tj}e_{tk}/Q_t), \qquad (j,k = 1, \dots, q; \; j \neq k).$$

It follows that

$$S_{tjk} = (n_0 + t)^{-1}\left(n_0 S_{0jk} + \sum_{r=1}^{t} e_{rj}e_{rk}/Q_r\right)$$

$$= (1 - \alpha_t)S_{0,jk} + \alpha_t c_{jk}(t),$$

where $\alpha_t = t/(n_0 + t)$ is a weight lying between 0 and 1, and

$$c_{jk}(t) = \frac{1}{t}\sum_{r=1}^{t} e_{rj}e_{rk}/Q_r.$$

This is just the *sample correlation* of the standardised, one-step ahead forecast errors observed up to time t. Thus S_{tjk} is a weighted average of the prior estimate $S_{0,jk}$ and the sample estimate $c_{jk}(t)$.

15.4.5 Further comments and extensions

Further theoretical and practical features of the analysis, and some extensions, deserve mention.

(a) Step ahead forecasting and filtering

Forecasting ahead to time $t + k$, the standard results apply to each of the q univariate series without alteration, giving Student T distributions

$$(Y_{t+k,j}|D_t) \sim T_{n_t}[f_{tj}(k), Q_t(k)S_{tj}].$$

The joint forecast distribution is multivariate T,

$$(\mathbf{Y}_{t+k}|D_t) \sim T_{n_t}[\mathbf{f}_t(k), Q_t(k)\mathbf{S}_t],$$

where $\mathbf{f}_t(k) = [f_{t1}(k), \dots, f_{tq}(k)]'$.

Similarly, retrospective time series analysis is based on filtered distributions for the state vectors, the usual results applying for each series. Thus, for $1 < k \le t$,

$$(\boldsymbol{\theta}_{t-k,j}|D_t) \sim T_{n_t}[\mathbf{a}_{tj}(-k), \mathbf{R}_t(-k)S_{tj}]$$

with the filtered mean vector $\mathbf{a}_{tj}(-k)$ and the common filtered variance matrix $\mathbf{R}_t(-k)$ calculated as in Theorem 4.4 of Chapter 4. The filtered distribution of the state matrix $\boldsymbol{\Theta}_{t-k}$ is matrix-variate T, being defined from the filtered joint distribution

$$(\boldsymbol{\Theta}_{t-k}, \boldsymbol{\Sigma}|D_t) \sim \mathrm{NW}_{n_t}^{-1}[\mathbf{a}_t(-k), \mathbf{R}_t(-k), \mathbf{S}_t],$$

where $\mathbf{a}_t(-k) = [\mathbf{a}_{t1}, \dots, \mathbf{a}_{tq}]$. In full matrix notation, the filtering equation for the matrix mean is analogous to that in Theorem 4.4 for each of its columns, given by

$$\mathbf{a}_t(-k) = \mathbf{m}_{t-k} + \mathbf{C}_{t-k}\mathbf{G}'_{t-k+1}\mathbf{R}^{-1}_{t-k+1}[\mathbf{a}_t(-k+1) - \mathbf{a}_{t-k+1}].$$

(b)   Correlation structure and principal components analysis

The covariance structure across the series can be explored by considering the inverse Wishart posterior for Σ at any time t. The matrices S_t provide estimates of the variances and covariances of the series. These provide obvious estimates of the correlations between series, that of the correlation σ_{jk}, $(j, k = 1, \dots, q; j \neq k)$, being given by

$$\frac{S_{tjk}}{(S_{tj}S_{tk})^{1/2}}.$$

These estimates can be derived as optimal Bayesian estimates of the actual correlations in a variety of ways, one such being developed in Quintana (1987), and noted in Quintana and West (1987).

   One common way of exploring joint structure is to subject an estimate of the covariance matrix to a principal components decomposition (Mardia, Kent and Bibby, 1979, Chapter 8; and Press, 1985, Chapter 9). Σ is non-negative definite, and usually positive definite, so it has q real-valued, distinct and non-negative eigenvalues with corresponding real-valued and orthogonal eigenvectors. The orthonormalised eigenvectors define the principal components of the matrix. Denote the eigenvalues of Σ by λ_j, $(j = 1, \dots, q)$, the corresponding orthonormal eigenvectors by η_j, $(j = 1, \dots, q)$, satisfying $\eta_j'\eta_j = 1$ and $\eta_i'\eta_j = 0$ for $i \neq j$. Suppose, without loss of generality, that they are given in order of decreasing eigenvalues, so that η_1 is the eigenvector corresponding to the largest eigenvalue λ_1, and so forth. Then the covariation of the elements of any random vector Y having variance matrix Σ is explained through the random quantities $X_j = \eta_j'Y$, $(j = 1, \dots, q)$, the principal components of Y. These X variates are uncorrelated and have variances $V[X_j] = \lambda_j$, decreasing as j increases. Total variation in Y is measured by $\lambda = \text{trace}(\Sigma) = \sum_{j=1}^{q} \lambda_j$, and so the jth principal component explains a proportion λ_j/λ of this total. Interpretation of such a principal components decomposition rests upon (i) identifying those components that contribute markedly to the variation, often just the first one or two; and (ii) interpreting the vectors η_j defining these important components.

   As in estimating correlations, Quintana (1987) shows that the eigenvalues and eigenvectors of the estimate S_t are, at time t, optimal Bayesian estimates of those of Σ, and so may be considered as summarising the covariance structure based on the posterior at t. The use of such estimated principal components is described in the next Section in the context of an application. For further applications and much further discussion, see Press, and Mardia, Kent and Bibby, as referenced.

(c)   Discount models

The use of discount factors to structure evolution variance matrices applies here directly as in the univariate DLM. $\mathbf{W}_t$ is common to each of the q univariate series, as are $\mathbf{C}_{t-1}$, $\mathbf{R}_t$, etc. Thus $\mathbf{W}_t$ may be based on $\mathbf{C}_{t-1}$ using discount factors that are common to each of the q univariate models.

(d)   Time varying Σ

The posterior for Σ increases in precision, measured by n_t, as t increases. In Section 10.8 of Chapter 10, consideration was given to the possibility that observational variances may vary stochastically over time, and to several arguments in favour of models at least allowing for such variation. Those arguments apply and extend here. One way of modelling such variation is to apply the discounting of variance learning described in univariate models in Section 10.8.2, this being noted here and applied in the following Sections. Full details simply follow those in Chapter 10, the key feature being that the degrees of freedom parameter n_t is discounted by a small amount between observations. With discount factor β, which will be very close to unity, the updating for Σ now takes the forms

$$n_t = \beta n_{t-1} + 1$$

and

$$\mathbf{S}_t = n_t^{-1}(\beta n_{t-1}\mathbf{S}_{t-1} + \mathbf{e}_t\mathbf{e}_t'/Q_t).$$

It follows that, for $0 < \beta < 1$, the posterior will not degenerate as t increases. Following Section 10.8.3 of Chapter 10, n_t converges to the finite limit $(1 - \beta)^{-1}$. Also, for large t, the posterior estimate $\mathbf{S}_t$ is given by

$$\mathbf{S}_t \approx (1 - \delta)^{-1}\sum_{r=0}^{t-1}\beta^r\mathbf{e}_{t-r}\mathbf{e}_{t-r}'/Q_{t-r}.$$

Hence the estimates of variances and covariances across series are exponentially weighted moving averages of past sample variances and sample covariances. For notation in the time-varying model, we have variance matrix Σ_t at time t. In filtering backwards in time, the retrospective filtered posteriors for Σ_{t-k} can be approximated as in Section 10.8.4 of Chapter 10 in the univariate case when Σ_t was scalar. Thus, for $k > 1$, the filtered estimate of Σ_{t-k} at time t is given by $S_t(-k)$, recursively calculated using

$$S_t(-k)^{-1} = S_t^{-1} + \beta[S_t(-k+1)^{-1} - S_t^{-1}].$$

Of course the static Σ model obtains when $\beta = 1$. Otherwise, β slightly less than unity at least allows the flexibility to explore a minor deviation away from constancy of Σ, that may be evidence of true changes in covariance structure, model misspecification, and so forth. See Section 10.8 for further discussion.

(e)   Matrix observations

The matrix model framework extends directly to matrix-valued time series, applying when $\mathbf{Y}_t$ is an $r \times q$ matrix of observations, comprising q separate time series of $r-$vector valued observations, in which the q series follow multivariate DLMs as in Section 15.2 but having the same $\mathbf{F}_t$ and $\mathbf{G}_t$ elements. Readers interested in further theoretical details of this, the other topics in this section and further related material, should refer to the earlier referenced works by Dawid, Quintana and West, and also the book by Press for general background reading. Related models are discussed, from different perspectives, in Harvey (1986), and Highfield (1984).

15.4.6   Application to exchange rate modelling

The model is illustrated here, following Quintana and West (1987). The interest is in analysing the joint variation over a period of years in monthly series of observations on international exchange rates. The data series considered are the exchange rates for pounds Sterling of the U.S. dollar, the Canadian dollar, the Japanese Yen, the Deutschmark, the French franc and the Italian lira. The original analysis in Quintana and West (1985) concerned data from January 1975 up to and including August 1984. Figures 15.1a and 15.1b provides plots of these six exchange rates, from the same starting point,

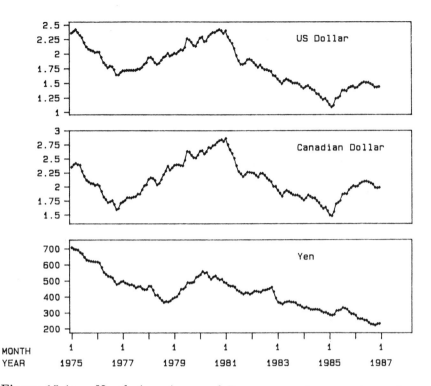

Figure 15.1a. North American and Japanese exchange rates series.

but including data up to the end of 1986. This full dataset is given in Table 15.1.[†]

From the graphs of the series, it can be seen that the trends may be approximately modelled as linear over rather short periods of time, the magnitudes and directions of the trends subject to frequent and often very marked change. After logarithmic transformation, the local linearity is still strongly evident, the marked changes no less important but the general appearance less volatile.

The model is based on the use of logarithmic transformations of each of the series, and variation in each of these transformed series is described as *locally linear*, using a second-order polynomial DLM, $\{\mathbf{E}_2, \mathbf{J}_2(1), ., .\}$. Note that this is *not* assumed to be a useful forecasting model for any other than the very short term (one-step ahead), but is used as a flexible, local trend description that has the

---

[†]Data source: U.K. Central Statistical Office databank, by permission of the Controller of H.M. Stationary Office

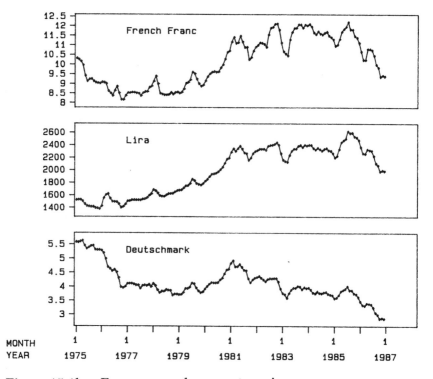

Figure 15.1b.   European exchange rate series.

ability to adapt to the series as time progresses, so long as the evolution variances are appropriately large. The focus is on identifying
the covariance structure across the series, and the basic DLM form
serves mainly to appropriately adapt to changes in trends over time.
Various analyses are reported by Quintana and West. Here slightly
different models are considered, the evolution variance sequence being based on the use of discount factors unlike those originally considered in by Quintana and West. The general results are rather
similar, however.

The model requires the specification of the observational variance
factors V_t, the evolution variances $\mathbf{W}_t$, the variance matrix discount
factor β and the initial prior (15.12). We report results from a model
in which $V_t = 1$ for all t, as is usually the case in the absence of information about additional sampling variation or recording errors
etc. The marked and frequent abrupt changes in trend would most
adequately be handled using some form of monitoring and intervention technique. Alternatively, a simpler approach to tracking the

Table 15.1. International exchange rate time series
(Source: CSO Databank)

Year	Month	$ U.S.A.	$ Canada	Yen	Franc	Lira	Mark
1975	1	2.36	2.35	707.8	10.31	1522	5.58
	2	2.39	2.40	698.3	10.24	1527	5.57
	3	2.42	2.42	695.1	10.15	1526	5.60
	4	2.37	2.40	692.4	9.95	1502	5.63
	5	2.32	2.39	676.3	9.41	1456	5.45
	6	2.28	2.28	668.9	9.13	1426	5.34
	7	2.19	2.19	647.5	9.23	1419	5.39
	8	2.12	2.12	630.1	9.25	1413	5.44
	9	2.08	2.09	624.8	9.13	1413	5.45
	10	2.06	2.06	621.9	9.06	1395	5.31
	11	2.05	2.05	620.2	9.03	1391	5.30
	12	2.02	2.02	618.0	9.01	1381	5.30
1976	1	2.03	2.04	617.8	9.08	1424	5.28
	2	2.03	2.01	611.0	9.06	1554	5.19
	3	1.94	1.92	584.2	8.99	1605	4.98
	4	1.85	1.82	552.3	8.62	1622	4.69
	5	1.81	1.77	540.5	8.50	1549	4.63
	6	1.76	1.72	527.9	8.36	1498	4.55
	7	1.79	1.74	526.4	8.64	1495	4.60
	8	1.78	1.76	518.6	8.86	1493	4.51
	9	1.73	1.69	497.1	8.51	1460	4.31
	10	1.64	1.60	477.1	8.16	1401	3.98
	11	1.64	1.61	483.0	8.16	1416	3.95
	12	1.68	1.71	494.3	8.37	1455	4.00
1977	1	1.71	1.73	498.3	8.52	1506	4.10
	2	1.71	1.76	487.2	8.52	1509	4.11
	3	1.72	1.81	481.7	8.55	1523	4.11
	4	1.72	1.81	473.5	8.54	1525	4.08
	5	1.72	1.80	476.7	8.51	1523	4.05
	6	1.72	1.82	468.6	8.49	1522	4.05
	7	1.72	1.83	456.1	8.36	1520	3.93
	8	1.74	1.87	463.8	8.52	1535	4.03
	9	1.74	1.87	465.2	8.58	1540	4.05
	10	1.77	1.95	450.9	8.60	1559	4.03
	11	1.82	2.02	444.8	8.81	1596	4.07
	12	1.85	2.04	446.6	8.88	1623	3.99

Table 15.1. (continued)

Year	Month	$ U.S.A.	$ Canada	Yen	Franc	Lira	Mark
1978	1	1.93	2.13	466.1	9.13	1686	4.10
	2	1.94	2.16	466.0	9.39	1668	4.03
	3	1.91	2.15	442.3	8.99	1633	3.88
	4	1.85	2.11	410.4	8.49	1589	3.78
	5	1.82	2.03	411.0	8.45	1582	3.83
	6	1.84	2.06	393.2	8.41	1580	3.83
	7	1.89	2.13	378.2	8.41	1604	3.89
	8	1.94	2.21	365.7	8.43	1623	3.87
	9	1.96	2.28	372.1	8.54	1625	3.86
	10	2.01	2.37	368.6	8.45	1628	3.69
	11	1.96	2.30	375.8	8.54	1651	3.73
	12	1.98	2.34	388.8	8.57	1671	3.73
1979	1	2.01	2.39	396.4	8.50	1677	3.71
	2	2.00	2.40	401.6	8.56	1683	3.72
	3	2.04	2.39	420.5	8.73	1714	3.79
	4	2.07	2.38	447.9	9.02	1748	3.93
	5	2.06	2.38	449.6	9.08	1752	3.93
	6	2.11	2.48	461.3	9.21	1785	3.98
	7	2.26	2.63	489.2	9.60	1854	4.12
	8	2.24	2.62	487.4	9.52	1833	4.10
	9	2.20	2.56	489.2	9.24	1785	3.95
	10	2.14	2.52	493.3	9.00	1770	3.84
	11	2.13	2.52	522.5	8.87	1760	3.78
	12	2.20	2.57	528.0	8.94	1785	3.81
1980	1	2.27	2.64	538.8	9.15	1823	3.91
	2	2.29	2.65	558.9	9.37	1854	4.00
	3	2.21	2.59	548.2	9.51	1896	4.08
	4	2.22	2.63	552.2	9.60	1934	4.14
	5	2.30	2.70	525.4	9.63	1942	4.13
	6	2.34	2.69	509.2	9.60	1951	4.13
	7	2.37	2.73	524.3	9.62	1973	4.14
	8	2.37	2.75	530.8	9.83	2008	4.24
	9	2.40	2.80	515.3	9.99	2045	4.30
	10	2.42	2.83	505.6	10.28	2111	4.45
	11	2.40	2.84	510.1	10.63	2177	4.59
	12	2.35	2.81	491.1	10.70	2194	4.62

Table 15.1. (continued)

Year	Month	$ U.S.A.	$ Canada	Yen	Franc	Lira	Mark
1981	1	2.40	2.86	485.9	11.16	2289	4.83
	2	2.29	2.75	471.8	11.41	2343	4.92
	3	2.23	2.66	466.6	11.09	2303	4.70
	4	2.18	2.59	467.9	11.13	2345	4.71
	5	2.09	2.51	461.0	11.47	2384	4.79
	6	1.97	2.38	442.3	11.16	2339	4.69
	7	1.88	2.27	435.1	10.87	2278	4.58
	8	1.82	2.23	425.2	10.89	2267	4.56
	9	1.82	2.18	416.8	10.25	2159	4.28
	10	1.84	2.22	426.4	10.36	2196	4.14
	11	1.90	2.26	424.8	10.69	2267	4.24
	12	1.91	2.26	416.5	10.89	2300	4.30
1982	1	1.89	2.25	423.8	10.99	2317	4.33
	2	1.85	2.24	435.1	11.12	2338	4.37
	3	1.81	2.21	435.6	11.10	2337	4.30
	4	1.77	2.17	431.9	11.05	2339	4.24
	5	1.81	2.23	428.8	10.89	2321	4.18
	6	1.76	2.24	441.2	11.54	2385	4.26
	7	1.73	2.20	442.1	11.89	2397	4.28
	8	1.73	2.15	446.6	11.94	2400	4.28
	9	1.71	2.12	450.4	12.11	2415	4.29
	10	1.70	2.09	460.1	12.13	2442	4.29
	11	1.63	2.00	431.9	11.79	2400	4.17
	12	1.62	2.00	392.4	11.11	2265	3.92
1983	1	1.57	1.93	366.0	10.66	2162	3.76
	2	1.53	1.88	361.5	10.54	2142	3.72
	3	1.49	1.83	354.9	10.44	2129	3.59
	4	1.54	1.90	366.2	11.28	2239	3.76
	5	1.57	1.93	369.2	11.67	2308	3.88
	6	1.55	1.91	371.8	11.87	2340	3.95
	7	1.53	1.88	367.3	11.89	2340	3.95
	8	1.50	1.85	367.0	12.08	2386	4.01
	9	1.50	1.85	363.3	12.08	2400	4.00
	10	1.50	1.85	348.6	11.90	2368	3.90
	11	1.48	1.83	347.3	12.06	2401	3.97
	12	1.44	1.79	336.4	12.02	2391	3.94

Table 15.1. (continued)

Year	Month	$ U.S.A.	$ Canada	Yen	Franc	Lira	Mark
1984	1	1.41	1.76	329.1	12.10	2403	3.96
	2	1.44	1.80	336.4	11.97	2402	3.89
	3	1.46	1.85	327.9	11.65	2350	3.78
	4	1.42	1.82	320.3	11.56	2327	3.76
	5	1.39	1.80	320.3	11.71	2357	3.82
	6	1.38	1.80	321.4	11.59	2333	3.77
	7	1.32	1.75	320.8	11.55	2311	3.76
	8	1.31	1.71	318.3	11.64	2338	3.79
	9	1.26	1.65	308.8	11.69	2352	3.81
	10	1.22	1.61	301.1	11.48	2317	3.74
	11	1.24	1.63	302.2	11.40	2309	3.71
	12	1.19	1.57	294.3	11.29	2270	3.69
1985	1	1.13	1.49	286.8	10.95	2199	3.58
	2	1.09	1.48	284.7	11.02	2229	3.61
	3	1.12	1.55	289.8	11.31	2336	3.70
	4	1.24	1.70	312.3	11.69	2446	3.83
	5	1.25	1.72	314.6	11.84	2476	3.88
	6	1.28	1.75	318.7	11.96	2502	3.92
	7	1.38	1.86	332.6	12.21	2620	4.01
	8	1.38	1.88	328.4	11.81	2591	3.87
	9	1.37	1.87	322.8	11.81	2597	3.87
	10	1.42	1.94	305.2	11.46	2537	3.76
	11	1.44	1.98	293.6	11.38	2523	3.73
	12	1.45	2.02	293.2	11.11	2478	3.63
1986	1	1.42	2.00	284.7	10.66	2368	3.47
	2	1.43	2.01	263.8	10.23	2269	3.34
	3	1.47	2.06	262.1	10.23	2263	3.32
	4	1.50	2.08	262.2	10.79	2332	3.40
	5	1.52	2.09	253.8	10.79	2322	3.39
	6	1.51	2.10	252.8	10.74	2312	3.37
	7	1.51	2.08	239.4	10.45	2229	3.25
	8	1.49	2.06	229.2	9.99	2111	3.07
	9	1.47	2.04	227.6	9.83	2075	3.00
	10	1.43	1.98	223.1	9.36	1980	2.86
	11	1.43	1.98	232.0	9.43	1996	2.88
	12	1.44	1.98	233.2	9.39	1984	2.86

changes in the series is to use a model allowing a reasonably large degree of variation in the trend component over all time. The model reported here is structured using a single discount factor $\delta = 0.7$ for the trend component, so that $\mathbf{W}_t = \mathbf{J}_2(1)\mathbf{C}_{t-1}\mathbf{J}_2(1)'(0.7^{-1} - 1)$, for all t. This very low discount factor is consistent with the view that there is to be a fair degree of change over time, particularly in the growth components of the series. The initial prior for the trend and growth parameters at $t = 0$ (January 1975) is very vague, being specified via $\mathbf{m}_0 = \mathbf{0}$, the 2×6 zero matrix, and

$$\mathbf{C}_0 = \begin{pmatrix} 100 & 0 \\ 0 & 4 \end{pmatrix}.$$

Recall that we are working on the logarithmic scale for each of the series.

For $\boldsymbol{\Sigma}$, the initial prior is also rather vague, based on $n_0 = 1$ and $\mathbf{S}_0 = \mathbf{I}$, the 6×6 identity matrix. Consistent with the volatility of the individual trends, it is apparent that the relationships amongst the series can be expected to vary noticeably over time. This is reinforced through consideration of changes in economic trading relationships between the various countries, and between the EEC, Japanese and North American sectors, over the long period of 12 years. Such changes are also evident from the figures, particularly, for example, in the changes from concordance to discordance, and vice versa, between the trends of the Yen and each of the North American currencies. Thus variance discounting is used to allow for adaptation to changes in $\boldsymbol{\Sigma}$ over time, and, again consistent with the marked nature of changes anticipated, a fairly low variance discount factor $\beta = 0.9$ is used. Note from the previous Section that the limiting value of the degrees of freedom parameter n_t in this case is $(1 - \beta)^{-1} = 10$, and that this limit is very rapidly approached (in this case, from below since $n_0 = 1$). At any time t, the estimate $\mathbf{S}_t$ is less heavily dependent on data in the past than in the static model, therefore more closely estimates local, or current, cross-sectional structure.

The estimate $\mathbf{S}_t$ may be analysed using a principal components decomposition as described in the previous Section. This provides inferences about $\boldsymbol{\Sigma}$ based on the data up to that time. On the basis of the model, the one-step ahead forecast variance matrix at t is simply proportional to $\mathbf{S}_t$, so that this is an analysis of the covariance structure of the forecast distribution. Having proceeded to the end of the analysis at $t = 144$, the revised, filtered estimates $S_{144}(-k)$ provide the raw material for such analyses.

Table 15.2 provides information from the principal components analysis of the estimated covariance matrix $S_{144}(-36)$, the filtered estimate of Σ at $t = 108$, corresponding to December 1983 — chosen to provide comparison with Quintana and West (1986) who consider time $t = 116$, albeit using a different model. Of the six eigenvalues of the estimate $S_{144}(-36)$, the first three account for over 97% of the total estimated variation in Σ. The proportions of total variation contributed by the first three components are approximately 73%, 18% and 6% respectively. The first principal component clearly dominates, explaining almost three quarters of total variation, although the second is also important, the third rather less so. Table 15.2 gives the corresponding three eigenvectors, defining the weightings of the elements of principal components $X_j = \lambda'_j Y_{108}$ (nb. for notational convenience, the earlier notation λ_j is used for the eigenvalues, although recall that they are estimates based on $S_{144}(-36)$).

The interpretation of the eigenvectors, and the first three principal component, is relatively straightforward and in line with the economic background. Bearing in mind that the vectors in the tables are uncertain estimates, the first vector implies that the first principle component is a weighted average of the six currencies, providing an average measure of the international value of the pound (as estimated for December 1983). In fact the compound measure defined by this component is very similar to U.K. official measures relative to "baskets" of currencies. Similar conclusions are found using a range of similar models. See Quintana (1987), Chapter 6, especially Figure 6.3, for further details. This average measure, and its development over time, provides an assessment of the buoyancy

Table 15.2. Weights of currencies in first three components (December 1983)

	λ_1	λ_2	λ_3
$ U.S.A.	0.27	−0.64	−0.13
$ Canada	0.27	−0.65	−0.09
Yen	0.50	0.06	0.86
Franc	0.49	0.25	−0.28
Lira	0.43	0.15	−0.23
Deutschmark	0.44	0.29	−0.31
% Variation	73	18	6

of Sterling on the international exchanges, and is obviously (and un-surprisingly) the major factor underlying the joint variation of the six exchange rates. The second principal component explains about 18% of the variation through a contrast of an average of the North American rates with an average of those of the EEC countries. The Yen receives little weight here. The contrast arises through the op-posite signs of the weights between the two sectors, and reflects the major effects on the value of Sterling of trading relationships between the two sectors. Note also the magnitudes of the weights. The U.S. and Canadian dollars are roughly equally weighted, being extremely highly correlated and so essentially indistinguishable. In the con-trasting sector, the Deutschmark and Franc are also roughly equally weighted, the Lira somewhat less so, refecting the dominance in the EEC of the effect of the German and French currencies on Sterling. Finally, the third component contrasts the Yen with the remaining currencies, though giving relatively little weight to the North Amer-ican currencies. This therefore essentially reflects the strength of the Yen relative to an aggregate measure of EEC countries.

Proceeding to the end of the series, $t = 144$ at December 1986, the estimate $S_{144} = S_{144}(0)$ may be similarly analysed. The conclusions are broadly similar to those above, although the differences are of interest. Table 15.3 provides details of the principal components analysis of S_{144} similar to those in Table 15.2.

The amount of variation explained by each of the components is similar to that at $t = 108$. The first component dominates at over 70%, the second is significant at 20%, the third less so at 9% and the other three negligible. The differences between this and the previous analysis lie in a more appropriate reflection the covariance structure in late 1986, the estimate of Σ more appropriately reflecting local

Table 15.3. Weights of currencies in first three components (December 1986)

	λ_1	λ_2	λ_3
$ U.S.A.	0.37	−0.51	0.29
$ Canada	0.33	−0.52	0.34
Yen	0.42	0.67	0.59
Franc	0.42	0.04	−0.53
Lira	0.47	0.06	−0.20
Deutschmark	0.42	0.10	−0.35
% Variation	71	20	9

conditions due to the discounting through $\beta = 0.9$ of past data. The first component is a similar basket of currencies, though slightly more weight is given to the North American currencies, and very slightly less to the Yen, in determining this weighted average. The second component is now seen to contribute rather more to total variation, accounting for over 20% in late 1986. This contrasts the North American currencies with the Yen and the EEC, although the EEC currencies are only very marginally weighted. The third component then explains a further 9% of total variation through a contrast between the EEC and non-EEC currencies. The dominance of the effect of the Yen is evident in a higher weighting relative to the North American currencies.

15.4.7 Application to compositional data analysis

A natural setting for the application of this class of models is in the analysis of time series of compositions, similar to the application in hierarchical forecasting mentioned in Section 15.2.3. One such application appears in Quintana (1987), and Quintana and West (1986), the series of interest there being U.K. monthly energy consumption figures according to energy generating source (gas, electricity, petroleum and nuclear). Although the models applied there do not directly address the compositional nature of the data, the application typifies the type of problem for which these models are suited. A common, univariate DLM form is appropriate for each of the series, and the focus is on analysis of covariance structure rather than forecasting.

With compositional (and other) series, the analysis of *relative* behaviour of series may often be simplified by converting from the original data scale to the corresponding time series of proportions. The idea is that if general environmental conditions (such as national economic policies, recessions and so forth) can be assumed to affect each of the series through a common multiplicative factor at each time, then converting to proportions removes such effects and leads to a simpler analysis on the transformed scale. Quintana and West (1988) describe the analysis of such series, the models introduced there being developed as follows.

Suppose that $\mathbf{z}_t$, $(t = 1, \dots)$, is a $q-$vector valued time series of *positive* quantities of similar nature, naturally defining q compositions of a total $\mathbf{1}'\mathbf{z}_t$ at time t. The time series of proportions is defined simply via

$$\mathbf{p}_t = (\mathbf{1}'\mathbf{z}_t)^{-1}\mathbf{z}_t = \left(\sum_{i=1}^{q} z_{ti}\right)^{-1} \mathbf{z}_t.$$

Having used the transformation to proportions in an attempt to remove the common affects of unidentified factors influencing the time series, we now face the technical problems of analysing the proportions themselves. A first problem is the restriction on the ranges of the elements of $\mathbf{p}_t$; our normal models of course assume each of the components to be essentially unrestricted and so further transformation may be needed. Aitchison (1986) describes the use of the logistic/log ratio transformation in the analysis of proportions, in a variety of application areas. Here this transformation may be applied to map the vector of proportions $\mathbf{p}_t$ into a vector of real-valued quantities $\mathbf{Y}_t$. A particular, symmetric version of the log ratio transformation is given by taking

$$Y_{tj} = \log(p_{tj}/p_t) = \log(p_{tj}) - \log(\hat{p}_t), \qquad (j = 1,\ldots,q),$$

where

$$\hat{p}_t = \prod_{j=1}^{q} p_{tj}^{1/q}$$

is the geometric mean of the p_{tj}. The inverse of this log ratio transformation is the logistic transformation

$$p_{ti} = \frac{\exp(Y_{ti})}{\sum\limits_{j=1}^{q} \exp(Y_{tj})}, \qquad (i = 1,\ldots,q).$$

Modelling $\mathbf{Y}_t$ with the matrix model of (15.11) implies a conditional multivariate normal structure. Thus the observational distribution of the proportions $\mathbf{p}_t$ is the multivariate logistic-normal distribution as defined in Aitchison and Shen (1980); see also Aitchison (1986). Clearly a sufficient condition for the vector $\mathbf{p}_t$ to have a logistic-normal distribution is that the series $\mathbf{z}_t$ is distributed as a multivariate log-normal, although the converse does not hold.

Suppose then that $\mathbf{Y}_t$ follows the model (15.11). Having proceeded from the proportions to the real-valued elements on the $\mathbf{Y}_t$ scale, we now face a second problem. It follows from the definition of Y_{tj}

that the elements sum to zero, $\mathbf{1}'\mathbf{Y}_t = 0$ for all t. This implies that $\boldsymbol{\Sigma}$ is singular, having rank $q - 1$. This, and other, singularities can be handled as follows. Suppose that we model $\mathbf{Y}_t$ ignoring the constraint, assuming as above that $\boldsymbol{\Sigma}$ is non-singular, its elements being unrestricted. Then the constraint can be imposed directly by transforming the series to $\mathbf{K}\mathbf{Y}_t$ where

$$\mathbf{K} = \mathbf{K}' = \mathbf{I} - q^{-1}\mathbf{1}\mathbf{1}', \qquad \mathbf{1} = (1,\dots,1)'.$$

Then, from equations (15.11), it follows that

$$\begin{aligned}
\mathbf{Y}'_t\mathbf{K} &= \mathbf{F}'_t\boldsymbol{\Psi}_t + (\mathbf{K}\boldsymbol{\nu}_t)', & \mathbf{K}\boldsymbol{\nu}_t &\sim \mathrm{N}[\mathbf{0}, V_t\boldsymbol{\Xi}^{-1}], \\
\boldsymbol{\Psi}_t &= \mathbf{G}_t\boldsymbol{\Psi}_{t-1} + \boldsymbol{\Omega}_t\mathbf{K}, & \boldsymbol{\Omega}_t\mathbf{K} &\sim \mathrm{N}[\mathbf{0}, \mathbf{W}_t, \boldsymbol{\Xi}^{-1}],
\end{aligned}$$

where $\boldsymbol{\Psi}_t = \boldsymbol{\Theta}_t\mathbf{K}$ and $\boldsymbol{\Xi} = \mathbf{K}\boldsymbol{\Sigma}\mathbf{K}$. Similarly, from (15.12) it follows that

$$(\boldsymbol{\Psi}_0, \boldsymbol{\Xi}|D_0) \sim \mathrm{NW}_{n_0}^{-1}[\mathbf{m}_0\mathbf{K}, \mathbf{C}_0, \mathbf{K}\mathbf{S}_0\mathbf{K}].$$

Thus the constrained series follows a matrix model and is the results in Theorem 15.4 apply. For a derivation of this, and related, result see Quintana (1987). Note that the quantities $\mathbf{F}_t$, V_t, $\mathbf{G}_t$, $\mathbf{W}_t$ and $\mathbf{C}_t$ are unaffected by such linear transforms, as is the use of discount factors to define $\mathbf{W}_t$.

With this in mind, initial priors for $\boldsymbol{\Theta}_0$ and $\boldsymbol{\Sigma}$ at time $t = 0$ can be specified in non-singular forms ignoring effects of the constraint, and then transformed as above.[†] Moreover, the unconstrained model is defined for the logarithms of $\mathbf{z}_t$, rather than the log-ratios, so that the unconstrained priors are those of meaningful quantities. Note that the transformation changes the correlation structure in obvious ways since it imposes a constraint. Whatever the correlation structure may be in the unconstrained $\boldsymbol{\Sigma}$, transformation will lead to negative correlations afterwards. This is most easily seen in the case $q = 2$ when the constraint leads to perfect negative correlation in $\boldsymbol{\Xi}$. Hence the joint structure may be interpreted through relative values of the resulting, constrained correlations.

Illustration of this model appears in Quintana and West (1988). The data series there are several years of monthly estimates of the value of imports into the Mexican economy. There are three series,

---

[†]This procedure is similar to that for setting a prior for a form-free seasonal DLM component subject to zero-sum constraints (Section 8.4, Chapter 8)

determining the composition of total exports according to end use, classified as Consumer, Intermediate and Capital. The data, running from January 1980 to December 1983 inclusive, appear in Table 15.4.

The time span of these series includes periods of marked change in world oil prices that seriously affected the Mexican economy, notably during 1981 and 1982. Rather marked changes are apparent in each of the three series. The log-ratio transformed series are plotted in Figure 15.2, where the compositional features are clearly apparent. Note, in particular the relative increase in imports of intermediate use goods at the expense of consumer goods during the latter two years. An important feature of the transformation is that it removes some of the common variation over time evident on the original scale, such as inflation effects and national economic cycles.

The model used for the transformed series is a second-order polynomial form as in the example of the previous Section. Again, as in that application, this is *not* a model to be seriously entertained for *forecasting* the series other than in the very short-term. However, as in many other applications, it serves here as an extremely flexible *local smoothing* model, that can track the changes in trend in the data and provide robust retrospective estimates of the time varying trend and cross-sectional correlation structure. Thus, for our purposes here, this simple, local trend description is adequate,

Table 15.4. Monthly Mexican import data (10^6)
(Source: Bank of México)

		Monthly data: A=Consumer, B=Intermediate, C= Capital										
		J	F	M	A	M	J	J	A	S	O	N	D
1980	A:	98.1	112.1	123.1	150.2	148.0	180.2	244.9	225.0	253.1	321.5	262.4	331.1
	B:	736.3	684.1	793.9	856.2	1020.9	986.9	1111.8	962.9	994.0	1049.2	954.0	1058.6
	C:	301.2	360.7	4290.9	358.7	354.6	419.8	438.4	445.2	470.3	514.8	483.4	606.1
1981	A:	246.5	135.4	183.4	178.2	194.4	189.2	226.6	254.8	298.9	362.2	270.2	273.0
	B:	1046.9	1051.6	1342.2	1208.0	1213.6	1291.1	1229.6	994.4	999.3	1120.4	977.1	1067.2
	C:	575.3	565.4	699.2	598.3	622.6	709.7	689.6	588.8	510.9	773.9	588.3	653.3
1982	A:	157.2	175.4	191.5	167.9	123.5	143.9	156.8	115.0	70.0	90.6	42.6	82.3
	B:	933.4	808.9	1029.2	894.3	811.8	866.1	762.0	631.0	476.6	470.0	314.9	419.5
	C:	514.5	684.7	568.4	436.0	312.6	388.7	367.1	330.1	232.9	270.1	220.8	176.5
1983	A:	52.3	44.5	42.9	39.2	55.7	56.6	41.9	60.5	40.2	33.1	39.5	47.4
	B:	257.0	318.6	457.2	440.3	506.4	529.6	488.1	548.3	502.0	474.9	392.4	431.7
	C:	85.2	83.2	126.7	131.3	144.4	165.5	159.5	137.4	128.2	190.7	268.8	198.1

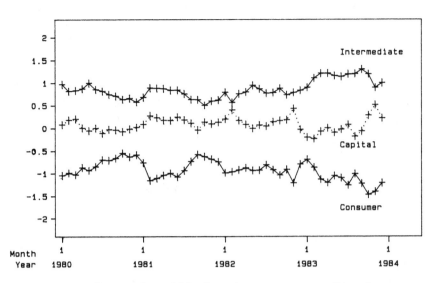

Figure 15.2.   Log-ratios of Mexican imports composition by use.

the basic, *locally linear* form of the model being supported by the graphs in Figure 15.2.

Quintana and West (1988) use a relatively uninformative initial prior that is used here for illustration (the problems involved in specifying an informative prior for Σ are well discussed by Dickey, Dawid and Kadane, 1986). For the unconstrained series,

$$\mathbf{m}_0 = \begin{pmatrix} -1 & 1 & 0 \\ 0 & 0 & 0 \end{pmatrix}, \qquad \mathbf{C}_0 = \begin{pmatrix} 1 & 0 \\ 0 & 1 \end{pmatrix}.$$

The prior for Σ has $\mathbf{S}_0 = 10^{-5}\mathbf{I}$ and $n_0 = 10^{-3}$, a very vague specification. These values are used to initialise the analysis, the corresponding posterior and filtered values for each time t then being transformed as noted above to impose the linear constraint. The evolution variance sequence is based on a single discount factor δ, so that $\mathbf{W}_t = \mathbf{G}\mathbf{C}_{t-1}\mathbf{G}'(\delta^{-1}-1)$, with the standard value of $\delta = 0.9$. In addition, the possibility of time-variation in Σ is allowed for through the use of a variance discount factor $\beta = 0.98$, small though sufficient to permit some degree of change.[†] In addition to this, one further modification is made to the model (and to each of those explored) to

---

[†]Of several models models with values in the *a priori* plausible ranges $0.8 \le \delta \le 1$ and $0.95 \le \beta \le 1$, that illustrated has essentially the highest likelihood. In particular, the static Σ models with $\beta = 1$ are very poor by comparison,

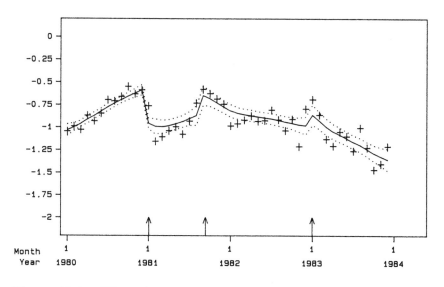

Figure 15.3.   Filtered trend in Consumer series.

cater by intervention for the marked changes in trend of the series that are essentially responses to major events in the international markets, particularly the plunge into recession in early 1981, for example. Interventions are made to allow for changes in model parameters that are greater than those anticipated through the standard discount factors δ and β. The points of intervention are January 1981, September 1981 and January 1983. At these points only, the standard discount factors are replaced by lower values $\delta = 0.1$ and $\beta = 0.9$ to model the possibility of greater *random* variation in both Θ_t and Σ, though not anticipating the direction of such changes.

Figures 15.3, 15.4 and 15.5 display the filtered estimates of the trends (the *fitted* trends) in the three imports series, with approximate 90% posterior intervals — these should not be confused with the associated *predictive* intervals that, though not displayed here, are roughly twice as wide. The times of intervention are indicated by arrows on the time axes, and the response of the model is apparent at points of abrupt change. Though the model could be refined, improvements are likely to be very minor, the fitted trends *smooth* the series rather well. The main interest is in correlation structure across

indicating support for time variation in Σ. The inferences illustrated are, of course, representative of inferences from models with discount factors near the chosen values.

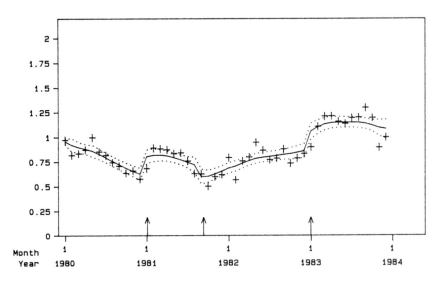

Figure 15.4.   Filtered trend in Intermediate series.

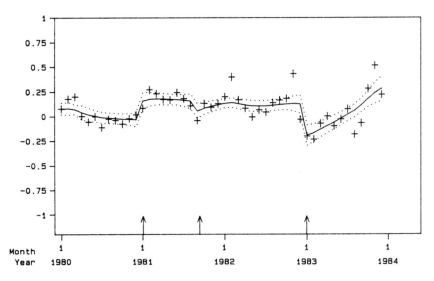

Figure 15.5.   Filtered trend in Capital series.

series, and Figure 15.6 clearly identifies the time variation in Ξ. The graph is of the correlations taken from the smoothed estimates as defined in Section 15.3.5. Note again that, though the series may move together at times suggesting positive values of the correlations in Σ,

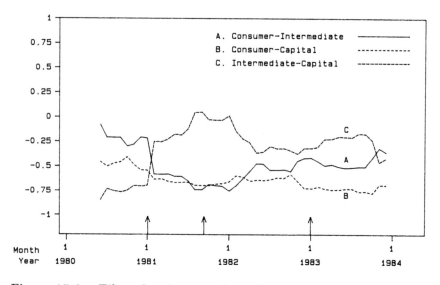

Figure 15.6.   Filtered estimates of correlations.

corresponding elements in $\boldsymbol{\Xi}$ will be negative due to the constraint. The estimates in the Figure bear this out. Slow, random changes are apparent here, accounting for changing economic conditions that affect the three import areas in different ways. A more marked change occurs at intervention prior to the onset of recession in early 1981. Cross-sectional inferences made in early 1981 based on this model would reflect these changes, which are quite marked, whereas those based on a static $\boldsymbol{\Sigma}$ model would be much more heavily based on the pre-1981 estimates of $\boldsymbol{\Sigma}$. Thus, for example, this model would suggest a correlation near -0.25 between the Intermediate and Capital series, whereas the pre-1981 based value is near -0.75.

Finally note that the estimates of $\boldsymbol{\Xi}$ over time may be subjected to principle components analyses just as in the unconstrained model of the previous Section. The fact that $\boldsymbol{\Xi}$ has rank $q - 1$ rather than full rank q simply implies that one of the eigenvalues is zero, the corresponding eigenvector being undefined and of no interest. Variation is explained by the remaining $q - 1$ principal components.

15.5   EXERCISES

(1) Consider the general multivariate DLM of Section 15.2. In Theorem 15.1, assume the prior distribution for $(\boldsymbol{\theta}_t | D_{t-1})$ of part

(b). Using Bayes' Theorem, show directly that $(\boldsymbol{\theta}_t|D_t)$ is normally distributed with moments given by the alternative expressions:

$$\mathbf{m}_t = \mathbf{C}_t[\mathbf{R}_t^{-1}\mathbf{a}_t + \mathbf{F}_t\mathbf{V}_t^{-1}\mathbf{Y}_t]$$

with

$$\mathbf{C}_t^{-1} = \mathbf{R}_t^{-1} + \mathbf{F}_t\mathbf{V}_t^{-1}\mathbf{F}_t'.$$

(2) Use the standard updating equations of Theorem 15.1 and the results of Exercise 1 above to deduce the following identities:

(a) $\quad \mathbf{A}_t = \mathbf{R}_t\mathbf{F}_t\mathbf{Q}_t^{-1} = \mathbf{C}_t\mathbf{F}_t\mathbf{V}_t^{-1};$

(b) $\quad \mathbf{C}_t = \mathbf{R}_t - \mathbf{A}_t\mathbf{Q}_t\mathbf{A}_t' = \mathbf{R}_t(\mathbf{I} - \mathbf{F}_t\mathbf{A}_t');$

(c) $\quad \mathbf{Q}_t = (\mathbf{I} - \mathbf{F}_t'\mathbf{A}_t)^{-1}\mathbf{V}_t;$

(d) $\quad \mathbf{F}_t'\mathbf{A}_t = \mathbf{I} - \mathbf{V}_t\mathbf{Q}_t^{-1}.$

(3) In the multivariate DLM of Section 15.2, consider the vector mean response function $\boldsymbol{\mu}_t = \mathbf{F}_t'\boldsymbol{\theta}_t$ at time t. The defining moments of the posterior for the mean response at time t,

$$(\boldsymbol{\mu}_t|D_t) \sim \mathrm{N}[\mathbf{f}_t(0), \mathbf{Q}_t(0)],$$

are easily seen to be given by

$$\mathrm{E}[\boldsymbol{\mu}_t|D_t] = \mathbf{f}_t(0) = \mathbf{F}_t'\mathbf{m}_t$$

and

$$\mathrm{V}[\boldsymbol{\mu}_t|D_t] = \mathbf{Q}_t(0) = \mathbf{F}_t'\mathbf{C}_t\mathbf{F}_t.$$

Use the updating equations for $\mathbf{m}_t$ and $\mathbf{C}_t$ to derive the following results.

(a) Show that the mean can be simply updated via

$$\mathrm{E}[\boldsymbol{\mu}_t|D_t] = \mathrm{E}[\boldsymbol{\mu}_t|D_{t-1}] + (\mathbf{I} - \mathbf{B}_t)\mathbf{e}_t,$$

or via the similar expression

$$\mathbf{f}_t(0) = (\mathbf{I} - \mathbf{B}_t)\mathbf{Y}_t + \mathbf{B}_t\mathbf{f}_t.$$

where $\mathbf{B}_t = \mathbf{V}_t\mathbf{Q}_t^{-1}$. Interpret the second expression here as a weighted average of two estimates of $\boldsymbol{\mu}_t$.

(b) Show that the variance can be similarly updated using

$$V[\boldsymbol{\mu}_t|D_t] = \mathbf{B}_t V[\boldsymbol{\mu}_t|D_{t-1}],$$

or, equivalently, using

$$\mathbf{Q}_t(0) = (\mathbf{I} - \mathbf{B}_t)\mathbf{V}_t.$$

(4) Prove the general step ahead forecasting results of Theorem 15.2 for the multivariate DLM.
(5) Prove the general, filtering results of Theorem 15.3 for the multivariate DLM.
(6) $\mathbf{Y}$ is an $n-$vector of sales of a large number n of competetive products. The forecast distribution is given by $\mathbf{Y} \sim N[\mathbf{f}, \mathbf{Q}]$ where

$$\mathbf{Q} = Q \begin{pmatrix} 1 & -\epsilon & -\epsilon & \cdots & -\epsilon \\ -\epsilon & 1 & -\epsilon & \cdots & -\epsilon \\ -\epsilon & -\epsilon & 1 & \cdots & -\epsilon \\ \vdots & \vdots & \vdots & \ddots & \vdots \\ -\epsilon & -\epsilon & -\epsilon & \cdots & 1 \end{pmatrix},$$

for some variance $Q > 0$ and scalar $\epsilon > 0$.
   (a) Calculate the implied forecast distribution for total sales $T = \mathbf{1}'\mathbf{Y}$ where $\mathbf{1} = (1, \dots, 1)'$.
   (b) Verify that $V[T] \geq 0$ if, and only if, $\epsilon \leq (n-1)^{-1}$, with $V[T] = 0$ when $\epsilon = (n-1)^{-1}$.
   (c) With n large, comment on the implications for aggregate forecasting.
(7) A series of n product sales $\mathbf{Y}_t$ is forecast using the constant model $\{\mathbf{I}, \mathbf{I}, \mathbf{V}, \mathbf{W}\}$ where all four, known matrices are $n \times n$. At time t, the prior for the state vector $\boldsymbol{\theta}_t$ is $(\boldsymbol{\theta}_t|D_{t-1}) \sim N[\mathbf{a}_t, \mathbf{R}_t]$ as usual. Let T_s be the total sales at any time s, $T_s = \mathbf{1}'\mathbf{Y}_s$.
   (a) Show, directly from normal theory or otherwise, that

$$V[T_{t+1}|\mathbf{Y}_t, D_{t-1}] = V[T_{t+1}|\mathbf{Y}_t, T_t, D_{t-1}] \leq V[T_{t+1}|T_t, D_{t-1}]$$

   with equality if, and only if,

$$C[\mathbf{Y}_t, T_{t+1}|T_t, D_{t-1}] = \mathbf{0}.$$

   (b) Calculate the joint normal distribution of T_{t+1}, T_t and $\mathbf{Y}_t$ given D_{t-1}.

(c) Using (b), prove that the required covariance in (a) is given by

$$\mathbf{r}_t - (\mathbf{r}_t + \mathbf{v})R_t/(R_t + W + V),$$

where $\mathbf{r}_t = \mathbf{R}_t\mathbf{1}$, $R_t = \mathbf{1}'\mathbf{r}_t$, $W = \mathbf{1}'\mathbf{W}\mathbf{1}$, $\mathbf{v} = \mathbf{V}\mathbf{1}$ and $V = \mathbf{1}'\mathbf{v}$. Deduce that the equality in (a) holds if, and only if,

$$\mathbf{r}_t = R_t\mathbf{v}/V.$$

(d) Comment on the relevance of the above results to two forecasters interested in T_{t+1} when $\mathbf{Y}_t$ becomes known. The first proposes to forecast T_{t+1} using $p(T_{t+1}|\mathbf{Y}_t, D_{t-1})$, the second proposes to use $p(T_{t+1}|T_t, D_{t-1})$,

(8) Verify the general results in parts (D) and (E) of Section 15.3.2. In particular, suppose that $\mathbf{Y} \sim N[\mathbf{f}, \mathbf{Q}]$ and $T = \mathbf{1}'\mathbf{Y}$. If the marginal distribution is revised to $(T|H) \sim N[f^*, Q^*]$ and Y is conditionally independent of H when T is known, show that $p(\mathbf{Y}|H)$ is as stated in (D), namely

$$(\mathbf{Y}|H) \sim N[\mathbf{f} + \mathbf{q}(f^* - f)/Q, \mathbf{Q} - \mathbf{q}\mathbf{q}'(Q - Q^*)/Q^2].$$

Extend this proof to verify the more general result in (E) when independent information H is used to revise the marginal forecast distribution of a vector of subaggregates, or more general linear combinations of the elements of $\mathbf{Y}$.

(9) In the framework of Section 15.3.2 (C), part (2), suppose that $a = 0$, $b = 1$ and $c = 1/\phi$ is uncertain, so that

$$(m|M, T, \phi) \sim N[T, M/\phi].$$

Suppose that uncertainty about the scale bias ϕ is represented by a gamma distribution with mean $E[\phi] = 1$ and n degrees of freedom,

$$\phi \sim G[n/2, n/2],$$

and that ϕ is considered to be independent of M and T.

(a) Show that, unconditional on ϕ,

$$(m|M, T) \sim T_n[T, M].$$

(b) Given the prior forecast distribution $T \sim N[f, Q]$, show that the posterior density is now

$$p(T|H) \propto \exp\{-0.5(T - f)^2/Q\}\{n + (T - m)^2/M\}^{-(n+1)/2}.$$

(c) Suppose that $f = 0$, $M = Q = 1$ and $n = 5$. For each value $m = 0, 2, 4$ and 6 plot the (unnormalised) posterior density in (b) as a function of T over the interval $-2 \leq T \leq m + 2$. Discuss the behaviour of the posterior in each cases, making comparison with the normal results in the case of known scale bias $c = 1$ (cf. Lindley, 1988).

(10) Consider the 3−dimensional DLM $\{I, I, \text{diag}(150, 300, 600), \text{diag}(25, 75, 150)\}$ applied to forecasting demand for three products. At time t, information D_t is summarised by

$$\mathbf{m}_t' = (100, 400, 600)$$

and

$$\mathbf{C}_t = \text{diag}(75, 125, 250).$$

(a) Calculate the one-step ahead forecast distribution $p(\mathbf{Y}_{t+1}|D_t)$.

(b) The total sales at time $t + 1$, $T_{t+1} = \mathbf{1}'\mathbf{Y}_{t+1}$, is targeted at $T_{t+1} = 960$. Calculate $p(\mathbf{Y}_{t+1}|T_{t+1}, D_t)$ for any T_{t+1} and deduce the joint forecast distribution for the three products conditional on the target being met.

(c) An independent forecasting method produces a point forecast m of the total. This is modelled as a random quantity such that

$$(m|T_{t+1}, D_t) \sim N[T_{t+1}, 750].$$

Calculate the revised forecast distribution $p(\mathbf{Y}_{t+1}|m, D_t)$ for any value of m, and deduce that when $m = 960$. Verify also that the revised forecast of the total is

$$(T_{t+1}|m, D_t) \sim N[1002, 525].$$

CHAPTER 16

APPENDIX: DISTRIBUTION THEORY
AND LINEAR ALGEBRA

16.1 DISTRIBUTION THEORY

16.1.1 Introduction

This Chapter contains a review and discussion of some of the basic results in distribution theory used throughout the book. Many of these results are stated without proofs and the reader is referred to a standard text on multivariate analysis, such as Mardia, Kent and Bibby (1979) for further details. The exceptions to this general rule are some of the results relating to Bayesian prior to posterior calculations in multivariate normal and joint normal/gamma models. Press (1985) is an excellent reference for such results in addition to standard, non-Bayesian multivariate theory. These results are discussed in detail as they are of some importance and should be clearly understood. For a comprehensive reference to all the distribution theory, see Johnson and Kotz (1972). For distribution theory specifically within the contexts of Bayesian analyses, see Aitchison and Dunsmore (1976), Box and Tiao (1973), and De Groot (1971).

It is assumed that the reader is familiar with standard univariate distribution theory.

Throughout this Chapter we shall be concerned with sets of scalar random variables such as $X_1, X_2, \ldots, (-\infty < X_i < \infty$, for each $i)$, that may be either discrete or continuous. For each i, the density of the distribution of X_i exists and is defined with respect to the natural counting or Lebesgue measure in the discrete and continuous cases respectively. In either case, the distribution function of X_i is denoted by $P_i(x)$, and the density by $p_i(x)$, $(-\infty < x < \infty)$. The density is only non-zero when x belongs to the range of X_i. In the discrete case

$$p_i(x) = \Pr(X_i = x) \qquad \text{and} \qquad P_i(x) = \sum_{y \le x} p_i(y),$$
$$(-\infty < x < \infty),$$

so that $p_i(x)$ is obtained from the differences of the distribution at consecutive values of x. In the continuous case

$$p_i(x) = \frac{dP_i(x)}{dx} \quad \text{and} \quad P_i(x) = \int_{-\infty}^{x} p_i(y)\, dy, \quad (-\infty < x < \infty).$$

From here on, the derivative and integral expressions for the continuous case will be applied throughout with the understanding that differences and summations are implied if X_i is discrete. This simplifies the exposition and unifies notation of the multivariate theory to follow.

16.1.2   Random vectors and distributions

$\mathbf{X} = (X_1, X_2, \dots, X_n)'$ is a random n-vector for any positive integer n. For each $\mathbf{x} = (x_1, x_2, \dots, x_n)'$, the distribution function of $\mathbf{x}$ is given by

$$P(\mathbf{x}) = \Pr(X_1 \le x_1 \cap \dots \cap X_n \le x_n).$$

$P(.)$ is also referred to as the joint distribution function of the random variables $X_1, \dots, X_n$ and is defined for all combinations of discrete and continuous variables. As in the univariate case we have:

(1) $P(\mathbf{x}) \ge 0$ and is non-decreasing in each x_i;
(2) as $x_i \to -\infty$ for all i, $P(\mathbf{x}) \to 0$;
(3) as $x_i \to \infty$ for all i, $P(\mathbf{x}) \to 1$.

MARGINAL DISTRIBUTIONS. If $\mathbf{X}'$ is partitioned into

$$\mathbf{X}_1' = (X_1, \dots, X_m) \quad \text{and} \quad \mathbf{X}_2' = (X_{m+1}, \dots, X_n),$$

the marginal distribution of $\mathbf{X}_1$ is simply the joint distribution of its elements $X_1, \dots, X_m$. Similarly, the marginal distribution of any subset of m of the X_i in $\mathbf{X}$ is defined by permuting these elements into the first m-subvector $\mathbf{X}_1$. Denote this marginal distribution by $P_1(\mathbf{x}_1)$ for any $\mathbf{x}_1$. Then

(1) $P_1(\mathbf{x}_1) = \lim P(\mathbf{x})$, where the limit is that obtained as $x_i \to \infty$ for each $i = m+1, \dots, n$;
(2) for $m = 1$, $P_1(x_1)$ is the univariate distribution of the scalar random variable X_1.

INDEPENDENCE. Given $P(\mathbf{x})$ and the marginal distributions $P_1(\mathbf{x}_1)$ and $P_2(\mathbf{x}_2)$, the vectors $\mathbf{X}_1$ and $\mathbf{X}_2$ are independent if and only if

$$P(\mathbf{x}) = P_1(\mathbf{x}_1)P_2(\mathbf{x}_2)$$

for all $\mathbf{x}$. More generally, for a partition of $\mathbf{X}'$ into $\mathbf{X}'_1, \ldots, \mathbf{X}'_k$, mutual independence of the subvectors is defined by the factorization

$$P(\mathbf{x}) = \prod_{i=1}^{k} P_i(\mathbf{x}_i)$$

where $P_i(\mathbf{x}_i)$ is the distribution function of $\mathbf{X}_i$.

DENSITY FUNCTIONS. Joint density functions are now defined and all further theory is based on density rather than distribution functions. $\mathbf{X}$ is a continuous random vector if and only if the joint distribution function has the form

$$P(\mathbf{x}) = \int_{-\infty}^{x_1} \cdots \int_{-\infty}^{x_n} p(\mathbf{y}) \, d\mathbf{y}$$

for some real and positive valued function $p(\mathbf{y})$. $p(\mathbf{x})$ is called the joint density of the X_i, $i = 1, \ldots, n$. The notation will be taken as including the cases that some, or all, of the X_i are discrete when the integrals are replaced by summations accordingly. In general, the density is given by

$$p(\mathbf{x}) = \frac{\partial^n P(\mathbf{x})}{\partial x_1 \ldots \partial x_n},$$

with derivatives being replaced by differences for discrete X_i. Note also that $p(\mathbf{x})$ is normalized:

$$\int_{-\infty}^{\infty} \cdots \int_{-\infty}^{\infty} p(\mathbf{x}) \, d\mathbf{x} = 1.$$

MARGINAL DENSITIES. Given the above partition of $\mathbf{X}'$ into $\mathbf{X}'_1$ and $\mathbf{X}'_2$, the marginal density of the marginal distribution of $\mathbf{X}_1$ is given by

$$p_1(\mathbf{x}_1) = \int_{-\infty}^{\infty} \cdots \int_{-\infty}^{\infty} p(\mathbf{x}) \, d\mathbf{x}_2,$$

the integration being taken over all $(n - m)$ elements of $\mathbf{X}_2 = \mathbf{x}_2$.

INDEPENDENCE. If $X_1, \ldots X_k$ are (mutually) independent, then

$$p(x) = \prod_{i=1}^{k} p_i(x_i).$$

CONDITIONAL DENSITIES. Conditional distributions are defined via conditional densities. The density of the m-vector X_1 conditional on a fixed value of the $(n - m)$-vector X_2 is defined, for all x_1 and x_2, by

$$p_1(x_1 \mid x_2) \propto p(x).$$

Since densities are normalized, $p_1(x_1 \mid x_2) = k(x_2) p(x)$ where the normalization constant is simply $k(x_2) = p_2(x_2)^{-1}$, assumed to exist. If $p_2(x_2) = 0$ then $p_1(x_1 \mid x_2)$ is undefined. Thus the conditional density of X_1 given X_2 is the ratio of the joint density of X_1 annd X_2 to the marginal density of X_2. Clearly then:

(1) $p(x) = p_1(x_1 \mid x_2) p_2(x_2) = p_1(x_1) p_2(x_2 \mid x_1)$;
(2) $p_1(x_1 \mid x_2) = p_1(x_1) p_2(x_2 \mid x_1) / p(x_2)$. This is the multivariate form of Bayes' Theorem;
(3) $p_1(x_1 \mid x_2) \equiv p_1(x_1)$ and $p_2(x_2 \mid x_1) \equiv p_2(x_2)$ if and only if X_1 and X_2 are independent.

16.1.3 Expectations

The expectation, or mean value, of a scalar funtion $g(X)$ is defined as

$$E[g(X)] = \int_{-\infty}^{\infty} \cdots \int_{-\infty}^{\infty} g(x) p(x) \, dx,$$

when the integral exists. A vector function $g(X)$ with i^{th} element $g_i(X)$ has expectation, or mean vector, $E[g(X)]$ whose i^{th} element is $E[g_i(X)]$. Similarly, the random matrix $G(X)$ with ij^{th} element $G_{ij}(X)$ has expectation given by the matrix with ij^{th} element $E[G_{ij}(X)]$.

MEAN VECTORS. The mean vector (or simply mean) of X is defined by $E[X] = (E[X_1], \ldots, E[X_n])'$

VARIANCE MATRICES. The variance (or variance-covariance) matrix of $\mathbf{X}$ is the $(n \times n)$ symmetric, non-negative definite matrix

$$V[\mathbf{X}] = E[(\mathbf{X} - E[\mathbf{X}])(\mathbf{X} - E[\mathbf{X}])'],$$

whose ij^{th} element is $E[(X_i - E[X_i])(X_j - E[X_j])]$. Thus the ith diagonal element is the variance of X_i and the ij^{th} off-diagonal element is the covariance between X_i and X_j.

Given the above partition of $\mathbf{X}'$ into the m-vector $\mathbf{X}_1'$ and the $(m-n)$-vector $\mathbf{X}_2'$, the corresponding partitions of the mean and variance matrix are

$$E[\mathbf{X}] = \begin{pmatrix} E[\mathbf{X}_1] \\ E[\mathbf{X}_2] \end{pmatrix} \quad \text{and} \quad V[\mathbf{X}] = \begin{pmatrix} V[\mathbf{X}_1] & C[\mathbf{X}_1, \mathbf{X}_2] \\ C[\mathbf{X}_2, \mathbf{X}_1] & V[\mathbf{X}_2] \end{pmatrix}$$

where $C[\mathbf{X}_1, \mathbf{X}_2] = E[(\mathbf{X}_1 - E[\mathbf{X}_1])(\mathbf{X}_2 - E[\mathbf{X}_2])'] = C[\mathbf{X}_2, \mathbf{X}_1]'$ is the $m \times (n - m)$ covariance matrix between $\mathbf{X}_1$ and $\mathbf{X}_2$. More generally, if $\mathbf{X}' = (\mathbf{X}_1', \ldots, \mathbf{X}_k')$, then $E[\mathbf{X}]' = (E[\mathbf{X}_1]', \ldots, E[\mathbf{X}_k]')$ and $V[\mathbf{X}]$ has diagonal blocks $V[\mathbf{X}_i]$, off-diagonal blocks $C[\mathbf{X}_i, \mathbf{X}_j] = C[\mathbf{X}_j, \mathbf{X}_i]'$, $(i = 1, \ldots, k; \ j = 1, \ldots, k)$. If $C[\mathbf{X}_i, \mathbf{X}_j] = \mathbf{0}$, then $\mathbf{X}_i$ and $\mathbf{X}_j$ are uncorrelated.

CONDITIONAL MEAN AND VARIANCE MATRIX. In general, the expectation of any function $g(\mathbf{X})$ may be calculated iteratively via

$$E[g(\mathbf{X})] = E[E[g(\mathbf{X}) \mid \mathbf{X}_2]].$$

The inner expectation is with respect to the conditional distribution of $(\mathbf{X}_1 \mid \mathbf{X}_2)$; the outer with respect to the marginal for $\mathbf{X}_2$. In particular this implies that:

(1) $E[\mathbf{X}_1] = E[\mathbf{m}_1(\mathbf{X}_2)]$ where $\mathbf{m}_1(\mathbf{X}_2) = E[\mathbf{X}_1 \mid \mathbf{X}_2]$;

(2) $V[\mathbf{X}_1] = V[\mathbf{m}_1(\mathbf{X}_2)] + E[V_1[\mathbf{X}_2]]$ where $V_1[\mathbf{X}_2] = V[\mathbf{X}_1 \mid \mathbf{X}_2]$.

INDEPENDENCE. If $\mathbf{X}_1$ and $\mathbf{X}_2$ are independent then they are uncorrelated so that

$$V[\mathbf{X}] = \begin{pmatrix} V[\mathbf{X}_1] & \mathbf{0} \\ \mathbf{0} & V[\mathbf{X}_2] \end{pmatrix},$$

a block diagonal variance matrix.

16.1.4   Linear functions of random vectors

LINEAR TRANSFORMATIONS. For constant matrices $\mathbf{A}$ and vectors $\mathbf{b}$ of suitable dimensions, if $\mathbf{Y} = \mathbf{A}\mathbf{X} + \mathbf{b}$ then:

$$E[\mathbf{Y}] = \mathbf{A}E[\mathbf{X}] + \mathbf{b} \qquad \text{and} \qquad V[\mathbf{Y}] = \mathbf{A}V[\mathbf{X}]\mathbf{A}'.$$

LINEAR FORMS. More generally, for suitable constant matrices $\mathbf{A}_1, \ldots, \mathbf{A}_k$ and vectors $\mathbf{b}$, if $\mathbf{Y} = \sum_{i=1}^{k} \mathbf{A}_i \mathbf{X}_i + \mathbf{b}$ then:

$$E[\mathbf{Y}] = \sum_{i=1}^{k} \mathbf{A}_i E[\mathbf{X}_i] + \mathbf{b}$$

and

$$V[\mathbf{Y}] = \sum_{i=1}^{k} \mathbf{A}_i V[\mathbf{X}_i] \mathbf{A}_i' + \sum_{i=1}^{k} \sum_{j \neq i}^{k} \mathbf{A}_i C[\mathbf{X}_i, \mathbf{X}_j] \mathbf{A}_j'.$$

In particular, if $\mathbf{Y} = \sum_{i=1}^{k} \mathbf{X}_i$ then

$$E[\mathbf{Y}] = \sum_{i=1}^{k} E[\mathbf{X}_i]$$

and

$$V[\mathbf{Y}] = \sum_{i=1}^{k} V[\mathbf{X}_i] + 2 \sum_{i=1}^{k} \sum_{j=1}^{i-1} C[\mathbf{X}_i, \mathbf{X}_j].$$

Finally, if $\mathbf{Y} = \sum_{i=1}^{k} \mathbf{A}_i \mathbf{X}_i + \mathbf{b}$ and $\mathbf{Z} = \sum_{j=1}^{h} \mathbf{B}_j \mathbf{X}_j + \mathbf{c}$, then:

$$C[\mathbf{Y}, \mathbf{Z}] = \sum_{i=1}^{k} \sum_{j=1}^{h} \mathbf{A}_i C[\mathbf{X}_i, \mathbf{X}_j] \mathbf{B}_j'.$$

16.2 THE MULTIVARIATE NORMAL DISTRIBUTION

16.2.1 The univariate normal

A real random variable X has a normal (or Gaussian) distribution with mean (mode and median) m, and variance V if and only if

$$p(X) = (2\pi V)^{-1/2} \exp\left[-(X-m)^2/2V\right], \qquad (-\infty < X < \infty).$$

In this case we write $X \sim N[m, V]$.

SUMS OF NORMAL RANDOM VARIABLES. If $X_i \sim N[m_i, V_i]$, $(i = 1, \ldots, n)$ have covariances $C[X_i, X_j] = c_{ij} = c_{ji}$, then $Y = \sum_{i=1}^{n} a_i X_i + b$ has a normal distribution with

$$E[Y] = \sum_{i=1}^{n} a_i m_i + b \qquad \text{and} \qquad V[Y] = \sum_{i=1}^{n} a_i^2 V_i + 2 \sum_{i=1}^{n} \sum_{j=1}^{i-1} a_i a_j c_{ij}$$

where a_i and b are constants. In particular, $c_{ij} = 0$ for all $i \neq j$ if, and only if, the X_i are independent normal in which case $V[Y] = \sum_{i=1}^{n} a_i^2 V_i$.

16.2.2 The multivariate normal

A random n-vector $\mathbf{X}$ has a multivariate normal distribution in n dimensions if and only if $Y = \sum_{i=1}^{n} a_i X_i$ is normal for all constant, non-zero vectors $\mathbf{a} = (a_1, \ldots, a_n)'$.

If $\mathbf{X}$ is multivariate normal, then $E[\mathbf{X}] = \mathbf{m}$ and $V[\mathbf{X}] = \mathbf{V}$ exist and we use the notation $\mathbf{X} \sim N[\mathbf{m}, \mathbf{V}]$. $\mathbf{m}$ and $\mathbf{V}$ completely define the distribution whose density is

$$p(\mathbf{X}) = \left[(2\pi)^n |\mathbf{V}|\right]^{-1/2} \exp\left[-(\mathbf{X}-\mathbf{m})'\mathbf{V}^{-1}(\mathbf{X}-\mathbf{m})/2\right].$$

The subvectors of $\mathbf{X}$ are independent if and only if they are uncorrelated. In particular, if $\mathbf{V}$ is block diagonal then the corresponding subvectors of $\mathbf{X}$ are mutually independent.

LINEAR TRANSFORMATIONS. For any constant $\mathbf{A}$ and $\mathbf{b}$ of suitable dimensions, if $\mathbf{Y} = \mathbf{AX} + \mathbf{b}$ then $\mathbf{Y} \sim N[\mathbf{Am}, \mathbf{AVA'}]$. If $\mathbf{AVA'}$ is diagonal, then the elements of $\mathbf{Y}$ are independent normal.

LINEAR FORMS. If $\mathbf{X}_i \sim N[\mathbf{m}_i, \mathbf{V}_i]$ independently, $i = 1, \ldots, k$, then, for constant $\mathbf{A}_1, \ldots, \mathbf{A}_k$ and $\mathbf{b}$ of suitable dimensions, $\mathbf{Y} = \sum_{i=1}^{k} \mathbf{A}_i \mathbf{X}_i + \mathbf{b}$ is multivariate normal with mean $\sum_{i=1}^{k} \mathbf{A}_i \mathbf{m}_i + \mathbf{b}$ and variance matrix $\sum_{i=1}^{k} \mathbf{A}_i \mathbf{V}_i \mathbf{A}_i'$.

MARGINAL DISTRIBUTIONS. Suppose that we have conformable partitions

$$\mathbf{X} = \begin{pmatrix} \mathbf{X}_1 \\ \mathbf{X}_2 \end{pmatrix}, \quad \mathbf{m} = \begin{pmatrix} \mathbf{m}_1 \\ \mathbf{m}_2 \end{pmatrix}, \quad \text{and} \quad \mathbf{V} = \begin{pmatrix} \mathbf{V}_{11} & \mathbf{V}_{12} \\ \mathbf{V}_{21} & \mathbf{V}_{22} \end{pmatrix}.$$

Then $\mathbf{X}_i \sim N[\mathbf{m}_i, \mathbf{V}_{ii}]$, $i = 1, 2$. In particular if X_i is univariate normal, $X_i \sim N[m_i, V_{ii}]$ for $i = 1, \ldots, n$.

BIVARIATE NORMAL. Any two elements X_i and X_j of $\mathbf{X}$ are bivariate normal with joint density

$$p_{ij}(x_i, x_j) = \left[(2\pi) \sqrt{V_{ii} V_{jj} (1 - \rho_{ij}^2)} \right]^{-1} \exp[-Q(x_1, x_2)/2]$$

where $Q(x_1, x_2)$ is the quadratic form

$$Q(x_1, x_2) = \frac{(x_i - m_i)^2}{V_{ii}} + \frac{(x_j - m_j)^2}{V_{jj}} - 2\rho_{ij} \frac{(x_i - m_i)}{\sqrt{V_{ii}}} \frac{(x_j - m_j)}{\sqrt{V_{jj}}}.$$

where $\rho_{ij} = \text{corr}(X_i, X_j) = V_{ij}/\sqrt{V_{ii} V_{jj}}$.

CONDITIONAL DISTRIBUTIONS. For the partition of $\mathbf{X}'$ into $\mathbf{X}_1'$ and $\mathbf{X}_2'$, we have:

(1) $(\mathbf{X}_1 \mid \mathbf{X}_2) \sim N[\mathbf{m}_1(\mathbf{X}_2), \mathbf{V}_1(\mathbf{X}_2)]$ where

$$\mathbf{m}_1(\mathbf{X}_2) = \mathbf{m}_1 + \mathbf{V}_{12} \mathbf{V}_{22}^{-1} (\mathbf{X}_2 - \mathbf{m}_2)$$

and

$$\mathbf{V}_1(\mathbf{X}_2) = \mathbf{V}_{11} - \mathbf{V}_{12} \mathbf{V}_{22}^{-1} \mathbf{V}_{21}.$$

The matrix $\mathbf{A}_1 = \mathbf{V}_{12} \mathbf{V}_{22}^{-1}$ is called the *regression matrix* of $\mathbf{X}_1$ on $\mathbf{X}_2$. The conditional moments are given, in terms of the regression matrix, by

$$\mathbf{m}_1(\mathbf{X}_2) = \mathbf{m}_1 + \mathbf{A}_1(\mathbf{X}_2 - \mathbf{m}_2) \quad \text{and} \quad \mathbf{V}_1(\mathbf{X}_2) = \mathbf{V}_{11} - \mathbf{A}_1 \mathbf{V}_{22} \mathbf{A}_1'.$$

(2) $(\mathbf{X}_2 \mid \mathbf{X}_1) \sim N[\mathbf{m}_2(\mathbf{X}_1), \mathbf{V}_2(\mathbf{X}_1)]$ where

$$\mathbf{m}_2(\mathbf{X}_1) = \mathbf{m}_2 + \mathbf{V}_{21}\mathbf{V}_{11}^{-1}(\mathbf{X}_1 - \mathbf{m}_1)$$

and

$$\mathbf{V}_2(\mathbf{X}_1) = \mathbf{V}_{22} - \mathbf{V}_{21}\mathbf{V}_{11}^{-1}\mathbf{V}_{12}.$$

The matrix $\mathbf{A}_2 = \mathbf{V}_{21}\mathbf{V}_{11}^{-1}$ is called the *regression matrix* of $\mathbf{X}_2$ on $\mathbf{X}_1$. The conditional moments are given, in terms of the regression matrix, by

$$\mathbf{m}_2(\mathbf{X}_1) = \mathbf{m}_2 + \mathbf{A}_2(\mathbf{X}_1 - \mathbf{m}_1) \quad \text{and} \quad \mathbf{V}_2(\mathbf{X}_1) = \mathbf{V}_{22} - \mathbf{A}_2\mathbf{V}_{11}\mathbf{A}_2'.$$

(3) In the special case of the bivariate normal in (1) above, the moments are all scalars, and the correlation between X_1 and X_2 is $\rho = \rho_{12} = V_{12}/\sqrt{V_{11}V_{22}}$. The regressions are determined by regression coefficients A_1 and A_2 given by

$$A_1 = \rho\sqrt{V_{11}/V_{22}} \quad \text{and} \quad A_2 = \rho\sqrt{V_{22}/V_{11}}.$$

Also

$$V[X_1 \mid X_2] = (1 - \rho^2)V_{11} \quad \text{and} \quad V[X_2 \mid X_1] = (1 - \rho^2)V_{22}.$$

16.2.3 Conditional normals and linear regression

Many of the important results in this book may be derived directly from the multivariate normal theory reviewed above. A particular regression model is reviewed here to provide the setting for those results.

Suppose that the p-vector $\mathbf{Y}$ and the n-vector $\boldsymbol{\theta}$ are related via the conditional distribution

$$(\mathbf{Y} \mid \boldsymbol{\theta}) \sim N[\mathbf{F}'\boldsymbol{\theta}, \mathbf{V}]$$

where the $(n \times p)$ matrix $\mathbf{F}$ and the $(p \times p)$ positive definite symmetric matrix $\mathbf{V}$ are constant. An equivalent statement is

$$\mathbf{Y} = \mathbf{F}'\boldsymbol{\theta} + \boldsymbol{\nu},$$

where $\nu \sim N[\mathbf{0}, \mathbf{V}]$. Suppose further that the marginal distribution of $\boldsymbol{\theta}$ is given by

$$\boldsymbol{\theta} \sim N[\mathbf{a}, \mathbf{R}]$$

where both $\mathbf{a}$ and $\mathbf{R}$ are constant, and that $\boldsymbol{\theta}$ is independent of ν. Equivalently,

$$\boldsymbol{\theta} = \mathbf{a} + \boldsymbol{\omega},$$

where $\boldsymbol{\omega} \sim N[\mathbf{0}, \mathbf{R}]$ independently of ν.

From these distributions it is possible to construct the joint distribution for $\mathbf{Y}$ and $\boldsymbol{\theta}$ and hence both the marginal for $\mathbf{Y}$ and the conditional for $(\boldsymbol{\theta} \mid \mathbf{Y})$.

MULTIVARIATE JOINT NORMAL DISTRIBUTION. Since $\boldsymbol{\theta} = \mathbf{a} + \boldsymbol{\omega}$ and $\mathbf{Y} = \mathbf{F}'\boldsymbol{\theta} + \nu = \mathbf{F}'\mathbf{a} + \mathbf{F}'\boldsymbol{\omega} + \nu$, then the vector $(\mathbf{Y}', \boldsymbol{\theta}')'$ is a linear transformation of $(\nu', \boldsymbol{\omega}')'$. By construction the latter has a multivariate normal distribution so that $\mathbf{Y}$ and $\boldsymbol{\theta}$ are jointly normal. Further:

(1) $E[\boldsymbol{\theta}] = \mathbf{a}$ and $V[\boldsymbol{\theta}] = \mathbf{R}$;

(2) $E[\mathbf{Y}] = E[\mathbf{F}'\boldsymbol{\theta} + \nu] = \mathbf{F}'E[\boldsymbol{\theta}] + E[\nu] = \mathbf{F}'\mathbf{a}$, and
$V[\mathbf{Y}] = V[\mathbf{F}'\boldsymbol{\theta} + \nu] = \mathbf{F}'V[\boldsymbol{\theta}]\mathbf{F} + V[\nu] = \mathbf{F}'\mathbf{R}\mathbf{F} + \mathbf{V}$;

(3) $C[\mathbf{Y}, \boldsymbol{\theta}] = C[\mathbf{F}'\boldsymbol{\theta} + \nu, \boldsymbol{\theta}] = \mathbf{F}'C[\boldsymbol{\theta}, \boldsymbol{\theta}] + C[\nu, \boldsymbol{\theta}] = \mathbf{F}'\mathbf{R}$.

It follows that

$$\begin{pmatrix} \mathbf{Y} \\ \boldsymbol{\theta} \end{pmatrix} \sim N\left[\begin{pmatrix} \mathbf{F}'\mathbf{a} \\ \mathbf{a} \end{pmatrix}, \begin{pmatrix} \mathbf{F}'\mathbf{R}\mathbf{F} + \mathbf{V} & \mathbf{F}'\mathbf{R} \\ \mathbf{R}\mathbf{F} & \mathbf{R} \end{pmatrix} \right].$$

Therefore, identifying $\mathbf{Y}$ with $\mathbf{X}_1$ and $\boldsymbol{\theta}$ with $\mathbf{X}_2$ in the partition of $\mathbf{X}$ in 16.2.2, we have

(4) $\mathbf{Y} \sim N[\mathbf{F}'\mathbf{a}, \mathbf{F}'\mathbf{R}\mathbf{F} + \mathbf{V}]$;

(5) $(\boldsymbol{\theta} \mid \mathbf{Y}) \sim N[\mathbf{m}, \mathbf{C}]$,
    where

$$\mathbf{m} = \mathbf{a} + \mathbf{R}\mathbf{F}[\mathbf{F}'\mathbf{R}\mathbf{F} + \mathbf{V}]^{-1}[\mathbf{Y} - \mathbf{F}'\mathbf{a}],$$

and

$$\mathbf{C} = \mathbf{R} - \mathbf{R}\mathbf{F}[\mathbf{F}'\mathbf{R}\mathbf{F} + \mathbf{V}]^{-1}\mathbf{F}'\mathbf{R}.$$

By defining $\mathbf{e} = \mathbf{Y} - \mathbf{F}'\mathbf{a}$, $\mathbf{Q} = \mathbf{F}'\mathbf{R}\mathbf{F} + \mathbf{V}$ and $\mathbf{A} = \mathbf{R}\mathbf{F}\mathbf{Q}^{-1}$, these equations become

$$\mathbf{m} = \mathbf{a} + \mathbf{A}\mathbf{e} \qquad \text{and} \qquad \mathbf{C} = \mathbf{R} - \mathbf{A}\mathbf{Q}\mathbf{A}'.$$

MULTIVARIATE BAYES' THEOREM. An alternative derivation of
the conditional distribution for $(\boldsymbol{\theta} \mid \mathbf{Y})$ via Bayes' Theorem provides
alternative expressions for $\mathbf{m}$ and $\mathbf{C}$. Note that:

$$p(\boldsymbol{\theta} \mid \mathbf{Y}) \propto p(\mathbf{Y} \mid \boldsymbol{\theta})p(\boldsymbol{\theta})$$

as a function of $\boldsymbol{\theta}$, so that

$$\ln[p(\boldsymbol{\theta} \mid \mathbf{Y})] = k + \ln[p(\mathbf{Y} \mid \boldsymbol{\theta})] + \ln[p(\boldsymbol{\theta})]$$

where k depends on $\mathbf{Y}$ but not on $\boldsymbol{\theta}$. Therefore

$$\ln[p(\boldsymbol{\theta} \mid \mathbf{Y})]$$
$$= k - \frac{1}{2}\left[(\mathbf{Y} - \mathbf{F}'\boldsymbol{\theta})'\mathbf{V}^{-1}(\mathbf{Y} - \mathbf{F}'\boldsymbol{\theta}) + (\boldsymbol{\theta} - \mathbf{a})'\mathbf{R}^{-1}(\boldsymbol{\theta} - \mathbf{a})\right].$$

The bracketed term here is simply

$$\mathbf{Y}'\mathbf{V}^{-1}\mathbf{Y} - 2\mathbf{Y}'\mathbf{V}^{-1}\mathbf{F}'\boldsymbol{\theta} + \boldsymbol{\theta}'\mathbf{F}\mathbf{V}^{-1}\mathbf{F}'\boldsymbol{\theta} + \boldsymbol{\theta}'\mathbf{R}^{-1}\boldsymbol{\theta} - 2\mathbf{a}'\mathbf{R}^{-1}\boldsymbol{\theta} + \mathbf{a}'\mathbf{R}^{-1}\mathbf{a}$$
$$= \boldsymbol{\theta}'\left\{\mathbf{F}\mathbf{V}^{-1}\mathbf{F}' + \mathbf{R}^{-1}\right\}\boldsymbol{\theta} - 2\left\{\mathbf{Y}'\mathbf{V}^{-1}\mathbf{F}' + \mathbf{a}'\mathbf{R}^{-1}\right\}\boldsymbol{\theta} + h$$

where h depends on $\mathbf{Y}$ but not on $\boldsymbol{\theta}$. Completing the quadratic form
gives
$$(\boldsymbol{\theta} - \mathbf{m})'\mathbf{C}^{-1}(\boldsymbol{\theta} - \mathbf{m}) + h^{*},$$

where, again, h^{*} does not involve $\boldsymbol{\theta}$,

$$\mathbf{C}^{-1} = \mathbf{R}^{-1} + \mathbf{F}\mathbf{V}^{-1}\mathbf{F}'$$

and

$$\mathbf{m} = \mathbf{C}\left\{\mathbf{F}\mathbf{V}^{-1}\mathbf{Y} + \mathbf{R}^{-1}\mathbf{a}\right\}.$$

Hence
$$p(\boldsymbol{\theta} \mid \mathbf{Y}) \propto \exp\left[-\frac{1}{2}(\boldsymbol{\theta} - \mathbf{m})'\mathbf{C}^{-1}(\boldsymbol{\theta} - \mathbf{m})\right]$$

as a function of $\boldsymbol{\theta}$, so that $(\boldsymbol{\theta} \mid \mathbf{Y}) \sim N[\mathbf{m}, \mathbf{C}]$, just as derived earlier.

Note that the two derivations give different expressions for **m** and **C** which provide, in particular, the matrix identity for **C**:

$$\mathbf{C} = \left[\mathbf{R}^{-1} + \mathbf{F}\mathbf{V}^{-1}\mathbf{F}'\right]^{-1} = \mathbf{R} - \mathbf{R}\mathbf{F}\left[\mathbf{F}'\mathbf{R}\mathbf{F} + \mathbf{V}\right]^{-1}\mathbf{F}'\mathbf{R}$$

which is easily verified once stated.

16.3   JOINT NORMAL/GAMMA DISTRIBUTIONS

16.3.1   The gamma distribution

A random variable $\phi > 0$ has a gamma distribution with parameters $n > 0$ and $d > 0$, denoted by $\phi \sim G[n, d]$, if and only if

$$p(\phi) \propto \phi^{n-1} \exp(-\phi d), \qquad\qquad (\phi > 0).$$

Normalization leads to $p(\phi) = d^n \Gamma(n)^{-1} \phi^{n-1} \exp(-\phi d)$ where $\Gamma(n)$ is the gamma function. Note that $E[\phi] = n/d$ and $V[\phi] = E[\phi]^2/n$.

Two special cases of interest are:

(1) $n = 1$, when ϕ has a (negative) exponential distribution with mean $1/d$;

(2) $\phi \sim G[n/2, d/2]$ when n is a positive integer. In this case $d\phi \sim \chi_n^2$, a chi-squared distribution with n degrees of freedom.

16.3.2   Univariate normal/gamma distribution

Let $\phi \sim G[n/2, d/2]$ for any $n > 0$ and $d > 0$, and suppose that the conditional distribution of a further random variable X given ϕ is normal $(X \mid \phi) \sim N[m, C\phi^{-1}]$, for some m and C. Note that $E[X] \equiv E[X \mid \phi] = m$ does not depend on ϕ. However $V[X \mid \phi] = C\phi^{-1}$. The joint distribution of X and ϕ is called (univariate) normal/gamma. Note that:

(1)

$$p(X, \phi) = \left(\frac{\phi}{2\pi C}\right)^{1/2} \exp\left[-\frac{\phi(X-m)^2}{2C}\right] \frac{d^{n/2}}{2^{n/2}\Gamma(\frac{n}{2})} \phi^{\frac{n}{2}-1} \exp\left[-\frac{\phi d}{2}\right],$$

$$\propto \phi^{(\frac{n+1}{2})-1} \exp\left[-\frac{\phi}{2}\left\{\frac{(X-m)^2}{C} + d\right\}\right],$$

as a function of ϕ and X.

(2)

$$p(\phi \mid X) \propto \phi^{(\frac{n+1}{2})-1} \exp\left[-\frac{\phi}{2}\left\{\frac{(X-m)^2}{C} + d\right\}\right]$$

so that

$$(\phi \mid X) \sim G\left[\frac{n^*}{2}, \frac{d^*}{2}\right].$$

where $n^* = n + 1$ and $d^* = d + (X - m)^2/C$.

(3)

$$p(X) = p(X, \phi)/p(\phi \mid X)$$

$$(-\infty < X < \infty)$$

$$\propto [n + (X-m)^2/R]^{-(n+1)/2},$$

where $R = C(d/n) = C/E[\phi]$. This is proportional to the density of Student T distribution with n degrees of freedom, mode m and scale R. Hence $X \sim T_n[m, R]$, or $(X - m)/R^{1/2} \sim T_n[0, 1]$, a standard Student T distribution with n degrees of freedom.

16.3.3   Multivariate normal/gamma distribution

As an important generalization, suppose that $\phi \sim G[n/2, d/2]$, and that the p-vector X is normally distributed conditional on ϕ, as $X \sim N[m, C\phi^{-1}]$. Here the p-vector m and the $(p \times p)$ symmetric positive definite matrix C are known. Thus each element of $V[X]$ is scaled by the common factor ϕ. The basic results are similar to 16.3.2, in that:

(1) $(\phi \mid X) \sim G[n^*/2, d^*/2]$, with parameters $n^* = n + p$ and $d^* = d + (X - m)'C^{-1}(X - m)$, (notice the degrees of freedom increases by p).

(2) X has a (marginal) multivariate T distribution in p dimensions with n degrees of freedom, mode m and scale matrix $R = C(d/n) = C/E[\phi]$, denoted by $X \sim T_n[m, R]$, with density

$$p(X) \propto [n + (X - m)'R^{-1}(X - m)]^{-(n+p)/2}.$$

In particular, if X_i is the ith element of X, m_i and C_{ii} the corresponding mean and diagonal element of C, then

$$X_i \sim T_n[m_i, R_{ii}]$$

where $R_{ii} = C_{ii}(d/n)$.

16.3.4 Simple regression model

The normal/gamma distribution plays a key role in providing closed form Bayesian analyses of linear models with unknown scale parameters. Details may be found in De Groot (1971) and Press (1985), for example. A particular regression setting is reviewed here for reference. The details follow from the above joint normal/gamma theory. Suppose that a scalar variable Y is related to the p-vector $\boldsymbol{\theta}$ and the scalar ϕ via

$$(Y \mid \boldsymbol{\theta}, \phi) \sim \mathrm{N}[\mathbf{F}'\boldsymbol{\theta}, k\phi^{-1}]$$

where the p-vector $\mathbf{F}$ and the variance multiple k are fixed constants. Suppose also that $(\boldsymbol{\theta}, \phi)$ have a joint normal/gamma distribution:

$$(\boldsymbol{\theta} \mid \phi) \sim \mathrm{N}[\mathbf{a}, \mathbf{R}\phi^{-1}]$$

and

$$\phi \sim \mathrm{G}[n/2, d/2]$$

for fixed scalars $n > 0$, $d > 0$, p-vector $\mathbf{a}$ and $(p \times p)$ variance matrix $\mathbf{R}$, and let $S = d/n = 1/\mathrm{E}[\phi]$. Then:

(1) $(Y \mid \phi) \sim \mathrm{N}[f, Q\phi^{-1}]$ where $f = \mathbf{F}'\mathbf{a}$ and $Q = \mathbf{F}'\mathbf{R}\mathbf{F} + k$;

(2) $Y \sim \mathrm{T}_n[f, QS]$, unconditional on $\boldsymbol{\theta}$ or ϕ;

(3) $(\boldsymbol{\theta}, \phi \mid Y)$ have a joint normal/gamma distribution. Specifically,

$$(\boldsymbol{\theta} \mid \phi, Y) \sim \mathrm{N}[\mathbf{m}, \mathbf{C}\phi^{-1}]$$

where, from 16.2.3, $\mathbf{m} = \mathbf{a} + \mathbf{R}\mathbf{F}(Y - f)/Q$ and $\mathbf{C} = \mathbf{R} - \mathbf{R}\mathbf{F}\mathbf{F}'\mathbf{R}/Q$, and

$$(\phi \mid Y) \sim \mathrm{G}[(n+1)/2, \{d + (Y - f)^2/Q\}/2];$$

(4) $(\boldsymbol{\theta} \mid Y)$ has a multivariate T distribution in p dimensions with $n + 1$ degrees of freedom, mode $\mathbf{m}$ and scale matrix $\mathbf{C}S$, denoted by $(\boldsymbol{\theta} \mid Y) \sim \mathrm{T}_{n+1}[\mathbf{m}, \mathbf{C}S]$. In particular, for the ith element θ_i of $\boldsymbol{\theta}$,

$$\theta_i \sim \mathrm{T}_{n+1}[m_i, C_{ii}S] \qquad \text{or} \qquad \frac{(\theta_i - m_i)}{\sqrt{C_{ii}S}} \sim \mathrm{T}_{n+1}[0, 1].$$

16.3.5 HPD regions

Consider any random vector $\mathbf{X}$ with (posterior) density $p(\mathbf{X})$. Density measures relative support for values of $\mathbf{X}$ in the sense that $\mathbf{X} = \mathbf{x}_0$ is better supported than $\mathbf{X} = \mathbf{x}_1$ if $p(\mathbf{x}_0) > p(\mathbf{x}_1)$. **Highest Posterior Density** regions (HPD regions) form the bases for inferences about $\mathbf{X}$ based on sets of values with high density. The basic notion is that the posterior probability of the set or region of values of $\mathbf{X}$ that have higher density than any chosen value $\mathbf{x}_0$ measures the support for, or against, $\mathbf{X} = \mathbf{x}_0$. Formally, the HPD region generated by the value $\mathbf{X} = \mathbf{x}_0$ is

$$\{\mathbf{X} \; : \; p(\mathbf{X}) \geq p(\mathbf{x}_0)\}.$$

The probability that $\mathbf{X}$ lies in this region is simply

$$\Pr[p(\mathbf{X}) \geq p(\mathbf{x}_0)].$$

If this probability is high, then $\mathbf{x}_0$ is poorly supported. In normal, linear models, HPD regions are always connected regions, or intervals, and these provide interval based inferences and tests of hypotheses through the use of posterior normal, T and F distributions, described below. Fuller theoretical details are provided by Box and Tiao (1973) and De Groot (1971).

MULTIVARIATE NORMAL POSTERIOR.

(1) Suppose that $\mathbf{X} = X$, a scalar, with posterior $X \sim N[m, C]$. Then, as is always the case with symmetric distributions, HPD regions are intervals symmetrically located about the median (here also the mode and mean) m. For any $k > 0$, the equal-tails interval

$$m - kC^{1/2} \leq X \leq m + kC^{1/2}$$

is the HPD region with posterior probability

$$\Pr[|X - m|/C^{1/2} \leq k] = 2\Phi(k) - 1$$

where $\Phi(.)$ is the standard normal cumulative distribution function. With $k = 1.645$ so that $\Phi(k) = 0.95$, this gives the 90% region $m \pm 1.645 C^{1/2}$. With $k = 1.96$, $\Phi(k) = 0.975$

and the 95% region is $m \pm 1.645C^{1/2}$. For any $k > 0$, the $100[2\Phi(k) - 1]\%$ HPD region for X is simply $m \pm kC^{1/2}$.

(2) Suppose that $\mathbf{X}$ is n–dimensional for some $n > 1$,

$$\mathbf{X} \sim \mathrm{N}[\mathbf{m}, \mathbf{C}]$$

for some mean vector $\mathbf{m}$ and covariance matrix $\mathbf{C}$. Then HPD regions are defined by elliptical shells centred at the mode $\mathbf{m}$, defined by the points $\mathbf{X}$ that lead to common values of the quadratic form in the density, namely

$$Q(\mathbf{X}) = (\mathbf{X} - \mathbf{m})'\mathbf{C}^{-1}(\mathbf{X} - \mathbf{m}).$$

For any $k > 0$, the region

$$\{\mathbf{X} \; : \; Q(\mathbf{X}) \leq k\}$$

is the HPD region with posterior probability

$$\Pr[Q(\mathbf{X}) \leq k] = \Pr[\kappa \leq k],$$

where κ is a gamma distributed random quantity,

$$\kappa \sim \mathrm{G}[n/2, 1/2].$$

When n is an integer, this gamma distribution is chi-squared with n degrees of freedom, and so

$$\Pr[Q(\mathbf{X}) \leq k] = \Pr[\chi_n^2 \leq k].$$

MULTIVARIATE T POSTERIORS. The results for T distribution parallel those for the normal, T distributions replace normal distributions and F distributions replace gamma.

(1) Suppose that $\mathbf{X} = X$, a scalar, with posterior $X \sim \mathrm{T}_r[m, C]$ for some degrees of freedom $r > 0$. Again HPD regions are intervals symmetrically located about the mode m. For any $k > 0$, the equal-tails interval

$$m - kC^{1/2} \leq X \leq m + kC^{1/2}$$

is the HPD region with posterior probability

$$\Pr[|X - m|/C^{1/2} \le k] = 2\Psi_r(k) - 1$$

where $\Psi_r(.)$ is the cumulative distribution function of the standard Student T distribution on r degrees of freedom. For any $k > 0$, the $100[2\Psi_r(k) - 1]\%$ HPD region for X is simply $m \pm kC^{1/2}$.

(2) Suppose that $\mathbf{X}$ is n−dimensional for some $n > 1$, $\mathbf{X} \sim T_r[\mathbf{m}, \mathbf{C}]$ for some mean vector $\mathbf{m}$, covariance matrix $\mathbf{C}$, and degrees of freedom $r > 0$. HPD regions are again defined by elliptical shells centred at the mode $\mathbf{m}$, identified by values of $\mathbf{X}$ having a common value of the quadratic form

$$Q(\mathbf{X}) = (\mathbf{X} - \mathbf{m})'\mathbf{C}^{-1}(\mathbf{X} - \mathbf{m}).$$

For any $k > 0$, the region

$$\{\mathbf{X} \; : \; Q(\mathbf{X}) \le k\}$$

is the HPD region with posterior probability

$$\Pr[Q(\mathbf{X}) \le k] = \Pr[\xi \le k/n],$$

where ξ is an F distributed random quantity, with n and r degrees of freedom,

$$\xi \sim F_{n,r}$$

(tabulated in Lindley and Scott, 1984, pages 50—55). Note that, when r is large, this is approximately a χ_n^2 distribution.

16.4 ELEMENTS OF MATRIX ALGEBRA

The reader is assumed to be familiar with basic concepts and results of matrix algebra including, for example: numerical operations on matrices; trace; determinant; transposition; linear and quadratic forms; singularity and inversion; symmetry; positive and non-negative definiteness; orthogonality; vector and matrix derivatives. In this Appendix we review some simple results concerning functions of square matrices and then explore in depth the eigenstructure and diagonalizability of such matrices. Foundational material, and more advanced theory, may be found in Bellman (1970) and Graybill (1969).

Thoughout the Appendix we consider square $(n \times n)$ matrices, such as $\mathbf{G}$, whose elements G_{ij}, $(i = 1, \ldots n; j = 1, \ldots, n)$ are real valued.

16.4.1 Powers and polynomials

POWERS. For any $k = 1, 2, \ldots,$ $\mathbf{G}^k = \mathbf{G}\mathbf{G}\ldots\mathbf{G} = \mathbf{G}\mathbf{G}^{k-1} = \mathbf{G}^{k-1}\mathbf{G}$. If $\mathbf{G}$ is non-singular, then $\mathbf{G}^{-k} = \left(\mathbf{G}^{-1}\right)^k$. Also, for all integers h and k, $\mathbf{G}^k\mathbf{G}^h = \mathbf{G}^{k+h}$ and $\left(\mathbf{G}^k\right)^h = \mathbf{G}^{kh}$.

POLYNOMIALS. If $p(\alpha)$ is a polynomial of degree m in α, $p(\alpha) = \sum_{r=0}^{m} p_r \alpha^r$, the matrix polynomial in $\mathbf{G}$, $p(\mathbf{G})$ is the $(n \times n)$ matrix defined by $p(\mathbf{G}) = \sum_{r=0}^{m} p_r \mathbf{G}^r$.

INFINITE SERIES. Suppose that, for all α with $|\alpha| \leq \max\{|G_{ij}|\}$, $p(\alpha)$ is a convergent power series $p(\alpha) = \sum_{r=0}^{\infty} p_r \alpha^r$. Then the matrix $p(\mathbf{G})$ defined as $\sum_{r=0}^{\infty} p_r \mathbf{G}^r$ exists with finite elements.

16.4.2 Eigenstructure of square matrices

The $(n \times n)$ matrix $\mathbf{G}$ has n eigenvalues $\lambda_1, \ldots, \lambda_n$ that are the roots of the polynomial of degree n in λ given by the determinant

$$p(\lambda) = |\mathbf{G} - \lambda\mathbf{I}|$$

where $\mathbf{I}$ is the $(n \times n)$ identity matrix. $p(\lambda)$ is the characteristic polynomial of $\mathbf{G}$ and the roots may be real or complex, occurring in pairs of complex conjugates in the latter case. Eigenvalues are sometimes referred to as the *characteristics values*, or *roots*, of the matrix.

The eigenvectors of $\mathbf{G}$ are $n-$vectors $\boldsymbol{\eta}$ satisfying

$$\mathbf{G}\boldsymbol{\eta} = \lambda_i\boldsymbol{\eta}, \qquad\qquad (i = 1, \ldots, n).$$

Eigenvectors are sometimes referred to as the *characteristics vectors* of the matrix.

This equation has at least one solution for each i. Various important properties are as follows:

(1) $|\mathbf{G}| = \prod_{i=1}^{n} \lambda_i$ and trace $(\mathbf{G}) = \sum_{i=1}^{n} \lambda_i$.

(2) The elements of η corresponding to λ_i are real valued if and only if λ_i is real valued.

(3) If the λ_i are distinct, then $\mathbf{G}\eta = \lambda_i\eta$ has a unique solution η_i (up to a constant scale factor of course) and the η_i are themselves distinct.

(4) If $\mathbf{G}$ is diagonal, then $\lambda_1, \ldots, \lambda_n$ are the diagonal elements.

(5) If $\mathbf{G}$ is symmetric then the λ_i are real. If, in addition, $\mathbf{G}$ is positive definite (or non-negative definite) then the λ_i are positive (or non-negative).

(6) If $\mathbf{G}$ has rank $p \le n$, then $n - p$ of the λ_i are zero.

(7) The eigenvalues of a polynomial or power series function $p(\mathbf{G})$ are given by $p(\lambda_i)$, $i = 1, \ldots, n$.

(8) If $p(\lambda) = |\mathbf{G} - \lambda\mathbf{I}|$ is the characteristic polynomial of $\mathbf{G}$ then

$$p(\mathbf{G}) = \mathbf{0}.$$

Since $p(\lambda) = \prod_{i=1}^{n}(\lambda - \lambda_i)$ then $\prod_{i=1}^{n}(\mathbf{G} - \lambda_i\mathbf{I}) = \mathbf{0}$. This result is known as the *Cayley-Hamilton Theorem*. It follows that, if $p(\lambda) = \sum_{i=0}^{n} p_i\lambda^i$, then $\sum_{i=0}^{n} p_i\mathbf{G}^i = 0$ and so

$$\mathbf{G}^n = -p_n^{-1}\sum_{i=0}^{n-1} p_i\mathbf{G}^i = q_0\mathbf{I} + q_1\mathbf{G} + \ldots + q_{n-1}\mathbf{G}^{n-1}, \text{ say.}$$

EIGENSTRUCTURE OF SYMMETRIC, POSITIVE DEFINITE MATRICES. The eigenstructure of a symmetric, positive definite matrix has particularly special features. Suppose that the $n \times n$ matrix $\mathbf{V}$ is symmetric and positive definite ie. a variance matrix. Then the n eigenvalues of $\mathbf{V}$ are real and positive. Without loss of generality, order the eigenvalues so that $\lambda_1 \ge \lambda_2 \ge \ldots \ge \lambda_n$. The corresponding eigenvectors are real-valued, and orthogonal so that $\eta_i'\eta_j = 0$, $(i, j = 1, \ldots, n; i \ne j)$ (Graybill, 1969, Chapter 3). The *orthonormalised* eigenvectors are defined by normalising the eigenvectors to have unit norm, so that, in addition to orthogonality, $\eta_j'\eta_j = 1$, $(j = 1, \ldots, n)$. These define define the **principal components** of the matrix $\mathbf{V}$. Suppose that $\mathbf{Y}$ is an $n \times 1$ random vector with variance matrix $\mathbf{V}$, so that $V[\mathbf{Y}] = \mathbf{V}$. Covariation of the elements of any random vector $\mathbf{Y}$ is explained through the random quantities $X_j = \eta_j'\mathbf{Y}$, $(j = 1, \ldots, n)$, the principal components of $\mathbf{Y}$. These X variates are uncorrelated and have variances $V[X_j] = \lambda_j$, decreasing as j increases. Total variation in $\mathbf{Y}$ is measured by $\lambda = \text{trace}(\mathbf{V}) = \sum_{j=1}^{n} \lambda_j$, and so the jth principal component explains a proportion λ_j/λ of this total.

16.4.3   Similarity of square matrices

Let $\mathbf{G}$ and $\mathbf{L}$ be $(n \times n)$ matrices and the eigenvalues of $\mathbf{G}$ be $\lambda_1, \ldots, \lambda_n$. Then $\mathbf{G}$ and $\mathbf{L}$ are similar (or $\mathbf{G}$ is similar to $\mathbf{L}$, $\mathbf{L}$ is similar to $\mathbf{G}$) if there exists a non-singular $(n \times n)$ matrix $\mathbf{H}$ such that

$$\mathbf{HGH}^{-1} = \mathbf{L}.$$

$\mathbf{H}$ is called the similarity matrix (or similarity transformation) and may be complex valued. Note that:

(1) $\mathbf{G} = \mathbf{H}^{-1}\mathbf{LH}$,

(2) $|\mathbf{G} - \lambda\mathbf{I}| = |\mathbf{H}^{-1}(\mathbf{L} - \lambda\mathbf{I})\mathbf{H}| = |\mathbf{H}|^{-1}|\mathbf{L} - \lambda\mathbf{I}||\mathbf{H}| = |\mathbf{L} - \lambda\mathbf{I}|$
so that the eigenvalues of $\mathbf{L}$ are also $\lambda_1, \ldots, \lambda_n$. Hence the traces of $\mathbf{G}$ and $\mathbf{L}$ coincide, as do their determinants.

(3) If $\mathbf{G}$ is similar to $\mathbf{L}$ and $\mathbf{L}$ is similar to $\mathbf{K}$, then $\mathbf{G}$ is similar to $\mathbf{K}$.

(4) Suppose that $\mathbf{G}_1, \ldots, \mathbf{G}_k$ are any k square matrices with the $(n_i \times n_i)$ matrix $\mathbf{G}_i$ similar to $\mathbf{L}_i$, $(i = 1, \ldots, k)$. Let $n = \sum_{i=1}^{k} n_i$. Then if $\mathbf{G}$ and $\mathbf{L}$ are the block diagonal matrices $\mathbf{G} = \text{block diag}[\mathbf{G}_1, \ldots, \mathbf{G}_k]$ and $\mathbf{L} = \text{block diag}[\mathbf{L}_1, \ldots, \mathbf{L}_k]$, it follows that $\mathbf{G}$ and $\mathbf{L}$ are similar. In particular, if the similarity matrix for $\mathbf{G}_i$ and $\mathbf{L}_i$ is $\mathbf{H}_i$ so that $\mathbf{H}_i\mathbf{G}_i\mathbf{H}_i^{-1} = \mathbf{L}_i$, the similarity matrix for $\mathbf{G}$ and $\mathbf{L}$ is $\mathbf{H} = \text{block diag}[\mathbf{H}_1, \ldots, \mathbf{H}_k]$.

By way of terminolgy, $\mathbf{G}$, $\mathbf{L}$ and $\mathbf{H}$ are said to be formed from the **superposition** of the matrices $\mathbf{G}_i$, $\mathbf{L}_i$ and $\mathbf{H}_i$, $(i = 1, \ldots, k)$, respectively.

DIAGONALIZATION. If $\mathbf{G}$ is similar to a diagonal matrix $\mathbf{L}$, then $\mathbf{G}$ is diagonalizable. In particular if $\mathbf{L} = \boldsymbol{\Lambda} = \text{diag}(\lambda_1, \ldots, \lambda_n)$, then, since

$$\mathbf{GH}^{-1} = \mathbf{H}^{-1}\boldsymbol{\Lambda},$$

the columns of $\mathbf{H}^{-1}$ are the eigenvectors of $\mathbf{G}$. It follows that:

(1) $\mathbf{G}$ is similar to $\boldsymbol{\Lambda}$ if the eigenvectors of $\mathbf{G}$ are linearly independent.

(2) If $\lambda_1, \ldots, \lambda_n$ are distinct, then $\mathbf{G}$ is diagonalizable.

(3) If $\lambda_1, \ldots, \lambda_n$ are not distinct, then $\mathbf{G}$ may not be diagonalizable. If $\mathbf{G}$ is symmetric or skew-symmetric, or orthogonal, then $\mathbf{G}$ is diagonalizable. In general, a different similar form for $\mathbf{G}$ is appropriate, as discussed in the next section.

DISTINCT EIGENVALUES. If $\lambda_1, \ldots, \lambda_n$ are distinct, then the eigenvectors are unique (up to a scalar constant) and linearly independent; they form the columns of $\mathbf{H}^{-1}$ in $\mathbf{HGH}^{-1} = \boldsymbol{\Lambda} = \mathrm{diag}(\lambda_1, \ldots, \lambda_n)$. In such a case it follows that:

(1) $\mathbf{G}^k = \left(\mathbf{H}^{-1}\boldsymbol{\Lambda}\mathbf{H}\right)^k = \left(\mathbf{H}^{-1}\boldsymbol{\Lambda}\mathbf{H}\right)\left(\mathbf{H}^{-1}\boldsymbol{\Lambda}\mathbf{H}\right)\ldots\left(\mathbf{H}^{-1}\boldsymbol{\Lambda}\mathbf{H}\right) = \mathbf{H}^{-1}\boldsymbol{\Lambda}^k\mathbf{H}$ for any integer k. Thus $\mathbf{G}^k$ is similar to $\boldsymbol{\Lambda}^k$, for each k, and the similarity matrix is $\mathbf{H}$. Note that $\boldsymbol{\Lambda}^k = \mathrm{diag}\left(\lambda_1^k, \ldots, \lambda_n^k\right)$ so that the eigenvalues of $\mathbf{G}^k$ are λ_i^k.

(2) For any n-vectors $\mathbf{a}$ and $\mathbf{b}$ not depending on k,

$$p(k) = \mathbf{a}'\mathbf{G}^k\mathbf{b} = \left(\mathbf{a}'\mathbf{H}^{-1}\right)\boldsymbol{\Lambda}^k\left(\mathbf{Hb}\right) = \sum_{i=1}^{n} c_i \lambda_i^k$$

for some constants $c_1, \ldots, c_n$, not depending on k.

In general the eigenvalues of $\mathbf{G}$ may be real or complex with the latter occuring in conjugate pairs. Suppose that $\mathbf{G}$ has p, $1 \leq p \leq n/2$, pairs of distinct complex conjugate eigenvalues ordered in pairs on the lower diagonal of $\boldsymbol{\Lambda}$. Thus $\lambda_1, \ldots, \lambda_{n-2p}$ are real and distinct, and

$$\left.\begin{array}{l} \lambda_j = r_j \exp\left(i\omega_j\right) \\ \lambda_{j+1} = r_j \exp\left(-i\omega_j\right) \end{array}\right\}, \qquad \begin{array}{l} j = n - 2(p-h) - 1 \\ \text{for } h = 1, \ldots, p. \end{array}$$

for some real, non-zero r_j and ω_j, (ω_j not an integer multiple of π), for each j.

Define the $(n \times n)$ matrix F as the block diagonal form

$$\mathbf{F} = \text{block diag}\left[\mathbf{I}, \begin{pmatrix} 1 & 1 \\ i & -i \end{pmatrix}, \ldots, \begin{pmatrix} 1 & 1 \\ i & -i \end{pmatrix}\right],$$

where $\mathbf{I}$ is the $(n - 2p)$ square identity matrix. By noting that

$$\begin{pmatrix} 1 & 1 \\ i & -i \end{pmatrix} \begin{pmatrix} \exp(i\omega_j) & 0 \\ 0 & \exp(-i\omega_j) \end{pmatrix} \begin{pmatrix} 1 & 1 \\ i & -i \end{pmatrix}^{-1}$$
$$= \begin{pmatrix} \cos(\omega_j) & \sin(\omega_j) \\ -\sin(\omega_j) & \cos(\omega_j) \end{pmatrix}$$

for each ω_j it is clear, by superposition, that $\mathbf{F}\boldsymbol{\Lambda}\mathbf{F}^{-1} = \boldsymbol{\Phi}$ where $\boldsymbol{\Phi}$ is the block diagonal matrix

$$\text{block diag}\left[\lambda_1, \ldots, \lambda_{n-2p}, \begin{pmatrix} r_1\cos(\omega_1) & r_1\sin(\omega_1) \\ -r_1\sin(\omega_1) & r_1\cos(\omega_1) \end{pmatrix};\right.$$
$$\left.\cdots; \begin{pmatrix} r_p\cos(\omega_p) & r_p\sin(\omega_p) \\ -r_p\sin(\omega_p) & r_p\cos(\omega_p) \end{pmatrix}\right].$$

Thus $\mathbf{\Lambda}$, and so $\mathbf{G}$, is similar to $\mathbf{\Phi}$. By contrast to $\mathbf{\Lambda}$, the block diagonal matrix $\mathbf{\Phi}$ is real-valued; $\mathbf{\Phi}$ is called the real canonical form of $\mathbf{G}$. Reducing $\mathbf{G}$ via the similarity transformation $\mathbf{H}$ to $\mathbf{\Lambda}$ simplifies the structure of the matrix but introduces complex entries. By contrast, the reduction to the "almost" diagonal $\mathbf{\Phi}$ remains real valued.

COMMON EIGENVALUES AND JORDAN FORMS. We now consider the case of replicated eigenvalues of $\mathbf{G}$. For any positive integer r, and any complex number λ, the Jordan block $\mathbf{J}_r(\lambda)$ is the $(r \times r)$ upper triangular matrix

$$
\mathbf{J}_r(\lambda) = \begin{pmatrix}
\lambda & 1 & 0 & \cdots & 0 \\
0 & \lambda & 1 & \cdots & 0 \\
0 & 0 & \lambda & \cdots & 0 \\
\vdots & \vdots & \vdots & \ddots & \vdots \\
0 & 0 & 0 & \cdots & \lambda
\end{pmatrix}.
$$

Thus the diagonal elements are all equal to λ, the super-diagonal elements are unity, and all other entries are zero.

Suppose that, in general, the $(n \times n)$ matrix $\mathbf{G}$ has s distinct eigenvalues $\lambda_1, \ldots, \lambda_s$ such that λ_i has multiplicity $r_i \geq 1$, $\sum_{i=1}^{s} r_i = n$. Thus the characteristic polynomial of $\mathbf{G}$ is simply $p(\lambda) = \prod_{i=1}^{s} (\lambda - \lambda_i)^{r_i}$. It can be shown that $\mathbf{G}$ is similar to the block diagonal Jordan form matrix $\mathbf{J}$ given by the superposition of the Jordan blocks

$$
\mathbf{J} = \text{block diag}\left[\mathbf{J}_{r_1}(\lambda_1), \ldots, \mathbf{J}_{r_s}(\lambda_s)\right].
$$

$\mathbf{J}$ is sometimes called the Jordan canonical form of $\mathbf{G}$. In this case it follows that, since $\mathbf{G} = \mathbf{H}^{-1}\mathbf{J}\mathbf{H}$ for some $\mathbf{H}$ then

$$
\mathbf{G}^k = \left(\mathbf{H}^{-1}\mathbf{J}\mathbf{H}\right)^k = \left(\mathbf{H}^{-1}\mathbf{J}\mathbf{H}\right)\left(\mathbf{H}^{-1}\mathbf{J}\mathbf{H}\right) \ldots \left(\mathbf{H}^{-1}\mathbf{J}\mathbf{H}\right) = \mathbf{H}^{-1}\mathbf{J}^k\mathbf{H}
$$

so that, for any integer k, $\mathbf{G}^k$ is similar to $\mathbf{J}^k$ and the similarity matrix is $\mathbf{H}$, for each k. The structure of $\mathbf{J}^k$ may be explored as follows:

(1) From the block diagonal form of $\mathbf{J}$,

$$
\mathbf{J}^k = \text{block diag}\left[\mathbf{J}_{r_1}(\lambda_1)^k, \ldots, \mathbf{J}_{r_s}(\lambda_s)^k\right],
$$

for each integer k.

(2) For a general Jordan block $\mathbf{J}_r(\lambda)$, let $\mathbf{J}_r(\lambda)^k$ have elements $m_{i,j}(k)$, $(i = 1, \ldots, r; \ j = 1, \ldots, r)$, for each integer k. Then, for $k \geq 1$,

$$m_{i,i+j}(k) = \begin{cases} \dbinom{k}{j} \lambda^{k-j} & \text{for } 0 \leq j \leq \min(k, r - i), \\ 0 & \text{otherwise .} \end{cases}$$

This may be proved by induction as follows. For $k = 1$, $m_{i,i}(1) = \lambda$ and $m_{i,i+1}(1) = 1$ are the only non-zero elements of $\mathbf{J}_r(\lambda)^1 = \mathbf{J}_r(\lambda)$ so the result holds. Assuming the result holds for a general k, we have $\mathbf{J}_r(\lambda)^{k+1} = \mathbf{J}_r(\lambda)^k \mathbf{J}_r(\lambda)$ so that

$$m_{i,i+j}(k + 1) = \sum_{h=1}^{r} m_{i,h}(k) m_{h,i+j}(1)$$

$$= \begin{cases} \lambda m_{i,i+j}(k) + m_{i,i+j-1}(k), & j \neq 0; \\ \lambda m_{i,i}(k), & j = 0. \end{cases}$$

Hence $m_{i,i}(k + 1) = \lambda m_{i,i}(k) = \lambda \lambda^k = \lambda^{k+1}$ as required. For $j < i$, both $m_{i,i+j}(k)$ and $m_{i,i+j-1}(k)$ are zero by hypothesis so that $m_{i,i+j}(k + 1) = 0$. Similarly, for $k + 1 < j \leq r$, $m_{i,i+j}(k + 1) = 0$. Finally, for $0 \leq j \leq k + 1$,

$$m_{i,i+j}(k + 1) = \lambda \binom{k}{j} \lambda^{k-j} + \binom{k}{j-1} \lambda^{k-j+1}$$

$$= \binom{k + 1}{j} \lambda^{k+1-j},$$

as required. Hence assuming the result true for any k implies it is true for $k + 1$. Since the result holds for $k = 1$, then it is true, in general, by induction.

(3) From (2) it follows that for any fixed r-vectors $\mathbf{a}_r$ and $\mathbf{b}_r$,

$$\mathbf{a}_r \mathbf{J}_r(\lambda)^k \mathbf{b}_r = \lambda^k p_r(k)$$

where $p_r(k) = p_{0r} + p_{1r} k + \ldots + p_{r-1\,r} k^{r-1}$ is a polynomial of degree $r - 1$ in k; the coefficients p_{ir} depend on $\mathbf{a}_r$, $\mathbf{b}_r$ and λ but not on k.

(4) From (1) and (2) it follows that, for any fixed n-vectors $\mathbf{a}$ and $\mathbf{b}$ and integer $k \geq 1$,

$$f(k) = \mathbf{a}'\mathbf{G}^k\mathbf{b} = \mathbf{a}'\mathbf{H}^{-1}\mathbf{J}^k\mathbf{H}\mathbf{b}$$

$$= \sum_{i=1}^{s} p_i(k)\lambda_i^k$$

where $p_i(k)$ is a polynomial function of k of degree $r_i - 1$, for each $i = 1, \ldots, s$.

When some of the eigenvalues of $\mathbf{G}$ are complex, a similar real canonical form is usually preferred to the complex valued Jordan form. In general, order the eigenvalues of $\mathbf{G}$ so that the complex eigenvalues occur in conjugate pairs. Thus if $\mathbf{G}$ has p distinct pairs of complex eigenvalues, then $\lambda_1, \ldots, \lambda_{s-2p}$ are real and distinct, and

$$\left.\begin{aligned}\lambda_j &= a_j\exp(i\omega_j) \\ \lambda_{j+1} &= a_j\exp(-i\omega)\end{aligned}\right\}, \qquad \begin{aligned} &j = s - 2(p-h) - 1, \\ &\text{for } h = 1, \ldots, p.\end{aligned}$$

Note that both λ_j and λ_{j+1} have multiplicity r_j $(= r_{j+1})$, for each such j. Now for each j, the $(2r_j \times 2r_j)$ Jordan form given by block $\mathrm{diag}[\mathbf{J}_{r_j}(\lambda_j), \mathbf{J}_{r_j}(\lambda_{j+1})]$ may be shown to be similar to the real valued matrix shown below.

Define the (2×2) real matrix $\mathbf{G}_j$, $(j = s - 2p + 1, \ldots, s - 1)$, by

$$\mathbf{G}_j = \begin{pmatrix} a_j\cos(\omega_j) & a_j\sin(\omega_j) \\ -a_j\sin(\omega_j) & a_j\cos(\omega_j) \end{pmatrix}$$

and let $\mathbf{I}$ be the (2×2) identity matrix. Then the similar matrix required is the $(2r_j \times 2r_j)$ (Jordan-form) matrix

$$\mathbf{L}_j = \begin{pmatrix} \mathbf{G}_j & \mathbf{I} & 0 & \cdots & 0 \\ 0 & \mathbf{G}_j & \mathbf{I} & \cdots & 0 \\ 0 & 0 & \mathbf{G}_j & \cdots & 0 \\ \vdots & \vdots & & \ddots & \vdots \\ 0 & 0 & 0 & \cdots & \mathbf{G}_j \end{pmatrix}.$$

Using this result it can be shown that the general $\mathbf{G}$ matrix is similar to the $(n \times n)$ block diagonal matrix whose upper diagonal block is the Jordan form corresponding to real eigenvalues and whose lower diagonal block is given by block $\mathrm{diag}[\mathbf{L}_1, \ldots, \mathbf{L}_p]$.

BIBLIOGRAPHY

Abraham, B., and Ledholter, A., 1983. *Statistical Methods for Fore-casting.* Wiley, New York.

Abramowitz, M., and Stegun, I.A., 1965. *Handbook of Mathematical Functions.* Dover, New York.

Aitchison, J., 1986. *The Statistical Analysis of Compositional Data.* Chapman-Hall, London.

Aitchison, J., and Brown, J.A.C., 1957. *The Lognormal Distribution.* Cambridge University Press, Cambridge.

Aitchison, J., and Dunsmore, I.R., 1975. *Statistical Prediction Analysis.* Cambridge University Press, Cambridge.

Aitchison, J., and Shen, S. M., 1980. Logistic-normal distributions: some properties and uses. *Biometrika* **67**, 261-272.

Alspach, D.L., and Sorenson, H.W., 1972. Nonlinear Bayesian estimation using Gaussian sum approximations. *IEEE Trans. Automatic Control* **17**, 439-448.

Amaral, M.A., and Dunsmore, I.R., 1980. Optimal estimates of predictive distributions. *Biometrika* **67**, 685-689.

Ameen, J.R.M., 1984. *Discount Bayesian models and forecasting.* Unpublished Ph.D. thesis, University of Warwick.

Ameen, J.R.M., and Harrison, P.J., 1984. Discount weighted estimation. *J. of Forecasting* **3**, 285-296.

Ameen, J.R.M., and Harrison, P.J., 1985a. Normal discount Bayesian models. In *Bayesian Statistics 2*, J.M. Bernardo, M.H. DeGroot, D.V. Lindley and A.F.M. Smith (Eds.). North-Holland, Amsterdam, and Valencia University Press.

Ameen, J.R.M., and Harrison, P.J., 1985b. Discount Bayesian multi-process modelling with cusums. In *Time Series Analysis: Theory and Practice 5*, O.D. Anderson (Ed.). North-Holland, Amsterdam.

Anderson, B.D.O., and Moore, J.B., 1979. *Optimal Filtering.* Prentice-Hall, New Jersey.

Ansley, C.F., 1979. An algorithm for the exact likelihood of a mixed autoregressive moving average process. *Biometrika* **66**, 59-65.

Azzalini, A., 1983. Approximate filtering of parameter driven processes. *J. Time Series Analysis* **3**, 219-224.

Baker, R.J., and Nelder, J.A., 1985. *GLIM Release 3.77.* Oxford University, Numerical Algorithms Group.

Barbosa, E., and Harrison, P.J., 1989. Variance estimation for multivariate DLMs. *Research Report* 160, Department of Statistics, University of Warwick.

Barnard, G.A., 1959. Control charts and stochastic processes (with discussion). *J. Roy. Statist. Soc.* (Ser. B) **21**, 239-271.

Barnett, V., and Lewis, T., 1978. *Outliers in Statistical Data.* Wiley, Chichester.

Bates, J.M., and Granger, C.W.J., 1969. The combination of forecasts. *Oper. Res. Quart.* **20**, 451-468.

Bellman, R., 1970. *Introduction to Matrix Analysis.* McGraw-Hill, New York.

Berger, J.O., 1985. *Statistical Decision Theory and Bayesian Analysis* (2nd edn.). Springer-Verlag, New York.

Berk, R.H., 1970. Consistency a posteriori. *Ann. Math. Stat.* **41**, 894-906.

Bernardo, J.M., 1979. Reference posterior distributions for Bayesian inference (with discussion). *J. Roy. Statist. Soc.* (Ser. B) **41**, 113-148.

Bordley, R.F., 1982. The combination of forecasts: a Bayesian approach. *J. Oper. Res. Soc.* **33**, 171-174.

Box, G.E.P., 1980. Sampling and Bayes' inference in scientific modelling and robustness. *J. Roy. Statist. Soc.* (Ser. A) **143**, 383-430.

Box, G.E.P., and Cox, D.R., 1964. An analysis of transformations (with discussion). *J. Roy. Statist. Soc.* (Ser. B) **26**, 241-252.

Box, G.E.P., and Draper, N.R., 1969. *Evolutionary Operation.* Wiley, New York.

Box, G.E.P., and Draper, N.R., 1987. *Empirical Model-Building and Response Surfaces.* Wiley, New York.

Box, G.E.P., and Jenkins, G.M, 1976. *Time Series Analysis: Forecasting and Control,* (2nd edn.). Holden-Day, San Francisco.

Box, G.E.P., and Tiao G.C., 1968. A Bayesian approach to some outlier problems. *Biometrika* **55**, 119-129.

Box, G.E.P., and Tiao G.C., 1973. *Bayesian Inference in Statistical Analysis.* Addison-Wesley, Massachusetts.

Broadbent, S., 1979. One way T.V. advertisements work. *J. Mkt. Res. Soc.* **21**, 139-165.

Broemeling, L.D. 1985. *Bayesian Analysis of Linear Models.* Marcel Dekker, New York.

Broemeling, L.D. and Tsurumi, H, 1987. *Econometrics and Structural Change.* Marcel Dekker, New York.

Brown, R.G., 1959. *Statistical Forecasting for Inventory Control.* McGraw-Hill, New York.

Brown, R.G., 1962. *Smoothing, Forecasting and Prediction of Discrete Time Series.* Prentice-Hall, Englewood Cliffs, NJ.

Brown, R.L., Durbin, J., and Evans, J.M., 1975. Techniques for testing the constancy of regression relationships over time. *J. Roy. Statist. Soc.* (Ser. B) **37**, 149-192.

Bunn, D.W., 1975. A Bayesian approach to the combination of forecasts. *J. Oper. Res. Soc.* **26**, 325-330.

Cleveland, W.S., 1974. Estimation of parameters in distributed lag models. In *Studies in Bayesian Econometrics and Statistics,* S.E. Fienberg and A. Zellner (Eds.). North-Holland, Amsterdam.

Colman, S., and Brown, G., 1983. Advertising tracking studies and sales effects. *J. Mkt. Res. Soc.* **25**, 165-183.

Davis, P.J., and Rabinowitz, P., 1967. *Numerical Integration.* Waltham, Massachusetts.

Dawid, A.P., 1981. Some matrix-variate distribution theory: notational considerations and a Bayesian application. *Biometrika* **68**, 265-274.

De Finetti, B., 1974, 1975. *Theory of Probability* (Vols. 1 and 2). Wiley, Chichester.

De Groot, M.H., 1971. *Optimal Statistical Decisions.* McGraw-Hill, New York.

Dickey, J.M., Dawid, A.P., and Kadane, J.B., 1986. Subjective-probability assessment methods for multivariate-t and matrix-t models. In *Bayesian inference and decision techniques: essays in honor of Bruno de Finetti,* P.K. Goel and A. Zellner (Eds.). North-Holland, Amsterdam.

Dickinson, J.P., 1975. Some statistical results in the combination of forecasts. *Oper. Res. Quart.* **26**, 253-260.

Duncan, D.B., and Horne, S.D., 1972. Linear dynamic regression from the viewpoint of regression analysis. *J. Amer. Statist. Ass.* **67**, 815-821.

Ewan, W.D., and Kemp, K.W., 1960. Sampling inspection of continuous processes with no autocorrelation between successive results. *Biometrika* **47**, 363-380.

Fox, A.J., 1972. Outliers in time series. *J. Roy. Statist. Soc.* (Ser. B) **34**, 350-363.

Fuller, W.A., 1976. *Introduction to Statistical Time Series.* Wiley, New York.

Gamerman, D., 1985. Dynamic Bayesian models for survival data. *Research Report* 75, Department of Statistics, University of Warwick.

Gamerman, D., 1987a. *Dynamic analysis of survival models and related processes.* Unpublished Ph.D. thesis, University of Warwick.

Gamerman, D., 1987b. Dynamic inference on survival functions. In *Probability and Bayesian Statistics*, R. Viertl (Ed.). Plenum, New York and London.

Gamerman, D., and West, M., 1987a. A time series application of dynamic survival models in unemployment studies. *The Statistician* **36**, 269-174.

Gamerman, D., and West, M., 1987b. Dynamic survival models in action. *Research Report* 111, Department of Statistics, University of Warwick.

Gamerman, D., West, M., and Pole, A., 1987. A guide to **SURVIVAL**: Bayesian analysis of survival data. *Research Report* 121, Department of Statistics, University of Warwick.

Gardner, G., Harvey, A.C. and Phillips, G.D.A., 1980. An algorithm for exact maximum likelihood estimation by means of Kalman filtering. *Applied Statistics* **29**, 311-322.

Gilchrist, W., 1976. *Statistical Forecasting.* Wiley, New York.

Godolphin, E.J., and Harrison, P.J., 1975. Equivalence theorems for polynomial projecting predictors. *J.R. Statist. Soc.* (Ser. B) **37**, 205-215.

Goldstein, M., 1976. Bayesian analysis of regression problems. *Biometrika* **63**, 51-58.

Good, I.J., 1985. Weights of evidence: a critical survey. In *Bayesian Statistics 2*, J.M. Bernardo, M.H. DeGroot, D.V. Lindley and

A.F.M. Smith (Eds.). North-Holland, Amsterdam, and Valencia University Press.

Granger, C.W.J. and Newbold, P, 1977. *Forecasting Economic Time Series*. Academic Press, New York.

Granger, C.W.J., and Ramanathan, R., 1984. Improved methods of combining forecasts. *J, Forecasting* **3**, 197-204.

Graybill, F.A., 1969. *Introduction to Matrices with Applications in Statistics*. Wadsworth, California.

Green, M, and Harrison, P.J., 1972. On aggregate forecasting. *Research Report* 2, Department of Statistics, University of Warwick.

Green, M, and Harrison, P.J., 1973. Fashion forecasting for a mail order company. *Oper. Res. Quart.* **24**, 193-205.

Harrison, P.J., 1965. Short-term sales forecasting. *Applied Statistics* **15**, 102-139.

Harrison, P.J., 1967. Exponential smoothing and short-term forecasting. *Man. Sci.* **13**, 821-842.

Harrison, P.J., 1973. Discussion of Box-Jenkins seasonal forecasting: a case study. *J. Roy. Statist. Soc.* (Ser. A) **136**, 319-324.

Harrison, P.J., 1985a. First-order constant dynamic models. *Research Report* 66, Department of Statistics, University of Warwick.

Harrison, P.J., 1985b. Convergence for dynamic linear models. *Research Report* 67, Department of Statistics, University of Warwick.

Harrison, P.J., 1985c. The aggregation and combination of forecasts. *Research Report* 68, Department of Statistics, University of Warwick.

Harrison, P.J., 1988. Bayesian forecasting in O.R.. In *Developments in Operational Research 1988*, N.B. Cook and A.M. Johnson (Eds.). Pergamon Press, Oxford.

Harrison, P.J., and Akram, M., 1983. Generalised exponentially weighted regression and parsimonious dynamic linear modelling. In *Time Series Analysis: Theory and Practice 3*, O.D. Anderson (Ed.). North Holland, Amsterdam.

Harrison, P.J., and Davies, O.L., 1964. The use of cumulative sum (CUSUM) techniques for the control of routine forecasts of produce demand. *Oper. Res.* **12**, 325-33.

Harrison, P.J., Leonard, T., and Gazard, T.N., 1977. An application of multivariate hierarchical forecasting. *Research Report* 15, Department of Statistics, University of Warwick.

Harrison, P.J., and Pearce, S.F., 1972. The use of trend curves as an aid to market forecasting. *Industrial Marketing Management* **2**, 149-170.

Harrison, P.J., and Quinn, M.P. 1978. A brief description of part of the work done to date concerning a view of agricultural and related systems. In *Econometric models presented to the Beef-Milk symposium*. Commission of European Communities, Agriculture EUR6101, March 1977, Brussels.

Harrison P.J., and Scott, F.A., 1965. A development system for use in short-term forecasting (original draft 1965, re-issued 1982). *Research Report* 26, Department of Statistics, University of Warwick.

Harrison, P.J., and Smith, J.Q., 1980. Discontinuity, decision and conflict (with discussion). In *Bayesian Statistics*, J.M. Bernardo, M.H. De Groot, D.V. Lindley and A.F.M. Smith (Eds.). University Press, Valencia.

Harrison P.J., and Stevens, C. F., 1971. A Bayesian approach to short-term forecasting. *Oper. Res. Quart.* **22**, 341-362.

Harrison P.J., and Stevens, C. F., 1976a. Bayesian forecasting (with discussion). *J. Roy. Statist. Soc.* (Ser. B) **38**, 205-247.

Harrison P.J., and Stevens, C. F., 1976b. Bayes forecasting in action: case studies. *Warwick Research Report* 14, Department of Statistics, University of Warwick.

Harrison, P.J., and West, M., 1986. Bayesian forecasting in practice. *Bayesian Statistics Study Year Report* 13, University of Warwick.

Harrison, P.J., and West, M., 1987. Practical Bayesian forecasting. *The Statistician* **36**, 115-125.

Harrison, P.J., West, M., and Pole, A., 1987. **FAB**, a training package for Bayesian forecasting. *Warwick Research Report* 122, Department of Statistics, University of Warwick.

Hartigan, J.A., 1969. Linear Bayesian methods. *J. Roy. Statist. Soc.* (Ser. B) **31**, 446-454.

Harvey, A.C., 1981. *Time Series Models*. Philip Allan, Oxford.

Harvey, A.C., 1984. A unified view of statistical forecasting procedures. *J. Forecasting* **3**, 245-275.

Harvey, A.C., 1986. Analysis and generalisation of a multivariate exponential smoothing model. *Management Science* **32**, 374-380.

Harvey, A.C., and Durbin, J., 1986. The effects of seat belt legislation on British road casualties: a case study in structural time series modelling (with discussion). *J. Roy. Statist. Soc.* (Ser. A) **149**, 187-227.

Heyde, C.C., and Johnstone, I.M., 1979. On asymptotic posterior normality for stochastic processes. *J. Roy. Statist. Soc.* (Ser. B) **41**, 184-189.

Highfield, R., 1984. Forecasting with Bayesian state space models. *Technical Report*, Graduate School of Business, University of Chicago.

Holt, C.C., 1957. Forecasting seasonals and trends by exponentially weighted moving averages. *O.N.R. Research Memo.* 52, Carnegie Institute of Technology.

Jazwinski, A.H., 1970. *Stochastic Processes and Filtering Theory.* Academic Press, New York.

Jeffreys, H., 1961. *Theory of Probability* (3rd edn.). Oxford University Press, London.

Johnston, F.R., and Harrison, P.J., 1980. An application of forecasting in the alcoholic drinks industry. *J. Oper. Res. Soc.* **31**, 699-709.

Johnston, F.R., Harrison, P.J., Marshall, A.S., and France, K.M., 1986. Modelling and the estimation of changing relationships. *The Statistician* **35**, 229-235.

Johnson, N.L., and Kotz, S., 1972. *Distributions in Statistics: Continuous Multivariate Distributions.* Wiley, New York.

Kalman, R.E., 1960. A new approach to linear filtering and prediction problems. *J. of Basic Engineering* **82**, 35-45.

Kalman, R.E., 1963. New methods in Wiener filtering theory. In *Proceedings of the First Symposium of Engineering Applications of Random Function Theory and Probability*, J.L. Bogdanoff and F. Kozin (Eds.). Wiley, New York.

Kitagawa, G., 1987. Non-Gaussian state-space modelling of nonstationary time series (with disussion). *J. Amer. Statist. Ass.* **82**, 1032-1063.

Kleiner, B., Martin, R.D., and Thompson, D.J., 1979. Robust estimation of power spectra (with discussion). *J. Roy. Statist. Soc.* (Ser. B) **3**, 313-351.

Kullback, S., 1983. Kullback information. In *Encyclopedia of Statistical Sciences* (Vol. 4), S. Kotz and N.L. Johnston (Eds.). Wiley, New York.

Kullback, S., and Leibler, R.A., 1951. On information and sufficiency. *Ann. Math. Statist.* **22**, 79-86.

Leamer, E.E., 1972. A class of information priors and distributed lag analysis. *Econometrica* **40**, 1059-1081.

Lindley, D.V., 1965. *Introduction to Probability and Statistics from a Bayesian viewpoint* (Parts 1 and 2). Cambridge University Press, Cambridge.

Lindley, D.V., 1983. Reconciliation of probability distributions. *Oper. Res.* **31**, 806-886.

Lindley, D.V., 1985. Reconciliation of discrete probability distributions. In *Bayesian Statistics 2*, J.M. Bernardo, M.H. DeGroot, D.V. Lindley and A.F.M. Smith (Eds.). North Holland, Amsterdam, and Valencia University Press.

Lindley, D.V., 1988. The use of probability statements. In *Accelerated life tests and experts' opinion in reliability*, C.A. Clarotti and D.V. Lindley (Eds.). In press.

Lindley, D.V., and Scott, W.F., 1984. *New Cambridge Elementary Statistical Tables*. Cambridge University Press, Cambridge.

Lindley, D.V., and Smith, A.F.M., 1972. Bayes' estimates for the linear model. *J. Roy. Statist. Soc.* (Ser. B) **34**, 1-41.

Lindley, D.V., Tversky, A., and Brown, R.V., 1979. On the reconciliation of probability assessments (with discussion). *J. Roy. Statist. Soc.* (Ser. A) **142**, 146-180.

Mardia, K.V., Kent, J.T., and Bibby, J.M., 1979. *Multivariate Analysis*. Academic Press, London.

Marriot, J., 1987. Bayesian numerical integration and graphical methods for Box-Jenkins time series. *The Statistician* **36**, 265-268.

Masreliez, C.J., 1975. Approximate non-Gaussian filtering with linear state and observation relations. *IEEE Trans. Aut. Con.* **20**, 107-110.

Masreliez, C.J., and Martin, R.D., 1977. Robust Bayesian estimation for the linear model and robustifying the Kalman filter. *IEEE Trans. Aut. Con.* **22**, 361-371.

Mazzuchi, T.A., and Soyer, R., 1987. A dynamic general linear model for inference from accelerated life tests. *Technical Report* 87/10, Institute of Reliability and Risk Analysis, George Washington University.

McCullagh, P., and Nelder, J.A., 1983. *Generalised Linear Models*. Chapman Hall, London and New York.

McKenzie, E., 1974. A comparison of standard forecasting systems with the Box-Jenkins approach. *The Statistician* **23**, 107-116.

McKenzie, E., 1976. An analysis of general exponential smoothing. *Oper. Res.* **24**, 131-140.

Migon, H.S., 1984. *An approach to non-linear Bayesian forecasting problems with applications.* Unpublished Ph.D. thesis, Department of Statistics, University of Warwick.

Migon, H.S., and Harrison, P.J., 1985. An application of non-linear Bayesian forecasting to television advertising. In *Bayesian Statistics 2*, J.M. Bernardo, M.H. DeGroot, D.V. Lindley and A.F.M. Smith (Eds.). North-Holland, Amsterdam, and Valencia University Press.

Morris, C.N., 1983. Natural exponential families with quadratic variance functions: statistical theory. *Ann. Statist.* **11**, 515-519.

Morris, P.A., 1983. An axiomatic approach to expert resolution. *Man. Sci.* **29**, 24-32.

Muth, J.F., 1960. Optimal properties of exponentially weighted forecasts. *J. Amer. Statist. Ass.* **55**, 299-306.

Naylor, J.C., and Smith, A.F.M., 1982. Applications of a method for the efficient computation of posterior distributions. *Applied Statistics* **31**, 214-225.

Nelder, J.A., and Wedderburn, R.W.M., 1972. Generalised linear models. *J. Roy. Statist. Soc.* (Ser. A) **135**, 370-384.

Nerlove, M., and Wage, S., 1964. On the optimality of adaptive forecasting. *Man. Sci.* **10**, 207-229.

Page, E.S., 1954. Continuous inspection schemes. *Biometrika* **41**, 100-114.

Pole, A., 1988a. Transfer response models: a numerical approach. In *Bayesian Statistics 3*, J.M. Bernardo, M.H. DeGroot, D.V. Lindley and A.F.M. Smith (Eds.). Oxford University Press.

Pole, A., 1988b. NOBS: Non-linear Bayesian forecasting software. *Warwick Research Report* 154, Department of Statistics, University of Warwick.

Pole, A., and Smith, A.F.M., 1983. A Bayesian analysis of some threshold switching models. *J. Econometrics* **29**, 97-119.

Pole, A., and West, M., 1988. Efficient numerical integration in dynamic models. *Warwick Research Report* 136, Department of Statistics, University of Warwick.

Pole, A., and West, M., 1989a. Reference analysis of the DLM. *J. Time Series Analysis*, (to appear.

Pole, A., and West, M., 1989b. Efficient Bayesian learning in non-linear dynamic models. *J. Forecasting*, (to appear).

Pole, A., West, M., Harrison, P.J., 1988. Non-normal and non-linear dynamic Bayesian modelling. In *Bayesian Analysis of Time Series and Dynamic Models*, J.C. Spall (Ed.). Marcel Dekker, New York.

Press, S.J., 1985. *Applied Multivariate Analysis: Using Bayesian and Frequentist Methods of Inference.* Krieger, California.

Priestley, M.B., 1980. System identification, Kalman filtering, and stochastic control. In *Directions in Time Series*, D.R. Brillinger and G.C.Tiao (Eds.). Institute of Mathematical Statistics.

Quintana, J.M., 1985. A dynamic linear matrix-variate regression model. *Research Report* 83, Department of Statistics, University of Warwick.

Quintana, J.M., 1987. *Multivariate Bayesian forecasting models.* Unpublished Ph.D. thesis, University of Warwick.

Quintana, J.M., and West, M., 1987. Multivariate time series analysis: new techniques applied to international exchange rate data. *The Statistician* **36**, 275-281.

Quintana, J.M., and West, M., 1988. Time series analysis of compositional data. In *Bayesian Statistics 3*, J.M. Bernardo, M.H. DeGroot, D.V. Lindley and A.F.M. Smith (Eds.). Oxford University Press.

Roberts, S.A., and Harrison, P.J., 1984. Parsimonious modelling and forecasting of seasonal time series. *European J. Oper. Res.* **16**, 365-377.

Russell, B., 1921. *The Analysis of Mind.* Allen and Unwin.

Salzer, H.E., Zucker, R., and Capuano, R., 1952. Tables of the zeroes and weight factors of the first twenty Hermite polynomials. *J. Research of the National Bureau of Standards* 48, 111-116.

Sage, A.P., and Melsa, J.L., 1971. *Estimation Theory with Applications to Communications and Control.* McGraw-Hill, New York.

Savage, L.J., 1954. *The Foundations of Inference.* Wiley, New York.

Schervish, M.J., and Tsay, R., 1988. Bayesian modelling and forecasting in autoregressive models. In *Bayesian Analysis of Time Series and Dynamic Models*, J.C. Spall (Ed.). Marcel Dekker, New York.

Schnatter, S., 1988. Bayesian forecasting of time series by Gaussian sum approximation. In *Bayesian Statistics 3*, J.M. Bernardo, M.H. De Groot, D.V. Lindley and A.F.M. Smith (Eds.). Oxford University Press.

Shaw, J.E.H., 1987. A strategy for reconstructing multivariate probability distributions. *Research Report* 123, Department of Statistics, University of Warwick.

Shaw, J.E.H., 1988. Aspects of numerical integration and summarisation. In *Bayesian Statistics 3*, J.M. Bernardo, M.H. DeGroot, D.V. Lindley and A.F.M. Smith (Eds.). Oxford University Press.

Smith, A.F.M., 1975. A Bayesian approach to inference about a change point in a sequence of random variables. *Biometrika* **63**, 407-416.

Smith, A.F.M., 1980. Change point problems: approaches and applications. In *Bayesian Statistics*, J.M. Bernardo, M.H. De Groot, D.V. Lindley and A.F.M. Smith (Eds.). University Press, Valencia.

Smith, A.F.M., and Cook, D.G., 1980. Switching straight lines: a Bayesian analysis of some renal transplant data. *Applied Statistics* **29**, 180-189.

Smith, A.F.M., and Pettit, L.I., 1985. Outliers and influential observations in linear models. In *Bayesian Statistics 2*, J.M. Bernardo, M.H. DeGroot, D.V. Lindley and A.F.M. Smith (Eds.). North-Holland, Amsterdam, and Valencia University Press.

Smith, A.F.M., Skene, A.M., Shaw, J.E.H., Naylor, J.C., and Dransfield, M., 1985. The implementation of the Bayesian paradigm. *Commun. Statist.: Theor. Meth.* **14**, 1079-1102.

Smith, A.F.M., Skene, A.M., Shaw, J.E.H., and Naylor, J.C., 1987. Progress with numerical and graphical methods for practical Bayesian statistics. *The Statistician* **36**, 75-82.

Smith, A.F.M., and West, M., 1983. Monitoring renal transplants: an application of the multi-process Kalman filter. *Biometrics* **39**, 867-878.

Smith, A.F.M., West, M., Gordan, K., Knapp, M.S., and Trimble, I., 1983. Monitoring kidney transplant patients. *The Statistician* **32**, 46-54.

Smith, J.Q., 1979. A generalisation of the Bayesian steady forecasting model. *J. Roy. Statist. Soc.* (Ser. B) **41**, 378-387.

Smith, R.L., and Miller, J.E., 1986. Predictive records. *J. Roy. Statist. Soc.* (Ser. B) **48**, 79-88.

Sorenson, H.W., and Alspach, D.L., 1971. Recursive Bayesian estimation using Gaussian sums. *Automatica* **7**, 465-479.

Souza, R.C., 1981. A Bayesian-entropy approach to forecasting: the multi-state model. In *Time Series Analysis*, O.D. Anderson (Ed.). North-Holland, Houston, Texas.

Stevens, C.F., 1974. On the variability of demand for families of items. *Oper. Res. Quart.* **25**, 411-420.

Sweeting, T.J., and Adekola, A.O., 1987. Asymptotic posterior normality for stochastic processes revisited. *J. Roy. Statist. Soc.* (Ser. B) **49**, 215-222.

Theil, H., 1971. *Principles of Econometrics*. Wiley, New York.

Thiel, H., and Wage, S., 1964. Some observations on adaptive forecasting. *Man. Sci.* **10**, 198-206.

Titterington, D.M., Smith, A.F.M., and Makov, U.E., 1985. *Statistical Analysis of Finite Mixture Distributions*. Wiley, Chichester.

Tong, H., 1983. *Threshold models in non-linear time series analysis*. Springer-Verlag, New York.

Tong, H., and Lim, K.S., 1980. Threshold autoregression, limit cycles and cyclical data (with discussion). *J. Roy. Statist. Soc.* (Ser. B) **42**, 245-292.

Trimble, I., West, M., Knapp, M.S., Pownall, R. and Smith, A.F.M., 1983. Detection of renal allograft rejection by computer. *British Medical Journal* **286**, 1695-1699.

Van Dijk, H.K., Hop, J.P., and Louter, A.S., 1987. Some algorithms for the computation of posterior moments and densities using Monte Carlo integration. *The Statistician* **36**, 83-90.

Van Duijn, J.J., 1983. *The Long Wave in Economic Life*. George Allen and Unwin, London.

West, M., 1981. Robust sequential approximate Bayesian estimation. *J. Roy. Statist. Soc.* (Ser. B) **43**, 157-166.

West, M., 1982. *Aspects of recursive Bayesian estimation*. Unpublished Ph.D. thesis, University of Nottingham.

West, M., 1984. Outlier models and prior distributions in Bayesian linear regression. *J. Roy. Statist. Soc.* (Ser. B) **46**, 431-439.

West, M., 1985a. Generalised linear models: outlier accomodation, scale parameters and prior distributions. In *Bayesian Statistics 2*, J.M. Bernardo, M.H. DeGroot, D.V. Lindley and A.F.M. Smith (Eds.). North-Holland, Amsterdam, and Valencia University Press.

West, M., 1985b. Combining probability forecasts. *Research Report* 67, Department of Statistics, University of Warwick.

West, M., 1985c. Assessment and control of probability forecasts. *Research Report* 69, Department of Statistics, University of Warwick.

West, M., 1986a. Bayesian model monitoring. *J. Roy. Statist. Soc.* (Ser. B) **48**, 70-78.

West, M., 1986b. Non-normal multi-process models. *Research Report* 81, Department of Statistics, University of Warwick.

West, M., 1987. Analysis of nasopharynx cancer data using dynamic Bayesian models. *Research Report* 109, Department of Statistics, University of Warwick (also Technical Report 7–1987 of the Department of Mathematics, II University of Rome).

West, M., 1988. Modelling expert opinion (with discussion). In *Bayesian Statistics 3*, J.M. Bernardo, M.H. De Groot, D.V. Lindley and A.F.M. Smith (Eds.). Oxford University Press.

West, M., 1989a. Bayesian analysis of agent opinion. Unpublished report, ISDS, Duke University.

West, M., 1989b. Modelling agent forecast distributions. Unpublished report, ISDS, Duke University.

West, M., and Harrison, P.J., 1986a. Monitoring and adaptation in Bayesian forecasting models. *J. Amer. Statist. Ass.* **81**, 741-750.

West, M., and Harrison, P.J., 1986b. Advertising awareness response model: micro-computer APL*PLUS/PC implementation. *Warwick Research Report* 118, Department of Statistics, University of Warwick.

West, M., and Harrison, P.J., 1989. Subjective intervention in formal models. *J. Forecasting* **8**, 33-53.

West, M., Harrison, P.J., and Migon, H.S., 1985. Dynamic generalised linear models and Bayesian forecasting (with discussion). *J. Amer. Statist. Ass.* **80**, 73-97.

West, M., Harrison, P.J., and Pole, A., 1987a. **BATS** : Bayesian Analysis of Time Series. *The Professional Statistician* **6**, 43-46.

West, M., Harrison, P.J., and Pole, A., 1987b. **BATS** : A user guide. *Warwick Research Report* 114, Department of Statistics, University of Warwick.

West, M., and Mortera, J., 1987. Bayesian models and methods for binary time series. In *Probability and Bayesian Statistics*, R. Viertl (Ed.). Plenum, New York and London.

Whittle, P., 1965. Recursive relations for predictors of non-stationary processes. *J. R. Statist. Soc.* (Ser. B) **27**, 523-532.

Winkler, R.L., 1981. Combining probability distributions from dependent information sources. *Management Science* **27**, 479-488.

Winters, P.R., 1960. Forecasting sales by exponentially weighted moving averages. *Man. Sci.* **6**, 324-342.

Woodward, R.H., and Goldsmith, P.L., 1964. *I.C.I. Monograph No. 3: Cumulative Sum Techniques.* Mathematical and Statistical Techniques for Industry, Oliver and Boyd, Edinburgh.

Young, A.S., 1977. A Bayesian approach to prediction using polynomials. *Biometrika* **64**, 309-317.

Young, A.S., 1983. A comparative analysis of prior families for distributed lags. *Empirical Econonomics* **8**, 215-227.

Young, P.C., 1984. *Recursive Estimation and Time Series Analysis.* Springer-Verlag, Berlin.

Zellner, A., 1971. *An Introduction to Bayesian Inference in Econometrics.* Wiley, New York.

AUTHOR INDEX

SUBJECT INDEX